Lecture Notes in Networks and Systems

Volume 341

The series "Lecture Notes in Networks and Systems" publishes the latest developments in Networks and Systems—quickly, informally and with high quality. Original research reported in proceedings and post-proceedings represents the core of LNNS.

Volumes published in LNNS embrace all aspects and subfields of, as well as new challenges in, Networks and Systems.

The series contains proceedings and edited volumes in systems and networks, spanning the areas of Cyber-Physical Systems, Autonomous Systems, Sensor Networks, Control Systems, Energy Systems, Automotive Systems, Biological Systems, Vehicular Networking and Connected Vehicles, Aerospace Systems, Automation, Manufacturing, Smart Grids, Nonlinear Systems, Power Systems, Robotics, Social Systems, Economic Systems and other. Of particular value to both the contributors and the readership are the short publication timeframe and the world-wide distribution and exposure which enable both a wide and rapid dissemination of research output.

The series covers the theory, applications, and perspectives on the state of the art and future developments relevant to systems and networks, decision making, control, complex processes and related areas, as embedded in the fields of interdisciplinary and applied sciences, engineering, computer science, physics, economics, social, and life sciences, as well as the paradigms and methodologies behind them.

Indexed by SCOPUS, INSPEC, WTI Frankfurt eG, zbMATH, SCImago.

All books published in the series are submitted for consideration in Web of Science.

More information about this series at https://link.springer.com/bookseries/15179

Rajendra Prasad Mahapatra ·
Sateesh Kumar Peddoju · Sudip Roy ·
Pritee Parwekar · Lavika Goel
Editors

Proceedings of International Conference on Recent Trends in Computing

ICRTC 2021

Springer

Editors
Rajendra Prasad Mahapatra
SRM Institute of Science and Technology
Ghaziabad, India

Sateesh Kumar Peddoju
Indian Institute of Technology Roorkee
Haridwar, India

Sudip Roy
Indian Institute of Technology Roorkee
Haridwar, India

Pritee Parwekar
SRM Institute of Science and Technology
Ghaziabad, India

Lavika Goel
Malaviya National Institute of Technology
Jaipur, India

ISSN 2367-3370 ISSN 2367-3389 (electronic)
Lecture Notes in Networks and Systems
ISBN 978-981-16-7117-3 ISBN 978-981-16-7118-0 (eBook)
https://doi.org/10.1007/978-981-16-7118-0

This Springer imprint is published by the registered company Springer Nature Singapore Pte Ltd.
The registered company address is: 152 Beach Road, #21-01/04 Gateway East, Singapore 189721, Singapore

Preface

This *Lecture Notes in Networks and Systems* (LNNS) volume contains the papers presented at the AICTE sponsored 9th International Conference on Intelligent Computing and Applications (ICRTC 2021) held during June 4–5, 2021, at SRM Institute of Science and Technology, Delhi-NCR Campus, Modinagar, Ghaziabad, India.

ICRTC 2021 is aiming at bringing together the researchers from academia and industry to report and review the latest progress in the cutting-edge research on various research areas of image processing, computer vision and pattern recognition, machine learning, data mining, big data and analytics, soft computing, mobile computing and applications, cloud computing, green IT and finally to create awareness about these domains to a wider audience of practitioners.

ICRTC 2021 received 310 paper submissions including two submissions from foreign countries. All the papers were peer-reviewed by the experts in the area from India and abroad and comments sent to the authors of accepted papers. Finally, 70 papers were accepted for online zoom presentation in the conference. This corresponds to an acceptance rate of 34% that is intended to maintain the high standards of the conference proceedings. The papers included in this Lecture Notes in Networks and Systems (LNNS) volume cover a wide range of topics in intelligent computing and algorithms and their real-time applications in problems from diverse domains of science and engineering.

The conference was inaugurated by Prof. Milan Tuba, Professor, Singidunum University, Serbia, on June 4, 2021. The conference featured distinguished keynote speakers as follows: Prof. Sheng-Lung Peng, National Tsing Hua University, Taiwan; Prof. Mufti Mahmud, Nottingham Trent University, Research Guidance by Shri. Aninda Bose, Senior Editor, Springer New Delhi, India; Address by Chief Guest Prof. Mike Hinchey, University of Limerick, Ireland.

We take this opportunity to thank the authors of the submitted papers for their hard work, adherence to the deadlines, and patience with the review process. The quality of a referred volume depends mainly on the expertise and dedication of the reviewers. We are indebted to the technical committee members, who produced excellent reviews in short time frames. First, we are indebted to the Hon'ble

Dr. T. R. Paari Vendhar, Member of Parliament (Lok Sabha), Founder-Chancellor, SRM Institute of Science and Technology, Shri. Ravi Pachamoothoo, Pro-Chancellor—Administration, SRM Institute of Science and Technology, Dr. P. Sathyanarayanan, Pro-Chancellor—Academics, SRM Institute of Science and Technology, Dr. R. Shivakumar, Vice President, SRM Institute of Science and Technology, Prof. C. Muthamizhchelvan, Vice Chancellor i/c, SRM Institute of Science and Technology for supporting our cause and encouraging us to organize the conference there. In particular, we would like to express our heartfelt thanks for providing us with the necessary financial support and infrastructural assistance to hold the conference. Our sincere thanks to Dr. D. K. Sharma, Professor and Dean; Dr. S. Viswanathan, Deputy Registrar; Dr. Navin Ahalawat, Professor and Dean (Campus Life), SRM Institute of Science and Technology, Delhi-NCR Campus, Modinagar, Ghaziabad, for their continuous support and guidance. We specially thank Dr. Dambarudhar Seth, Professor, Dr. Pritee Parwekar, Associate Professor, and Dr. Veena Khandelwal, Associate Professor, Co-Converners-ICRTC 2021, SRM Institute of Science and Technology, Delhi-NCR Campus, of this conference for their excellent support and arrangements. Without them it is beyond imagination to conduct this conference. We thank the international advisory committee members for providing valuable guidelines and inspiration to overcome various difficulties in the process of organizing this conference. We would also like to thank the participants of this conference. The faculty members and students of SRM Institute of Science and Technology, Delhi-NCR Campus, Modinagar, Ghaziabad, deserve special thanks. Without their involvement, we would not have been able to face the challenges of our responsibilities. Finally, we thank all the volunteers who made great efforts in meeting the deadlines and arranging every detail to make sure that the conference could run smoothly. We hope the readers of these proceedings find the papers inspiring and enjoyable.

Ghaziabad, India — Dr. Rajendra Prasad Mahapatra
Haridwar, India — Dr. Sateesh Kumar Peddoju
Haridwar, India — Dr. Sudip Roy
Ghaziabad, India — Dr. Pritee Parwekar
Jaipur, India — Dr. Lavika Goel
June 2021

Contents

Hybrid Precoding Based on Machine Learning for Millimeter Wave System 1
Divya Singh

Applying Predictive Analysis Methods for Detection of Driver Drowsiness 9
Abra Shafiq Siddiqi, Sherin Zafar, M. Afshar Alam, Samia Khan, Nida Iftekhar, and Siddhartha Sankar Biswas

Comparative Analysis of Various Supervised Machine Learning Techniques Used for Sentiment Analysis on Tourism Reviews 19
Manoj Kumar Sahu and Smita Selot

An Enhanced Machine Learning Classification System to Investigate the Status of Micronutrients in Rural Women 51
V. V. R. Maheswara Rao, N. Silpa, Gadiraju Mahesh, and Shiva Shankar Reddy

Advanced Hydroponics Fodder System with Temperature Control Using Arduino 61
Ahbab Rizvi, Aditi Raj, Anishrawa Sharma, and Vinesh Kumar

Multicriteria-Based Trustworthiness Testing of Web Sites 73
Aanjey Mani Tripathi, Arunima, and Anjali Anand

Design and Implementation of e-learning Platform Using Data Analysis 81
Sunil Kumar, Aanjey Mani Tripathi, Hansika Bhatia, Gurneet Kaur, Daksh Aggarwal, and Divyansh Chauhan

Prediction and Detection of COVID-19 Using Machine Learning 91
Diksha Goel, Mani Vats, Ayush, Priyanshi Baliyan, and Punit Mittal

Music Therapy for Transforming Human Negative Emotions: Deep Learning Approach ... 99
S. G. Shaila, T. M. Rajesh, S. Lavanya, K. G. Abhishek, and V. Suma

Interest-Based News Feed ... 111
Utkarsh Gupta, Ayush Saini, and Ankush Gupta

Gesture Decoder for Communication Using Deep Learning ... 121
Mukul Dayal, Mohak Goel, Apoorv Garg, and Ajay Kumar Singh

Vehicle Owner Identification from Number Plate ... 131
Harsh Vashishtha, Gopal Sharma, Abhishek Malik, Shweta Paliwal, and Aanjey Mani Tripathi

Smart Monitoring of Drunk Driver Using IOT and Machine Learning Based Anomaly Detection ... 139
Sujatha Karimisetty, A. Arjuna Rao, Y. Dinesh Kumar, and P. Kumudini Devi

An Effective and Innovative Framework with Authentication Schema and Key Generation in Internet of Things ... 147
S. Karthick and N. Gomathi

Sarcastic Sentiment Detection and Polarity Classification of Tweets Using Supervised Learning ... 161
M. S. M. Prasanna, S. G. Shaila, A. Vadivel, Mahima Mittal, K. Hithyshi, and H. R. Pooja Shree

Visualization of Online Social Dynamics for Forensic Investigation of User's Behavior ... 173
Govind Verma, Manash Sarkar, and Dambarudhar Seth

Bi-CRS: Bio-Inspired Cluster-Based Routing Scheme for D2D Communication in IoT ... 187
M. B. Yashoda and Vrinda Shivashetty

A Cloud-based Trusted Framework for Industrial Connected Vehicles ... 201
M. Deva Priya, G. Amirthavarsini, S. Angu Kaushika, and K. Deeptheshanmathie

Classification of COVID-19 Tweets Using Deep Learning Classifiers ... 213
M. Deva Priya, M. Saranya, N. Sharaha, and S. Tamizharasi

Neural Network-Based Method to Predict PCOS in Women ... 227
Asis Kaur Baweja and Megha Gupta

Integrated IoT Blockchain-Based Smart Agriculture System ... 237
M. Deva Priya, P. Anantha Prabha, K. Gokulakrishnan, A. Joe Raymond, and T. Karthickraja

Feedback Investigation on Twitter Dataset Using Classification Approaches 251
Yogesh Kumar, Sameeka Saini, Harendra Sharma, Ritu Payal, and Arpit Mishra

Detecting the Change in Microgrid Using Pattern Recognition and Machine Learning 263
S. Ramana Kumar Joga, Amit Kumar, Pampa Sinha, and Manoj Kumar Maharana

Electroencephalogram Signal Processing with Independent Component Analysis and Cognitive Stress Classification Using Convolutional Neural Networks 275
Venkatakrishnan Sutharsan, Alagappan Swaminathan, Saisrinivasan Ramachandran, Madan Kumar Lakshmanan, and Balaji Mahadevan

Perceiving Machine Learning Algorithms to Analyze COVID-19 Radiographs 293
Shilpa Hari Prakash, Adithya Narayan K. A., Gouri S. Nair, and Sandhya Harikumar

Identification of Anomalies in Images Using CNN and Autoencoders Techniques 307
V. Navya and Kavita Avinash Patil

Big Data Privacy Preserved of END to END Delivery Solution with TRIZ Mechanism 317
Ashutosh Dixit, Nidhi Tyagi, and Rajeev Kumar

Heart Disease Prediction Using Various Machine Learning Algorithms 325
Arshdeep Singh Chudhey, Aditya Sharma, and Mrityunjay Singh

Home Security Using Facial Recognition System 337
Vibhu Dixit, Srishti Agarwal, and Nidhi Tyagi

Mechanism for Big Data Security Related to GMPLS/MPLS Networks 347
Rajani Singh and Ashutosh Dixit

Deep Learning Algorithms based Vehicle Mobility Prediction from Satellite Imagery During Pandemic 359
M. Deva Priya, A. Sahaya Gebin, S. Selva Kumar, and R. G. Vipin

Textual Posts Analysis Using Voting Classifier 371
Naman Goel, Yash Rastogi, Yash Verma, and Rajesh Kumar Singh

People Monitoring and Social Distancing Using Computer Vision 381
Nidhi Tyagi, Riya Upadhyay, Vatsal Gupta, and Shekhar Chauhan

Radiological Image Synthesis Using Cycle-Consistent Generative Adversarial Network 391
Rahul Nehra, Abhisikta Pal, and B. Baranidharan

Mouse Assistance for Motor-Disabled People Using Computer Vision 403
P. Anantha Prabha, K. Srinivash, S. Vigneshwar, and E. Viswa

Two-Way Economical Smart Device Control and Power Consumption Prediction System 415
P. Anantha Prabha, N. Arjun, J. Gogul, and S. Divya Prasanth

Image Colorization Using Deep Convolution Autoencoder 431
Vaibhav Singh, Deepak Arora, and Puneet Sharma

Binary Tree-Based Asymmetric Moderate Algorithm for Secured DNS 443
Puneet Sharma, Deepak Arora, and Anil Kumar

Multimodal Analysis of TED Talks Toward Designing a Recommendation System 455
M. N. Renuka Devi and Gowri Srinivasa

Effective Resource Allocation in Network Slicing Using Artificial Intelligence 469
Anushka Svara and Ramkumar Jayaraman

Pulmonary Image Analysis with Computer Vision Using Binary Classification Method for COVID-19 Detection 483
S. V. N. S. Vaibhav Ram and Samiappan Dhanalakshmi

Bone Fracture Detection Using Image Processing Methods 493
M. Uma Devi, Sweta Nayak, and Aman Gupta

An Inventory Model with Pareto Distribution Deterioration with Weibull Demand Rate and Time-Dependent Shortages 503
S. Sindhuja and P. Arathi

Performance Evaluation of Full Adder Using Magnetic Tunnel Junction 517
Jyoti Garg and Subodh Wairya

Intrusion Detection System and Vulnerability Identification Using Various Machine Learning Algorithms 527
Kishor P. Jadhav and Mohit Gangwar

Multi-entity Topic Modeling and Aspect-Based Sentiment Classification Using Machine Learning Approach 537
Ganesh N. Jorvekar and Mohit Gangwar

Demographic Filtering for Movie Recommendation System Using Machine Learning 549
Sanket Rastogi, Divyanshu Agarwal, Jatin Jain, and K. P. Arjun

Kubernetes Continuous Development 559
Sakshi Sharma, Ayush Shankar Pathak, Harshit Malhotra, Soumalya Ghosh, and Garima Pandey

Movie Recommendation and Mood Prediction System—Leisures 569
Poornima Dave, Shashi Bhushan Singh, Divyansh Chaurasia, and Avneesh Kumar

Brain Tumor Classification Using Convolution Neural Network 579
Preeti Sharma and Anand Prakash Shukla

Perceptual Energy Weight Matrix Based Adaptive Block Compressed Sensing for Marine Image Compression 589
R. Monika, Samiappan Dhanalakshmi, and R. Kumar

Self-Adaptive Cuckoo Search-Based Cluster Head Selection for Maximizing Network Lifetime in Wireless Sensor Networks 599
G . Rajeswarappa and S. Vasundra

State-Wise Analysis and Prediction of Covid-19 in India 613
Kritika Malhotra and Amit Prakash Singh

Generation of One-Time Password for the Authentication of Software Requirements Using Secure Hash Algorithms 627
Javed Ahmad, Chaudhary Wali Mohammad, and Mohd. Sadiq

QAudious: A Question Answering System for Academic Lecture Recordings 641
Dipti Pawade, Avani Sakhapara, Swati Pandey, Harsh Vasa, Kajal Shethia, and Sana Pagarkar

Summarization of Tweets Related to Disaster 651
Avani Sakhapara, Dipti Pawade, Baiju Dodhia, Jigar Jain, Omkar Bhosale, and Onkar Chakrawar

Analysing Effectiveness of ML Algorithms Used to Predict Diabetes Mellitus 667
Archit Sharma, Ashwani Raj, and Ramkumar Jayaraman

Enhancing the Decentralized Application (Dapp) for E-commerce by Using the Ethereum Blockchain 679
P. Shamili and B. Muruganantham

Efficient Machine Learning-Based Diagnosis System for Breast Cancer 695
M. Deva Priya, J. Sangeetha Priya, S. Sam Peter, Viraja Ravi, S. Karthik, and R. Karthik

Deep Learning-based Stock Market Prediction 709
A. Christy Jeba Malar, M. Deva Priya, M. Kavin Kumar,
S. Mangala Arunsankar, K. V. Bilal, and S. Karthik

Neural Machine Translation Using Attention Mechanism 717
Sai Yashwanth Velpuri, Sonakshi Karanwal, and R. Anita

Implementation of Automatic Detection of Text from Complex Images and Converting to Semantic Speech 731
Gorli Santoshi, Y. Srinivas, and Mullapudi Sri Harsha

Various Security Problems and Its Solving for Future Dynamic IoT-Based Smart Home Automation 739
S. Karthick, N. Gomathi, R. P. Mahapatra, Anitha Rajakumari,
and Pritee Parwekar

Machine Vision for Intelligent Vehicles in India 761
Vatsal Verma, Srishti Ahlawat, and Darpan Khanna

An Efficient Contour Detection Approach for Extracting Rim from Wheel Images 771
D. Karthik, P. Mirunalini, R. Priyadharsini, and T. T. Mirnalinee

Design of GaN HEMT Based Down-Convertor Passive Mixer from S Band to X Band 781
Manoj Kumar Vishnoi and Satya Sai Srikant

Psychological Impacts of Covid-19 on Human Health 787
Md Adnan Baig and Ihtiram Raza Khan

Improved Symbiosis Organism Search Algorithm-Based Clustering Scheme for Enhancing Longevity in Wireless Sensor Networks (WSNs) 799
Sengathir Janakiraman and Bandi Rambabu

A Novel 3D Facial Recognition for Digital Payments 809
K. C. Prabu Shankar and M. Hema

Denoising of Thermal Images Using Deep Neural Network 827
Ruchika Thukral, A. S. Arora, Ashwani Kumar, and Gulshan

Author Index 835

Editors and Contributors

About the Editors

Prof. Rajendra Prasad Mahapatra is currently working at SRM Institute of Science & Technology, DELHI-NCR Campus as Professor and Head—Computer Science & Engineering and Dean Admissions. Prof. R. P. Mahapatra has vast experience of 20 years as an academician, researcher and administrator. During these 20 years, he has worked India and abroad. He has been associated with Mekelle University, Ethopia for more than 2 years. Six candidates have successfully completed their Ph.D. under his supervision and 8 students are perusing their research under his guidance. Prof. R. P. Mahapatra has authored more than 70 research papers which are published in international journals like Inderscience, Emerald, Elsevier, IEEE and Springer. He has authored three books, five book chapters and two granted patents. He is a fellow member of I.E. (India), Senior member of IACSIT, Singapore, Life member of ISTE, Member of IEEE and many more reputed bodies. He is a recipient of the Institution Award from Institution of Engineers (India), Calcutta on 56th Annual Technical Session on 15 February 2015, Madhusudan Memorial Award from Institution of Engineers (India), Calcutta on 57th Annual Technical Session on 14 February 2016, Certificate of Excellence from SRM University, NCR Campus on 9 February 2015.

Sateesh Kumar Peddoju is an Associate Professor at Indian Institute of Technology Roorkee, India. He is the Senior Member of ACM and IEEE, and recipient of Cloud Ambassador Award from AWS Educate, IBM SUR Award, Microsoft Educate Award, university merit scholarship, best teacher award in his previous employment, best paper/presentation awards, and travel grants. He has publications in reputed journals like IEEE TIFS, IEEE Access, IEEE Potentials, MTAP, WPC, IJIS, and PPNA, and conferences like ACM MobiCom, IEEE TrustCom, IEEE MASS, ACM/IEEE ICDCN, and ISC. He is the co-author of the book *Security and Storage Issues in the Cloud Environment* and co-editor of the book *Cloud Computing Systems and Applications in Healthcare*. Sateesh is serving as a reviewer for IEEE TCC, IEEE

TSC, IEEE Access, MTAP, COSE, COMNET, JNCA, and CLUS. He is the Program Chair for IEEE MASS 2020 and founding Steering Committee Chair for SLICE 2018. He is on board for IEEE SLICE, IEEE DSEA, IEEE MASS, IEEE ATC, IEEE SmartComp, IEEE iNIS, and IoTSMS. He has received grants from NMHS, MEITY, Railtel, MHRD, DST, IBM, Samsung, CSI, Microsoft. He is guiding many interns, bachelor's, master's, doctoral students. His research interests include Cloud Computing (Cloud, Edge, and Fog), Ubiquitous Computing (Mobile, Pervasive, and IoT), and Security (Network, Cloud, IoT, Mobile, and Blockchain).

Dr. Sudip Roy is an Assistant Professor in the Department of Computer Science and Engineering of Indian Institute of Technology (IIT), Roorkee, India since July 2014. He is also an associated faculty member of the Centre of Excellence in Disaster Mitigation and Management (CoEDMM) in IIT Roorkee, India since April 2015. He is a JSPS fellow (long-term) in the College of Information Science and Engineering, Ritsumeikan University, Japan, during April 2021 to January 2022. He has authored one book, one book chapter, two granted U.S. patents including 25 research articles in international peer-reviewed journals and 40 papers in international peer-reviewed conference proceedings. His current research interests include computer-aided-design for digital systems, electronic design automation (EDA) for microfluidic lab-on-a-chips, algorithm design, optimization techniques, information and communication technologies (ICT) for disaster risk reduction (DRR). He is a recipient of JSPS Invitational Fellowship Award (Long-Term) from the Japan Society for the Promotion of Science (JSPS), Govt. of Japan in 2021, Early-Career Research Award from Department of Science and Technology, Govt. of India in 2017 and Microsoft Research India Ph.D. Fellowship Award in 2010. He is a member of IEEE and ACM.

Dr. Pritee Parwekar is an Associate Professor in Department of Computer Science Engineering, Faculty of Engineering and Technology. She has been an academician from last 20 years. She has been awarded Ph.D. in the area of Wireless Sensor Networks. She holds a life membership of IEEE, ACM, CSI, ISTE. She has also contributed as Computer Society of India (CSI) State Student Coordinator for two years. Her research interests include Internet of Things, Cloud Computing, Machine Learning, Wireless Sensor Networks, Software Engineering, Information Retrieval Systems, Social Media, Data Mining. She has published more than 50 research papers in peer-reviewed journals with SCI, SSCI, Scopus Index, and in conference proceeding of Springer and IEEE. She has been a resource person to many Workshops, FDPs and International Conferences. She has been program chair and organizing chair of many international conferences. She has been invited to many workshops and international conferences as organizer, session chair, and member in Advisory/Program committees. She has peer-reviewed IEEE Network Magazine, Springer, IGI Global, Inderscience, Evolutionary Intelligence (EVIN) Journals, Personal and Ubiquitous Computing, Multimedia Tools and Applications, Expert Systems and many more international journals. She has guided many M.Tech. Thesis, B.Tech. Major and Minor Projects. Presently she is guiding four Ph.D. scholars.

Dr. Lavika Goel is currently an Assistant Professor in the Department of Computer Science and Engineering at Malaviya National Institute of Technology (NIT), Jaipur, Rajasthan. She earlier worked at Birla Institute of Technology and Science, Pilani for 4.5 years during July 2015–January 2020. She holds a corporate experience working at Oracle India Private Ltd. She is Ph.D., M.E. and B.Tech. with distinction in Computer Science and Engineering from Delhi Technological University. She was awarded the prestigious 'Young Scientist award' by VIFRA International Foundation in 2015. She has been endorsed by Prof. Ajith Abraham, who is the pioneering author of hybrid intelligent Systems and Director, Machine Intelligence & Research (MIR) Labs, USA on Researchgate. She also coauthored a paper with him which Prof. Abraham himself presented in Mexico in 2012. She is offered to write two books currently for Cambridge University Press and Wiley India, one from her Ph.D. research work and the other which will be a text book on Artificial Intelligence for undergraduate students. She recently received in May 2018, approval of external funding of Rs. 144.32 lakhs as Principal Investigator from Department of Science and Technology (DST), NRDMS division and of Rs. 40 lakhs as Co-Principal Investigator from Department of Science and Technology (DST), ICPS division for her novel ideas of using nature inspired heuristics for effectively developing a recommender system for two different applications, one for designing a question answering system and other for crop selection in Rajasthan, India. Based on her background assessment as outstanding researcher and current performance, she was awarded the key role as the Head of Research Initiatives and projects at BITS Pilani, for coordinating the award of sponsored projects from government agencies. In addition to this, she is a true team leader and she is one of the founding members of a popularly growing research group, called as the Computational Intelligence and Research Group (CiRG), now a registered national society of importance where she serves as the Quality Assurance Head and Technical Core Committee member. She is also the Director and founding member of Sports Engineering Association (SEA) of India where she is focusing on improving sports through AI. She has published around 60 research papers in SCI indexed International conferences, journals and book chapters including Elsevier, IEEE, Taylor & Francis, Springer, etc.

Contributors

K. G. Abhishek Dayananda Sagar University, Bangalore, Karnataka, India

K. A. Adithya Narayan Department of Computer Science and Engineering, Amrita Vishwa Vidyapeetham, Amritapuri, India

Divyanshu Agarwal Computer Science and Engineering, Galgotias University Greater, Noida, Uttar Pradesh, India

Srishti Agarwal Meerut Institute of Engineering & Technology, Meerut, India

Daksh Aggarwal Department of Computer Science and Engineering, Meerut Institute of Engineering and Technology, Meerut, India

Srishti Ahlawat Department of Computer Science and Engineering, SRM Institute of Science and Technology, Modinagar, India

Javed Ahmad Department of Applied Sciences and Humanities, Faculty of Engineering and Technology, Jamia Millia Islamia (A Central University), New Delhi, India

M. Afshar Alam Jamia Hamdard, New Delhi, India

G. Amirthavarsini Department of Computer Science and Engineering, Sri Krishna College of Technology, Coimbatore, Tamilnadu, India

Anjali Anand Department of Computer Science and Engineering, Meerut Institute of Engineering and Technology, Meerut, UP, India

P. Anantha Prabha Department of Computer Science and Engineering, Sri Krishna College of Technology, Coimbatore, Tamilnadu, India

S. Angu Kaushika Department of Computer Science and Engineering, Sri Krishna College of Technology, Coimbatore, Tamilnadu, India

R. Anita Department of Computer Science and Engineering, SRM Institute of Science and Technology, Kattankulathur, Chennai, Tamil Nadu, India

P. Arathi Department of Mathematics and Actuarial Science, B.S. Abdur Rahman Crescent Institute of Science and Technology, Chennai, India

K. P. Arjun Computer Science and Engineering, Galgotias University Greater, Noida, Uttar Pradesh, India

N. Arjun Sri Krishna College of Technology, Coimbatore, Tamil Nadu, India

A. Arjuna Rao Miracle Educational Society Group of Institutions, Vizianagram, India

A. S. Arora Department of Electrical and Instrumentation Engineering, Sant Longowal Institute of Engineering and Technology, Longowal, Punjab, India

Deepak Arora Department of Computer Science and Engineering, Amity University, Lucknow, Uttar Pradesh, India

Arunima Department of Computer Science and Engineering, Meerut Institute of Engineering and Technology, Meerut, UP, India

Ayush Department of Computer Science and Engineering, MIET, Meerut, India

Md Adnan Baig Department of Computer Science, Jamia Hamdard, Delhi, India

Priyanshi Baliyan Department of Computer Science and Engineering, MIET, Meerut, India

B. Baranidharan SRM Institute of Science and Technology, Kattankulathur, Chennai, India

Asis Kaur Baweja Department of CSE, SRM KTR, Chennai, India

Hansika Bhatia Department of Computer Science and Engineering, Meerut Institute of Engineering and Technology, Meerut, India

Omkar Bhosale Department of Information Technology, K. J. Somaiya College of Engineering, Vidyavihar, Mumbai, India

Shashi Bhushan Singh Computer Science and Engineering, Galgotias University, Uttar Pradesh, Greater Noida, India

K. V. Bilal Department of Information Technology, Sri Krishna College of Technology, Coimbatore, Tamilnadu, India

Siddhartha Sankar Biswas Jamia Hamdard, New Delhi, India

Onkar Chakrawar Department of Information Technology, K. J. Somaiya College of Engineering, Vidyavihar, Mumbai, India

Divyansh Chauhan Department of Computer Science and Engineering, Meerut Institute of Engineering and Technology, Meerut, India

Shekhar Chauhan Student, Meerut Institute of Engineering & Technology, Meerut, India

Divyansh Chaurasia Computer Science and Engineering, Galgotias University, Uttar Pradesh, Greater Noida, India

A. Christy Jeba Malar Department of Information Technology, Sri Krishna College of Technology, Coimbatore, Tamilnadu, India

Arshdeep Singh Chudhey Department of Computer Science and Engineering, Jaypee University Of Information Technology, Waknaghat, India

Poornima Dave Computer Science and Engineering, Galgotias University, Uttar Pradesh, Greater Noida, India

Mukul Dayal Department of Computer Science and Engineering, MIET, Meerut, India

K. Deeptheshanmathie Department of Computer Science and Engineering, Sri Krishna College of Technology, Coimbatore, Tamilnadu, India

M. Deva Priya Department of Computer Science and Engineering, Sri Krishna College of Technology, Coimbatore, Tamil Nadu, India

Samiappan Dhanalakshmi Department of Electronics and Communications Engineering, Faculty of Engineering and Technology, SRM Institute of Science and Technology, College of Engineering and Technology, Kancheepuram, Tamil Nadu, India; Department of ECE, College of Engineering and Technology, Faculty of Engineering and Technology, SRM Institute of Science and Technology, Kattankulathur, Tamil Nadu, India

Y. Dinesh Kumar Department of CSE, Satya Institute of Technology and Management, Vizianagram, India

S. Divya Prasanth Sri Krishna College of Technology, Coimbatore, Tamil Nadu, India

Ashutosh Dixit Bhagwant University, Ajmer, Rajasthan, India;
IIMT College of Management Greater Noida, Noida, U.P, India

Vibhu Dixit Meerut Institute of Engineering & Technology, Meerut, India

Baiju Dodhia Department of Information Technology, K. J. Somaiya College of Engineering, Vidyavihar, Mumbai, India

Mohit Gangwar Engineering, Bhabha University, Bhopal, Madhya Pradesh, India

Apoorv Garg Department of Computer Science and Engineering, MIET, Meerut, India

Jyoti Garg Department of Electronics Engineering, Dr. A.P.J. Abdul Kalam Technical University, Lucknow, India

Soumalya Ghosh School of Computer Science and Engineering, Galgotias University, Greater Noida, India

Diksha Goel Department of Computer Science and Engineering, MIET, Meerut, India

Mohak Goel Department of Computer Science and Engineering, MIET, Meerut, India

Naman Goel Department of Computer Science Engineering, Meerut Institute of Engineering and Technology (MIET), Meerut, U.P, India

J. Gogul Sri Krishna College of Technology, Coimbatore, Tamil Nadu, India

K. Gokulakrishnan Department of Computer Science and Engineering, Sri Krishna College of Technology, Coimbatore, Tamilnadu, India

N. Gomathi Department of Computer Science and Engineering, Vel Tech Rangarajan Dr.Sagunthala R&D Institute of Science and Technology, Chennai, Tamilnadu, India

Gulshan Department of Electrical and Instrumentation Engineering, Sant Longowal Institute of Engineering and Technology, Longowal, Punjab, India

Aman Gupta Department of Computer Science and Engineering, SRM Institute of Science and Technology, Chennai, Tamil Nadu, India

Ankush Gupta Department of Computer Science and Engineering, M.I.E.T, Meerut, India

Megha Gupta Department of CS, MSCW, University of Delhi, Delhi, India

Utkarsh Gupta Department of Computer Science and Engineering, M.I.E.T, Meerut, India

Vatsal Gupta Student, Meerut Institute of Engineering & Technology, Meerut, India

Shilpa Hari Prakash Department of Computer Science and Engineering, Amrita Vishwa Vidyapeetham, Amritapuri, India

Sandhya Harikumar Department of Computer Science and Engineering, Amrita Vishwa Vidyapeetham, Amritapuri, India

Mullapudi Sri Harsha ANITS Engineering College, Vishakhapatnam, India

M. Hema Department of Computer Science and Engineering, SRM Institute of Science and Technology, Chennai, Tamil Nadu, India

K. Hithyshi Dayananda Sagar University, Bangalore, Karnataka, India

Nida Iftekhar Jamia Hamdard, New Delhi, India

Kishor P. Jadhav Department of Computer Science and Engineering, Bhabha University, Bhopal, Madhya Pradesh, India

Jatin Jain Computer Science and Engineering, Galgotias University Greater, Noida, Uttar Pradesh, India

Jigar Jain Department of Information Technology, K. J. Somaiya College of Engineering, Vidyavihar, Mumbai, India

Sengathir Janakiraman Hyderabad, India

Ramkumar Jayaraman Department of Computing Technologies, SRM Institute of Science and Technology, Kattankulathur, Chennai, India

A. Joe Raymond Department of Computer Science and Engineering, Sri Krishna College of Technology, Coimbatore, Tamilnadu, India

S. Ramana Kumar Joga School of Electrical Engineering, Kalinga Institute of Industrial Technology, Bhubaneswar, India

Ganesh N. Jorvekar Department of Computer Science and Engineering, Bhabha University, Bhopal, Madhya Pradesh, India

Sonakshi Karanwal Department of Computer Science and Engineering, SRM Institute of Science and Technology, Kattankulathur, Chennai, Tamil Nadu, India

Sujatha Karimisetty Department of CSE, Dadi Institute of Engineering and Technology, Visakhapatnam, India

S. Karthick Department of Computer Science and Engineering, Vel Tech Rangarajan Dr.Sagunthala R&D Institute of Science and Technology, Chennai, Tamilnadu, India

T. Karthickraja Department of Computer Science and Engineering, Sri Krishna College of Technology, Coimbatore, Tamilnadu, India

D. Karthik Sri Sivasubramaniya Nadar College of Engineering, Chennai, Tamil Nadu, India

R. Karthik Department of Computer Science and Engineering, Sri Krishna College of Technology, Coimbatore, Tamil Nadu, India

S. Karthik Department of Computer Science and Engineering, Delhi-NCR Campus, SRM Institute of Science and Technology, Ghaziabad, India

Gurneet Kaur Department of Computer Science and Engineering, Meerut Institute of Engineering and Technology, Meerut, India

M. Kavin Kumar Department of Information Technology, Sri Krishna College of Technology, Coimbatore, Tamilnadu, India

Ihtiram Raza Khan Department of Computer Science, Jamia Hamdard, Delhi, India

Samia Khan Jamia Hamdard, New Delhi, India

Darpan Khanna Department of Computer Science and Engineering, SRM Institute of Science and Technology, Modinagar, India

Amit Kumar School of Electrical Engineering, Kalinga Institute of Industrial Technology, Bhubaneswar, India

Anil Kumar Department of Computer Science and Engineering, Amity University, Uttar Pradesh, India

Ashwani Kumar Department of Electrical and Instrumentation Engineering, Sant Longowal Institute of Engineering and Technology, Longowal, Punjab, India

Avneesh Kumar Computer Science and Engineering, Galgotias University, Uttar Pradesh, Greater Noida, India

R. Kumar Department of ECE, College of Engineering and Technology, Faculty of Engineering and Technology, SRM Institute of Science and Technology, Kattankulathur, Tamil Nadu, India

Rajeev Kumar Teerthanker Mahaveer University, Moradabad, India

Sunil Kumar Department of Computer Science and Engineering, Meerut Institute of Engineering and Technology, Meerut, India

Vinesh Kumar Department of Computer Science and Engineering, Meerut Institute of Engineering and Technology, Meerut, UP, India

Yogesh Kumar Dev Bhoomi Institute of Technology, Dehradun, India

P. Kumudini Devi Department of CSE, Dadi Institute of Engineering and Technology, Visakhapatnam, India

Madan Kumar Lakshmanan CSIR—Central Electronics Engineering Research Institute, Pilani, Rajasthan, India

S. Lavanya Dayananda Sagar University, Bangalore, Karnataka, India

Balaji Mahadevan Department of Electrical and Electronics Engineering, SSN College of Engineering, Chennai, India

R. P. Mahapatra Department of Computer Science and Engineering, SRM Institute of Science and Technology, Delhi-NCR Campus, Ghaziabad, India

Manoj Kumar Maharana School of Electrical Engineering, Kalinga Institute of Industrial Technology, Bhubaneswar, India

Gadiraju Mahesh Department of Computer Science and Engineering, S.R.K.R. Engineering College, Bhimavaram, Andhra Pradesh, India

V. V. R. Maheswara Rao Department of Computer Science and Engineering, Shri Vishnu Engineering College for Women(A), Bhimavaram, Andhra Pradesh, India

Harshit Malhotra Bachelor of Computer Science and Engineering, Galgotias University, Greater Noida, India

Kritika Malhotra University School of Information, Communication and Technology, Guru Gobind Singh Indraprastha University, Delhi, India

Abhishek Malik Department of Computer Science and Engineering, Meerut Institute of Engineering and Technology, Meerut, India

S. Mangala Arunsankar Department of Information Technology, Sri Krishna College of Technology, Coimbatore, Tamilnadu, India

Aanjey Mani Tripathi Department of Computer Science and Engineering, Meerut Institute of Engineering and Technology, Meerut, UP, India

T. T. Mirnalinee Sri Sivasubramaniya Nadar College of Engineering, Chennai, Tamil Nadu, India

P. Mirunalini Sri Sivasubramaniya Nadar College of Engineering, Chennai, Tamil Nadu, India

Arpit Mishra Dev Bhoomi Institute of Technology, Dehradun, India

Mahima Mittal Dayananda Sagar University, Bangalore, Karnataka, India

Punit Mittal Department of Computer Science and Engineering, MIET, Meerut, India

Chaudhary Wali Mohammad Department of Applied Sciences and Humanities, Faculty of Engineering and Technology, Jamia Millia Islamia (A Central University), New Delhi, India

R. Monika Department of ECE, College of Engineering and Technology, Faculty of Engineering and Technology, SRM Institute of Science and Technology, Kattankulathur, Tamil Nadu, India

B. Muruganantham Department of Engineering and Technology, SRM Institute of Science and Technology, Chennai, India

Gouri S. Nair Department of Computer Science and Engineering, Amrita Vishwa Vidyapeetham, Amritapuri, India

V. Navya Department of Electronics and Communication, East Point College of Engineering and Technology, Avalahalli, Bengaluru, Karnataka, India

Sweta Nayak Department of Computer Science and Engineering, SRM Institute of Science and Technology, Chennai, Tamil Nadu, India

Rahul Nehra SRM Institute of Science and Technology, Kattankulathur, Chennai, India

Sana Pagarkar Department of Information Technology, K. J. Somaiya College of Engineering, Vidyavihar, Mumbai, India

Abhisikta Pal SRM Institute of Science and Technology, Kattankulathur, Chennai, India

Shweta Paliwal Department of Computer Science and Engineering, Meerut Institute of Engineering and Technology, Meerut, India

Garima Pandey School of Computer Science and Engineering, Galgotias University, Greater Noida, India

Swati Pandey Department of Information Technology, K. J. Somaiya College of Engineering, Vidyavihar, Mumbai, India

Pritee Parwekar Department of Computer Science and Engineering, SRM Institute of Science and Technology, Delhi-NCR Campus, Ghaziabad, India

Ayush Shankar Pathak Bachelor of Computer Science and Engineering, Galgotias University, Greater Noida, India

Kavita Avinash Patil Department of Electronics and Communication, New Horizon College of Engineering, Marathahalli, Bengaluru, Karnataka, India

Dipti Pawade Department of Information Technology, K. J. Somaiya College of Engineering, Vidyavihar, Mumbai, India

Ritu Payal Dev Bhoomi Institute of Technology, Dehradun, India

H. R. Pooja Shree Dayananda Sagar University, Bangalore, Karnataka, India

K. C. Prabu Shankar Department of Computer Science and Engineering, SRM Institute of Science and Technology, Chennai, Tamil Nadu, India

M. S. M. Prasanna Dayananda Sagar University, Bangalore, Karnataka, India

R. Priyadharsini Sri Sivasubramaniya Nadar College of Engineering, Chennai, Tamil Nadu, India

Aditi Raj Department of Computer Science and Engineering, Meerut Institute of Engineering and Technology, Meerut, UP, India

Ashwani Raj Department of Computing Technologies, SRM Institute of Science and Technology, Kattankulathur, Chennai, India

Anitha Rajakumari Department of Computer Science and Engineering, SRM Institute of Science and Technology, Delhi-NCR Campus, Ghaziabad, India

T. M. Rajesh Dayananda Sagar University, Bangalore, Karnataka, India

G . Rajeswarappa Department of CSE, JNTUA, Ananthapuramu, India

Saisrinivasan Ramachandran Department of Electrical and Electronics Engineering, SSN College of Engineering, Chennai, India

Bandi Rambabu Hyderabad, India

Sanket Rastogi Computer Science and Engineering, Galgotias University Greater, Noida, Uttar Pradesh, India

Yash Rastogi Department of Computer Science Engineering, Meerut Institute of Engineering and Technology (MIET), Meerut, U.P, India

Viraja Ravi Department of Computer Science and Engineering, Sri Krishna College of Technology, Coimbatore, Tamil Nadu, India

Shiva Shankar Reddy Department of Computer Science and Engineering, S.R.K.R. Engineering College, Bhimavaram, Andhra Pradesh, India

M. N. Renuka Devi Dayanada Sagar University, Bengaluru, Karnataka, India

Ahbab Rizvi Department of Computer Science and Engineering, Meerut Institute of Engineering and Technology, Meerut, UP, India

Mohd. Sadiq Department of Computer Science and Automation, Indian Institute of Science Bangalore, Bangalore, Karnataka, India

A. Sahaya Gebin Department of Computer Science and Engineering, Sri Krishna College of Technology, Coimbatore, Tamilnadu, India

Manoj Kumar Sahu Shri Sankaracharya Technical Campus, Bhilai, Chhattisgarh, India

Ayush Saini Department of Computer Science and Engineering, M.I.E.T, Meerut, India

Sameeka Saini Dev Bhoomi Institute of Technology, Dehradun, India

Avani Sakhapara Department of Information Technology, K. J. Somaiya College of Engineering, Vidyavihar, Mumbai, India

S. Sam Peter Department of Computer Science and Engineering, Sri Krishna College of Technology, Coimbatore, Tamil Nadu, India

J. Sangeetha Priya Department of Computer Science and Engineering, Sri Krishna College of Technology, Coimbatore, Tamil Nadu, India

Gorli Santoshi ANITS Engineering College, Vishakhapatnam, India

M. Saranya Department of Computer Science and Engineering, Sri Krishna College of Technology, Coimbatore, Tamilnadu, India

Manash Sarkar Atria Institute of Technology, Bengaluru, Karnataka, India

Smita Selot Shri Sankaracharya Technical Campus, Bhilai, Chhattisgarh, India

S. Selva Kumar Department of Computer Science and Engineering, Sri Krishna College of Technology, Coimbatore, Tamilnadu, India

Dambarudhar Seth Department of Computer Science and Engineering, SRMIST, Ghaziabad, UP, India

S. G. Shaila Dayananda Sagar University, Bangalore, Karnataka, India

P. Shamili Department of Engineering and Technology, SRM Institute of Science and Technology, Chennai, India

N. Sharaha Department of Computer Science and Engineering, Sri Krishna College of Technology, Coimbatore, Tamilnadu, India

Aditya Sharma Department of Computer Science and Engineering, Jaypee University Of Information Technology, Waknaghat, India

Anishrawa Sharma Department of Computer Science and Engineering, Meerut Institute of Engineering and Technology, Meerut, UP, India

Archit Sharma Department of Computing Technologies, SRM Institute of Science and Technology, Kattankulathur, Chennai, India

Gopal Sharma Department of Computer Science and Engineering, Meerut Institute of Engineering and Technology, Meerut, India

Harendra Sharma SRM Institute of Science and Technology, Modinagar, Ghaziabad, India

Preeti Sharma Department of Computer Science and Engineering, KIET Group of Institutions, Ghaziabad, U.P., India

Puneet Sharma Department of Computer Science and Engineering, Amity University, Lucknow, Uttar Pradesh, India

Sakshi Sharma Bachelor of Computer Science and Engineering, Galgotias University, Greater Noida, India

Kajal Shethia Department of Information Technology, K. J. Somaiya College of Engineering, Vidyavihar, Mumbai, India

Vrinda Shivashetty Sai Vidya Institute of Technology, Bangalore, India;
VTU Belgaum, Belgaum, India

Anand Prakash Shukla Department of Computer Science and Engineering, KIET Group of Institutions, Ghaziabad, U.P., India

Abra Shafiq Siddiqi Jamia Hamdard, New Delhi, India

N. Silpa Department of Computer Science and Engineering, Shri Vishnu Engineering College for Women(A), Bhimavaram, Andhra Pradesh, India

S. Sindhuja Department of Mathematics and Actuarial Science, B.S. Abdur Rahman Crescent Institute of Science and Technology, Chennai, India

Ajay Kumar Singh Department of Computer Science and Engineering, MIET, Meerut, India

Amit Prakash Singh University School of Information, Communication and Technology, Guru Gobind Singh Indraprastha University, Delhi, India

Divya Singh Department of Electronics and Communication, GLA University, Mathura, India

Mrityunjay Singh Department of Computer Science and Engineering, Jaypee University Of Information Technology, Waknaghat, India

Rajani Singh IIMT College of Engineering Greater Noida, Noida, U.P, India

Rajesh Kumar Singh Department of Computer Science Engineering, Meerut Institute of Engineering and Technology (MIET), Meerut, U.P, India

Vaibhav Singh Department of Computer Science and Engineering, Amity University, Lucknow, India

Pampa Sinha School of Electrical Engineering, Kalinga Institute of Industrial Technology, Bhubaneswar, India

Satya Sai Srikant SRM Institute of Science and Technology, Delhi-NCR Campus, Modinagar, Ghaziabad, India

Y. Srinivas GITAM University, Vishakhapatnam, India

Gowri Srinivasa Department of Computer Science and Engineering, PES University, Bengaluru, Karnataka, India

K. Srinivash Sri Krishna College of Technology, Coimbatore, Tamil Nadu, India

V. Suma Dayananda Sagar University, Bangalore, Karnataka, India

Venkatakrishnan Sutharsan Department of Electrical and Electronics Engineering, SSN College of Engineering, Chennai, India

Anushka Svara Department of Computing Technologies, SRM Institute of Science and Technology, Kattankulathur, Chennai, India

Alagappan Swaminathan Department of Electrical and Electronics Engineering, SSN College of Engineering, Chennai, India

S. Tamizharasi Department of Computer Science and Engineering, Sri Krishna College of Technology, Coimbatore, Tamilnadu, India

Ruchika Thukral Department of Electrical and Instrumentation Engineering, Sant Longowal Institute of Engineering and Technology, Longowal, Punjab, India

Aanjey Mani Tripathi Department of Computer Science and Engineering, Meerut Institute of Engineering and Technology, Meerut, India

Nidhi Tyagi MIET, Meerut, U.P, India;
Shobhit Institute of Engineering & Technology, Meerut, India

M. Uma Devi Department of Computer Science and Engineering, SRM Institute of Science and Technology, Chennai, Tamil Nadu, India

Riya Upadhyay Student, Meerut Institute of Engineering & Technology, Meerut, India

A. Vadivel ICFAI University, ICFAITECH, FST Hyderabad, Hyderabad, Telangana, India

S. V. N. S. Vaibhav Ram Department of Electronics and Communications Engineering, Faculty of Engineering and Technology, SRM Institute of Science and Technology, College of Engineering and Technology, Kancheepuram, Tamil Nadu, India

Harsh Vasa Department of Information Technology, K. J. Somaiya College of Engineering, Vidyavihar, Mumbai, India

Harsh Vashishtha Department of Computer Science and Engineering, Meerut Institute of Engineering and Technology, Meerut, India

S. Vasundra Department of CSE, JNTUACEA, Ananthapuramu, India

Mani Vats Department of Computer Science and Engineering, MIET, Meerut, India

Sai Yashwanth Velpuri Department of Computer Science and Engineering, SRM Institute of Science and Technology, Kattankulathur, Chennai, Tamil Nadu, India

Govind Verma Department of Computer Science and Engineering, SRMIST, Ghaziabad, UP, India

Vatsal Verma Department of Computer Science and Engineering, SRM Institute of Science and Technology, Modinagar, India

Yash Verma Department of Computer Science Engineering, Meerut Institute of Engineering and Technology (MIET), Meerut, U.P, India

S. Vigneshwar Sri Krishna College of Technology, Coimbatore, Tamil Nadu, India

R. G. Vipin Department of Computer Science and Engineering, Sri Krishna College of Technology, Coimbatore, Tamilnadu, India

Manoj Kumar Vishnoi SRM Institute of Science and Technology, Delhi-NCR Campus, Modinagar, Ghaziabad, India

E. Viswa Sri Krishna College of Technology, Coimbatore, Tamil Nadu, India

Subodh Wairya Department of Electronics and Communication Engineering, IET, Uttar Pradesh, Lucknow, India

M. B. Yashoda Sai Vidya Institute of Technology, Bangalore, India;
VTU Belgaum, Belgaum, India

Sherin Zafar Jamia Hamdard, New Delhi, India

Hybrid Precoding Based on Machine Learning for Millimeter Wave System

Divya Singh

Abstract Millimeter wave (mmWave) large output input (MIMO) is considered to be the emergent key for the subsequent generation of wireless communication, where hybrid precoding is an important way to reduce cost complexity as well as power consumption linked with diverged signal components. Though, the basic restriction of current hybrid systems is that they fail to make full use of location data due to high computer complexity. A hybrid design is proposed in this paper for power generation, in which a lesser number of inverters and switches is available in the analog component as a replacement for a large extent of phase shifters. Subsequently, stimulated by cross-entropy performance in machine learning, the hybrid precoding scheme of this new structure is based on a systematic CE (ACE) system. It seeks to fundamentally revitalize the material distribution in the analog/digital precoder thereby reducing cross-entropy that can deliver a result with a substantially high probability close to the ideal one. The results of the simulation confirm that the proposed scheme will achieve an average value and energy efficiency far lower than that of conventional schemes.

Keywords Millimeter wave · Machine learning · Cross entropy · Hybrid precoding · 5G

1 Introduction

It is predicted that wireless data transmission would boost by 1000 times by 2020 and could increase by more than 10,000 by 2030 [1], therefore advancing the employment of the 5G theory to cope with the explosive data boom. The use of vast amounts of less widely used materials in ultra-high-frequency spectrum, such as millimeter-wave, is one of the most powerful ways to meet the needs of 5G [2]. One advancement of mmWave communications is that there is a tenfold rise in network traffic relative to

D. Singh (✉)
Department of Electronics and Communication, GLA University, Mathura, India
e-mail: divya.singh@gla.ac.in

R. P. Mahapatra et al. (eds.), *Proceedings of International Conference on Recent Trends in Computing*, Lecture Notes in Networks and Systems 341,
https://doi.org/10.1007/978-981-16-7118-0_1

existing wireless networks. In other words, mmWave signals bring space-free path loss [3] to order-of-magnitude enhancement.

However, most current hybrid precoding techniques involve a complex network shifter network in which all antennas with high-precision phase shifts are connected to each RF series [4, 5]. Although this technology is responsible for high design freedom to attain near-term efficiency, at high hardware costs and energy consumption, it needs many hundreds of high-grade transformation points [4]. There have recently been two types of schemes proposed to fix this problem. The first step is to explicitly use the shifts of the restricted resolution process instead of the upper phase adjustments [6, 7]. Consumed power can be reduced by adopting a huge number of phase shifters in phase shifter network deprived of apparent operational losses. The second trade is to install a phase shift network by using switch network [7–9]. It can considerably decrease the cost of hardware and energy use at the cost of operating losses.

The use of a switch-inverter (SI)—a built-in hybrid design—is implemented in this paper with very low hardware costs and power consumption. The analog portion of the projected properties is obtained with a small amount of switching and power consumption. Subsequently, performance analysis is provided to quantify the performance gap between traditional construction structures and proposed hybrid structure. Subsequently, to encourage CE performance in machine learning [10], a systematic CE (ACE) system based on the hybrid precoding is implemented scheme of this new structure. Specifically, with the distribution of material possibilities in the hybrid precoder, this program periodically begins to produce multiple hybrid precoder. Thereafter, it accurately balances these selected hybrid layers according to their relative values and filters the potential distribution of material in the precoder hybrid by reducing the CE. Repeating such a process, it may eventually be able to produce a hybrid precoder that is closer to the one with high potential [11–14]. The simulation results confirm that our scheme can achieve a much closer average value and energy efficiency than traditional schemes.

2 System and Model

Consider the standard millimeter wave system, where one BS with the same line (ULA) for *Nt* antennas and a user with available antennas is considered. *Ns* data streams are sent by the BS to the user that are independent to each other, and it is also assumed that no information is available on communication links. Furthermore, it is supposed that user and BS have identical RF chains (N_{RF}) that fulfill the requirements of Ns.

$$Ns \leq N_{\mathrm{RF}} \leq N\mathrm{t} \quad \text{and} \quad N\mathrm{t} \leq N_{\mathrm{RF}} \leq Nr$$

[3].

Figure 1a demonstrates the traditional architectures for precoding, the one with phase shifters of finite resolution (PS based), i.e., architecture) [7]. Figure 1a shows

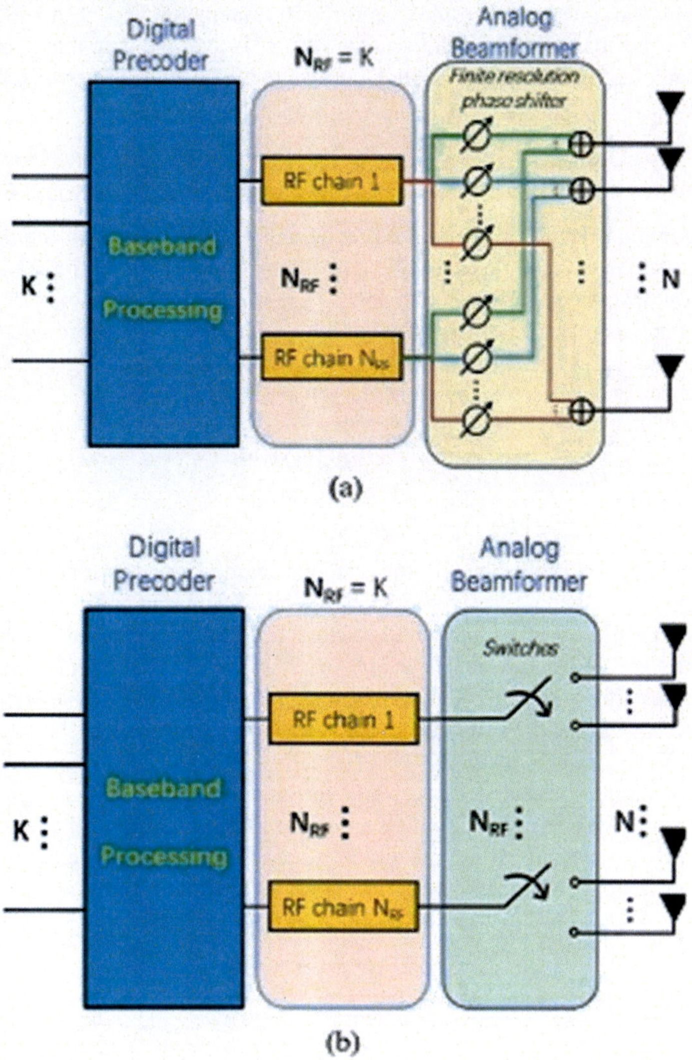

Fig. 1 **a** Phase shifter-based architecture. **b** Switch inverter-based architecture

conventional architecture based on PS that requires a complex phase shifter network. Even though the PS based design enjoys elevated freedom of concept to attain the near-optimum performance [7], a big number is needed of shifters in process. Furthermore, the energy consumption of the finite stage shifter is too important. These still make the conventional architecture based on PS suffer from high-energy ingestion.

The signal at receiver side received by the user is expressed by

$$y = gHx + n \tag{1}$$

where received signal y is a complex vector with dimension $Nr \times 1$. $n \sim \mathcal{CN}(0, \rho n)$ is a complex thermal AWGN with variance ρn. 'g' is average gain that is calculated for the channel. The transmitted signal x in vector form with dimension $Nt \times 1$ is specified by:

$$x = F_A F_D s \tag{2}$$

The baseband precoder F_D and RF precoder F_A are defined with the dimension $N_{\text{RF}} \times Ns$ and $Nt \times N_{\text{RF}}$, respectively. s of dimension $Ns \times 1$ is the signal before transmission designated by

The channel matrix H with L propagation path $Nr \times Nt$ for crowded channel model is shown by:

$$H = \sqrt{\frac{N_t N_r}{L}} \sum_{l=1}^{L} \alpha_l a_r(\theta^r_{l,} \emptyset^r_l) a_t^H(\theta^t_l, \emptyset^t_l) \tag{3}$$

To solve the issues that conventional architecture faces, an architecture based on SI is proposed, as shown in Fig. 1b, that can be regarded as a better trade-off among the near-optimal and energy-efficient PS based and SW based architecture oriented. Specifically, in the SI-based proposed, each RF chain is linked to a sub-antenna architecture. Instead of all N antennas [15], each and every RF chain is connected to a sub-antenna array with $M = N/N_{\text{RF}}$ antennas. In addition, each RF chain is linked via only one inverter and M switches instead of N phase shifters to the sub-antenna array.

To do the same, it is needed to clarify the hardware limitations caused by the SI-based architecture suggested, which are different from those of the conventional ones. The first restriction is that, instead of a complete matrix, the analog precoder matrix F_{RF} should be a block diagonally matrix as a whole matrix [16, 17].

where the M to 1 analog precoder on the nth sub antenna array is f_n^{RF}. The second restriction is that due to the use of only inverters and switches, all of the F_{RF}'s N nonzero components should belong to $\frac{1}{\sqrt{N}}\{-1, +1\}$.

3 Experimental Results

Spectral and energy efficiency simulation results are used to evaluate the performance of the hybrid precoding scheme based on ACE. Simulation parameters in this paper are taken as: antenna spacing in UPA-d1 = d2 = λ/2 is assumed to be used by the BS. For the kth user, it can be assumed that: (1) Lk = 3; (2) $\alpha^{(l)}{}_k \sim \text{CN}(0, 1)$ for $1 \leq l \leq L_k$; and (3) $\phi(l)\,k$ and $\theta(l)\,k$ follow the uniform distribution $U(-\pi, \pi)$ for $1 \leq l \leq Lk$ [10]. The signal-to-noise ratio is well defined as ρ/σ^2.

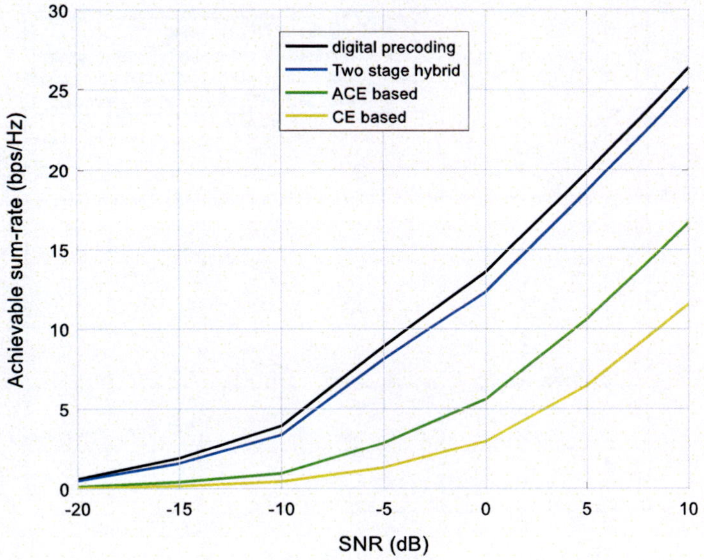

Fig. 2 Achievable sum rate

The achievable sum-rate relation, where the green line is the proposed ACE based hybrid precoding with inverter-based switch architecture, is shown in Fig. 2. Next, it can be observed that the ACE algorithm beats the standard CE algorithm, which is the purple line, with a slight increase in complexity.

In addition, it is also evident that a device can attain better efficiency than the switch-based architecture of AS hybrid precoding, which is the green line, and similar to the phase shifter-based architecture of two-stage hybrid precoding, which is the blue line.

Energy efficiency is seen in Fig. 3. The model for energy usage and the principles adopted are described here. The red line is the energy efficiency of the proposed analog/digital precoding based on ACE. It is noted that the highest energy efficiency can be achieved through our scheme.

By comparison, when K is thin, the two-stage hybrid precoding, which is the blue line, can only perform better than the completely digital precoding, which is the black line.

4 Conclusion

SI-based analog/digital precoding structural design is suggested in this paper where analog portion consists a lesser number of inverters and switches. This architecture is an energy efficient. Simulation results reveal that the performance fissure appears to be constant and narrow between the digital one and SI-based architecture. Then,

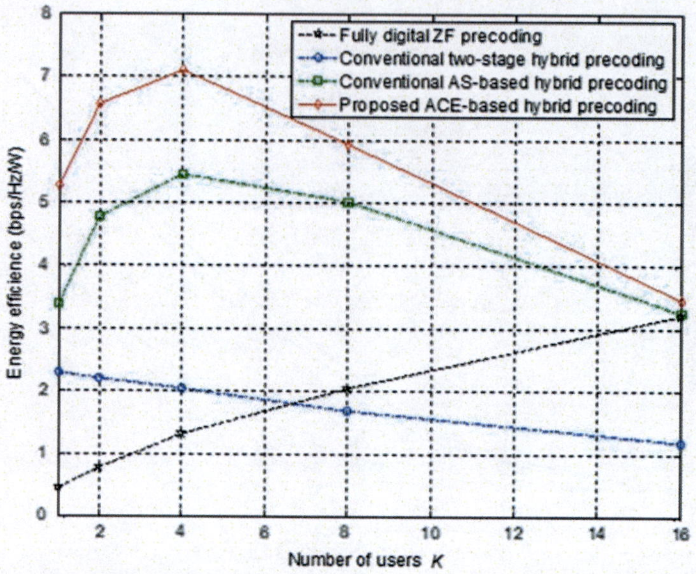

Fig. 3 Energy efficiency

analog/digital precoding built on ACE with low complexity for switch inverter-based architecture can be further analyzed by using the CE optimization theory in machine learning. The simulation results check that energy efficiency and fair average achievable rate performance can be attained by this scheme much higher than customary schemes.

References

1. Swindle hurst AL, Ayanoglu E, Heydari P, Capolino F (2014) Millimeter-wave massive MIMO: the next wireless revolution?. IEEE Commun Mag 52(9):56–62
2. Rappaport TS, Sun S, Mayzus R, Zhao H, Azar Y, Wang K, Wong GN, Schulz JK, Samimi M, Gutierrez F (2013) Millimeter wave mobile communications for 5G cellular: It will work! IEEE Access 1:335–349
3. Marzetta TL (2010) Noncooperative cellular wireless with unlimited numbers of base station antennas. IEEE Trans Wireless Commun 9(11):3590–3600
4. Heath RW, Gonzalez-Prelcic N, Rangan S, Roh W, Sayeed A (2016) An overview of signal processing techniques for millimeter wave MIMO systems. IEEE J Sel Top Signal Process 10(3):436–453
5. El Ayach O, Rajagopal S, Abu-Surra S, Pi Z, Heath RW (2014) Spatially sparse precoding in millimeter wave MIMO systems. IEEE Trans Wireless Commun 13(3):1499–1513
6. Sohrabi F, Yu W (2015) Hybrid beamforming with finite-resolution phase shifters for large-scale MIMO systems. In: Proceedings IEEE SPAWC workshops, July 2015, pp 136–140
7. Alkhateeb A, Leus G, Heath RW (2015) Limited feedback hybrid precoding for multi-user millimeter wave systems. IEEE Trans Wireless Commun 14(11):6481–6494

8. Alkhateeb A, Nam Y-H, Zhang J, Heath RW (2016) Massive MIMO combining with switches. IEEE Wireless Commun Lett 5(3):232–235
9. Mendez-Rial R, Rusu C, Alkhateeb A, Gonz´alez-Prelcic N, Heath RW (2015) Channel estimation and hybrid combining for mmWave: phase shifters or switches?. In: Proceedings ITA workshops, Febrauary 2015, pp 90–97
10. Sayeed A, Brady J (2013) Beamspace MIMO for high-dimensional multiuser communication at millimeter-wave frequencies. In: Proceedings IEEE GLOBECOM, December 2013, pp 3679–3684
11. Xie H, Gao F, Zhang S, Jin S (2017) A unified transmission strategy for TDD/FDD massive MIMO systems with spatial basis expansion model. To appear in IEEE Trans Veh Technol
12. Xie H, Gao F, Jin S (2016) An overview of low-rank channel estimation for massive MIMO systems. IEEE Access 4:7313–7321
13. Rubinstein RY, Kroese DP (2013) The cross-entropy method: a unified approach to combinatorial optimization. Monte-Carlo simulation and machine learning. Springer Science and Business Media
14. Han S, Chih-Lin I, Xu Z, Rowell C (2015) Large-scale antenna systems with hybrid precoding analog and digital beamforming for millimeter wave 5G. IEEE Commun Mag 53(1):186–194
15. Gao X, Dai L, Han S, Chih-Lin I, Heath RW (2016) Energy-efficient hybrid analog and digital precoding for mmWave MIMO systems with large antenna arrays. IEEE J Sel Areas Commun 34(4):998–1009
16. M´endez-Rial R, Rusu C, Gonz´alez-Prelcic N, Alkhateeb A, Heath RW (2016) Hybrid MIMO architectures for millimeter wave communications: phase shifters or switches?. IEEE Access 4:247–267
17. Chen J-C, Wen C-K, Wong K-K (2016) An efficient sensor node selection algorithm for sidelobe control in collaborative beamforming. IEEE Trans Veh Technol 65(8):5984–5994

Applying Predictive Analysis Methods for Detection of Driver Drowsiness

Abra Shafiq Siddiqi, Sherin Zafar, M. Afshar Alam, Samia Khan, Nida Iftekhar, and Siddhartha Sankar Biswas

Abstract Life is the most precious asset, and thousands of times due to terrible car accidents lives are lost. However, real-time sleep detectors used in cars can significantly prevent these accidents and save precious lives around the world. The main reason is the driver's inattention, which is mainly called driver drowsiness. The driver's drowsiness monitoring system is used in conjunction with a high-frequency detection system. The system uses the input video stream from the driver to target the driver's natural visual changes, such as constantly closing the eyes and using artificial intelligence and scientific drowsiness to slow down the rate of changes in facial expressions and detection measures. The proposed work focuses on the different monitoring systems used in drowsiness detection and the process of the detection system. It is recommended to use a driver drowsiness detector connected to a key predictive analytics system. This research work focuses on the analysis of various sleepiness systems in order to perform a better predictive analysis. Machine learning is a very important example of predictive analytics, so this article focuses on various machine learning techniques and their effectiveness in detecting drowsiness in various systems.

Keywords Drowsiness detection · Driver drowsiness detector · Facial expression · Predictive analysis · Machine learning algorithms

---

A. S. Siddiqi (✉) · S. Zafar · M. A. Alam · S. Khan · N. Iftekhar · S. S. Biswas
Jamia Hamdard, New Delhi, India

S. Zafar
e-mail: sherin.zafar@jamiahamdard.ac.in

M. A. Alam
e-mail: aalam@jamiahamdard.ac.in

N. Iftekhar
e-mail: nida.iftekhar@jamiahamdard.ac.in

S. S. Biswas
e-mail: ssbiswas@jamiahamdard.ac.in

R. P. Mahapatra et al. (eds.), *Proceedings of International Conference on Recent Trends in Computing*, Lecture Notes in Networks and Systems 341,
https://doi.org/10.1007/978-981-16-7118-0_2

1 Introduction

Statistical data show that more than 20% of people have been killed, and 76,000 people prescribe to 76,000 people with a zero file due to traffic accidents [1]. Statistics emphasize even more than as most of the road accidents are caused by driving drowsiness. When the driver falls asleep, there is a loss of administration for the clean final driving vehicle, and most of them cause a serious shock to another car in operation or any position. The term “drowsiness” literally means a trend to fall asleep. Somnolence is asleep as a dream. In addition, the sleeping stage can be divided as a wake out (not sleep), the dream of the ocular movement not fast and the fast-moving sleep. It can be further subdivided even more into the following three stages of [2].

Step I: Sleep archery changes (single)

Stage II: Light

Step III: Deep sleep

Mainly in drowsiness, we evaluate the various shocks generated that have several characteristics and features [3] including execution of individual vehicles running on highway, operator generally young men, legal level of alcohol in blood, and numerous numbers of skate or brake signals related to these functions. However, the police database uses the following indicators to determine the drowsiness accident [4].

- If the vehicle has any mechanical failure or defect
- No slip
- Insufficient sleep
- Run away from the appropriate road, or run behind another movement and permanent car
- Excluded. Possible potential causes, too close to the vehicles near the front vehicle, are generally the statistical data obtained using these methods generally for several complexity numbers and cannot be completely verified for road accidents caused by drowsiness of the driver. Therefore, an accident that is recognized by driving drowsiness can be more sonic and catastrophic than these statistical data. Therefore, to avoid an accident, it is more important to measure and effectively detect driver’s driving when the driver is alert. The second half of the dissertation of research focuses on the process of detecting drowsiness and progress in the field of driving drowsiness and future advances detection systems in this area.

Sections 2 and 3 discuss the detection and conclusions of driving drowsiness, respectively, with the end of the methodology, research and purpose of the driver. The dissertation of research finally explains the conclusion in Sect. 4. The resources used during the construction of references and research work are presented in last section.

2 Methodology

The behavioral dimension strategies correctly reveal the degrees of drowsiness via use of high-definition set up cameras constant in an automobile such that the alternate in facial expressions inclusive of yawning, eye state, blinking fee, sluggish eye closure, head motion, etc. The clinical tactics often exercise precise method to process motive force capabilities of face recorded from the furnished digital digicam as depicted in Fig. 1. After amassing the above-noted capabilities, the further, paintings are counted to reveal drowsiness degrees. This is mainly acquired through use of technology of machine getting to know which regularly include, convolution neural networks or hidden Markov models and support vector machines (SVM) [5]. These methods but are carried out by means of precise traits in addition to labeled information matrices with the intention to construct structures detecting drowsiness and obviate deadly hazards. However, the maximum difficult component in detection procedure is to locate large dataset that includes featured versions at some stage in specific direction of race in addition to pores and skin pigments. This being a sizeable difficulty due to myriad of protection in addition to discretionary troubles which could definitely stand up for the duration of the complete procedure of publishing those datasets for instructional in addition to business purpose [5].

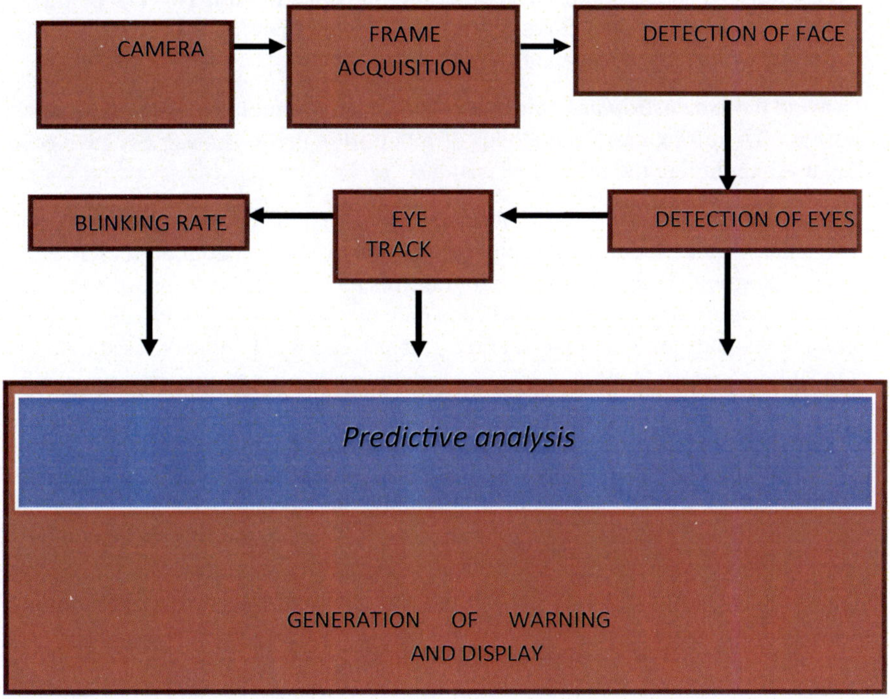

Fig. 1 Drowsiness detection process

The following expressions noted underneath are marked from the facial capabilities that are mainly won from a face of motive force:

1. Changing facial features analysis: Here the method combines multiple facial features such that the motive force drowsiness is detected. This encompasses expressions including wrinkles discovered mainly at the brow, additionally the top poses which can be occasionally extreme [6].
2. Rate of eye blink: This method detects the fee of blinking by thinking about the frequency with which the eye blinks such that the drowsiness is measured. Since, the regular blinking fee of a human eye is 10 per sec [7], the drowsiness is detected if the blinking fee falls the given number.
3. Closed eyes analysis: This expression is very important and is often used to specifically measure the driver's drowsiness. This includes technologies such as percent closed eyes (PERCLOS) and eye height-to-width ratio (EAR). EAR is identified as the ratio of the height to the width of the eye. Instead, PERCLOS is calculated as the percentage of eye closure within a certain period. However, the main difference between PERCLOS and EAR is that the latter will distinguish the proportion of eyes, the latter will decrease, and the former will detect whether the eyes are wide enough, or the eyes are closed.
4. Analyze yawning: Yawning is often observed when you are tired or bored. It exists mainly in drivers, which defines the premise for them to fall asleep during the journey. The method used can measure the open mouth. This can be traced using the driver's yawn function to observe the shape of the mouth and the position of the corners of the lips [8].

To avoid these catastrophic accidents, it is very important to measure the driver's drowsiness. Therefore, various methods of monitoring driver drowsiness have been widely used, including the following:

i. Physiological measures—Defines the relationship between physiological signals, such as electromyography (EMG), electroencephalogram (EEG), electrocardiogram (ECG), electrooculogram (EOG), and drivers. These measurements produce highly efficient and accurate measurement data, but due to some real flaws, they are not universal on a global scale. 4444 vehicle-based metrics, including various metrics such as accelerator pedal pressure, braking mode, steering wheel movement, out-of-lane position. It is continuously evaluated, including any changes in these indicators that exceed certain thresholds. Therefore, this indicates that the driver's possibility of drowsiness is greatly increased. Of all vehicle-based measurement methods, the steering wheel produces the most accurate results. However, because the vehicle-based method is noninvasive, it may not be a reliable and effective determination technique in the long run, mainly because it only depends on the nature of the road and the driving skills of the driver.
ii. Angle behavioral or computer vision measurement: This method is much more stable than the vehicle-based method because it focuses on the person himself, rather than the vehicle-based surveillance system. Focus on the driver's natural

reaction and use the camera to gather information to detect small changes in the driver's facial features, including monitoring yawning, eye closing, blinking, head posture changes, and more of these symptoms of drowsiness are detected, the driver will be alerted. This noninvasive method has gained great popularity in determining drowsiness. When conducting experimental work in the real atmosphere, the positive detection rate is greatly reduced. It is given in the following table. Table1 lists the work of several authors mentioned in this work, who used behavioral methods to detect driver drowsiness. There are many processes and techniques for face detection and drawing and determining features collected from video data. However, most of the work is close

Table 1 Previous works list on driver drowsiness detection using behavioral measures

Sensor used	Drowsiness measure	Detection techniques	Feature extraction	Classification	Positive detection rate (%)
CCD camera	Yawning	Gray projection and gravity center template	Gabor wavelets	LDA	91.97
Simple camera	Eye blinking	Algorithm diamond searching for tracing the face and Cascaded classifiers	Frequency of continuous blinks, frequency of eye blink, duration of eyelid closure,	Region mark algorithm	98
CCD micro camera with infrared illuminator	Pupil	AdaBoost	Red eye effect, detection method for texture	Eye height and eye width ratio	92
Camera	Multi-scale dynamic featuring	Gabor filter	LBP	AdaBoost	98.33
Digital video camera	Facial action	Gabor filter	Decomposing wavelet	SVM	96
IR camera	Eye state	Gabor filter	Condensation algorithm	SVM	93
Fire wire camera and Webcam	Eye closure frequency and duration of eye closure	Hough transform	Transforming discrete wavelet	Neural classifier	95
IR illuminator attached to camera	PERCLOS	Detecting face by Haar algorithm	Kalman filter algorithm unscented	SVM	99

to various databases, and various algorithms may follow these databases. The main reason for this situation is that the system data set is insufficient and cannot be used mainly as a benchmark for research. Therefore, it is difficult to distinguish these methods only by determining the efficiency of the reported data. Techniques from machine learning research are mainly used to determine the different stages of sleepiness, which will be listed in the following discussion. Similarly, it includes analysis of various measures and methods that are sufficient to constitute a driver's drowsiness detection system.

3 Predictive Analysis Techniques for Drowsiness Detection

There exist numerous procedures and techniques for detection of face as well as to draw and determine features collected from the video data. However, majority of these works conducted approach various databases which could follow the individual algorithms. This exists mainly because of insufficiently systematic datasets which could be mainly applied as a point of references in the study. Thus, it is difficult to distinguish methods by merely determining the reported data efficiencies. The techniques of machine learning studies are mainly used to determine the different stages of drowsiness that are now listed along with their discussions below. Also, including the analysis of various measures and methods that sufficiently form a drowsiness detection system for drivers.

a. Hidden Markov Model (HMM): This is basically a statistical data models and methods applied to predict and determine about the hidden states based on the actual perceived states which are defined by the probability methods. HMMs were extracted during the late 1960s period and early 1970s period (Deng and Wu [9]). In this era, these methods have a global application in various technologies such as modeling sequence errors, DNA gene annotation, and face expression recognition. Table 2 shows the varied range of features as well as perspectives which are used by hidden Markov model-based driver drowsiness detectors. The authors have beautifully enumerated various fresh facial characteristics using changing wrinkles that are differentiated by calculating the edge potencies from the driver's facial features by using a camera of infrared nature to exclude the

Table 2 Driver drowsiness detection based on HMM technique

Metric %	Classifiers	Frame per second(fps)	Accuracy
State of eye and position of head	HMM	16–20	N/A*
State of eye	HMM	N/A*	95.9
Blink of eye	SVM and HMM	61	90.99
Eye state	HMM	3	99.7
Closure of eye and other varied features	HMM and SVM	20	97
Blinking eye	HMM	25	95.7

variations in illumination such that it permits the proper functioning both in day and night conditions. Howbeit, the deeper and strong wrinkles formed on faces of the older people may provide the false results generated by this system. In contrast, the HMM techniques implied for the eye tracking movements which are based on the varied geometrical features and color of eyes. For eliminating the illumination, authors have used a 2-stage Lloyd–Max quantization method which is calculated to be sufficient to eliminate changes in illumination. Unfortunately, the system suggested is drafted for the inside conditions which fails to display the facial features when the driver is actually not forward facing. Table 2 enlists the HMM technique below used for driver drowsiness detection [9].

b. Support Vector Machines (SVMs): SVMs are specialized, supervised, and efficient learning techniques used as regression and classification system. These specialized methods focused to find a hyperplane to discontinue data of training into predefined classes. To analyze the varied states of the driver drowsiness obtained from the labeled data SVMs are initially used. A significant work on SVM done to mark its capabilities for drowsiness detection. In numerous measures and techniques have been put forward to determine a significant amount of drowsiness in driver's using SVMs. This technique further, attained a 100% accuracy for detection of face howbeit, a potential downfall still exists in the approach that is the minute rate of mount attainable which possibly could end in insufficient and inappropriate expressions obtained from facial. Table 3 represent the list of authors who worked on driver drowsiness detection using SVMs technology including their measures and data [10].
c. Convolutional Neural Network (CNN)

This technique is very similar to a neural network commonly used that is also compiled of the neurons which further consists of the learnable weights. CNN's uses varied spatial convolutions layers that are convenient enough for drawing images used to display the strong spatial relations. The methods have shown success in certain areas of image recognition, classification, and video analysis. The images were trimmed to the dimensions of (48*48 square) which were incorporated into

Table 3 Detection of driver drowsiness using SVM

Measure	Classifiers	Frame per second(fps)	Accuracy %
State of eye	HOG and SVM	5	91.6
Eye closure and yawning	Binary SVM including Linear kernel	15	94.58
Closure of eye	Haar features with SVM	60	99.74
State of eye	SVM	25	98.4
State of eye	SVM	15	93.5

Table 4 Driver drowsiness detection using the CNN technique

Metric	Methods	Classifiers	Accuracy %
Gaze of eye	Algorithm of Jones and Viola	CNN	98.32
State of Eye	PERCLOS, AdaBoost and LBF	CNN	95.18
State of eye	Eye state and mouth	MTCNN and DDDN	91.6
Visual features	Algorithm of Jones and Viola	Softmax layer CNN	78

the initial layer of the network that comprises basically of 20 filters. The overall network consists of 2-layer systems. CNN provided a data which preceded to one-layer softmax for the purpose of classification. CNN system, however, could not contemplate the change in the head pose thus, can result in a fail. Neural network has been utilized for attaining more efficient results. In this system, the face feature followed by a combined filter for correlation encompassed with a Kalman filter system to track face robustly. The regions of face are drawn out and passed directly to a 3D-CNN system further to a gradient boosting machine for classification in an optimum manner. The advantage added to this system is that it also works well with the change in head position. Table 4 presents the list which worked on the driver drowsiness detection using CNN technique [11].

4 Conclusion and Future Scope

As mentioned in the research paper, there are in fact different technologies and methods to detect and determine driver drowsiness fatigue. The research work attempts to evaluate various technologies under development to devise the best way to avoid death caused by collapsing sleepiness. This research article reviews and discusses various methods that can be used to detect driver drowsiness. However, there is no accepted definition of drowsiness, but the research work certainly contains various systems and methods for obtaining drowsiness in a simulated environment. The various methods used to detect drowsiness mainly include subjective, behavioral measures, vehicle-based methods, and physiological methods, and are discussed in detail, including their advantages and disadvantages. Although the accuracy and efficiency of sleepiness detection are high when physiological measures are used, it is very detrimental. However, this invasiveness of physiological measures can be resolved with noncontact electrode placement. Therefore, it is worth developing a hybrid system that uses vehicle-based measurements and physiological measurements such as ECG and behavioral measurements to develop a suitable and effective sleepiness detection system. Similarly, it is also very important to maintain and focus on the driving environment for the best suitable results. The further purpose of the author is to find out the data set and apply better machine learning algorithms for better results. This is a problem that is not considered in this research. This will be our post implementation work. Random forest and decision tree are algorithms proposed by the author for the data set of identity verification, aiming to establish a highly accurate analysis model.

References

1. Aboalayon K, Faezipour M (2019) Single channel EEG for near real-time sleep stage detection. In: Proceedings of the international conference on computational science and computational intelligence. Yogyakarta, pp 641–645
2. George A, Routray A (2016) Real-time eye gaze direction classification using convolutional neural network. In: International conference signal processing communications, pp 1–5
3. Manu N (2017) Facial features monitoring for real time drowsiness detection. In: Proceedings 201612th international conference innovations information technology IIT 2016, pp 78–81
4. Choi IH, Jeong CH, Kim YG (2016) Tracking a driver's face against extreme head poses and inference of drowsiness using a hidden Markov model. Appl Sci 6(5)
5. Huynh P, Kim YG (2017) Detection of driver drowsiness using 3D deep neuralnetwork and semi-supervised gradient boosting machine. vol 10116
6. Parkhi OM, Vedaldi A, Zisserman A (2015) Deep face recognition. In: BMVC, vol. 1. (2015) pp 6
7. Shakeel, Bajwa M (2019) Detecting driver drowsiness in real time through deep learning based object detection. Adv Comput Intell 283–296
8. Zhang J, Su L, Geng, Xiao Z (2017) Driver fatigue detection based on eye staterecognition. In: Proceedings—2017 international conference machine vision information technology C. pp 105–110
9. Deng W, Wu R (2019) Real-time driver-drowsiness detection system using facial features. IEEE Access 7:118727–118738
10. Ren S, Cao X, Wei Y, Sun J (2014) Face alignment at 3000 FPS via regressing local binary features, In: Proceedings IEEE conference computing vision pattern recognition, pp 1685–1692
11. Dwivedi K, Biswaranjan K, Sethi A (2014) Drowsy driver detection using representation learning. In: Souvenir 2014 IEEE international advanced computing conference IACC 2014, pp 995–999
12. Cech J, Soukupova T (2016) Real-time eye blink detection using facial landmarks. In: 21st computing vision winter working
13. AL-Anizy J, Nordin MJ, Razooq MM (2015) Automatic driver drowsiness detection using harr algorithm and support vector machine techniques. Asian J Appl Sci

Comparative Analysis of Various Supervised Machine Learning Techniques Used for Sentiment Analysis on Tourism Reviews

Manoj Kumar Sahu and Smita Selot

Abstract On various social media sites, many tourists share their impressions and opinions in the form of text, photographs, and videos. To conduct sentiment analysis, these tourist reviews are collected and preprocessed. Text-based sentiment analysis currently relies on the creation of dictionaries and machine learning models to extract sentiments from text data. This form of sentiment analysis is commonly used in tourist satisfaction, tour recommendation, tourist place ranking, and decision-making. Traditional techniques in sentiment analysis in tourism domains are examined in this study. Using the Weka Tool, the experiment ran various machine learning algorithms on the tourist data collected from various websites and social media platforms. Lowercasing, lemmatization, stop word elimination, and nonword removal were all used as preprocessing techniques. The efficiency of classifiers was significantly improved as a result of these techniques. Sentiment processing was carried out using a supervised machine learning method in this experiment. This experiment's results are compared to one another. The outcomes are examined. As compared to J48, random forest, Logistic, and Naïve Bayes classifiers, LibSVM obtained the highest accuracy (92.5%) from the tourism data in this analysis.

Keywords Sentiment analysis · Machine learning · Opinion mining

1 Introduction

Millions of travelers and tourists share different types of content for mutual gain on common platforms like YouTube, Facebook, Twitter, TripAdvisor, Instagram, and other pages. Many studies of sentiment analysis have discovered that online feedback and opinions have a significant impact on tourist conduct and decision-making. Textual reviews are usually where you will find this stuff. Sentiment analysis creates an atmosphere in which user feedback can be collected, analyzed, and classified

M. K. Sahu (✉) · S. Selot
Shri Sankaracharya Technical Campus, Bhilai, Chhattisgarh, India

R. P. Mahapatra et al. (eds.), *Proceedings of International Conference on Recent Trends in Computing*, Lecture Notes in Networks and Systems 341,
https://doi.org/10.1007/978-981-16-7118-0_3

based on the polarity meaning in the reviews. We may conclude that sentiment analysis is used to assess a tourist's or writer's feelings about a specific subject or tourist destination. Sentiment analysis can be used in a variety of areas. The accuracy and other features of major supervised machine learning techniques that can be used in sentiment analysis on tourism feedback were collected and observed in this study.

The computational study of people's sentiment by feelings, thoughts, behaviors, events, concerns, and individual subjects is known as sentiment analysis. It is a mental mechanism for assessing a user's emotions and thoughts. Via social media articles, users can freely share their experiences, views, thoughts, and feelings about various trending events, topics, and so on. Via images, text, audio and video messages, and posts, reviewers may share their feelings, mental states, and moments; and express their opinions on specific social, tourist destination, national, and international issues. These reviews are analyzed to see how much emotion is expressed in these articles. Sentiment analysis is a fresh and effective method that allows viewers to extract pertinent details and summarize the overall emotions of the feedback. In recent years, a variety of approaches to completing this challenge have emerged. The following are examples of different types of sentiment analysis.

1.1 Lexicon-Based

Lexicon-based approach used natural language processing concepts to assess the polarity of reviews/tweets based on sentiment lexicon of reviews and matching them with data. The appropriate sentiment values provided for the reviews, describing how each review is classified as neutral, positive or negative words from the dictionary. To evaluate opinion at the aspect, sentence, expression, and document levels a variety of natural language processing techniques are used, including bag-of-words, part of speech, Hidden Markov Model, N-gram algorithms, parsing techniques, and massive emotion lexicon acquisition.

1.2 Dictionary-Based

This approach is based on the collection and annotation of terms (seeds) manually. This collection was created by looking up synonyms and antonyms for dictionary words. WordNet is one such dictionary, and it is currently working on a thesaurus called SentiWordNet.

1.3 *Corpus-Based*

The aim of the corpus-based approach is to provide dictionaries for a specific domain like tourism, hotel reviews, movie reviews, etc. These dictionaries start with a collection of seed opinion expressions, which are then expanded using semantic or computational methods to identify words with related meanings.

2 Literature Reviews

This segment gives a quick overview of relevant works for sentiment analysis in tourism reviews. To begin, we take a quick look at the supervised machine learning approach, which is based on classification methods in a variety of datasets. Many scholars have recently proposed various approaches to sentiment analysis. Total 25 research papers are reviewed and most similar and related 17 papers are listed as follows (Table 1).

3 Methodology

The proposed approach is demonstrated in this section. The first step is to build a tourism review dataset, which will then be used for sentiment analysis. The proposed experiment is divided into three stages (dataset preparation, feature extraction, and evaluation), which are listed below (Fig. 1).

3.1 *Data Collection Source*

This experiment is based on TripAdvisor, HolidayIQ, Twitter, Facebook and Google English reviews of tourist places in Chhattisgarh and Maharashtra, India. Therefore, a datasets with reviews until June 2020 are collected a total of 11,000 (5500 positive and 5500 negative) reviews. Manual sentiment labeling (positive and negative) is accomplished.

3.2 *Data Preprocessing*

Preprocessing data entails extracting key words from a collection of data in order to produce useful or analytical information. Spaces, stop words, special symbols (!, @, *), hash tags, redundant words, and emojies are all examples of data that should be

Table 1 Literature reviews of similar research

Authors	Data	Result
Albusaidi et al [4]	Twitter data	The outcome of this paper is focused on measuring the satisfaction of tourists of the tourism services provided to tourists by measuring the sentiments of tourists in the Twitter hash tags
Floris and Campagna [10]	Booking.com, TripAdvisor.com	According to the findings of this report, a tourist destination's success is determined not only by the tourist sector offer, but also by the destination's territorial climate, which includes the location's natural, physical, and cultural characteristics, as well as the facilities and services accessible
Kirelli and Arslankaya [17]	Twitter data	The combination of the N-gram technique and the KNN classification algorithm generated the highest effective rate of 74.63%. The KNN classification technique with the combination of N-gram technique provides the highest success rate with 74.63
Cieliebak et al [9]	10,000 German tweets	In all but one instance, CNN outperforms SVM (SB10k-DAI). Surprisingly, SVM performs better on SB10k when trained on MGS than when trained on SB10k (60.50 instead of 56.98), although the classifier gains in all other cases when trained on the same corpus
Kamel and Gayar [14]	Hong Kong tourism	According to their findings, GRNN was the only method that generated three highly accurate forecasts, one decent forecast, and one fair forecast. KNN, MLP, GP, and RBF did not achieve any forecasting accuracy
Zheng et al [42]	Twitter, Facebook, Flickr, Microblogs	They get a 4% higher precision and a 70% reduction in computational complexity when they use the most important subset of 20 features as their final random forests input, opposed to when they use the whole range of features. Random forests outperformed LambdaMART, Ranking SVM, ListNet, and RankBoost in a comparison of five different machine learning algorithms
Al-Azani and El-Alfy [2]	ASTD, ArTwitter, QCRI, Syria	Models focused on LSTM and GRU outperform other classifiers by a small margin in the experiments, with bidirectional GRU providing the strongest results
Jianqiang et al [11]	Twitter data	The model of GloVe CNN achieves a high degree of sentiment classification precision, as shown by the experimental results
Roshanfekr et al [28]	www.digikala.com	Deep learning methods outperformed NBSVM in terms of F-score. One explanation may be the unsupervised implementation of word vector representations

(continued)

Table 1 (continued)

Authors	Data	Result
Vateekul and Koomsubha [36]	Thai Twitter data	As per the study DCNN is the highest classifier, led by LSTM. Traditional classification techniques such as SVM and NB have significantly lower accuracy than both of these techniques, but not maximum entropy
Valdivia et al [34]	TripAdvisor.com	Although some sentences have the opposite polarity, the users usually rate their overall experience favorably. As a consequence, this ranking is inapplicable to all sentences. About 47% of matches are based on detecting three polarities (+ve, Neutral, and -ve)
Ahirwal [1]	TripAdvisor.com	The simulation results show that KNN has an accuracy of around 80%, SVM has an accuracy of around 89% , and RF has an accuracy of around 93 percent
Windasari and Eridani [39]	TripAdvisor.com	A precision of 85 percent was obtained from the test results. The recall value for positive sentiment is 100%, while the recall value for negative sentiment is 62.5% . Positive sentiment has an accuracy rate of 80%, whereas negative sentiment has an accuracy rate of 100%
Kim et al [16]	TripAdvisor, Yelp, and Booking.com	Due to the limited size of the intent corpus, SVM outperforms deep neural network; in reality, DNN's precision score is 3% higher than SVM's, but SVM beats DNN's recall score by 6%, resulting in a higher F-measure
Cankurt and Subasi [8]	www.turizm.gov.tr, www.die.gov.tr, www.tursab.org.tr, http://evds.tcmb.gov.tr	These models were investigated and evaluated, and it was discovered that- (1) The accuracy of the ANN model using the second dataset is R = 0.9879, R.A.E. = 14.41%, and R.R.S.E. = 16.27%, which is higher than the R = 0.9821, R.A.E. = 16.96%, and R.R.S.E. = 19.11% accuracy of the ANN model using the first dataset (2) The accuracy of the SVR model with R = 0.9932, R.A.E. = 10.36%, and R.R.S.E. = 11.66% is higher than the accuracy of the SVR model with R = 0.9875, R.A.E. = 14.57%, and R.R.S.E. = 15.84% (3) With R = 0.9932, R.A.E. = 10.36%, and R.R.S.E. = 11.66% , the SVR model with the second dataset has the highest accuracy
Biba and Mane [6]	Albanian political news	HyperPipes is the best-performing classifier on average. The best performing algorithm for each subject, however, varies depending on the topic
Salur [29]	Twitter data	Among the multilayer perception, Naive Bayes, and random forest algorithms, random forest had the best classification precision

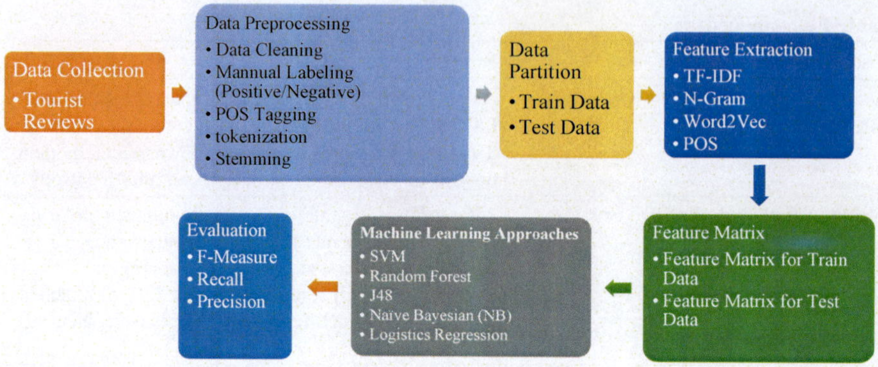

Fig. 1 Sentiment analysis methodology

preprocessed. Data cleaning is essential because tweets and reviews contain many syntactical terms that are useless for research. Once the data has been wiped, the data is processed. To prepare reviews for feature extraction, major tasks such as splitting a summary into sentences, splitting a sentence into words, tokenization, stop word scanning, part-of-speech marking, stemming, and case transformation are performed [31].

3.3 *Feature Extraction*

In sentiment analysis, feature extraction is a crucial step. It includes converting a piece of text into a feature vector. Any of the most commonly used sentiment analysis features can be found in this section.

3.3.1 The Term Frequency-Inverse Document Frequency (TF-IDF)

Although the TF-IDF algorithm is simple to use and extremely efficient; it does have some drawbacks. In today's world of big data, modern data processing techniques are needed before research can begin. A variety of researchers have suggested the adaptive TF-IDF algorithm, which is a superior version of the TF-IDF algorithm [25]. It is a mathematical approach for determining the value of terms based on their frequency across all documents in a dataset. The TF-IDF can be used to identify key factors that travelers used to convey satisfaction and expectations in online reviews. The TF-IDF algorithm, on the other hand, would result in dimensional catastrophe [20]. Frequency refers to the number of times a phrase appears in a piece of document. It is considered important in standard text classification activities. In sentiment analysis, however, term presence has been shown to be more significant than term frequency since the presence of a single term will also reverse the polarity of the whole sentence.

3.3.2 N-Gram Features

N-grams are commonly used to define meaning in natural language processing tasks. It is unclear, however, whether higher order N-grams work better than lower order N-grams.

3.3.3 Word2Vec

Word2Vec is a well-known similarity analysis tool for determining the degree of similarity between two text datasets. Mikolov, Sutskever, Chen, Corrado, and Dean suggested the Word2Vec algorithm based on Google's deep learning platform [19]. Word2Vec's key premise is that terms that are semantically and syntactically identical appear in the same ways with a high likelihood.

3.3.4 Part Of Speech (POS)

A POS is a set of words with similar grammatical properties (syntax, morphology) in English [3]. Nouns, verbs, adjectives, adverbs, pronouns, prepositions, conjunctions, interjections, and numerals, articles, and determiners are all used.

3.4 Data Partition

Separate the training and testing data first. Data training accounts for 80% of the total, while data testing accounts for 20%. The remaining 20% of the data from the 11,000 feedback were used as test data, with the remaining 80% being used as train data.

3.5 Machine Learning Approach

In sentiment analysis, a machine learning method specifies that an algorithm is trained on a testing dataset before being applied to the actual dataset. Before allowing the algorithm to work with new, unfamiliar results, machine learning techniques train it with unique inputs and known outputs [35]. The explanation for this is that we want to train our algorithm so that it can later be used to identify new unknown inputs. For sentiment analysis, a number of machine learning techniques are available.

3.5.1 Support Vector Machine (SVM)

Vapnik is the first to suggest the support vector machine (SVM) method. Many studies have shown that the SVM can outperform other classification algorithms in terms of accuracy [35]. Image classification, text categorization, object identification, data science, and data classification are only a few of the real-world problems where SVM has been used. SVM has been shown to be more effective than other supervised learning approaches on a consistent basis. However, the accuracy of SVM is highly dependent on how the cost parameter and kernel parameters are set for certain datasets [22]. It is a supervised learning technique that creates a mapping function from a collection of training data. SVMs are commonly used for classification and nonlinear regression problems because they can differentiate between linear and nonlinear effects [38]. A mapping function that sorts the data in the categorized dataset may be a classification function. SVM is very useful for text classification since it separates them linearly [12].

To help distinguish the different groups, SVM tries to find the best hyperplane in N-dimensional space. It finds the ideal hyperplane by maximizing the margin distance between the groups' observations using the Hinge loss function. The number of input features determines the height of the hyperplane. The hyperplane's dimension is N-1 if the number of features is N.

The training set for an SVM classifier is (a_1, b_1), (a_2, b_2),..., (a_n, b_n), where a_i is an input vector and b_i is the symbol. The hyperplane of partition can be defined as follows [32, 35]:

$$\lambda \cdot a + m = 0 \tag{1}$$

where m is the hyperplane's offset; λ is the normal vector of the partition hyperplane. To create the bilateral blank field, a partition hyperplane is used, i.e., $2/\parallel \lambda \parallel$. The partition hyperplane must be as far away from the point in the training dataset as possible, which can be described as follows [32]

$$\text{Minimize}\,\phi(\lambda) = \frac{1}{2}\|\lambda\|^2 \tag{2}$$

A constraint condition must be met, which is defined as follows [35]

$$b_i(\lambda \cdot a_1 + m) \geq 1 \tag{3}$$

The Lagrange function can be defined as follows [35]:

$$L(\lambda, m, \alpha) = \frac{1}{2}(\lambda \cdot \lambda) - \sum_{i=1}^{n} \alpha_i(y_i(\lambda \cdot a_i + m) - 1) \tag{4}$$

Subject to the following two conditions, i.e., $\sum_{i=1}^{n} y_{i\alpha_i} = 0$ and $\alpha_i \geq 0$, then the following formula can be used to find the Lagrange function's minimum [32]

$$\max\ Q(\alpha) = \sum_{i=1}^{n} \alpha_i - \frac{1}{2} \sum_{i,\, j=1}^{n} \alpha_i \alpha_j b_i b_j \left(a_i a_j\right) \tag{5}$$

The optimal class function can be defined as follows [37]

$$f(x) = \operatorname{sgn}\left(\left(\lambda^* \cdot a\right) + m^*\right) = \operatorname{sgn}\left(\sum_{i=1}^{n} \alpha_i^* b_i \left(a_i \cdot a\right) + m^*\right) \tag{6}$$

3.5.2 J48

To classify and perform correct classification results, the J48 algorithm is used. The J48 algorithm is best machine learning algorithms for evaluating results continuously and categorically [30]. It takes up too much memory space when used for a fast reason, but it performs well and accurately in classifying medical data [15]. A basic C4.5 classification decision tree is used in the J48 classifier. It dynamically generates a binary tree for you. Using this classifier a tree is constructed to model the classification process. The tree is then applied to each tuple in the database, resulting in a label for each [23]. Data is stored in documents of the following format:

$$(x, Y) = (x1,\ x2,\ x3, \ldots xk,\ Y) \tag{7}$$

The target variable, Y, is the one we are trying to understand, explain, or generalize, and the dependent variable is the one we are trying to understand, describe, or generalize. The vector x, which is used for that reason, is made up of the input variables $x1$, $x2$, $x3$, and so on.

3.5.3 Logistic Regression

The odds ratio is calculated using logistic regression when there are many explanatory variables. This is somewhat similar to multiple linear regressions since the response variable is binomial [24]. The end result is the influence of each vector on the odds ratio of the observed occurrence of interest. The key benefit is that it eliminates conflicting effects by examining the relationship between both variables [33]. The analysis of categorical outcome variables is well adapted to logistic regression, which is a subset of regression models.

We use a cost function, also known as Log Loss, instead of mean squared error. This can be broken down into two cost functions: one for $a = 1$ and one for $a = 0$.

$$\text{Cost}\ (h_\theta(a),\ b) = \begin{cases} -\log(h_\theta(a)) & \text{if } b = 1 \\ -\log(1 - (h_\theta(a)) & \text{if } b = 0 \end{cases} \tag{8}$$

Cost Function of Logistic Regression

The function converts every real number into a number between 0 and 1. We use sigmoid to map projections to probabilities in machine learning.

$$P = \frac{1}{1 + e^{-(A_0 + A_1 \cdot X_1 + A_2 \cdot X_2 + A_{3.} X_3 + \ldots + A_n \cdot X_n)}} \tag{9}$$

Logistic regression equation.

3.5.4 Naïve Bayesian (NB) Classifier

The Naive Bayesian classifier is based on the Bayes principle, which states that an attribute value's effect on a given class is independent of other attributes. This assumption is known as class conditional independence. The assumptions that different NB classifiers make about the distribution of features are what distinguishes them the most. The NB classifier's assumptions on feature distribution are known as event models [13]. Multinomial or Bernoulli distributions are common for discrete features. These assumptions result in two distinct models, which are often misunderstood [18, 40]. A Gaussian distribution is a common assumption when dealing with continuous functions. Bayesian classifiers are important because they have a theoretical foundation for classifiers that do not explicitly employ Bayes' theorem [41]. The expectation of class conditional freedom, which is one of NB's main shortcomings, is closely correlated with its predictive accuracy [21]. This presumption makes calculation easier. However, in fact, there can be dependencies between variables.

$$P(c|x) = \frac{P(x|c) \cdot P(c)}{P(x)} \tag{10}$$

where P represents posterior probability, c represents class, X represents predictor, $P(x|c)$ is likelihood, $P(c)$ is class prior probability, and $P(x)$ represents predictor prior probability.

$$P(c|X) = P(x1|c) \times P(x2|c) \times \ldots P(xn|c) \times P(c) \tag{11}$$

The posterior likelihood for each class is determined in this equation. It requires a frequency table for all predictors versus groups, as well as the ability to quantify the probability of all predictors.

3.5.5 Random Forest

Leo Breiman and Adele Cutler provides the random forest algorithm, which incorporates the Bootstrap aggregating and random subspace method approaches to construct a series of decision trees that are then categorized using the decision tree set [27]. In random forest, which contains multiple decision tree classifiers, the performance categories are calculated using the mode of decision tree classification results [7]. The random forest algorithm employs two random selection methods to construct a single decision tree: the first is the random selection of training samples, and the second is the random selection of sample characteristics attributes [26]. After all of the decision trees have been created, the final classification result is decided using an equal weight voting method [27]. Because of the precision of its classifications, this classifier has become common in data analysis. By selecting features at random, the association between trees is reduced, resulting in improved prediction power and performance [5].

3.6 Evaluation

To get the evaluation result, we used performance indicators like Accuracy, F-Measure, Recall, and Precision, as well as training time, true positive rate, false positive rate, test timing, and area under the ROC curve.

3.6.1 Accuracy

Accuracy is defined as the percentage of correct predictions for the test data. The accuracy rate is determined as the number of accurate predictions divided by the total number of predictions. Divide the total number of possible predictions by the number of valid predictions to get the response.

$$\text{Accuracy} = \frac{\text{Number of Correct Predictions find}}{\text{Total Number of Predictions made}} \tag{12}$$

3.6.2 Precision

Precision is measured by the percentages of positive instances out of the total estimated positive instances.

$$\text{Precision} = \frac{\text{True Positive (TP)}}{\text{True Positive(TP)} + \text{False Positive(FP)}} \tag{13}$$

3.6.3 Recall

The percentages of positive instances out of the total number of positive instances are used to measure recall. The percentage of True Positive and False Negative is the actual number of positive instances.

$$\text{Precision} = \frac{\text{True Positive (TP)}}{\text{True Positive(TP)} + \text{False Negative(FN)}} \tag{14}$$

3.6.4 F1-Score

The harmonic mean of precision and recall is the F1-score. The F1-score is a number that varies from 0 to 1. It provides you with information about your classifier's precision and robustness.

$$\text{F1} = \frac{2 * \text{precision} * \text{recall}}{\text{precision} * \text{recall}} \tag{15}$$

3.6.5 Receiver Operating Characteristics (ROC) Curve

One of the most widely used assessment criteria for evaluating the results of a classification model is the ROC curve. The true positive rate (TPR) and false positive rate (FPR) are plotted on the curve. The ROC curve is a metric for determining a classification model's consistency.

4 Experiment and Result

Separate the training and testing data first. Data training accounts for 80% of the total, while data testing accounts for 20%. Then, using the algorithm chosen, classify the training data. The training data's classification results are then added to the testing data. Perform a performance assessment after the data has been classified (Table 2)..

Classification Result and Performance Evaluation of Naïve Bayes Classifier

According to the findings in the table above, data with a total of 11,000 reviews/tweets using the Weka Tool given an accuracy of 81.9545% for Naive Bayes. As a consequence of the findings, the average of Precision, Recall, and F1-Measure is 0.820. The time it took to construct the model was 4.14 s, and it took 1.17 s to test it on the test split.

ROC Curve for Naïve Bayes Classifier

Table 2 Evaluation of Naïve Bayes classifier

Method	Correctly classified instance	Incorrectly classified instance	Rate of TP	Rate of FP	Precision	Recall	F1-Measure	MCC	ROC area	PRC area
Naïve Bayes	81.9545%	18.0455%	0.820	0.181	0.820	0.820	0.820	0.639	0.902	0.898

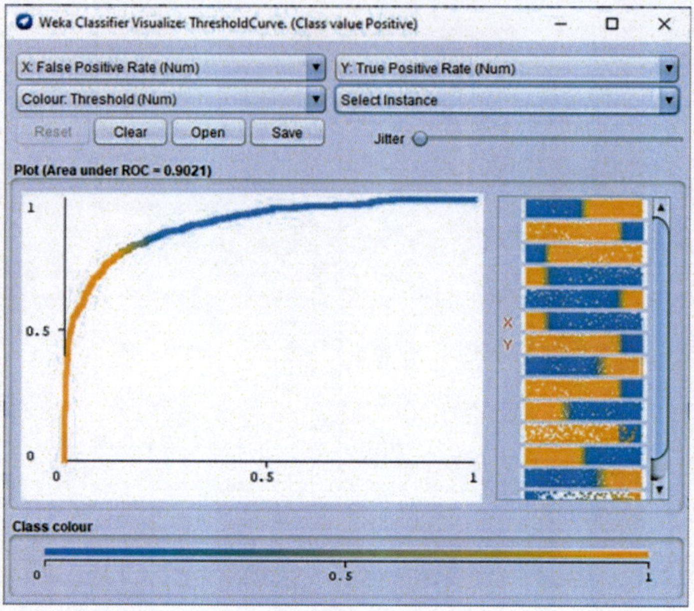

Fig. 2 ROC for Naïve Bayes classifier (class value positive)

The ROC curve for the Naive Bayes Classifier between False Positive Rate and True Positive Rate as shown in Fig. 2 (Class Value Positive). The ROC curve for the Naive Bayes Classifier between False Positive Rate and True Positive Rate as seen in Fig. 3. (Negative Class Value).

Precision Recall Curve for Naïve Bayes Classifier -

Figure 4 shows the PRC curve between Recall and Precision for Naïve Bayes Classifier (Class Value Positive) similarly Fig. 5 shows the PRC curve between Recall and Precision for Naïve Bayes Classifier (Class Value Negative).

Classification Result and Performance Evaluation of LibSVM Classifier

According to the findings in the table above, LibSVM had an accuracy of 92.5% using data from 11,000 reviews/tweets and Weka Tools. As a consequence of the results, the average of Precision, Recall, and F1-Measure is 0.925. The time it took to develop the model was 21.29 s, and the time it took to test it on the test split was 3.89 s (Table 3).

ROC Curve for LibSVM Classifier -

Figure 6 shows the LibSVM Classifier's ROC curve for False Positive Rate and True Positive Rate. Similarly, (Class Value Positive) Fig. 7 displays the ROC curve for the LibSVM Classifier between False Positive Rate and True Positive Rate (Class Value Negative).

Precision Recall Curve for LibSVM Classifier

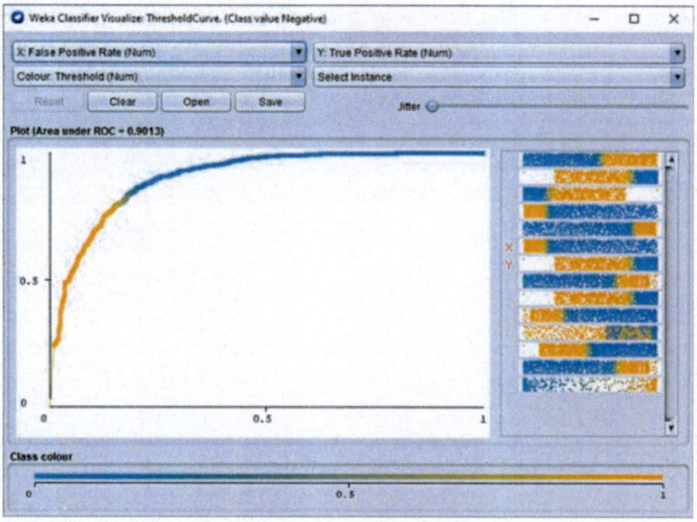

Fig. 3 ROC for Naïve Bayes classifier (class value negative)

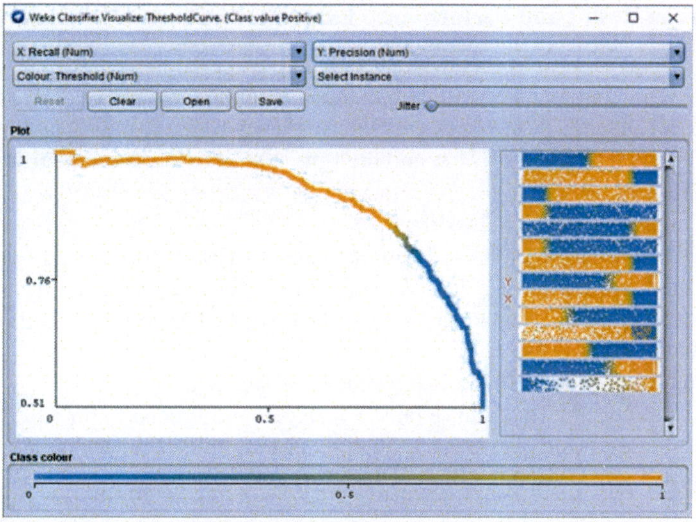

Fig. 4 PRC for Naïve Bayes classifier (class value positive)

Figure 8 shows the PRC curve between Recall and Precision for LibSVM Classifier (Class Value Positive) similarly Fig. 9 shows the PRC curve between Recall and Precision for LibSVM Classifier (Class Value Negative).

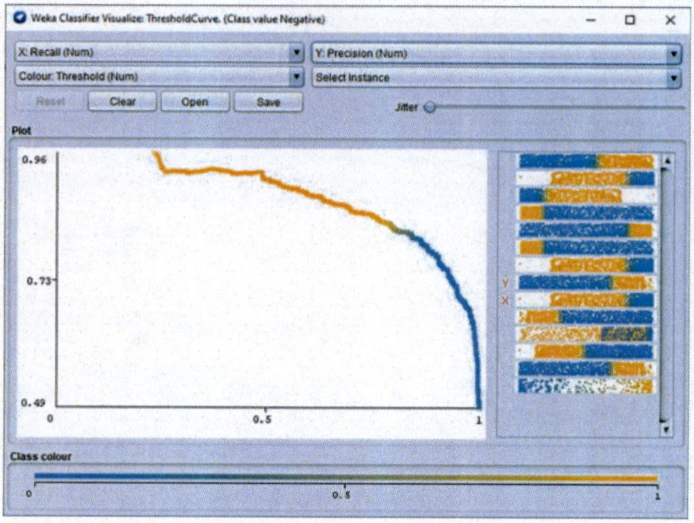

Fig. 5 PRC for Naïve Bayes classifier (class value negative)

Classification Result and Performance Evaluation of Logistic Classifier

Data with a total of 11,000 reviews/tweets using Weka Tools provided an accuracy of 86.5909% for Logistic, according to the results in the table above. The average of Precision, Recall, and F1-Measure is 0.866 as a result of the results. The time it took to create the model was 1023.71 s, and the time it took to test it on the test split was 0.13 s (Table 4).

ROC Curve for Logistic Classifier

The ROC curve between False Positive Rate and True Positive Rate for Logistic Classifier as shown in Fig. 10 (Class Value Positive) likewise The ROC curve between False Positive Rate and True Positive Rate for Logistic Classifier as seen in Fig. 11 (Class Value Negative).

Precision Recall Curve for Logistic Classifier

Figure 12 shows the PRC curve between Recall and Precision for Logistic Classifier (Class Value Positive) similarly Fig. 13 shows the PRC curve between Recall and Precision for Logistic Classifier (Class Value Negative).

Classification Result and Performance Evaluation of Random Forest Classifier

Data with a total of 11,000 reviews/tweets using Weka Tools provided an accuracy of 90.0455% for random forest, according to the findings in the table above. The average of Precision, Recall, and F1-Measure is 0.900 as results. The time it took to develop the model was 74.61 s, and the time it took to validate it on the test split was 0.77 s (Table 5).

ROC Curve for Random Forest Classifier

The ROC curve between False Positive Rate and True Positive Rate for Random Forest Classifier (Class Value Positive) is shown in Fig. 14, and the ROC curve

Table 3 Evaluation of LibSVM classifier

Method	Correctly classified instance	Incorrectly classified instance	Rate of TP	Rate of FP	Precision	Recall	F1-Measure	MCC	ROC area	PRC area
LibSVM	92.5%	7.5%	0.925	0.076	0.926	0.925	0.925	0.851	0.924	0.893

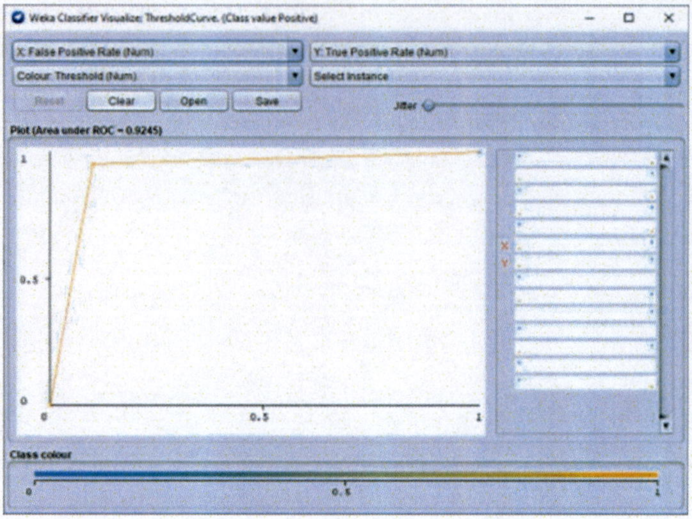

Fig. 6 ROC for LibSVM classifier (class value positive)

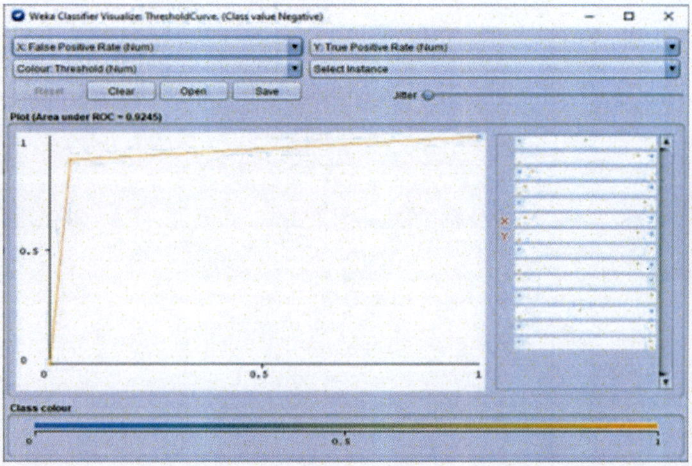

Fig. 7 ROC for LibSVM classifier (class value negative)

between False Positive Rate and True Positive Rate for Random Forest Classifier (Class Value Positive) is shown in Fig. 15 (Class Value Negative).

Precision Recall Curve for Random Forest Classifier

Figure 16 shows the PRC curve between Recall and Precision for Random Forest Classifier (Class Value Positive) similarly Fig. 17 shows the PRC curve between Recall and Precision for Random Forest Classifier (Class Value Negative).

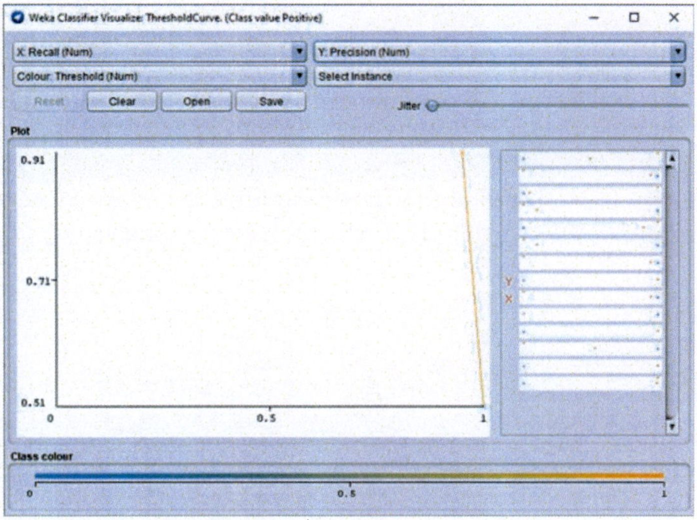

Fig. 8 PRC for LibSVM classifier (class value positive)

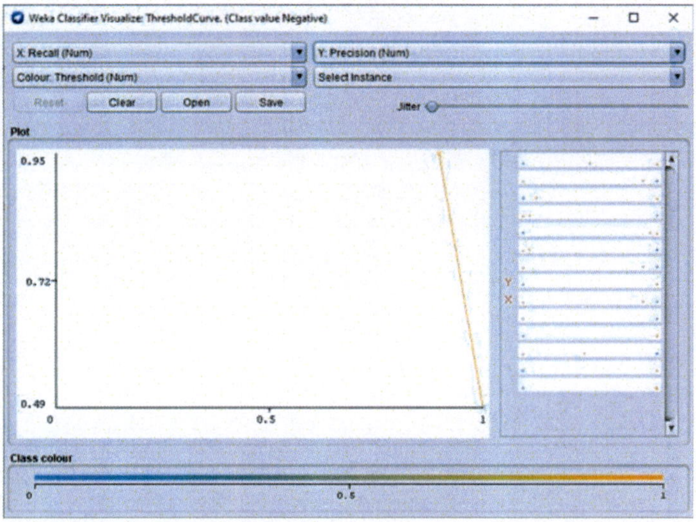

Fig. 9 PRC for LibSVM classifier (class value negative)

Classification Result and Performance Evaluation of J48 Classifier

Data with a total of 11,000 reviews/tweets using Weka provided an accuracy of 83.0455% for J48, based on the findings in the table above. The average of Precision, Recall, and F1-Measure is 0.830 as a result of the results. The tree has 586 leaves

Table 4 Evaluation of logistic classifier

Method	Correctly classified instance	Incorrectly classified instance	Rate of TP	Rate of FP	Precision	Recall	F1-Measure	MCC	ROC area	PRC area
Logistic	86.5909%	13.4091%	0.866	0.134	0.866	0.866	0.866	0.732	0.911	0.871

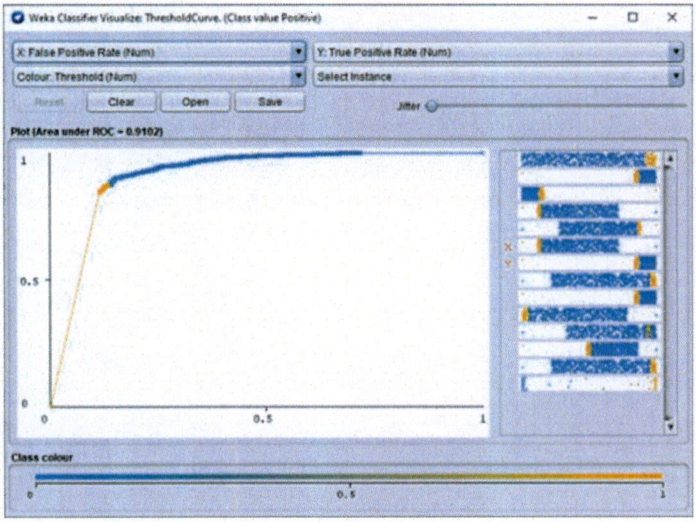

Fig. 10 ROC for logistic classifier (class value positive)

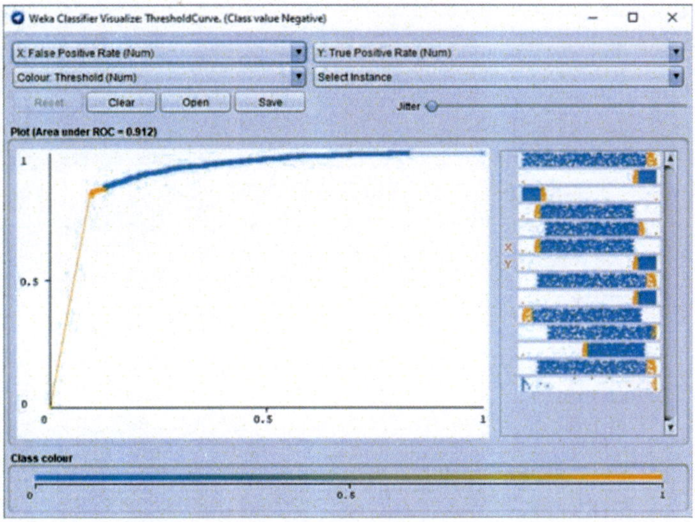

Fig. 11 ROC for logistic classifier (class value negative)

and is 1171 feet tall. The time it took to construct the model was 216.63 s, and it took 0.06 s to validate it on the test break (Table 6).

ROC Curve for J48 Classifier

The ROC curve between False Positive Rate and True Positive Rate for J48 Classifier is shown in Fig. 18 (Class Value Positive) likewise The ROC curve between

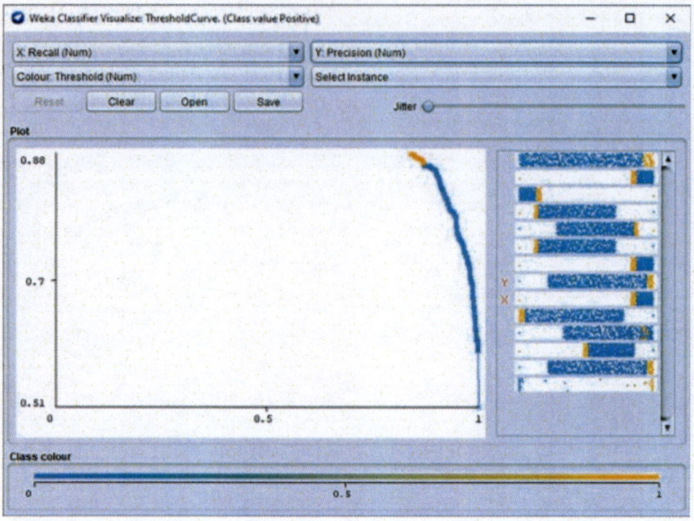

Fig. 12 PRC for logistic classifier (class value positive)

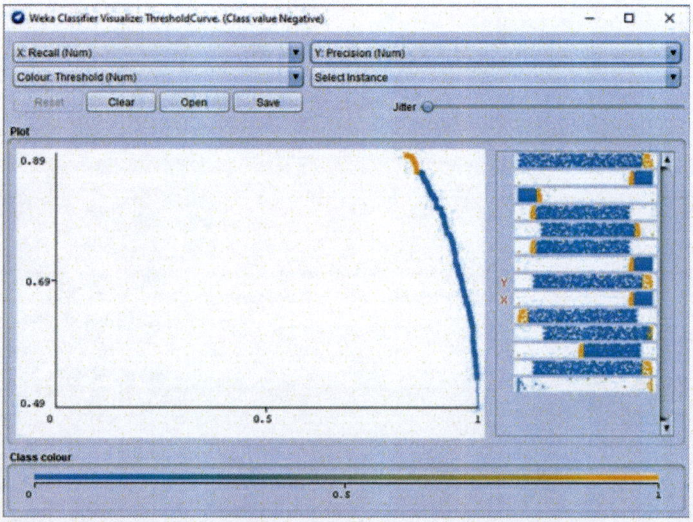

Fig. 13 PRC for logistic classifier (class value negative)

False Positive Rate and True Positive Rate for J48 Classifier is seen in Fig. 19 (Class Value Negative).

Precision Recall Curve (PRC) for J48 Classifier

Figure 20 shows the PRC curve between Recall and Precision for J48 Classifier (Class Value Positive) similarly Fig. 21 shows the PRC curve between Recall and Precision for J48 Classifier (Class Value Negative).

Table 5 Evaluation of random forest classifier

Method	Correctly classified instance	Incorrectly classified instance	Rate of TP	Rate of FP	Precision	Recall	F1-measure	MCC	ROC area	PRC area
Random forest	90.0455%	9.9545%	0.900	0.100	0.901	0.900	0.900	0.801	0.963	0.961

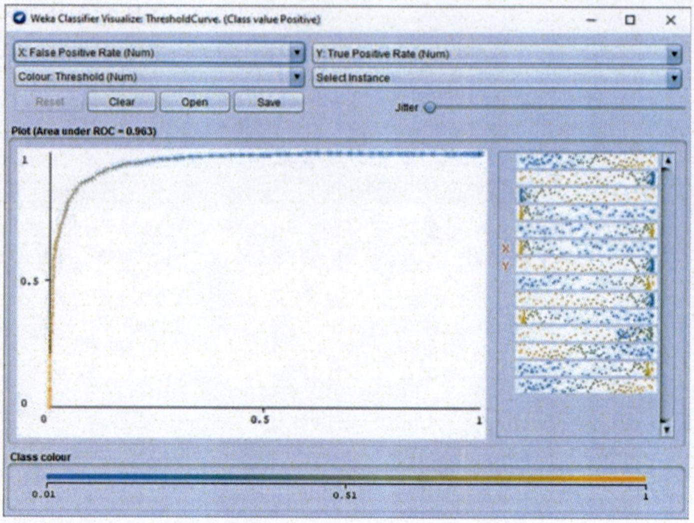

Fig. 14 ROC for random forest classifier (class value positive)

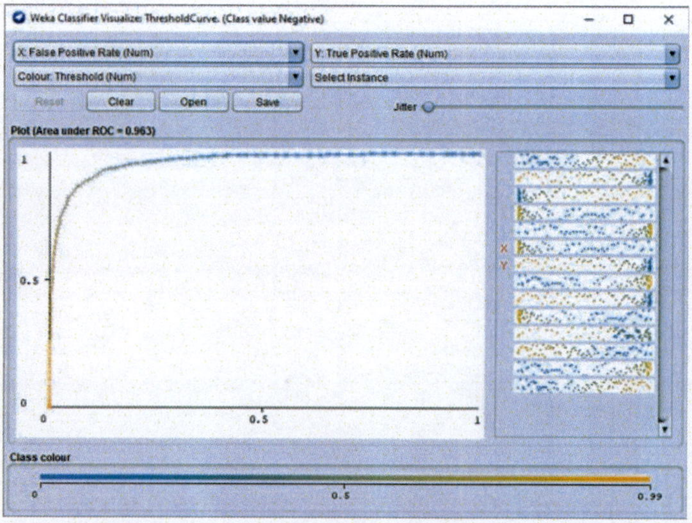

Fig. 15 ROC for random forest classifier (class value negative)

According to the accuracy results, LibSVM classifier produced the highest accuracy of 92.5%, while random forest produced the highest accuracy of 90.0455% as compared to others. As a result, in the case of tourism data obtained, LibSVM and random forest algorithm are very well used to perform sentiment analysis. The support vector machine classifier was found to be the most reliable and complete classifier in the experiment, with highest values for all four measurements (Accuracy,

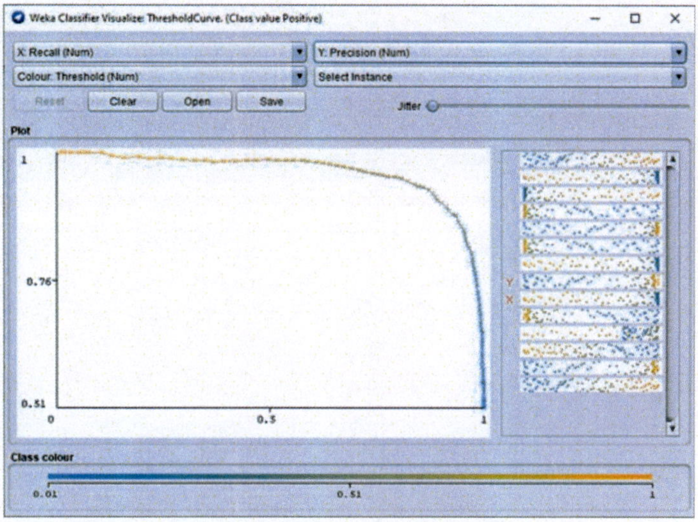

Fig. 16 PRC for random forest classifier (class value positive)

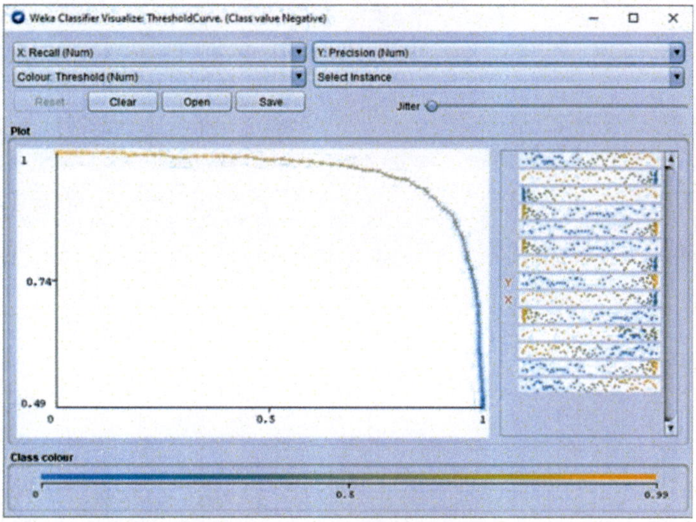

Fig. 17 PRC for random forest classifier (class value negative)

Precision, Recall, and F1-score). The results show that random forest is a classifier worth considering, as it can achieve high values on all of the measures while only marginally outperforming the support vector machine classifier (Fig. 22 and Table 7).

Table 6 Evaluation of J48 classifier

Method	Correctly classified instance	Incorrectly classified instance	Rate of TP	Rate of FP	Precision	Recall	F1-Measure	MCC	ROC area	PRC area
J48	83.0455%	16.9545%	0.830	0.170	0.830	0.830	0.830	0.661	0.836	0.797

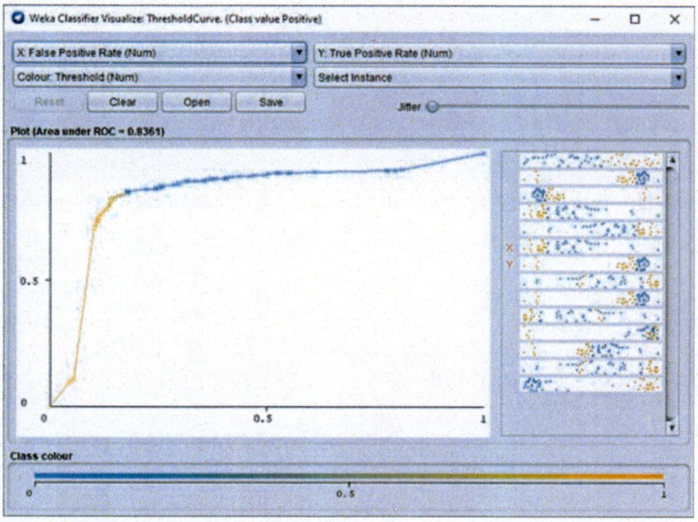

Fig. 18 ROC for J48 classifier (Class value positive)

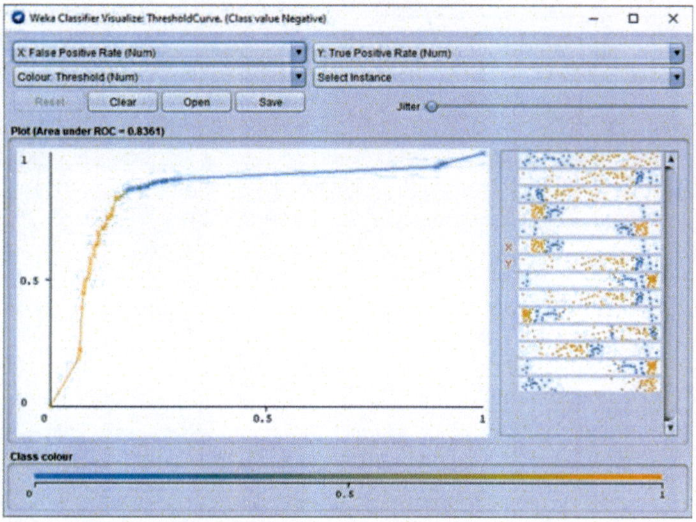

Fig. 19 ROC for J48 classifier (Class value negative)

5 Conclusion and Implications for the Future

In this analysis, performance of five different sentiment classification methods was analyzed on tourism dataset of 11,000 (5500 positive and 5500 negative) reviews. It is observed that SVM with linear kernel outperforms all other models under identical

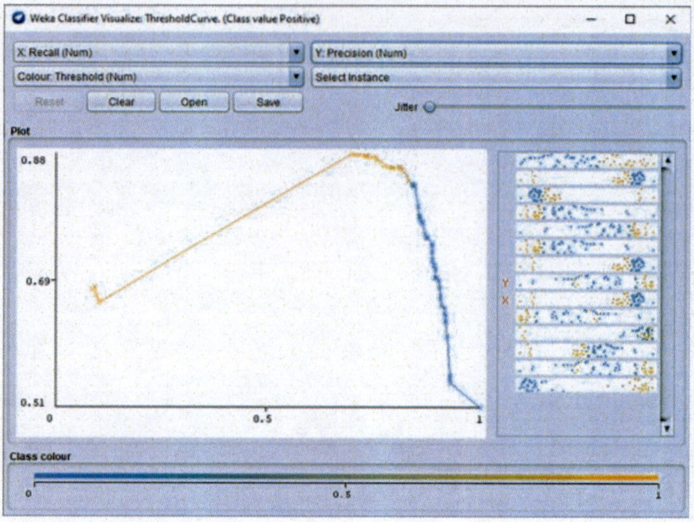

Fig. 20 PRC for J48 classifier (class value positive)

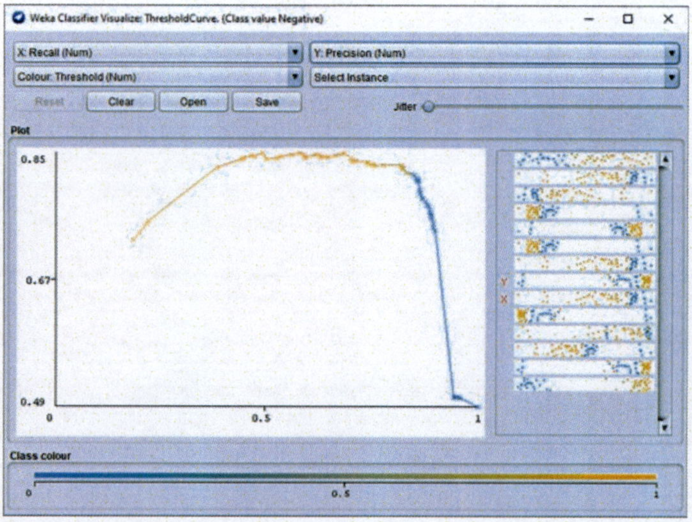

Fig. 21 PRC for J48 classifier (class value negative)

feature set. Experiments were conducted on different feature set and it is concluded that linear SVM gives highest accuracy for binary classification of reviews as it is able to generate optimum hyperplane with linear kernel. This study can be further enhanced by incorporating multilevel classification of reviews and finding sentiments on different aspects of tourism domain. Further, we can perform these analysis using sarcasm detection, double negation, compound and multiple sentences, and complex

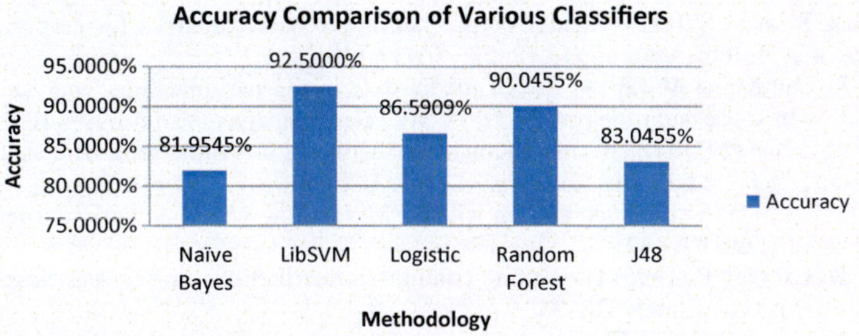

Fig. 22 Accuracy comparison of various classifiers

Table 7 Comparison of various classifiers

Methodology	Accuracy (%)
Naïve Bayes	81.9545
LibSVM	92.5000
Logistic	86.5909
Random forest	90.0455
J48	83.0455

reviews for enhanced study. Deep learning models can raise accuracy in multilevel aspect based sentiment classification.

Acknowledgements The authors are grateful to the Government of India's Technical Education Quality Improvement Programme, Phase-III (TEQIP-III) for financially supporting the relevant research.

References

1. Ahirwal PRR (2018) Tourism review sentiment analysis using lexicon features and machine learning approach. 5(4)
2. Al-Azani S, El-Alfy E (2018) Emojis-based sentiment classification of arabic microblogs using deep recurrent neural networks. In: 2018 international conference on computing sciences and engineering (ICCSE). pp 1–6. https://doi.org/10.1109/ICCSE1.2018.8374211
3. Alaei AR et al (2017) Sentiment analysis in tourism:capitalizing on bi data. https://doi.org/10.1177/0047287517747753
4. Albusaidi HS et al (2017) Twitter data streaming and apturing for tourism dashboard application. 5(2):63–66
5. Ali J et al (2012) Random forests and decision trees. December 2013
6. Biba M, Mane M (2014) Sentiment analysis through machine learning : an experimental evaluation. 195–203. https://doi.org/10.1007/978-3-319-01778-5
7. Breiman LEO (2001) Random forests. 5–32
8. Cankurt S, Subasi A (2015) Developing tourism demand forecasting models using machine learning techniques with trend , seasonal , and cyclic components. 3(1)

9. Cieliebak M et al (2017) A twitter corpus and benchmark resources for german sentiment analysis. 45–51. https://doi.org/10.18653/v1/w17-1106
10. Floris R, Campagna M (2014) Social media data in tourism planning: analysing tourists' satisfaction in space and time roberta floris. Michele Campagna. 8(May):997–1003
11. Jianqiang Z et al (2018) Deep convolution neural networks for twitter sentiment analysis. IEEE Access 6:23253–23260. https://doi.org/10.1109/ACCESS.2017.2776930
12. Joachims T (2005) Text categorization with SVM: learning with many relevant features. Eur Conf Mach Learn 137–142
13. John GH, Langley P (1995) Estiamting continuous distributions in Bayesan classifiers. Elev Conf Uncertain Artif Intell 338–345
14. Kamel N, Gayar N (2008) El: tourism demand foreacsting using machine learning methods
15. Kaur G (2014) Improved J48 classification algorithm for the prediction of diabetes improved J48 classification algorithm for the prediction of diabetes. https://doi.org/10.5120/17314-7433
16. Kim ZM et al (2016) Classif Travel-related Intents in Textual Data 3(1):96–101
17. Kirelli Y, Arslankaya S (2020) Sentiment analysis of shared tweets on global warming on twitter with data mining methods: a case study on Turkish language. Comput Intell Neurosci. https://doi.org/10.1155/2020/1904172
18. Metsis V et al (2006) Spam filtering with Naive Bayes—which Naive Bayes? In: 3rd conference email anti-spam—proceedings, CEAS 2006
19. Mikolov T et al (2013) Distributed representations of words and phrases and their compositionality. In: Burges CJC et al (eds) Advances in neural information processing systems. vol 26. Curran Associates, Inc, pp 3111–3119
20. Minai A (2018) ScienceDirect 47:2005–2007
21. Niloy NH, Navid MAI (2018) Naïve bayesian classifier and classification trees for the predictive accuracy of probability of default credit card clients. 3(1):1–12. https://doi.org/10.11648/j.ajdmkd.20180301.11
22. Patel R (2017) Sentiment analysis on twitter data using machine learning by Ravikumar Patel A thesis submitted in partial fulfillment of the requirements for the degree of MSc Computational Sciences The Faculty of Graduate Studies
23. Patil TR, Sherekar SS (2013) Performance analysis of Naive Bayes and J48 classification algorithm for data classification. 6(2)
24. Peng CJ et al (2002) The Use Interpretation of Logistic Regre Higher Educ J 43(3):1988–1999
25. Qaiser S, Ali R (2018) Text mining: use of TF-IDF to examine the relevance of words to documents. Int J Comput Appl 181(1):25–29. https://doi.org/10.5120/ijca2018917395
26. Ren Q et al (2017) Research on machine learning framework based on random forest algorithm. In: AIP conference proceedings. vol 1820. https://doi.org/10.1063/1.4977376
27. Ren Q et al (2017) Research on machine learning framework based on random forest algorithm research on machine learning framework based on random forest algorithm. vol 080020. https://doi.org/10.1063/1.4977376
28. Roshanfekr B et al (2017) Sentiment analysis using deep learning on Persian texts. In: 2017 Iranian conference on electrical engineering (ICEE). pp 1503–1508. https://doi.org/10.1109/IranianCEE.2017.7985281
29. Salur MU (2019) SmartSenti : a twitter-based sentiment analysis system for the smart tourism in Turkey
30. Saravanan N, Gayathri V (2018) Performance and classification evaluation of J48 algorithm and Kendall ' s based J48 algorithm (KNJ48). 59(2):73–80
31. Schmunk S et al (2013) Information and communication technologies in tourism 2014. Inf Commun Technol Tour (2013). https://doi.org/10.1007/978-3-319-03973-2
32. Shi L et al (2010) Financial data mining based on support vector machines and ensemble learning. In: 2010 international conference intelligence computing technology automation ICICTA 2010. vol 2. 313–314. https://doi.org/10.1109/ICICTA.2010.787
33. Sperandei S (2014) Lessons in biostatistics understanding logistic regression analysis
34. Valdivia A et al (2019) Neurocomputing Inconsistencies on TripAdvisor reviews : a unified index between users and sentiment analysis methods. Neurocomputing 353:3–16. https://doi.org/10.1016/j.neucom.2018.09.096

35. Vapnik VN (1995) The nature of statistical learning theory
36. Vateekul P, Koomsubha T (2016) A study of sentiment analysis using deep learning techniques on Thai Twitter data. In: 2016 13th international joint conference on computer science and software engineering (JCSSE). pp 1–6. https://doi.org/10.1109/JCSSE.2016.7748849
37. Wang H et al (2010) Web service classification using support vector machine. In: Proceedings international conference tools with artificial intelligence ICTAI. vol 1. pp 3–6. https://doi.org/10.1109/ICTAI.2010.9
38. Wang SJ et al (2009) Empirical analysis of support vector machine ensemble classifiers. Expert Syst Appl 36 (3 PART 2):6466–6476. https://doi.org/10.1016/j.eswa.2008.07.041
39. Windasari IP, Eridani D (2017) Sentiment analysis on travel destination in Indonesia. 276–279
40. Xu S (2018) Bayesian Naïve Bayes classifiers to text classification. J Inf Sci 44(1):48–59. https://doi.org/10.1177/0165551516677946
41. Yeh IC, Lien CH (2009) The comparisons of data mining techniques for the predictive accuracy of probability of default of credit card clients. Expert Syst Appl 36(2 PART 1):2473–2480. https://doi.org/10.1016/j.eswa.2007.12.020
42. Zheng S et al (2016) A machine learning-based tourist path prediction. 0–4

An Enhanced Machine Learning Classification System to Investigate the Status of Micronutrients in Rural Women

V. V. R. Maheswara Rao, N. Silpa, Gadiraju Mahesh, and Shiva Shankar Reddy

Abstract The health and nutrition status of marginalized women with constraint economic resource in rural setting remain a great challenge. In addition, poor nutritional knowledge, attitudes, and dietary practices significantly contribute to the rapid growth of malnutrition among the rural women. In an attempt to re-architect the present scenario, it is necessary to employ the interventions of science and technology in rural women community. To adapt the interventions of science and technology practically, it is very much required to collect, analyze, and interpret the data on health profile, family history, dietary habits, and blood profile of an individual. Naturally, this data are huge in volume, potentially high dimensional, absolutely diversified, and structurally complex. To extract and study the insights of such data and segment the rural women efficiently, it is necessary to employ a capable data mining technique. Toward this goal, the investigators in the present paper present an intelligent data classification system (IDCS). The IDCS is designed with four stages on the basis of classification technique with a comprehensive approach. Initially, the first stage of IDCS focusses on collecting the data from various resources on nutritional knowledge, attitudes, and dietary practices of the rural women systematically. Later, it rightly preprocesses the data with the standard practices and prepare in suitable structure to apply classification technique. Subsequently, an intelligent classifier is modeled using learning algorithm which can classify the rural women in desired number of groups with respect to nutritional levels. Finally, this system identifies suitable interventions of science and technology like creating awareness about nutritional knowledge, arranging training workshops to prepare the food without losing nutrients and ready to cook nutritious complementary food, promoting homestead gardens and backyard poultry, etc., appropriately to the respective rural women groups. This system helps to identify the under nutritional women accurately and

V. V. R. Maheswara Rao (✉) · N. Silpa
Department of Computer Science and Engineering, Shri Vishnu Engineering College for Women(A), Bhimavaram, Andhra Pradesh, India

G. Mahesh · S. S. Reddy
Department of Computer Science and Engineering, S.R.K.R. Engineering College, Bhimavaram, Andhra Pradesh, India

R. P. Mahapatra et al. (eds.), *Proceedings of International Conference on Recent Trends in Computing*, Lecture Notes in Networks and Systems 341,
https://doi.org/10.1007/978-981-16-7118-0_4

efficiently. This model decreases the rate of fault and demonstrates substantial efficiency. To authenticate the proposed research study, a number of experiments are done and results are presented in this paper.

Keywords Classification · Data mining applications · Machine learning · Intelligent systems · Micronutrient analysis etc

1 Introduction

The rapid advancement in the past decades has witnessed an unprecedented change in the lifestyle of rural women affecting the knowledge on nutrition, food attitudes, eating practices, and health consciousness led to negative impact on the health of rural women. On the other hand, the modern agricultural practices [1] resulted in producing fewer nutrients food. The rural women are consuming unbalanced diet, nutrition transition consequences of occurrence of chronic diseases like increased blood pressure, glucose, abnormal blood lipids, and obesity at exponential rate. To study the complete insights of such data and segment the rural women, it is essential to employ a capable data mining technique.

The rural women from 30 villages of 3 mandals in the West Godavari district of Andhra Pradesh, India, are selected for the present study with the support of DST-SEED division. The data are collected using the standard survey form designed under the supervision of nutrition expert, pertaining series of questions to retrieve the information regarding the nutritional knowledge, attitudes, dietary practices, family history, and present health status of an individual rural women. To understand the insights of this data, the present investigators are triggered and motivated toward the advancements of data mining techniques [2, 3] in order to study and explore the needs of different people.

Among all techniques of data mining [4–6], the present problem endorses the requirement of applying the classification technique to segment the rural women in desired groups. A classification is a systematic approach with a task of handover data objects to one of numerous pre-defined classes. Each classification technique builds on a learning algorithm to decide the fitness of relationship between the data object and the class. As conventional classification techniques require more domain experts' intervention, thus, the investigators in the present work pay attention toward learning classification techniques which classify the desired rural women groups with reduced human intervention.

With this background, the present investigators reported in this paper, the necessity of learning algorithm [7–11] to classify the rich nutrition group and poor nutrition group among the rural women. It helps to provide right interventions of science and technology like creating awareness about nutritional knowledge, arranging training workshops to prepare the food without losing nutrients and ready to cook nutritious complementary food, promoting homestead gardens and backyard poultry, etc., among the rural women.

The remaining paper is ordered as follows: Initially, it gives a brief related work. Then, the proposed intelligent data classification system is presented. Subsequently, it provides mathematical model in detail. Later, the experimental analysis is showcased. Finally, the conclusions are made.

2 Related Work

In the literature, many authors [4, 9–12] expressed that applying of the learning algorithms are more suitable for dynamically growing data. The experimental results of classification techniques [5, 7, 13] have been proposed to improve the overall success rate in segmenting the health care data and its management. Some more authors [2, 6–8, 14] also expressed the relevance of the machine learning approach in the field of health care, analysis of steady growth of related health survey data, and predict human deceases. In the literature [15, 16], the classification technique is used on the health data sets having more features for the diagnosis of heart deceases and they also expressed in their future work, learning classification techniques lead to better performance. The research provided by some more authors [15–18] in the literature found the presentations of data mining which are growing in analyzing health care data for improved health plan creation and finding of decease outbreak with root causes. They also mentioned the relevance of learning system that classifies the data accurately and alerts the people about the dietary habits promptly.

3 Proposed Intelligent Data Classification System (IDCS)

The micro nutrient deficiency analysis of rural women is a great socio economic importance, responsible not only in India across the world. This is not always clinically apparent but more associated with food consumption habits and family history data. To understand the significant knowledge in the insights embedded in nutritional knowledge, attitudes and practices of rural women and segmentation of this data accurately become a central problem. In addition, handling of such data lies not only high volume, potentially high dimensional, absolutely diversified and also structurally complex is one of the main motivation for the present investigators. To adapt the right interventions of science and technology for improving, the nutritional levels of right rural women recognized as a straightforward motivation factor for the current central problem of a classification.

In order to overcome the inherited challenges of segmenting rural women, the investigators in the present paper propose an intelligent data classification system as shown in Fig. 1, which follows fundamentally a comprehensive approach and modeled as a four stage model. The IDCS helps to classify the rural women with respect to the nutritional status accurately and efficiently. Initially, the first stage of IDCS concentrates on collecting the data through various possible sources like

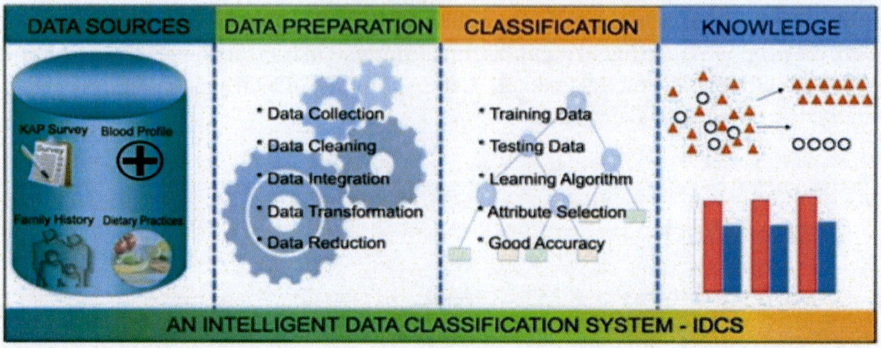

Fig. 1 Architecture of IDCS

nutritional knowledge, dietary, and practices of rural women under the guidance of nutritional expert. In the second stage, the IDCS mainly focuses on the preparation of data by employing necessary preprocessing techniques suitable to classification algorithms. The IDCS, in third stage, classifies the rural women, namely poor and rich nutritional groups by adapting an intelligent classifier which is modeled using a learning algorithm. The final stage of IDCS identifies right interventions of science and technology to the right groups based on classification of the previous stage under the supervision of nutritionist.

3.1 Data Sources

In order to accurately classify the rural women with respect to nutritional levels, it is highly dependent on the selection of data resources. Toward this, the investigators in the present work elect survey form as a primary resource to estimate the nutritional levels of rural women. The questions in the survey form are designed under the guidance of nutritional expert. The questions intrinsically retrieve the facts about nutritional knowledge, attitudes, dietary practices, and family history of rural women. To adapt interventions of science and technology and classification of rural women accurately, the investigators select the blood sample as a secondary resource.

3.2 Data Preparation

IDCS data preparation includes all milestones, including data processing, data cleaning, data integration, data transformation, and data reduction, in order to prepare data suitable for the original raw data classification algorithm. Data collection work is conducted on the basis of inputs drained from rural women's nutritional status. The possible ways of collecting the data of rural women are knowledge, attitude and practices (KAP) survey, family history, dietary habits, and blood profile under the supervision of nutritional expert. The whole data preparation operation increases the importance of the data to the basic techniques of classification. The resultant processed information feeds to the next stage of IDCS as an input to successfully classify the rural women based on their nutritional status.

3.3 Classification of Rural Women

The data collected and prepared from various resources are potentially high diversified and having poor nutritional rural women data as well as rich nutritional rural women data. To adapt interventions of science and technology, the data of rural women having rich nutrition is not required. Thus, it is essential to classify the rural women with respect to nutritional levels into two groups as rich nutritional women group (RNWG) and poor nutritional women group (PNWG). To perform this task, the investigators in the present paper propose intelligent data classification system (IDCS) that acquires the knowledge from the facts of standard nutritional data. Together with this intelligence, the IDCS classifies RNWG and PNWG accurately. In addition, it helps in implementing the interventions of science and technology for the successful implementation of overall proposed work.

The IDCS takes the prepared data and classifies RNWG and PNWG intelligently with less time. The prepared data collected from various resources include dietary habits data, health data, and family history data. Toward this, the IDCS algorithm is intended that it acquires intelligence from the related and standard nutritional facts. For the proposed IDCS derived attributes reserved as set of related facts, which are specific in nature and classify the set of classes as RNWG and PNWG. These derived attributes from the original collected data used as training data to the IDCS. The complete collected and prepared data become testing data for IDCS.

In order to classify any new data in the testing data either RNWG or PNWG, IDCS using learning intelligence approach by probe a sequence of queries about the features of newly entered data. In each time, IDCS receives the answer and trail up query is probed until IDCS reaches a deduction about the class name of new entry. The sequence of inquiries and their one-to-one responses can be structured in form of a tree consisting of nodes and edges. A root node is a node with no edges inward and two or more edges outward. Any inner node properly has one inner edge and two or more outer edges. The process of IDCS tree development is done by developing start and internal nodes with the nutritional knowledge, attitude, and practices of rural women and each end node represents as a class labels. The labels of class are either RNWG or PNWG.

To classify the rural women data accurately, one can use the intelligent systems to reduce the time for overall processing of proposed work. There are numerous tree algorithms in theory that could be developed from an accepted number of attributes. But discovery the learning tree algorithm, which is right feasible and capable of acquiring intelligence for large size of rural women data is highly complicated. These algorithms specifically utilize a sensible line of attack to induce the intelligence. One such algorithm is IDCS, which is customized and is suitable for typical nature of rural women data. As stated earlier, due to the complex nature of data, one has to focus on creation of the related facts-base of derived attributes. The IDCS design consists of data-training and data-testing. Finally, IDCS generates the learning tree structure based on a set of rules and which consists of nodes and class labels.

The input for IDCS is obtained from the data prepared from the survey forms of rural women, which is having highly diverse and voluminous entries. Each entry corresponds to a details of individual rural women consists of numerous attributes of nutritional knowledge, attitude, dietary practices, and details of health status. For the proposed IDCS, all these numerous attributes treated as testing data (TA), which consists of entries with unknown labels.

Among all the attributes, some of the attributes are useful to classify the rural women. These attributes are considered as training data or derived attributes. Based on the notable features of tree, these attributes are identified, which are discrete, continues, and specific in nature. The entries in training data have known labels. The set of derived attributes $DA = \{DA\ 1,\ DA\ 2,\ \ldots,\ DAn\ \}$, which forms knowledge base for IDCS. This knowledge base makes the IDCS to acquire enough learning to classify the women automatically and reduces the processing time significantly. The proposed IDCS derives many attributes from details of nutritional knowledge, nutritional attitudes, dietary practices, and health status of rural women collected and prepared from various resources to classify as RNWG and PNWG. A sample of derived attributes is shown in Table 1, and a sample characteristic of rich and poor nutritional rural women is showcased in Tables 2 and 3, respectively.

Algorithm of IDCS

TreeCreate(DA, TA)

Table 1 Sample derived attributes

Derived attribute	Description
Adequacy of food	Each critical nutrient, fiber, and energy are adequately given by foods
Balanced diet	Food options at the cost of fiber should not emphasize nutrients or food
Calorie count	Foods have the amount of energy required to maintain the optimum weight
Equability of food	Excess sugar, sodium, glucose, or other unnecessary ingredients are not produced by foods
Diversity of food	Food varies from day to day
Health status	Details of present health condition
Hereditary complaints	Details of a diseases or disorders that are inherited genetically
Life style habits	Details of common lifestyle habits that cause diseases
Vitamin deficiency symptoms	Details of signs and symptoms of vitamin deficiencies

Table 2 Sample characteristics of rich nutritional rural women

Rich nutritional women group (RNWG):
• Take adequate food that is essential nutrient, fiber, and energy • Maintain balanced diet choices • Have the knowledge of calorie control on food • Do not have excess weight, sodium, glucose, or other unnecessary components if you are conscious of the food • Consume multiple types of foods regularly • Good condition of present health • No diseases or disorders that are inherited genetically • Good lifestyle habits • No signs and symptoms of vitamin deficiencies

Table 3 Sample characteristics of poor nutritional rural women

Poor nutritional women group (PNWG):
• Take less nutritious, fiber, or high energy critical foods • Unable to maintain balanced diet choices • No knowledge of calorie control on food • Less conscious on the nutrients of the food • Eat same kind of foods daily • Suffer from health conditions • Have diseases or disorders that are inherited genetically • Bad lifestyle habits and addictions • Observe signs and symptoms of vitamin deficiencies

Step 01: If (ConditionEnd(DA, TA) = True) then
Step 02: LastNode = CreateNewNode( )
Step 03: LastNode.Label = GiveLabel(DA)
Step 04: Return LastNode
Step 05: Else
Step 06: Root = CreateNewNode( )
Step 07: Root.ConditionCheck = DeriveBestAttribute(DA, TA)
Step 08: PO ={po /po, possible outputs, ConditionCheck()}
Step 09: For every po Є PO perform
Step 10: DApo ={da / Root.ConditionCheck(da) = po and d Є DA}
Step 11: Succ = TreeCreate(DApo, TA)
Step 12: Add Succ as successor of parent, label it as po
Step 13: close for loop
Step 14: stop if
Step 15: Return parent

The learning algorithm works by selecting DeriveBestAttribute() recursively (step 07) and extending the tree leaf nodes (step 11 and 12) until ConditionEnd() is satisfied (step 01). The details of methods used in the learning algorithm are follows,

CreateNewNode(): The IDCS makes use of this function to create the tree with an extension of newly created node. Either a test condition or a class label is given to a new node in the IDCS tree.

ConditionCheck(): An attribute check requirement must be selected for every recursive stage of IDCS TreeCreate() to be divided into two subgroups as RNWG and PNWG. To execute this stage, the IDCS algorithm uses a ConditionCheck() method to calculate the quality of each condition of an attribute.

ConditionEnd(DA, TA): By checking if all records have either the identical class label or even the same attribute content, IDCS adopts this feature to stop the IDCS tree expansion. A further approach to terminate the feature is to check if the number of records has dropped underneath the lower limit.

GiveLabel(): To evaluate the class label to be allocated to the very last node, the IDCS uses this function. The last node is labeled in most situations to a category with a larger set of training data.

DeriveBestAttribute(): This role is processed by the IDCS to find out which derived attribute must be chosen as a testing condition for the training to be extracted.

4 Experimental Analysis

Learning performance of IDCS: The sample rural women data collected across 30 villages in 3 mandals are considered for experimentation with one year data. The experiments prove that IDCS reduces the human intervention to classify the data. In IDCS, the error margin between the testing dataset and testing dataset is almost reduced and is estimated to be an amount of 0.2 presented in Fig. 2.

Computing Performance of IDCS: In a typical setting with regard to the processing time of IDCS, many experiments are performed. The findings demonstrated that, as shown in Fig. 3, IDCS consistently takes minimized execution time.

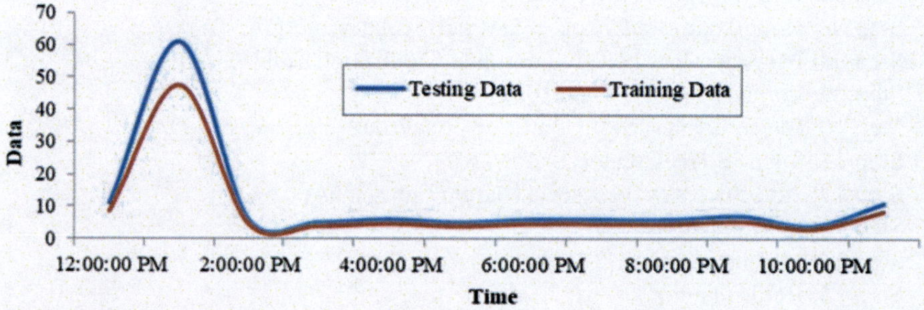

Fig. 2 Error rate between training and testing data

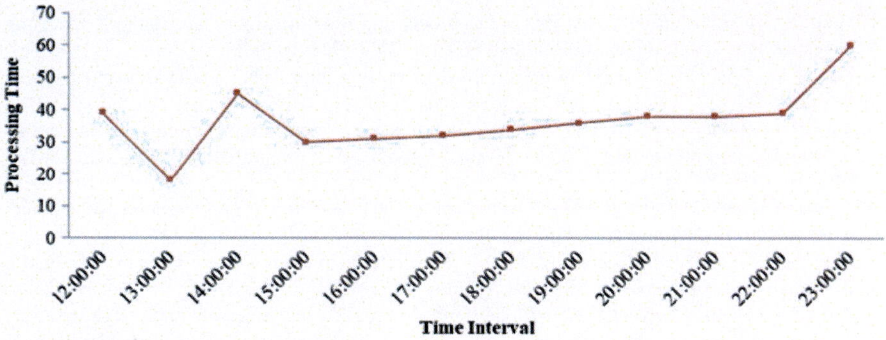

Fig. 3 Processing performance of IDCS

5 Conclusion

Classification is a promising research area in the era of data mining to study the insights of the rural women data and segment the rural women with respect to nutritional levels. The investigators in the present paper outlined the importance of all the stages, namely selection of suitable sources of data, data preparation, classification, and knowledge interpretation. The investigators introduced an intelligent learning algorithm IDCS to separates the rich nutritional women group and poor nutritional women group efficiently. The IDCS reduces the human intervention and classifies the said groups automatically in less time. Also, mathematical model is presented to ensure the goodness of error rate. Finally, the proposed system helped to adapt right interventions of science and technology like creating awareness about nutritional knowledge, arranging training workshops to prepare the food without losing nutrients and ready to cook nutritious complementary food, promoting homestead gardens and backyard poultry, etc., among the less nutritional rural women.

References

1. Acham H, Oldewage-Theron WH et al (2012) Dietary diversity, micro nutrient intake and their variation among black women in informal settlements in South Africa: a cross-sectional study. IJNAM 4(2):24–39
2. Hian CK, Gerald T (2012) Data mining applications in healthcare. J Healthcare Info Managem 19(2):64–72
3. Tipawan S, Kulthida T (2012) Data mining and its applications for knowledge management: a literature review from 2007 to 2012. IJDKP 2(5):13–24
4. Health Catalyst (2014) It all starts with a data warehouse
5. Juan A (2014) Lara, David Lizcano, "A general framework for time series data mining based on event analysis: application to the medical domains of electroencephalography and stabilometry." J Biomed Inform 51:219–241
6. Olegas N, Olga K (2010) Data mining applications in healthcare: research versus practice. Institute of Mathematics and Informatics, Vilnius University, pp 58–70

7. Hamidi H, Daraei A (2016) Analysis of pre-processing and post-processing methods and using data mining to diagnose heart diseases. IJE Trans 29(7):921–930
8. Tomar D, Agarwal S (2013) A survey on data mining approaches for healthcare. IJBSBT 5(5):241–266
9. Santos RS, Malheiros SMF et al (2013) A data mining system for providing analytical information on brain tumors to public health decision makers. Comput Methods Programs in Bio Med 109:269–282
10. Senthil Kumar D, Sathyadevi G, Sivanesh S (2011) Decision support system for medical diagnosis using data mining. IJCSI 8((3)1):147–153
11. Mohit K, Rayid G, Zhu-Song M (2010) Data mining to predict and prevent errors in health insurance claims processing. KDD'10, Washington, DC, USA, pp 65–73
12. Anguera A, Barreiro JM et al (2016) Applying data mining techniques to medical time series: an empirical case study in electroencephalography and stabilometry. Comput Struct Biotechnol J 185–199
13. Olegas N (2015) Development and application of data mining methods in medical diagnostics and healthcare management. Doctoral Dissertation, Vilnius University
14. Patel S, Patel H (2016) Survey of data mining techniques used in healthcare domain. Int J Info Sci Techniques 6(1/2):53–60
15. Alzahani SM, Althopity A et al (2014) An overview of data mining techniques applied for heart disease diagnosis and prediction. Lecture Notes on Information Theory 2(4):310–315
16. Frantzidis CA, Bratsas C et al (2010) On the classification of emotional bio signals evoked while viewing affective pictures: an integrated data-mining-based approach for healthcare applications. IEEE Trans Inf Technol Biomed 14(2):309–318
17. Shubpreet K, Bawa RK (2015) Future trends of data mining in predicting the various diseases in medical healthcare system. IJEIC 6(4):17–34
18. Abdullah AA, Mohammed GA (2013) Application of data mining: diabetes health care in young and old patients. J King Saud Univ—Comput Info Sci 25:127–136

Advanced Hydroponics Fodder System with Temperature Control Using Arduino

Ahbab Rizvi, Aditi Raj, Anishrawa Sharma, and Vinesh Kumar

Abstract In hydroponics [1–3] fodder [4] system, hydroponics are the plants which are grown in the absence of soil and pesticides or fertilizers, i.e., only using hydro(water). Fodder is the part of plants that is being fed to the cattle and livestock. This part is produced using temperature sensor DS18B20 which measures the temperature and give readings on the LCD display underneath the roots. The code is uploaded using USB cable to Arduino UNO. These plants require low maintenance and are low cost. The plants can be developed fully in 8–10 days. For making it conventional for traditional farmers, use of nutrient solution is not necessary, and tap water can also be used. In a country like India, wheat is the most liked and chosen grain for traditional farming, and hence, wheat is widely used to produce fodder. The fodder produced looks like 25 cm high green compromising of the whole grown plant, i.e., seeds, plants, roots, and stems. For production of about 1 kg of good quality wheat fodder, about 2–3 L of water is required, and it yields about 5–6 times of good quality fodder on regular basis. The fodder which is produced using automation hydroponics technique is more digestive and nutrient rich which also provides other dietary benefits to the cattle. From cost point of view, the price of seed compromises of about 86% of the total cost of fodder. A cow requires about 5–10 kg of dietary supplement everyday, and hence, hydroponics wheat fodder is capable of providing organic, fresh, and nutritious fodder to the cattle without any chemicals and harmful supplements. Feeding of hydroponics fodder will increase the cattle and livestocks' digestability which will in-turn result in increased production of by-product (milk) as

A. Rizvi (✉) · A. Raj · A. Sharma · V. Kumar
Department of Computer Science and Engineering, Meerut Institute of Engineering and Technology, Meerut, UP 250001, India
e-mail: ahbab.rizvi.cse.2017@miet.ac.in

A. Raj
e-mail: aditi.raj.cse.2017@miet.ac.in

A. Sharma
e-mail: anishrawa.sharma.cse.2017@miet.ac.in

V. Kumar
e-mail: vinesh.kumar@miet.ac.in

R. P. Mahapatra et al. (eds.), *Proceedings of International Conference on Recent Trends in Computing*, Lecture Notes in Networks and Systems 341,
https://doi.org/10.1007/978-981-16-7118-0_5

compared to conventional/traditional feeding methods. Therefore, fodder produced using this automated system could be used by the farmers to feed their livestock in a cost-effective manner.

Keywords Arduino UNO · Fodder · Hydroponics · Temperature · Livestock

1 Introduction

Hydroponic [5] fodder systems for dairies, livestock, or poultry are developing as the traditional farmers are attracted by the idea of putting one kg of seed into a hydroponic system yielding five times its weight in fodder. Research analysis is positive in fodder production or animal health benefit. The main objective of this advancement is to direct farmers weigh the high production costs of fodder relative to any comparative dietary benefits obtained from feeding hydroponics fodder. The seeds are grown in trays on regular basis (Fig. 1), and water is dropped at timely intervals. Within eight days, the plants attain a height of 20–22 cm (Fig. 3); they are then taken out from the trays, and then, the fodder is ready to be fed to the livestock. Hence, it is the best way to overcome demand for quality livestock feed, especially for dairy farmers in developed areas having less land to produce the required amount of fodder for the livestock (Fig. 2 and 3).

Fig.1 Wheat at day 1

Fig. 2 Wheat at day 4

Fig.3 Wheat at day 8

2 Methodology

For growing better quality fodder, we need to maintain the moisture and temperature. The hydroponic fodder should be grown under controlled environment with the temperature between 16 and 30 °C and moisture between 78 and 85%. For increasing fodder quantity demand, we can build a hydroponic fodder system, this system requires a little space, we need to use 10 × 10 ft shade to grow hydroponic fodder, and it is better to choose shady location near the livestock (Figs.4, 5, 6, 7, 8 and 9).

For proper ventilation, some open space between the roof and side walls is left as it ensures that when the shed is well ventilated and airy, temperature and humidity

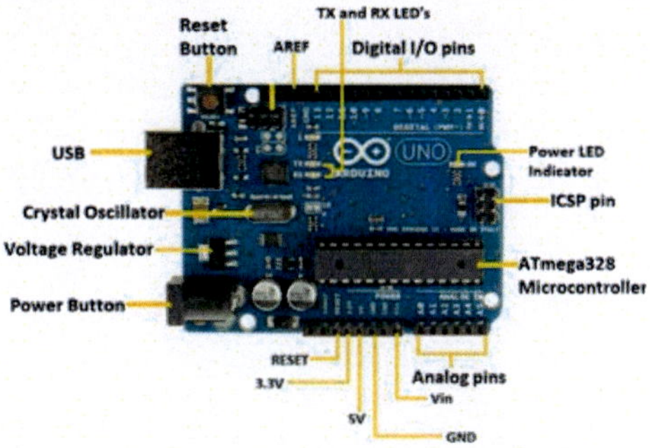

Fig. 4 Arduino UNO

Fig. 5 16X2 I2C LCD display

Fig. 6 Relay board

Fig. 7 DS18B20 temperature sensor

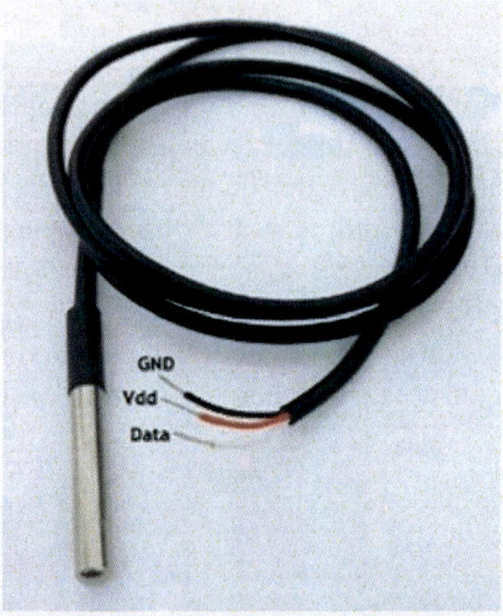

Fig. 8 Growing cabinet

Fig. 9 Rack for trays

Fig. 10 Trays

are easily maintained to grow hydroponic fodder, a medium size tray Fig. 10 of about 1 × 3 Ft made up of good plastic and strong enough to hold the weight of the fodder. The seeds are kept moist, so metal trays are avoided because they can easily corrode. A total of 15–20 small size holes are made for draining any excess water. Maximum three to four layers Fig. 9 of rack should be kept, but we have to ensure that rack should not be too high because it would be difficult to spray water. Enough space between two layers should be kept, so water can be easily sprayed, the seeds also create slope for each layer at one side of the rack for easy and quick water drain. Small drainage line Fig. 12 under the slope side of a rack should be created to properly drain out water.

3 Impact on Farmers

Hydroponics is just a single unit which require less space like a chamber (Fig. 8) with arrangements like temperature, humidity, and light intensity for extreme sprouting for the growth of fodder seeds like oat, barley, wheat, etc. To grow 8–9 kg of green fodder in 7–9 days, we need a moderate temperature of 30–35 °C temperature, 70–75% R humidity, and 60% shed of 1 kg wheat which is needed.

According to many researchers, it found that during summer season 2 L of water is enough for per kg of seeds to increase the amount of seeds up to 6–8 times. By choosing this method of farming, one can grow very healthy, sublime, and economically viable fodder for goat and dairy farming. Farmers can build their own hydroponics system (Fig. 9) as it is very easy to fabricate and all components are easily available in the market, and there is no need to purchase those hydroponics systems from market. It will reduce the cost and become more efficient for the farmers.

4 System Components

1. Arduino UNO
2. 16×2 I2C LCD display, relay board
3. DS18B20 temperature sensor (1 wire, waterproof)
4. 12 V adapter (power supply), 4.7 K (ohm) resistor, bread board, connecting wires
5. Growing cabinet, rack for trays, Trays (with holes)
6. Seeds
7. Watering system (drip pipes), PVC pipes (hard)
8. Water tanks, water pump.

4.1 Arduino UNO

Arduino UNO (Fig. 4) is a microcontroller in which we upload code with the help of USB cable, and in our project [5], the input will be taken in accordance with the temperature sensor readings and will give the respective output [6–8].

4.2 16×2 I2C LCD Display

It is used to display the extracted data to make the process easier and interactive, and also it will show the output, i.e., our readings of temperature sensor, the suitable temperature, etc., and I2C enables easier circuit connection with Arduino UNO.

4.3 (i) Relay Board

Relay board Fig. 6 is used to bifurcate the power supply from one pump to another. Hence, we can choose when and how to operate multiple pumps [9].

4.4 DS18B20 Temperature Sensor (1 Wire, Waterproof)

DS18B20 (Fig. 7) is a wired contact-based temperature sensor which can be used in water also. Thus, it is the most suitable choice of temperature sensor in hydroponics fodder system [10, 11].

4.5 Growing Cabinet

Most suitable cabinets Fig. 8 used in hydroponics are in square shape because they are easy to fabricate and easy to use. We can use any type of material (mild steel pipes, UPVC, or bamboo), and size of the cabinets is based on the number of trays used. These cabinets should be covered with 50% of green shed net of polyhouse film for conserving humidity and redundancy in light intensity [12].

4.6 (i) Rack for Trays

These racks Fig. 9 can be designed or fabricated with mild steel, UPVC, or bamboo. Tray rack height depends on the day cycle of fodder that is if we use 2–3 racks so for that height should be 6 inch, above three racks height should be 10–12 inch, and for more higher level, height can be increased up to 15 inch. For avoiding fungal growth and water lodging, a gentle slope is a plus point.

4.7 (ii) Trays

There are various kind of trays Fig. 10 available in the market which differ in their stuff and durability. Specialized hydroponics trays are costly (2*2 feet tray can cost upto 350). Generally, we use 1.5 feet*1 feet trays which have holes at the bottom, and each holes are at equal distance. These holes are used to avoid fungal infection and any kind of water lodging. For safety reasons after every use, tray needs to be sanitized by diluted hydrogen peroxide and sun drying.

Fig. 11 Seeds (wheat)

Fig. 12 Drip line

4.8 Seeds

Wheat Fig. 11, oat, maize, and barely are highly used for hydroponics. According to the needs of farmers, they can choose the seeds on the basis of rate and availability in the market. To improve the quality of the seeds, broken seeds are removed by using brine water treatment. To make sure that seeds should be free from bacterial and fungal infection, seeds are soaked in potassium permanganate for some hours.

For hydroponics fodder cultivation, 0.5 kg/Ft2 amount of seed is sufficient.

4.9 Drip Line

Generally, 16 mm HDEP drip Fig. 10 line is the best-suited method for watering in hydroponics fodder system. In this, the drip is connected to each tray in the cabinet and hence irrigating it uniformly [13, 14].

4.10 Pump

This pump Fig. 13 is used for pumping water for irrigation that will be supplied through drip lines. This is an AC operated water pump basically used in water coolers.

Fig. 13 Pump

5 Precautions

1. The seeds should not overlap each other.
2. All seeds should be germinated before being placed in the tray.
3. The setup should never be kept in sunlight.
4. The tray must contain tiny holes to sustain water requirements.
5. Suitable temperature should be set according to seasons.

6 Circuit Diagram of Setup

See Fig. 14

7 Conclusion

In future, there will be growing demand for land, i.e., grazing areas or arable land which produces animal feed needed to balance livestock production. Presently, about 17% of the domestic requirements for cereals, 21% starchy foods, and 4% oilseeds obtained from livestock production. Additionally, the part of production is used as animal feed and also exported in many countries. Big rise in the demand for livestock fodder may occur in Eastern and North African parts of the world. As there is less productive land or sufficient water in these regions to increase the food and feed productivity to meet the demand of fodder, this may sustain much larger feed production.

At present the graph of permanent grassland is changing gradually, it is observed that there has been a decrease of about 8.5 million hectares per year in the world over

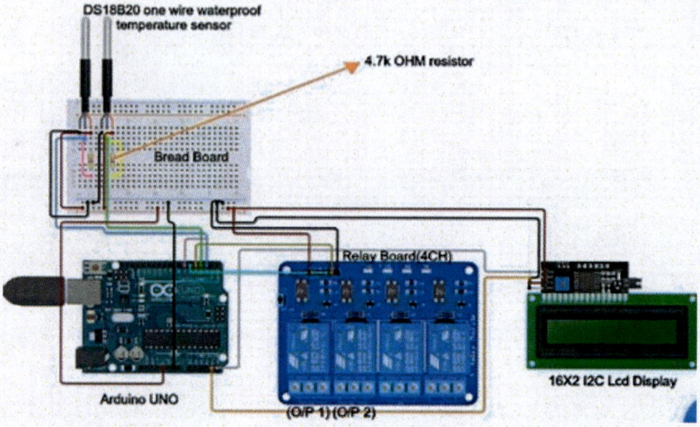

Fig. 14 Circuit diagram

the past thirty years. If this situation persists in the future, the demand of land for livestock production will come from feed and fodder production. Livestock fodder production is more nutritious and energetic than conventional production. As a result of rapid growth of the population of cattle species and the more efficiently usage of nitrogen by the animals, the level of increase in ammonia (NH3) and nitrogen oxide (N2O) emissions is less that is expected on the basis of growth of total fodder production.

In nutshell, the productivity of animal fodder provides the most rational way for using a considerable part of agricultural by-product. A affluence of knowledge has gathered in the last 15–20 years pointing the practicality of the operation and pinpointing their risk. The technology used sometime might be at large demand and need institute but are not conventional for rural areas. Judicial step is straggle behind feasible execution, but under the illumination of enlarge economical stress, enormous development may be anticipate in the coming years.

References

1. Aliac CJG, Maravillas E (2018) IOT hydroponics management system. In: IEEE 10th international conference on humanoid, nanotechnology, information technology, communication and control, environment and manangement (HNICEM)
2. Bharti NK, Dongargaonkar MD, Kudkar IB, Das S, Kenia M (2019) Hydroponics system for soilless farming integrated with android application by internet of things and MQTT broker. In: IEEE Pune section international conference (PuneCon)
3. Saraswathi D, Manibharathy P, Gokulnath R, Sureshkumar E, Karthikeyan K (2018) Automation of hydroponics green house farming using IOT. In: IEEE International conference on system, computation, automation and networking (ICSCA)
4. Tryhuba A, Tryhuba I, Ftoma O, Boyarchuk O (2019) Method of quantitative evaluation of the risk of benefits for investors of fodder producing cooperatives. In: IEEE 14th international

conference on computer sciences and information technologies (CSIT)
5. Ali MF, Thakur P, Mendiratta P, Gupta N (2019) IoT-based solar hydroponics farming. In: 6th international conference on computing for sustainable global development (INDIACom)
6. Eridani D, Wardhani O, Widianto ED (2017) Designing and implementing the arduino-based nutrition feeding automation system of a prototype scaled nutrient film technique(NFT hydroponics using total dissolved solids (TDS) sensor. In: 4th international conference on information technology, computer, and electrical engineering (ICITACEE)
7. Karmokar C, Hasan J, Khan SA, Alam MII (2018) In: Arduino UNO based smart irrigation system using GSM module, soil moisture sensor, sun tracking system and inverter. In: International conference on innovations in science, engineering and technology (ICISET)
8. Gopinath A, Arun C, Hanumanthaiah A, Murugan R (2020) An analogy of the datalogger implementation in Hyd UNO and PSoC5LP*
9. Lin H, Rui-Qi M, Shi-Gang C, Yong-Li Z, Xmg-Li W (2018) Design of intelligent plant growth cabinet environment monitoring and control system. In: Chinese control and decision conference (CCDC)
10. Ping L, Yucai Z, Zeng X, Ting-fang Y (2007) A design of the temperature test system based on grouping DS18B20. In: 2nd IEEE conference on industrial electronics and applications
11. Darwish HA, Taalab AI, Assal H (2001) A novel overcurrent relay with universal characteristics. In: IEEE/PES transmission and distribution conference and exposition. Developing New Perspectives (Cat. No.01CH37294)
12. Qiu Y-F, Meng G (2013) The effect of water saving and production increment by drip irrigation schedules. In: Third international conference on intelligent system design and engineering applications
13. Anand S, Guha SK (1988) On-line diagnosis of problems in drip infusion system by impedance and flow measurement. In: Proceedings of the annual international conference of the IEEE engineering in medicine and biology society
14. Skakun S, Franch B, Roger J-C, Vermote E, Becker-Reshef I, Justice C, Santamaria Artigas A (2016) Incorporating yearly derived winter w maps into winter wheat yield forecasting model
15. Xiong F (2015) Wireless temperature sensor network based on DS18B20, CC2420, MCU AT89S52. In: IEEE international conference on communication software and networks (ICCSN)

Multicriteria-Based Trustworthiness Testing of Web Sites

Aanjey Mani Tripathi, Arunima, and Anjali Anand

Abstract In this age, Internet being affordable and fast has become the foremost popular means to retrieve any sort of information. Presently, people depend upon the Web for everything. The gathering of data from the Internet witnessed to be a standard phenomenon within the present technological advanced era since any information can be easily accessed and available to everyone. In fact, Internet has been used as a source for spreading awareness about anything among the public. People often visit online sources to accumulate details about something. Most people tend to believe that the information available online is reliable. When any user searches for something, he gets the results within a flash. But the question to ponder over is, "Are the results obtained be trustable?". Unfortunately, there is no surety for accuracy of results because of contrasting data provided by distinct sites about the equivalent thing. So, the matter is, "How does a user ensure which Web site is reliable?". In the current scenario, where the majority of the population relies on Web sources for any information, the necessity to test the correctness of data available online requires utmost attention. Hence, this paper proposed a scheme to ensure the trustworthiness of Web sites. In this scheme, a tool to check and display genuine Web sites for a particular keyword is made.

Keywords Trustworthiness · Authentication · Popularity · Domain age · Filtration · Security · Web links

A. Mani Tripathi (✉) · Arunima · A. Anand
Department of Computer Science and Engineering, Meerut Institute of Engineering and Technology, Meerut, UP 250005, India
e-mail: aanjey.mani@miet.ac.in

Arunima
e-mail: arunima.g.cse.2017@miet.ac.in

A. Anand
e-mail: anjali.anand.cse.2017@miet.ac.in

R. P. Mahapatra et al. (eds.), *Proceedings of International Conference on Recent Trends in Computing*, Lecture Notes in Networks and Systems 341,
https://doi.org/10.1007/978-981-16-7118-0_6

1 Introduction

People look for all kinds of information online day-to-day starting from general topics like latest songs, current affairs to certain data including facts and figures. When searched for a specific data, the sole thing got to be done is to write down the query in any of Google, Yahoo, Bing, or other search engines. The results are provided instantly. However, surety about the correctness of information cannot be guaranteed as it comes from distinct sources and of varied standard. What does trust mean exactly? For Web site effectiveness, trust is that object can be estimated by the eagerness of visitors to associate with it somehow or another. At the point, when the object is a Web page, that implies not just look at the page, but accepting the data introduced, or following up on it. Considering an example, user gives the input as "Total Land area of Uttar Pradesh." For the given input, britannica.com says "243,286 square kilometers," villageinfo.in says "Total area of Uttar Pradesh is 2,40,928 km^{2}″ while Sap.ipni.net says "24.09 million hectares." This shows that the results are not matched and which successively not acceptable too. Then, which one should be trusted? This left the user with a choice to pass each Web site himself so as to realize required results. This not only waste time in unwanted searching but also a slow and useless process.

Trustworthiness means the power to be relied on as truthful or sincere. But due to the contrasting information from various sources, it becomes difficult for users to make a decision which one among the Web sites provides more trustworthy data. This arises the necessity to supply some mechanism through which the trustworthiness of Web sites can be checked.

Hence, this paper contains one such scheme that helps in ensuring the same. This scheme contains the creation of a tool named SEARCH BIRD for checking and displaying the genuine Web sites for a particular keyword. There are a lot of parameters using which user can determine trustworthiness in content provided online. For this scheme, three factors have taken into consideration namely authentication, domain age, and popularity for checking trustworthiness.

2 Literature Review

While retrieving any information, accuracy of data is important. What proportion of relevant result set is provided by search engines defines its utility. A particular word may be contained in many of web pages, but some pages may be more related and popular than others. Different search engines employ different methods to rank a page in their search results. PageRank is one of the popular algorithms used for ranking a web page. Google uses PageRank algorithm to rank a webpage in their search results that stress on counting the amount and quality of links to roughly estimate the importance of a web page. PageRank algorithm features a drawback that the Web site visited higher has a higher page rank regardless of what the data is in the content, and this suggests that a Web site having higher page rank may contain wrong data. This creates a problem among users, and they often get confused or believed wrong data. They have to anticipate truth based on rank of a page. Thus, observation is that wrong information may be present in a page with higher rank. Hyperlink-Induced Topic Search algorithm (HITS) also called hubs and authorities is a link analysis algorithm that is used to web pages. This scheme defines two outcomes for every page, i.e., authority value predicts about how much worthy the content of page is and a hub value that evaluates the value of its link to other pages. HITS is an iterative algorithm that ranks web pages based on the in-links and out-links of web pages on the Web. Both of these approaches are unable to resolve veracity problem (i.e., conformity to facts/accuracy) as these are identifying most important web pages. But only popularity cannot assure the accuracy of information.

In the [1], author proposed a solution to improve Google's ranking capability using a heuristic approach based on the use of differential weights to back links depending upon the link location in the Web directory space. This approach was accomplished using DiffRank, a modified PageRank algorithm.

Panchal [2] presented a similar work of TRUTHFINDER algorithm that aims at resolving out conflicting facts from numerous Web sites and finding true facts among them. It uses two parameters, i.e., confidence of facts and trustworthiness of Web sites. But it has restriction that it assumes initial Web site trustworthiness to be 0.9, even if it is popular, not trustworthy or authoritative Web site. It assumes that a provider either provide good facts or bad facts for every object. Secondly, recalculating the trustworthiness of Web sites given by user reduce system performance.

A semi-supervised approach for veracity problem has proposed in [3], in which a confidence score is assign to each fact, so that true facts have higher scores than false facts. Some training data is used in this approach for improving accuracy. After comparing with the ground truth fact dataset (contain highly confident fact), ranked Web sites are displayed. Confidence of facts and trustworthiness of data sources are inferred using ground truth data. In this approach, additional keywords need to be provided with the normal search keyword.

3 Proposed Method

Trust is not something that can be assured only by seeing or reading the content of the web pages. This cannot be done that if it seems to be trustworthy, then one can trust otherwise not. Web site homepage is not only the factor that influences trustworthiness, and in addition, there are a lot of factors too that affect how user decides to trust in the content provided by the online sources.

Hence, proposed method contains the creation of a tool to check and display genuine Web sites for particular keyword. There are multiple criteria used to check the trustworthiness of the Web sites.

3.1 Factors Included for Checking Trustworthiness

3.1.1 Authentication/Security

This factor is used to check security of the Web site. It checks whether the site is authorized or not, i.e., whether it has proper security or not. There are two protocols, namely HTTP and HTTPS. HTTPS is a secure version of HTTP. Nowadays, HTTPS is essential for every Web site. For Web sites, to switch from HTTP to HTTPS (more secure), they need to get SSL certified. To keep user data secure, preventing attackers from designing a fake copy of the site, and verifying ownership of sites, a Web site needs to get SSL certified. Being SSL certified ensures that Web site is actually is what it claims to be. HTTPS provides confidentiality, integrity, and add trust.

3.1.2 Popularity

This factor is used to check the popularity of the Web site. Place it holds in the search engine's list or position. Popularity plays a major role in any search engine result set. Popularity is not measured manually. Rather than, the search engines employ mathematical algorithm to filter out relevance and then rank in order of popularity. Search engines use ranking process to determine where a particular content should be displayed on a Search Engine Result Page (SERP).

3.1.3 Domain Age

It is the age of domain, i.e., from the creation of Web site for how long it has existed. It an important attribute to be considered in SEO [4, 5]. More older domain names will be, more authority they will have than newer domain names. Importance to SEO will be more valued for a long-lasting and well-kept domain name. In the system, Last modified domain date will be checked. The age of the domain will be checked in this.

The proposed method contains the creation of the tool SEARCH BIRD to check and display genuine Web sites for particular keyword. The methodology is as follow:

The user will search a topic on SEARCH BIRD [6–8]. When getting the topic in request, it will be extracted and checked whether it already exists in database or not. If exits, links from database extracted and showed to the user. If not found in database, then, first of all, top ten Web sites from different search engine (i.e., Google, Yahoo, and Bing) will be fetched. It will give the first ten popular Web sites for particular keyword from all the three search engines. Then, the authentication / security of all those top ten sites will be checked. Unsecured sites get removed from the list. Now only all secured popular sites remained. Next, the domain age of all those sites will be checked. Then, they are sorted in terms of domain age. Now, three lists of Web sites (popular secured and age sorted lists) of Google, Yahoo, Bing are obtained. Thereby, a final list is made by following priorities:

- First priority to those sites which are present in all three search engines.
- Second priority to Google's list remaining sites
- Third to Yahoo
- Fourth to Bing

Next, the keyword and links will be added to database. Finally, final list of all trusted sites for the keyword searched generated and showed to the user (Fig. 1).

4 Implementation

The implementation of the proposed method is as follows:

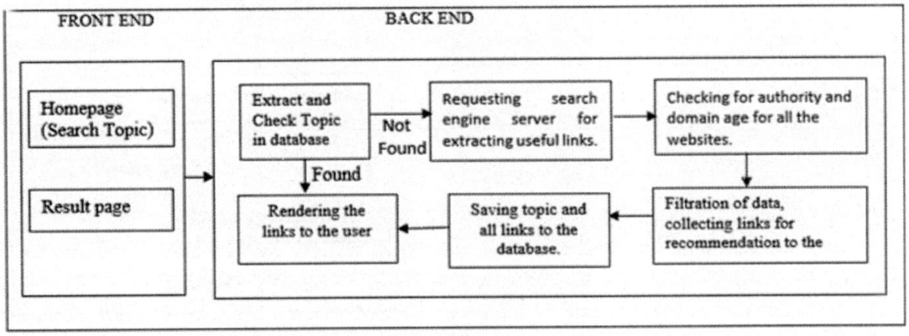

5 Algorithm

Input: Topic to be searched.
Output: List of Trustworthy Websites.
Step 1: Render one html page to input a search topic in it.
Step 2: When get the topic in request, extracting and checking it whether it is already exists in database or not. If it is existing, then returning True else returning False.
Step 3: If topic is found in Database, extracting links from Database and showing to the user.
Step 4: If the topic is not found in database, then requesting to all search engine server for extracting useful links.

google link, google text = (result)google.

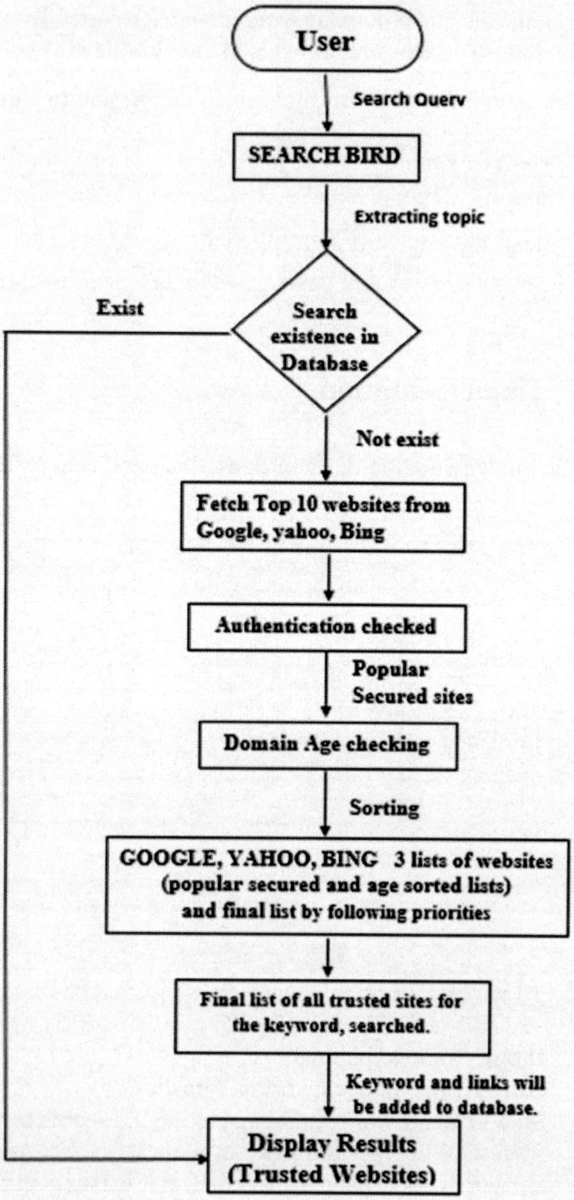

Fig. 1 Work flow of proposed method

yahoo link, yahoo text = (result)yahoo.
bing link, bing text = (result) bing.

Step 5: After getting links from all search engines, checking for authority and domain age for all the websites

websites = authority Check for (google_link).
websites = authority Check for (yahoo_link).
websites = authority Check for (bing_link).
then filtration of the data is conducted.

Step 6: After successful of filtration of data, collecting links for recommendation to the user.
Step 7: After collection of recommendation links, saving topic and all links to the database.
Step 8: After Saving links to database, rendering the links to the user.

6 Conclusion

World Wide Web is known to be the mostly desired and significant origin of information. However, due to of contrasting in nature and of about varying from top standard to cheap standard, surety about the accuracy of the information provided by the Web sources cannot be guaranteed. The proposed method contains the creation of tool for checking and displaying trustworthy Web sites. Because, this system gives trustworthy Web site as outcome, large amount of time wasted in unwanted searching would be saved, and the process would be more efficient than manual searching as well.

7 Future Scope

By adding new features, this scheme can be easily implemented when required. Reusability is also possible.

1. Bandwidth is a crucial factor that affects the Web sites' performance for high traffic Web sites. Higher the bandwidth, faster the data can be extracted and transferred. So, using higher Internet bandwidth, the performance of the proposed tool can be increased.
2. Reusability reduces design, coding, and data structure. Reducing the amount of code also simplifies understanding, which in turn increases the performance of the proposed tool.

8 Result Analysis

During search at the starting, it is taking time due to depth search on three different search engines. Later, it provides the recommended Web sites which is the common Web site of all the three search engines. It also suggested other Web sites for that keyword in each search engines. The recommended Web sites are based on different parameters, i.e., Authentication/Security, Popularity, and Domain Age. This work has been assigned higher priority to those Web sites which are recommended Web sites.

References

1. Kumar S, Krishna CR, Solanki AK (2017) Time efficient public key cryptography for enhancing confidentiality and integrity in a wireless sensor network. Int J Comput Sci Netw Secur (IJCSNS) 17(1):81
2. Panchal V, Pillai S, Singh A (2012) Truth finder algorithm for multiple conflicting information providers on the web. Int J Comput Appl 5:1–4
3. Sharma PR, Patil ME (2014) Extracting trustworthy data from multiple conflicting information using semi-supervised approach. Int J Adv Electron Comput Eng 2(11):271–275
4. Shahriar H, Zulkernine M (2012) Trustworthiness testing of phishing websites: a behavior model-based approach. Futur Gener Comput Syst 28(8):1258–1271
5. Parwekar P (2020) SGO a new approach for energy efficient clustering in WSN. In: Sensor technology: concepts, methodologies, tools, and applications. IGI Global, pp 716–734
6. Ramachandran S, Sujaya P, Sharon J, Vetriselvi R (2009) Enhanced trustworthy and high-quality information retrieval system for web search engines. arXiv preprint arXiv:0911.0914
7. Singal H, Kohli S (2016) Trust necessitated through metrics: estimating the trustworthiness of websites. Proc Comput Sci 85:133–140
8. Ansari S, Gadge J (2012) Architecture for checking trustworthiness of websites. Int J Comput Appl 44(14):22–26

Design and Implementation of e-learning Platform Using Data Analysis

Sunil Kumar, Aanjey Mani Tripathi, Hansika Bhatia, Gurneet Kaur, Daksh Aggarwal, and Divyansh Chauhan

Abstract Digitalization is growing like never and e-learning is a major part of it. Over the last decade, a significant growth has been seen in e-learning. The advancements and innovations in technology have created the option of distance learning for people which incorporates different mediums of learning, e.g., video tutorials, slideshows, etc. Moreover, ever since the COVID-19 pandemic has started, the concept of online classes has become an integral part of many people. This coronavirus has compelled to cease the normal functioning of schools, colleges, and universities, and thus, the use of online learning has enhanced. The objective of this work is to develop a robust virtual platform for the students which would work as an alternative for the offline college classes. Many analysis techniques have been incorporated for teachers to evaluate the students' performance and also to classify them according to the special skills and abilities they possess. We have used J2EE technologies and Bootstrap framework while developing the system. MySQL is used for the database requirement. This system would help the students and teachers to manage their daily study pattern. Most importantly, it gives the benefit of studying at your own pace, anytime, and anywhere.

S. Kumar · A. M. Tripathi (✉) · H. Bhatia · G. Kaur · D. Aggarwal · D. Chauhan
Department of Computer Science and Engineering, Meerut Institute of Engineering and Technology, Meerut, India
e-mail: aanjey.mani@miet.ac.in

S. Kumar
e-mail: sunil.kumar@miet.ac.in

H. Bhatia
e-mail: hansika.bhatia.cse.2017@miet.ac.in

G. Kaur
e-mail: gurneet.kaur.cse.2017@miet.ac.in

D. Aggarwal
e-mail: daksh.aggarwal.cse.2017@miet.ac.in

D. Chauhan
e-mail: divyansh.chauhan.cse.2017@miet.ac.in

R. P. Mahapatra et al. (eds.), *Proceedings of International Conference on Recent Trends in Computing*, Lecture Notes in Networks and Systems 341,
https://doi.org/10.1007/978-981-16-7118-0_7

Keywords Online learning · Virtual classes · Analysis · Classification

1 Introduction

Online learning is the process of learning with the help of the Internet. It is often popularly mentioned as "e-learning." It is basically the process of studying without having to physically attend classes or lectures. The Web today is flooded with the Websites offering numerous online learning options. With the rapid advancement in science and technology, these digital learning platforms have evolved a lot and have played a very significant role in people's life. Although, there exist some issues that also plague e-learning, it has appeared as an outstanding alternative for the continuation of studies, when the corona pandemic had locked everybody inside their homes. The transition from offline to online classes became a new normal. The Web learning platforms are being seen as a boon in disguise, providing aid to fight with sudden closure in these testing situations [1–3].

In this project, a completely digitalized Web platform has been created for a college. It includes a complete shift from offline to online mode of education. In addition of providing the basic perks of e-learning like anytime-anywhere learning, doubt-resolving support, attending live classes, accessing learning material, etc., there are certain other useful features as well. Since estimating the level of participation of students in their respective classes and measuring their progress rate is quite a big challenge for teachers, the analysis feature has been included for this. It analyzes the students' performance based on their attendance and their assignment submissions. Furthermore, for giving the teachers an insight about the special skills of their students, a remarkable feature of clustering has been added that groups or classifies students according to similar skills [4, 5].

The project basically consists of three modules: the admin, the teacher, and the student and have their own options and authorizations.

1. **Admin:** The admin has the control over all the activities. He can add, remove, or update students, teachers, or classes and can post informative notices and view the analysis reports of students [6].
2. **Teacher:** Teachers, on the other hand, can send study material, post notices and can answer a query. They can share links for live classes whenever scheduled. Assignments are posted that have to be completed before the due date and time. The performance of enrolled students can be analyzed based on their attendance and the assignments submissions. Grouping of students on similar skills is yet another option available to teachers [7].
3. **Student:** The students can access the notes, posted notices, etc. They have the option of resolving their queries whenever needed. They can also visualize their own performance within a selected period of time [8].

For implementing the analysis feature quite interactively, the graph APIs have been used that give us an attractive visual experience to the viewer and which are easy to understand. It is a nice-looking way of understanding the patterns of fall and rise in attendance and the timely submission of assignments. After analyzing the performance and participation of students, the required actions can also be taken [9, 10].

For the clustering of students based on similar skill set, various data analysis and text classification concepts are put into use. The students are made to enter their skills and achievements of various types like cultural, educational, sports, etc. The teachers using the clustering feature, can get the information of all those students who possess a particular skill, that the teacher is intended to find. For that, Apache Lucene and the Java library are used [11, 12].

2 Methodology

This project of developing a full-fledged robust Web application for the purpose of e-learning has been implemented by using the Java Enterprise Edition. The Bootstrap framework is employed for the responsiveness. For handling the database, MySQL has been put into use [13].

(1) With the effective use of the Java Enterprise Edition as the backend technology, the enrolled students have been given a medium to access the syllabus description, chapter notes, videos lectures, informative links, and posted notices with the ability to ask queries whenever required. Students have access to live classes as per the schedule which is implemented by embedding the Live Streaming APIs.

(2) They can view and submit assignments and view their performance by looking at the grades. The interactive feature of performance analysis by graphs and charts (using the graph APIs) has been added for the ease of students that helps them to analyze their performance.

(3) The admins have been given the privilege to manage the users of the application. We have employed a robust database for managing the storage of all kinds of information (using JDBC and MySQL). He can post notices regarding any holidays, updated schedules, etc. He can analyze the performance of each student.

(4) Teachers can post notices, add content, assign assignments with due dates to the students of their respective classes. By evaluating the solutions, teachers can award grades on the students' work as well as manage the queries of his/her students.

(5) Teachers can arrange live classes to give the students a better experience to study and clarify their doubts more efficiently. Both the students and the teachers are provided with a user-friendly interface using the CSS and Bootstrap technology which keeps them away from boredom and makes learning fun.

(6) Teachers and admin have a great feature of clustering or grouping the students based on similar skill set. Apache Lucene library has been used to extract

useful information from free-form text which in turn is used to find students who have similar interest. This is quite a useful feature as it gives better insights for organizing a competition or fest, etc.

Implementation: Process Description

The general architecture of this project of a virtual learning platform is shown in Fig. 1. It fully describes the features and functionalities provided to each of the modules. To access the Web application, the users have to first login through their respective credentials. The three modules are

(1) admin, (2) teacher, and (3) student.

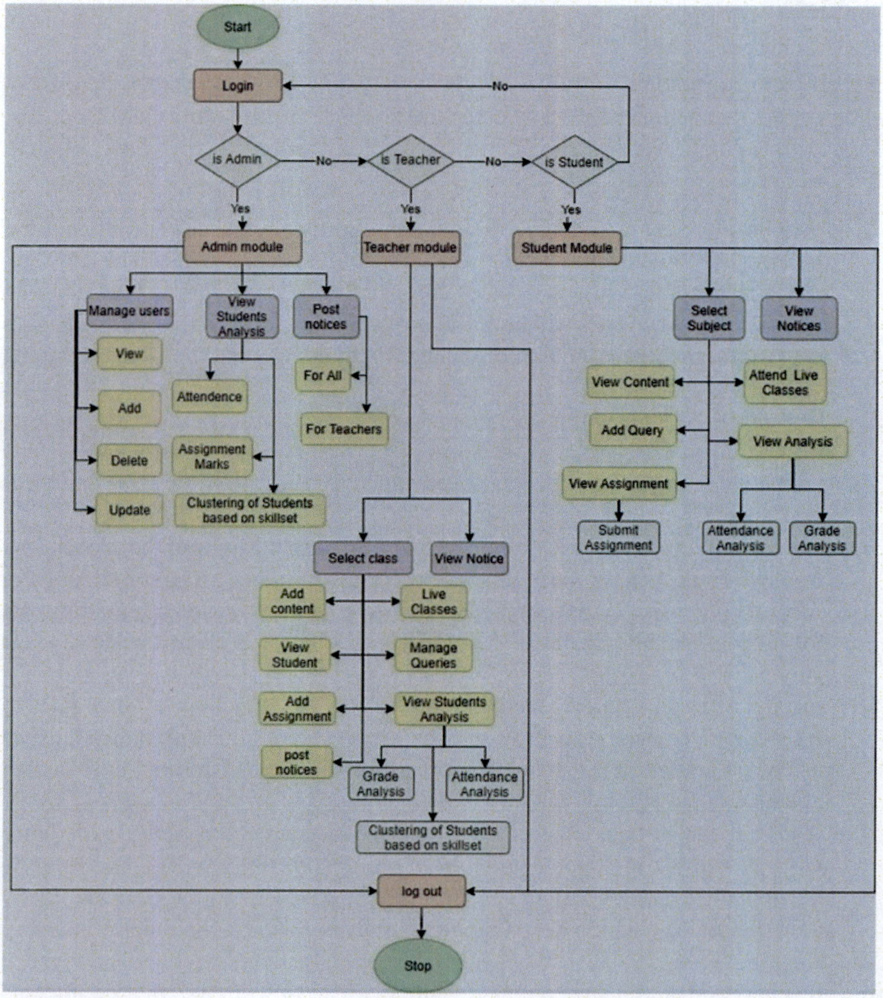

Fig. 1 General architecture

Extraction of information from the text (terminology and process)

Terminology

For the purpose of clustering using text analysis, Apache Lucene library is used. In this Java library, various APIs are there which perform the extraction of keywords. Given below is the description of the API's major functionalities.

(1) **Tokenizer:** Tokenizer breaks up the sentences into pieces. Tokenizers are of many types and we obtain varying token streams (sequences of fragments of text) as output based on the tokenizer we had used.
(2) **Stemmers:** We use stemmers to obtain the root of the word in the string. The language used also plays an essential part. For example, the word "merge" is the base word for the words like "merge," "merged," "merging," etc. It comes out to be quite useful if we have fewer words which are the basic ones while retrieving information, because the noise gets reduced and the main intent of the document still persists. Porter stemmer is one of the popular algorithms for this.
(3) **TokenFilter:** For normalizing or filtering the tokens, we apply a TokenFilter which accepts the stream of tokens. For example, StopFilter eliminates frequently occurring words and words which are of no use. Also, TokenText is normalized to lowercase with the help of LowerCaseFilter. The language has major role to play here as well. In English language, the stops words include "is," "or," "am," "to," "had," etc.
(4) **Analyzer:** An analyzer is the higher-level class that uses tokenizers to supply tokens from the input, uses stemmers to scale back the token, uses filters to suppress/normalize the tokens. This is the category that glues the opposite three main parts. Different combinations of filters and tokenizers are used by different analyzers. For example, StandardAnalyzer uses StandardTokenizer to extract tokens from the string, pass that through LowerCaseFilter to convert tokens into the lowercase then pass the stream of tokens through StopFilter to urge obviate most ordinarily used English words. Stemming is not performed by default. By mixing and matching tokenizer and token filters, one can develop a customized analyzer according to the need.

Process

(a) **Filtering the words:** When a teacher wishes to search for the names of students that are good in an entered skill or activity, the profiles of the students of a given class are scanned, i.e., the text is read word by word and the unwanted words are taken away. In this process of filtering, all commonly used English words are removed. Moreover, the censor rules can also be applied if required.
(b) **Carry out stemming on words:** Words like "merges" or "merged" or "merging" which all mean "merge." Stemming is basically the process of minimizing a word to its root. So, if a student's profile consists of words like "dances," "danced" or "dancing," the root word which would come out of it will be "dance."

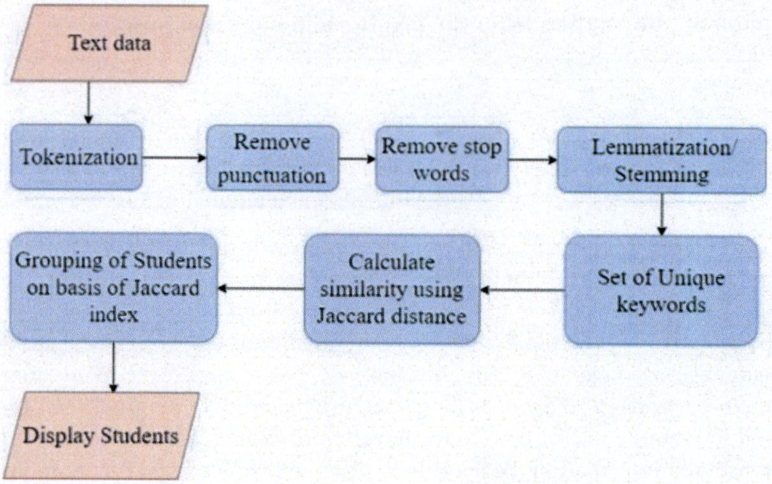

Fig. 2 Process of similarity calculation

(c) **Evaluating extent of similarity:** Finally, when the above two processes would be completed, we will end up having the original text (student's profile) represented by a group of keywords. These keywords will be treated as a set of unique words. While calculating the extent of similarity between two students' profile or between a profile and the searched skill, similarity can be better represented as a numerical value for calculation purposes which shows extent of similarity. There is a scale ranging from 0 to 1, where 0 indicates no similarity and 1 symbolizes complete similarity. For calculating the similarity or the diversity between two words, we are using Jaccard index (Fig. 2).

The Jaccard similarity coefficient or better known as Jaccard similarity index makes a comparison between the sentences and checks the resemblance. The higher the value, the more similarity between the documents. The results may be imprecise in case of small datasets with incomplete observations.

Jaccard index $J(X, Y) = |X \cap Y|/|X \bigcup Y|$ where X and Y are sets and $J(X, Y)$ lies between 0 and 1.

The above process is repeated for all the profiles of students. If the Jaccard index results to be greater than 0, the compared texts are said to be similar and put into a common set and displayed in decreasing order of the resulting Jaccard index.

3 Result Analysis

The availability of education regardless of time and place is an obvious and most essential benefit of an e-learning application. The results of the features like analysis and clustering are quite satisfactory. With the use of graph API, the analysis of students' performance has become interactive and user-friendly. The results of skill search can be improved by increasing the textual content. As it is clear from the above table (Fig. 3), the students' profiles are compared with the searched skill entered by the teacher and the similarity index is expressed in terms of the Jaccard index. The more the value, the more similarity between the documents.

As shown in graph (Fig. 4), the student with fifth profile will be found in the result

	football	coding	annual fest	volleyball	cricket
I am good at football. I also like to play volleyball.	0.142	0.0	0.0	0.142	0.0
I won first prize in a coding competition and has also participated in various debate competitions	0.0	0.090	0.0	0.0	0.0
I participated in state level cricket tournament and won third prize in robo champs	0.0	0.0	0.0	0.0	0.090
I volunteered in the annual fest . I also like to play football	0.125	0.0	0.250	0.0	0.0
I helped to code the college fest website. I play for the college cricket team	0.0	0.111	0.100	0.0	0.111

Fig. 3 Table showing Jaccard index of different students' profiles against searched skill

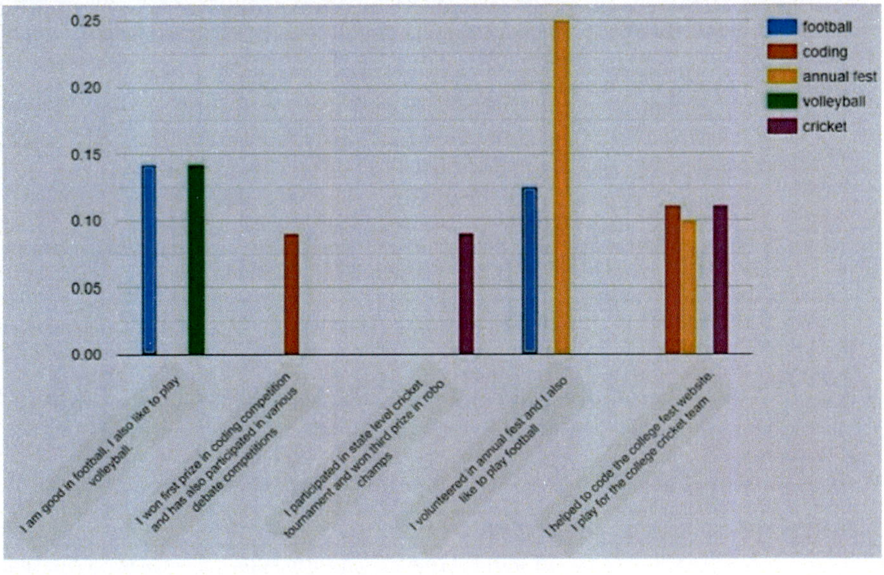

Fig. 4 Graphical representation of Jaccard index

when teacher will search for students good at coding, cricket or organizing fests, and events. When teachers search for students good at football, the students with first and fourth profile would be seen in results.

4 Conclusion

The main objective of creating this e-learning platform is to provide a successful online alternative for regular offline classes that were discontinued because of the outbreak of COVID-19.

The design and implementation of this project will make sure that the Website can be consumed by the teachers as well as the students with ease. The class notes, assignments, assistance for doubts, attendance, and other activities of an offline environment are taken into consideration and are implemented in the best possible way. The analysis and clustering features also have increased the usefulness of the platform.

References

1. Muhammad ABA, Nina KM, Juliana J, Norlida B (2017) Design and implementation of an online education and evaluation system. In: The IEEE 9th international conference on engineering education (ICEED), Kanazawa, Japan, November 9–10
2. Abdallah M, Mohammad NI, Ali BN, Hanan L, Abdallah S (2018) E-Learning: challenges and research opportunities using machine learning and data analytics. IEEE Access 6:39117–39138
3. Juan AL, Shadi A, Sonia P (2019) Special issue on the current trends in E-learning assessment. In: Springer Science+Business Media, LLC, part of Springer Nature
4. Jaccard index https://en.wikipedia.org/wiki/Jaccard_in
5. Tripathi A, Singh S (2017) A literature review on algorithms for the load balancing in cloud computing environments and their future trends. Math Comput Model 21(1):64–73
6. Kumar V, Kumar R, Singh J (2017) An efficient and secure RSA-based certificateless signature scheme for wireless sensor networks. Int J Trust Managem Comput Commun 4(1):17–35
7. Kumar V, Kumar R (2015). A cooperative black hole node detection and mitigation approach for MANETs. In: International conference for information technology and communications. Springer, Cham, pp 171–183
8. Tripathi AM, Singh S (2017) EEEMRP: extended energy efficient multicast routing protocol for MANET. In: International conference on information and communication technology for intelligent systems. Springer, Cham, pp 473–481
9. Kumar S, Krishna CR, Solanki AK (2017) A survey on security architecture and key management systems in a wireless sensor network. Int J Comput Sci Netw Secur (IJCSNS) 17(4):263
10. Singh B, Bansal A, Kumar S, Garg A (2010) Analysis for power control and security architecture for wireless ad-hoc network. In: International conference on contemporary computing. Springer, Berlin, Heidelberg, pp 205–216
11. Kumar S, Krishna CR, Solanki AK (2017) Error prone transmission system to resist data loss in a wireless sensor network. Int J Comput Netw Inform Secur (IJCNIS) I(11):17–26

12. Kumar S, Krishna CR, Solanki AK (2018) A technique to resolve data integrity and confidentiality issues in a wireless sensor network. In: 2018 8th international conference on cloud computing, data science and engineering (Confluence). IEEE, pp 183–188
13. Kumar V, Kumar R (2015) An adaptive approach for detection of blackhole attack in mobile ad hoc network. Proc Comput Sci 48:472–479
14. Java class library https://en.wikipedia.org/wiki/Java_Class_Library
15. MySQL client library driver for Oracle. https://docs.oracle.com/database/121/DRDAA/mysql_driver.htm#DRDAA29191
16. Apache Lucene Core libray https://lucene.apache.org/core/8_7_0/core/index.html

Prediction and Detection of COVID-19 Using Machine Learning

Diksha Goel, Mani Vats, Ayush, Priyanshi Baliyan, and Punit Mittal

Abstract Presently, the discovery of COVID infection 2019 (Coronavirus) is one of the fundamental difficulties on the planet, given the fast spread of the illness. Ongoing insights show that the quantity of individuals determined to have Coronavirus is expanding dramatically, with more than 1.6 million affirmed cases; the sickness is spreading to numerous nations across the world. In this investigation, we dissect the frequency of Coronavirus appropriation across the world. We present a computerized reasoning strategy dependent on a profound convolutional neural organization (CNN) to distinguish COVID-19 patients utilizing genuine world datasets. Our framework inspects chest X-beam pictures to recognize such patients. Our discoveries demonstrate that such an investigation is important in Coronavirus conclusion as X-beams are helpfully accessible rapidly and at low expenses. Experimental discoveries acquired from 1000 X-beam pictures of genuine patients affirmed that our proposed framework is valuable in recognizing Coronavirus and accomplishes a F-measure scope of 95–99%. Our proposed framework can essentially help distinguish the most tainted urban communities, and it has uncovered that waterfront territories are intensely affected by the Coronavirus spread as the quantity of cases is fundamentally higher in those regions than in non-seaside zone.

Keywords X-ray · Convolutional neural network · FBprophet · COVID-19 · Machine learning

1 Introduction

The COVID ailment is an overall sickness which has being initiated before Chinese specialist in Wuhan, the capital city of Hubei territory landscape China, in Dec 2019

D. Goel (✉) · M. Vats · Ayush · P. Baliyan · P. Mittal
Department of Computer Science and Engineering, MIET, Meerut, India

P. Mittal
e-mail: punit.mittal@miet.ac.in

R. P. Mahapatra et al. (eds.), *Proceedings of International Conference on Recent Trends in Computing*, Lecture Notes in Networks and Systems 341,
https://doi.org/10.1007/978-981-16-7118-0_8

As of now, there is no embraced mortal vaccination for battling it. In the meantime, fever and hack are the most broadly perceived contamination manifestations. Different side effects may happen, including chest uneasiness, sputum advancement, and an irritated throat. COVID-19 may advance to viral pneumonia which has a 5.8% mortality hazard. The demise pace of Coronavirus is identical to 5% of the passing pace of the 1918 Spanish influenza pandemic.

The portable applications propose customers to self-separate and alarm the concerned prosperity experts when someone corrupted by the contamination. Additionally, they screen tainted individuals, and the last people that they had contact with. Later it was initial point by point, the disorder has feast drastically across the creation and has develop an overall apprehension. An assessment drove revealed downfall speed of Coronavirus is 4.5% across the universe. The passing speed of patients in the age extent of 70–79 years is 8.0%, however, that of patients more than 80 years is 14.8%. The creators additionally affirmed that patients over the age of 50 years with constant diseases are at the most elevated danger and ought to in this way play it safe. One of the fundamental dangers of Coronavirus is its quick spread, with an expected 1.5–3.5 individuals getting contaminated by the sickness upon contact with a tainted individual. This suggests that if 10 individuals are Coronavirus positive, they are bound to contaminate 15–35 others. Thusly, Coronavirus can contaminate an extremely enormous number of individuals in a couple of days except if intercession measures are actualized [1] .

2 Contribution of This Study

Coming up next are the center commitments of this investigation:

- We propose a mechanized astute framework for recognizing Coronavirus patients from non-patients based on chest X-beam pictures. Our framework immediately peruses the design of a chest X-beam picture, uses concealed examples to recognize Coronavirus patients, and diminishes the requirement for manual pre-preparing steps.
- Empirical discoveries got from 1000 chest X-beam pictures of patients affirmed that our proposed framework can recognize Coronavirus patients with a precision of 95–99%.
- We give an insightful expectation framework to foreseeing the quantity of patients affirmed to have gotten the sickness, recuperated from the infection, and passed on from the illness over the course of the following 7 days utilizing three estimating techniques. Our proposed framework has been prepared and tried on datasets created from certifiable cases and has anticipated the quantities of Coronavirus affirmations, recuperations, and passing in Australia and Jordan with a normal precision of 94.80 and 88.43%, individually.
- We feature the most influenced zones and show that beach front territories are vigorously affected by Coronavirus contamination and spread as the quantity of

cases in those regions is fundamentally higher than that in other non-waterfront zones.

2.1 Related Work

The examination and identification of Coronavirus have been broadly researched over the most recent couple of months. The initial segment of this segment delivers issues identified with Coronavirus identification dependent on profound learning approaches utilizing CT outputs and chest X-ray pictures. The subsequent part audits the connected written works to evaluate future appraisals of the quantity of Coronavirus affirmations, recuperations, and passings.

Coronavirus has now gotten a worldwide pandemic inferable from its quick spread. It is trying to identify uncovered people since they do not show infection manifestations right away. Accordingly, it is important to discover a strategy for assessing the quantity of conceivably contaminated people consistently to embrace the fitting measures. Manmade intelligence can be utilized to look at an individual for Coronavirus as an option in contrast to conventional tedious and costly strategies. In spite of the fact that there are a few examinations on Coronavirus, this investigation zeroed in on the utilization of simulated intelligence in estimating Coronavirus cases and diagnosing patients for Coronavirus disease through chest X-ray pictures [2] .

Manmade intelligence can likewise be utilized for anticipating through existing proof. Hence, foreseeing potential outcomes in the short term can assist specialists with receiving the vital estimates zeroed in on two principle ideas. The principal idea included examinations identified with the analysis of Coronavirus, and the second elaborate investigations identified with the expectation of the quantity of individuals who will be tainted in the coming days. The examination investigation kept up that the vast majority of the current models are poor and one-sided. The creators proposed that examination based Coronavirus information ought to be freely accessible to support the selection of all the more explicitly planned discovery and expectation models [3].

2.2 System Design

Our proposed profound learning-based Coronavirus identification includes a few stages. The stages are summed up in the accompanying five stages:

Stage 1: Gather the chest X-beam pictures for the dataset from Coronavirus patients and sound people.

Stage 2: Create 1000 chest X-beam pictures utilizing information expansion.

Stage 3: Address the pictures in a component space and apply profound learning.

Stage 4: Fragment the data file into two sets: a preparation and an approval lay.

Stage 5: Assess the presentation of the finder on the approval dataset.

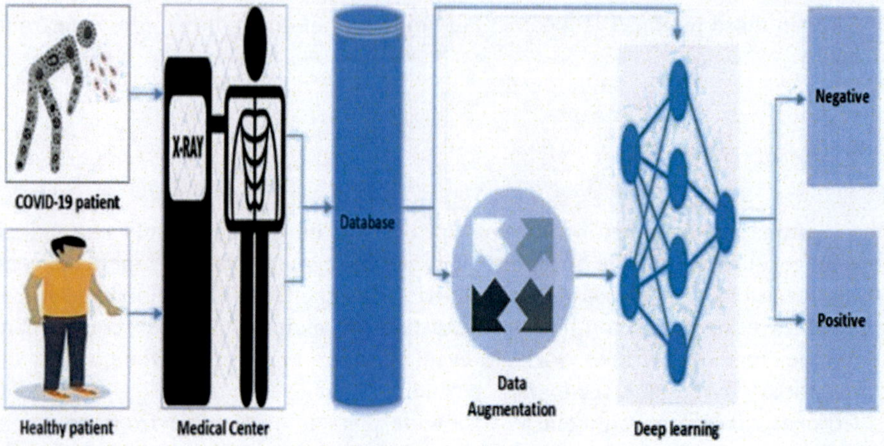

Fig. 1 Architecture of the proposed system

A. **Dataset**

Two sorts of datasets were utilized in the assessment, the first dataset (without increase) and the enlarged dataset. The dataset contained the accompanying: (a) a solid dataset containing chest X-beam pictures of sound people and (b) a Coronavirus dataset containing chest X-beam pictures of COVID-19 patients. The first dataset was acquired from the Kaggle information base, and its absolute number of pictures (Fig. 1).

Attributable to the restricted accessibility of chest X-ray pictures, we produced our dataset utilizing information expansion. Information enlargement is an artificial intelligence technique for expanding the size and the variety of named preparing sets by creating various cycles of the examples in a dataset. Information enlargement strategies are usually utilized in ML to address class irregularity issues, diminish overfitting in profound learning, and improve intermingling, which at last adds to better outcomes. The all out number of pictures in the dataset got 1000 subsequent to applying enlargement [4].

2.3 Methodology

There are two phases in this project:-

1. Prediction
2. Detection

Along these lines, there are two models. First model is prediction, we have some information, and we partition the information in two type of 60 and 40%, so 60%

for preparing and 40% for testing. At that point, our model is prepared by utilizing FBprophet [1].

FBprophet utilizes time as a regressor and attempts to fit a few straight and non-direct elements of time as a segments. As a matter of course, FBprophet will fit the information utilizing a direct model; however, it tends to be changed to the nonlinear model (coordinations development) from its contentions, a reasonable number of missing or exception information [5].

Also, the subsequent model is detection by utilizing Convolutional Neural Organization (CNN) Move Learning. We arrange the individual is crown positive or negative. There are numerous pictures of chest X-beams which is crown positive, negative, and nonpartisan. It is utilized to distinguish if the individual is contaminated by Coronavirus by move learning calculation for disease by means of chest X-beam.

Subsequently, we can use an association arranged on unessential classes in a huge dataset (commonly Imagenet) and apply it to our own anxiety considering the way that there are comprehensive, low-level features split between pictures. The photographs in the Caltech 101 dataset are actually similar to those in the Imagenet dataset, and the data a model learns on Imagenet should conveniently move to this endeavor.

3 Experimental Results

Right off the bat, we anticipated the quantity of Coronavirus affirmations, recuperations, and passings throughout the following n days utilizing FBprophet.

Besides, we inspected whether chest X-beam pictures can be utilized to create refined characterization models for Coronavirus expectation. We present the use of a profound learning calculation on two datasets. Exact discoveries demonstrated that our proposed framework is solid in distinguishing Coronavirus and has a F-measure scope of 95–99%, as uncovered [6].

This examination built up a CNN-based Coronavirus location model that was tried with both the first and the enlarged datasets. All the chest X-beam pictures utilized were resized to 224 × 224 pixels while disregarding the angle proportion. Figure 2a

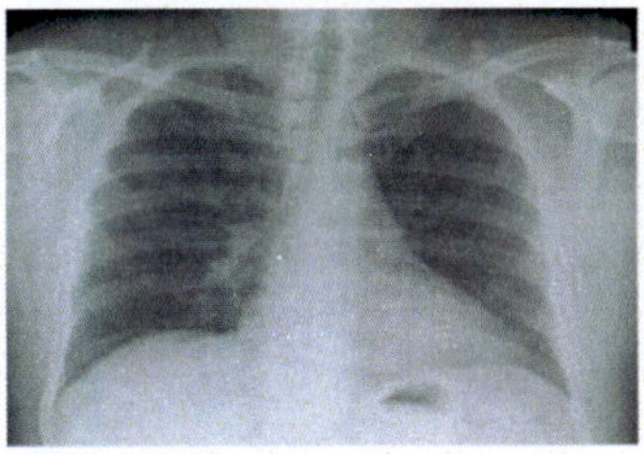

(a) Healthy

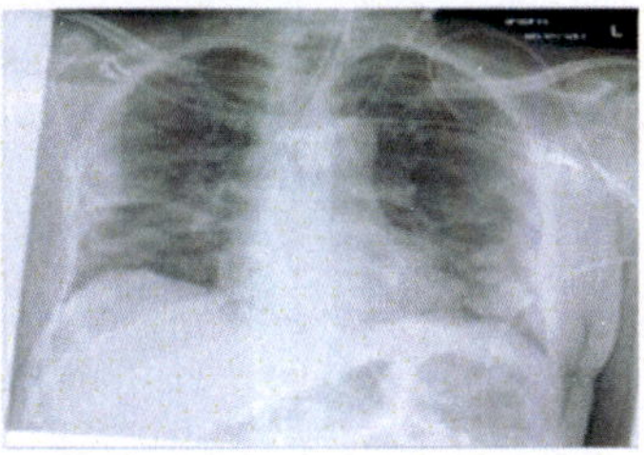

(b) Coronavirus Infected

Fig. 2 Chest X-ray images

and b present the chest X-beam pictures of sound and Coronavirus contaminated patients, individually. The gathered dataset was arbitrarily part into a preparation information subset and a testing information subset [7].

The Coronavirus identifier was prepared and tried on the gathered dataset, 80% of which was utilized for preparing, and the excess 20% was utilized for testing. The loads of the CNN were haphazardly initialized, and the cluster size was differed up to 25 and exactly set to 25 to evade overfitting and to accomplish the most elevated preparing precision. Moreover, the study figure have being at first put 0.1. It subtleties the exactness of Coronavirus locator and its misfortune [3] esteems for the actualized finder with growth and subtleties the precision of similar indicator with growth and its misfortune esteems [8] (Figs. 3 and 4).

Extraordinary consideration should be paid to the evasion of overfitting in the un expanded dataset, particularly while expanding the ages as the approval gradually improves to start with and afterward quits improving when the ages are expanded,. At the point when the enlarged dataset is utilized, the hole between the preparation and approval decreases after a couple ages, as demonstrated in figure. Hence, a more prominent improvement is accomplished in the preparation cycle, and a more summed up and strong Coronavirus locator is accomplished utilizing the CNN models while actualizing information growth on the dataset. Figures show the increased chest X-ray pictures of Coronavirus patients and solid individual separately [9].

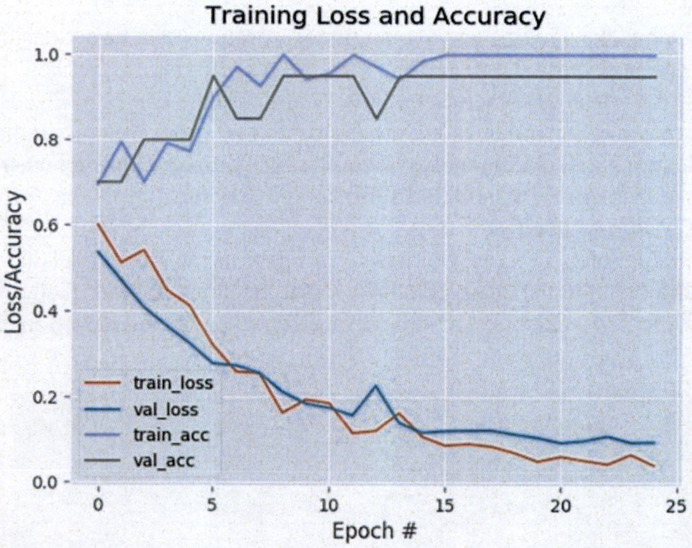

Fig. 3 Performance of CNN models in the original dataset

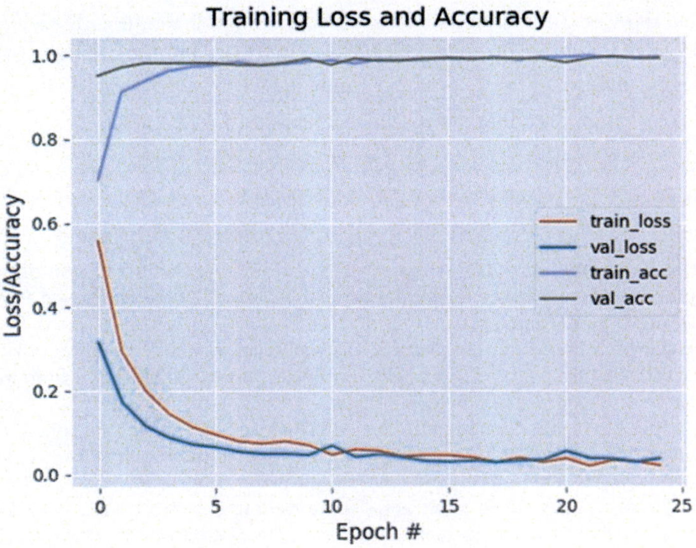

Fig. 4 Performance of CNN models in the augmented dataset

4 Conclusion

The fast spread of COVID-19 across the world and the expanding number of passings require dire activities from all areas. Future expectation of potential diseases will empower specialists to handle the results adequately. Moreover, it is important to stay aware of the quantity of tainted individuals by performing normal registration, and it is frequently imperative to isolate contaminated individuals and receive clinical measures. Also, consideration ought to be given to a few different variables to control the spread of COVID-19, for example, the natural impacts and the similitudes among the most influenced regions, and cautious measures ought to be received. In this paper, AI-based procedures were proposed for the expectation and conclusion of COVID-19: [3].

Prediction models, for example, the PA, ARIMA, and LSTM calculations were utilized to anticipate the quantity of COVID-19 affirmations, recuperations, and passings over the course of the following seven days. Dad conveyed the best execution. It anticipated the quantity of COVID-19 affirmations, recuperations, and passings in Australia and acquired expectation correctness of 99.94, 90.29, and 94.18%, separately. It likewise anticipated the quantity of COVID19 affirmations, recuperations, and passings in Jordan and acquired forecast exactness of 99.08, 79.39, and 86.82%, separately. Then, examination concerning more refined estimating and expectation strategies is a subject of a future work [10].

A determination model utilizing VGG16 was proposed to distinguish COVID-19 utilizing chest X-beam pictures. The model permits the quick and dependable location of COVID-19, empowering it to accomplish a F-proportion of 99% utilizing an expanded dataset. In a future report, we will consider diagnosing COVID-19 in chest CT examine pictures utilizing the VGG-XX forms and look at their exhibitions utilizing bigger datasets [11].

References

1. World Health Organization (2020) Research facility testing for Covid infection 2019 (COVID-19) in presumed human cases: break direction, 2 March 2020. World Health Organization, World Health Organization
2. Wu Z, McGoogan JM (2019) Attributes of and significant exercises from the Covid infection 2019 (COVID-19) flare-up in China: rundown of a report of 72 314 cases from the Chinese center for disease control and prevention. Jama
3. Maghdid HS, Asaad AT, Ghafoor KZ, Sadiq AS, Khan MK (2020) Diagnosing COVID-19 Pneumonia from X-Ray and CT Images utilizing deep learning and transfer learning algorithms. arXiv preprint arXiv:2004.00038
4. Chen A (2020) China's Covid application could have unintended results. MIT Technology Review. Accessible https://www.technologyreview.com/2020/02/13/844805/Covidchina-applicationclose-contactsurveillance-Coronavirusinnovation
5. Feng C, Elazab A, Yang P, Wang T, Zhou F, Hu H et al (2019) Profound learning framework for Alzheimer's disease diagnosis by means of 3D-CNN and FSBi-LSTM. IEEE Access 7:63605–63618
6. Worldometers (2020). Covid Cases. Accessible https://www.worldometers.info/Covid/
7. Santosh K (2020) Simulated intelligence driven tools for coronavirus episode: need of active learning and cross-populace train/test models on multitudinal/multimodal data. J Med Syst 44:1–5
8. Ogunleye A, Qing-Guo W (2019) XGBoost model for persistent kidney sickness determination. IEEE/ACM Exchanges on Comput Sci Bioinform
9. Yin H, Mukadam B, Dai X, Jha N (2019) DiabDeep: pervasive diabetes diagnosis dependent on wearable medical sensors and efficient neural networks. IEEE Trans Emerg Topics Comput
10. Artificial intelligence T, Yang Z, Hou H, Zhan C, Chen C, Lv W et al (2020) Relationship of chest CT and RT-PCR testing in Covid infection 2019 (COVID-19) in China: a report of 1014 cases. Radiology 200642
11. Narin A, Kaya C, Pamuk Z (2020) Programmed detection of Coronavirus disease (COVID-19) using X-beam images and deep convolutional neural networks. arXiv preprint arXiv:2003.10849

Music Therapy for Transforming Human Negative Emotions: Deep Learning Approach

S. G. Shaila, T. M. Rajesh, S. Lavanya, K. G. Abhishek, and V. Suma

Abstract The paper presents the music therapy approach on controlling and transforming the human emotions. The proposed approach considers six basic emotions such as *happy, sad, angry, fear, depression,* and *surprise* along with *neutral* expression. Here, *happy, surprise, and neutral* are considered as positive emotions and *sad, angry, fear, and depression* are considered as negative emotions. Initially, the proposed approach used FER-(2013) dataset for training Convolution Neural Network (CNN) for facial expression recognition. Later, the proposed approach uses real-time video clips to analyze the emotions. In the first stage, the zeroth frame of the video clip is considered and preprocessed. Later the image frames are proceded for CNN for feature extraction and facial expression classification. For the negative emotions related facial expresssions, related category of music is played for certain time. It is noticed that facial deformation with respect to features such as eye-eyebrows, nose, and lips will change that exhibits emotion transformation. These changes are analyzed, and when music stops, the proposed approach extracts (n + T)th frame and proceeded for deep learning using CNN for facial expression classification. The experiments are performed and noticed that the proposed approach of music therapy has greater impact on transforming internal state of human emotions and supports in avoiding the future disasters.

Keywords Music · Negative emotions · Facial expression · Classification · CNN

1 Introduction

Nowadays, the younger generation is more subjected to negative emotions such as *depression, sad, and angry*. There are various reasons behind these emotional changes

S. G. Shaila (✉) · T. M. Rajesh · S. Lavanya · K. G. Abhishek · V. Suma
Dayananda Sagar University, Bangalore, Karnataka, India
e-mail: shaila-cse@dsu.edu.in

T. M. Rajesh
e-mail: rajesh-cse@dsu.edu.in

R. P. Mahapatra et al. (eds.), *Proceedings of International Conference on Recent Trends in Computing*, Lecture Notes in Networks and Systems 341,
https://doi.org/10.1007/978-981-16-7118-0_9

which are exposed to the stress due to workload, studies, exams, etc. These issues lead to increase in suicides and this need to be addressed in the early stage. In general, emotions represent the internal state of human mind which is accompanied by chemical changes associated with feelings, thoughts, and degree of pleasure/displeasure. Emotions play a big role during communication among human beings. It is observed that emotions are controlled by the surrounding people and events. As face is the index of mind, human express their emotions in terms of facial expressions while communicating with others. Facial components such as eyes, eyebrows, mouth, nose, and chin plays a vital role in expressing emotions. Ekman [1] proposed six basic expressions such as *happy, sad, angry, depression, surprise, and fear*. Among these *sad, angry, depression, and fear* are considered are negative emotions. Suppose the negative emotions are not recognized and handled in the early stages, it can have adverse effect. For instance, the sadness of a person is not recognized and taken care that may lead to depression in later stages, and this stage can take an individual to an extent of committing suicide or any mental health related issue without their knowledge.

According to the recent report published by World Health Organization (WHO), India stands first among the depression countries. Approximately 8,00,000 people die due to suicide every year among which 80% are youth of age 15 to 19 years. India, as a biggest nation of youth population, would soon become highest number of suicidal place due to depression which needs to be addressed early. Recently lot of research is done on this domain, and some of the literatures are presented here. Sourina has proposed two algorithms such as Information acquisition algorithm and music therapy algorithm [2] for real-time EEG-based emotion recognition for music therapy. The authors distinguish different type of therapy for pain related issues. Gilda used CNNs for emotion recognition [3] in which the emotional module used two different sections dataset description and model description. Srivastava et al. [4], proposed an approach of identifying emotions using a video where face expression is detected and extracted using Viola–Jones algorithm and classification of features based on support vector machine. Suja et al. [5], proposed two approaches and applied on cropped face and whole face for extracting features from the image taken from database Cohn-Kanade and JAFFE. Dual-tree complex wavelet transform (DT-CWT) and Gabor wavelet transform are used to make feature vector with neural network and K-nearest neighbor which act as classifiers to identify emotions. The output shows that cropped face approach performs better than whole face approach. Hosen et al. [6], the author describes about playing music and changing wallpaper of personal computer based on identifying the mood of person if sad, then music will be played to make person happy and the wallpaper will be displayed as any happy image to change the mood of person. HAAR features are used to extract facial features and principal component analysis algorithm act as classifier. Kumar et al. [7], describes about recognizing six basic emotion in 4D video of BU4DFE database with the help of geometric-based method. A video which express emotion consists of frames which have neutral, onset, apex and offset of that emotion. Here apex frame is identified in a video sequence and then Euclidean distance is calculated between apex, and neutral frame to make feature vector. The feature vector is passed to Random forest and support vector machine for classifying the emotion based on music. Thus, above-discussed literatures represent few of the issues such as such as manual selection of music, random selection may not match with the mood of user, and manual music classification is tedious.

The paper presents the effect of music on controlling the negative emotions of human. The approach analyzes the initial frames of the video clips and proceeded to recognize the emotions. The approach uses the CNN for extracting facial features and expression classification. If it results in negative emotions, related class of music is played. Facial expression changes are observed and analyzed. The proposed approach analyses the nth frame when deformation is observed in facial expression and CNN classifies the emotions. The experiments are performed on real time and noticed that the proposed approach of music physiotherapy has big impact on human internal state of emotions, thereby avoiding the disasters that happen due to negative emotions of human being. The rest of the paper is organized as follows. In Sect. 2, we present the proposed approach, and in Sect. 3, 4, experimental results are presented and conclude the paper in the last section.

2 Proposed Methodology

2.1 Overview Architecture

The proposed approach overview is represented in Fig. 1. This is a real-time application. The proposed approach considers input from the camera that records live videos. In general, the video clips are in the form of image frames.

The initial image is considered as zeroth frame, and image preprocessing will be done. In preprocessing stage, the noise will be removed, and region of interest (ROI)

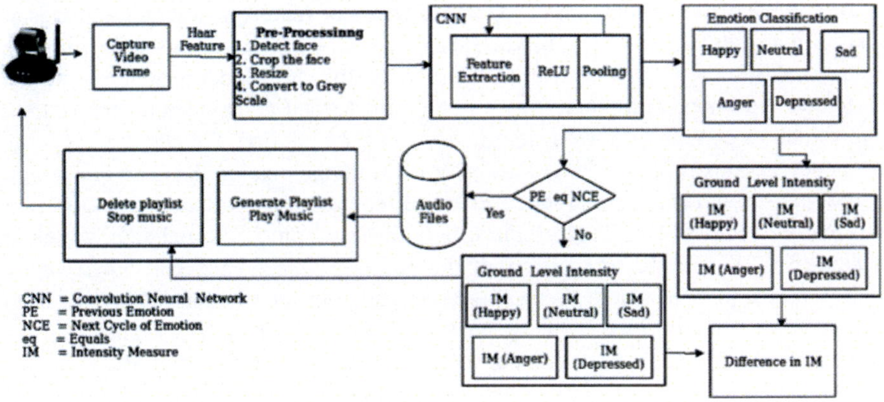

Fig. 1 Architecture of the proposed approach

is considered to process the image data fast. This is done using "Haar-Feature" which considers face region alone and crops the unnecessary area. The cropped face is further converted from color scale to gray scale image which reduces three channels to one channel. Later, the image is resized to process fast. This process is continues for the all the images frames of the videos. Further, deep learning is used for feature extraction and learning. The preprocessed sample of the initial image frame is processed to the CNN. CNN extracts the facial features such as eyes, eyebrows, nose, and lips. The action units and salient points are considered as landmarks. These feature vectors are processed through many intermediate layers before resulting in facial expressions classification. Once the expressions are classified, its intensity is visualized using histogram. If the classification results in negative emotions, those individuals are proceeded for experimentation with a music that calm downs their internal state. In the initial level, various music playlist is created for different categories of emotions and stored in database. Based on the type of negative emotion a person holds, the music is automatically selected from the playlist and played for 15 min. The impact of music is analyzed by the deformation in facial features such as lips, nose, eyebrows, and eyes in terms of facial expression. After certain time stamp, the nth frame of the video clip is captured and analyzed for the facial expression classification by CNN. CNN classifies the facial expression of the nth frame. It is observed that there exist polarity change in the emotions from negative to positive. This can be done by measuring the facial deformations in zeroth and nth frame of the same video clip. The intensity of the facial expression in the nth frame of the video clip is visualized using histogram.

2.2 Feature Extraction and Classification

The proposed approach use CNN, which is a deep neural network for feature extraction and classification.

The detailed working process of CNN in terms of feature extraction and classification is presented here. The input layer of the CNN receives the input samples and proceeds to the hidden layer of CNN. Here, numbers of hidden layers are added based on the problem complexity and computation level. Each hidden layer has three layers such as convolution layer, relu layer, and max pooling. Convolution layer is the most crucial part of the CNN. This layer takes image pixel in the form of a tensor matrix (tensors). The tensors related to facial features are extracted and converted into feature vectors. The feature matric is built based on the kernel size, which is of size (3×3). The feature matrix of defined size where each cell has a value. Kernel of size (3×3) where (0, 1, 2) are row and column numbers. Values of each cell are automatically filled by using backpropagation method. The kernel helps in convolving the image matrix. The convolution of image after every layer is learnt. As layer goes deeper more feature it starts learning. This is depicted in Fig. 2.

Later, at certain point, image shrinks and at a stage where it is not possible to be convolved more and it may not learn many important features. After every convolve,

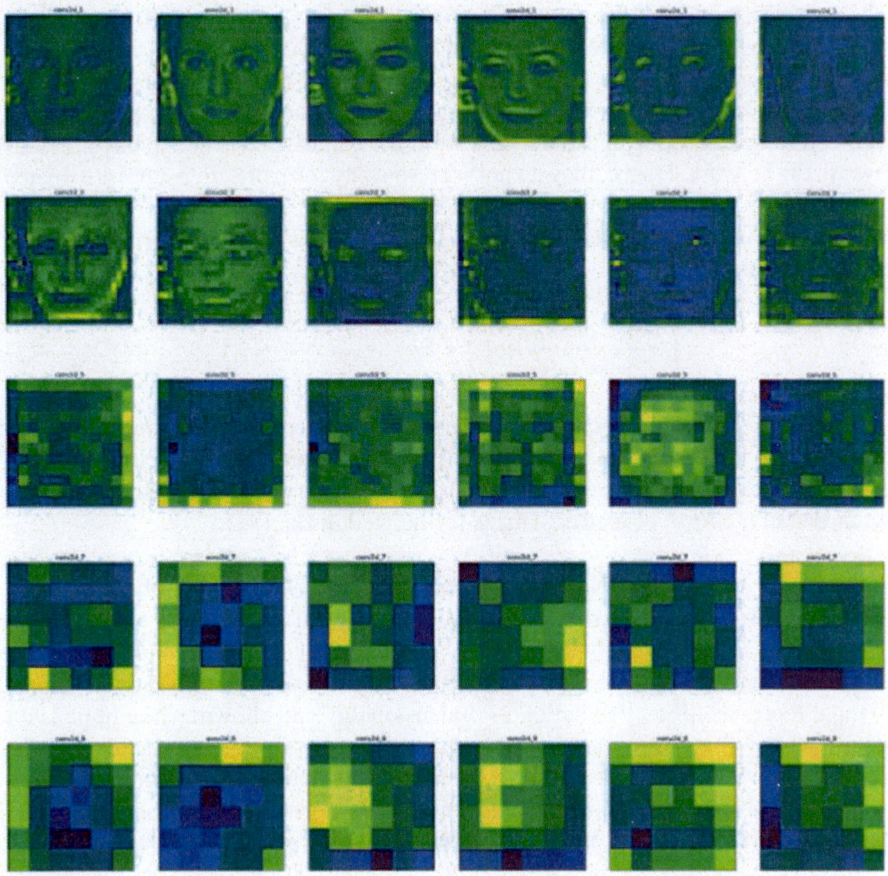

Fig. 2 Features at each layer

the resultant matrix reduces its size as shown in Eq. (1).

$$(n - f + 1) * (n - f + 1) \tag{1}$$

where 'n' is Image size and 'f' is filter size.

To overcome this, the padding process is used in convolution layer. The padding adds the specified number of cells around the image. The proposed approach used the common padding "padding = same." This maintains the same image size as the input image and learns many more corner features. Equation (2) represents the resultant output with padding.

$$(n + 2p - f + 1) * (n + 2p - f + 1) \tag{2}$$

where 'p' represents padding.

The framework used filter size 'f' that represents the size of the matrix which supports in convolving. The cells in filter matrix are parameters obtained from back-propagation. An extra neuron included with each pre-output layer is represented by Bias and stores the value of 1 for each action. Bias units are not tied to any previous layer in the network. Hence, they do not represent any form of activity, but are treated the same as any other weight. This is depicted in Eq. (3). The framework uses stride 's' to decide how many shifts of the row and column must be made horizontally and vertically in matrix.

$$\left(\frac{n+2p-f}{s}+1\right)*\left(\frac{n+2p-f}{s}+1\right) \tag{3}$$

Relu layer involves in activation function and in general does not require any parameter to learn. It takes input from the convolution layer and eliminates the negative values and replace with 0. The positive values are retained and generates the output in the form of matrix. This is formulated in Eq. (4)

$$\begin{Bmatrix} \text{No change} & \text{if pixel value} > 0 \\ 0 & \text{if pixel value} \le 0 \end{Bmatrix} \tag{4}$$

Max pooling is a sample-based discretization process wherein it down-samples an input representation by reducing its dimensionality and allowing assumptions on features that is binned in various sub-regions. This is performed to achieve over-fitting by providing an abstracted form of the representation. Also, it minimizes the computational cost by reducing the number of parameters to learn and provides basic translation invariance to the internal representation. Max pooling is done by applying a *max filter* to non-overlapping sub-regions of the initial representation. Further the output of max pooling is fed to the output layer of CNN. This layer returns the predicted value stored in the form of array. This value is matched with the trained model to classify the facial expression. The proposed approach use adaptive moment estimation as an optimizer with the learning rate of 0.0001. Further, to validate how well the model is being trained, predictions are done. This approach of CNN helps in deciding about parameter and the optimizer to be retained or changed. It helps in finding whether model has over-fitted or under-fitted during the defined epoch. In general, epoch decides number of times the complete dataset to be passed forward and backward. During this process, model learns different features from the batch of training dataset. The proposed approach used 55 epochs before the model starts over-fitting. After successful training of the mode, the processed features are restored. Only successful predictions are saved which contains numerical value of learned feature. After every epoch, loss function is calculated and the graph is plotted for every epoch which helps in deciding whether the model is under fitting or over-fitting.

2.3 Emotion Transformation Based on Music

The CNN classifies the facial expressions present in the zeroth frame into six basic emotions such as *happy, sad, angry, fear, depression,* and *surprise*. As discussed earlier, *sad, angry, fear,* and *depression* are considered as negative emotions, and *happy, surprise,* and *neutral* are considered as positive emotions. Suppose CNN classifies the facial expressions present in the zeroth frame as negative emotion and automatically the music from the related playlist is played. The music library is classified into classical, happy, relaxing, chill, feel good, energy booster, and mythological tune categories. Each category contains many more related music. Effect of these libraries is noticed after plotting the emotion intensity.

3 Results and Discussions

Initially, for classifier learning, the proposed approach used FER-(2013) dataset that has size of 65 Gb. The dataset is cross validated into testing and training set. Table 1 gives the details. The dataset is processed to pre-processing and then fed to CNN for learning and classification.

Once CNN learns to classify the facial expressions as positive and negative in terms of six basic emotions with static images, the proposed approach uses real-time video dataset recorded by camera. The implementation is carried out in Python3.6. The facial expression recognition is carried out with dynamic images. The images were captured using in-built camera. The detected faces are then adapted to train the face. Here, noise and blurriness is removed from the image and resized. The proposed approach achieved accuracy of 76% over the real-time dataset, when CNN with five layers with learning rate 0.0001 is used. Here, model goes over-fitting after four epoch as shown in Fig. 3.

When learning rate is reduced to 0.00001, with the same channel, accuracy of model increases to 86% with the model being trained for 40 cycles which is represented in Fig. 4.

The CNN classifies *sad* emotion when processed with the real-time video clip. This is represented in Fig. 5a and b, the histogram represents *sad* emotions for the input subject (Image). Since CNN classifies it as negative emotion, the music from

Table 1 Number of images in training and testing set

Training set		Testing set	
Emotions	Count	Emotions	Count
Happy	7367	Happy	1774
Angry	2361	Angry	957
Sad	4939	Sad	1247
Neutral	4965	Neutral	1233
Fear	3206	Fear	831

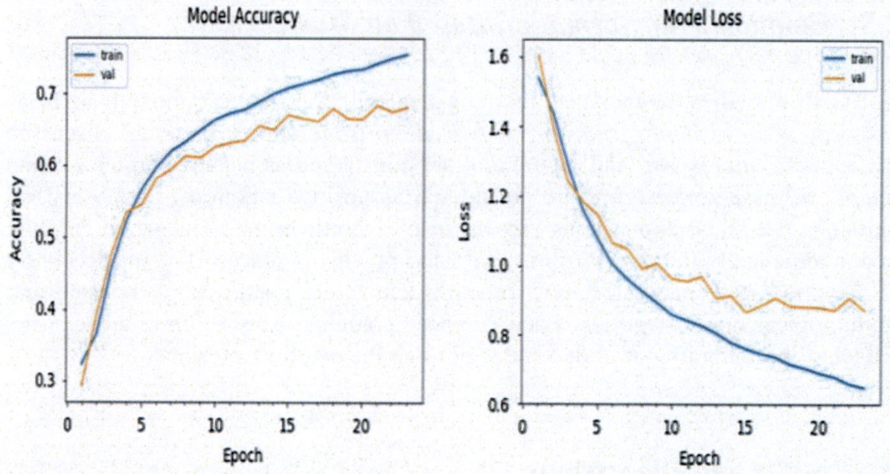

Fig. 3 Accuracy and loss graph for learning rate 0.0001

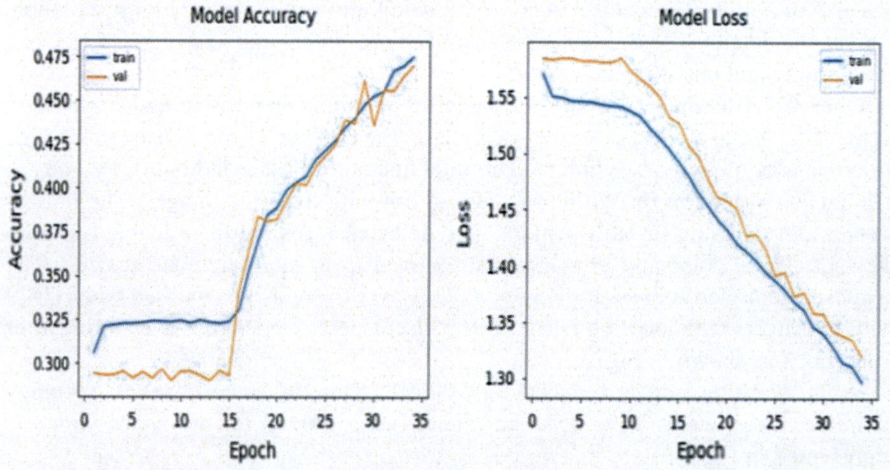

Fig. 4 Accuracy and loss graph for learning rate of 0.00001

the library is played for certain time stamp (15 min). The proposed approach automatically selected the 'Relaxing' music category and further proceeded with automatic selection of music from the playlist. It is noticed after (n + T)th frame, that there exist certain deformation in facial expression that represents slow transformation from *sad* to *neutral*. Each facial feature is responsible for these deformations and is represented in Fig. 6a and b.

The histogram represents *neutral* emotion for the subject after (n + T)th (Image) frame. Here, T represents the time stamp (min). The concurrent image frame, when the time stamp of 15 min get completed will be considered. Further, the proposed

Before: Music

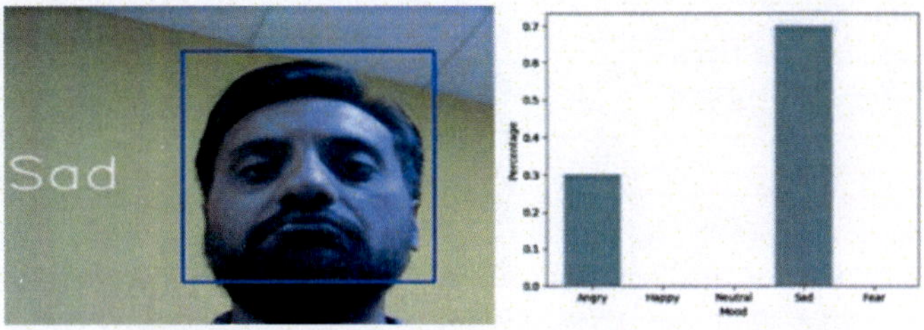

(a) Input image (b) CNN classifies Emotion: *sad*

Fig. 5 **a** Input image **b** CNN classifies emotion: *sad*

After: MUSIC

(Music category: Relaxing)

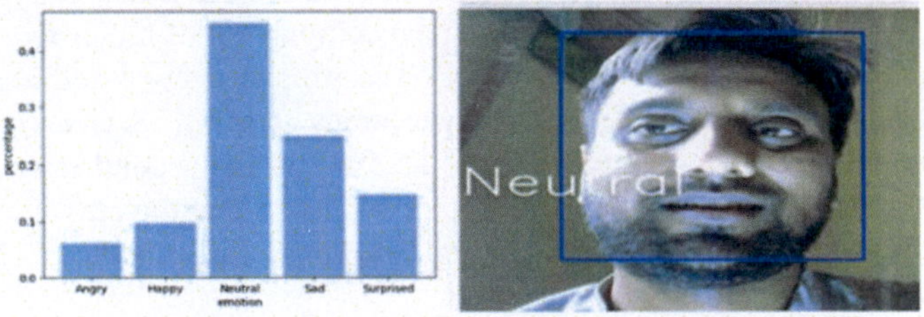

(a) Input image (b) CNN classifies Emotion: *Neutral*

Fig. 6 **a** Input image **b** CNN classifies emotion: *Neutral*

approach automatically selects the music category: 'Energy Booster' and proceeded with automatic selection of music from the playlist. It is noticed after (n + 2 T)th frame, that there exist certain facial deformation in expression that represents slow transformation from *Neutral* to *happy* emotion. This is represented in Fig. 7a and b. The histogram represents *happy* emotion for the subject after (n + 2 T)th (Image) frame. Here, 2 T represents the time stamp of 30 min.

Continue: MUSIC

(Music category: Energy Booster)

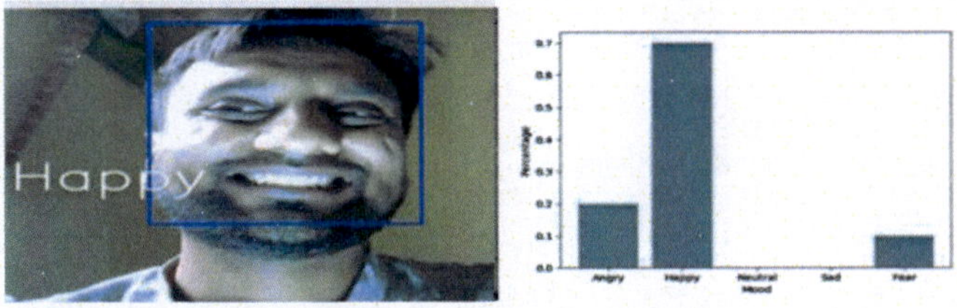

(a) Input image (b) CNN classifies Emotion: *happy*

Fig. 7 **a** Input image **b** CNN classifies emotion: *happy*

Table 2 Performance evaluation of CNN for FER—(2013) dataset

No of subjects: 15			
Negative emotion categories	Estimated prediction	Classifier prediction	Classification accuracy (%)
Sad	1247	1010	80
Neutral	1233	1037	84
Happy	831	735	88

Table 2 represents the performance evaluation of proposed approach In terms of classification accuracy for FER-(2013 dataset), Table 3 shows the impact of music in transformation the negative emotions for 15 subjects.

Table 4 presents the performance comparison of the proposed CNN model with other state-of-the-art approaches. The proposed approach achieves better performance with average classification accuracy as 92.7%. Srinivas et al. [4] have achieved lowest classification accuracy of 85.5% and the second lowest is 88.9% by Gilda et al. [3]. It is observed from the above results that the classification accuracy of the proposed CNN model is encouraging compared to other state-of-the-art approaches.

Table 3 Impact of music in transformation the negative emotions

No of subjects: 15								
Video frames	*Subjects with negative emotions*							
	Sad		Depression		Anger		Fear	
Zeroth frame	8		3		2		2	
	Subjects with positive emotions							
	Happy	Neutral	Happy	Neutral	Happy	Neutral	Happy	Neutral
(n + T)th frame	3	5	0	3	1	1	0	2
(n + 2 T)th frame	6	2	2	1	1	1	1	1

Table 4 Performance evaluation of proposed model with other state-of-the-arts approaches

Approaches	Accuracy (%)
Proposed approach	**94.7**
Srinivas et al. [4]	85.5
Gilda et al. [3]	88.9

4 Conclusion

The paper presents the effect of music on controlling the negative emotions of human. The proposed approach considers six basic emotions represented such as *happy, sad, angry, fear, depression,* and *surprise* along with *neutral* expression. The approach analyzes the initial frames of the video clips and automatically recognizes the emotions. The approach uses the deep neural network for extracting features and learning facial expressions. If the emotions are recognized as negative, related music from the music category is automatically selected and played for certain time. The changes in the facial deformation with respect to emotions are analyzed. The proposed approach classifies the emotion for $(n + T)$th frame when music stops. Initially, the experiments are performed on benchmark dataset FER-(2013) to learn the facial expressions. Later, real-time video frames are considered and noticed that the proposed approach of music therapy has big impact on human internal state of emotions, thereby avoiding the disasters that can happen in future.

References

1. Ekman P, Friesen WV (1978) Facial action coding system: a technique for the measurement of facial movement. Consulting Psychologists Press, Palo Alto
2. Olga S, Liu Y, Nguyen MK (2012) Real-time EEG-based emotion recognition for music therapy. J Multimodal User Interf. 5:27–35 Springer
3. Gilda S, Zafar H, Soni C, Waghrdekar K (2017) Smart music player ıntegration facial emotion recognition and music mood recommendation. IEEE WISPNET
4. Srivastava S (2012) Real time facial expression recognition. In: Meghanathan N, Chaki N, Nagamalai D (eds) Advances in computer science and ınformation technology. Computer science and ınformation technology (CCSIT 2012). Lecture notes of the ınstitute for computer sciences, social ınformatics and telecommunications engineering, vol 86. Springer, Berlin, Heidelberg, pp 124–133. https://doi.org/10.1007/978-3-642-27317-9_13
5. Suja P, Tripathi S, Deepthy J (2014) Emotion recognition from facial expressions using frequency domain techniques. In: Thampi S, Gelbukh A, Mukhopadhyay J (eds) Advances in signal processing and ıntelligent recognition systems. Advances in ıntelligent systems and computing, vol 264. Springer, Cham, pp 299–310. https://doi.org/10.1007/978-3-319-04960-1_27
6. Hosen P, Himel N, Adil M, Moon NN, Nur FN (2019) music playing and wallpaper changing system based on emotion from facial expression. In: Emerging technologies in data mining and ınformation security. Springer, Singapore, pp 79–87
7. Kalyan Kumar VP, Suja P, Tripathi S (2016) Emotion recognition from facial expressions for 4D videos using geometric approach. In: Thampi S, Bandyopadhyay S, Krishnan S, Li KC, Mosin S, Ma M (eds) Advances in signal processing and ıntelligent recognition systems. Advances in ıntelligent systems and computing, vol 425. Springer, Cham, pp 3–14. https://doi.org/10.1007/978-3-319-28658-7_1

Interest-Based News Feed

Utkarsh Gupta, Ayush Saini, and Ankush Gupta

Abstract There exist various news providing Web applications (e.g., Hindustan Times, Aninews, Beebom, etc.) that provide news articles of various domains (like technology, sports, business, etc.) to the user. However, in today's world of busy life, one may not get enough time to go through different news Web sites to read various long news articles. This paper provides an approach to construct a Web-based platform for the users where a user can get daily news from different news Web sites that belong to various domains at a single place only and also in a very concise manner. Moreover, the user can select news categories and the news Web sites from which he or she wants news from. For example: One may want to get news articles that are only related to sports and technology and that too from ANI News and Beebom news Web sites. This paper proposes an idea of creating a platform that ensures customized and concise news feed for the user.

Keywords Web scrapping · Personalized news · Information retrieval · Text summarization · Natural language processing

1 Introduction

News is found to have a large impact on a person's opinion formation regarding different concerns. Unlike conventional newspapers, the Internet has made a great change in the information flow and the way it reaches a person. Now, there exist many news providing Web sites like ANI News, Beebom, Times Of India, etc., that provides numerous news articles on various domains to the user. However, it is

U. Gupta (✉) · A. Saini · A. Gupta
Department of Computer Science and Engineering, M.I.E.T, Meerut, India
e-mail: utkarsh.gupta.cse.2017@miet.ac.in

A. Saini
e-mail: ayush.saini.cse.2017@miet.ac.in

A. Gupta
e-mail: ankush.gupta@miet.ac.in

R. P. Mahapatra et al. (eds.), *Proceedings of International Conference on Recent Trends in Computing*, Lecture Notes in Networks and Systems 341,
https://doi.org/10.1007/978-981-16-7118-0_10

important that a user shall be given liberty to subscribe to the feeds (news sources) they are generally intrigued by and also the type of news they are interested to read.

Our research in this paper is based on aggregation and integration of various different technological aspects like Web scrapping, information retrieval, text summarization and server management in order to provide a user with a valuable experience of reading news. The proposed architecture of the system is somewhat unique in its own respect in handling of real-time information fetching and turns out to be effective when tested to run in real-world environment like cloud.

2 Methodology

News Feed Preference: User can customize his preference any time and can select his preferred news source per category. By default, automated recommendations will be made for a new user based on the popularity of a Web site by checking which Web site has the most number of users subscribed to it for a particular category.

Management of Servers: Since scrapping different Web sites and performing news summarization are time taking tasks; therefore, we cannot deploy just one server for web scrapping [1] and user interaction as it will downgrade the performance of our application. As a solution to this, our project involves maintaining of two servers, one will perform Web scrapping for collecting news articles from several news Web sites and the other one will be the application server that will be interacting with the users.

2.1 Web Scrapper Server

This server will be fetching latest news articles from various Web sites listed by the developers. Firstly, it will collect the links (URLs) of all the latest articles. Then, It will check which article is newly added article by checking if it is already present in our database or not. If the article is already present, it will avoid fetching the same article again thus saving time, and if the article is not present then it will fetch the article. After fetching the article, it will perform preprocessing tasks on the article (such as new categorization, making summary, and storing date in general format). Finally, it will construct article object and save it to [2] database.

(1) Scrap Web sites at regular intervals

Set a timer function that will automatically call fetchNews() method to fetch latest new articles from certain set of Web sites. Suppose the news will be fetched at an interval of 20 min and the Web sites are scrapped at a gap of 10 min.

(2) Steps to fetch links of latest articles from a Web site:

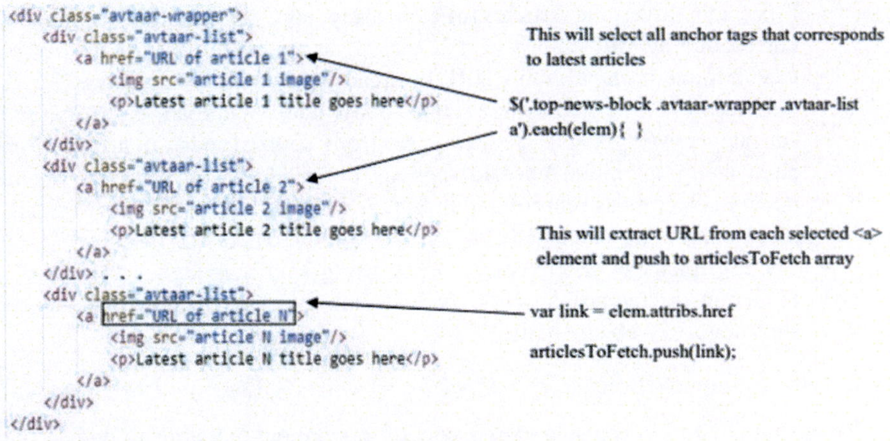

Fig. 1 Simplified HTML structure of ANI News Web site

Let us take the example of scrapping ANI News Web site using Node.js [3] and Selenium Web driver [4] to understand the scrapping methodology:

- $('.top-news-block.avtaar-wrapper.avtaar-list a').each(elem){} will select all anchor tags that belong to latest articles as depicted in Fig. 1
- Next step would be to loop over these tags and extract URL from each tag: var link = elem.attribs.href
- News category and subcategory will be taken from URL of the article by splitting it on '/' as shown in Fig. 2, e.g., var category =
- News category and subcategory will be taken from URL of the article by splitting it on '/', e.g., var category = link.split("/") [4], subcategory = link.split("/") [5];
- Push each link into articlesToFetch list
- Finally, articlesToFetch list will contain URL of latest articles. This list will be used to fetch individual article.

(3) Steps to fetch article corresponding to the link of the article.

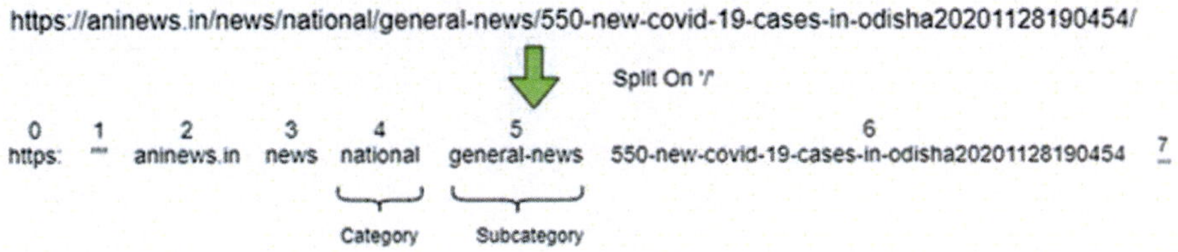

Fig. 2 Link structure and link categorization example

- First, loop over the articlesToFetch array and for each link fetch the corresponding article.
- Check if an article with this URL is already present in the database.
- Article properties like article's title, body, date of upload, category, and subcategory (if exists) are extracted from source code of article Web page using selenium, cheerio (node.js library for Web scrapping), and jQuery selectors as represented in Fig. 2, e.g., var title = $("#news-detail-block.content h1").text();
- Then summary of article body is made using TextRank algo.
- Final constructed article: {title, body, summary, date, thumbnail, category, subcategory, Web site name}

(4) News Summarization:

News summarization is also a vital aspect of this project that aims to deliver the best and precise news articles to the user. Among various applications of Natural Language Processing [5], text summarization has a vast impression on our lives. As the digital media and online publishing is growing, no one has time to read long news articles and to determine whether they are helpful or not. Use of TextRank algorithm [6–8] for text summarization is discussed below.

TextRank is an extractive and unsupervised algorithm for text summarization [9]. Figure 4 depicts the flow of the TextRank algorithm.

Steps for making summary of news article:

- First split the news article body into individual sentences. Figure 3 describes how to tokenize article body into individual sentences by making a split on period that are outside quotes and are also not a part of a decimal number
- Then, for each sentence, vector representation (word embeddings) is made.
- Then, similarities between sentence vectors are evaluated and saved in a matrix

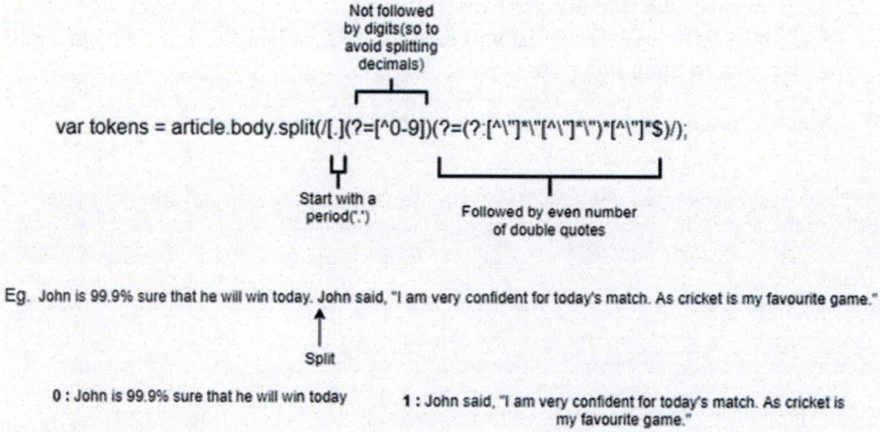

Fig. 3 Sentence tokenization

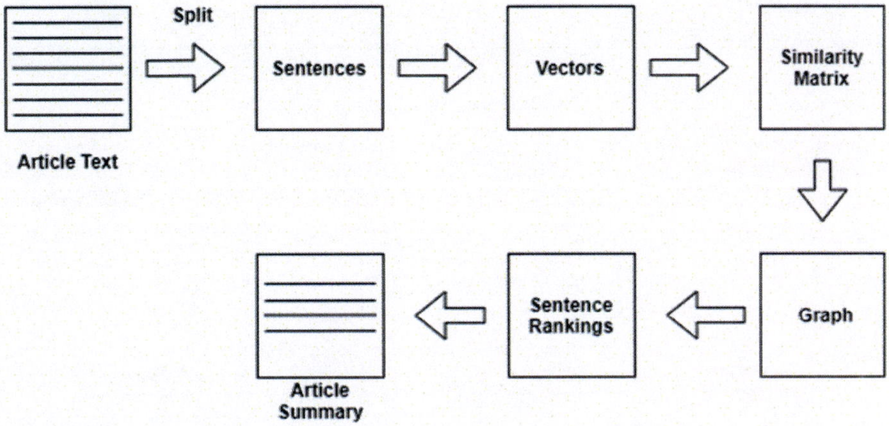

Fig. 4 Text summarization flow

- Then, for sentence rank calculation, a graph is made from similarity matrix with sentences as vertices and similarity scores as edges.
- In the final step, a specific number of top-ranked sentences are used to construct the final news summary.

2.2 *Application Server*

This server would be interacting with the users. It will provide users with an interface to select their news preferences like what type of articles they would like to read daily and from which particular news Web sites. Once the user is done with saving his or her news preference, it will populate user's home page with his personalized news.

(1) Personalized News Feed: Personalized news feed of the user will be constructed based on which category of news user wants to read and from which Web sites. User can change his or her news preferences anytime. It is also important to understand that the user may not always get time to open the Web site and then read his personalized news feed. As a solution to this, we propose sending emails to the user that will contain news template populated with user's personalized news feed.

(2) Custom News Search: Custom news search will search articles based on the search keyword or search phrase. Firstly, the articles will be searched in our own database, and if in case no such related articles are present in our database, then it will do a dynamic search on external Web site. This will be implemented by making text indexes in the database, e.g., In Mongodb [10]: db.articles.createIndex({title:'text',body:'text'});

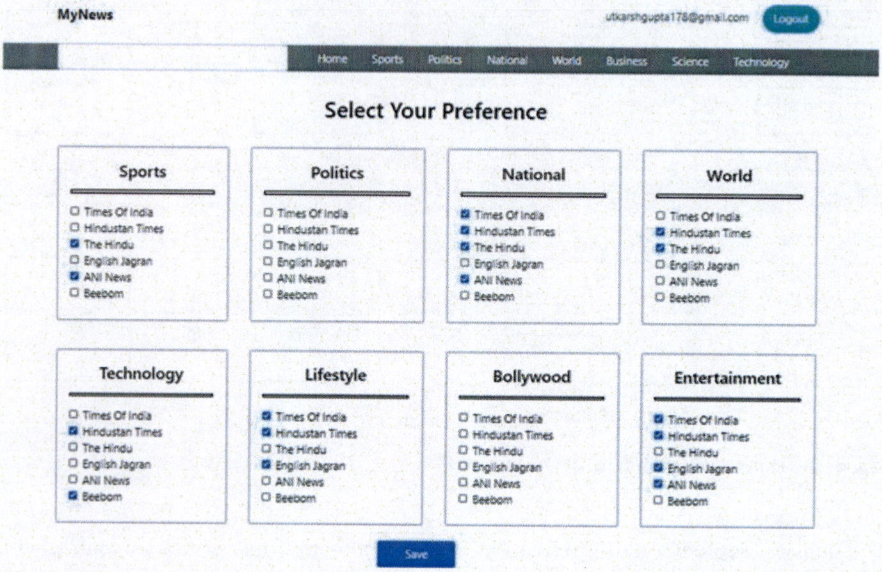

Fig. 5 Snapshot of user's preference page

3 Output Screen

User's news feed will be updated automatically on the Web portal and can also be delivered to the users through digital mediums (like emails and Web push notifications). Some of the project functionalities along with their screenshots are mentioned below:

1. Choose Preference—Fig. 5 shows the interface that will take user's news preferences, so to ensure that user will only see the news articles that are of his or her interest.
2. News Summary—Fig. 6 shows the comparison between the original article that is taken from www.beebom.com and the summarized article generated by our project.

4 Performance Analysis

A. Web Scrapper Server Performance:

Graph in Fig. 7 shows the CPU utilization of our web scrapper server hosted on AWS [11] ec2-instance with Intel(R) Xeon(R) CPU E5-2676 v3 @ 2.40 GHz 2.40 GHz and 1 GB Ram. Point A marks the start of the instance. Point B marks the start of the server. The region between B and C shows the timeout before the scrapping of first Web site is started. Point C marks the start of scrapping

Original Article	Summarized Article
Samsung Will Allow You to Use Bixby Voice to Unlock the Galaxy S21  Smartphones have gone from using PIN and patterns to fingerprint sensors and face unlock to safeguard your personal data. Biometric authentication is now a must-have feature as it simplifies unlocking the smartphone. Samsung is reportedly planning to introduce a new biometric unlocking into the mix with the Galaxy S21 series. As per the latest SamMobile report, Samsung is not looking to kill its in-house voice assistant, Bixby, and replace it with Google Assistant. Instead, it **intends to offer a voice unlock feature** via Bixby Voice on the upcoming Galaxy S21 series. The voice unlock feature will sit alongside the fingerprint, face unlock, and other smart unlock options in Android. The voice recognition feature, powered by Bixby Voice, will be **part of OneUI 3.1** on the Galaxy S21 series. It will initially be limited to the company's latest flagship series. But, it will eventually trickle down to older Samsung flagships and various Galaxy A-series models as well. The exact details about how this feature will work are still unknown. Will it be more secure than face unlock or fingerprint recognition? This will be the first time that Samsung will roll out "voice recognition unlocking" on its smartphones. I'm excited to try it out but I'm also skeptical about how good it will be. We are all aware of the shortcomings of Bixby. Add to this the fear of safeguarding your data with voice, which might easily be over-ridden with a recording. It will be interesting to see what measures Samsung has in place to avoid its misuse and offer true voice unlock functionality to users with the Galaxy S21 series.	Samsung Will Allow You to Use Bixby Voice to Unlock the Galaxy S21  The voice recognition feature, powered by Bixby Voice, will be part of OneUI 3.1 on the Galaxy S21 series.It will be interesting to see what measures Samsung has in place to avoid its misuse and offer true voice unlock functionality to users with the Galaxy S21 series.The voice unlock feature will sit alongside the fingerprint, face unlock, and other smart unlock options in Android. **Uploaded On :** Thu Nov 26 2020 05:47:00 GMT+0000 (Coordinated Universal Time) Read **Original Article** from beebom Share On Whatsapp
Words Count: 269	Words Count: 64

Fig. 6 Comparison between original article and summarized article

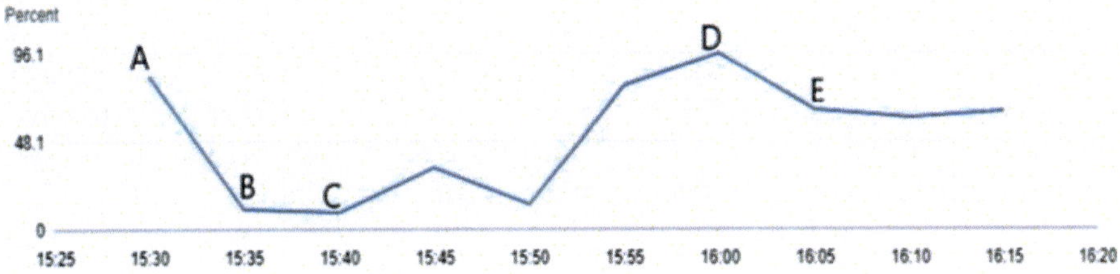

Fig. 7 CPU utilization of web scrapper

intervals for first news Web site and so on. The curve between C and D is elevated because latest news articles have been added to our database within this duration. However after a while, the curve becomes steady at point E (and onwards) because most of the latest articles have already been fetched.

B. Application Server Performance:

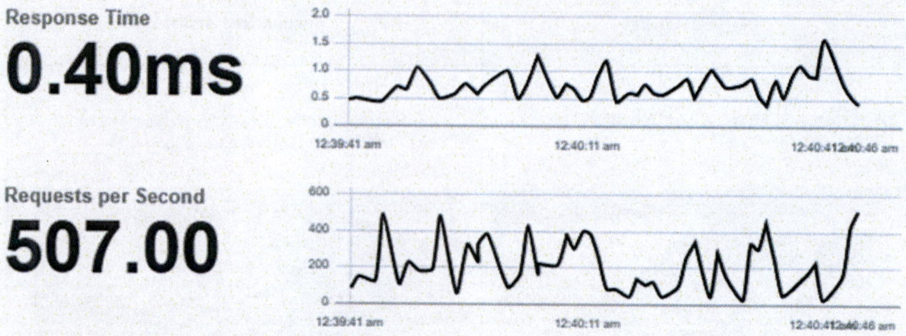

Fig. 8 Response time versus requests per second

Graph in Fig. 8 represents the result of a load test [12] that was conducted on our application server. Curve of the graph represents response time of our Web site at varying number of requests received per second

5 Conclusion and Future Works

In this paper, we talked about the idea of performing two level of personalization on news articles and the concept of news summarization to ensure a superior news reading experience for a user. Ability to empower user to specify news sources of their own choice gives this project an extra edge above the existing news applications. We have also discussed our methodology to scrap and gather information from disparate sources and storing their summary into our database. In the last section, we have analyzed the performance of Web scrapper server and measured the response time of application server against varying number of requests received.

This paper is confined only to English news articles, thus targeting only English understanding audience. However, to make this project of more use and cover all types of people, it can be extended to provide news in the reader's native language also. In addition to this, the algorithm used here for text summarization is an extractive and unsupervised algorithm, so to make a more effective summary in future it could also be swapped with a supervised text summarization algorithm.

References

1. Anand VS, Kedar GP, Shweta AG (2018) An overview on web scraping techniques and tools. Int J Future Revol Comput Sci Commun Eng 4(4). ISSN: 2454–4248
2. Erdinç U (2020) A novel web scraping approach using the additional information obtained from web page. IEEE Access

3. Bangare SL, Gupta S, Dalal M, Inamdar A (2016) Using node. Js to build high speed and scalable backend database serve. Int J Res Advent Technol (E-ISSN: 2321–9637) Special Issue National Conference "NCPCI-2016", 19 March 2016
4. Jyoti D, Kirti B, Rohini S (2017) A study on functioning of selenium automation testing structure. Int J Adv Res Comput Sci Softw Eng 7(5)
5. Sainath TN, Mohamed A–R, Kingsbury B, Ramabhadran B (2015) Deep convolutional neural networks for LVCSR. In: IEEE international conference on Ac
6. Mihalcea R, Paul T (2004) TextRank: bringing order into texts. In: Proceedings of EMNLP. vol 4(4)
7. Mihalcea (2004) Graph-based ranking algorithms for sentence extraction, applied to text summarization. In: Proceedings of the 42nd annual meeting of the association for computational Li
8. Srivastava N, Hinton G, Krizhevsky A, Sutskever I, Salakhutdinov R (2014) Dropout: a simple way to prevent neural networks from overfitting. J Mach Learn Res 15:1929–1958
9. Adhiraj C, Monalisa D (2020) Chapter 23 design and implementation of an automatic summarizer using extractive and abstractive methods. Springer Science and Business Media LLC
10. Hema K, Sudheep Elayidom M, Santhanakrishnan T (2016) MongoDB—a comparison with NoSQL databases. Int J Scientif Eng Res 7(5):1035. ISSN 2229–5518
11. Mukherjee S (2019) Benefits of AWS in modern cloud (March 6, 2019). Available at SSRN https://ssrn.com/abstract=3415956 or https://doi.org/10.2139/ssrn.3415956
12. Sharmila S, Ramadevi E (2014) Analysis of performance testing on web applications. Int J Adv Res Comput Commun Eng 3(3)

Gesture Decoder for Communication Using Deep Learning

Mukul Dayal, Mohak Goel, Apoorv Garg, and Ajay Kumar Singh

Abstract The project focuses on building a Deep Learning/Machine Learning model that can interpret the sign language, i.e., the hand and finger gestures of a person. The model will convert the signs into corresponding words and characters. So that an easy communication can be developed between the people who are not able to speak or hear properly. The model will be trained on images consisting of the hand gestures for different characters used by deaf and dumb persons to communicate. The model will be tested on the image dataset as well as on real world images. For the model, we have considered various Deep Learning approaches such as ResNet-50, VGG-16, Inception Network, Transfer Learning and compared their accuracy on the dataset. Using Transfer Learning on the VGG-16 architecture by training it on the dataset containing the images of the hand signs, we obtained highest accuracy, so we used this approach. The pre-training of the VGG-16 Network is done on various different real world images which helps the model to understand different patterns of the image easily.

Keywords CNN · Convolutional Neural Network · VGG · Visual Geometry Group · DL · Deep Learning · TL · Transfer Learning

1 Introduction

Communication is an important aspect of human life. It helps us to transfer information or our thoughts to other people. Speech is a frequently used medium of communication used by people. The people having problems related to hearing or speaking face a lot of issues in communicating with everyone and this gives rise to a huge communication gap. There are several means such as visual aids as a solution but they are both costly and complicated. One of the solutions is to use sign language.

M. Dayal · M. Goel · A. Garg · A. K. Singh (✉)
Department of Computer Science and Engineering, MIET, Meerut, India

M. Dayal
e-mail: mukul.dayal.cse.2017@miet.ac.in

R. P. Mahapatra et al. (eds.), *Proceedings of International Conference on Recent Trends in Computing*, Lecture Notes in Networks and Systems 341,
https://doi.org/10.1007/978-981-16-7118-0_11

Sign language involves continuously combining various shapes of hand, their direction and motion, body and arms movement to convey the speaker's message and ideas to the receiver.

Sign language consists of finger and hand gestures, which signify words as symbols, links between those words conveying the meaning of the signs using hand gestures. Spelling words by the aid of fingers are essential, as it helps in conveying words that do not depict a meaning in association by word level like names, addresses, etc. Due to which, fingerspelling is not generally preferred in a large scope as it is complicated to understand or use. Furthermore, very few people know the universal sign language, which is why it is difficult for dumb people to communicate with others.

Therefore, an algorithm that can differentiate between different signs in sign language can work as a solution for the above discussed problem. We have compared a plethora of machine learning algorithms that can interpret sign language and assessed their accuracies.

2 Related Work

Sanil Jain and KV Sameer Raja made a Sign Language Recognition System with the help of colored datasets. They used various visual words, Histogram of Gradients (HoG) and Gaussian random for extraction of features. Support Vector Machine was trained using three subjects, and an accuracy of 54.63% was achieved on a new user [1].

3 Preliminaries

3.1 Dataset

For this project, ISL dataset [2] is used. The dataset contains RGB images of 200 × 200 pixels. The size of the dataset is 87,000. Dataset contains 29 different label classes from which 26 are for alphabetic characters A–Z and remaining 3 classes are for space characters, deleting the last image and nothing if the model does not detect anything. These labels will help in real-time applications and classification as there can be many incidents where the gesture is not recorded perfectly.

The test dataset contains some part of the training dataset images which we have kept for testing and also real world images, so that the model can be applied on the real world images easily.

Indian Sign Language (ISL) Dataset

The gestures are in the form of RGB images with a resolution of (200 × 200 × 3) pixels. To which we will first apply some preprocessing (explained in Sect. 4.1) to transform the dataset into the required format for training.

3.2 Algorithms Used

3.2.1 VGG-16 as Base Model

VGG-16 was proposed by Karen Simonyan and Andrew Zisserman in the paper "VERY DEEP CONVOLUTIONAL NETWORKS FOR LARGE-SCALE IMAGE RECOGNITION."

The Network takes RGB Images of size (224*224*3) as the input containing various convolution layers of (3*3) and (2*2) filters with max pooling layers, the model outputs a vector of size 1000. For our model, we have used this network as a base and removed the final output layer and built our further model on top of it. We have use the second last layer which has the size of (2*2*512) to build further.

3.2.2 CNN for Further Improvements

Convolutional Neural Networks (CNN) [3] is a neural networks architecture which contains convolution layers to capture the features from the images, with pooling and many fully connected layers which are used for processing data. A CNN model is formed using these three layers: Convolution, followed by a nonlinearity function, pooling layers and an output layer that can be a classification or regression layer.

Convolution: In convolution, we extract the main features from the input image using matrix multiplication operation, and it finds the relationship between the pixels and identifies different patterns in images. It generates a feature map as an output. Further on this, a nonlinearity function is applied which is also known as Activation Function, which can be a ReLU(Rectified Linear Unit), Sigmoid or Hyperbolic tangent function.

Activation Function is an element-wise operation that introduces nonlinearity in a convolution network.

Pooling: Pooling also called Down Sampling is used to capture the important features from the feature map and reduce the size of the data. We can use many types of pooling layers such as Max Pooling, Mean Pooling, Min Pooling, etc. [3].

Fully connected layer: It is a multi-unit layer where each node of one layer is connected to each node of another layer, and these are used as hidden units in neural networks. [3].

A CNN model is formed using the combination of above stated layers (Fig. 1).

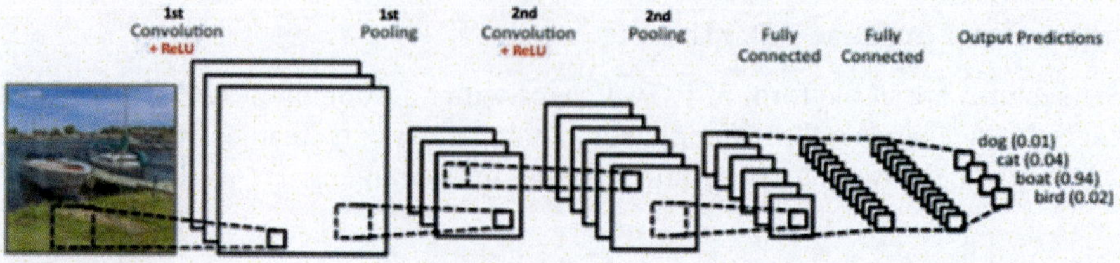

Fig. 1 CNN

3.2.3 Transfer Learning

In transfer learning, another model which is trained on a large dataset but for a separate task is used. We transfer the learning of that model to build a new model on top of the initial learning of that model. This process will tend to work under two conditions: first the input of the new model should be in the same format as the pre-trained model and second the initial model should be trained on more data then the new one. The low level features of the pre-trained model will be helpful in learning for the new model.

We have used Transfer learning [4] for our model, in which we used a pre-trained VGG-16 architecture model that was trained on a dataset consisting of large and discrete images. The prior knowledge gained by the model earlier can be utilized by the new model. The pre-trained model weights are saved in a HDF5 file which is loaded later on to acquire the understanding of the pre-trained model.

4 Methodology

4.1 Data Preprocessing

The RGB images in the dataset are of size (200*200*3). So, to use the dataset for training we first resize them to (224*224*3) array of pixels by padding them from the corners using the openCV library and normalizing the pixel values by dividing from 255.

4.2 Model Used

The Architecture of the Model Used:

	Layers used	Output/filter size	Units	Param #
1	Initial Input Layer	224 * 224 * 3	0	0
2	Convolution Layer 1.1	11 * 11 (Same padding)	64	1792
3	Convolution Layer 1.2	5 * 5 (Same padding)	64	36,928
4	Max Pooling Layer 1	2 * 2 (Valid padding)	0	0
5	Convolution Layer 2.1	11 * 11 (Same padding)	128	73,856
6	Convolution Layer 2.2	5 * 5 (Same padding)	128	147,584
7	Max Pooling Layer 2	2 * 2 (Valid padding)	0	0
8	Convolution Layer 3.1	3 * 3 (Same padding)	256	295,168
9	Convolution Layer 3.2	3 * 3 (Same padding)	256	590,080
10	Convolution Layer 3.3	3 * 3 (Same padding)	256	590,080
11	Max Pooling Layer 3	2 * 2 (Valid padding)	0	0
12	Convolution Layer 4.1	3 * 3 (Same padding)	512	1,180,160
13	Convolution Layer 4.2	3 * 3 (Same padding)	512	2,359,808
14	Convolution Layer 4.3	3 * 3 (Same padding)	512	2,359,808
15	Max Pooling Layer 4	2 * 2 (Valid padding)	0	0
16	Convolution Layer 5.1	3 * 3 (Same padding)	512	2,359,808
17	Convolution Layer 5.2	3 * 3 (Same padding)	512	2,359,808
18	Convolution Layer 5.3	3 * 3 (Same padding)	512	2,359,808
19	Max Pooling Layer 5	2 * 2 (Valid padding)	0	0
20	Flattening Layer	0	0	0
21	Fully Connected Layer 1	100%	512	12,845,568
22	Dropout Layer 1	20%	0	0
23	Fully Connected Layer 2	100%	128	65,664
24	Output Layer	100%	29	3741

Total number of parameters in the model are 27,629,661 from which 12,914,973 are trainable and the rest are non-trainable. We have used Adam Optimizer to compile the model which is present in the Keras Optimizers Library with Sparse Categorical Cross Entropy as the loss function.

The metric used for training the model is F1 score, which is defined as

Precision = Actual Positive Label that are Predicted Positive/(Actual Positive Label that are Predicted Positive + Wrongly predicted Positive Label)
Recall = Actual Positive Label that are Predicted Positive/(Actual Positive Label that are Predicted Positive + Wrongly predicted Negative Label)
F1_score = 2 * (Precision * Recall)/(Precision + Recall).

4.3 Pre-training on the Model

A pre-trained VGG-16 architecture model is used in our project. The VGG model is trained on 14 million images dataset that has about 1000 classes, we used these pre-trained weights and remove the last layer of the model and add our own layers on top of it, we first Flatten the second last layer and add a hidden layer of 512 units with a dropout of 0.2, further adding another fully connected layer of 128 units and output layer in the end for prediction.

4.4 Training Process

First, we train the layers added on top of the VGG network using the ISL dataset.

After that once the model is trained enough then we fine tune the model by setting the trainability of the last few layers of the VGG network to True, and training the model again for a few epochs for improving the final accuracy.

4.5 Application Overview

Our application will consist of a front end part that will take the input image either through the user's webcam or any camera device. This image will then be passed to the corresponding python script that will apply the required preprocessing on it after which the image is passed to our trained model which will classify the image as one of the character classes. Finally, we will map the corresponding character according to the class output by the model and return the result back to the user (Fig. 2).

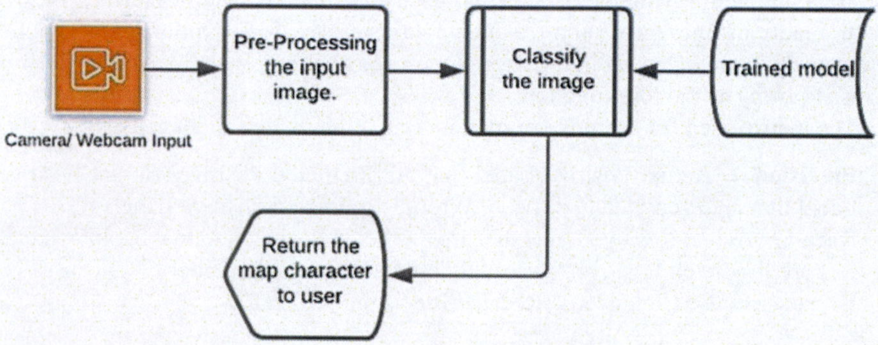

Fig. 2 Implementation flow chart

5 Results

5.1 *Confusion Matrix*

The confusion matrix [6] is the summary of output results. The rows of the matrix show the actual label and the columns represent the label predicted by the model, and the leading diagonal of the matrix gives the correct predictions of the model. Using the confusion matrix, we can check the accuracy of the model as well as its precision and recall, which helps in finding the ways by which the model can be improved.

In our model, the confusion matrix is used to check correctness of each sign separately, so that we can see to which specific sign A–Z, Space, Nothing, Delete, the model is not performing correctly and we can improve the accuracy by increasing the number of images for that particular sign (Fig. 3).

5.2 *Model Accuracy and Loss Graph*

This graph gives us the learning of the model on corresponding epochs. Model accuracy graph displays the accuracy of the model after a particular epoch so that we can monitor the learning. Also the loss graph gives the improvement in the model as the loss reduces with epochs. The model gives the accuracy of 96% on the training dataset with the F1 score of 0.92 (Fig. 4).

6 Conclusion and Future Work

After analyzing and implementing various Deep Learning approaches on the dataset such as ResNet-50, VGG-16, Inception Network, Transfer Learning. We conclude that Convolutional Neural Network using transfer learning over VGG-16 architecture has achieved the highest accuracy. The result shows that the model performs well from previous approaches. Therefore, the model can be used for Sign Language Interpretation. The accuracy can further be improved by fine tuning the final model with more epochs.

The application can further be improved so that it can automatically construct the sentences based on the characters predicted by the model. This can be done by using a RNN or a python script containing sentence constructing methods.

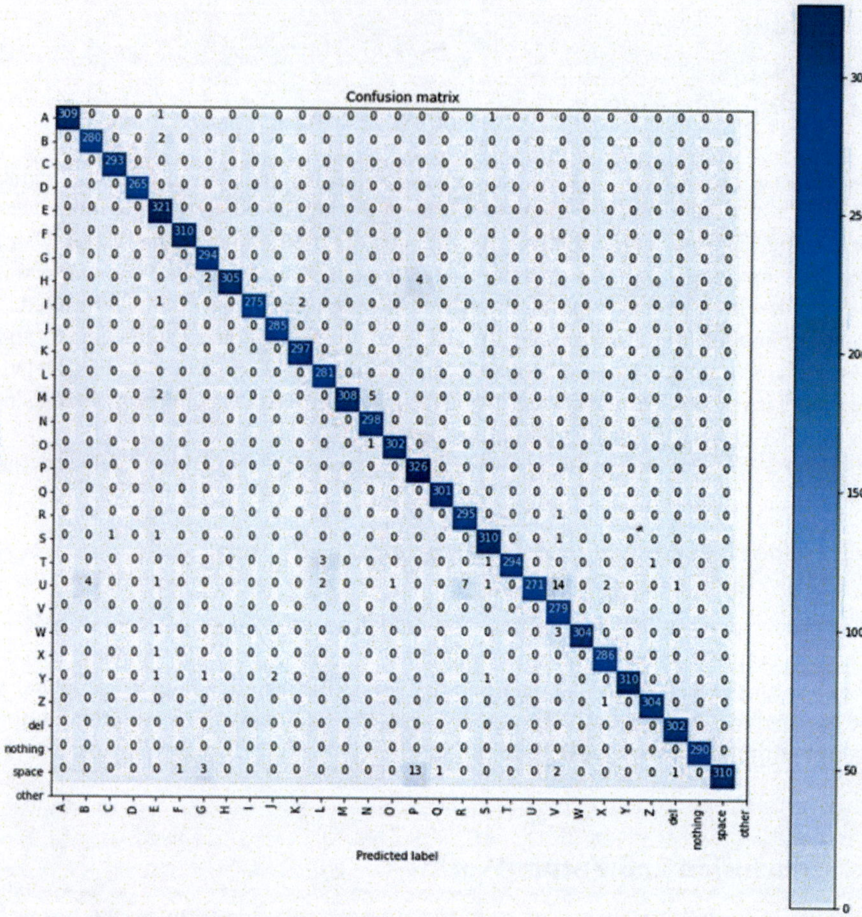

Fig. 3 Confusion matrix of model

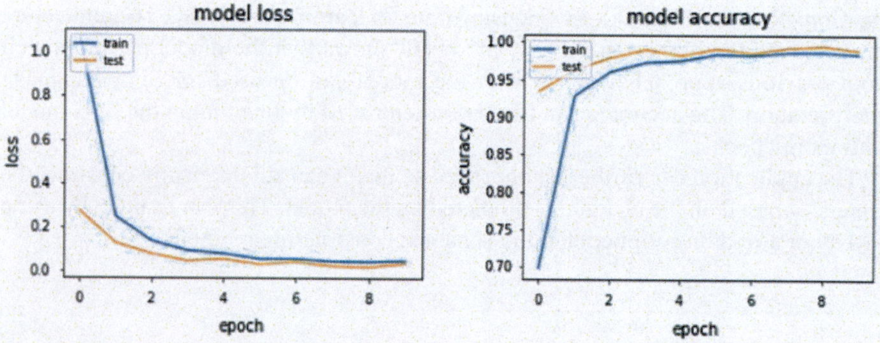

Fig. 4 Accuracy and loss versus epoch curve

References

1. Jain S, Raja KVS Indian sign language recognition . Indian Institute of Technology, Kanpur Course Project-CS365A
2. Padmavathi S, Saipreethy MSV (2013) Indian sign language character recognition using neural networks. IJCA Special Issue Rec Trends Pattern Recogn Image Anal RTPRIA 40–45
3. O'Shea K, Nash R (2015) An introduction to convolutional neural networks, vol 9. [Submitted on 26 Nov 2015 (v1), last revised 2 Dec 2015 (this version, v2)]
4. Huang J-T, Li J, Yu D, Deng L, Gong Y (2013) Cross-language knowledge transfer using multilingual deep neural networks with shared hidden layers. In: 2013 IEEE international conference on acoustics, speech and signal processing (ICASSP). IEEE, pp 7304–7308
5. Simonyan K, Zisserman A (2015) Very deep convolutional networks for large-scale image recognition. Published as a conference paper at ICLR 2015
6. Visa S, Ramsay B, Ralescu AL, Van Der Knaap E Confusion matrix-based feature selection 13:1–298. MAICS, 2011—researchgate.net

Vehicle Owner Identification from Number Plate

Harsh Vashishtha, Gopal Sharma, Abhishek Malik, Shweta Paliwal, and Aanjey Mani Tripathi

Abstract Vehicle management and owner identification have become a significant drawback for all countries. Generally, it's troublesome to spot a vehicle owner who breaks the traffic rules and drives too quickly. Therefore, it's uphill to catch and penalize those people because the speed of the vehicle is too fast for the person to retrieve the number from that vehicle. Therefore, there's a requirement to enhance the Automated Number Plate Recognition (ANPR) system. There are several ANPR programs available nowadays. These systems are made up of different methodologies; however, it is still quite difficult to retrieve the license number from a moving vehicle as a number of the factors like high speed of the automotive, different license plate on a different vehicle, the language of number (Hindi, English, etc.) and varied lighting conditions will considerably have an effect on the recognition. Most programs work under these limits.

Keywords ANPR · Vehicle Identification · Frame Processing

H. Vashishtha · G. Sharma · A. Malik · S. Paliwal · A. M. Tripathi (✉)
Department of Computer Science and Engineering, Meerut Institute of Engineering and Technology, Meerut, India
e-mail: Aanjey.mani@miet.ac.in

H. Vashishtha
e-mail: harsh.vashishtha.cse.2017@miet.ac.in

G. Sharma
e-mail: gopal.sharma.cse.2017@miet.ac.in

A. Malik
e-mail: abhishek.malik.cse.2017@miet.ac.in

S. Paliwal
e-mail: shweta.paliwal@miet.ac.in

R. P. Mahapatra et al. (eds.), *Proceedings of International Conference on Recent Trends in Computing*, Lecture Notes in Networks and Systems 341,
https://doi.org/10.1007/978-981-16-7118-0_12

1 Introduction

Vehicle owner identification systems are an important technology that is based on the ANPR technology. This system can be used for various purposes like owner identification, vehicle identification, law enforcement, collecting toll money and vehicle parking. This system uses a number plate for the identification of vehicle and owner from both video and an image [1–3]. In today's scenario, vehicle identification system is used in the area of traffic control and at the same time deals with the issues related to access control and tracking. The researches illustrated that detection of features from vehicle images is a difficult task and different techniques have been proposed in order to perform this detection better. Vehicles are considered important source of information as they are capable of serving different purposes such as Vehicle Surveillance, Traffic Management and Vehicle Verification. The working of the system is classified into four parts: Video or image of a car is used as an input parameter, second task is to perform the extraction of number plate, and third task is the extraction of characters and fourth is character recognition [4, 5].

2 Methodology

First step is to find out the correct moving image of the car which seems to be very easy but a difficult task. Because it is very difficult to click an image of the fast moving vehicle on roads so that any part of the vehicle is not left and most importantly the Vehicles Number Plate because it is the only way to find out the owners identity and the identity is correct only when then program detects the position of the number plate correctly and read the characters correctly written on the Number Plate [4, 5]. These programs follow a variety of ways to obtain a car number plate from a car and extract the car number from the image.

There are various steps involved in the vehicle owner identification system. One of the major steps in this process is to identify the location of the number plate in the image or video. Then, we can move forward to identify characters in that number plate and save the output in a string with all special characters removed from that string. String should only contain alphanumeric characters. After that we can pass that string to DBMS and then we can look for the owner's details in a database containing all the information about the owner of all the registered license plates.

2.1 *Steps Required in Vehicle Owner Identification System*

1. **Selection for photo or video**: In the first step, we take input from the user whether the user wants to detect the number plate from a video or photo. If user chooses to detect number plate from a video.

```
C:\Python\Python385\python.exe C:/Users/Acer/Desktop/Programs/ANPR_Both/npr.py
License Plate Recognition
1.Video
2.Photo
Enter your choice:
```

Fig. 1 Selection of input

2. **Frame Division**: First step in number detection from a video is dividing video into frames.
3. **Frame processing**: After dividing video into frames, the program will start processing frames one by one to detect the number. After detecting the number in any frame, program will stop processing more frames.
4. **Image/frame processing**: Image Processing basically deals with performance of operations on an image in order to get the enhanced version of the image.

There are 4 steps in image processing–Rescale the size of image and grayscale it use bilateral filter converting grayscale image into edged image Search Contours in the Obtained image (Figs. 1, 2, 3 and 4).

```
Enter video title:1.mp4
Creating..../data/frame0.jpg
Creating..../data/frame1.jpg
Creating..../data/frame2.jpg
Creating..../data/frame3.jpg
Creating..../data/frame4.jpg
```

Fig. 2 Division of frames

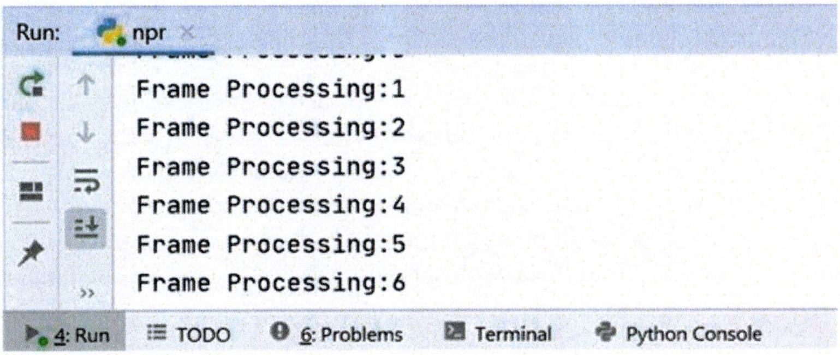

Fig. 3 Frame processing

```
img = cv2.resize(img, (600, 400))

gray = cv2.cvtColor(img, cv2.COLOR_BGR2GRAY)
```

Fig. 4 Algorithm for resize and grayscale

3 Steps for Image Conversion

Step 1: Rescale the size of image and grayscale it: Rescaling the size of the image will help us to deal with images of different sizes, but we have to make sure that number plate is in the frame after resizing. Gray scaling is one of the most common steps in all image processing algorithms. This helps us to improve the accuracy of the algorithm as we do not need to deal with color details, while processing the image. This step boosts the speed of algorithms and helps to reduce time complexity. Result after this step will look like the sample image below.

Step 2: Use Bilateral filter: All input images will contain some important and futile information but here we only need the information about the number plate, so all the other information is waste which is known as "Noise." We use filters to remove this waste information from the system (Fig. 5).

Step 3: Converting grayscale image into edged image: Now in this step, we are going to extract the edges from the image. We have used OpenCv to extract the edges. In OpenCv, the canny edge method is the most efficient method to extract the edges. Canny edge methods extract those edges which have a gradient ranges between a minimum and maximum value.

Step 4: Search contours in the obtained image: Firstly, we find the contours in our image then we are going to consider the first 15 resulting contour images and sort them from largest to the smallest. The contours will find all the shapes with a closed area. So, we can find the number plate using contours as the number plate also has a closed area (Fig. 6).

Detection of Number plate: In this step we will detect the location of the number plate from the input image. For this purpose, we will have to once we have the right contours, we save them with a variable called screen Cnt and draw a Rectangle around it, so that we can be sure that we detect the number plate right.

As we have the correct position of the number plate, so the outer image and other information are useless. Therefore, we start hiding the whole image outside the area

Fig. 5 Algorithm for Bilateral filter

```
gray = cv2.bilateralFilter(gray, 13, 15, 15)
```

```
contours=cv2.findContours(edged.copy(),
cv2.RETR_TREE, cv2.CHAIN_APPROX_SIMPLE)
contours = imutils.grab_contours(contours)
contours = sorted(contours,
key=cv2.contourArea, reverse=True)[:15]
```

Fig. 6 Algorithm for contours detection

where the number plate is. After hiding the image, we crop the detected number plate.

Character Segmentation: After identifying the position of the number plate, we should crop the number plate from the image and save the cropped number plate as a new image. Then we can use this image to find the characters present in the number plate. In the new image, we can use edge detection technique or gray scale the image. This helps the program to identify characters more efficiently. Although we can use the original image too and it will also produce very effective results. Character segmentation is a method that looks to decompose the image of the sequence of characters in the sub images of certain symbols. It is one of the decision-making processes in the OCR (optical character recognition).

Character Identification: Now the program is going to extract alpha-numerical characters from the obtained segmented image using OCR (optical character recognition). The segmented image contains some alphanumeric characters. We are going to use pytesseract to extract the alphanumeric characters from the segmented image (Fig. 7).

Owner Identification: The final step in this project is to identify the owner of the vehicle from the number which is extracted from the number plate. For this, we have to create a Database of all registered vehicles and then search the extracted number in the database.

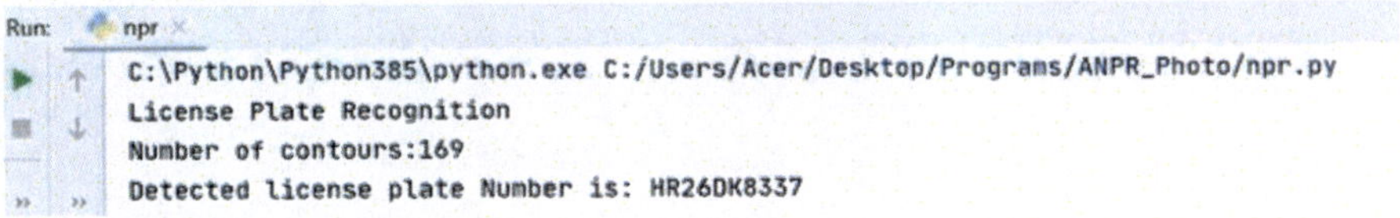

Fig. 7 Identification of characters

4 Result

After successful execution of this program, we get the details (such as Name, Mobile number and Address) of the owner of the vehicle from the input image or video. In most of the cases, the detected number is correct, and we get the correct owner details. It is a fast way to identify the owner of the vehicle. It accurately finds the number of vehicles from most of the videos and images.

Below is the example of the correctly detected Number plate and Number of the Vehicle (Figs. 8, 9 and 10).

But in a few cases where the number is not clearly visible or the car is very far away in the photo or video, the program will be unable to detect the correct number, and

Fig. 8 Input image of vehicle

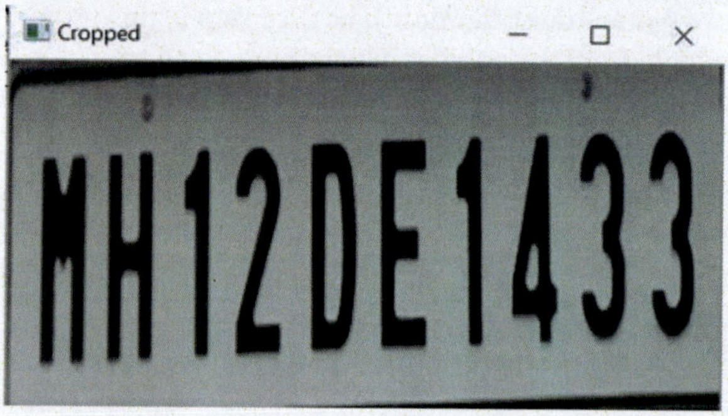

Fig. 9 Detected number plate

```
C:\Python\Python385\python.exe C:/Users/Acer/Desktop/Programs/ANPR_Photo/npr.py
License Plate Recognition
Number of contours:239
Detected license plate Number is: MH12DE1433
```

Fig. 10 Correct detected number

hence, the owner details are also wrong. Sometimes the detected number is incorrect due to the OCR package (in this case pytesseract) we have used in the program. So, to overcome this problem, we have to train more data. Here is an example in which the detected number is incorrect due to the OCR package.

As we can see in this example program correctly detected the location of the number plate but the number detected is not completely accurate.

Correct Number—MH 12FN 7866 Detected number—NH 12FN 7865

In this project, we have observed that there are multiple factors responsible for identification of number plate such as image quality, distance b/w camera and vehicle, number of vehicles in the photo or video, etc. So these factors can bring a lot of change in the results. So, we should try to input only those images which have a number plate in the center of image and the size of the number plate should be large enough to be identified by the program (Fig. 11 and Table 1).

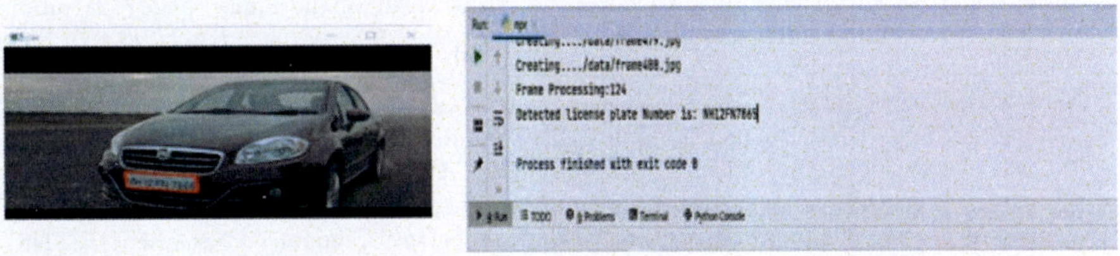

Fig. 11 Wrong detected number

Table 1 Result analysis from different image

Actual number in image or video	Total number of characters	Number plate identified	Detected number by program	Correctly identified characters	Accuracy (%)
MH 12 DE 1433	10	Yes	MH 12 DE 1433	10	100
MH 12 FN 7866	10	Yes	MH 12 FN 7865	8	80
MH 20 DV 2363	10	Yes	WH 20 OY 2363	8	80
RIP LSI	6	Yes	RIP LSI	6	100
KL 40 L 5577	9	Yes	KL 40 L 5577	9	100

5 Conclusion

Vehicle owner identification system is a very useful software for number plate detection and identification of vehicle's owner. This software saves time by removing the need to enter the number manually. This system has various applications like vehicle identification, owner identification, law enforcement, collecting toll and parking systems for vehicles. In future, this technology can be improved by implementing some rules such as only simplified number plates should be used without any special symbol or any design on the number plate and by implementing proper imaging setups so that the number plate is clearly visible in the image or video.

References

1. Abdelwahab ABMM, Thabet AM, Abdelsadek AM (2011) Automatic number plate recognition system. Annals of the University of Craiova. Math Comput Sci Ser 38(1):62–71. ISSN 1223-6934
2. Karwal H, Girdhar A (2015) Vehicle number plate detection system for Indian vehicles. IEEE
3. Anagnostopoulos CNE, Anagnostopoulos IE, Psoroulas ID, Loumos V, Kayafas E (2008) License plate recognition from still images and video sequences: a survey. 9(3):377–391
4. Kumar S, Krishna CR, Solanki AK (2017) Time efficient public key cryptography for enhancing confidentiality and integrity in a wireless sensor network. Int J Comput Sci Netw Sec (IJCSNS) 17(1):81
5. Tripathi AM, Singh S (2017, March) EEEMRP: extended energy efficient multicast routing protocol for MANET. In: International conference on information and communication technology for intelligent systems. Springer, Cham, pp 473–481
6. Saif N, Ahmmed N, Pasha S, Shahrin MSK, Hasan MM, Islam S, Jameel ASMM (2019) Automatic license plate recognition system for Bangla license plates using convolutional neural network. In: TENCON 2019–2019 IEEE region 10 conference (TENCON). IEEE, pp 925–930
7. Sengur A, Guo Y (2011) Color texture image segmentation based on neutrosophic set and wavelet transformation. Comput Vis Image Underst 115(8):1134–1144
8. Rawat M Design and comparison of agglomerative hierarchical clustering. Int J Comput Appl 975:8887
9. Tripathi A, Singh S (2017) A literature review on algorithms for the load balancing in cloud computing environments and their future trends. Math Comput Model 21(1):64–73
10. Kumar S, Krishna CR, Solanki AK (2018, February) A technique to analyze cyclomatic complexity and risk in a wireless sensor network. In: 2018 5th international conference on signal processing and integrated networks (SPIN). IEEE, pp 602–607

Smart Monitoring of Drunk Driver Using IOT and Machine Learning Based Anomaly Detection

Sujatha Karimisetty, A. Arjuna Rao, Y. Dinesh Kumar, and P. Kumudini Devi

Abstract Now-a-days, many road accidents are occurring due to the alcohol consumption of the person who is driving the vehicle. Though drunken driving is a threatening issue, detection of drunk driver has become a cumbersome issue. Hence, a stable system is required with continuous monitoring of drunk driver. The Smart Monitoring of Drunk Driver using IOT and Machine Learning helpful to the owner of the vehicle if he handovers the vehicle to the driver. The owner of the vehicle can continuously monitor the vehicle movements by using the features such as GPS and can track the driver mannerism by using the IOT sensors that sense both alcohol and abnormal movements of the vehicle. This requires a two-step procedure, first using Internet of Things Drunk Driver Detection (IOTDDD) and next adapting Support Vector Machine Learning Based Anomaly Detection (SVMAD). This system helps police in easily identifying the drunk drivers in heavy traffic.

Keywords Smart Monitoring of Drunk Driver using IOT and Machine Learning · Internet of Things Drunk Driver Detection(IOTDDD) · Support Vector Machine Learning Based Anomaly Detection (SVMAD)

1 Introduction

Drivers who consumed alcohol have many times lost control over the vehicle and loose the road sense and hence make dangerous accidents. The threatening reason behind all road accidents that is common to all countries in the whole world is drunk driving. Metropolitan cities are facing major issue and higher count of accidents are

---

S. Karimisetty (✉) · P. Kumudini Devi
Department of CSE, Dadi Institute of Engineering and Technology, Visakhapatnam, India

A. Arjuna Rao
Miracle Educational Society Group of Institutions, Vizianagram, India

Y. Dinesh Kumar
Department of CSE, Satya Institute of Technology and Management, Vizianagram, India

R. P. Mahapatra et al. (eds.), *Proceedings of International Conference on Recent Trends in Computing*, Lecture Notes in Networks and Systems 341,
https://doi.org/10.1007/978-981-16-7118-0_13

on highways at the city outskirts which are due to drunken drivers and clearly no special method is adapted to eradicate these accidents [1].

Andhra Pradesh has been placed third on the list of states with the highest accident rates in 2017.A.P. witnessed 30,045 road accidents with 8,210 deaths which makes 7.52% of the country's total. The below table shows the accident rate and cause for the accidents in A.P. Out of these 30,045 accidents, 20% are only due to drunken driving.

Vehicle Driving needs quick and smart decisions, actions that are reflexive, ultimate concentration in order to avoid major accidents. The function of the brain will be improper and that results in unwanted incidents on road if any person consumes alcohol beyond safety limit. This is a problem for driver, fellow-passenger and also the other people who are pedestrians or driving other vehicles [2].

The drunk drivers lose their concentration and they do common mistakes like moving with high speed, jumping signal and damage the vehicle by hitting static objects such as trees, buildings. In present scenario, traffic-cops use breath analyzers for detecting drunken persons which is difficult to check in modern-day traffic. The disadvantages of the present system are becoming a challenging task to police in checking vehicles and there is a chance for drunk drivers to escape from the police in the traffic. So, there is a need for an effective system to check every drunk driver. Hence, this is becoming a issue for government as well as police [3].

2 Smart Monitoring of Drunk Driver Using IOT and Machine Learning (SMDDIM)

Smart Monitoring of Drunk Driver using IOT and Machine Learning involves two step process where the first module uses IOT and next module uses Machine Learning techniques to analyze the behavior of the driver. When a person enters the vehicle to ignite the engine by turning the key the system starts working activating the sensor and detects the level of alcohol content. When the alcohol consumed by driver is crossing the extreme limit, immediately the owner and nearby police station server get the notification.

The objectives of the SMDD are as follows

- This system can be used in the various vehicles for detecting the drunken driver. As the project is based on IOT an alcohol sensor is to detect.
- It can immensely decrease the rate of drunken driving cases which often leads to accidental deaths.
- An android application is used by both police and the owner of the vehicle in order to get notified when the driver was drunk.
- Through GPS, the vehicle can be tracked by the police as well as the owner.
- Severe punishment will be given to the drunk driver by the police so the driver may not repeat drunk driving in future. Repetition of this activity is known that may indulge to permanent seize of driver's license.

- Machine learning technique Support Vector Machine-Based Anomaly Detection is used for detecting the abnormal behavior of the driver when he is drunk.

2.1 Internet of Things Drunk Driver Detection (IOTDDD)

The IOT kit contains MQ3 Alcohol sensor, NodeMCU ESP8266, Transformer, Bridge rectifier, Voltage Regulator, HC-05 Bluetooth module. The architecture of the IOTDDD is shown in Fig. 1.

- Alcohol concentration level is detected by MQ3 alcohol sensor.
- The NodeMcu consists of ESP8266 Wi-Fi module which provides internet connectivity.
- Step down transformers are used to decrease the voltage from 230 to 12 V
- Alternating Current (AC) to Direct Current (DC) is done by bridge rectifier.
- Voltage Regulator 12 V DC received from bridge rectifier splits into 5 V each to Node MCU, and MQ3 sensor.
- Serial connection for wireless connection is done by HC-05 Bluetooth Module with SPP.

In this system, in order to notify the owner and police Mobile App is used. The block diagram is as shown in Fig. 1.

This kit is placed in the vehicle by the owner who handovers the vehicles to driver. However, it is required to analyze the behavior of the driver as the owner is accused even if the mistake is done by the driver and he has to bear all the loss on concerned vehicle.

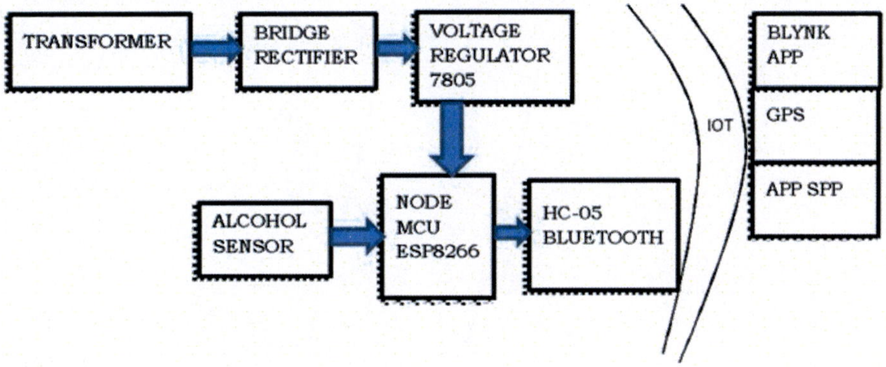

Fig. 1 Block diagram of IOTDDD

2.2 *Support Vector Machine Learning Based Anomaly Detection (SVMAD)*

Anomaly detection plays key role in machine learning and this is used to detect the abnormal behavior of the driver. One class Support Vector Machine abbreviated as SVM is utilized for Anomaly Detection on unsupervised problem where unlabeled training data is used. This algorithm detects the unusual behavior of the driver by analyzing the vehicle movements and informs the concerned to make a check of drunk driver. Sensors are used to identify the vehicle abnormal motion analysis [4].

SVM is the classifier that uses a hyperplane that uses maximization of margin that separates two classes as shown in Fig. 2. This is based on statistical learning method over the training data set. SVM follows an assumption that in training set all samples used are distributed independently. The SVM training algorithm makes the decision surface to deviate from position optimal in features [5] (Fig. 2).

The algorithm is as follows

1. First the margin is maximized as an optimal hyperplane is defined.
2. Trace the misclassifications by extending to non-linear separable problems.
3. Problem reformulation is done by mapping data to high dimensional space that makes easy to classify.

Training set that is given as input is collected from the sensor used to predict the sudden variations such as frequent sudden brakes, sharp cuts and so on. The sensor originally delivers the unstructured data and then the data is converted to structured and given as input to SVM. The procedure is shown in Fig. 3 [6].

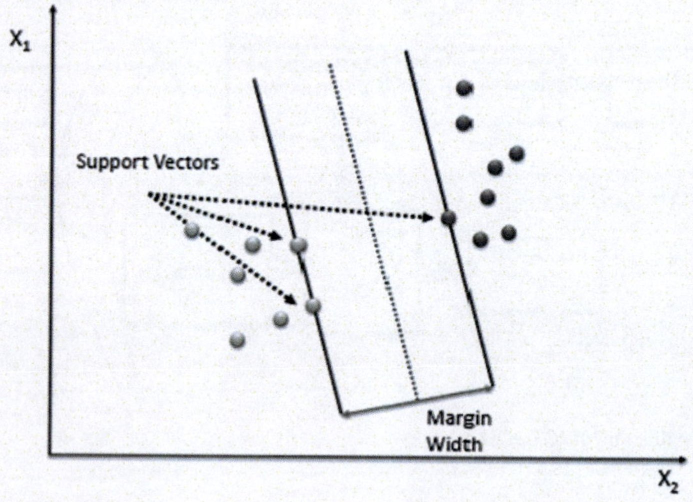

Fig. 2 SVM classification

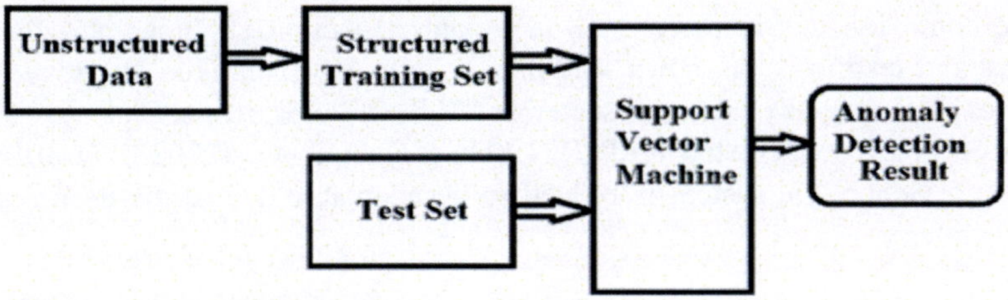

Fig. 3 Anomaly detection using SVM

3 Results and Discussion

3.1 Experimental Setup of IOTDDD

The Internet of Things Drunk Driver Detection is assembled as shown in Fig. 4, testing is being done by exposing the sensor to alcohol, the exact proportion of alcohol in air is being sensed, and the data is examined for analysis.

Fig. 4 IOTDDD simulating model

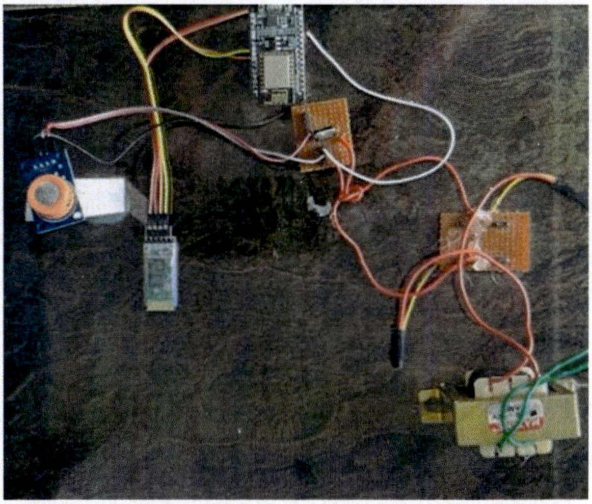

Table 1 Confusion matrix

No of samples = 10,000	Predicted No	Predicted Yes	
Actual NO	TN = 7160	FP = 40	7200
Actual YES	FN = 8	TP = 2792	2800
	7168	2832	

The confusion matrix performed on a dataset using 10,000 samples for both training and testing is as shown in Table 1. True Positive, True Negative, False Positive, False Negative, are represented as TP, TN,FP, FN, respectively.

The accuracy is calculated as (TP/TN)/(No of Samples) = (2792 + 7160)/10000 = 0.9952. Hence, the system is proved to be adaptable due to the accuracy and Positive rate.

4 Conclusion

Major problem observed in accidents is due to Drunk and Driving everywhere. The threat is not only for drivers but to innocent tress passers. Mainly children and adults having less awareness on vehicle speed estimation and road crossing rules are affected. Some are deceased and others have to lead their life handicapped. This will effect economically the entire family due to medical expenses. In this paper, real-time model is presented that can reduce the accidents due to drunken driving. Therefore, using this toolkit, alcohol-related road accidents can be reduced, and hence, this kind of detectors has greater importance in future. Automating this IOT system using machine learning will increase the probability in detection and sends timely alert. Through this Smart Monitoring of Drunk Driver using IOT and Machine Learning the owner of the vehicle or police can do continuous monitor of alcohol detector and preventive device of accidents. The future scope is to implement Multicore parallel processing based SVM using parallel processing proposed by the authors in Multicore Parallel Processing Concepts [7].

Acknowledgements The authors would like to thank the Chairman of Dadi Institute of Engineering and Technology, Visakhapatnam, Sri Dadi Ratnakar for his support in establishing the lab and constant motivation for completing the work successfully.

References

1. Sahabiswas S et al (2016) Drunken driving detection and prevention models using Internet of Things. In: 2016 IEEE 7th annual information technology, electronics and mobile communication conference (IEMCON), Vancouver, BC, pp 1–4
2. Charniya NN, Nair VR (2017)Drunk driving and drowsiness detection. In: 2017 international conference on intelligent computing and control (I2C2), Coimbatore, pp 1–6
3. Sandeep K, Ravikumar P, Ranjith S (2017) Novel drunken driving detection and prevention models using Internet of Things. In: 2017 international conference on recent trends in electrical, electronics and computing technologies (ICRTEECT), Warangal, pp 145–149
4. Kishore B, Satyanarayana MRS, Sujatha K (2016) Efficient fault detection using support vector machine based hybrid expert system. Int J Syst Assur Eng Manag 7(Supplement 1):34–40
5. Rao AA, Sujatha K, Saragada N, Sastry MS, Prasad KR (2015)Automation of metal charge calculations using support vector machine. In: 2015 international conference on man and machine interfacing (MAMI), Bhubaneswar, pp 1–5

6. Cristianini N, Shawe-Taylor J (2000) An introduction to support vector machines and other Kernel-based learning methods. Cambridge University Press, Cambridge
7. Sujatha K, Nageswara Rao PV, Rao AA, Sastry VG (2015) Multicore parallel processing concepts for effective sorting and searching. In: 2015 international conference signal processing and communication engineering systems (SPACES), 2–3 January 2015, pp 162–166

An Effective and Innovative Framework with Authentication Schema and Key Generation in Internet of Things

S. Karthick and N. Gomathi

Abstract Internet of Things (IoT) plays vital role on heterogeneous wireless network communications especially on wireless sensor networks (WSN). WSN communication are generally concentrating low energy consuming and distributed based interconnected sensor nodes. Hence, in such network, access security is much more essential and the user authentication scheme is one of popular security topics in WSN. Previous authentication works usually focus on one user to one sensor accessing scenario. However, for future IoT applications, such as smart-home, there plenty of self-define sensor nodes in WSN architecture, where one user usually wants to control multiple sensor devices in a short time or at the same time. In such network phenomenon, we call it as multiple accesses scenario. Accordingly, this research work proposes an authentication and key agreement scheme which enables a remote user to efficiently complete multiple authentication processes at a time in the multiple access scenario. This proposed authentication scheme is suitable for the resource-constrained WSN architecture. Further, our scheme also considers the security flaws of two-factor authentication and designs a stronger security protection. In our security feature and performance evaluation, our proposed scheme achieves several security goals and, meanwhile, ensures the efficiency.

Keywords WSN · IoT · Authentication · Data security · Key-generation · Sensor nodes · Password protection · Identification

S. Karthick (✉) · N. Gomathi
Department of Computer Science and Engineering, Vel Tech Rangarajan Dr.Sagunthala R&D Institute of Science and Technology, Chennai, Tamilnadu, India

N. Gomathi
e-mail: gomathin@veltech.edu.in

R. P. Mahapatra et al. (eds.), *Proceedings of International Conference on Recent Trends in Computing*, Lecture Notes in Networks and Systems 341,
https://doi.org/10.1007/978-981-16-7118-0_14

1 Introduction

The applications in WSN [1] cover the wide life fields, including traffic monitor, healthcare monitor, landslide detection, assert tracking, etc. In WSN, access security is much more essential for WSN. If without any security defense, malicious users could disclose secret data and personal privacy easily. Xuei et al. [13], the fifth model seems to be appropriate for the communication model in IoT environment. Turkanoviç and Hölbl [10] also apply this model. Their novel authentication scheme allows one user to access the desired sensor node directly in a secure without the aid of GWN. However, for some applications, like smart-home concept, there are dozens of sensor nodes in WSN architecture and the user may want to control more than one sensor devices in a short time or at the same time. In such network phenomenon, we call it as multiple accesses scenario. According to previous research, in such a circumstance, the authentication sessions should be initiated again and again in a very short time. Some unnecessary security risks will arise and it seems not an efficient authentication manner. In our authentication model, the user just needs to connect with one of multiple sensor nodes and pass authentication message to it. Then, the chosen sensor node, namely Representative Sensor Node (RSN), delegates authentication task to corresponding GWN. After that, GWN transfers the authentication and key agreement message to other sensor nodes and these sensor nodes can obtain the session key securely.

Those sensor nodes send back their keys to GWN and GWN can respond those secure session keys to the User. Via our proposed algorithm, for authentication of multiple accesses, the user just needs to implement authentication and key agreement algorithm once. Hence, our proposed algorithm saves much communication resource and achieves effectively.

2 Literature Review

All over the internet, thesis, research work, assignment work etc. provide various definition of IoT. M. Syafiq Mispan, Mark Zwolinski and Basel Halak define IoT network consists of plenty of inter-connected network communication nodes (example: sensors nodes, actuators, connectors, processors and some other network communication based things). Network system would like to provide more detailed definition of IoT as 'Internet of Things' is the realization of concept that all the things over the world are connected to a heterogeneous network. IoT include Sensors, internet enabled wireless devices, internet enabled wired devices, servers, protocols which enable smooth flow of information, security software and devices, Database system/cloud system, and web portals.

2.1 Sensor

A sensor are the physical devices which can measure some value from its surrounding environment. For example, a thermometer could measure the current. Speedometer could measure the speed. There is various type of sensor. In general term sensor are physical device which sense the surrounding environment. Sensors communication network nodes are the end point in the IoT network.

2.2 Internet Enabled Devices

These are the devices which have the capability to transferred information which is collected through various sensors or from other means, to the internet whether on the other hand server and a database system is connected.

2.3 Servers

Servers are computer system which have vast processing power. Server could be a single computer or as set of inter connected computers. These have sufficient memory, processing power and power supply. Most of the IoT server run all the time. Main task of these server is to process the data and provide the information whenever required.

2.4 Cloud System/database System

A cloud or database system is the place where retrieved data and processed data is stored for the further use. Cloud system have trillions of exabyte storage spaces. It has its own authentication, verification, validation, and other security protocols to keep data safe.

2.5 Security Software

There is so many security software which are continuously running the keep data safe from various attack from intruders and malicious programs.

2.6 Physically Unclonable Function

For data retrieval we need to find answer of few queries: Important characteristics of data retrieved? Is data being static, user information such as "user id, name, gender"; or dynamic data such as user's tweet and its network? Which kind of data is important for analysis? How it will be processed? What is actual size of data collected? It is very easier to keep finding of certain keyword associated with any hashtag rather than keyword which is not associated.

2.7 Encryption and Decryption

Encryption is a technique or process by which we convert plain text or data into code form which is called cypher text. This cypher text looks meaningless. If someone receive this cypher text then he can't receive any information from this without decoding it. Decryption is a technique or process by which we convert coded cypher text into plain text.

3 Proposed Methodology

3.1 One New Authentication and Key Agreement Model

This paper proposes a new authentication model composed of four relationship models, such as User-Representative sensor (U-R), Representative sensor-GWN (R-G), GWN-Multi-sensor node (G-M) and GWN-User (G-U) relation models as shown in Fig. 1.

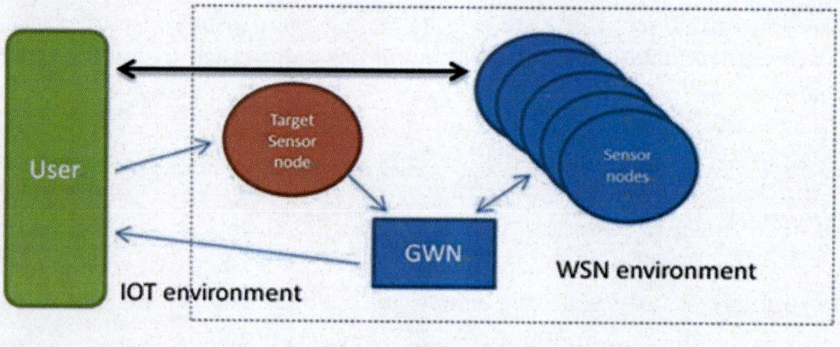

Fig. 1 The new authentication model for one bunch of accesses scenario

Table 1 Notations used in our proposed authentication algorithm

Parameters	Definitions
S_j	Sensor node *j*
U_i	The User
SID_j	The identification of the sensor node *j*
UPW_i	The user password assigned by the User *i*
MPK_i	The reformed user password of U_i
T	Delay of network transmission
UID_i	Identification of the U_i
q_i	A randomly number generated by *SC*
Tri	The current timestamp sending message out from the user
TSj	The current timestamp sending message out from the sensor node
TCG	Current timestamp in GWN
ISj	Information about $S_j$in GWN
IUi	The information about the User *i* in THE GWN
RSN	Representative Sensor Node
SID_{Sa}	The *SID* set of these nodes
S_a	The set of all sensor nodes expected to be controlled by the User
K_i	Key-generation by U_i
K_j	Key-generation by S_j

3.1.1 An Effective and Measurable Key Generation and Authentication

Five phases of authentication algorithm:

1. Preprocessing or Pre-deployment
2. Sign up or Registration
3. Sign in or Login
4. Verification and Authentication
5. Verification key generation and Response phases.

Table 1 shows the definition of parameters used in the following explanation of this algorithm.

3.2 *Preprocessing or Predeployment*

Deployment and pre-definition in WSN are main work items in pre-deployment phase. The service provider should provide Smart Cards with unique UID_i to the users and these UIDs are stored in GWN meanwhile. Each sensor nodes are defined with SID_j and has a secret key KGWN-S_j shared by GWN. Initially, GWN stores SID_j, KGWN S_j for sensor node *j* and UID_i for the User *i*. The User should randomly select one of sensor node in WSN to be the RSN. While the User i wants to access WSN, the user firstly registers with GWN and so does the Representative network communication nodes.

3.3 Sign up or Registration

Sign up or registration process combination of user first data share and receives login credential. Following data table will give information about registration phase.

Step 1: In the user end, User i prepares four parameters, UID_i, UPW_i, q_i and the current timestamp T_{R1}, to initiate sign up phase (registration).
Step 2: The User i finds a modified password, $MPK_i = H(q_i UPW_i)$.
Step 3: The user computes masked password $RUK_i = H\ (T_{R1} || MPK_i)$.
Step 4: The user obtains RU_i via XORing UID_i and RUK_i. Then the user sends RU_i, RUP_i, T_{R1} to the GWN.
Step 5: The GWN computes $RUK_i = RU_i\ UID_i$. And then it also generates random secret key associating with the User i, P_{GUi}.
Step 6: With UID_i, P_{GWN} and P_{GUi}, the GWN computes $r_i = H(UID_i || P_{GWN} || P_{GUi})$ and $u_i = H(RUK_i || P_{GUi} || r_i)$.
Step 7: After obtaining r_i and u_i, the GWN computes $w_i = r_i\ u_i$.
Step 8: GWN personalizes the user network card with following parameter set $\{w_i, r_i, P_{GUi}, T_{RI}\}$ and then sends it to the network communication User i.
Step 9: Network users stores $\{w_i, r_i, P_{GUi}, T_{RI}\}$ into smart card in the end of this phase. After the U-G registration phase, the user uses MPK_i instead of UPW_i thereafter.

The procedure of t S-G registration phase is depicted with following figure (Fig. 2).

Step 1: At the beginning, the sensor node j prepares random value q_j and the timestamp T_{S1}.
Step 2: Network sensor node j computes masked and distributed secret key, $RSK_j = H(P_{GSj} || q_j || T_{SI})$, with the use of P_{GSj}, q_j, T_{S1}.
Step 3: Similar to above step, the sensor node j computes $RSP_j = H(SID_j || RSK_j || P_{GSj})$, with the use of SID_j, RSK_j, P_{GSj}.
Step 4: The sensor node j computes $RS_j = RSK_j\ RSP_j$. After these initial computations, the sensor node j sends the parameter set $\{SID_j, RS_j\ RSK_j, T_{SI}\}$ to the GWN.
Step 5: Upon receiving parameters from the sensor node j, the GWN checks if $T_{CG} - T_{S1} > T$. Then GWN computes $*RSP_j$ with the use of RS_j, RSK_j. (Note: the * sign in the front of the notation denotes this notation is candidate notation needed to be verified).
Step 6: With the use of SID_j, RSK_j and P_{GSj}, the GWN computes another RSP_j. The result of comparison between both RSP_j determines the validation of this registration message. If both RSP_j are different, this registration terminates.

User (U_i) (the known information: UID_i, UPW_i)	GWN (the known information: P_{GWN}, I_{Sj} $=\{SID_j, P_{GSj}\}$, $I_{Ui}=\{UID_i\}$)
Generate a random q_i, and current time-stamp T_{R1} MPK_i=H ($q_i UPW_i$) RUK_i=H ($T_{R1}\|\|MPK_i$) RU_i= UID_i RUK_i	
$\{RU_i, UID_i, T_{R1}\}$ →	
	Checks if T_{CG}- T_{R1}>T? RUK_i= $RU_i UID_i$ Generate random P_{GUi}associating with the user i r_i=H ($UID_i\|\|P_{GWN}\|\|$ P_{GUi}) u_i=H ($RUK_i\|\|$ $P_{GUi}\|\|$ r_i) w_i= r_i u_i personalizes user's smart card USC $USC=\{w_i,r_i,P_{GUi}\}$
← $\{w_i,r_i,P_{GUi},T_{R1}\}$	
Store $\{w_i,r_i,P_{GUi},T_{R1}\}$ into smart card	

Fig. 2 U-G registration procedure

Step 7: The GWN computes $r_j = H(*RSP_j||P_{GWN})$ and $u_j = H(RSK_j||P_{GSj}||P_{GWN})$.
Step 8: The GWN adds RSP_j into the I_{Sj} and then computes $w_j = r_j$ *compare* u_j. Then, the GWN sends information set $\{w_j, r_j, T_{G1}\}$ to S_j.
Step 9: On receipt of information set $\{w_j, r_j, T_{G1}\}$, the sensor node j checks if $T_{CS} - T_{G1} > T$. If no, the sensor node stores w_j, r_j or it rejects this message (Fig. 3).

3.4 Sign in or Login

In Login Phase, the user randomly chooses a Representative Sensor Node (*RSN*) in S_a and tries to login it. Figure 4 shows U-S login procedure. The details of this sign in or login are as following,

Step 1: Terminal exploits T_{R1} and MPK_i to compute $*RUK_i = H(T_{R1}||MPK_i)$ and then it employs the secret information in smart card, $\{r_i, P_{GUi}\}$, to compute $*u_i = H(RUK_i||P_{GUi}||r_i)$.
Step 2: After computing u_i, the terminal computes $*w_i = r_i{*}u_i$.
Step 3: The terminal checks if computed $*w_i$ is equal to w_i receives in network card and if result holds, go to next Step, or this login process is terminated.
Step 4: The terminal calculates network mask with *UID*, $RUP_i = H(T_{R2}||UID_i)$.

GWN (the known information: P_{GWN}, $I_{Sj}=\{SID_j, P_{GSj}\}$, $I_{Ui}=\{UID_i\}$)	Sensor node (S_j) (the known information: SID_j, P_{GSj})
	chooses random $q_j$and timestamp T_{S1} RSK_j=H (P_{GSj}|| q_j || T_{S1}) RSP_j= H (SID_j ||RSK_j ||P_{GSj}) RS_j= RSK_j RSP_j
$\{SID_j, RS_j, RSK_j, T_{S1}\}$	
Checks if T_{CG}– T_{S1}>T? *RSP_j= RS_j RSK_j RSP_j=H (SID_j ||RSK_j ||P_{GSj}) Check if *RSP_j =RSP_j? r_j=H (*RSP_j||P_{GWN}) u_j=H (RSK_j|| P_{GSj}|| P_{GWN}) GWN adds RSP_jinto the I_{Sj} w_j= r_j u_j	
$\{w_j, r_j, T_{G1}\}$ →	
	Checks if T_{CS}-T_{G1}>T? Stores w_j,r_j

Fig. 3 S-G registration procedure

Step 5: The terminal computes $M_i = H(u_i || P_{GUi} || T_{R2} || r_i)$ with the use of (u_i, P_{GUi}, T_{R2}, r_i).

Step 6: The terminal chooses random nonce K_i and computes D_i with the use of K_i, M_i.

Step 7: The terminal sends the parameter set $\{RUP_i, w_i, D_i, M_i, T_{R2}, SID_{Sa}\}$ to RSN. SID_{Sa} denotes the SID sequence in S_a. The first one in SID_{Sa} is SID of the RSN.

Step 8: The RSN receives the login message and checks if $T_{CS} - T_{R2} > T$. If the inequality holds, the RSN rejects this login request.

3.5 Authentication Phase

This phase can be divided into RSN-GWN authentication phase and GWN-MSN authentication phase. Figure 5 shows RSN-GWN procedure.

Step 1: RSN computes u_j, using the formula as defined previously, i.e. $u_j = w_j$ compare r_j.

Step 2: Then, it conducts L_j, i.e. $L_j = H(P_{GSj} || T_{R2} || T_{S2} || SID_j)$ u_j.

User (U_i) (the known information: UID_i, MPK_i)	RSN (S_j) (the known information: SID_j, P_{GSj}, w_j, r_j)
SC login Input password MPK_i $*RUK_i = H\ (T_{R1} \| \| MPK_i)$ $*u_i = H\ (*RUK_i \| \| P_{GUi} \| \| i)$ $*w_i = r_i * u_i$ Check if $*w_i = w_i$? $RUP_i = H\ (T_{R2} \| \| UID_i)$ $M_i = H\ (u_i \| \| P_{GUi} \| \| T_{R2} \| \| r_i)$ Select random nonce K_i $D_i = K_i \aleph\ M_i$ Choose the RSN from S_a, and SID_{Sa}denotes the SID set of S_a. $\{RUP_i, w_i, D_i, M_i, T_{R2}, SID_{Sa}\}$ →	Checks if T_{CS}- $T_{R2} > T$?

Fig. 4 User-sensor login procedure

Step 3: The sensor node sends $\{RUP_i, w_i, D_i, M_i, T_{R2}, T_{S2}, w_j, L_j, SID_{Sa}\}$ to GWN.
Step 5: The candidate $*u_j = w_j$ compare r_j, hence, can be obtained with the use of w_j, r_j.
Step 6: GWN can also compute $u_j = L_j\ H(P_{GSj} \| T_{R2} \| T_{S2} \| SID_j)$. Then, GWN compares both u_j and if they are the same, mutual authentication between sensor node and GWN is successful. Or this authentication process terminates.
Step 7: In order to verify the validation of received RUP_i, the GWN computes $*RUP_i = H(T_{R1} \| UID_i)$ with each UIDs in its database. If there is no adequate UID for this received RUP_i, the GWN reject this registration message and the authentication process ends.
Step 8: GWN computes $r_i = H(UID_i \| P_{GWN} \| P_{GUi})$.
RSN-GWN authentication sub-phase
Step 9: GWN computes $*u_i = w_i$ compate r_i.
Step 10: GWN computes $*M_i = H(*u_i \| P_{GUi} \| T_{R2} \| r_i)$ with the use of $*u_i$, *PGUi*, *TR2*, *ri*.

The RSN (S_j) (the known information: SID_j, P_{GSj},w_j,r_j)	GWN (the known information: P_{GWN},I_{Sj} =\{SID_j, RSP_j, P_{GSj}\}, I_{Ui}=\{P_{GUi},UID_i\})
u_j= $w_j$compare r_j L_j=H (P_{GSj}|| T_{R2}|| T_{S2}||SID_j) u_j \{RUP_i, w_i, D_i, M_i, T_{R2}, T_{S2}, w_j, L_j, SID_{Sa}\} →	
	Checks if T_{CG}– T_{S2}>T? r_j=H (RSP_j||P_{GWN}) *u_j= w_j comparer_j u_j= L_jH (P_{GSj}|| T_{R2}|| T_{S2}||SID_j) checks if *u_j=u_j? *For each UID do* *RUP_i=H (T_{R2}|| UID_i) Checks if any *RUP_i= RUP_i? if yes, continue process, or authentication fails. r_i=H (UID_i||P_{GWN}|| P_{GUi}) *u_i= w_i compare r_i *M_i=H (*u_i|| P_{GUi}|| T_{R2}|| r_i) check if *M_i=M_i? K_i=D_i *M_i

Fig. 5 RSN-GWN authentication procedure

Step 11: GWN checks if $*M_i$ is equal to received M_i. If not, this authentication process terminates, or GWN computes K_i with use of received D_i and $*M_i$.

After verifying authentication message between RSN-GWN, GWN can retrieve the session key K_i and then sends it to all sensor nodes in S_a. Each sensor node j receives, pair key (K_i, K_j) back to the user via GWN. Figure 6 shows this authentication GWN-MSN eprocedure.

GWN-MSN authentication sub-phase preprocessing: For each sensor node S_j in S_a, GWN should run the following steps.

Step 1: GWN computes OSP_j and OSK$_j$ with the formulas, $OSP_j = H(r_j \| SID_j \| T_{G3} \| T_{S2} \| T_{R2})$ and OSK$_j$ $= H(r_j \| P_{GSj} \| T_{G3} \| T_{S2} \| T_{R2})$.
Step 2: Then, GWN computes $OS_j = OSP_j\ OSK_j$.
Step 3: GWN hides the user's key, K_i, in $GC_{ij} = H(OSP_j \| T_{G3} \| T_{S2} \| T_{R2})\ K_i$.
Step 5: In sensor node S_j's end, it checks if $T_{CS} - T_{G3} > T$ once it receives the messages from GWN. If inequality holds, the authentication and key agreement process terminates, or S_j computes $*OSP_j$.
Step 6: S_j computes another OSP_j with the use of its known information and the received message.
Step 7: S_j checks if $*OSP_j$ is equal to OSP_j. If it is true, S_j can obtain the user's key, $K_i = GC_{ij}\ H(OSP_j \| T_{G3} \| T_{S2} \| T_{R2})$. Or, this process terminates. Keys, the session key pair, SK_{ij}, is generated.
Step 8: S_j chooses a random nonce K_j to be its key. By the combination of both.

GWN (the known information: P_{GWN}, I_{Sj} $=\{SID_j, RSP_j, P_{GSj}\}$, $I_{Ui}=\{RUP_i, P_{GUi}, UID_i\}$)	Sensor node (S_j) (the known information: SID_j, P_{GSj}, w_j, r_j)

For each sensor node S_j in S_a $OSP_j= (r_j||SID_j|| T_{G3}||T_{S2}|| T_{R2})$. $OSK_j=$ H $(r_j|| P_{GSj}|| T_{G3}||T_{S2}|| T_{R2})$ $OS_j=OSP_j OSK_j$
$GC_{ij}=$ H $(OSP_j|| T_{G3}||T_{S2}|| T_{R2})$ K_i

$\{OS_j, T_{G3}, T_{S2}, T_{R2}, OSK_j, GC_{ij}\}$

Checks if T_{CS}- $T_{G3}>T$?
$*OSP_j= OS_j OSK_j$
$OSP_j=$ H $(r_j||SID_j|| T_{G3}||T_{S2}|| T_{R2})$
Check if $*OSP_j = OSP_j$?
$K_i=GC_{ij}$H $(OSP_j|| T_{G3}||T_{S2}|| T_{R2})$
Choose random nonce K_j
SK_{ij}=H $(K_i K_j)$
OC_{ij}=H (H (r_j) || $T_{G3}||T_{S2}|| T_{R2}|| T_{S3})$ K_j

$\{OC_{ij}, T_{G3}, T_{S2}, T_{R2}, T_{S3}, OSP_j\}$
←

Check if T_{CG}- $T_{S3}>T$?
Check if OSP_j=H $(r_j||SID_j|| T_{G3}||T_{S2}|| T_{R2})$?
$K_j= OC_{ij}$H (H (r_j) || $T_{G3}||T_{S2}|| T_{R2}|| T_{S3})$
OU_{ij}=H (H (r_i) || $T_{G3}||T_{S2}|| T_{R2}|| T_{S3})$ K_j

Fig. 6 GWN-MSN authentication procedure

Step 9: S_j hides its key K_j in OC_{ij}, where $OC_{ij} = H(H(r_j)||T_{G3}||T_{S2}||T_{R2}||T_{S3})\ K_j$. Then it sends $\{OC_{ij}, T_{G3}, T_{S2}, T_{R2}, T_{S3}, OSP_j\}$ to GWN.
Step 10: S_j hides its key K_j in OC_{ij}, where $OC_{ij} = H(H(r_j)||T_{G3}||T_{S2}||T_{R2}||T_{S3})$ K_j. Then it sends $\{OC_{ij}, T_{G3}, T_{S2}, T_{R2}, T_{S3}, OSP_j\}$ to GWN.
Step 11: GWN checks if $T_{CG} - T_{S3} > T$, then if inequality holds, this process terminates, or continues next step.
Step 12: GWN checks if OSP_j is equal to $H(r_j||SID_j||T_{G3}||T_{S2}||T_{R2})$ for S_j in S_a. If the inequality holds, this process terminates, or continues next step.
Step 13: GWN recovers the key, K_j, via XORing OC_{ij} and $H(H(r_j)||T_{G3}||T_{S2}||T_{R2}||T_{S3})$.
Step 14: GWN hides the corresponding key of $S_j$in OU_{ij}, where $OU_{ij} = H(H(r_i)||T_{G3}||T_{S2}||T_{R2}||T_{S3})K_j$.

3.6 Key Response Phase

Until now, each session key pair has been conducted by each sensor node in S_a and then, GWN sends back the corresponding key of the sensor nodes in S_a to the User. Figure 7 shows this procedure.

User (U_i) (the known information: UID_i, RUK_i)	GWN (the known information: P_{GWN}, I_{Sj} ={SID_j, RSP_j, P_{GSj}}, I_{Ui}={ P_{GUi}, UID_i})
	V_{ij}=H ($UID_i \| r_i \|$ $T_{G3} \| T_{S2} \|$ $T_{R2} \|$ T_{S3}) OV_{ij}= OU_{ij} V_{ij}
{ V_{ij},OV_{ij},T_{G3},T_{S2}, T_{R2}, T_{S3}, T_{G4}} ←	
Checks if T_{CU}- T_{G4}>□T ? Checks if V_{ij}= H ($UID_i \| r_i \|$ $T_{G3} \| T_{S2} \|$ $T_{R2} \|$ T_{S3}) $OUij$= $OVijVij$ K_j= OU_{ij} H (H (r_i) || $T_{G3} \| T_{S2} \|$ $T_{R2} \| T_{S3}$) SK_{ij}=H(K_iK_j)	

Fig. 7 The procedure of key response phase

Step 1: For assurance of legitimacy of the user, GWN computes V_{ij}, containing the user information UID_i and r_i, for verification.
Step 2: GWN computes OV_{ij} which contains the S_j's key and the validation information, V_{ij}. Then GWN sends $\{V_{ij}, OU_{ij}, T_{G3}, T_{S2}, T_{R2}, T_{S3}, T_{G4}\}$ to the network users. Meanwhile computes iteration UID$_i$ and r_i values.
Step 3: If receives acknowledgment of the authentication message, the User checks if $T_{CU} - T_{G4} > \Box T$ and if inequality holds, the process terminates, or continues next step.
Step 4: The User checks if V_{ij} is equal to $H(UID_i \| r_i \| T_{G3} \| T_{S2} \| T_{R2} \| T_{S3})$ with the use of its smart card information. If it is true, the user has verified V_{ij}.
Step 5: The user gets OU_{ij} via XORing OV_{ij} and V_{ij}.
Step 6: The user recovers the key of a sensor node in S_a with the formula $K_j = OU_{ij}\ H(H(r_i) \| T_{G3} \| T_{S2} \| T_{R2} \| T_{S3})$.

Finally, sensor node, S_j, in S_a have the session keys $\{SK_{ij}\}$. The communication between them, therefore, is secure.

Security and performance analysis our proposed scheme can support the following features:

1. Secure key agreement: The user's key, K_i, and K_j, are all transferred by secure authentication process.
2. Mutual authentication: In our protocol, each message delivery should be verified its legitimacy through authentication message.
3. Strong password protection: With the use of $MPK_i = H(q_i UPW_i)$. Thereafter, this modified password substitutes for the weak password.
4. User anonymity: For ID protection, in our proposed protocol, the user uses masked ID, namely $RUP_i = H(T_{RI} \| UID_i)$, to conceal real UID in general context.

Next, to evaluate the performance of our proposed key-generation and authentication protocol. Here, we compare our authentication scheme with recent another SC-based scheme. Firstly, we can discover recent SC-based schemes support mutual authentication.

The key agreement is also supported in recent years. In order to add sensor node dynamically. Our scheme can resist weak-password-based attacks. For the first security metrics, our scheme uses randomly-reformed password and ID is user legitimate verification and hence can resist all attacks arising from this security metric and our scheme outperforms other schemes. Next, our scheme can resist reply attack and most schemes can resist this attack, except to Yeh's scheme [16]. For Stolen-verifier and privileged in-sider attacks, our scheme does not use password table and we also reform our weak password. Hence, our authentication scheme resists from reply attack. For Denial of service (DoS) attacks, our propose scheme, the schemes of Das et al. [2] do not resist this attack. In our scheme, authentication message must be handled by GWN and transmitted to all other sensors and the user. Hence, the malicious user cannot use GWN bypassing attack to intrude our scheme. Further, GWN in our scheme just uses $14T_h/k$ to complete k authentication requests. Hence, for one user to multiple sensor nodes accessing scenario, our scheme outperforms than other schemes.

4 Conclusion

Through this propose authentication schema and key-generation protocol in WSN based IoT. We propose a hybrid authentication model suitable to one user to multiple sensor nodes accesses scenario. Further, we also adopt the suggestions of previous study to design strong password protection, which enables our scheme against various attacks arising from weak password. Besides, our scheme also achieves several securities features, such as mutual authentication with communication nodes, secure key-generation protocol and password protection schema. This research work give a cryptanalysis of our scheme to show the robust of our authentication and key agreement scheme.

References

1. Akyildiz IF, Su W, Sankarasubramaniam Y, Cayirci E (2002) Wireless sensor networks: a survey. Comput Netw 38:393–422
2. Das AK, Sharma P, Chatterjee S, SingJ K (2012) A dynamic password-based userauthentication scheme for hierarchical wireless sensor networks. J Comput Netw 35:1646–1656
3. Farash MS, Turkanovic´ M, Kumari S, Holbl M (2015) An efficient user authentication and key agreement scheme for heterogeneous wireless sensor tailored for the Internet of Thing environment. Ad Hoc Netw 36:152–176

4. Kumar P, Lee H (2011) Cryptanalysis on two user authentication protocols using smart card for wireless sensor networks. IEEE Wireless Advanced, London
5. Lee CC, Li CT, Chen SD (2011) Two attacks on a two-factor use authentication in wireless sensor networks. Parallel Process Lett 21:21–26
6. Li CT, Weng CY, Lee CC (2013) An advanced temporal credential-based security scheme with mutual authentication and key agreement for wireless sensor networks. Sensors 13:9589–9603
7. Amin R, Biswas GP (2016) A secure light weight scheme for user authentication and key agreement in multi-gateway based wireless sensor networks. Ad Hoc Netw 36:58–80
8. Sun D, Li J, Feng Z, Cao Z, Xu G (2013) On the security and improvement of a two-factor user authentication scheme in wireless sensor networks. Personal Ubiquitous Comput 17:895–905
9. Turkanovic´ M, Brumen B, Hölbl M (2014) A novel user authentication and key agreement scheme for heterogeneous ad hoc wireless sensor networks, based on the Internet of Things notion. Ad Hoc Netw 20:96–112
10. Turkanovic´ M, Hölbl M (2013) Notes on a temporal-credential-based mutual authentication and key agreement scheme for wireless sensor networks. Wirel Pers Commun 77:1–16
11. Wang D, Ma CG (2012) On the (in)security of some smart-card-based password authentication schemes for WSN. Published in IACR Cryptology ePrint Archive
12. Wang D, Wang P (2014) Understanding security failures of two-factor authentication schemes for real-time applications in hierarchical wireless sensor networks. Ad Hoc Netw 20:1–15
13. Xue K, Ma C, Hong P, Ding R (2013) A temporal-credential-based mutual authentication and key agreement scheme for wireless sensor networks. J Comput Netw 36:316–323
14. Yuan J, Jiang C, Jiang Z (2010) A bio-metric-based user authentication for wireless sensor networks. Wuhan Univ J Nat Sci 15:272–276
15. Yang GM, Wong DS, Wang HX, Deng XT (2008) Two-factor mutual authentication based on smart cards and passwords. J Comput Syst Sci 74:1160–1172
16. Yeh H, Chen T, Liu P, Kim T, Wei H (2011) A secured authentication protocol for wireless sensor networks using elliptic curves cryptography. Sensors 11:4767–4779

Sarcastic Sentiment Detection and Polarity Classification of Tweets Using Supervised Learning

M. S. M. Prasanna, S. G. Shaila, A. Vadivel, Mahima Mittal, K. Hithyshi, and H. R. Pooja Shree

Abstract Nowadays, due to the rapid growth in web technologies and internet usage, there is an enormous amount of data available in WWW and also growing progressively. The data includes online learning, exchanging ideas and opinions, etc. People are using social networking websites such as Twitter, Facebook, Instagram, Google+ as a common platform for all these. These platforms nowadays have become very popular as they permit people to share and exchange their views about any topics/events and have considered by the researchers for sentiment analysis, opinion mining, text summarization, question answering, etc. In this paper, we proposed the approach to classify the tweets into sarcastic and non-sarcastic sentences. Initially, Twitter data is preprocessed and feature extraction is done that considers n-grams, POS tags, term position and frequencies. Further, the approach use machine learning and deep learning algorithms for identifying sarcasm in tweets. Further, the classification algorithms such as Naive Bayes, Decision Tree, Random Forest and CNNs used to classify the polarities of non-sarcastic sentences. The experimental results are promising with good classification accuracy.

Keywords Sarcasm · Sentiment analysis · Polarity · Classification · Tweets · CNN

1 Introduction

Nowadays, sentiment analysis plays a vital role in social media networks. It is the process of understanding an opinion about a given topic from the individual perspective. Sentiment analysis is also referred to opinion mining as it analyzes people opinion. User opinions can be positive, negative or sarcastic based on the product or event. It is through sentiment analysis one can analyze the attitudes and emotions

M. S. M. Prasanna · S. G. Shaila (✉) · M. Mittal · K. Hithyshi · H. R. Pooja Shree
Dayananda Sagar University, Bangalore, Karnataka, India
e-mail: shaila-cse@dsu.edu.in

A. Vadivel
ICFAI University, ICFAITECH, FST Hyderabad, Hyderabad, Telangana, India

R. P. Mahapatra et al. (eds.), *Proceedings of International Conference on Recent Trends in Computing*, Lecture Notes in Networks and Systems 341,
https://doi.org/10.1007/978-981-16-7118-0_15

of an individual. Sentiment analysis plays important role in improving customer service, crisis management, alter marketing strategy. Sentiment analysis helps practitioners and researchers who are working with Human–Computer Interaction (HCI) data in the field of political science, marketing, advertising, psychology, sociology, economics. Also most of the industries atomize the entire process through sentiment analysis. It enables an individual to analyze the emotions and impressions of the customer when they contact, prior to receiving support, and how effective the service is at increasing satisfaction. In general, sentiment analysis is done from coarse level to fine level. Coarse sentiment analysis deals with calculating the sentiments of a document and fine sentiment analysis deals with attribute level. However, analyzing sentiments is challenging in micro-blogging content coming from Twitter because of the sort of the language used in them include short forms, memes, and emotions to express sentiments. These sentiments can be non-sarcastic or sarcastic tweets. Non-sarcastic tweets are simple and straight forward to analyze without any sarcasm, for instance "*He is a handsome boy.*" Sarcastic tweets are hard to analyze as they include sarcasm for instance "*Have you wondered how many ghosts you have hit with your car.*" Sentiment analysis and polarity classification in non-sarcastic sentences are easy when compared with the sarcastic tweets. This is hard and tedious because of the length of the tweet, unstructured and include short forms, slang words and misspellings in tweets, emotions, special characters to express sentiments. The approach in [1] proposed captured real time tweets using Hadoop framework and used machine learning algorithms to extract Sarcastic sentiments. The approach in [2] designed sentiment classification framework by extracting and integrating many features such as question mark symbol, word gazetteer, emoticons, exclamation, n-grams, etc. The authors in [3] have gathered Twitter data using Twitter API. Further, the Dictionary based approach is used to analyze tweets posted by various users. The approach classified the user opinions into positive, negative or sarcastic tweets. The approach in [4] used neural network based semantic model for detecting sarcasm in text. The approach used Support Vector Machine (SVM) that is labeled with semantic and syntactic information. The proposed model combines Convolution Neural Network (CNN), Long Short Term Memory (LSTM) network along with a Deep Neural Network (DNN). The authors in [5] used Twitter dataset for addressing the issues of sentiment analysis. Initially, the approach used machine learning and deep learning techniques to analyze and classify sentiments. Further, the approach built ensemble method based on majority vote and achieved classification accuracy of 83.58%. The authors in [6] proposed the model that extracts four sets of features which covers various types of Sarcasm. The authors analyzed the importance of each set of features and emphasized the pattern-based features for sarcasm detection in tweets. The model achieved the classification accuracy of 83.1%. The approach in [7] used collected tweets on electronic products from the Twitter micro-blogging site and used preprocessing on tweets to classify sarcasm. The approach used unsupervised learning algorithm to determine the positive, negative and neutral polarities. The authors in [8] proposed an approach that classifies tweets. The approach used Hadoop Ecosystem to improve its efficiency and enhance scalability. Hadoop ecosystem use Map Reduce parallel processing paradigm on distributed processing platform. The authors in [9] used various information analysis methods such as different hashtags analysis, Twitters network-topology, and various events over the network, influence impact for sentiment analysis. The above mentioned approaches considered less features for classifying polarities for non-sarcastic sentences and lack in achieving good classification accuracy. Sarcastic sentences hold figurative language which includes sarcasm, irony and metaphor, can be defined as universal aspect of communication from classic religious texts to modern micro-texts. Analysis of this type of sarcastic sentences as positive or negative opinions is a hard problem. Sarcasm detection is demanding as the literal meaning of a context is subjective. In real time applications, to analyze and detect irony is completely dependent on the social factors such as belief, intentions and sentiments. Due to this reason, one cannot train classifier consistently. In this paper, the approach presents two stages. Thus, the objective of the proposed approach lies in:

- Classification of sarcastic and non-sarcastic tweets
- Classification of the non-sarcastic tweets into positive polarity and negative polarity.

The rest of the paper is organized as follows. The related works are presented in the next section. In Sect. 3, we present the proposed approach and in Sect. 4, experimental results are presented and conclude the paper in the last section.

2 Proposed Methodology

Sentiment analysis is considered as pattern classification problem. Figure 1 represents the overall architecture that is mainly divided into three stages such as preprocessing, feature extraction and classification phase. The tweets from Twitter dataset are given

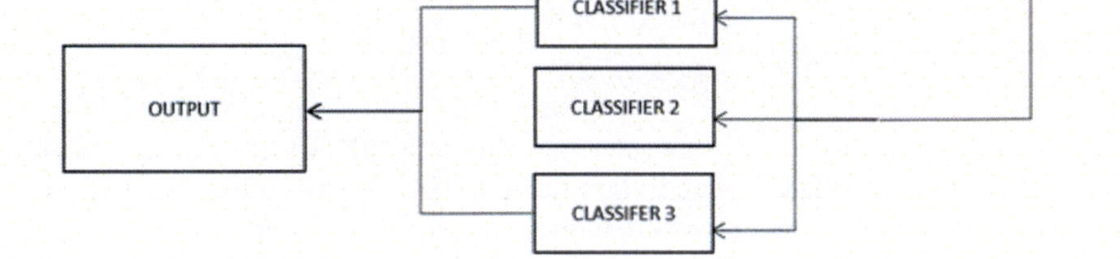

Fig. 1 Overview architecture of the sarcasm detection and sentiment analysis

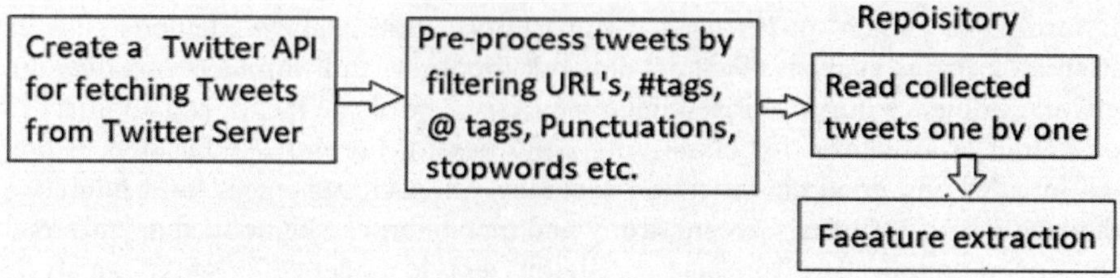

Fig. 2 Preprocessing of tweets

as the input to the preprocessing. In preprocessing, the data is cleaned, noise is removed along with special characters. Further, features are extracted such as terms that influence on positive and negative polarities, POS tagging is done and text patterns are analyzed. Further, the text is proceeded to classify sarcastic and non-sarcastic tweets. If the given tweet is sarcastic in nature then it is processed as 1 and non-sarcastic, it is 0. Later, the non-sarcastic tweets are given as inputs to the supervised learning for classifying their polarities. Classifiers determine whether the tweets are positive or negative or neutral. Below subsections explain in detail.

2.1 *Preprocessing*

The preprocessing is the primary step in sentiment analysis. The proposed approach used Twitter API for collecting tweets. The dataset use Tweepy-a python library built as a wrapper on the Twitter API to extract Tweet details such as text, date-time and the author name. Further, the Twitter data base is preprocessed. The approach used filtering technique to remove quotes, re-tweets, spam, tweets of other languages, and tweets with URLs. Thus, the dataset now consists of unique Tweet IDs assigned by Twitter. Further, the tweets are stored in repository either in text or csv files. In this stage, the approach checks for null values. Next, the approach proceeds for cleaning the data by eliminating the special symbols that occur in text or headlines column. This can be done using regular expressions. This is represented in Fig. 2. Further, the Twitter data processed for next stage which is feature extraction stage and is explained in below subsection in detail.

2.2 *Feature Extraction*

In feature extraction stage, the proposed approach uses CNN, which is a Deep Neural Network for feature extraction and classification. The approach extracts N-grams such as unigram, bigram models along with their frequency counts as features. This is important to consider as they convey sentiments via phrases. Further, Parts Of Speech (POS) Tagging such as adjectives, adverbs, verbs and nouns are considered to understand the text pattern as they are good indicators of subjectivity and sentiments. This in general, creates syntactic dependency patterns. The term position is important feature as it affects the overall sentiment of the text. Negation is considered as an important feature as it alters the polarity of the opinions. In this work, Convolution Neural Network (CNN) is used to recognize the sarcastic or non-sarcastic tweets.

2.3 Classification

This is the third stage in sentiment analysis. Here, the proposed approach performs classification in two stages. In the first stage of classification, CNN classifies tweets into sarcastic and non-sarcastic tweets. In second stage, classifying the non-sarcastic tweets into positive, negative and neutral polarities are performed.

2.3.1 Classification of Sarcastic and Non-Sarcastic Tweets

In general, tweets are represented as n words. The tweet is presented in a vector wherein each word gets replaced with its dictionary index. Input is fed to a neural network layer which in turn converts every word into a D dimensional distributional vector. In order to solve varying lengths of input, padding is done to the tweet vector that is represented as a matrix. The detailed working process of CNN in terms of feature extraction and classification is presented here. The input layer of the CNN receives the input samples and proceeds to the hidden layer of CNN. Here, numbers of hidden layers are added based on the problem complexity and computation level. Each hidden layer has three layers such as Convolution Layer, Relu layer and Max Pooling. Convolution layer is the most crucial part of the CNN. The objective of a CNN is to minimize frequency changes by convolutional filters and obtaining word sequences which discriminate as a feature map for the Long Short Term Memory (LSTM) layer. Convolutional filter having the same D dimensional input matrix performs element-wise product between the column dimension of the input matrix and the filter matrix creating a vector component which in turn gives feature map as output. The power of semantic modeling is displayed by Recurrent Neural Network (RNN) where it incorporates feedback cycles in the networks design. A temporal component included in the RNN allows to store the related information being present in the model. The RNN is not able to find long term dependencies, rather than it introduces LSTM. LSTM is able to present long term dependencies by a group of gates defined in each memory cell, where d is called the memory dimension of state that is hidden in LSTM, and it does not suffer from diminishing or increasing gradient, while back propagation is performed through time. The result of LSTM is given as an input to the connected DNN layer, wherein higher order feature is produced depending on the output of LSTM. By adding a Softmax layer, the training of the networks is carried out where binary cross entropy error is minimized and accuracy is maximized in the classification of sarcastic and non-sarcastic tweets. This is depicted in Fig. 3.

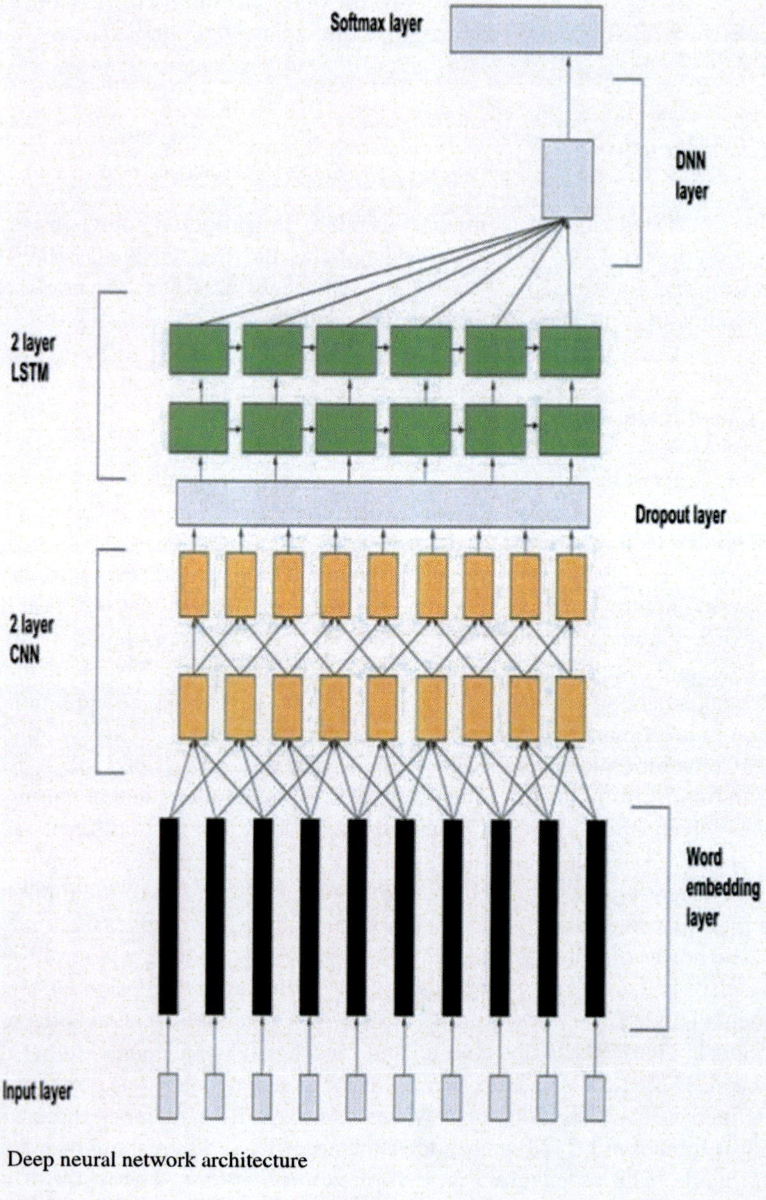

Fig. 3 Deep neural network architecture

2.3.2 Classification of Sarcastic and Non-Sarcastic Tweets

This section discusses about the classification of polarities in non-sarcastic tweets using various Machine Learning algorithms. Here, the approach uses Decision Tree, Naïve Bayes, Random Forest and CNN algorithms. The Naïve Bayes model assigns the sentences using joint probabilities for words and classes. Suppose, for the feature vector (v, …, v_n) and the class C_k is given, the Bayes theorem is represented mathematically as shown in Eq. (1).

$$P(C_k|v_1, v_2, \ldots, v_n) = \frac{P(C_k)P(v_1, v_2, \ldots, v_n|C_k)}{P(v_1, v_2, \ldots, v_n)} \tag{1}$$

Naive conditional independence assumes that each feature of vector v_i is conditionally independent of every other feature of vector v_j for $i \neq j$ for the given class C_k. This is depicted in Eq. (2).

$$P(v_1|C_k, v_1, \ldots, v_n) = P(v_1|C_k) \tag{2}$$

Thus, the above mathematical relation can be simplified as shown in Eq. (3).

$$P(C_k|v_1, \ldots, v_n) = \frac{P(C_k)\prod_{i=1}^{n} P(v_1|C_k)}{P(v_1, \ldots, v_n)} \tag{3}$$

As $P(v_1 \ldots v_n)$ is constant, the classification rule depicted in Eqs. (4) and (5) can be used based on the feature variables values.

$$P(C_k|v_1, \ldots, v_n) \alpha P(C_k) \prod_{i=1}^{n} P(v_1|C_k) \tag{4}$$

$$\hat{y} = \arg_k \max P(C_k) \prod_{i=1}^{n} P(v_1|C_k) \tag{5}$$

Further, the approach use Decision tree model for classification. The tree has a set of nodes that represents the attribute of the data set, and its results are represented by a child node. The decision tree is a supervised model which takes data with existing labels. For every node in the decision tree the best decision has to be taken. Next, the approach use Random Forest classifier for classifying non-sarcastic sentences in polarity wise. Random Forest uses ensemble approach with decision trees and classifies based on the aggregated decision of those trees. For a set of tweets (t_1, t_2, …, t_n) and their respective sentiment labels (l_1, l_2, …, l_n), bagging repeatedly picks a random instance (T_b, L_b) including replacement. Each classification tree is trained using a different random sample (T_b, L_b) where b ranges from 1 … B. Finally, a majority vote is taken for predictions of these B trees. The performance evaluation of these classifiers is represented in Sect. 4.

3 Results and Discussions

The proposed approach used the Twitter dataset. The dataset is cross validated into testing and training set. The details of Twitter dataset are depicted in Table1. The approach had undergone preprocessing with the raw data to make it more convenient to analyze. Here, the semantic analysis is performed and features are extracted. Various machine learning techniques are used on the training dataset to learn and to classify the polarities of the dataset.

3.1 *Performance Evaluation of CNN in Classifying Sarcasm and Non-Sarcasm*

The Twitter dataset is cross validated into training and testing set and preprocessed. The preprocessed dataset is fed to CNN for learning and classification. The CNN extracts the features and classifies the sarcastic and non-sarcastic tweets. The proposed approach achieved accuracy of 91% over the training dataset, when CNN with 5 layers with learning rate 0.00001 is used. Here, model being trained for 40 Epoch as shown in Fig. 4.

Table 1 Details of the dataset

Training dataset	4500	Testing dataset	3500
Negative	2324	Negative	2123
Positive	2176	Positive	1377

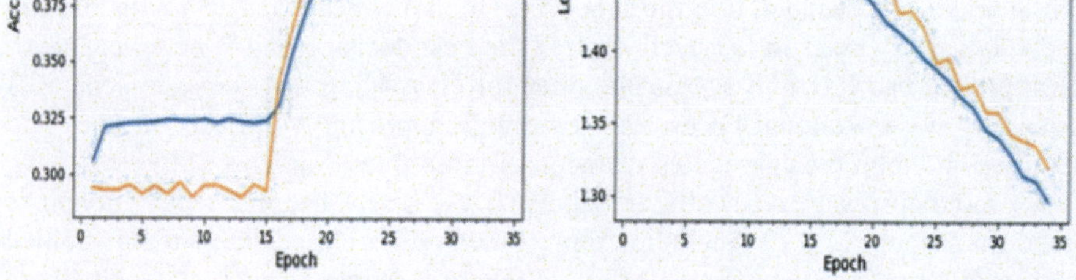

Fig. 4 Accuracy and loss graph for learning rate of 0.00001

Table 2 Performance evaluation of CNN in classifying sarcasm and non-sarcasm

Tweets	Desired output	Estimated iutput
i love when people text back	1	1
A lot of classes are useless, privileged to be a student	1	1
i love working in Sydney river it makes me want to go to work so much more !	1	1
i love that thing you do when you read my text and do not reply	1	0

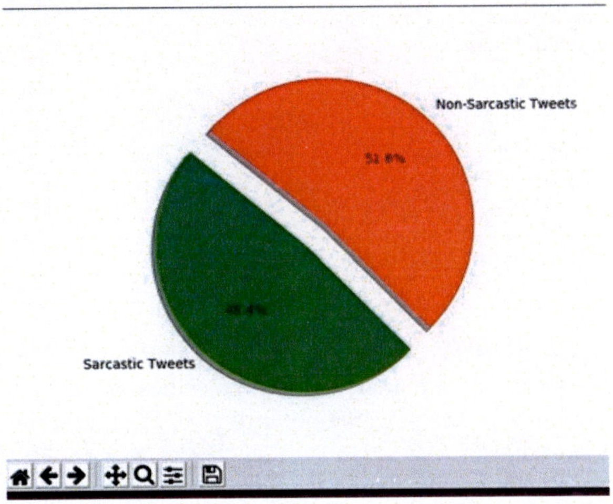

Fig. 5 Pie Chart representation if classification

Table 2 represents the sample tweets and CNN classifies into sarcastic and non-sarcastic sentences. The human annotation is represented as desired output and classifier output is the estimated output. If the text contains sarcasm, it will be denoted as 1 which represents sarcastic tweet and 0 represents non-sarcastic tweet, meaning no sarcasm in text.

Figure 5 represents the CNN classification of tweets into sarcastic and non-sarcastic for validated dataset.

3.2 *Performance Evaluation in Polarity Classification Using Machine Learning*

Table 3 represents the sample tweets of non-sarcastic sentences. The proposed approach uses machine learning algorithms such as Naïve Bayes, Decision Tree, Random Forest and CNN algorithms to classify the non-sarcastic tweets in polarities. Based on the extracted feature vectors such as POS tagging, term position,

Table 3 Performance evaluation of classifiers in classifying polarities of non-sarcastic tweets

Tweets	Desired output	Nave Bayes	Decision tree	Random forest	CNN
		Classifier output	Classifier output	Classifier output	Classifier output
I received a call from someone in Miami office for a blog. http://tumblr.com/xcn21w6o7	1	1	1	1	1
I love you guys so much that it hurts	0	1	1	1	1
im sick 'cough cough'	0	0	0	0	0
My new car was stolen….by my mother who wanted to go pose at church	0	0	0	0	0

presence/absence of negations, Term Frequency that are discussed in Sect. 3, the tweet patterns are analyzed and polarities are classified. The approach represents positive as 1 and negative polarity as 0. Figure 6 represents the CNN classification of polarities.

Table 4 represents the performance evaluation of Naïve Bayes, Decision Tree, Random Forest and CNN classifiers in terms of classification accuracy. The table shows the accuracy obtained for different sizes of dataset. It is observed that as the

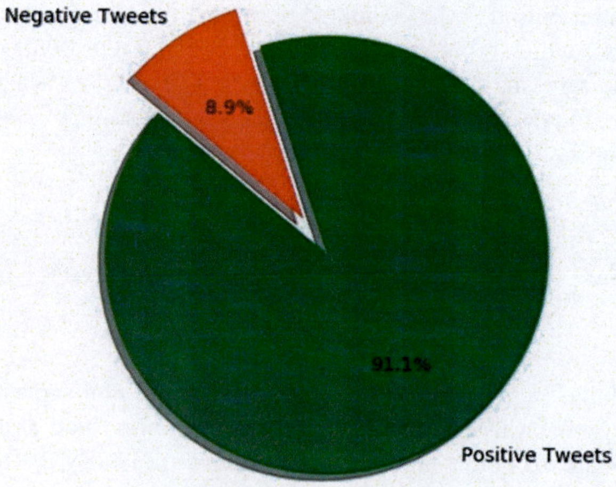

Fig. 6 Pie chart representation of CNN polarity classification

Table 4 Classification accuracy of classifiers in classifying polarities of non-sarcastic tweets

Dataset	Classifier accuracy			
	Naïve Bayes	Decision tree	Random forest	CNN
10	0.52	0.54	0.55	0.77
50	0.58	0.58	0.61	0.79
100	0.59	0.60	0.65	0.83
500	0.64	0.61	0.73	0.88
1000	0.67	0.65	0.75	0.91

dataset grows and the learning rate grows, and the classification accuracy is improved to the greater extent.

4 Conclusion and Future Works

In this paper, we consider Twitter data mainly for analyzing sentiments. Initially, Twitter data is preprocessed. Later, features such as n-grams, POS tags, Term position and frequency have been extracted and further preceded with classification. CNN is used to classify the tweets to sarcastic and non-sarcastic sentences. Later, machine learning algorithms such as Naive Bayes, Decision Tree, Random Forest and CNNs used to classify the polarities of non-sarcastic sentences. The experimental results are promising with good classification accuracy. Further, we consider to classify the polarities of sarcastic tweets as a future work.

References

1. Bharti SK, Vachha B, Pradhan RK, Babu KS, Jena SK (2016) Sarcastic sentiment detection in tweets streamed in real time: a big data approach. Dig Commun Netw 2(3):108–121. ISSN 2352-8648, https://doi.org/10.1016/j.dcan.2016.06.002
2. Deshwal A, Sharma SK (2016) Twitter sentiment analysis using various classification algorithms. In: 5th ınternational conference on reliability, ınfocom technologies and optimization (Trends and Future Directions) (ICRITO'2016), Noida, India, pp 251–257. https://doi.org/10.1109/ICRITO.2016.7784960
3. Mishra P, Rajnish R, Kumar P (2016) Sentiment analysis of Twitter data: case study on digital India. In: International conference on ınformation technology (InCITe'2016)—the next generation IT summit on the theme—Internet of Things: connect your worlds, Noida, pp 148–153. https://doi.org/10.1109/INCITE.2016.7857607
4. Ghosh A, Veale T (2016) Fracking Sarcasm using neural network. In: 7th workshop on computational approaches to subjectivity, sentiment and social media analysis, pp 161–169
5. Ansari AF, Seenivasan A, Anandan A, Lakshmanan R (2017) Twitter sentiment analysis. github.com/abdulfatir/Twitter-sentiment-analysis/blob/master/docs/report
6. Bouazizi M, Otsuki OT (2016) A pattern-based approach for sarcasm detection on Twitter. IEEE Access 4:5477–5488. https://doi.org/10.1109/ACCESS.2016.2594194

7. Patel A, Patel AA, Butani SG, Sawant PB (2017) Literature survey on sentiment analysis of Twitter Data using machine learning approaches. Int J Innov Res Sci Technol 3:10
8. Sheela LJ (2016) A review of sentiment analysis in Twitter data using Hadoop. Int J Database Theor Appl 9(1):77–86
9. Anber H, Salah A et al (2016) A literature review on Twitter data analysis. J Comput 8(3):241–249. https://doi.org/10.17706/IJCEE.2016.8.3.241-249

Visualization of Online Social Dynamics for Forensic Investigation of User's Behavior

Govind Verma, Manash Sarkar, and Dambarudhar Seth

Abstract Online social media platforms represent an intriguing issue: how to best describe the various classes of user conduct. Generally, user conduct portrayal techniques, in view of user singular highlights, are not proper networking websites. In these conditions, users cooperate with the site and with different users through a progression of various interfaces that let them to transfer and view content, pick companions, rank most loved substance, buy in to users and do numerous different associations. Diverse communication examples can be noticed for various gatherings of users. Get the measure of trust and distrust between users in online social network is an essential research problem. In our paper, we propose a technique for portraying and distinguishing user practices in social networks. We focus on structured and comprehensive overview of different features of distingue user behavior in social network. To start with, we gathered information from YouTube, Facebook, Twitter, LinkedIn and utilized a bunching calculation to aggregate users that share comparable standard of conduct. Moreover, we have demonstrated that credits that come from the user social collaborations, as opposed to ascribes comparative with every individual user, are acceptable discriminators and permit the distinguishing proof of significant user practices. Subsequently, we introduce and talk about exploratory after effects of the utilization of proposed approach. An array of valuable profiles, obtained from the investigation of the YouTube, Facebook, Twitter, LinkedIn test is introduced. The ID of various classes of user conduct could possibly improve, to epitomize, proposal frameworks for commercials in online social communities.

Keywords Network environment · Social networks · Social networking site · User behaviors · Security

G. Verma (✉) · D. Seth
Department of Computer Science and Engineering, SRMIST, Ghaziabad, UP, India
e-mail: govindv@srmist.edu.in

M. Sarkar
Atria Institute of Technology, Bengaluru, Karnataka, India

R. P. Mahapatra et al. (eds.), *Proceedings of International Conference on Recent Trends in Computing*, Lecture Notes in Networks and Systems 341,
https://doi.org/10.1007/978-981-16-7118-0_16

1 Introduction

Scholastic researchers focus on security issues in huge critical networks, similar to mix frameworks, the overall of things, cloud organizations, etc., since, in such a case that these networks have security issues, it'll cause incredible monetary shortfall. Conversely, barely any scientists represent considerable authority in individual security issues. Despite the fact that these occurrences caused extraordinary repercussions, they neglected to cause incredible misfortunes in an extremely brief timeframe. In contrast to occurrences in mechanical organizations, the dangers of such episodes are steady and extensive. Wilcox and Bhattacharya [7] Right now, OSNs (online interpersonal organizations) turned into a basic a piece of human existence and furthermore the fundamental gratitude to acquire and share individual data. Also, administrators can utilize users' very own data discretionarily for a few purposes, the partner editorial manager planning the survey of this original copy and favoring. It could be deciphered in a few settings and viewpoints in controls like law, wellbeing science, sociologies, and PC and information science. Security assurance in various ends isn't general. The strategies in develop modern industrial networks don't appear to be reasonable for social networking. For the most part, security will be denied in light of the fact that the option to be distant from everyone else and to possess independence from impedance or interruption. For moreover, in social networks, confronting a particularly extravagant and huge organization climate, homogeneity makes it difficult to ponder security assurance from just a private viewpoint; the organization climate during which are discovered furthermore revealed personal data. Additionally, the affirmation of security spillage of different users, like community users and normal users, changes extensively [8]. The common strategies for ensuring users' protection in interpersonal organizations incorporate namelessness, decentralization, encryption, data security guidelines, protection setting, access control and improving users' security attention and protection behavior, Pan et al. [10]. The rest four techniques rely upon the administrators of social networking sites to be executed, which have demonstrated to be questionable. At the indistinguishable time, complex security settings bargain user's experience. Studies have over and again indented users on the grounds that the most easily broken connection in security, however, come together of studies utilized or zeroed in on this thought and directed exploration during this work. In this paper, we have used the last strategy to develop the user's security mindfulness and security improvisation by protection estimation to highlight the matter of protection spillage. This technique analyzes such a large number of users. With respect to the minuscule world principle, any two users can arrive at each other inside six stages. Only one user's connection can arrive at many millions of different users. With this enormous and convoluted network structure, it becomes difficult for conventional algorithms to oblige this model successfully. At a certain time, there are an outsized number of envision accounts, zombie accounts, damage accounts, public accounts, etc. These records encompass the security of common users. In different investigations, this perception was not entertained; they just broke down all the pals around a user.

During the time spent picking users who are immovably associated with the security spillage of target users, we don't that there exist advantageousness get away from provisions, which suggests that after successfully killing the redundant users referred to above, Jain et al. [9] we found that there have been as yet a couple of users who have a genuinely cozy relationship with the target user inside the graph structure yet who hold information about the target user that has lost its insurance regard. These users are near the goal user, however, at this point they're irrelevant. In various systems, we decided to use the user's direct follow to loosen up this issue. Li [14] in the wake of attempting numerous strategies, we chose to follow the user's behavior to unwind this issue. In past examinations, specialists considered executing user behavior follows to live security; be that as it may, their utility in barring repetitive users was let alone Yan et al. [12]. In past assessments, experts considered executing user lead follows to live security; nevertheless, their utility in accepting redundant users was not to mention. As of here, we have maintained our underlying work on individual assurance scoring system also we loosen up the procedure from individuals to all or any users inside the entire association structure graph and join the direct of users to compensate for the shortcomings of past existing examinations. Our responsibilities during this paper are listed below:

- We propose the possibility of essential comparable structure and end the matter advantageousness inside the traditional hub centrality situating algorithm.
- We propose direct social closeness to feature lead user conduct qualities.
- We settle the matter of reasonableness and dispose of repetitive with abundance users.

Major doubt in these assessments is that all the users in online media all make substance, need to share the substance by making it easy to find, or banter with anticipated target swarm. To improve correspondence and allot resources effectively for better open organizations, government needs to grasp assorted online media content sorts and better separation of inhabitants' social media practices of information sharing and dispersal. With everything taken into account, online media uses by governments should grasp the various sorts of users and their practices the extent that the substance types they make, recurrence of making content easier to arrange by others Fu et al. [13]. The perception of the user practices will have the alternative to zero in on some particular subset of user types. In this examination, we intend to perceive and arrange different sorts of online media users by user media content sorts and frequencies, and to look at social differentiations by user's sorts. We will zero in on Facebook online media to understand the user practices in e-government setting. We present a social media user conduct to investigate the above requests. The user conduct perceives the social media users into different kinds, and licenses to inspect the social media based data creation, plotting and zeroing in on practices by different user types. Social networking sites have an assortment of choices and applications that make them appealing to a wide crowd. Facebook has made it feasible for people to meet on the web and has filled colossally in ubiquity as of late. Facebook offers an easy method to quickly relate with companions. However, when considered in component, there are numerous issues of long range social networking can present, for example, reliance, protection and security issues. To encourage answer these inquiry, this examination dissects four recommendations. The principal proposal expresses that users who look at security as a significant factor in an interpersonal interaction site more are probably going to change their settings on in any event a month to month premise Yadav et al. [5]. The subsequent recommendation states users who have perceived wholesale fraud is as a significant security concern are bound to change their settings on at any rate a yearly premise. The third proposal expresses that users of Facebook who have left their security on a default setting have hard to an infection or malware assault. At long last, the forward proposal states users of Facebook who have their protection set to an "everyday practice" setting haven't got an assault on their profile.

2 Literature Survey

In social networking, Facebook is solitary of the leading sites; with there were 39.17 million Facebook users in India in December 2020, which represented 28% of its entire general population. Basically of them were men—75%. People developed 25–34 were the biggest user (14.70 million). The most striking difference among individuals occurs inside people developed 25–34, where men lead by 7.30 million. The site works by getting users to associate with one another dependent on their experience or shared interests. It moreover permits them to join packages that have practically identical tendencies. Every user seeks after an online profile which contains particular information on the users, for example, their name and email address. Part of being on Facebook fuses users posting sees which illuminate others about the thing they are doing. These updates by then show up on their amigos' newsfeeds comparatively as on the user's channel. These information are accessible to anybody and are viewed as in the public region. Because of such a data posted, it is viable for an assailant to amass and target users dependent on the individual data they share. The planner of Facebook has the past conveyed that security isn't as huge as the value that the site offers. Customized benefits and zeroed in on advancing on Facebook rely upon users' near and dear data. Fitting organizations reliant on near and dear information licenses associations to area likely users and advance their items. Past examinations have focused in on the use instances of school understudies on facebook and didn't dissect the security issues looked by these understudies on facebook. In this assessment, we include the online security gives that users of facebook experience, and we propose how these issues may be mitigated. Regardless of the security outrages, the examinations, the reports that commitment on its fundamental stage is in decay, Facebook has by and by posted an expansion in active users in its 2019 update. Most importantly, on users-Facebook currently serves 2.37 billion monthly active users, an expansion of 55 million on the past quarter.

In the past time of protection appraisal research, inspectors zeroed in on the information revealed by users' profiles. In imaginatively utilized profile data to evaluate the affectability and see limit of characteristics, by then surveyed the user's security status through a Bayesian model kept up their examination, it's give a more ordinary, considerably more numerically sensible strategy to decide user security scores in Online Social Networks by consolidating affectability and see limit with IRT (Item Response Theory); this assessment exceptionally advanced the examination of confirmation evaluation. In coming about examination, Almabdy et al. [8] masterminded an arrangement to give an affirmation settings wizard to users to control them to choose sensible profile settings. Jain and Katarya [9] masterminded a straightforward condition to discover the affectability of profile things, by then saw as both the affectability and see limit of the data inside the user's profile and arranged a record an inspiration on this explanation. Pan et al. [10] found the key sections affecting users' self-disclosure of individual data. Yun et al. [11] evaluated users' profiles spread on various informal organizations and got the users' security divulgence status; they are the first to propose protection appraisal in different social affiliations. Malicious users in relational association stages are presumably going to show individual direct principles that not equivalent to should be normal users, because their goals in intensifying their own necessities and purposes (e.g., advance a particular thing or certain political points of view or theory). User direct investigation isn't only helpful in getting a start to finish cognizance of user objective, yet it is in like manner basic to the disclosure of toxic social bots' records in online casual networks. User lead presumably changes under different conditions. Fu et al. [13] proposed that condition study is associated with programming administration prerequisite examination, which may empower the investigation of any change in user's necessities. Such an investigation is important to grasp the dynamic necessities of an item organization atmosphere. Li et al. [14] acquainted a structure with the advancement of user individual direct norm in media video proposal organizations on online casual networks. Their structure relies upon social setting and explores the changes in user need for various social conditions. Such user lead data will be jumped if we have permission to the user's logs [15] or user's snap streams (e.g., recorded by relational association stages). The qualification in user direct is gotten, for example, by examining the image search logs of users to overview the chase objective of vacillated users [16], and this technique can support improvement of web records. Shi et al. [17] used user click stream data to fabricate a tick stream graph model to address user direct and perceive unmistakable user social affairs, so on distinguish poisonous records. There have also been distinctive researches that exhibit user objective and unusual records could similarly be settled through direct investigation, and social condition in empowering the appreciation of users' dynamic lead. Wang et al. [18] constructed a novel convolutional neural detail maintained user direct, program substance and setting information to fabricate a tick model and be encouraged the user's gobble tendencies to enhance up chase quality. Li et al. [19] accumulated an outsized proportion of user information on the Twitter and YouTube, around 13 million channel activities, exploring and perceiving atypical practices that veer off basically from enormous extension subtleties through user lead in two relational associations Yan et al. [12].

3 Fundamental of Social Network in Graph Theory Perspective

The idea of graph theory is broadly utilized in social media. As a usual here the users or the individuals included are considered as the nodes or vertices Wang et al. [18]. What's more, any connection between the users because of regular preferences or shared friendship is considered as edges.

- **Graph Theory in Facebook**: Majority know about Facebook nowadays. You can click 'like' in the event that you discover something amiable, "tag" your companions in different "posts," put remarks in posts and above all become friends with somebody whom you know and furthermore somebody whom you don't have the foggiest idea. The idea of graph theory is utilized in Facebook with every individual as nodes and each like, share, remark, and tag as edges.
- **Graph Theory in Twitter**: Here, the people are considered as nodes and in the event that one individual follows another, at that point which is considered as the edge between the two users. Graph theory in various fields:
- **Mathematical and other Geometrical investigations**: If we don't think about the length of the edge and vertical points of a polygon, at that point it tends to be securely expected as a graph with its vertices and edges. This reality probably won't be that helpful in the investigation of polygons; however, this hypothesis is generally utilized in the investigation of surfaces and articles with higher measurement. Graph is additionally utilized in polynomial math.
- **Designing transportation networks**: Architects and organizers utilize the idea of graph prior to building expressway and bridges. They consider the different urban areas as nodes and the expressway associating them as edges. It is utilized locally inside the urban areas too for example while developing a bus station. Here, the bus station goes about as nodes and the streets interfacing them are edges. Here, the distance between the bus stations for example loads or estimations of the edges are considered in light of the fact that that will show the time taken to cross the distance.
- **Communication networks**: A PC network whether concentrated or dispersed structures a graph. The web directing framework and other information and packet routing frameworks in a PC network represents to a graphical structure with the different PCs or device as nodes and the routing way between them as edges.
- **WWW**: WWW in like manner addresses a tremendous diagram. Various pages in the web are considered as hubs and if there is any hyperlink between two pages, by then that is the edge between those two pages.
- **Social network**: Regardless, we need to have some idea with respect to Network. There are a couple of various methods of formally describing an organization, dependent upon the piece of science used. The most well-known and versatile definition is gotten from chart speculation, an interpersonal organization is conceptualized as a diagram, that is, a bunch of vertices (or hubs, centers) addressing social substances or objects and a bunch of lines addressing in any event one

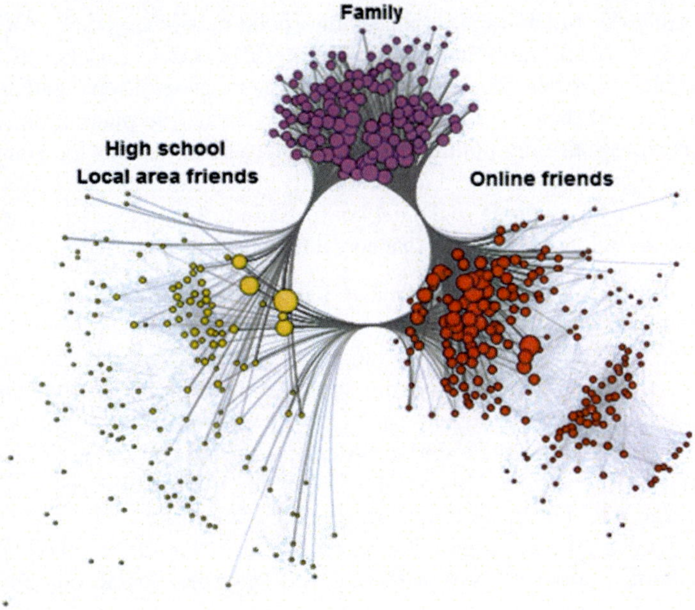

Fig. 1 Network of friends

social relations among them. An organization, nevertheless, is more than a chart since it contains additional information on the vertices and lines.

- **Brief thought on Social Network**: When we need to represent to any type of relations in the general public as connections, it tends to be named as Social Network. The example of interdependency between every person (nodes) can be founded on various perspectives, viz.—friendship, interconnection between families, basic interest, monetary trade, despise, sexual relationships, or connections of convictions, information or renown (Fig. 1).

Figure shows the relationship of a specific gathering of school companions. Every vertex represents to a critical individual in the friendship, and the edges indicate an association by relationship and connection of any sort that has been meant by four different colors.

4 Taxonomy of Privacy and Security Problems in Online Social Network Complete Description of Privacy and Security Threat in OSN

Social media is a sort of correspondence between the information generator and end user for serving the purpose of online interchanges that are helpful in creation of virtual networks which are utilizing online social networks (OSN). An informal community is a social graph that depicts an interface between users, associations, and their social exercises. The nodes represent users, associations, gatherings and the edges represent their connections. An OSN is an online place utilized by last users to make informal communities or associations with others that have comparable perspectives, interests, exercises, and additionally genuine associations. In the current online space, many long range social networking administrations are accessible. The basic highlights in long range informal social networking sites are:

- Every one of existing online social network interpersonal communication services are internet based, utilizing an Internet association. Substances are put away on distributed stockpiling through a brought together access the board framework. This substance can be gotten to from anyplace utilizing an Internet association and internet browsers.
- OSN users need to make a public profile for informal community objections as per their predefined stage. This profile information is basically used for the check cycle to sign into the person to person communication site.
- Approximately, all current individual to individual long range social communication administrations support users in developing their social relationship with various users by partner a user's profile with others having practically identical profile information.
- Individual fascinating component of the current OSNs is to facilitate substances on these sites are user produced, while OSNs utilize these substances for dealing purposes.

The essential target of OSNs is to surrender the output to the most noteworthy users. For certain activities, users use OSNs, for instance, Facebook, Twitter, and LinkedIn. Here, Online Social Network users share information about themselves and their exercises Radha et al. [21]. In this part of appropriated data, parts of the revealed output are kept private and should not be altered in any way shape or form. The users share their regular daily existence routine through notification or by sharing photographs and accounts. Special OSN users use PDAs to take pictures and make accounts for distributed through OSNs, Bagretsov et al. [4]. This data can have both region information and metadata into it. OSN expert suppliers accumulate an extent of data about their users to offer altered administrations, yet it might be used for business purposes. Furthermore, users' data may in like manner be given to pariahs, which lead to security spillages. This information can allow pernicious users to utilize and assault the insurance of a person. Information recuperation and data security are two creating territories in software engineering disciplines that have different targets. Information recuperation offers procedures to data extraction. It similarly offers a bunch of techniques to a relationship for data examination and making decisions subject to this recuperated information. Data security shield information from unapproved and poisonous access that discloses, changes, attacks, or obliterates the data set aside or shared on the web.

5 Perspective of Social Network User Behavior

We recognized highlights that are identified with user behavior from summed up within given it. The portrayal of these highlights is removed and their setting is clarified. There are highlights that don't straightforwardly in user behavior; notwithstanding, these are contribute in framing user behavior, for example, character characteristics. These highlights can't be named as a type of user behavior; notwithstanding, along with ONS use, they help to separate user behaviors.

5.1 Behavior Recognition

Social media user behavior lead examination is a huge zone of exploration that active different attributes of users to be contemplated. The social behavior understands and estimates of user assumption toward explicit things help commercial houses and enterprises find the aim part of areas to zero in on. The assumption for user objective relies upon the coordinated efforts inside a site, which is huge for retargeting. Generally, E-business site and commercial showcase networks screen track the inquiry of users to fathom their points and practices. Besides, comprehension the interest and making pertinent appropriate information present for online users on the friendship of the temperamental development of information on the Internet actualizes investigation and showing of the web course behavior of online users. Web mining methods have gotten advantageous to investigate and isolate important material from web information, and they are arranged into web data mining, web usage mining and web structure mining. Essentially, web records are removed by web information mining, users perusing exercises are examined by web use mining, and the actual association structure of locales is dissected by web structure mining Shi et al. [17]. Web logging, which control user webpage course information, for instance IP address, user/user id, date, time, strategy, status code and size of the article, are used by web use information mining to envision user behavior. The sign is an assortment of user trades that is invigorated each time the user gets to locales. The web sign data are then set up by a gathering of activities, for instance Data Cleaning, User Identification, Session Identification, Enhanced Pattern Tree Construction, and Pattern Recognition, to recuperate the use illustration of each customer of the webpage subject to his/her activities during the examining, for instance, clicks, visited locales, re-visitations of, etc. Many delicate registering and information mining calculations have been proposed by experts to perceive accommodating examples in the user's web profile.

5.2 *Behavior Prediction*

As referenced already, there are highlights that can't straightforwardly be characterized as user behavior; notwithstanding, they impact of user behavior though regard ONS. They characterized these highlights and depiction for each classify is given prediction on user behavior.

- **Simplicity of use**: Skill or exertion needed to work ONS and usability.
- **Gratifications**: Sentiments of joy got from utilizing ONS. Stream insight while utilizing ONS. It is terms of gratifications, unwinding, gluttonous, delight, and diversion.
- **Character Traits**: Individual characteristics that structure user direct on ONS and separate behavior from others activist or unfavorable demeanor for ONS that impacts its habit. This terms Attitude, character qualities Colossal Five sociability, reasonableness, self-importance, neuroticism, and unwavering quality.
- **Confidence**: Frequent ONS venture for gaining affirmation and certification to fabricate confidence, human being in a social context present in ONS participations and activities to facilitate make vibes of value.
- **Social impact**: Relatives, partners, companion sway to facilitate impacts ONS use. The users are ones gathering of companions that choose and influence ONS usage. This terms social effect, social shackle, relative's pressure, and acknowledged practice.

6 Types of User Behavior

We separated persons that go under the group of user behavior. As various conditions were used to address a forlorn idea, we assembled the critical terms and assigned them with identifiers that are more sensible to address the idea. Terms are the features used in examinations of user behavior. Identifiers are the imprints consigned by us to store up these terms. We moreover gave a minimal portrayal of all of these identifiers. A couple of identifiers are taken from the wording of social brain research, where appropriate. The order of various user behaviors is given.

- **Social Investigation**: To investigate, search, and look at exercises of elective user on ONS. ONS is that the terms of Surveillance, social examination, social surfing, perusing.
- **Social relationship**: Relation with others like companions, family, and friends on ONS, creating and keeping up new ties. Partner with the group of friends on ONS related inclination a fundamental a piece of it. Assets got from ONS bonds and associations. ONS is that the conditions of social association, social correspondence, social association, social alliance and social capital, more grounded and more vulnerable ties.

- **Recurrence of utilization**: Time spent on ONS and furthermore the degree of exercises performed on it. ONS is the terms of Frequency of visits, recurrence of progress statuses.
- **Data Control**: Managing data through access the board choices on ONS and predominant the degree of information unveiled on ONS. ONS is simply the terms of confidentiality interface highlights, restriction, mindfulness, self-efficacy, self-revelation and self-projection.
- **Self-compass reading**: Creating ideal impacts on ONS for introducing the necessary picture of one-self. ONS is that the conditions of self-introduction, personality the board, self-direction.
- **Correspondence**: teaming up on ONS inside a similar recurrence as their ONS contacts or companions.
- **Social confidence**: uncertainty in contributive and partaking on ONS of boldness.

7 Types of Attacks

The security issue and protection concern are the significant necessities of the extensive variety interpersonal social sites. Nonetheless, there are various deadliest attacks suffers in all these person to person social networking sites and protecting the anticipated users from these hostile assault have been the troublesome task of various social agent and engineers. The crucial security assaults are assembled into different classes.

- **Active Attacks**: Active attacks form the new hubs normally and endeavoring to connect with the associated hubs and gain the permission to various hubs.
- **Passive Attacks**: This is absolutely mysterious and imperceptible.
- **Security Breach**: Find interface among nodes, edges and potentially distinguish the connection between them. Here are away from of various assaults in online social media organization and offered the possible response for how to manage the assaults safely.
- **Online Social Networking Infrastructure attacks**: S-murf IP Attack, UDP Flood Attack, TCP SYN Deluge Attack.
- **Malware Attacks**: Spy-ware, Adware, Crime-product, Browser Ruffians, Downloader, and Tool Bars.
- **Phishing Attacks**: Misleading phishing (E-mail, messages), Search program phishing, Malware base phishing, Key loggers.
- **Pernicious identical twin Attacks**: Collective planning attack.
- **Uniqueness Extensive scam Attack**: Use complex passwords, avoid mystery key reuse.
- **Physical Attack**: Impersonation, Harassment through messages.
- **Cyber (Computer-generated) bullying**: Don't perceive the messages that are proposed to dangers or damage.

8 Data Preparation

The present arising innovation and patterns have brought about various frameworks and stages through which social elements collaborate and speak with each other, for example, email, Text Messaging, Telephone/Mobile organizations and above all OSNs like Facebook, Instagram, Twitter, and YouTube. These frameworks empower their users to hitch, and share thoughts, data, and interests through different assortments of connections upheld by them. Also, OSNs permit users to publish their own data, extra sight and sound substance, and connection to different users whom they identify with. The correspondence and collaboration administrations gave by these frameworks empower to uncover the basic social networks of their users and consequently they speak to a novel occasion to survey and get them Li et al. [19]. A top to bottom analysis of social networks structure and development can bring about a vastly improved plan of future interpersonal organization based frameworks. Online social networks offer numerous helpful properties that reflect certifiable social networks attributes, which incorporate little world behavior, huge neighborhood bunching, and the presence of colossal emphatically associated segment and arrangement of very close gatherings or networks. The wide prominence of OSNs and their basic access have likewise come about inside the abuse of their administrations. Other than the trouble of safeguarding user security, OSNs face the test of adapting to bother some users and their noxious exercises inside the social networks Siddula et al. [22]. The first normal kind of pernicious movement recognized in OSNs is spamming which includes noxious users (spammers) to communicate immaterial data inside the sort of messages, IMs, remarks, text messages and presents on as sizable measure of genuine users as could be expected under the circumstances. Spamming is done generally with a point of advancing items, viral advertising, spreading crazes, and now and again would be never really authentic users of an OSN in order to diminish their trust inside the specific help. Long range interpersonal communication alludes to gathering of individuals and associations together by means of some medium, in order to share contemplations, interests, and exercises. There are a few online informal organization administrations are accessible like Facebook, Twitter, LinkedIn, Google+ and so on which give simple to utilize and intuitive interface to append with individuals with inside the nation a distant too. There additionally are a few portable based long range informal communication benefits sure of applications like WhatsApp and so on.

9 Conclusion

This study utilizes both quantitative and abstract techniques to research the fortunate combination between long range social network use and dangers concerns. It shows that the enjoyments of using Facebook will by and large surpass the clear risks to assurance. The most notable strategy for security confirmation attenuation profile deceivability through keeping induction to associates is similarly an uncommonly weak part; a helpful arrangement rather than a proficient method to manage guaranteeing insurance. Most users don't seem to comprehend that restricting permission to their data doesn't enough address the danger coming about in view of the entirety, quality and dauntlessness of the data they give. Taking everything into account, restricting profile deceivability to "buddy just" fundamentally infers binding it inside the conspicuous small piece of the chilly crowd. Anyway long users feed the impalpable bit of the cold mass with wide individual data that they update purposefully and perpetually, and their security is in threat. Given the attention in on age get-togethers, the strong interest of casual network environment, and the way that snitch, baiting, hacking, phishing, data mining, and utilization of individual data by outsiders are a reality in these associations and not just a theoretical possibility, and this paper shows that active adults should be instructed about threat to their security in a way that truly changes their lead. The makers believe that the revelations of this examination may add to this pattern of tutoring in PC education.

References

1. Khanday HA, Ganai AH, Hashmy R (2018) A comparative analysis of identifying influential users in online social networks. In: 2018 international conference on soft-computing and network security (ICSNS), Coimbatore, pp 1–6. https://doi.org/10.1109/ICSNS.2018.8573634
2. Rahman M, Karim R (2016) Comparative study of different methods of social network analysis and visualization. In: 2016 international conference on networking systems and security (NSysS), Dhaka, pp 1–7. https://doi.org/10.1109/NSysS.2016.7400702
3. Suleimanov A, Abramov M, Tulupyev A (2018) Modelling of the social engineering attacks based on social graph of employees communications analysis. In: 2018 IEEE industrial cyber-physical systems (ICPS), St. Petersburg, pp 801–805. https://doi.org/10.1109/ICPHYS.2018.8390809
4. Bagretsov GI, Shindarev NA, Abramov MV, Tulupyeva TV (2017) Approaches to development of models for text analysis of information in social network profiles in order to evaluate user's vulnerabilities profile. In: 2017 XX IEEE international conference on soft computing and measurements (SCM), St. Petersburg, pp 93–95. https://doi.org/10.1109/SCM.2017.7970505
5. Yadav AK, Johari R, Dahiya R (2019) Identification of centrality measures in social network using network science. In: 2019 international conference on computing, communication, and intelligent systems (ICCCIS), Greater Noida, India, pp 229–234. https://doi.org/10.1109/ICCCIS48478.2019.8974553
6. Nakerekanti M, Narasimha VB (2019) Analysis on malware issues in online social networking sites (SNS). In: 2019 5th international conference on advanced computing and communication systems (ICACCS), Coimbatore, India, pp 335–338. https://doi.org/10.1109/ICACCS.2019.8728536
7. Wilcox H, Bhattacharya M (2016) A framework to mitigate social engineering through social media within the enterprise. In: 2016 IEEE 11th conference on industrial electronics and applications (ICIEA), Hefei, pp 1039–1044. https://doi.org/10.1109/ICIEA.2016.7603735
8. Almabdy S (2018) Comparative analysis of relational and graph databases for social networks. In: 2018 1st international conference on computer applications and information security (ICCAIS), Riyadh, pp 1–4. https://doi.org/10.1109/CAIS.2018.8441982

9. Jain L, Katarya R (2018) A systematic survey of opinion leader in online social network. In: 2018 international conference on soft-computing and network security (ICSNS), Coimbatore, pp 1–5. https://doi.org/10.1109/ICSNS.2018.8573639
10. Pan T et al (2017) Threat from being social: vulnerability analysis of social network coupled smart grid. IEEE Access 5:16774–16783. https://doi.org/10.1109/ACCESS.2017.2738565
11. Yun X, Li S, Zhang Y (2015) SMS worm propagation over contact social networks: modeling and validation. IEEE Trans Inf Forensics Secur 10(11):2365–2380. https://doi.org/10.1109/TIFS.2015.2455413
12. Yan Z, Wang M (2017) Protect pervasive social networking based on two-dimensional trust levels. IEEE Syst J 11(1):207–218. https://doi.org/10.1109/JSYST.2014.2347259
13. Fu T, Liu P, Ding Y, Zhang Y (2018) Secure and efficient large content broadcasting in mobile social networks. IEEE Access 6:42108–42118. https://doi.org/10.1109/ACCESS.2018.2858237
14. Li X, Xin Y, Zhao C, Yang Y, Luo S, Chen Y (2020) Using user behavior to measure privacy on online social networks. IEEE Access 8:108387–108401. https://doi.org/10.1109/ACCESS.2020.3000780
15. Guo L, Zhang C, Fang Y (2015) A trust-based privacy-preserving friend recommendation scheme for online social networks. In: IEEE transactions on dependable and secure computing, vol 12(4), pp 413–427, 1 July–Aug. 2015. https://doi.org/10.1109/TDSC.2014.2355824
16. Abbas F, Rajput U, Oh H (2016) PRISM: PRivacy-aware interest sharing and matching in mobile social networks. IEEE Access 4:2594–2603. https://doi.org/10.1109/ACCESS.2016.2564985
17. Shi P, Zhang Z, Choo KR (2019) Detecting malicious social bots based on clickstream sequences. IEEE Access 7:28855–28862. https://doi.org/10.1109/ACCESS.2019.2901864
18. Wang H, Hu W, Qiu Z, Du B (2017) Nodes' evolution diversity and link prediction in social networks. IEEE Trans Knowl Data Eng 29(10):2263–2274. https://doi.org/10.1109/TKDE.2017.2728527.
19. Li R et al (2017) Perturbation-based private profile matching in social networks. IEEE Access 5:19720–19732. https://doi.org/10.1109/ACCESS.2017.2748958
20. Abawajy JH, Ninggal MIH, Herawan T (2016) Privacy preserving social network data publication. IEEE Commun Surv Tutorials 18(3):1974–1997. https://doi.org/10.1109/COMST.2016.2533668
21. Radha D, Kulkarni S (2017) A social network analysis of world cities network. In: 2017 2nd international conference on computational systems and information technology for sustainable solution (CSITSS), Bangalore, pp 1–6. https://doi.org/10.1109/CSITSS.2017.8447571
22. Siddula M, Li L, Li Y (2018) An empirical study on the privacy preservation of online social networks. IEEE Access 6:19912–19922. https://doi.org/10.1109/ACCESS.2018.2822693

Bi-CRS: Bio-Inspired Cluster-Based Routing Scheme for D2D Communication in IoT

M. B. Yashoda and Vrinda Shivashetty

Abstract Internet of Things (IoT) is the quickly evolving network of interconnected objects that are able to collect and exchange data using embedded sensors. IoT is a very promising research field due to its significance in many real world applications. Device-to-Device (D2D) communication is the key technologies to inflate the performance of future generation network. D2D communication is communication between two or more entities without going through a central base station. Key technologies to strengthen the future IoT networks are multihop routing and relaying. This paper reviews several multihop routing protocols, but with the internet's evolution and the rise of the internet of things, current routing protocols can no longer accommodate the large increase in nodes. Management of communication network and routing is becoming very difficult due to the growing size, topology changes and complexity of the network. A new bio-inspired cluster based routing scheme (Bi-CRS) based on Particle swarm Optimization (PSO) is developed to find optimal routing path for D2D communication in IOT. The performance of Bi-CRS is compared with Bee-Inspired Routing algorithm (BIRA). The proposed algorithm out performs in terms of convergence, delay, energy consumption and packet transmission.

Keywords IoT · SI · CH · D2D communication · Multihop routing · 5G · PSO

1 Introduction

IoT is a cutting-edge concept that is gaining momentum in the next wave of technological networks, bringing the concept of linking countless physical objects to the internet. The vast amount of internet-connected "stuff" poses a number of challenges and research opportunities. D2D communication captivated a substantial attention

M. B. Yashoda (✉) · V. Shivashetty
Sai Vidya Institute of Technology, Bangalore, India

V. Shivashetty
e-mail: vrinda.shetty@saividya.ac.in

VTU Belgaum, Belgaum, India

R. P. Mahapatra et al. (eds.), *Proceedings of International Conference on Recent Trends in Computing*, Lecture Notes in Networks and Systems 341,
https://doi.org/10.1007/978-981-16-7118-0_17

and considered as favorable technology for IOT networks due to its high-energy efficiency, high-data rate and lower delay. On the basis of spectrum usage, D2D communication is categorized into inband (or licensed) and outband (or unlicensed) communications. Underlay and overlay are two categories of inband communication while outband communication is categorized into controlled and autonomous.

Authors in [1], describes five architectural and component disruptive design changes for 5th generation (5G) cellular networks, include: device-centric architecture, massive MIMO, millimeter wave, smart devices and Machine-to-Machine communication. To reach the demands of 5G in terms of high data rate, low latency and high-spectral efficiency, it is more important to concentrate on smartness of the devices and device-centric architecture that is D2D communication [2]. Devices can interact directly with objects in its proximity using D2D communication without traversing through the base station. The main advantage of D2D communication is high-data rates and low latency because of the shorter signal travel path.

Apart from the advantages, there are several challenges that are still open for research. One of the major challenges is to find optimal route for data transmission from source node to the target node in D2D communication. Routing is key function to achieve efficient communication in a network. Architecture of D2D communication consists of communication between heterogeneous devices with different computing capabilities. Since the computing devices are energy constrained, multihop routing and relaying is most desirable [3]. Due to the device constraints like limited battery and memory size, the UEs or computing devices use low power IoT networks like Wireless Fidelity, ZigBee and Bluetooth. A routing protocol which addresses all the limitations of existing protocols is needed to efficiently deliver the data in D2D network. Main objective is to survey the routing techniques introduced in wireless sensor networks, cellular networks and IoT, and to develop a novel routing scheme to find optimal route for D2D communication in Internet of Things (IoT) with required intelligence.

The major concern of wireless networks is positioning the nodes, clustering and data gathering. The bio-inspired algorithms drive to put together these routing issues as optimization problems. The conventional optimization techniques use high energy, as the network size increases this will become a difficult problem. Thus in resource constrained networks bio-inspired algorithms give desirable results, as it needs less memory and computing resources. Bio-inspired algorithms are metaheuristic optimization algorithms used to obtain optimal solution to a given problem. These algorithms mimic the dynamics of biological evolution. Bio-inspired algorithms are categorized as: Evolutionary based, Swarm intelligence and Ecology based. Swarm intelligence, has many properties that are useful in numerous engineering applications, like network routing. A basic feature of future design paradigms will be emerging intelligence-swarm agents interact locally and with the environment, agents follow simple rules, and give rise to intelligent, self-organized, collective global behavior.

Outline of the paper: Sect. 2 discusses the related work of D2D communication in WSNs, IoT and cellular network. Then, we present the proposed model for D2D communication in IoT along with the performance analysis in Sect. 3, andthe conclusion of this study is presented in Sect. 4.

2 Related Work

The routing of data from source to target is a basic part of any large scale network like IoT. Routing protocols falls under three main categories: Proactive, Reactive and Hybrid routing protocols. This classification is based on the way by which routing decisions are made. Proactive protocols are table driven; in which each node maintain a table containing routes for all possible destinations. These protocols do not work well with large networks. In reactive protocols network nodes discover routes on-demand. Routes are discovered using flooding has mainly two phases: route discovery and route maintenance. A hybrid protocol integrates reactive and proactive protocols.

Several routing protocols for communication in WSNs, MANETs, cellular networks and IoT are studied. Performance of Ad hoc protocols like DSDV, AODV and DSR is compared in [4]. Performance of DSDV is very poor compared to other two, and DSR performed better than AODV. But the performance of all three protocols degrades as node mobility increases. AOMDV-IOT proposed in [5] uses four messages: RREQ, RREP, RERR and HELO in the communication process. To handle the routing, each node keeps track of two tables namely internet connecting table and routing table. This increases the overhead on the memory of the nodes. AOMDV does not optimize the routing path, since it does not have the knowledge of available bandwidth of routing paths. By taking AOMDV as the base authors in [6] developed a new routing algorithm Load Balancing Based selective AOMDV (LBS-AOMDV). The simulation results show the enhancement in QoS of D2D communication.

In [7], authors proposed that in cellular networks, D2D communication is possible using an energy efficient multihop routing. They invented a system for selecting the next hop that takes into account the congestion levels of neighbor nodes in order to expedite data distribution, while conserving node transmission capacity. The simulation results show that the proposed algorithm will reduce the end-to-end data transmission delay and increase the likelihood of successfully transmitting a packet in a D2D network, while conserving node resources.

A routing algorithm to enhance throughput of D2D communication under interference constraint is proposed in [8]. The routing algorithm first finds the computing power of each D2D node, which helps to determine the outage probability of the links. The rate matrix is then constructed using this knowledge, allowing Dijkstra's algorithm to fine-tune the best data transmission path. Data transmission is based on Sequential Link Activation (SLA), which stick to the outage probability constraint. Simulation results showed that multihop routing provide greater throughput as compared to single hop D2D communications.

A routing algorithm for D2D communication in 5G cellular networks is presented in [9]. Author's main goal is to widen the base station coverage. The authors use the LBS-AOMDV algorithm as a model for implementing the Network Assisted Routing (NAR) algorithm in 5G networks for D2D communication. It is clear from the results that there is 35% improvement in energy saving and 5% increase in total number of packets sent.

Hanan H. Hussein et al., have discussed about D2D scenarios and applications in 5G cellular networks [10]. Machines are placed next to each other in M2M communication applications. In such cases, it is feasible to apply D2D communication techniques, which raises network efficiency, reduces delay and machines can have long battery life which are necessary conditions for future communication networks. RPL was developed by the Internet Engineering Task Force (IETF) as a routing protocol for low power and lossy networks (LLNs). RPL could be used for different applications like Healthcare, Smart environments, Transport, Industry and military by making changes in few of the configuration parameters depending upon application requirements [11]. But its performance is not good for dynamic environments.

The main barriers for D2D communication to find the optimal route from source to sink are energy constraints, network stability, congestion, and link failure of the nodes. These barriers results in packet loss, rapid energy usage, and longer delay, which leads to considerable fall in the network performance. To resolve the challenges and improve network efficiency, Valmik Tilwari et al. proposed the EMBLR method [12]. The proposed routing scheme has overcome several challenges of selection of an optimal route. Network performance degradation is noticed, because of the interference caused by spectrum sharing. Table 1 summarizes the routing protocols for D2D communication.

Many researchers try to apply SI based routing algorithms for routing in D2D IoT communication. Authors in [13] designed a framework to use D2D communication for disaster management in 5G using ACO algorithm. SDN controller assistance is taken by the base station to maintain reliable connectivity to the computing nodes in the emergency zone. The framework finds the isolated nodes using D2D in disaster zone. D2D is used for critical communication so that damage and loss of life is reduced. FINDER extends the network life time and improves the probability of message delivery.

To improve the efficiency of IoT network, some intelligent algorithm is required to route the data not only inside the network but also outside the network [14]. The authors used the powerful IoT communications with Ant System (EICAntS) routing algorithm to send healthcare data between three towns, each of which has its own internet of things network. Proposed algorithm shows less latency and better path identification compared to the base method. Improved version of Bee Colony algorithm BIRA is proposed to meet D2D communication in IoT network in [16]. Each computing nodes uses two media access control layer protocols. In terms of end-to-end delay, packet transmission ratio, and energy consumption, BIRA outperforms AODV.

3 Proposed Model

D2D communication is a state-of-the-art data transmission mechanism developed to improve the effectiveness of next generation network. D2D communication exhibits some challenges like complex architecture, heterogeneous devices, energy

Table 1 Summary of the conventional and SI based routing protocols for D2D communication in WSN and IoT

Reference	Year	Algorithm	Findings
[4]	2010	AODV, DSR, DSDV	AODV and DSR performs better then DSDV, and as the mobility of node increases the performance of all three protocols decreases
[5]	2010	AOMDV-IOT	Does not optimize the routing paths because it does not understand the context
[1]	2016	Proximity services	Architectural upgrades to the EPC core network are needed to support the coexistence of D2D and cellular links. Despite the fact that the ProSe standard was developed years ago, most vendors did not produce chipsets that supported it
[7]	2016	Multihop routing	When choosing the next hop in multihop routing, the congestion level of the neighbor nodes is taken into account. This speeds up data delivery while conserving node transmission capacity
[8]	2017	Multihop routing with SLA based link selection	Routing algorithm which maximizes the given source-sink D2D pair's throughput. Other link activation technique could be used instead of Sequential Link Activation to reduce system idle probability
[11]	2019	RPL	RPL is accepted as a routing protocol in IOT. But its performance is not good for dynamic environments
[12]	2020	EMBLR	Routing scheme has overcome several challenges of selection of an optimal route. Network performance degradation is noticed, because of the interference caused by spectrum sharing
[13]	2018	FINDER: ACO	Ant colony optimization (ACO) based routing algorithm for critical communication. It is clustering and relay based to communicate with the base station. The algorithm elevates individual nodes energy and the network's lifetime

(continued)

Table 1 (continued)

Reference	Year	Algorithm	Findings
[14]	2018	EICAntS	A variation of ACO is used for transmission of data between three IoT networks. The new algorithm is more reliable compared to basic ACO
[15]	2018	HACO-ABC-CHS	Ant colony and Bee colony optimization based algorithm is used for the cluster head selection. The premature convergence problem of ACO is solved using bee colony optimization
[16]	2019	BIRA	Bee-inspired routing algorithm for D2D IoT communication, shows better performance compared to AODV in terms of delay, packet transmission and energy consumption
[17]	2019	Elastic PSO	The A* algorithm is used to design the paths. The Particle Swarm Optimization (PSO)-based scheme employs shrinking optimization to determine the overall best route generated by the local fittest nodes, allowing for faster convergence

constraints, and autonomous and distributed infrastructure. We assert that bio-inspired intelligent algorithms can successfully handle these challenges because of their unique properties like Decentralized, Scalability, Interoperability, Fault tolerance, and Resource management.

IoT network devices (i.e., sensor nodes) are power-constrained, have limited capacity, and have limited processing power. In an IoT network, devices constantly sense data and send it to the destination. This may lead to congestion at the destination node, as a result sensor node will fail to route the data to the base station [18]. To solve this problem, the network can be divided into clusters and fittest node in each cluster is considered as the cluster head (CH). The choice of the best CH would increase the lifespan of the IoT network. The proposed system model is shown in Fig. 1.

3.1 Initialize the Nodes

The number of nodes in the search space is set and the parameters of nodes in the IoT network are initialized in this phase. Since the nodes of IoT network are energy constrained, preserving the battery life is very crucial to increase the lifetime of the network. The parameters of nodes like, initial energy, communication range, energy

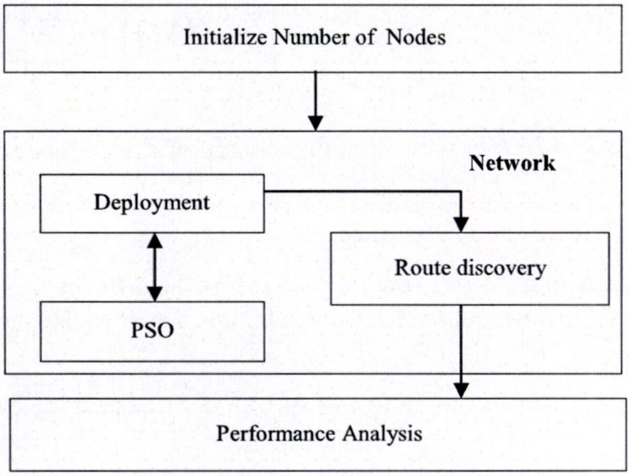

Fig. 1 System model

of node, initial load at each node and temperature are set. The initial simulation parameters for D2D communication are also configured.

3.2 *Network Deployment*

In order to avoid energy depletion of nodes and congestion at the BS, network is divided into several regions. Each region or cluster has an optimal cluster head. The life of the network and the selection of the optimal path to the destination from the source depend on the fittest cluster head. The CHs are picked depending on a variety of factors; including node's left over energy (N_{energy}), congestion at the node (N_{load}), temperature of the node (N_{temp}), delay (N_{delay}) in moving nodes to the destination, and the distance to the destination from the chosen CH (N_{dist}). The cluster head is selected using following Eq. (1).

$$\text{NF}_i = \upsilon * N_{\text{temp}} + \phi * N_{\text{load}} + \chi * N_{\text{energy}} + \varphi * (1 - N_{\text{dist}}) + \omega * \left(1 - N_{\text{delay}}\right) \tag{1}$$

where ν, ϕ, χ, φ and ω are weighted parameters, when added the total must be equal to 1.

The CH selection parameters are calculated using following equations:

Energy Computation

The left over energy at the CH node plays a major role in extending the network's existence and improving its efficiency. Energy at each node changes following data forwarding and receipt. The information is transferred before the node's energy is depleted. Energy consumption can be calculated using Eq. (2).

$$N_{\text{energy}} = \frac{1}{n}\left\{\sum_{x=1}^{n} E\left(N_s^x\right)\right\} + \frac{1}{T_{\text{ch}}}\left\{\sum_{t=1}^{\text{ch}} E\left(E_{\text{ch}}^t\right)\right\} \tag{2}$$

where $E\left(N_s^x\right)$ is the draining energy of the xth node after the transmission of data and $E\left(E_{ch}^t\right)$ draining energy of tth CH.

Calculation of Distance

Equation (3) gives the formula to find the distance from source node to CH and from CH to destination. The node which is nearer to the destination node is chose as CH.

$$N_{\text{dist}} = \sum_{x=1}^{n}\sum_{t=1}^{\text{ch}} \frac{\left\|d_s^x - d_{\text{ch}}^t\right\| + \left\|d_{\text{ch}}^t - d_{\text{dest}}\right\|}{m * m} \tag{3}$$

$d_s^x - d_{\text{ch}}^t$: The distance between source xth node to the respective CH.
$d_{\text{ch}}^t - d_{\text{dest}}$: The distance from tth CH to the destination.
m: sensing area.

Delay Calculation

Network performance is enhanced by sending data packets from source to destination in a short period of time. Equation (4) shows the formula to calculate the time taken to send data from CH to destination and n denotes number of nodes.

$$N_{\text{delay}} = \frac{\text{Max} \sum_{t=1}^{\text{ch}} \text{CH}_t}{n} \tag{4}$$

3.3 *The PSO Algorithm*

PSO in inspired by flock of birds and school of fishes. Consider a flock of birds circling a food-scavenging area. The particle (i.e., bird in search space) closest to the food source chirps loudly and flock as a whole follow its direction. This continues till one of the birds reaches the food source. Each particle has fitness value which is calculated by optimized function.

Three variables are maintained by the algorithm:

- Desired result
- Best of all (GlBest) optimal position found by particle swarm search
- A value that specifies when the algorithm can come to a halt if the desired isn't found.

Each particle has the following characteristics:

Step 1: Initialization

1.1: for each particle in the population P:

- initialize X_i randomly
- initialize V_i randomly
- evaluate the fitness $f(X_i)$
- initialize $PBest_i$ with a copy of X_i

1.2: Initialize $GlBest_i$ with a copy of X_i with the best fitness

Step 2: Repeat until the termination criterion is satisfied:

2.1: for each particle i:

- update X_i and V_i according to equations (5) and (6)
- evaluate the fitness $f(X_i)$
- if $f(PBest_i) < f(X_i)$ then $PBest_i \leftarrow f(X_i)$
- if $f(GlBest_i) < f(X_i)$ then GlBesti $\leftarrow f(X_i)$

Fig. 2 The PSO algorithm

- Information about a potential solution
- Particle velocity
- A personal best (PBest) optimal position found by particle *i*.

PSO algorithm is given in Fig. 2. Formula to update particle velocity and position are given in Eqs. (5) and (6), where c_1 and c_2 are acceleration constants and ω is inertia constant.

$$V_i = \omega V_i + c_1 * \text{random}(0, 1) * (\text{PBest}_i - X_i) + c_2 * \text{random}(0, 1) * (\text{GlBest} - X_i) \tag{5}$$

$$X_i = X_i + V_i \tag{6}$$

3.4 Route Discovery

Every node in the proposed PSO algorithm emits a periodic hello packet to keep its neighbor informed about the active link. The hello packet consists of the node identifier (ID), and it's left over energy. This information will help other nodes to select efficient neighbor in optimal path selection. Each node maintains a routing table with entries for source ID, destination ID, path ID, hop count and path cost. When a node wants to send data, it first looks in its routing table to see if an optimal path to the destination is available. If node could not find a valid path in its routing table, it initiates the route discovery process. When it finds a best optimal path, it

Table 2 Simulation parameters

Parameter	Value
Simulation range	900 m × 900 m
Payload size	512 bytes
Initial Energy	100 J
Number of nodes	70
Simulation time	600 s
Propagation range	40 m
Radio propagation model	Two Ray Ground
Wireless network	Wi-Fi IEEE 802.11
MAC layer protocols	IEEE 802.11a, IEEE 802.11b
Network Interface Type	WirelessPhy

updates its routing table. Once the optimal path is found, the node transmits the data packet to the destination.

3.5 *Performance Evaluation*

Simulation is carried out and the results of energy efficient PSO based routing algorithm is compared with BIRA. On the Ubuntu 16.04 Linux operating system, the proposed routing algorithm is simulated with ns^{-2}. IoT network nodes are spread across a 900 × 900 m^2 region. Two MAC layer protocols, IEEE 802.11a and IEEE 802.11b, are introduced at each node to preserve network heterogeneity. The parameters considered in simulation are listed in Table 2.

Experimental results

End-to-End delay: Figure 3 shows how the proposed routing algorithm is quicker than BIRA at delivering data packets to their destination. The graph depicts how the delay increases as the size of the network grows. This is due to a rise in the number

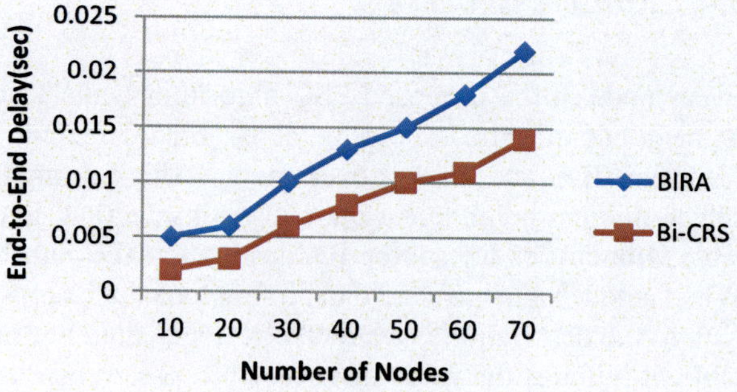

Fig. 3 End-to-End latency at different number of nodes

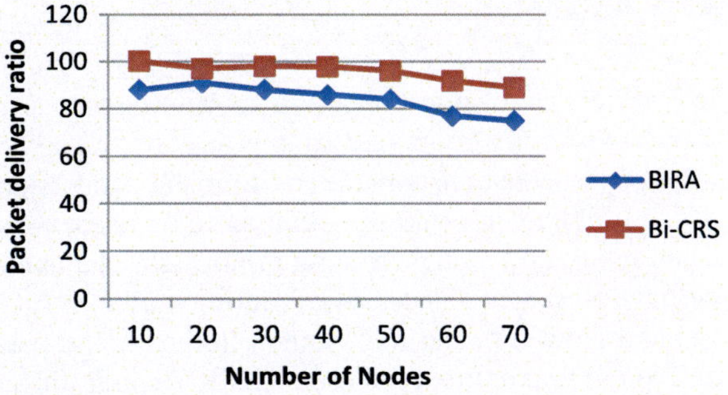

Fig. 4 Ratio of packet transmission at different number of nodes

of hops in the routing path from the source to the destination node. When comparing the average delays of both algorithms, BIRA's end-to-end delay is 0.089, whereas Bi-CRS' is 0.054.

Packet delivery ratio: Bi-CRS outperforms BIRA in terms of packet transmission ratio at various numbers of nodes, as shown in Fig. 4. Bi-CRS transmits packets with a performance rate of 87–100%, while BIRA's is 75–91%. In comparison with the bee-inspired routing algorithm, on an average Bi-CRS shows a 9% improvement in packet distribution ratio.

Energy consumption: Proposed algorithm takes several parameters in to considerations while selecting the CH. One of the important parameters is the left over energy at the node. CH is chosen as the node with the most remaining energy and is closest to the destination. As shown in Fig. 5, Bi-CRS uses less energy than BIRA and increases system reliability. That's because in Bi-CRS, instead of looking for the best qualified neighbor as in BIRA, the source node often chooses the CH to route the packet to the destination.

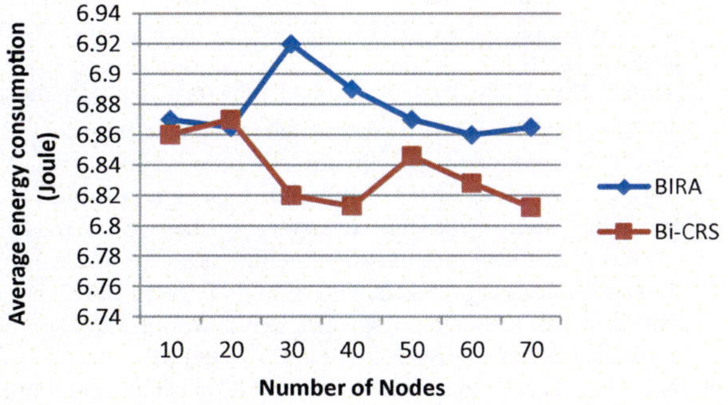

Fig. 5 Average energy consumption at various node counts

4 Conclusion

IoT system is a large collection of heterogeneous resource constrained devices. Device-to-Device communication is an essential part of IoT system to design, establish and perpetuate sustainable environment. Understanding the routing of sensor data is one of the main challenges that an IoT system would face. Building energy efficient, robust and reliable routing protocols is challenging. The first part of the paper, we have presented a summary of routing protocols for D2D communication in WSN, Cellular Networks and IoT system. In the second part, we proposed Bi-CRS based on PSO to find optimal route in D2D IoT communication. Selection of CH is made intelligently using several parameters including node's left over energy, congestion at the node, temperature of the node, delay in moving nodes to the destination, and the distance to the destination from the chosen CH. Since the data transmission is through the efficient CH, Bi-CRS shows faster convergence, energy efficiency, and increase in the network performance as compared to BIRA.

References

1. Hicham M, Abghour N, Ouzzif M (2016) Device-To-Device (D2d) Communication under Lte-advanced networks. Int J Wirel Mob Netw (IJWMN) 8(1)
2. Ahmed S, Rashid M, Alam F, Fakhruddin B (2019) A disaster response framework based on IoT and D2D communication under 5G network technology. In: 2019 29th ınternational telecommunication networks and applications conference (ITNAC), 2019 Nov 27. IEEE, pp 1–6
3. Hamadani MK, Mahdi AH (2019) Centralized multi-hop routing for device-to-device communication: simulation and results. In: 2019 11th international conference on electronics, computers and artificial intelligence (ECAI), Pitesti, Romania, pp 1–6. https://doi.org/10.1109/ECAI46879.2019.9042024
4. Tuteja A, Gujral R, Thalia S (2010) Comparative performance analysis of DSDV, AODV and DSR routing protocols in MANET using NS2. In: 2010 International conference on advances in computer engineering. 978-0-7695-4058-0/10 $26.00 © 2010 IEEE. https://doi.org/10.1109/ACE.2010.16
5. Tian Y, Hou R (2010) An improved AOMDV routing protocol for Internet of Things. In: 2010 International conference on computational intelligence and software engineering, Wuhan, pp 1–4. https://doi.org/10.1109/CISE.2010.5676940
6. Tata C, Kadoch M (2014) Multipath routing algorithm for device-to-device communications for public safety over LTE Heterogeneous networks. In: 2014 1st international conference on information and communication technologies for disaster management (ICT-DM), Algiers, pp 1–7. https://doi.org/10.1109/ICT-DM.2014.6918583
7. Park J (2016) Fast and energy efficient multihop D2D routing scheme. Hindawi Publ Corp Int J Distrib Sens Netw 2016. Article ID 2148734
8. Madabhushi S, Gopal GR, Murthy CR (2017) Optimal routing and data transmission for multi-hop D2D communications under stochastic interference constraints. In: 2017 twenty-third national conference on communications (NCC), Chennai, pp 1–6. https://doi.org/10.1109/NCC.2017.8077104
9. Bastos AV, Silva CM, da Silva DC (2018) Assisted routing algorithm for D2D communication in 5G wireless networks. In: 2018 wireless days (WD), Dubai, pp 28–30.https://doi.org/10.1109/WD.2018.8361688

10. Hussein HH, Elsayed HA, Abd El-kader SM (2020) Intensive benchmarking of D2D communication over 5G cellular networks: prototype, integrated features, challenges, and main applications. Wirel Netw 26:3183–3202. https://doi.org/10.1007/s11276-019-02131-2
11. Kharrufa H, Al-Kashoash HAA, Kemp AH RPL-based routing protocols in IoT applications: a review. IEEE Sens J 19(15):5952–5967. ISSN 1530-437X
12. Tilwari V, Dimyati K, Hindia MHDN, Izam TFBTMN, Amiri IS (2020) EMBLR: a high-performance optimal routing approach for D2D communications in large-scale IoT 5G network. Symmetry 12:438. https://doi.org/10.3390/sym12030438
13. Thomas A, Raja G (2019) FINDER: a D2D based critical communications framework for disaster management in 5G. Peer-to-Peer Netw Appl 12:912–923. https://doi.org/10.1007/s12083-018-0689-2
14. Priyanka JK (2018) Ant colony optimization based routing in IoT for healthcare services. In: Proceedings of the second international conference on intelligent computing and control systems (ICICCS 2018). IEEE Xplore Compliant Part Number: CFP18K74-ART; ISBN 978-1-5386-2842-3
15. Janakiraman S (2018) A hybrid ant colony and artificial bee colony optimization algorithm-based cluster head selection for IoT. In: 8th international conference on advances in computing and communication (ICACC-2018). 1877-0509 © 2018
16. Almazmoomi AM, Monowar MM (2019) On designing bee inspired routing algorithm for device-to-device communication in the Internet of Things. (IJACSA) Int J Adv Comput Sci Appl 10(11)
17. Wang H, Zhou Z (2019) A heuristic elastic particle swarm optimization algorithm for robot path planning. Information 10(3):99
18. Reddy PK, Rajasekhara Babu M (2017) Energy efficient cluster head selection for Internet of Things. New Rev Inf Netw 22:54–70

A Cloud-based Trusted Framework for Industrial Connected Vehicles

M. Deva Priya, G. Amirthavarsini, S. Angu Kaushika, and K. Deeptheshanmathie

Abstract To add a member in a distributed network, Vehicular Clouds (VCs) offer better scope for exploration. Incorporation of VCs aids in developing vehicular frameworks with acute reception. Reliability should be ensured in the VC for delivering additional processing capabilities and managing the statistics collected within the scope. In this paper, a secure evaluation framework that ensures protection of statistics, trustworthiness of data and accessibility of assets is proposed. It is essential to offer reliability of the framework that is based on VCs. A three-level secure structure that is sustainable and appropriate for vehicular organizations to detect reckless vehicles is designed. Cooperative Driving Performance Rating (CDPR) element is used to certainly apprehend rashly driven cars based on Machine Learning (ML) based computations. To generate alarms, a three-layered framework is formed for existing vehicular organizations which incorporate vehicles, Road Side Units (RSUs) and cloud users. The scheme provides paths to the vehicles by involving reduced transmission overhead of the CDPR element. The vehicles transfer the statistics of the user movements to the RSUs. Based on the gathered statistics, the vehicles are charged. Support Vector Machine (SVM) and Decision Tree (DT) are used to carry out the process. The proposed CDPR element can be incorporated into a vehicular organization and a simulated framework is designed using Simulation of Urban Mobility (SUMO) to assess the performance. The proposed system offers a secured alarming framework to remove rashly driven cars and rightly deliver cautions to the organizations.

M. Deva Priya (✉) · G. Amirthavarsini · S. Angu Kaushika · K. Deeptheshanmathie
Department of Computer Science and Engineering, Sri Krishna College of Technology, Coimbatore, Tamilnadu, India
e-mail: m.devapriya@skct.edu.in

G. Amirthavarsini
e-mail: 17tucs009@skct.edu.in

S. Angu Kaushika
e-mail: 17tucs010@skct.edu.in

K. Deeptheshanmathie
e-mail: 17tucs026@skct.edu.in

R. P. Mahapatra et al. (eds.), *Proceedings of International Conference on Recent Trends in Computing*, Lecture Notes in Networks and Systems 341,
https://doi.org/10.1007/978-981-16-7118-0_18

Keywords Trustworthy · Evaluation framework · Road side units · Support vector machine

1 Introduction

Cloud computing includes resources that offer storage and computing capability without user intervention for effective management. These facilities enable the users to work at ease. Larger clouds which are dominant allow functions to be distributed from central servers to multiple locations. Clouds can be available to only one organization (private) or limited to various organizations (public).

Cloud computing allows us to share resources to achieve economic needs and requirements. Public and hybrid cloud computing aid in reducing IT building and development costs. Proponents also hope that it helps in business, trading and transaction processes, making everything faster. According to business administrators, it aids in the improvement of the sales rate and communication among customers. It meets the management demands and involves reduced maintenance costs. It plays a major role in the development of IT by offering dynamic resources that meet all the demands. It offers a facility of 'pay as you go.' Though it is very useful to the public, it may lead to unwanted operation costs for the administrators if they are not aware of the cloud payments and structures. The demand for cost-effective computers, secondary devices and high-capacity networks, evaluation of service-oriented architecture, utility computing, autonomic computing and hardware virtualization have increased the use of cloud computing.

Traffic simulations facilitate the assessment of infrastructure adjustments in addition to coverage adjustments. Initially, they were applied on the streets. The effectiveness of environmental zones may be examined and optimized through a simulation before it is deployed in the actual world. Simulation of Urban Mobility (SUMO) is a free, open, microscopic and non-stop street side visitor's simulation suite designed to deal with big street networks. It permits the modeling of intermodal site visitor structures inclusive of street vehicles, public shipping and pedestrians. It includes a number of assisting tools which manage responsibilities together with course exploring, visualizing, community imports and emission evaluation. It is more suitable for customizing fashions and also gives diverse Application Programming Interfaces (APIs) to remotely manage the simulation.

As the consequences of reckless driving are more, the traits are examined after which their outcomes are hypothetically examined through a survey. To circumvent reckless driving, a protection method is applied to the current day motors as more fatalities are common these days. Any behavior not adhering to the protection scheme is considered as reckless.

Due to drastic consequences of reckless use, a cloud computing based protective scheme that generates alarms for rash driving vehicles is proposed. Cooperative Driving Performance Rating (CDPR) scheme is proposed to mechanically overcome the stumbling of rash driving vehicles, in which the consequences will be dispatched to the motorist close by. These features are incorporated using a three-tier machine structure. CDPR method is implemented through a signaling process.

2 Related Work

Hafeez et al. [1] have proposed a logical structure based on the dependable and dedicated short-range communication control channel to work with safe applications in Vehicular ad hoc networks.

Al-Sultan et al. [2] have developed Dedicated Short-Range Communications (DSRC) for Vehicle Ad hoc NETworks (VANETs) which allows only vehicles that are closer to converse with each other, else to communicate with wayside requirements. Saiprasert and Pattara-Atikom [3] have designed a rash driving report scheme based on a smartphone platform. Data streams including the time of speed profile for the current journey are collected through a built-in GPS receiver. Codeca et al. [4] have focused on various research groups differing from mechanization to congestion engineering that deals with the issues based on collision among vehicles, automation shipping systems and moving designs based on data gathered from diverse sensors.

Zhang et al. [5] have dealt with several approaches to offer incentives to inspiring users who contribute for mobile crowd detecting applications. These economic incentive schemes provide users intuitive incentives. Monetary schemes are more common than those of entertainment and service as the latter classes of schemes are designed as applications. Kato et al. [6] have propounded suitable input and output descriptions of assorted network traffic along with a supervised Deep Neural Network (DNN) system that varies from customary Neural Networks (NNs). The authors have compared the performance with Open Shortest Path First (OSPF) in terms of signaling overhead, throughput and delay.

Luvizon et al. [7] have propounded a non-intrusive video dependent system for measuring vehicle speeds in urban roadways. This system makes use of an improved motion and text indicators to proficiently trace vehicle license plates in regions with motion. Distinct features on the license plate regions are selected, pursued across multiple frames and resolved for perspective alteration.

Yu et al. [8] have propounded a fine-grained irregular driving behavior detection and identification system (D3) to deal with the real-time monitoring of precise abnormal driving using smartphone sensors. Coarse and fine-grained identifications are conducted which include weaving, swerving, side slipping, fast U-turns, turning with a broad radius and rapid braking.

Mahmud et al. [9] have presented a review of the contemporary proximal substitute safety signs. It offers an inclusive investigation on ideologies, applications, merits, demerits and application appropriateness for varied types of conflict assessment using proximal substitute signs.

Fadlullahet et al. [10] have proposed a network traffic control system that is based on Internet core and wired/wireless assorted backbone networks. The prevailing network strategies are not capable of coping with the constantly changing network conditions due to incredible traffic growth. Deep learning with the contemporary revolution in the ML area seems to be a feasible method for the network operators to design and take control of their networks in a quicker and independent fashion.

Martinez et al. [11] have dealt with the driving style that plays a predominant role in controlling the vehicle energy and ensuring driving safety. The authors have focused on offering trust for developing innovative driver assistance systems in vehicle automation. It is a study on driving style classification and recognition reviewing a variation of algorithms with specific prominence on ML methods based on present and upcoming trends.

Matousek et al. [12] have focused on the driving behavior that acts as a main issue in ensuring traffic safety. Irregular behavior like exceptionally violent or inert driving can threaten the driver and traffic members. The schemes used for analyzing driving behaviors are based on classification that demand labeled data for normal and irregular behavior. Anomalous driving forms are identified based on outlier detection which does not necessitate such data. In addition to the necessary dataset, prevailing schemes face issues with the varying nature that overlaps with usual behavior.

Gao et al. [13] have dealt with network efficacy maximization problems that are solved by assuming the availability of data that is not locally accessible. Incentive schemes for honest reporting in optimization of network with local information are studied, and the influence of data and incentive on the results is exposed. Wang et al. [14] have proposed Light Detection and Ranging, radar and camera making use of the automation feature in sensors. Nevertheless, due to the inherent restrictions of the sensors, autonomous vehicles are susceptible to taking inaccurate conclusions leading to severe catastrophes. Networking and communication technologies can be used to deal with these scarcities and seem to be more dependable, realistic and effective to stimulate information communication.

Memos et al. [15] have considered Internet of Things (IoT) as the new technology that aims to connect the physical objects to the Internet making a vast universal network of things that is capable of sharing information and finish planned tasks, offering substantial benefits to users and corporates of a Smart City (SC) by using Efficient Algorithm for Media-based Surveillance System (EAMSuS).

Li et al. [16] have dealt with visual object tracing and deep learning as trackers have attained state-of-the-art performance on several benchmarks. Nevertheless, these trackers can scarcely offer better performance at better speed. Siamese Region Proposal Network (Siamese-RPN) an end-to-end trained image is proposed for a wide variety of image sets. Image reconstruction along with objects focusses on feeding into the framework and the complete organization is trained from end-to-end.

Sikander and Anwar [17] have focused on driver fatigue that leads to accidents. Exhaustion based accidents causes increased number of death and leads to more harm when compared to drivers who are watchful. Physical factors include driver qualities, timing, driving period and remote Photo Plethysmo Graphy (rPPG) algorithms that offer better results when incorporated into the reliable framework.

Zhao et al. [18] have dealt with developing smart cities that are based on diverse advanced technologies including ultra-dense networking, mobile edge computing and Software-Defined Networking (SDN). A smart routing scheme is very much needed to deal with increased traffic and massive growth of the complex networking environment.

Wang et al. [19] have focused on the emerging mobile devices to improvise transmission which leads to the amplification of many challenging and computation-intensive mobile applications. The outcomes reveal that in contrast to the classic Open Shortest Path First (OSPF) approach under several states, the suggested Deep Reinforcement Learning based Resource Allocation (DRLRA) offers better results.

Luo et al. [20] have dealt with autonomous driving and many new technologies for automatic cars. Security threats of localization and locating technologies for developing autonomous driving are explored along with some countermeasures. It is seen that the target drivers can be spoofed to arrive at a wrong and dangerous destination, and be tracked down in real time [20].

3 Proposed Trusted Framework for Industrial Connected Vehicles Using Cloud Computing

The proposed trusted framework for industrial connected vehicles using cloud computing is a strong reliable framework of VC for ensuring safety. A three-level security framework is proposed to ensure viability and dependability in VCs. In this framework, security and privacy are ensured by detecting reckless driving involving the following steps (a) VC security evaluation, (b) VC privacy evaluation, (c) Multi-criteria group Decision-Making (DM) Fuzzy set theory modeling, (d) Adaptive evaluation.

3.1 VC Security Evaluation

The proposed trusted framework includes a series of processes that come with well-defined inputs and outcomes. A suitable method for developing trust demands in a particular pattern of VCs is developed. The major processes include estimation, checking for constraint violation and building of framework. The target (proper variables) and instances that are concepts for estimation are explored. The basic concepts of VC security evaluation are to check the condition variable of the target.

3.2 VC Privacy Evaluation

The trusted framework includes three major components to meet the security requirements of a classic VC environment. The privacy evaluation includes the aggregation of the security levels like Security in Cloud Computing (SCC) and Security in Cloud

System (SCS). The security level evaluation is based on the singular perception framework. To receive a reasonable solution, Group Decision Makers (GDMs), the most suitable approach for knowledge synthesis is used. It supports ensuring privacy in the network. The cryptographic trusts are serially provided by Road Traffic Accidents (RTA) and the common trusts are transferred to each RSU in the service regions. Every trust Public Key Cryptography (PKC) is sent to all the vehicles by RSU, but the private trust Secret Key Cryptography (SKC) is sent only to RTA.

3.3 Multi-criteria Group DM

When several components are used to check safety solutions in a VC framework, it is essential to verify all the criteria. Any system cannot work properly by considering all the evaluation criteria and hence the tradeoff that exists among them should be analyzed. Estimating and choosing better safety standards are the goals of the framework. A multi-decision algorithm that feeds the fuzzy set ranking model of the proposed estimation framework is proposed. The Vehicle-to-Road Side Unit (V2R) and Road Side Unit-to-Vehicle (R2V) authentications are provided for ensuring V2V authentication which comprises of the following steps. One RSU and a number of vehicles are considered.

Step 1: The RSU periodically broadcasts information that is used for providing V2R and R2V authentication. Hence, the vehicles within the range of transmission can acquire the information at the RSU (ID_{RSU}, T_{Curr}, PKC_{Trust}, AD, Nonce, $SIG_{RSU}(ID_{RSU}||T)$).

where,

ID_{RSU}—ID of the RSU involved in broadcasting

T_{Curr}—Time stamp

PKC_{Trust}—Public Trust of PKC allotted by the RTA that is used in the present time interval

AD—Advertisement that is the invite of V2R authentication in the ensuing step. Nonce is used for freshness

$SIG_{RSU}(ID_{RSU}||T)$—Identity Based Signature (IBS) for R2V authentication that is produced from the 'ID_{RSU}' and the time stamp 'T_{Curr}'.

Step 2: A vehicle forwards a message to the conforming RSU by using IBS for providing V2R authentication in the following scenarios.

- A vehicle demands a new or updated pseudonym to provide authentication and communication in the VANET
- A vehicle obtains a new 'ID_{RSU}' from the broadcast from RSU.

The vehicle unicasts the current pseudonym to the RSU (ID_{RSU}) which includes P_{Veh}, T_{Curr}, $Join_{Req}$, $SIG_{Veh}(ID_{Veh}||T_{Curr})$.

where,

ID_{RSU}—RSU that is the destination

P_{Veh}—Currently produced pseudonym
$Join_{Req}$—Join request message
$SIG_{Veh}(ID_{Veh}||T)$—Digital signature produced from 'P_{Veh}' and 'T_{Curr}'.

Step 3: On receiving the '$Join_{Req}$' from a vehicle, the RSU confirms the signature and admits it only if the message is authentic. The RSU updates the 'P_{Veh}' and informs the RTA. The RSU produces offline signatures $SIG_{Offline-Veh}(P_{Veh})$ from 'P_{Veh}'. It then broadcasts an allocation message to the vehicles in the area of coverage to support V2V authentication and IBS for R2V authentication. The message incorporates a '$P_{Veh}|SIG_{Offline-Veh}|ID_{RSU}$' set in the form, $SIG_{Offline-Veh}(P_{Veh}) = ID_{RSU}$ including a nonce yet to be combined with the digital signature $SIG_{RSU}(ID_{RSU}||T)$. Here, 'ID_{RSU}' refers to the conforming RSU, where '$P_{Veh}|SIG_{Offline-Veh}|ID_{RSU}$' set is produced. The vehicles in the transmission range of the RSU obtain the message and admit it if the signature is valid. Concerning the suitable set based on verification, the vehicle saves and updates them. Else, drops it.

3.4 Fuzzy Set Theory Modeling

In consideration of the safety and privacy, the estimation for the safety level of industrial CV is difficult and challenging. In this situation, a Decision-Making (DM) tree algorithm is required for every computation. A fuzzy set is preferred as the best solution in many areas. It is used to manage the efficiency of the linguistic values in decision-making algorithms such as high and low values that can be replaced with numeric values between binary set values (1–0) with the help of fuzzy set modeling. This helps in making the decision-making approach more efficient.

3.5 Adaptive Evaluation

Adaptive evaluation schemes deal with the Simple Additive Weighting (SAW) method which is preferred for optimal evaluation. Due to its simplicity and ability to spot the undeveloped spaces of alternatives, SAW is the most familiar DM tree algorithm. Also, V2V authentication is also done in inner RSU which is preferred for protective transmission among vehicles. It receives the '$P_{Veh}|SIG_{Offline-Veh}|ID_{RSU}$' sets for authentication and verification. The acceptor vehicles can make use of the online signature for V2V evidence. If the vehicle is willing to connect and transmit to another vehicle within the range, it uses the SIGnature (SIG) when offline vehicle $Veh(P_{Veh}, T_{Curr}, SIG_{Veh}(SIG_{Offline-Veh}(P_{Veh}||T_{Curr})$ receives the authentication message from the vehicle. Then, the online signature is checked, and the set is stored.

4 Experimental Investigation and Results

The performance of the proposed trusted vehicular evaluation framework is investigated based on simulation experiments conducted using SUMO, an open-source simulator that helps in traffic analysis to attain incessant estimation for specific vehicles with real-world topologies of traffic. In contrast to the existing SVM based framework, the suggested framework identifies reckless driving vehicles. The simulation area for implementing the proposed evaluation framework is identified. The updated data available on the cloud will be considered for simulation. A simulation time of 300 s with a pause time of 20 s is taken.

Figure 1a shows the proposed evaluation framework which makes use of a DM algorithm along with the fuzzy set modeling to provide accurate binary values including time period and type of vehicle based on weight. This proposed system offers 8% better accuracy in contrast to SVM based approach. Figure 1b shows the potential of the proposed trustworthy vehicular evaluation framework in terms of number of passing vehicles and the corresponding time. The RSU based detection of the proposed evaluation framework provides the information about the count of Light Motor Vehicle (LMV), Heavy Motor Vehicle (HMV), Heavy Passenger Motor Vehicle (HPMV) in varying time periods (in hours). The count differs from 15 HMV at 7 PM, 7 HMV at 8 PM, 9 HMV at 9 PM, 10 HMV at 10 PM, 18 HMV at 11 PM. This predominant enhancement is mainly due to the capability of the Fuzzy set modeling in exploring the vehicle counts that helps in RSU based traffic congestion detection.

Figure 2a shows the efficient Intelligent Transportation System (ITS) provided by the cloud server which allows choosing the origin and destination at the specified time. The ITS provides details of accurate level of traffic in various routes along with parameters such as proximity between the vehicles, best route, distance, speed, estimated time to reach and the fuel consumption details. Figure 2b shows the potential of the proposed evaluation framework for location tracking using DM tree. The vehicle location tracking allows us to track the vehicle using vehicle ID.

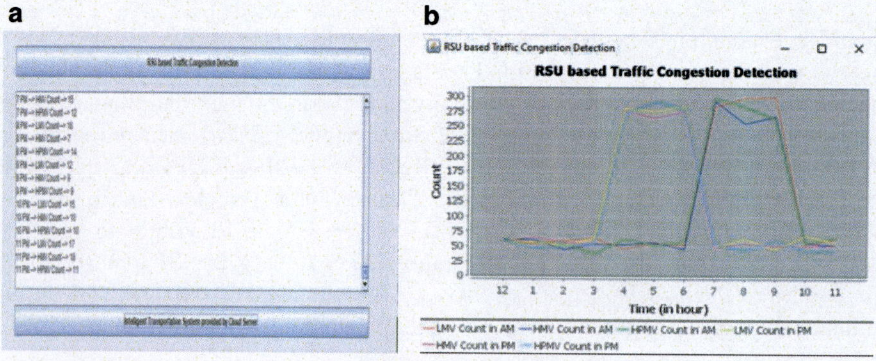

Fig. 1 **a** Fuzzy set theory modeling, **b** RSU based traffic congestion detection

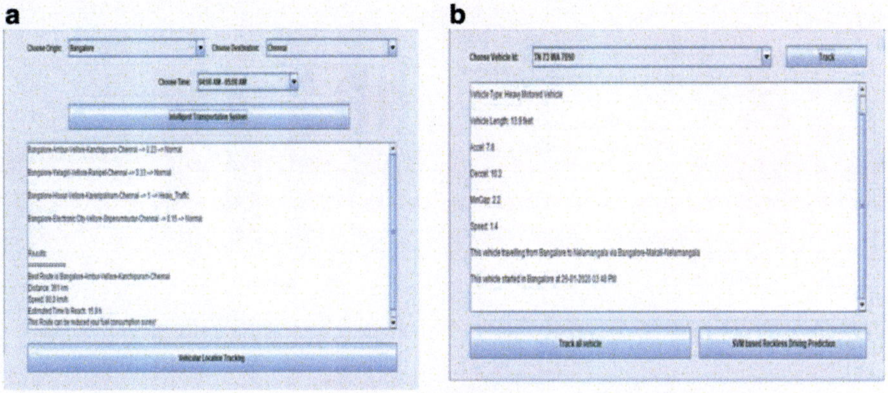

Fig. 2 **a** Intelligent transportation system provided by cloud server, **b** Vehicular location tracking

It also provides information such as vehicle type, vehicle length, type of fuel used, minimum gap and speed.

Figure 3a shows the SVM based adaptive prediction of vehicle details such as RSU and state of the vehicle. It also provides the details of RMSE, Precision, Recall, F-score, Accuracy of the vehicles. If the target information is not considered precisely, less than 50% of the dataset is only evaluated. On the other hand, sometimes SVM is better when compared to DM trees as it segregates the classes with conditions. The DM tree is an effective and optimal method for regression and is also used in classification. Figure 3c shows the DM tree based privacy and reckless driving prediction which provides the details of vehicles such as RSUs and state of vehicles. RMSE, Precision, Recall, F-score, Accuracy of the vehicles are measured. In contrast to SVM, the DM tree based privacy method processes the details more accurately and also provides information about reckless driving vehicles. It is more secure and

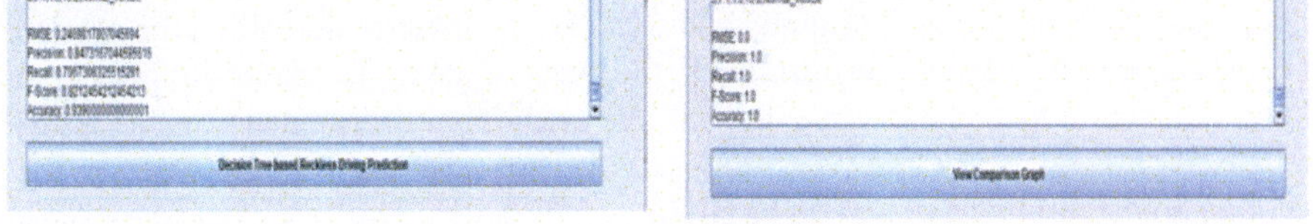

Fig. 3 **a** SVM based adaptive prediction, **b** Decision tree based privacy and reckless driving prediction

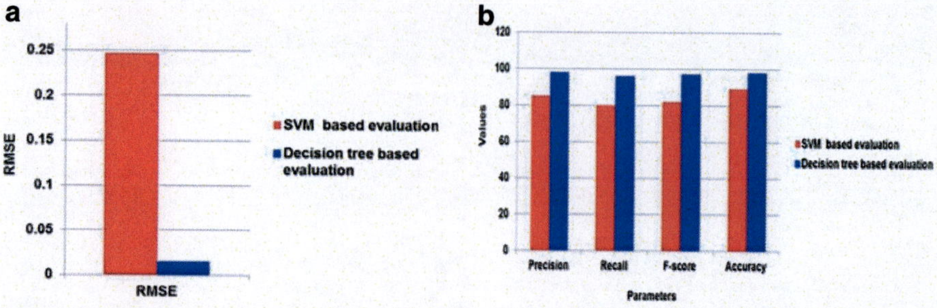

Fig. 4 **a** RMSE, **b** Precision, Recall, F-score and Accuracy of SVM and DM tree based approaches

private when compared to the SVM approach. Reckless driving vehicle authentication gives the vehicle information that is in a bad state for privacy preservation with the ability to trace for non-repudiation. IBS method is used for V2V and R2V authentication.

Figure 4 shows the comparison between the SVM approach and the DM tree approach. When compared to the SVM approach, the DM tree approach is far better and gives an optimal solution. The factors in the DM tree influence it to be better and optimal than the SVM algorithm. One among the factors is the capability to characterize and split the attributes and parameters to a particular class clearly. The estimated time period for calculation in the DM tree is faster than the SVM approach. At the same time, the DM tree provides optimal and accurate solution. DM tree is useful for handling regression based problems. This estimation regarding RMSE in the DM tree approach is 0.01, while the SVM approach involves RMSE of 0.25. The accuracy of the DM tree scheme is 98%, while SVM approach offers an accuracy of 89%.

5 Conclusion

As the impact of reckless driving increases, a secured framework that identifies the impact of dangers is developed to stop vehicles from moving toward wild vehicles. To access the likelihood of accidents, reckless driving is assessed through a security and privacy evaluation framework. By using the observed data from neighboring vehicles, CDPR subsequently identifies reckless vehicles. A three-level framework is designed with present vehicular organizations. Additionally, the information about the motion of reckless vehicles can be transferred. On the basis of the collected information, the cloud users worldwide can assess every vehicle's driving performance by using SVM and DM tree based approach. The proposed evaluation framework by utilizing the DM tree algorithm is simulated using SUMO test system. The proposed framework precisely identifies reckless vehicles and offers better accuracy, security and privacy.

References

1. Hafeez KA, Zhao L, Ma B, Mark JW (2013) Performance analysis and enhancement of the DSRC for VANET's safety applications. IEEE Trans Veh Technol 62(7):3069–3083
2. Al-Sultan S, Al-Bayatti AH, Zedan H (2013) Context-aware driver behavior detection system in intelligent transportation systems. IEEE Trans Veh Technol 62(9):4264–4275
3. Saiprasert C, Pattara-Atikom W (2013, January) Smartphone-enabled dangerous driving report system. In: 2013 46th Hawaii international conference on system sciences. IEEE, pp 1231–1237
4. Codeca L, Frank R, Engel T (2015, December) Luxembourg sumo traffic (lust) scenario: 24 hours of mobility for vehicular networking research. In: 2015 IEEE vehicular networking conference (VNC). IEEE, pp 1–8
5. Zhang X, Yang Z, Sun W, Liu Y, Tang S, Xing K, Mao X (2015) Incentives for mobile crowdsensing: a survey. IEEE Commun Surv Tutorials 18(1):54–67
6. Kato N, Fadlullah ZM, Mao B, Tang F, Akashi O, Inoue T, Mizutani K (2016) The deep learning vision for heterogeneous network traffic control: proposal, challenges, and future perspective. IEEE Wirel Commun 24(3):146–153
7. Luvizon DC, Nassu BT, Minetto R (2016) A video-based system for vehicle speed measurement in urban roadways. IEEE Trans Intell Transp Syst 18(6):1393–1404
8. Yu J, Chen Z, Zhu Y, Chen Y, Kong L, Li M (2016) Fine-grained abnormal driving behaviors detection and identification with smartphones. IEEE Trans Mob Comput 16(8):2198–2212
9. Mahmud SS, Ferreira L, Hoque MS, Tavassoli A (2017) Application of proximal surrogate indicators for safety evaluation: a review of recent developments and research needs. IATSS Res 41(4):153–163
10. Fadlullah ZM, Tang F, Mao B, Kato N, Akashi O, Inoue T, Mizutani K (2017) State-of-the-art deep learning: evolving machine intelligence toward tomorrow's intelligent network traffic control systems. IEEE Commun Surv Tutorials 19(4):2432–2455
11. Martinez CM, Heucke M, Wang FY, Gao B, Cao D (2017) Driving style recognition for intelligent vehicle control and advanced driver assistance: a survey. IEEE Trans Intell Transp Syst 19(3):666–676
12. Matousek M, Yassin M, van der Heijden R, Kargl F (2018, June) Robust detection of anomalous driving behavior. In: 2018 IEEE 87th vehicular technology conference (VTC Spring). IEEE, pp 1–5
13. Gao J, Zhao L, Shen X (2018) Network utility maximization based on an incentive mechanism for truthful reporting of local information. IEEE Trans Veh Technol 67(8):7523–7537
14. Wang J, Liu J, Kato N (2018) Networking and communications in autonomous driving: a survey. IEEE Commun Surv Tutorials 21(2):1243–1274
15. Memos VA, Psannis KE, Ishibashi Y, Kim BG, Gupta BB (2018) An efficient algorithm for media-based surveillance system (EAMSuS) in IoT smart city framework. Fut Gener Comput Syst 83:619–628
16. Li B, Yan J, Wu W, Zhu Z, Hu X (2018) High-performance visual tracking with siamese region proposal network. In: Proceedings of the IEEE conference on computer vision and pattern recognition, pp 8971–8980
17. Sikander G, Anwar S (2018) Driver fatigue detection systems: a review. IEEE Trans Intell Transp Syst 20(6):2339–2352
18. Zhao L, Wang J, Liu J, Kato N (2019) Routing for crowd management in smart cities: a deep reinforcement learning perspective. IEEE Commun Mag 57(4):88–93
19. Wang J, Zhao L, Liu J, Kato N (2019) Smart resource allocation for mobile edge computing: a deep reinforcement learning approach. IEEE Trans Emerg Topics Comput
20. Luo Q, Cao Y, Liu J, Benslimane A (2019) Localization and navigation in autonomous driving: threats and countermeasures. IEEE Wirel Commun 26(4):38–45

Classification of COVID-19 Tweets Using Deep Learning Classifiers

M. Deva Priya, M. Saranya, N. Sharaha, and S. Tamizharasi

Abstract Coronavirus (COVID-19) is a major threat to the country. People get scared on seeing the posts related to COVID-19 on the social media. This paper focusses on this issue and applies Natural Language Processing (NLP) techniques on Sina Weibo data to classify COVID-19-related information into seven categories including warning and advice, notice and action, contribution of money, products or services, emotional support, looking for help, expressing and assessing doubts and counter-rumours based on situational information. Only situational intelligence is of use to the public and the authorities to respond to the outbreak. It is, therefore, necessary to recognise them and comprehend how it is being distributed on social media, so that proper information publication techniques can be updated about the COVID-19 outbreak. The informative efficiency of word embedding is tested and compared with a Bag-of-Words (BoW) approach based on classification accuracy. The initial findings indicate the guaranteed capability of the textual data in the overview of various stages of the outbreak. Emergency response reviews from social media after the epidemic offer useful insights about effective public health conditions. On Sina Weibo, the influential Chinese social media, posts with negative feelings are useful in analysing public issues. About 999,978 randomly chosen COVID-19 based Weibo articles are reviewed from 1 January 2020 to 18 February 2020. From the results, it is evident that the proposed system generates accurate information about COVID-19 so that people need not rely on social media.

M. Deva Priya (✉) · M. Saranya · N. Sharaha · S. Tamizharasi
Department of Computer Science and Engineering, Sri Krishna College of Technology, Coimbatore, Tamilnadu, India
e-mail: m.devapriya@skct.edu.in

M. Saranya
e-mail: 17tucs205@skct.edu.in

N. Sharaha
e-mail: 17tucs217@skct.edu.in

S. Tamizharasi
e-mail: 17tucs244@skct.edu.in

R. P. Mahapatra et al. (eds.), *Proceedings of International Conference on Recent Trends in Computing*, Lecture Notes in Networks and Systems 341,
https://doi.org/10.1007/978-981-16-7118-0_19

Keywords COVID-19 · Medical diagnosis · Decision support systems

1 Introduction

The COVID-19 epidemic in Wuhan, China, is a worldwide public health hazard. It is the reason for 48% number of deaths and is estimated to cause 25 million deaths per year by 2030. In a situation like COVID-19 outbreak, people continue to use social media platforms to provide the necessary facts and exchange their views regarding the situation. The crucial problem in social sensing applications lies in acknowledging the exactness of the reported observations and the reliability of data sources. An emotional-aware truth detection scheme that obviously includes emotional information of human reported data is presented in this paper. The new truth discovery scheme solves a maximum likelihood estimation problem to determine correctness. In this paper, the Italian discussion on COVID-19 crisis on Twitter is considered. The dataset contains tweets from Italy which was most affected by the epidemic before and after the national closure (Decree of the Presidency of the Council of Ministers of 9 March 2020). In particular, sentiment analysis has proven to be a valuable tool for identifying the polarity of texts on social media, more broadly for analysing user-generated content created during a disruptive case. The theory is confirmed by using unique indices, and Random Forest (RF) and K-Nearest Neighbour (KNN) classification models. To the best of our understanding, this is the first attempt that highlights the study of Italian texts from the social media website on the outbreak of COVID-19.

Current studies have not focussed on the interpretation of the situation details. Some studies have classified help demands and aid as situational information but have ignored other kinds of situational information like public views which expose their anxieties, emotional sustenance that show their sympathy to victims. However, it is important to classify situational details in a wide range of ways. For example understanding criticism-related facts may allow the authorities to understand the people's key issues and come up with appropriate responses. Exploring the emotional support may help the authorities learn about the social support behaviours of social media users and make greater use of those voluntary services.

Situational information is a post which provides strategic, actionable information that can allow people to make decisions, educate on how to access specific information from diverse sources, or offer instant post-influence support to those affected by the tragedy. The sharing of this intelligence demands law enforcement and persons associated to know the requirements and control their intentions. Based on the interpretation, this study categorises information linked to COVID-19 as situational information based on the following: (1) warning and advice (2) notice and action (3) Contribution of money, products or services (4) emotional support (5) looking for help (6) expressing and assessing doubts and (7) counter-rumours.

In addition, questioning and criticising information also address the socio-political origins and consequences and responsibilities for the outbreak. Discussions allow people to assure for the authenticity of information or to improve their understanding of the epidemic. Counter-rumour awareness also allows the public to discover the facts and alleviate the uncertainty created by misinformation. Given the fact that exchanging these different categories of information supports the effectiveness of disaster/crisis relief systems, they are classified into situational information.

2 Related Work

Tsang and Larson [1] have stated that US photographs may be demoralised by direct visualisation and via comment processes that have useful qualitative and quantitative functionality. In this sense, both the 2D automated cardiac movement prediction and the related strain measurements are shown to be effective instruments for the diagnosis of cardiovascular disease.

Huang et al. [2] have developed a time-sensitive truth discovery mechanism that clearly includes source receptiveness and claim lifetime as a demanding analytical framework. This scheme deals with the maximum likelihood approximation problem to find the claim accuracy and the source dependability.

Marshall and Wang [3] have developed a framework to deal with an emotion-based truth detection in social sensing applications which has become dominant in Cyber-Physical Systems (CPS) with social sensors to instinctively report observations about the environment. An emotion-based truth detection scheme that obviously includes expressive information in an analytical framework is designed. Maximum possibility estimation to prove claim correctness is performed.

Kwon et al. [4] have analysed the dissemination of misinformation via Twitter and user friendship. They have observed that misinformation spreads by a few false information diffusers who consistently make false claims. They have also mentioned that many of the articles referencing false news include hearing or guessing. They have dealt with collective diffusion patterns of rumours and following the exact modifications in prognostic powers across features. Statistical analysis aids structural and temporal features to segregate rumours from non-rumours over an enduring window, but they are not accessible during the preliminary dissemination stage. User and linguistic features, on the other hand, are available during the early dissemination stage. A rumour classification algorithm is proposed over short- and long-time windows.

Farajtabar et al. [5] have developed a social reinforcement learning scheme to contest the dispersion of fake news. Precisely, an intervention model that promotes the dispersion of true news in a social network to alleviate the influence of fake news is designed. News dispersion is modelled as a Multivariate Hawkes Process (MHP) to make interferences that are studied using policy optimisation. Feature illustration of states that defines mitigation activities and building reward functions that measure efficiency are chosen. The problem of fake news mitigation is mapped into a reinforcement learning framework. A policy iteration scheme is developed that is distinctive to multivariate networked point process with the aim of enhancing the actions for offering better reward under budget restrictions.

Thota et al. [6], as COVID-19 is the principal reason of death in developing countries, it is essential to develop assessment techniques, leading to the analysis and treatment of diseases. There are many approaches that are used to evaluate the fitness. Medical imaging can be used to check the physical activity along with Magnetic Resonance Imaging (MRI) and Ultrasound Imaging (UI). However, due to the eminent accuracy, the UI is suited for rapid sequence of heart. The collected ultrasound pictures offer information that is important for assessing cardiac activity.

Liu and Wu [7] have proposed a model for primary identification of fake news on social media by categorising propagation paths. The propagation path is a multivariate time series wherein the tuples include a vector demonstrating features of a user who are involved in dispersing the news. A time series classifier that includes recurrent and convolutional networks is included which considers the global and local disparities of user features along the path correspondingly to identify bogus news.

Tschiatschek et al. [8] have designed an algorithm that involves Bayesian inference for finding forged news and collectively learns about users' abating accuracy over time. It uses posterior sampling to dynamically establish a trade-off exploitation (choosing news that improves the objective value at a given time) and exploration (choosing news that increases the value of data in learning about users' failing accuracy).

Shu et al. [9] have built real-world datasets to measure users' trust level on false news and choose representative groups of skilled users who are capable of identifying recognising forged news items as false and inexperienced users who are more probable to trust fake news. A relative analysis of explicit and implicit profile features of these user groups which shows their capability to distinguish fake news is done.

Satoh et al. [10] have propounded a solution to identify fake news without considering any other subtasks. It is based on an article's textual and visual features. Precisely, language models are used to learn text characteristics, and image-related features are learnt from VGG-19 that is pre-trained on ImageNet dataset. The Twitter and Weibo datasets are considered.

Shu et al. [11] have designed a FakeNewsTracker that involves a method for perception and identification of misinformation. It will finally obtain information for news items and the relative factors which might allow study on recognition and forecast of misrepresentation using advanced visual elements.

Maakoul et al. [12] have focussed on the issue of fake news related to COVID-19 as a universal pandemic with a substantial influence on several sectors. An aggregation system that detects and analyses fake news associated with COVID-19 epidemic in the Moroccan environment-based datasets from Facebook is proposed.

3 Problem Statement

The widespread propagation of false information is possible through social media platforms. In this context, an increase in false news about COVID-19 is seen on social media, and it has become a major source of concern in the society. They are most common in the comments section. News verification is a critical step in identifying deliberately false information found in posts and comments. The detection of fake news, on the other hand is an uphill task, and interpreting different dialects used in posts and comments is the major challenge. The data collection strategy used in this paper employs a platform for gathering tweets and comments from Twitter. A machine learning-based approach is propounded for detecting false news.

4 Proposed Work

The proposed model ensures tranquillity among the citizens during epidemic situations. The media dataset can be utilised to make characterisation models which can get familiar with the information that is deceitful. Twitter medical dataset about the disease is taken to predict the correct COVID-19 signs and maintain serenity. This system uses deep learning algorithms for prediction.

4.1 Deep Learning for Prediction

Deep learning is the sub-field of Machine Learning (ML) that emphases on information at several levels of participation and definition, and each level contains various processing units for multiple dispensations amid input and output layers. It operates based on the concept of function-based structure, whereby a hierarchy is maintained by lower-level job functions. It extends the primary tasks of Neural Network (NN) which is being applied in the area where Boltzmann machines and autoencoder-decoder techniques are utilised. This approach can be used in image processing, layer-wise pre-training and Natural Language Processing (NLP). Sequential COVID-19 disease assumption is supported in the model where Gated Recurrent Unit (GRU) is utilised to obtain higher precision.

The Recurrent Neural Network (RNN) is considered to be suitable for serial features and sequential data based on analytical study. Long Short-Term Memory (LSTM) was suggested by Hochreiter and Schmidhuber [13]. The result is very promising for sequence-based tasks. GRU, another contemporary LSTM approach is easier than LSTM; however, the effect is very remarkable. Deep learning process includes data collection, data pre-processing (noisy and unrelated data are removed) and data visualisation phases. Further, KNN and RF classification algorithms are used for categorising the posts. The diagram in Fig. 1 provides an outline of the model.

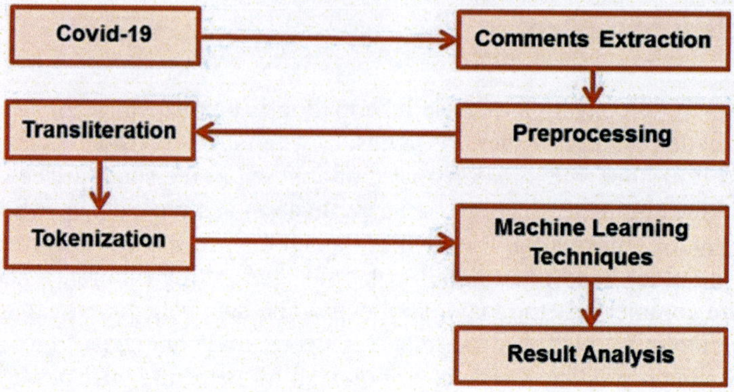

Fig. 1 Block diagram of working model

The proposed system deals with analysing the veracity of information distributed in Twitter. Figure 1 gives an overview of the proposed system. The method comprises of collecting, sorting, categorising public feedback on the post and measuring the percentage of each collection and then concluding whether the post is false in or not. The main functions include data mining, pre-processing, transformation, tokenisation, application of ML algorithms and performance analysis.

4.2 Data Collection

Data collection deals with analysing and crawling data to gain information from diverse data sources in diverse structures and forms. Twitter dataset is used in this work where users connect, discuss about themselves and post about various areas.

4.3 Data Pre-processing

There are various data pre-processing techniques. Data cleaning deals with removing noise and correct discrepancies in the data. Data amalgamation focusses on merging data from diverse sources into an intelligible data store. Data reduction deals with reducing the data size by combining, eradicating redundant features or clustering.

- **Outliers:** If the value of an attribute diverges from the predicted value, it is considered as an outlier. They are treated using an InterQuartile Range (IQR). It detects outliers by defining a range of information in the dataset. It is a better choice if the dataset is nearly symmetric, wherein the median equals its midrange.
- **Missing values:** Many tuples have no values for numerous attributes like COVID-19 affected patient's blood group. This is solved by ignoring the tuples when the

class label is not found in COVID-19 health dataset. This method is efficient only when the tuple includes numerous attributes with omitted values. By leaving the tuple, remaining attribute values in the tuple can be used.

- **Discretisation:** It focusses on reducing the amount of values for constant attributes. This is made possible by breaking the range of continuous attribute into separate intervals. It lessens the time required for building the prediction model and improving the prediction results. From the COVID-19 affected people, the attributes like age, blood group, oxygen level are discretised.
- **Sampling:** The dataset consisting of 17 million tweets is used in this study. The common metric used to assess ML schemes is accuracy. This measure demands that the data is balanced. The prediction problem demands a better rate of recognition of patients who are highly affected. An imbalanced dataset can be converted to a balanced one using two methods. Oversampling the minority class (patients with the disease) and undersampling the majority class. Undersampling and oversampling are used to handle the problem of misbalancing. Synthetic Minority Oversampling TEchnique (SMOTE) is used wherein the minority class is oversampled by generating synthetic samples instead of oversampling with replacement. It chooses minority samples and creates synthetic examples along the same line segment combining some or all KNNs fitting in the minority class.

4.4 Data Classification

Data classification systems are designed to deal with the unstructured or semi-structured data in contrast to database applications which use only structured data.

Prediction is done using K-Nearest Neighbour (KNN) algorithm and Random Forest (RF) algorithms. The social media content is classified into true and fake. KNN and RF supervised ML algorithms are used to solve classification and regression problems.

Regression logistics algorithm is found to be more accurate when compared to Naive Bayes (NB) algorithm. But, it has some drawback; i.e. by default, compilation ends excluding the residual 'responses'. It may be that half of the comments collected are either 'douaa' ads or others, and data extraction from images is difficult as more space and time are consumed [12]. System is improved by adding the ability to verify the information published in the image form by working on text extraction and recognition from images.

K-Nearest Neighbour (KNN)

K-Nearest Neighbours (KNN) is one of the most basic supervised ML algorithms, yet very essential. A supervised ML algorithm is dependent on labelled input data in order to learn a function which is capable of producing an output whenever a new unlabelled data is given as input. KNN is based on labelled data to generate a function that gives the appropriate output in case unlabelled data is given as input. It is accustomed to indicate the label as 1, −1 or 0. Mathematical calculations are not performed as it would be pointless. It is assumed that related objects are in proximity. KNN makes predictions by finding the likeness between the sample taken as input and every training instance. Data can be rescaled by using normalisation when using KNN. The merits of KNN for classification include simple implementation, robust nature based on the search space, as classes need not be linearly separable.

Random Forest (RF)

Random Forest (RF) enables measuring the relative significance of every feature. Prediction is done by examining the symptoms of corona-affected people before and after COVID-19. One of the biggest advantages of RF is its versatility. It is a predictive modelling tool and not a descriptive tool. It can be used for both regression and classification tasks, and it is also easy to view the relative importance it assigns to the input features. RF is also a very handy algorithm because the default hyperparameters are used often to produce a good prediction result. Understanding the hyperparameters is straightforward. One of the biggest problems in ML is overfitting, and it is highly overcome by using the RF classifier. If there are enough trees in the forest, the classifier does not overfit the model. The main limitation is that, increasing the number of trees can make the algorithm too slow and ineffective for real-time predictions. In general, these algorithms are fast to train, but quite slow to create predictions once they are trained. A more accurate prediction requires more trees, which results in a slower model. In most real-world applications, the RF algorithm is fast enough.

```
Algorithm
Input: Posts, comments
Output: Decision
Start
For each post(Post, Comment)
  For each comment
    If Comment=='fake'
      Post[fake]+=1
    If Comment=='real'
      Post[real]+=1
    End if
    If (Post[fake]/ count(Comment)*100) > (Post[real]/count(Comment)*100)
      Conclude Post as 'fake'
    else
      Conclude Post as 'Real'
    End if
  End for
End For
End
```

5 Implementation

Health dataset consists of parameters like name, smoking, overweight, bad cholesterol, hereditary, blood sugar of a COVID-19-affected people. This dataset is used to find the real symptoms of COVID-19. Population dataset includes a classification of 18 years, 18–24 years, 25–44 years, 45–64 years and 65 years and more. Tweet analysis consists of parameters like user id, system id and feedback. This dataset consists of COVID-19-related tweets. Figures 2, 3 and 4 show the feature extraction in health, population and Twitter analysis dataset.

From Fig. 5, it is seen that the number of positive reviews is 799,982 and negative reviews is 199,995 out of 999,977. KNN offers 4, 16, 14 and 3.4% better Precision, Recall, *F*-score and Accuracy when compared to RF algorithm (Fig. 6).

Similarly, KNN offers 1.7 times lesser RMSE when compared to RF (Fig. 7). Null value imputation using classification algorithm is made. KNN is faster in finding

Name	Smoking	Overweight	Heriditary	BadCholostrol	BloodSugar
Gobi	Never	Yes	No	140	60
Ganesh	Current	Yes	Yes	340	160
Gokul	Past	Yes	No	240	200
Karthi	Never	No	No	200	220
Yuvaraj	Current	No	No	123	80
Syril	Never	Yes	No	156	223
Rajesh	Past	Yes	Yes	200	256
Syed	Never	Yes	No	260	360

Fig. 2 Feature extraction of health dataset

Under 18 years	18 to 24 years	25 to 44 years	45 to 64 years	65 years and over
	9.9	29	22.8	13
30.4	9.1	32.5	22.3	5.7
26.6	10	29.5	20.9	13
25.4	9.8	28.1	22.7	14
27.3	9.9	31.6	20.5	10.6
25.6	10	32.6	22.2	9.7
24.7	8	30.3	23.2	13.8
24.8	9.6	30.2	22.4	13
20.1	12.7	33.1	21.9	12.2
22.8	8.3	28.6	22.7	17.6
26.5	10.2	32.4	21.3	9.6
24.4	9.5	29.9	22.9	13.3

Fig. 3 Feature extraction of population dataset

User ID	System ID	Feedback
1	1	Govt plans to raise covid from 7% to 9%...
2	1	covid collection seems to have established
3	3	covid revenue collections for the month ...
4	2	covid revenue collections at Rs 863.18 b...
5	5	Tamilnadu raise day by day covid disease ...
1	4	10 percentage tax on LTCG for covid-19 ...
6	3	Tweets for educational purposes only an...
3	2	Official twitter handle of the GoI for quer...
4	1	Official twitter handle for IT related queri...
7	2	Tweets for educational purposes only an...
9	2	Offline mechanism too for delayed covid r...
11	2	10 percentage tax on LTCG

Fig. 4 Feature extraction of tweet analysis dataset

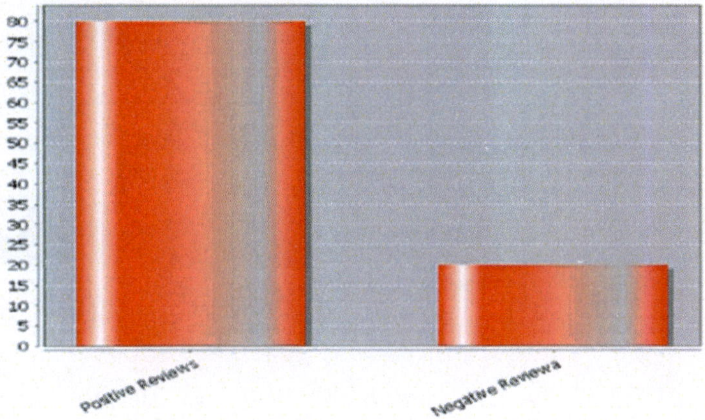

Fig. 5 Classification report based on tweet analysis

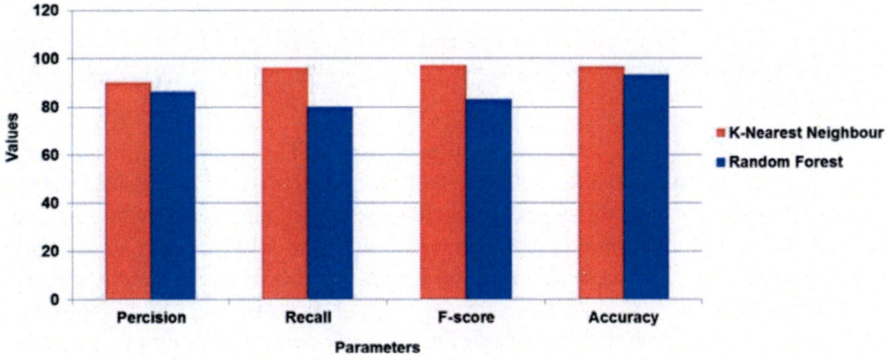

Fig. 6 Precision, recall, *F*-score and accuracy of KNN and RF

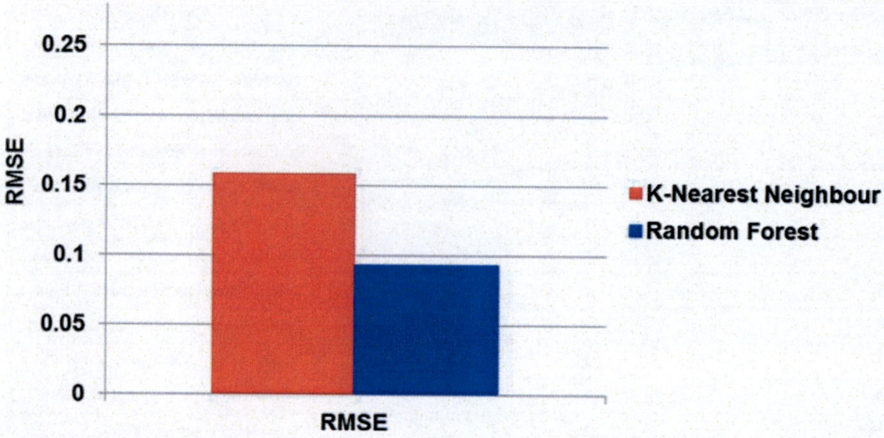

Fig. 7 RMSE of KNN and RF

Fig. 8 Categorisation of posts and comments

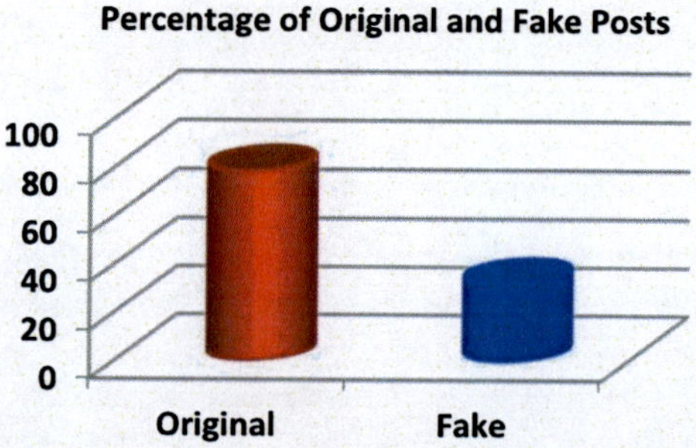

the missing values when compared to the RF algorithms. Therefore, accuracy of KNN algorithm is high.

Figure 8 shows the percentage of posts classified into original and fake.

6 Conclusion

COVID-19 is the key health hazard in human society. This paper summarises the state-of-the-art approaches and methods available to predict this epidemic. Deep learning and ML such as KNN and RF offer accurate medical diagnosis. In this paper, Twitter posts and comments are analysed in order to detect fake news using ML approaches. RF offers 93% Accuracy, whereas KNN offers 96.4% Accuracy. In the future, it is planned to detect the amount of fake news in posts and comments in other languages like Arabic, Chinese, Hindi and process the fake news content in video.

References

1. Tsang A, Larson K (2014) Opinion dynamics of skeptical agents. In: Proceedings of the 2014 international conference on autonomous agents and multi-agent systems, pp 277–284
2. Huang C, Wang D, Chawla N (2015) Towards time-sensitive truth discovery in social sensing applications. In: IEEE 12th international conference on mobile Ad Hoc and sensor systems. IEEE, pp 154–162
3. Marshall J, Wang D (2016) Towards emotional-aware truth discovery in social sensing applications. In: IEEE international conference on smart computing (SMARTCOMP). IEEE, pp 1–8
4. Kwon S, Cha M, Jung K (2017) Rumor detection over varying time windows. PloS one 12(1):e0168344
5. Farajtabar M, Yang J, Ye X, Xu H, Trivedi R, Khalil E, ... Zha H (2017) Fake news mitigation via point process based intervention. In: International conference on machine learning. PMLR, pp 1097–1106
6. Thota A, Tilak P, Ahluwalia S, Lohia N (2018) Fake news detection: a deep learning approach. SMU Data Sci Rev 1(3):10
7. Liu Y, Wu YF (2018) Early detection of fake news on social media through propagation path classification with recurrent and convolutional networks. In: Proceedings of the AAAI conference on artificial intelligence, vol 32, no 1
8. Tschiatschek S, Singla A, Gomez Rodriguez M, Merchant A, Krause A (2018) Fake news detection in social networks via crowd signals. In: Companion proceedings of the web conference, pp 517–524
9. Shu K, Wang S, Liu H (2018) Understanding user profiles on social media for fake news detection. In: IEEE conference on multimedia information processing and retrieval (MIPR). IEEE, pp 430–435
10. Singhal S, Shah RR, Chakraborty T, Kumaraguru P, Satoh SI (2019) Spotfake: a multi-modal framework for fake news detection. In: IEEE fifth international conference on multimedia big data (BigMM). IEEE, pp 39–47
11. Shu K, Mahudeswaran D, Liu H (2019) FakeNewsTracker: a tool for fake news collection, detection, and visualization. Comput Math Organ Theory 25(1):60–71
12. Maakoul O, Boucht S, El Hachimi K, Azzouzi S (2020) Towards evaluating the COVID'19 related fake news problem: case of morocco. In: IEEE 2nd international conference on electronics, control, optimization and computer science (ICECOCS). IEEE, pp 1–6
13. Hochreiter S, Schmidhuber J (1997) Long short-term memory. Neural Comput 9(8):1735–1780

Neural Network-Based Method to Predict PCOS in Women

Asis Kaur Baweja and Megha Gupta

Abstract Polycystic ovary syndrome is a hormonal disorder that causes enlarged ovaries with small cysts on the outer edges. Due to the formation of numerous collections of fluids (follicles) on ovaries, they fail to release eggs regularly. This syndrome is common among women of reproductive age. The major issue with this is that a majority of women who have this condition are not aware about it due to lack of awareness and understanding of this syndrome. The traditional methods used for its detection are time-consuming, unreliable as well as costly. In order to rectify this problem, this paper presents an ideal model of neural network that predicts PCOS with an accuracy of 93% on the basis of some rudimentary features. This predictor is free of cost and gives instant results. Machine learning techniques like K-nearest neighbors and logistic regression have also been used in the past year papers to predict this syndrome. The features used to train these model include the results of several blood tests and ultrasound that are hard to procure. Once analyzed, these are enough for detection of PCOS, thereby making the inclusion of other features as well as the design of predictor highly dispensable. The neural network-based model presented in this paper not only provides better results but also holds the potential of eliminating costly as well as tedious medical steps like ultrasound and blood test, thereby making the detection process more feasible.

Keywords Neural network · PCOS · MLP

1 Introduction

Polycystic ovary syndrome (PCOS) [1] is a syndrome that only affects women and is seldom characterized by an imbalance of sex hormones. The precise cause of

A. K. Baweja (✉)
Department of CSE, SRM KTR, Chennai, India
e-mail: ab8161@srmist.edu.in

M. Gupta
Department of CS, MSCW, University of Delhi, Delhi, India

R. P. Mahapatra et al. (eds.), *Proceedings of International Conference on Recent Trends in Computing*, Lecture Notes in Networks and Systems 341,
https://doi.org/10.1007/978-981-16-7118-0_20

this syndrome is not known, but as mentioned in Fig. 1, plethora of symptoms and repercussions are entailed with it. Irregular periods, excessive weight gain, acne, hirsutism are some of its major recurrent symptoms. If not taken care of at the right stage, this can lead to long-term complications like depression, anxiety, type two diabetes, cardiovascular diseases, breast cancer, ovarian cancer as well as trouble in conceiving.

The most vital problem with PCOS today is that it has a drastically high number of undetected patients [2] (roundabout 75% non-detectability rate). Women having frequent access to doctors keep paying visits to get the comorbidities and symptoms of this syndrome treated, yet being unaware of its presence throughout their lives. Studies show that 20% of women who have miscarriages are due to PCOS [3], and every one in five women in India happens to have this syndrome. Detecting PCOS is a lengthy process which involves ultrasound and blood tests, due to the high cost and lack of awareness only few women bother taking these tests. The most eminent bottleneck in the traditional methods of PCOS detection is: The process is time-consuming, costly as well as the results obtained are not reliable. The model presented in this paper aims to eliminate all these issues.

Artificial neural network [4] is a collection of nodes that are mapped on to one another with the help of mathematical functions. The idea behind it is to create an artificial system that perceives information akin to human brain that functions with help of neurons. As neural networks are emulators of human brains, they can tackle high-dimensionality classification problems with ease. In this work, to solve the rectification problem, a multilayer perception (MLP) feedforward artificial neural

Fig.1 Symptoms of PCOS

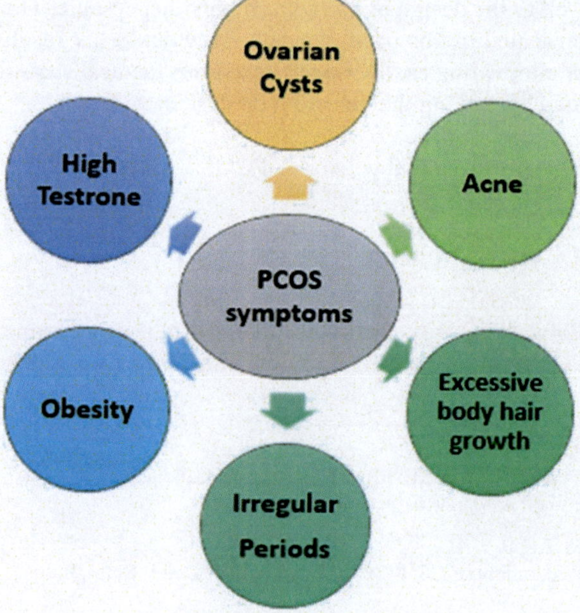

network [5] is used. What distinguishes MLP from normal linear preceptors is its usage of a nonlinear activation function in every neuron other than the input nodes. This makes it capable of classifying data that is otherwise not linearly separable.

2 Literature Survey

In work [2], an intricate study has been performed that highlights the extent and severity of this condition. According to statistics, 30% infertility cases occur due to PCOS although the etiology of the syndrome cannot be determined. In [3], author has discussed about how the lack of diagnostic criteria leads to a high degree of variability as well as uncertainty in detection.

In polycystic ovarian syndrome: diagnosis and management [6], it is clearly stated that the traditional methods of PCOS detection are incompetent. Women with irregular cycles are diagnosed with this even though 20% women who have PCOS have normal cycles. When a patient is clinically evaluated for the possibility of PCOS, it is important to search for signs of IR. Upper-body obesity is a key component of the IR syndrome. However, obesity is not required for the diagnosis of PCOS, only 35–50% of PCOS patients are obese. Pelvic ultrasonography is helpful in evaluation as well, but polycystic ovaries are not specific for PCOS with over 20% of "normal" women having this finding.

A cross-sectional study of polycystic ovarian syndrome among adolescent and young girls in Mumbai, India, [7] states that 10.7% PCOS-tested patients had high androgen level, hyperinsulinemia (serum insulin >15 μlU/mL) was present among 19.2% of diagnosed PCOS cases, and 71.8% patients were not obese. From this, we can reason that none of these tests are 100% reliable, and there is a severe need of obliterating traditional methods of detection.

İn work [8], author has mentioned that PCOS is now viewed as an eminent metabolic and reproductive disorder which leads to substantial risk of acquiring type two diabetes. It is a genetic disease which is complex in nature; after replicating and mapping several PCOS genetic susceptibility loci, it has been insinuated that this syndrome is an ancient trait. İn [9], PCOS has been projected as morphological or predominantly biochemical disorder. Women seeking help from health care professionals to resolve issues of amenorrhea, excessive hair growth, obesity, acne, and infertility often receive a diagnosis of PCOS, making it the most common endocrine abnormality among women of reproductive age in the U.S.

The binary classification problem of determining whether one is having PCOS or not is tackled using two approaches [10]. A comparative study is presented after applying K-nearest neighbor and logistic regression to solve the problem. Best results were obtained by logistic regression which gave the accuracy of 92 after inclusion of features like number of follicles which cannot be determined without medical assistance.

After analyzing these papers, conclusion can be drawn that PCOS is an issue that has been neglected for long and now requires undeviated attention; thus, turning a blind eye on it is not advisable.

3 Proposed Approach

3.1 Dataset Description and Feature Selection

Dataset [polycystic ovary syndrome (PCOS)] procured from Kaggle [8] was used to test and train the proposed neural network model. It consists information of 540 patients from 10 hospitals of Kerala, India. The dataset consisted of 39 features which are often associated and are believed to be either symptoms or causes of this syndrome. Some of the prominent features out of these 39 are the ones that can only be detected through medical assistance; for instance, detection of follicles on ovaries takes place with help of ultrasound. Similarly, anti-Mullerian hormone test (AMH test) which checks the ability of producing those eggs in women that holds the potential of getting fertilized; although an eminent feature, this can only be tested by taking a patient's blood sample.

Women only undergo the hassle of taking these tests when they have trouble of conceiving; also, once these tests are done, a doctor can easily take a look at them and without much analysis can tell if a person has PCOS or not; hence, the problem of lack of awareness as well as detection remains unsolved; moreover, it makes the predictors using this features for detection quite redundant. In the past research papers, people have come up with efficient methods that predict PCOS with remarkable accuracy on the basis of these features [10].

In this work, these features have been knowingly excluded and toiled and succeeded in creating a model that not only gives better accuracy but predicts this syndrome on the basis of elementary binary questions which any woman is capable of answering on her own.

A questioner is built out of these features which will require the user to answer some simple questions like their age, if their menstrual cycle is regular or not, if they have excessive hair fall, ten such questions in totality. This questionnaire will be connected to the proposed neural network model which will immediately detect the probability of a woman having PCOS. This approach will solve the problem of generating awareness because of the easy accessibility of this form and prompt detection, eliminating the tedious intermediator medical steps.

Figure 2 provides a complete list of the 10 selected features out of the 39 given features of the data, mean, standard deviation, count, along with other necessary statistical information. It can be seen from the figure that there is no missing information in the data. This implies there is no need of dropping any row from the dataset.

	count	mean	std	min	25%	50%	75%	max
PCOS	540.0	0.366667	0.482341	0.000000	0.000000	0.000000	1.000000	1.0
Age (yrs)	540.0	0.652508	0.119121	0.020833	0.562500	0.645833	0.729167	1.0
BMI	540.0	0.089762	0.042012	0.048246	0.078572	0.087469	0.096244	1.0
Cycle(R/I)	540.0	0.648148	0.228523	0.500000	0.500000	0.500000	1.000000	1.0
Weight gain(Y/N)	540.0	0.411111	0.492492	0.000000	0.000000	0.000000	1.000000	1.0
hairgrowth(Y/N)	540.0	0.344444	0.475627	0.000000	0.000000	0.000000	1.000000	1.0
Skin darkening (Y/N)	540.0	0.350000	0.477412	0.000000	0.000000	0.000000	1.000000	1.0
Hair loss(Y/N)	540.0	0.427778	0.495215	0.000000	0.000000	0.000000	1.000000	1.0
Pimples(Y/N)	540.0	0.388889	0.487950	0.000000	0.000000	0.000000	1.000000	1.0
Fast food (Y/N)	540.0	0.414815	0.493147	0.000000	0.000000	0.000000	1.000000	1.0
Reg.Exercise(Y/N)	540.0	0.246296	0.431253	0.000000	0.000000	0.000000	0.000000	1.0

Fig. 2 List of selected features

Figures 3 and 4 help us to visualize the data properly. A heat map is square matrix that denotes the co-relativity or similarity between two features. From Fig. 3, it is evident which features play a major role in determining PCOS. The first row of the heat map, being the most important, tells co-relativity of this syndrome with all other features. Higher color intensity corresponds to higher rate of similarity. Thus, it can be concluded that consumption of fast food, hair fall, presence of pimples, darkened

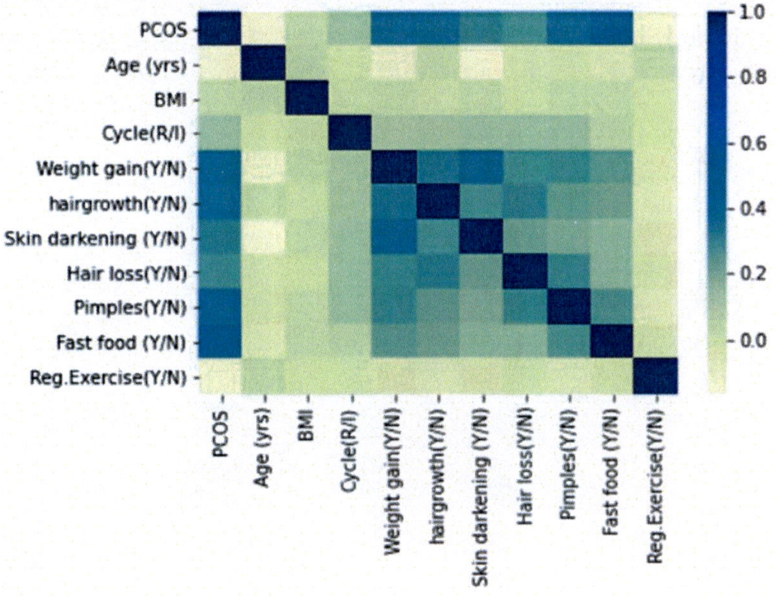

Fig. 3 Features heat map

Fig. 4 PCOS count

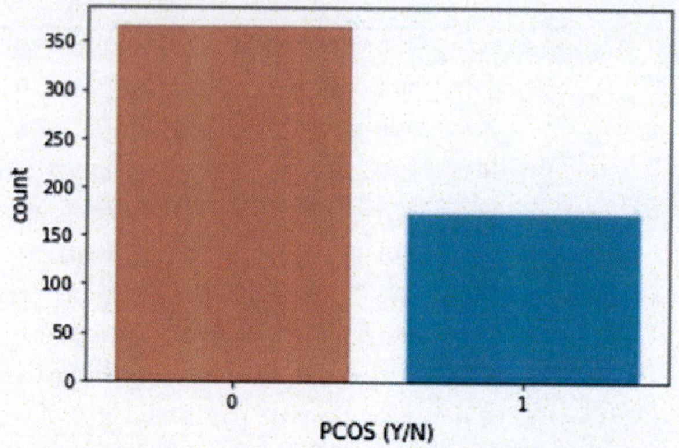

skin patches are more deterministic features than others. The distribution of weights and biases at the time of training of the model takes place in accordance with the similarity rate of the features.

Figure 4 tells us that around 350 people did not have PCOS, while 150 from the tested gave positive result. From this, it can be inferred that dataset is not skewed, and without any further manipulations, it can be used to train any model.

3.2 *MLP Feedforward Artificial Neural Network*

The architecture of the neural network used to train the proposed model is shown in Fig. 5. It leads to the optimum results. The neural network consists of one input

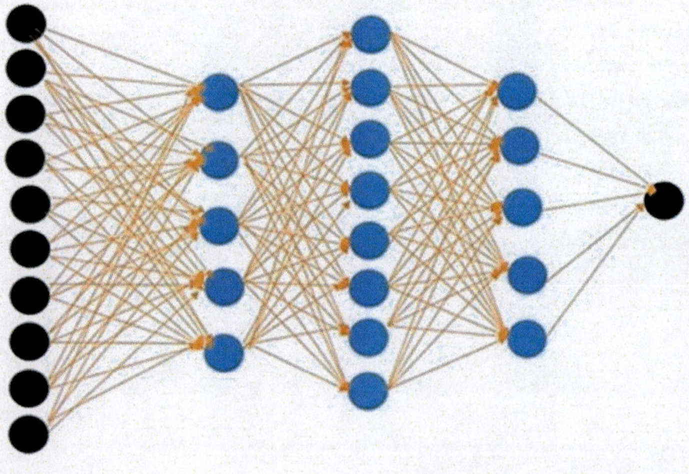

Fig. 5 ANN architecture

layer with ten nodes, three hidden layers, each constituting of five, eight, and five neurons, respectively, followed by a single output neuron. All the layers used in the neural network are dense in nature, i.e., every neuron in a particular layer maps to all the neurons present in consecutive layer.

It is necessary to choose the number of hidden layers as well as number of neurons in each layer with extreme care in order to obtain accurate results. On one hand, there is not any need of introducing hidden layers incase the data is linearly separable, while, on the other hand, excessive number of hidden layers increases the complexity of the model making it more prone to over fitting. The next daunting task is to select the number of neurons in hidden layer. Although there is no set criteria of selecting number of neurons for each layer, it is ideal to keep the number between the sizes of input as well as output layer. One can keep on decreasing the number of neurons in subsequent layers to get more and more close to the target class. After keeping the above-mentioned factors in mind and after a series of trial and error, the proposed approach has come up with the above architecture.

$$y_i = f\left(\sum w_{ij}x_i + b\right) \tag{1}$$

In accordance to Eq. (1), each neuron receives input from neurons of previous layer and further ends up calculating the output signal on the basis of the same. In order to optimize the cost function, the weights and bias in Eq. (1) keep on updating through backpropagation.

A neural network which uses $f(x) = x$ as an activation function and is linear in nature will not add any nonlinearity to the network thereby making the existence of multilayers and neurons redundant. In order to introduce nonlinearity to any network, selecting the correct activation function is very crucial. The proposed model did not use the sigmoid activation function which maps every input to a value that lies between 0 and 1, because at the time of backpropagation, the weights are updated with the help of derivative of sigmoid function. Such small derivative will give rise to vanishing gradient problem. In order overcome this situation, the proposed model has used ReLU as the activation function [11] which is:

$$f(x) = \max(0, x)$$

whose derivative will either be 0 or be 1.

The next step in building the model is selecting an optimizer that will minimize the lose function by updating weights and biases. For the proposed model, Adam optimizer [12] has been selected as it combines the advantages of RMSprop as well as stochastic gradient decent with momentum. v_t and s_t calculated using Eqs. (2) and (3), respectively, are the estimates of first and second moment of the gradient. The weights used to train the proposed model are updated using Eqs. (4) and (5).

$$v_t = \beta_1 * v_{t-1} - (1 - \beta_1) * g_t \tag{2}$$

$$s_t = \beta_2 * s_{t-1} - (1 - \beta_2) * g_t^2 \quad (3)$$

$$\Delta\omega_t = -\eta \frac{v_t}{\sqrt{s_t + \epsilon}} * g_t \quad (4)$$

$$\omega_{t+1} = \omega_t + \Delta\omega_t \quad (5)$$

η initial learning rate
g_t gradient at time t along ω_j
v_t exponential average of gradients along ω_j
s_t exponential average of squares of gradients along ω_j
β_1, β_2 hyperparameters.

4 Results and Discussions

The following results have been obtained using MLP classifier of Scikit library in Python Jupyter notebook.

In this model, following parameters values are used (Table 1):

Dataset is split into two parts; 80% was used to train the model, while 20% was used to test it. For result, parameters used are precision, recall, support, and $F1$ score.

Precision can be seen as the measure of exactness of a classifier. Precision is the ratio of true positive to the sum of true and false positive.

$$\text{Precision} = \frac{\text{True positive}}{\text{True positive} + \text{False positive}} \quad (6)$$

Recall shows us the ability of a classifier to correctly identify all positive instances. For all instances that were actually positive, what percent was classified correctly? The answer to this will give us recall. In simpler terms, recall is defined as the ratio of true positive to the sum of true positive and false negative.

$$\text{Recall} = \frac{\text{True positive}}{(\text{True positive} + \text{False negative})} \quad (7)$$

Support is the number of actual occurrences of the class in a specified model.

Table 1 Parameters used in results

Parameters	Value
η: initial learning rate	0.001
$\beta1$	0.9
$\beta2$	0.999
ϵ	1e−08

Fig. 6 Results

	precision	recall	f1-score	support
0	0.91	0.97	0.94	64
1	0.95	0.86	0.90	44
accuracy			0.93	108
macro avg	0.93	0.92	0.92	108
weighted avg	0.93	0.93	0.93	108

One of the best methods for comparing classification models is through $F1$-score. $F1$-score is the harmonic mean of precision and recall, i.e.,

$$F1 = \frac{2 * (\text{Precision} * \text{Recall})}{(\text{Precision} + \text{Recall})} \tag{8}$$

Macro-average is not a weight average and is calculated by the following formula:

$$\text{Macro - average} = F1_{\text{class1}} + F1_{\text{class2}} \cdots + F1_{\text{class}N} \tag{9}$$

$$\text{Weighted average} = F1_{\text{class1}} * W1 + F1_{\text{class2}} * W2 \cdots + F1_{\text{class}N} * WN \tag{10}$$

The weight in the above formula depends on number of true labels of each class.

It is evident from Fig. 6 that proposed model has performed satisfactorily on all above parameters. Model was trained successfully, and accuracy of 93% was obtained. Cumulative $F1$-score of 92, precision of 93, and recall 91.5 were also obtained.

5 Conclusion and Future Work

Polycystic ovarian syndrome is not a disease; it is a condition. Although this has no ultimate cure, its early detection along with few healthy changes in life style can help in overcoming sever repercussions. The proposed model is an attempt to solve this persisting problem in India by generating awareness through a test that is easily assessable, does not include any medical hassle, and can therefore be taken by every women. This work is providing a solution to polycystic ovary syndrome problem by building a PCOS detector with help of a neural network that predicts the probability of a woman having this with 93% accuracy on the basis of some rudimentary features. Due to the simplicity of the proposed approach, it is feasible to deploy this model into a real-time application, and the only cost associated with this project will be that of deployment and maintenance of the software. This one-time cost when compared to the money that each patient spends in acquiring blood test, insulin test, follicle detection test, and numerous other unreliable medical tests is very minimal.

In the future, this deep learning model will be deployed and made accessible in form of a Web site for easy usage.

References

1. Rotterdam ESHRE/ASRM-sponsored PCOS consensus workshop group (2004) Revised 2003 consensus on diagnostic criteria and long-term health risks related to polycystic ovary syndrome (PCOS). Hum Reprod 19(1):41–47
2. Barthelmess EK, Naz RK (2014) Polycystic ovary syndrome: current status and future perspective. J Front Biosci-Elite 6(1):104–119
3. Wolf W, Wattick R, Kinkade O, Olfert M (2018) Geographical prevalence of polycystic ovary syndrome as determined by region and race/ethnicity. Int J Environ Res Public Health 15:2589
4. Mishra M, Srivastava M (2014) A view of artificial neural network. In: International conference on advances in engineering and technology research, Unnao, India, pp 1–3
5. Bebis G, Georgiopoulos M (1994) Feed-forward neural networks. IEEE Potentials 13(4):27–31
6. Sheehan MT (2004) Polycystic ovarian syndrome: diagnosis and management. Clin Med Res 2(1):13–27. https://doi.org/10.3121/cmr.2.1.1
7. Joshi B, Mukherjee S, Patil A, Purandare A, Chauhan S, Vaidya R (2014) A cross-sectional study of polycystic ovarian syndrome among adolescent and young girls in Mumbai, India. Indian J Endocr Metab 18(3):317–324. https://doi.org/10.4103/2230-8210.131162
8. Polycystic ovary syndrome (PCOS) dataset, kaagle.com. https://www.kaggle.com/prasoonkottarathil/polycystic-ovary-syndrome-pcos
9. Diamanti-Kandarakis E, Dunaif A (2012) Insulin resistance and the polycystic ovary syndrome revisited: an update on mechanisms and implications. Endocr Rev 33(6):981–1030
10. Tanwani N (2020) Detecting PCOS using machine leraning. Int J Mod Trends Eng Sci 7(1):15–20
11. Nwankpa C, Ijomah W, Gachagan A, Marshall S (2020) Activation functions: comparison of trends in practice and research for deep learning
12. Kingma D, Ba J (2017) Adam: a method for stochastic optimization. In: International conference on learning representations. arXiv (2017)

Integrated IoT Blockchain-Based Smart Agriculture System

M. Deva Priya, P. Anantha Prabha, K. Gokulakrishnan, A. Joe Raymond, and T. Karthickraja

Abstract Smart irrigation and automated crop field cultivation are innovative applications in the field of IoT. In this paper, an IoT-based agricultural information monitoring system involving IoT devices is designed. The controller unit collects data related to crops by connecting with temperature, humidity and soil dampness sensors. It is customized with limit estimations of temperature and moisture content to improve the agricultural yields. Challenges faced in the design of such architectures are addressed in this paper. The data collected from the sensors are transferred to the ThingSpeak cloud platform, where it is processed. The network is trained using Linear Regression (LR) model to guide the farmers for increased yield in the upcoming years.

Keywords Internet of things · Smart agriculture · Arduino · Cloud · Linear regression

1 Introduction

Agriculture is the backbone of our nation. Farmers are in need of appropriate information related to cultivation and agriculture-related activities making it more effective. As diverse technologies are available, it is predominant to apply it to the field of

M. Deva Priya · P. Anantha Prabha (✉) · K. Gokulakrishnan · A. Joe Raymond · T. Karthickraja
Department of Computer Science and Engineering, Sri Krishna College of Technology, Coimbatore, Tamilnadu, India

M. Deva Priya
e-mail: m.devapriya@skct.edu.in

K. Gokulakrishnan
e-mail: 17tucs039@skct.edu.in

A. Joe Raymond
e-mail: 17tucs054@skct.edu.in

T. Karthickraja
e-mail: 17tucs056@skct.edu.in

R. P. Mahapatra et al. (eds.), *Proceedings of International Conference on Recent Trends in Computing*, Lecture Notes in Networks and Systems 341,
https://doi.org/10.1007/978-981-16-7118-0_21

agriculture. IoT plays dominant role in smart agriculture. Sensors gather information from agriculture fields. Farmers look into soil preparation, and the assumptions are made to cultivate based on the efficacy of the parcel of land to grow diverse types of crops. They find it difficult to forecast the moisture in soil, water level and predominantly environmental conditions that are essential for cultivating crops and maintaining the farm. It takes more time to examine these records, and hence to overcome this, it is essential to act based on the climatic conditions to maintain the crop yield. Nowadays, farmers apply different technologies to improve the crop yield and maintain the field. Among the technologies developed for monitoring the agricultural field, the user can view the data and maintain the field by regular checking using IoT-based system. Farmers should be capable of checking the live changes occurring in the field like soil moisture, humidity records and temperature of the environment for irrigation. More advanced techniques like storage of data, regular monitoring of the atmospheric conditions to improve the level of maintenance and productivity in the field and to retrieve the data by using storage techniques are necessary. To implement these kinds of techniques, a system is designed.

Many farm-related development activities are based on assumptions. Atomization and dynamic nature of atmospheric conditions are becoming more dominating to achieve this goal. Utilization of coherent systems in the field of agribusiness can bring out essential changes leading to better yields offering better profit by incorporating diverse developing strategies [1]. A remote sensor-based network that gathers data from diverse sensors deployed in the field and forwards it to the end-user is proposed. A coordinated framework that guarantees better efficiency and yield by observing all phases of deployment, design, and gathering is needed [2]. A smart cultivation system that is based on mechanization and IoT is designed. Efficient horticulture, smart watering system, soil and greenhouse monitoring, and cultivation measures in terms of yields are some of the key applications [3]. IoT demands features like interoperability, heterogeneity, memory optimized hardware components and offers security. In agribusiness, IoT finds its place in diverse applications. For instance, it can be used to analyze the soil supplements for envisioning the suitable crop that would increase the yield. A smart watering system can be designed based on soil dampness and environmental conditions [4].

Wireless Sensor Networks (WSNs) enable monitoring of greenhouse parameters in agriculture. As rain water is not evenly distributed, it is very vital for farmers to observe and support uniform dissemination of water to crops based on the type of crops. The parameters demand a complete analysis to select the right method. With the development in WSN technologies and smaller sensor devices, it is likely to use them for monitoring and controlling the environment parameters.

IoT system helps in fulfilling the basic requirements of the end-user who is engaged in monitoring crops in the field. Advanced techniques like automation in monitoring and precision of field management are to be incorporated. Here, data collected from sensors are stored on the cloud and secured using blockchain technology. For example, distributed data, monitoring records of field and more related data can be accessed from anywhere, anytime. This kind of technique mainly aids in live checking and storing of data.

2 Literature Survey

Gutiérrez et al. [5] have focused on water distribution for improving agricultural yields. An algorithm is designed to deal with the estimations of temperature and soil dampness and sent to a microcontroller-based gateway to regulate the water quantity. The framework is controlled by photovoltaic boards and supports duplex communication link that depends on a cellular-Internet interface permitting data monitoring and irrigation through a web page. Due to the energy independence and reduced cost, the system is capable of using it in water inadequate areas.

Devi and Kumari [6] have dealt with the fieldbus concept, where data transmission is mostly controlled by appropriate wired communication, which can be interchanged with a hybrid system including both wired and wireless communication to enjoy both and improve system performance and throughput. ZigBee protocol (IEEE 802.15.4) is used. Ethernet IEEE 802.3 is used to monitor the atmospheric conditions. Partial Root Zone Drying Process (PRZDP) is executed to conserve water to the maximum extent possible. Interaction with the farmers gives them knowledge about diverse techniques and enables them to implement it efficiently in farms to get better yield.

Shakthipriya [7] has proposed a WSN-based solution for agriculture, wherein sensor motes with numerous external sensors specifically leaf and soil moisture, soil pH and pressure sensors are employed. The mote activates the water sprinkler during water insufficiency. The value of soil pH sensor is forwarded to the Base Station (BS), which in turn informs the farmer by sending a Short Messaging Service (SMS) through a Global System for Mobile communications (GSM) modem. The farmer can choose the essential fertilizer and crop for the ensuing season. Rice crop observation module using WSN is propounded to afford a helping hand to farmers in real-time, enabling smart agriculture and consequently improving crop production.

Getu and Attia [8] have designed an electronic system for automatic monitoring of water pumps used in fields based on the level of soil moisture. The perceived signal from the sensor that measures the soil moisture is handled by a conditional comparator circuit conforming to various levels of soil moisture. The logic circuit with the output signals is used to trigger relays that take control of the motors circuit used for water pumping. The motor speed is varied based on the level of soil moisture, which is high when the soil is dry. A timer circuit is designed based on the anticipated watering time to determine the period for which water is to be pumped.

Channe et al. [9] have designed a multi-disciplinary model for smart agriculture involving IoT, sensors, cloud computing, mobile computing and big data analysis. The details including farmer details, soil properties, vendors and marketing agencies, e-governance schemes and existing environment are stored on the agro cloud storage. Big data analysis is carried out for fertilizer demands, crop sequence investigation, whole production and present stock and market necessities. The proposed model is focused on increasing the production and cost control of products.

Patil and Kale [10] have dealt with the climatic changes and rainfall. Smart agriculture is based on information technology combined with IoT. This Remote Monitoring

System (RMS) is designed to gather real-time data that offers easy access for facilities through SMS and recommendations on weather outline, crop to be cultivated, etc.,

Prathibha et al. [11] have dealt with the integration of IoT and agriculture. Observing environmental features is the foremost factor to increase the yield of crops. The temperature and humidity are monitored using CC3200 single chip. Camera is interfaced with CC3200 to take images and send them through MMS to farmers using Wi-Fi.

Suciu et al. [12] have focused on the increasing need for quality and quantity in the food sector. Empirical measurements from IoT devices using remote telemetry applications are dealt. Big data and decentralized cloud can deal with the agricultural demands. The architecture uses Platform as a Service (PaaS), ADCON Remote Telemetry Units (RTUs), Software as a Service (SaaS) as signified by Grafana and influxDB database.

Demestichas et al. [13] have presented an outline of the prevailing and possible threats in agriculture. The progress of ICT solutions and how they can be used are discussed. A wide review on the usage of ICT in agriculture on the related emergent threats and susceptibilities are discussed. The ICT innovations, methods, aids, threats and alleviation measures are discussed by focusing on the likely impacts they could have on agriculture.

Vangala et al. [14] have focused on IoT and blockchain technologies that are the speedily developing fields involved in improving the state of food chain today. A detailed review to examine the schemes that offer security using blockchain is done. General blockchain-based security design is proposed. A thorough cost-related study is done along with a comparative analysis to know the drawbacks in the prevailing research. The security goals are also listed.

3 Blockchain Technology

Some predominant features of blockchain are discussed below [15].

- **Decentralization:** In a distributed system, there exists numerous trusted agencies which collaborate to preserve data in a blockchain [16]
- **Persistence:** Data is resilient to tampers as data are endorsed by numerous nodes and duplicated at all nodes. Any variation in a copy can be recognized when the ledger copies vary. They are made tamper-proof by forbidding removal of blocks.
- **Anonymity:** Each node can have several pseudonyms or addresses to guarantee privacy conservation of transactions [17].
- **Auditability:** To validate a transaction, a user may take a node and find the transactions. The transactions are validated before recording them during mining. This verification guarantees traceability, and non-denial of transactions is also attained [18].

The challenges confronted in blockchain are listed below.

- **Scalability:** Blocks can be appended to a blockchain but not deleted; the quantity of space is the major challenge. Numerous solutions include restraining the number of transactions handled per unit time, the trade-off that exists between small and large blocks, trivial clients and segregated blocks [19, 20].
- **Privacy Leakage:** The transactions are openly accessible, and hence transactional privacy is not ensured [21, 22]. Diverse techniques are designed to connect users with their pseudonyms [23, 24].
- **Selfish Mining:** They are vulnerable to conspiracy from nodes if more than 50% of nodes are not true. Such nodes would append blocks to the chain inversing the present transaction that is formerly endorsed [25, 26].

4 Proposed System

To improve the efficiency of the product thereby supporting the farmer and the country, the improvement which assesses the idea of gathering data and giving suggestions is incorporated. IoT and cloud computing in agriculture result in improved production of crops by regulating the cost, observing performance and maintenance, thereby helping the farmers. The blockchain concept of storing data is used for managing crop related details, where the real-time sensed data is coupled with IoT and cloud computing technologies helping to build a sustainable smart agriculture. An IoT network involving less power and cost for smart agriculture is developed. It is seen that the lifetime of the network is extended due to less power consumption. The proposed system helps in collection of data based on conditions like environment, temperature and efficiency of soil. Remote sensor frameworks are used for checking the farm conditions, and small-scale controllers are used to control and maintain the productivity of the field. The data are secured using blockchain technology, and only authenticated members can view the values. The automated systems enable viewing and gathering of data, thus diminishing human error and expenses to a greater extent. It efficiently guides the farmer to change crop cycles. Dampness, temperature, soil moisture and location are sent to the mobile application of the client. If the water content in the soil is less than the threshold, then an alert is sent and the motor is turned on.

The proposed architecture is shown in Fig. 1 which includes several components. This keeps the farmer updated about the status of cultivation.

- The temperature sensor is used to measure the surrounding temperature
- The humidity sensor is used to measure the relative humidity and analyze patterns and forecast weather
- Soil dampness sensor is used to find the degree of dampness in the soil
- Arduino microcontroller is employed to collect information from sensors and send to the LED display as well as to the cloud where it is processed
- A network is trained at the cloud which is used to take decisions in the future
- The data are secured by completely authenticating the process
- The information about crops can be known remotely on a smartphone or laptop.

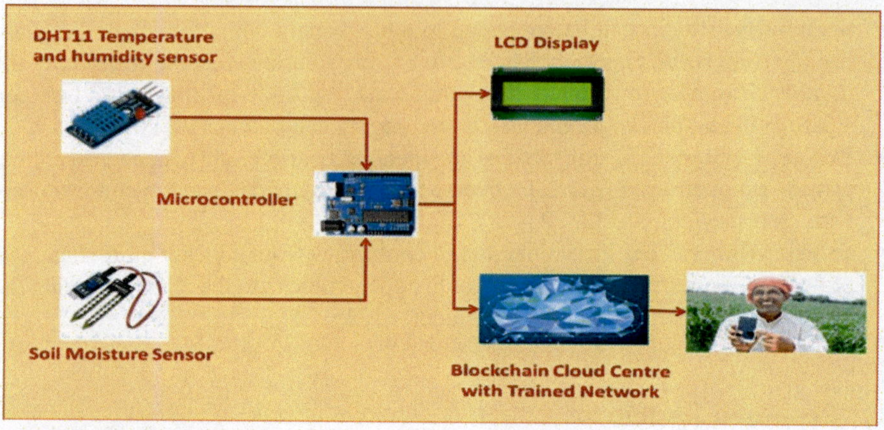

Fig.1 Architectural diagram

A blockchain-based framework for smart agriculture is shown in Fig. 2, and the timing diagram is shown Fig. 3.

To ensure security, all the entities undergo a registration stage with a trusted authority (TA), before endorsing them for access to the available data. During registration, the TA assigns unique identity to the smart sensors, micro-controllers, and LED with credentials containing secret keys. The sensors positioned in the field should make a secure communication with other neighboring sensors and smart devices. A shared authentication and key settlement system is highly needed. The

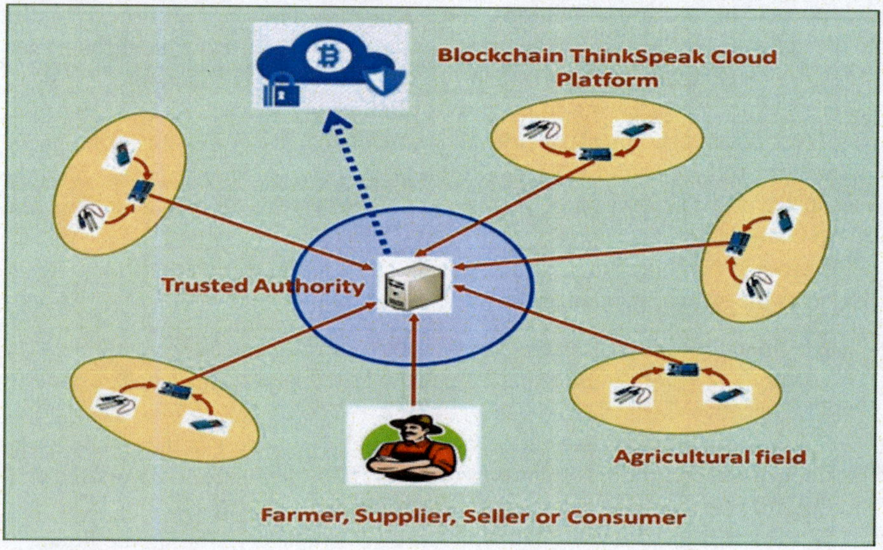

Fig. 2 Blockchain-based framework

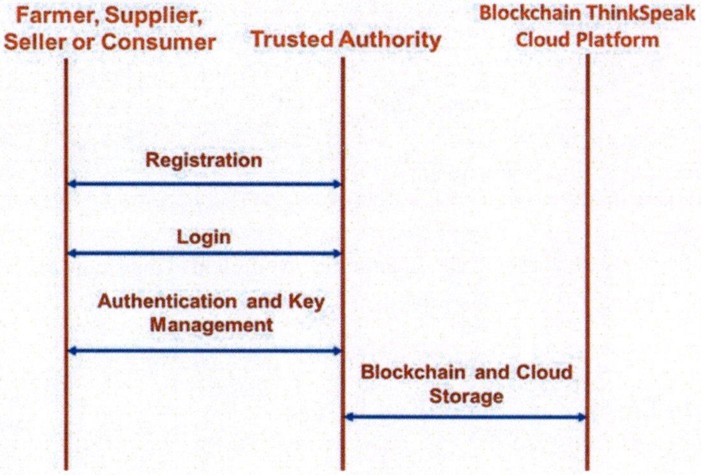

Fig. 3 Timing diagram

user may be a farmer, supplier, seller, or consumer. An external user may also be given admittance to real-time data from the installed devices. Nevertheless, user authentication is wanted, wherein a user establishes session keys with the retrieved smart devices such that they equally authenticate each other. The information gathered forms numerous transactions which are then used to form blocks. As the information is reserved and intimate, private blockchain can be considered in the smart agriculture environment. The owner's public key is used to encrypt the transactions and is stored in blocks. These blocks are extracted using effectual consensus algorithm. Once the blocks are validated by all the nodes in the peer-to-peer (P2P) network, the blocks are appended to the blockchain preserved by the cloud center.

5 Implementation

The proposed system includes the following phases: data collection, training, and classification.

5.1 Data Collection

Smart cultivation monitoring is a substantial sign in contemporary agriculture. The factors which have impact on the agricultural environment are obtained, and the whole fields are monitored by fitting a sensor and smart control system. Based on the parameters, the infrastructures including irrigation and heat conservation system

Table 1 Input parameters table

Field name	Attribute	Type	Size
ID	Primary key	Integer	10
Temperature	Null	Integer	12
Humidity	Null	Integer	12
Moisture	Null	Integer	12

are controlled for efficient cultivation and improved yield. This smart analyzer helps in obtaining high yield and better quality (Table 1).

5.2 Data Storage on the Cloud Server

Cloud computing includes the following features.

- An IoT network generates more amount of data from numerous devices and performs separation, analysis and storage. Much storage is required with robust development and influential parallel computing capability on a huge scale
- Several modifications are carried out on the resource load
- IoT offers increased computing power based on PaaS

Data including temperature, humidity and moisture of soil are stored on the cloud, and authenticated users can have access to them by logging in. ThingSpeak cloud is used in this project. ThingSpeak enables sensors, devices and Websites to forward and store data on the cloud. The data can be monitored, analyzed, modified and interacted with web services and devices. ThingSpeak is an IoT web platform which allows one to collect, visualize analyze data any time and react according to it. It is free of cost and was designed by ioBridge in 2010. It helps one to build IoT-based network without the need of setting up extra servers. The collection of data is done by REpresentational State Transfer (REST) or Message Queuing Telemetry Transport (MQTT). Data survey and visualization are done by MATLAB analytics. There are options to include various plugins that allow users to display Google gauge and other custom visualization and controls in private view. The main component of ThingSpeak is its channel which stores data received from various devices. Each channel can save upto eight fields along with device location, URL, etc. The channel can be made public which can be seen by other users or private which needs the API key to view the data. The private channel can be shared for some specific users.

ThingSpeak supports accessing MATLAB to aid in sensing data. It aids in combining and calculating data and run planned calculations at particular intervals. Plotting functions that aid visual understanding can be built. More refined examination by merging data from numerous channels is made possible (Fig. 4). Once the device is connected to the Internet, data can be retrieved from the kit and stored in the ThingSpeak cloud storage. The values can be seen on the LCD display.

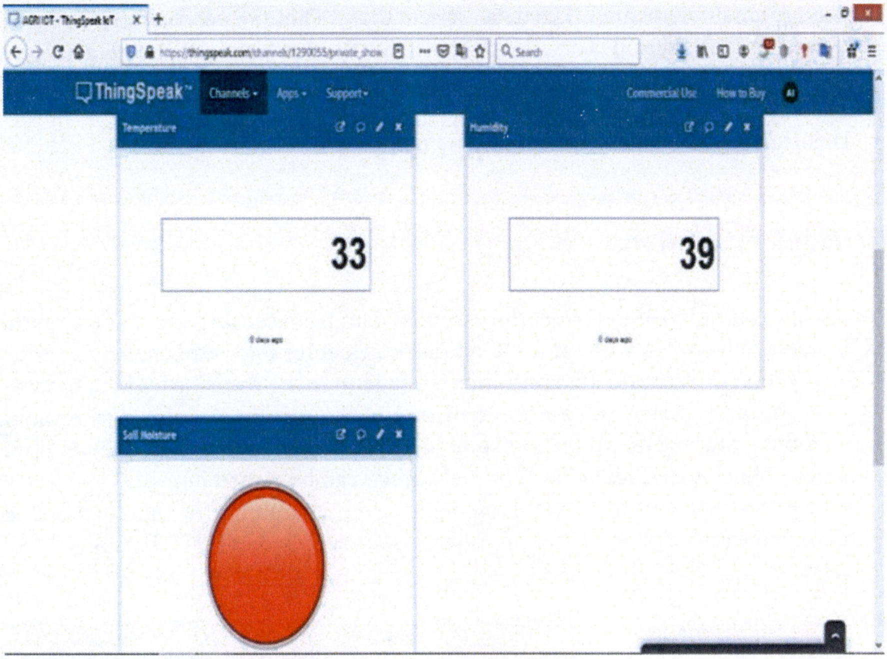

Fig.4 ThingSpeak data

5.3 *Training*

Regression and classification algorithms are applied on the data to design models that are capable of predicting periodical irrigation plans. Linear Regression (LR) model is chosen due to its dominance and capability to interpret. As simple LR models are not very successful, these models that focus on apprehending non-linear associations between independent and dependent variable are also developed.

5.4 *Data Classification*

Classification involves the following steps.

- The moisture level in the soil is sensed and read by the microcontroller through General Purpose Input Output (GPIO 21) pin and forwarded through Analog to Digital Converter (ADC)
- The values are analyzed. If the value = 1, then the soil is considered to be dry. If value = 0, then the soil is considered to be wet
- The data are then made available to the MQTT server on ThingSpeak platform depending on the value detected. The alarms are generated, and the conforming

triggers are implemented. The data are sent to the ThingSpeak database. In case it is observed that the soil is dry, then Simple Notification Service (SNS) is activated, and SNS mailer notifies the user (farmer, supplier, seller or consumer) through mail

- The decisions are taken accordingly by the farmer.

6 Implementation

An IoT kit is developed to collect the essential data from sensors and display on the LCD display integrated with the kit. Once a specific threshold is touched by a sensor or a collection of sensors, the controller is made to make decisions which activate the conforming actuator. In case the moisture in the soil drops below a threshold, then a motor is turned on till the moisture level is restored to the desired level. If the amount of light is insufficient, an external source can be turned on.

Figure 5 shows the hardware components and assembling of the components including moisture sensor, microcontroller, power supply and LCD display.

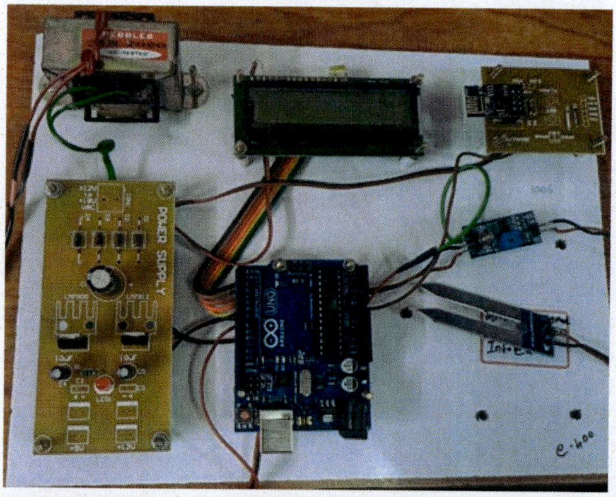

Fig. 5 Working model

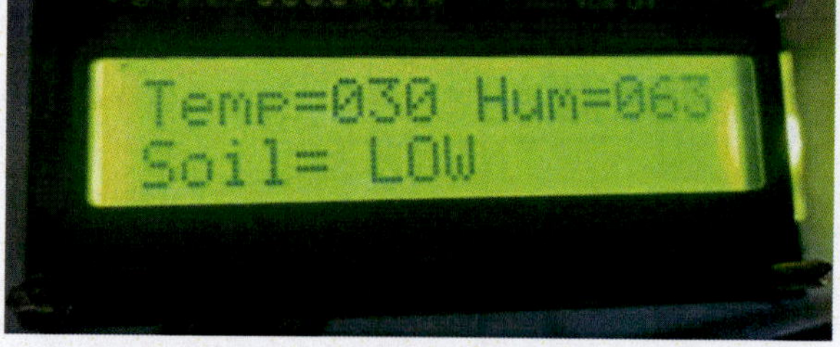

Fig. 6 Kit output

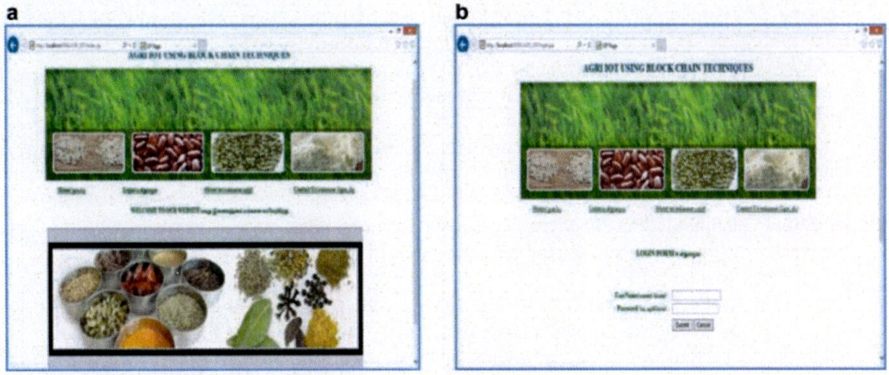

Fig. 7 **a** UI output **b** Login page

Figure 6 shows the temperature and humidity in degrees (Fahrenheit) and moisture level as either high or low. The data are passed to the ThingSpeak cloud server, where it is analyzed and used to train the network. The results are displayed on any smart device, either a smart phone or a laptop.

Figure 7a shows the home page of the Website. It also supports Tamil. Figure 7b shows the login page where only the registered users can access the Website and view the required data.

7 Conclusion and Future Enhancement

The proposed design describes the use of IoT in the farming sector. This model focuses on improving the yield by anticipating the soil and moisture conditions of the soil. IoT has permitted agricultural crop monitoring in a simple and skilled way to enhance the yield. As power loss and improper methods of processing play a major role, automation is the need of the hour. Farmers stay connected and are aware of the nature of the field from anywhere, anytime. As the data are stored on the cloud and processed, resource scarcity is highly avoided. Linear Regression (LR) is used for training the network. This enables the farmers to take decisions quickly, and the whole process is automated. The system is capable of self-learning and is capable of taking independent decisions.

References

1. Kim Y, Evans RG, Iversen WM (2008) Remote sensing and control of an irrigation system using a distributed wireless sensor network. IEEE Trans Instrum Meas 57(7):1379–1387

2. Sharma BB, Kumar N (2021) IoT-based intelligent irrigation system for paddy crop using an internet-controlled water pump. Int J Agric Environ Inform Syst (IJAEIS) 12(1):21–36
3. Lakhiar IA, Jianmin G, Syed TN, Chandio FA, Buttar NA, Qureshi WA (2018) Monitoring and control systems in agriculture using intelligent sensor techniques: a review of the aeroponic system. J Sens
4. Siva KN, Bagubali A, Krishnan KV (2019) Smart watering of plants. In: International conference on vision towards emerging trends in communication and networking (ViTECoN). IEEE, pp 1–4
5. Gutiérrez J, Villa-Medina JF, Nieto-Garibay A, Porta-Gándara MA (2013) Automated irrigation system using a wireless sensor network and GPRS module. IEEE Trans Instrum Measur 63(1):166–176
6. Devi DVV, Kumari GM (2013) Real-time automation and monitoring system for modernized agriculture. Int J Rev Res Appl Sci Eng (IJRRASE) 3(1):7–12
7. Sakthipriya N (2014) An effective method for crop monitoring using wireless sensor network. Middle-East J Sci Res 20(9):1127–1132
8. Getu BN, Attia HA (2015) Automatic control of agricultural pumps based on soil moisture sensing. In: AFRICON. IEEE, pp 1–5
9. Channe H, Kothari S, Kadam D (2015) Multidisciplinary model for smart agriculture using internet-of-things (IoT), sensors, cloud-computing, mobile-computing and big-data analysis. Int J Comput Technol Appl 6(3):374–382
10. Patil KA, Kale NR (2016) A model for smart agriculture using IoT. In: International conference on global trends in signal processing, information computing and communication (ICGTSPICC). IEEE, pp 543–545
11. Prathibha SR, Hongal A, Jyothi MP (2017) IoT based monitoring system in smart agriculture. In: International conference on recent advances in electronics and communication technology (ICRAECT). IEEE, pp 81–84
12. Suciu G, Istrate CI, Diţu MC (2019) Secure smart agriculture monitoring technique through isolation. In: Global IoT summit (GIoTS). IEEE, pp 1–5
13. Demestichas K, Peppes N, Alexakis T (2020) Survey on security threats in agricultural IoT and smart farming. Sensors 20(22):6458
14. Vangala A, Das AK, Kumar N, Alazab M (2020) Smart secure sensing for IoT-based agriculture: blockchain perspective. IEEE Sens J
15. Zheng Z, Xie S, Dai HN, Chen X, Wang H (2018) Blockchain challenges and opportunities: a survey. Int J Web Grid Serv 14(4):352–375
16. Jindal A, Aujla GS, Kumar N (2019) Survivor: a blockchain based edge-as-a-service framework for secure energy trading in SDN-enabled vehicle-to-grid environment. Comput Netw 153:36–48
17. Tanwar S, Bhatia Q, Patel P, Kumari A, Singh PK, Hong WC (2019) Machine learning adoption in blockchain-based smart applications: the challenges, and a way forward. IEEE Access 8:474–488
18. Sharma PK, Kumar N, Park JH (2018) Blockchain-based distributed framework for automotive industry in a smart city. IEEE Trans Ind Inf 15(7):4197–4205
19. Malavolta G, Moreno-Sanchez P, Schneidewind C, Kate A, Maffei M (2019) Anonymous multi-hop locks for blockchain scalability and interoperability. In: 26th Annual network and distributed system security symposium, NDSS
20. Kim S, Kwon Y, Cho S (2018) A survey of scalability solutions on blockchain. In: International conference on information and communication technology convergence (ICTC). IEEE, pp 1204–1207
21. Meiklejohn S, Pomarole M, Jordan G, Levchenko K, McCoy D, Voelker GM, Savage S (2013) A fistful of bitcoins: characterizing payments among men with no names. In: Proceedings of the 2013 conference on internet measurement conference, pp 127–140
22. Kosba A, Miller A, Shi E, Wen Z, Papamanthou C (2016) Hawk: the blockchain model of cryptography and privacy-preserving smart contracts. In: IEEE symposium on security and privacy (SP). IEEE, pp 839–858

23. Mastan ID, Paul S (2017) A new approach to deanonymization of unreachable bitcoin nodes. In: International conference on cryptology and network security. Springer, Cham, pp 277–298
24. ShenTu Q, Yu J (2015) Research on anonymization and de-anonymization in the bitcoin system. arXiv preprint arXiv:1510.07782
25. Bai Q, Zhou X, Wang X, Xu Y, Wang X, Kong Q (2019) A deep dive into blockchain selfish mining. In: ICC 2019–2019 IEEE international conference on communications (ICC). IEEE, pp 1–6
26. Sapirshtein A, Sompolinsky Y, Zohar A (2016) Optimal selfish mining strategies in bitcoin. In: International conference on financial cryptography and data security. Springer, Berlin, pp 515–532

Feedback Investigation on Twitter Dataset Using Classification Approaches

Yogesh Kumar, Sameeka Saini, Harendra Sharma, Ritu Payal, and Arpit Mishra

Abstract The public's views and feelings about individuals or brands have shifted significantly as a result of social media. Twitter, a free social media network that allows users to broadcast happenings from their daily lives, is one of the many social media platforms available. The sentiment analysis concept is deployed on tweets data to predict feedback using machine learning. Prediction of sentiment analysis is a revolutionary methodology for evaluating public emotions with their feedback uploaded to their personal media. In this study, we analyzed the performance of Naive Bayes, logistic regression, and random forest on airlines dataset concerning the accuracy, precision, recall, etc. After sentiment analysis, the dataset is classified into negative, positive, and neutral classes and compares the accuracy of the algorithm along their execution time.

Keywords Logistic regression · Naïve Bayes · Random forest · Precision · Sentiment analysis

1 Introduction

The sentiments are something that can be measured through different parameters and sentiment analysis is a technique by which we can identify the feedbacks of any organization, social website, e-commerce website and other websites. By using a machine learning algorithm, we are extracting people's emotions in the form of subjective comments and applying text analysis. Natural language processing is defined as the way in which, we are getting user as positive, negative, and neutral regarding the particular organization. The main purpose of doing sentiment analysis is to get people's thoughts on a particular service provider. Here without analyzing the name

Y. Kumar (✉) · S. Saini · R. Payal · A. Mishra
Dev Bhoomi Institute of Technology, Dehradun, India
e-mail: cse.yogeshkumar@dbgidoon.ac.in

H. Sharma
SRM Institute of Science and Technology, Modinagar, Ghaziabad, India
e-mail: harendrs1@srmist.edu.in

R. P. Mahapatra et al. (eds.), *Proceedings of International Conference on Recent Trends in Computing*, Lecture Notes in Networks and Systems 341,
https://doi.org/10.1007/978-981-16-7118-0_22

of the user, we are only processing comments for finding reviews. Public opinion is very important to know the actual feedback of the people. Nowadays number of papers is available on sentiment analysis, and lots of work have been done on this. Nowadays e-commerce Website is very famous for shopping before that product feedback is important for buying product. Now sentiment analysis is a text analyzing technique by which one can decide particular product comments that whether it is positive or negative. This text analysis is useful in evaluating the trainer using students feedback [1]. Some of the previous research conducted in this area include, SA performing on feedback given by the user on open-source multimedia contents for checking the percentage of user satisfaction [2]. Automated twitter SA system is dealing to predict user's online activity which provides an opportunity to identify negative and positive emotion rate by ML algorithm [3]. Most people have some disease in their body and taking drugs regularly but the main problem is how to decide our self which drugs will give better results than others. Nowadays people take medicine online so there is a platform in which users can identify which drugs contain what merits and demerits [4]. Some researchers have done on checking which algorithm is giving the best efficiency on a particular problem [5]. There is no model that can determine whether a society or group of people is distinct based on tweets (posts/messages) with identical textual material, tweets based on the same place, or tweets based on the invariant time for the subscriber timeline. To build an insightful, smart model that helps to provide online advice and tips for clients and merchants both see in [6]. The importance of automatic sentiment analysis (SA) for polling the opinions of a manufacturing article is expanded. This study aims to determine the potential of various approaches of sentiment analysis combined with star ratings, for identifying the genuine conclusion of the survey [7]. Assessment mining idea is deployed to predict the teacher evaluation with student feedback. The student's written feedback can be analyzed and conveyed as positive or negative. In this paper, the writer used a fusion-based neural network classifier to predict the feedback of the student is positive or negative [8]. Twitter is a social media network that is used to track the performance of public opinion on a variety of topics, as well as stock time series analysis [9, 10], to analyze health [11], classification of movies or analysis of box office [12, 13], prediction of crime [14], sports [15], movie, and election forecasting [16, 17].

The paper consists of 5 sections, and it is structured as follows:

1. This section presents the introductory part and literature review of the paper.
2. In this section, the algorithm used in this study is given and explained.
3. This section describes the procedure of sentiment analysis in the form of an algorithm.
4. This section presents practical results and evaluation.
5. This section draws the conclusion.

2 Proposed Methodology

The design of an optimization algorithm for emotional assessment will be the primary goal of the study. As input data, Twitter reviews will be used. At first, stop words will be omitted and the input data will be lemmatized. We performed normalization on the tweets done by the reviewer. For high accuracy, an optimization method gathers features and categorizes feelings using machine learning algorithms. The steps taken are below.

- Collect the data.
- Preprocessing of the data.
- Feature selection.
- Sentiment analysis was done by calculating polarity of the text.
- Applied machine learning algorithm to predict the model and calculate recall, *f*-measure, accuracy, and precision for the different classes.

Figure 1 shows the proposed working flowchart.

2.1 Data Collection

Using the API, the input data is obtained from the Twitter data source. The input data used in this work has been collected from twitter data source with the help of Application program interface. It is used to gather input data. It is an user interface that resides between source and the user. Twitter Website is used as a source that contains the tweets of the user. In this paper, airline data is taken. It contains attributes like airline_sentiment, tweet_id, airline, negative reason, retweet_count, negative_reason_confidence, airline_sentiment_gold, name, text, etc. Figure 2 shows the word cloud of tweets.

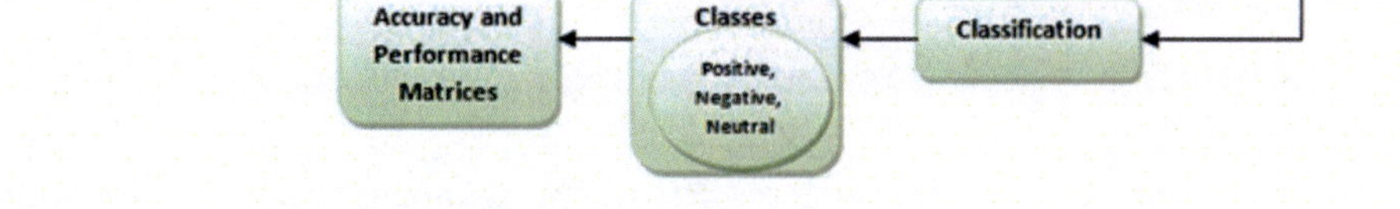

Fig. 1 Flowchart of the proposed working process

Fig. 2 Word cloud of tweets

tweet_id	sentiment	tweets
1	neutral	@VirginAmerica What @dhepburn said.
2	positive	@VirginAmerica plus you've added commercials t...
3	neutral	@VirginAmerica I didn't today... Must mean I n...
4	negative	@VirginAmerica it's really aggressive to blast...
5	negative	@VirginAmerica and it's a really big bad thing...
...	...	...
14636	positive	@AmericanAir thank you we got on a different f...
14637	negative	@AmericanAir leaving over 20 minutes Late Flig...
14638	neutral	@AmericanAir Please bring American Airlines to...
14639	negative	@AmericanAir you have my money, you change my ...
14640	neutral	@AmericanAir we have 8 ppl so we need 2 know h...

Fig. 3 Before preprocessing

2.2 Data Preprocessing

Preprocessing data is a type of data mining that is used to convert raw data into a usable and effective shape. Tweets contain many specific symbols with comments. Such symbols are treated as noisy and thus excluded from the dataset as irrelevant. To avoid noisy, inconsistent, incomplete data, preprocessing of data is necessary.

Figures 3 and 4 show the before preprocessing and after preprocessing comments respectively.

2.3 Removing of URL, Special Characters and Stopwords

Normally, we are not considered the URL in the sentiment analysis. Uniform Resource Locators are used for reference to a location on the Web page and Website but do not provide information related to people's sentiments. So, we have to remove

tweet_id	sentiment	tweets
1	neutral	said
2	positive	plus add commercials experience tacky
3	neutral	today must mean need take another trip
4	negative	really aggressive blast obnoxious entertainme...
5	negative	really big bad thing
...	...	...
14636	positive	thank get different flight chicago
14637	negative	leave minutes late flight wam communication ...
14638	neutral	please bring american airlines
14639	negative	money change flight answer phones suggestio...
14640	neutral	ppl need know many seat next flight plz put s...

Fig. 4 After preprocessing

this. No meaning is provided by special characters such as (hyphen) or/(slash), so we have to also remove them. We delete the $ or any currency sign because the currency does not play a role (for instance in sentiment analysis). Also, terms like I, me, you, he, and others increase text data size without improving efficiency, therefore removing them is a good idea.

2.4 Sentiment Analysis

After collection of the data in.csv file, we have done sentiment analysis and create a new column sentiment. Sentiment analysis was done by calculating polarity of the text. If the value of polarity is greater than zero then the text is classified into positive, for less than zero text classified into negative, and for zero text classified into neutral.

Algorithm for sentiment analysis

```
Input:   text
Output: sentiment
STEP1: sentiment = []
Using TextBlob will calculate the Polarity of whole tweets which
is having an in-text column:
STEP2: a1 ["polarity"] = getpolarity(text)
STEP 3: polarity = list(numpy.array(a1["polarity"]))
STEP 4: for i=0 to text.length() do
STEP 5:      if polarity[i]>0 then
STEP 6:           sentiment.append(positive)
STEP 7:      elseif  polarity[i] <0 then
STEP 8:           sentiment.append(negative)
STEP 9:    else
STEP 10:          sentiment.append(neutral)
STEP 11:    endif
STEP 12: endfor
STEP 13: return sentiment
```

3 SGD Classifier with Algorithm

SGD classifier is a linear classifier that has been tuned for SGD (SVM, logistic regression, and so on). There are two distinct thoughts here. SGD is a method of optimization, whereas logistic regression, also known as linear support vector machine, is a method of machine learning. The optimization method seeks to minimize or maximize the machine learning model's loss function. We will use the one vs. all approach to convert it to a multiclass classifier [as we know, logistic regression is a binary classifier]. In this method, when we work with a class, that class is designated by 1 and the rest of the class is denoted by 0. We use the gradient descent approach to solve this problem. For implementing logistic regression, we have used SGD classifier (loss = "log") which is based on the gradient descent technique. Gradient descent is the most widely used optimization algorithm.

$$h = \frac{1}{1 + e^{-Z}} \tag{1}$$

Z: input feature multiply by random initialize value denoted by theta

$$Z = \theta X \tag{2}$$

Here, X use the input feature

Cost function and gradient descent

$$J = -\frac{1}{m}\left[\sum y^i \log h^i + (1 - y)\log\left(1 - h^i\right)\right] \quad (3)$$

M—the number of training data, *y*—label on the original output, *h*—hypothesis or expected output.

This is the gradient descent equation, which we will use to update theta values in each iteration.

$$\theta = \theta - \alpha \sum \left(h^i - y^i\right)x^i_j \quad (4)$$

Shuffle the data set at random to ensure that the parameters are trained evenly for each category of data.

Step1: Let (x^i, y^i) be the training example

Step2: $Cost\left(\theta, \left(x^{(i)}, y^{(i)}\right)\right) = \frac{1}{2}\Sigma(h\theta(x^{(i)} - y^i)^2$

Step3: $J_{train}(\theta) = \frac{1}{m}\Sigma\, Cost(\theta, (x^{(i)}, y^{(i)}))$

Step4: **for** i=1 to m **do**

Step5: $\theta_j = \theta_j - \alpha \times \Sigma(h_\theta(x^i - y^i)x^i_j)$

Step6: For every j=0……. n

Step7: **endfor**

Step8: Repeat the steps 4 to 6 till trained the model

Step9: Calculate precision, accuracy, f-measure, recall, CPU time for the obtained classes.

Output: Precision, accuracy, f-measure, and recall.

3.1 *Algorithm for Proposed Work*

This section provides the algorithm used in this paper.

Input: set of tweets

Output: accuracy, prediction, recall, *f*-measure, and CPU time.

Step 1 Divide data into training and testing data set in 67:33 ratios.

Step 2 Train the model using Naive Bayes, logistic regression (using SGD classifier()), random forest classifier using fit().

Step 3 Now, using the predict() method, the dataset predicted.

Step 4 Classify the data into three categories: neutral, positive, and negative.

Step 5 Calculate precision, accuracy, *f*-measure, and recall for the different classes.

4 Experimental Setup

This section compares the prediction performance of three feedback analysis machine learning algorithms to show the experimental results. Using the Twitter API, the data for the input is taken from the twitter data source. The tweets are preprocessed and sorted into several categories. Best features are collected from preprocessed data using the logistic method, and then an ensemble of machine learning algorithms is used to classify tweets into different classes. The tweets are mostly divided into three categories. There are three types of emotions represented in the data: negative, positive, and neutral.

Figure 5 displays there are three categories for the overall number of tweets. Out of 14,640 tweets, 2363 are positive, 9178 are negative, and 3099 are neutral. Figures 6 and 7 show the classification of tweets as positive, negative, and neutral.

Table 1 represents the performance matrix of the different algorithms used in this paper. It shows the accuracy, CPU time, precision, recall, and *F*-measure of the used algorithm on positive, negative, and neutral tweets data.

Performance matrices can be defined as follows

$$\text{Accuracy} = \frac{(\text{TP} + \text{TN})}{(\text{TP} + \text{FP} + \text{FN} + \text{TN})} \tag{5}$$

where TN = true negative, TP = true positive, FP = false positive, and FN = false negative.

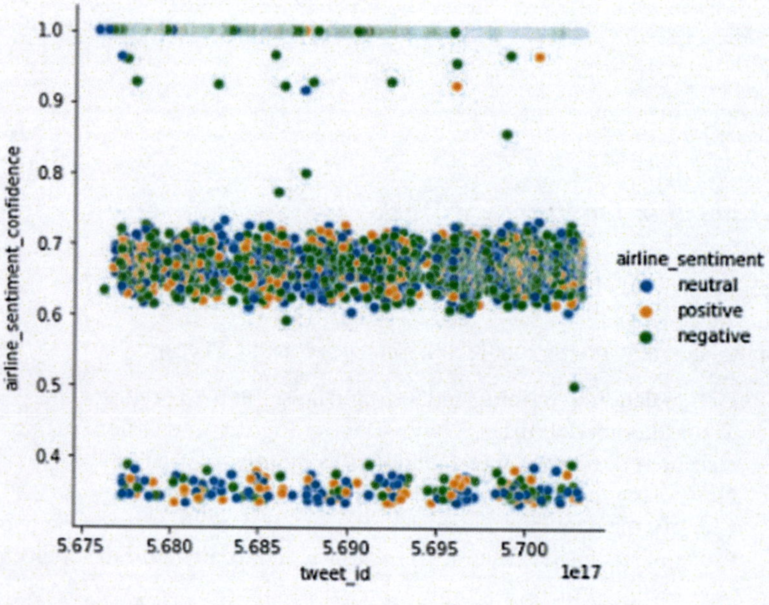

Fig. 5 Graphical representation of tweets as neutral, positive, and negative

Fig. 6 Classification of tweets

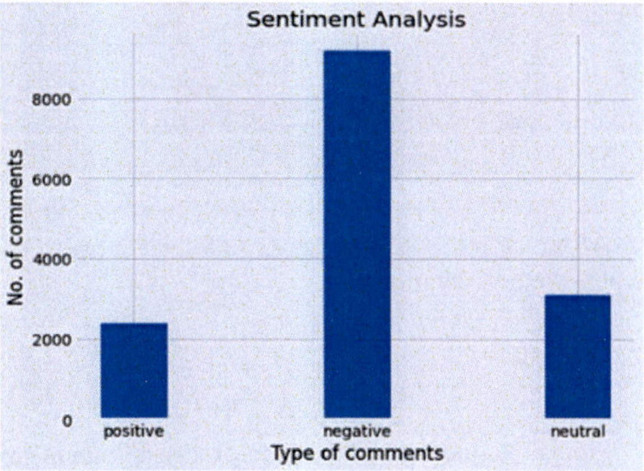

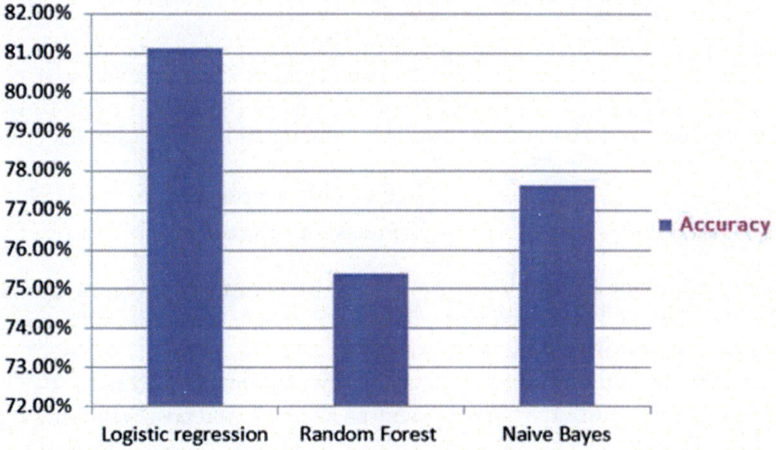

Fig. 7 Graph representation of accuracy of logistic regression, random forest and Naïve Bayes algorithm

Table 1 Algorithm performance metrics

Classifiers	Classes	Accuracy (%)	Precision	Recall	*F*-measure	CPU time (sec)
Random forest	Negative	75.04	0.76	0.96	0.85	16.5
	Neutral		0.69	0.35	0.47	
	Positive		0.77	0.43	0.55	
Logistic regression	Negative	81.11	0.86	0.91	0.88	16.9
	Neutral		0.67	0.61	0.64	
	Positive		0.77	0.68	0.72	
Naïve Bayes	Negative	77.29	0.80	0.94	0.86	245
	Neutral		0.66	0.44	0.53	
	Positive		0.76	0.58	0.66	

Table 2 Tweets classification

Sentiment class	Total tweets	Classified
Positive	4832	692
Negative	4832	3242
Neutral	4832	898

Accuracy refers to the number of correct predictions made by the model for all types of categorization tasks.

$$\text{Precision} = \frac{(\text{TP})}{(\text{TP} + \text{FP})} \tag{6}$$

Precision is defined as the ratio of appropriate instances to all recovered favorable instances.

$$\text{Recall} = \frac{(\text{TP})}{(\text{TP} + \text{FN})} \tag{7}$$

It denotes the fraction of retrieved samples among all relevant samples

$$F1 - \text{Score} = 2*\frac{(\text{Precisoin} * \text{Recall})}{(\text{Precisoin} + \text{Recall})} \tag{8}$$

Logistic regression algorithm has 81.11% highest accuracy while, random forest has 75.38%, and Naive Bayes has 77.62%. Figure 7 represents the graphical representation of these three algorithms.

In this work, for training and testing, data is divided into a 67:33 ratio. Tweets are categorized into three different classes: neutral, negative, and positive. Using logistic regression, 692 people are classified as positive, 3242 as negative, and 898 as neutral. Table 2 gives a summary of the classification of tweets.

The algorithm is applied in python. In this proposed method three algorithms applied random forest, logistic regression, and Naive Bayes on the airline dataset. From the graph, it is seen that logistic regression was given a high performance on the taken data. This method can be applied in different fields such as.

In the digital market, we can observe the feedback and can improve productivity. This can do a monthly or yearly basis. This work helps to know customer feedback for travels like airline services.

5 Conclusion

To evaluate public sentiment for travels such as the airline's Twitters API allows approved users to gain detailed information from tweets. The practical result shows negative, positive, and neutral thoughts of the public for an American airline. This kind of opinion analysis could provide important input to the company and assist them in identifying bad tweets that are rotating in the viewer's mind. Early detection of bad tweets can help a corporation improve its public relations. The given model is evaluated in three stages. In the first stage remove the noisy data, and the accuracy is calculated in the second stage using an optimization technique. In the third stage training set data is categorized into neutral, negative, and positive. Out of three algorithms, logistic regression has given the highest accuracy 81.11%. In the future work, authors can apply different machine learning techniques for better accuracy of our construct model to distinguish into strongly positive, positive, strongly negative, negative, or neutral opinions mining.

References

1. Selvapandian D, Thamba Meshach W (2020) An efficient sentiment analysis on feedback assessment from student to provide better education. In: Proceedings of the fourth international conference on I-SMAC (IoT in social, mobile, analytics and cloud) (I-SMAC). pp 1293–1300
2. Woldemariam Y (2016) Sentiment analysis in a cross-media analysis framework. In: IEEE international conference on big data analysis (ICBDA). Hangzhou, China, pp 1–5. https://doi.org/10.1109/ICBDA.2016.7509790
3. Sharma P, Sharma AK (2020) Experimental investigation of automated system for twitter sentiment analysis to predict the public emotions using machine learning algorithms. Mater Today: Proc. 2214–7853. https://doi.org/10.1016/j.matpr.2020.09.351.LNCS
4. Colón-Ruiz C, Segura-Bedmar I (2020) Comparing deep learning architectures for sentiment analysis on drug reviews. J Biomed Inform **110**. https://doi.org/10.1016/j.jbi.2020.103539
5. Naresh A, Venkata Krishna P (2020) An efficient approach for sentiment analysis using machine learning algorithm. Evol Intel. https://doi.org/10.1007/s12065-020-00429-1
6. Perti A, Trivedi MC, Sinha A (2020) Development of intelligent model for twitter sentiment analysis. Mater Today: Proc. 33(7):4515–4519. ISSN 2214-7853. https://doi.org/10.1016/j.matpr.2020.08.004
7. Al-Natour S, Turetken O (2020) A comparative assessment of sentiment analysis and star ratings for consumer reviews. Int J Inform Manage. 54. https://doi.org/10.1016/j.ijinfomgt.2020.102132
8. Selvapandian D (2020) An efficient sentiment analysis on feedback assessment from student to provide better education. In: Proceedings of the fourth international conference on I-SMAC (IoT in social, mobile, analytics and cloud) (I-SMAC) IEEE Xplore Part Number: CFP20OSV-ART; ISBN: 978-1-7281-5464-0
9. Vu T-T (2012) An experiment in integrating sentiment features for tech stock prediction in twitter. In: Workshop on information extraction and entity analytics on social media data, Mumbai, The COLING 2012 Organizing Committee, pp 23–38
10. Ohmura M, Kakusho K, Okadome T (2014) Stock market prediction by regression model with social moods. Int J Soc Behav Educ Econ Manage Eng **8**(10)
11. Larsen ME, Boonstra TW, Batterham PJ, O'Dea B, Paris C, Christensen H (2015) We feel: mapping emotion on Twitter. IEEE J Biomed Health Inform
12. Lu Y (2014) Integrating predictive analytics and social media. In: IEEE conference on visual analytics science and technology (VAST). IEEE, Paris, France
13. Pak A, Patrick P (2010) Twitter as a corpus for sentiment analysis and opinion mining. LREC 10
14. Wang X, Gerber MS, Brown DE (2012) Automatic crime prediction using events extracted from twitter posts. In: Social computing, behavioral-cultural modeling and prediction. Springer Berlin, pp 231–238

15. Barnaghi P, Ghaffari P, Breslin JG (2015) Text analysis and sentiment polarity on FIFA world cup 2014 Tweets. In: Conference CM SIGKDD'15, Sydney, Australia, workshop on large-scale sports analytics (LSSA)
16. Kagan V, Stevens A, Subrahmanian VS (2015) Using Twitter sentiment to forecast the 2013 Pakistani election and the 2014 Indian election. IEEE Intell Syst 1
17. Gayo-Avello D (2012) I wanted to predict elections with Twitter and all I got was this Lousy paper—a balanced survey on election prediction using Twitter data. arXiv preprint arXiv:1204.6441. University of Oviedo Spain

Detecting the Change in Microgrid Using Pattern Recognition and Machine Learning

S. Ramana Kumar Joga, Amit Kumar, Pampa Sinha, and Manoj Kumar Maharana

Abstract Microgrid is efficient and low-cost local grid. It consists of both local sources and loads. With the integration of more renewable energy sources, the chances of fault occurrence also increased. Fault is the abnormal condition that happen in microgrid, and it leads to change in the behavior of the system. These changes should be detected as fast as possible to avoid property damage and human loss. In this method, different machine algorithm-based classifiers are used to detect these changes in microgrid. Discrete wavelet transforms (DWT) based signal processing technique is used to analyze the transient signals pattern recognition method is used to classify the fault and non-fault data. An IEEE 9 bus test microgrid system is used to verify the proposed algorithm. The proposed methodology results are collated with dual-tree complex wavelet transform (DTCWT) based signal processing technique. Necessary simulation work is done in the MATLAB.

Keywords Artificial intelligence · Classifiers · Microgrid · Signal processing

1 Introduction

Microgrids are the local grid with renewable sources, residential and industrial loads. The control of microgrid is easier than the regular distribution system. Basically, microgrids categorized in to three types, they are DC microgrid, AC microgrid and hybrid microgrid. Fault is an abnormal condition of analysis of electrical equipment which is defined as the imperfection of the electrical network, from this it causes more or less current and deflected from the intended path, which damaging the electrical equipment and make it abnormal flow of electric current. The flow of current in the

S. R. K. Joga (✉) · A. Kumar · P. Sinha · M. K. Maharana
School of Electrical Engineering, Kalinga Institute of Industrial Technology, Bhubaneswar, India

P. Sinha
e-mail: pampa.sinhafel@kiit.ac.in

M. K. Maharana
e-mail: mkmfel@kiit.ac.in

R. P. Mahapatra et al. (eds.), *Proceedings of International Conference on Recent Trends in Computing*, Lecture Notes in Networks and Systems 341,
https://doi.org/10.1007/978-981-16-7118-0_23

intended path will be disturb from the fault [1]. The main reason for the change in current flow is the voltage misbalancing, overvoltage due to switching surges, caused by aging of the conductor, high flow of wind and atmospheric temperature, falling trees on the distribution line, falling wire on the road [2]. The effect of the fault in the system is to make abnormal the flow of electric current in the electric system, damaging the electrical equipment and make it work less and even cause death of human, birds and animals. Basically, the main effect of fault in the system is to make it abnormal to get distinguish results from the system. The main abnormalities that occur in the microgrid are open circuit fault, short circuit faults, arc faults and irregularities in the renewable sources like partial shading in PV source, irregular winds in wind turbine power generation. It is very essential to detect these faults as fast as possible. In the literature, already some methods are proposed to detect the faults in microgrid. Mohammad Amin Jarrahi et al. proposed a fault classification scheme along change detection through squared three phase currents, and it is discussed in [3]. Subhransu R. Samantaray et al. proposed a microgrid protection scheme using integrated impedance angle method assisted by PMU, and it is more discussed in [4]. Renxi Gong et al. proposed a method to identify the power quality disturbances in microgrid using convolution neural networks and it is more discussed in [5]. James J. Q. Yu et al. proposed a new fault detection method using wavelet transform and DNN, and it is more explained in [6]. Kartika Dubey et al. proposed a differential protection scheme for microgrid using impedance angle method and it is discussed in [7]. Rudranarayan Pradhan et al. proposed a method to detect change in islanded microgrid using positive power sequence component, and it is more discussed in [8]. Hamid Reza Baghaee et al. proposed an anti-islanding method to protect microgrid consisting of solar charged electric vehicle batter charge stations from faults using support vector machine, and it is more discussed in [9]. Moinul Haque proposed a fault detection scheme in microgrid consists of electrical vehicle using deep learning and it is more discussed in [10]. Animesh Sahoo et al. proposed a method to detect fault in AC microgrid based on Teager-Huang, and it is discussed in [11].

In this paper, a wavelet and sum of squares of detail coefficients of voltage signal is considered for to detect the changes in the microgrid. Classifiers of machine learning are used to classify the fault. The accuracy of the machine learning classifiers also improved through this method.

2 Proposed Methodology

Machine learning is one of the applications of artificial intelligence. It reduces the human interference in solving problem. Change detection and fault classification are considered as classification problem. These classification problems are solved by machine learning classifiers. These machine learning classifiers are classified in to three types, and they are supervised learning classifier, unsupervised learning classifier and reinforcement learning classifier. In this paper, supervised learning-based machine learning classifiers are used, and they are given in Table 1.

Table 1 List of supervised machine learning classifiers

Machine learning classifier	Problem type	Learning type	Advantages	Disadvantages
Decision tree [12]	Regression and classification	Supervised	Requires less effort for data preparation No normalization of data is required initially	It is very sensitive even a small change in data effect the accuracy
Naive Bayes [13]	Probabilistic classifier	Supervised	It is very fast algorithm	It takes that all predictors (or features) are independent, rarely happening in real life
Support vector machine (SVM) [14]	Regression and classification	Supervised	It is is relatively memory efficient	Not suitable for large data sets
k-nearest neighbors' algorithm (KNN) [15]	Regression and classification	Supervised	Very easy to implement for multi-class	It doesn't perform well on imbalanced data

The basic fault detection method involves three stages, they are (a) measuring voltage, current and phasor parameters (b) extraction of features and pattern recognition and (c) fault classification through machine learning classifiers. The flow chart of proposed algorithm is given in Fig. 1.

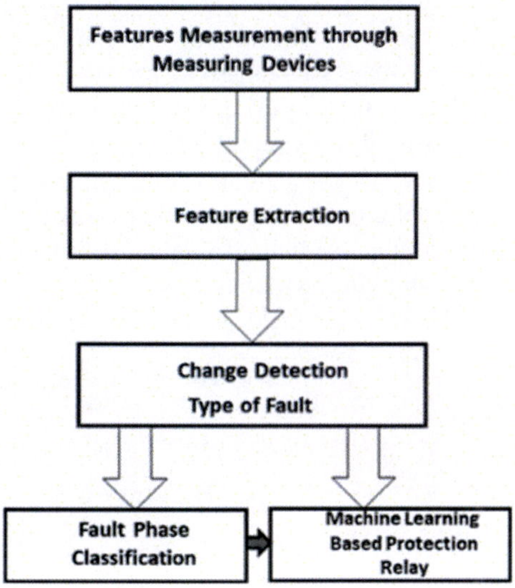

Fig. 1 Flowchart of proposed algorithm

Feature extraction is done through discrete wavelet transform based signal process technique and results are compared with dual-tree complex wavelet transform. Proposed methodology starts with measuring the parameters like voltage, current and phasor. These parameters are extracted further for selecting good features. Classifier classify these good features by training themselves with supervision. The accuracy of classification problem is evaluated through confusion matrix. Confusion matrix is a data science tool used to evaluate the performance of the classification problem.

3 Results and Discussions

3.1 Type of Fault Detection

IEEE 9 bus system with distributed generation is uses as test system to verify the proposed methodology. The main source is coming from the Grid of 18 kV, and it is connected to bus 2. A distributed generation named diesel generator of 13.8 kV rating is connected at the end of bus number 3. A Solar Array of 16.5 kV with MPPT is connected at the end of bus number 1. Residential Load, Industrial Loads are connected at bus number 5, 6 and bus 8. Various faults are given to the microgrid at bus number 7 to detect the change in the system. The block diagram of IEEE9 bus microgrid is given in Fig. 2

The MATLAB Simulink is used to simulate the proposed test system. The proposed algorithm should detect the change in the microgrid for various fault condition. Short circuit (Line-Ground) and arc faults are given and tested to detect the change in the system. A short circuit fault and arc fault are given at time interval from 0.6 to 0.8 s applied at line bus 5–7. The voltage response before fault and after fault is plotted, and it is shown in Fig. 3

The voltage signals are measured at bus number 7. The process uses DWT and DTCWT based signal processing technique to decompose the signal. A 6-Level DWT is used to extract the features of the signals. Detail coefficients are collected at every level, they are V_{D_o} is detail coefficient of original voltage signal, $V_{D_{\mathrm{LG}}}$ is detail coefficient of signal during short circuit fault, and $V_{D_{\mathrm{Arc}}}$ is detail coefficient of signal during arc fault. To detect the change in the deviation, sum of the square of all deviation is calculated.

$$\text{For Short Circuit Fault, } \sum_{0}^{6} V_{D_{\mathrm{LG}}}^2$$

$$\text{For Arc Circuit Fault, } \sum_{0}^{6} V_{D_{\mathrm{Arc}}}^2$$

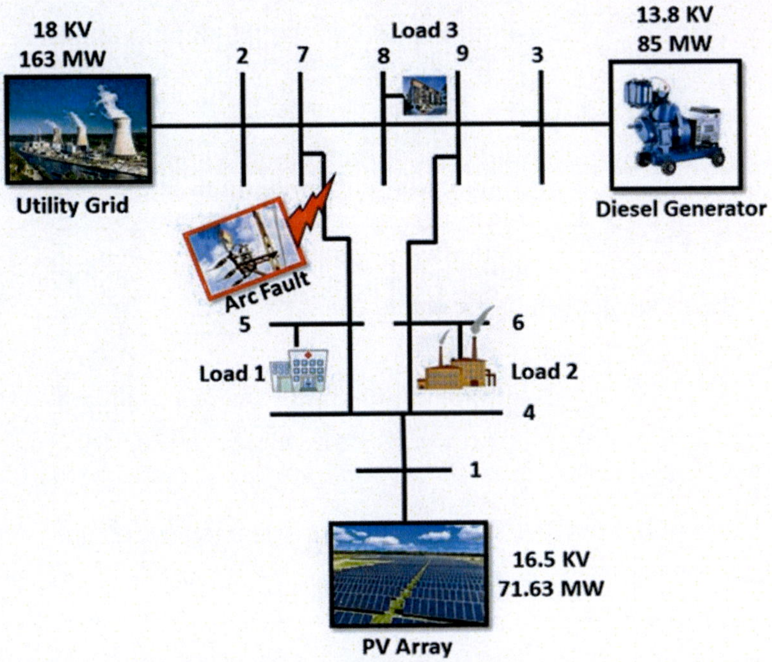

Fig. 2 Proposed TEST microgrid system layout

These sum of the squares of details coefficients of voltage signal during short circuit fault and arc fault is measured, it is tabulated with chart, and it is shown in Fig. 4

The change in the microgrid is detected during fault condition. To make this process without any human interference, machine learning algorithms are used to detect the fault type that occurred in microgrid. The window length considered is from 0.61 to 0.62 s. Total 402 signals, 201 signals during arc fault and 201 signals during LG fault are used for the pattern recognition. The total training time is 0.764 s and prediction speed is 27,000 obs/sec. All these data are trained by the classifiers and the performance is assessed through confusion matrix. The confusion matrix obtained after training the data in various machine learning algorithms are given in Fig. 5.

Considering DT based classifier, 402 signals are considered for training. In those, 198 signals are mapped themselves as true fault signals and 3 signals as False LG fault and out of 201 LG fault signals 200 signals mapped themselves as True LG fault and 1 signal grouped as False Arc Fault. The overall accuracy of the classification is 99.0%. In support vector machine and K-Nearest Neighbor based Machine Learning Classifiers, 402 signals are trained and 201 signals mapped themselves as true arc fault and 0 signals as False short circuit fault and out of 201 short circuit fault signals 201 signals grouped as True short circuit fault and 0 signal grouped as False Arc Fault. The overall accuracy of the classification is 100.0%. In Naive Bayes based

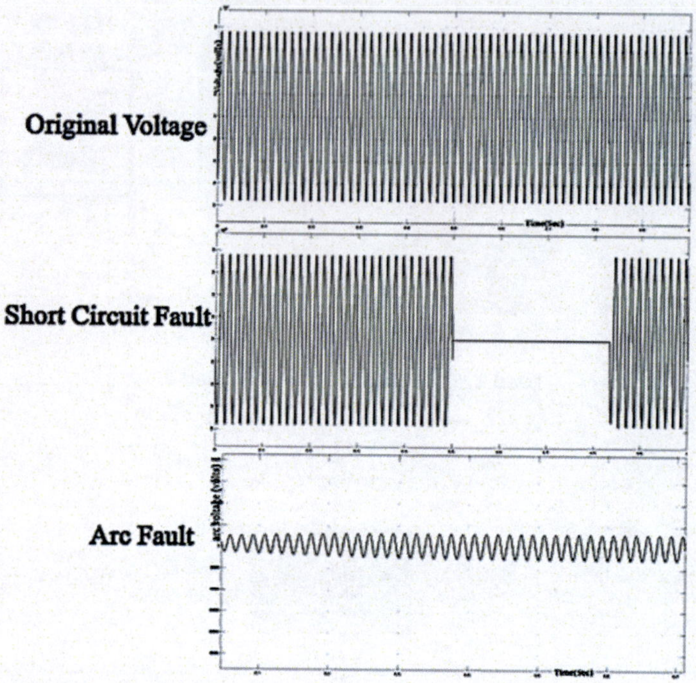

Fig. 3 Voltage response at bus: before fault, short circuit fault and arc fault

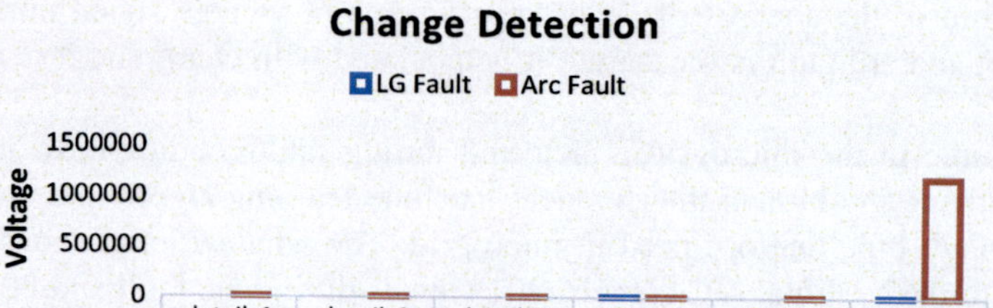

	details1	details2	details3	details4	details5	details6
LG Fault	0.0044	0.009	0.0491	2.33	0.0344	68.6792
Arc Fault	77.6473	16.1836	872.6583	4112.9	516.8008	1183900

Fig. 4 Change detection: sum of squares of detail coefficients of signal during arc fault and LG fault

Machine Learning Classifier, out of 201 arc fault signals 201 signals grouped as true arc fault and 0 signals as False LG fault and out of 201 LG fault signals 197 signals grouped as True LG fault and 4 signals grouped as False Arc Fault. The overall accuracy of the classification is 99.0%.

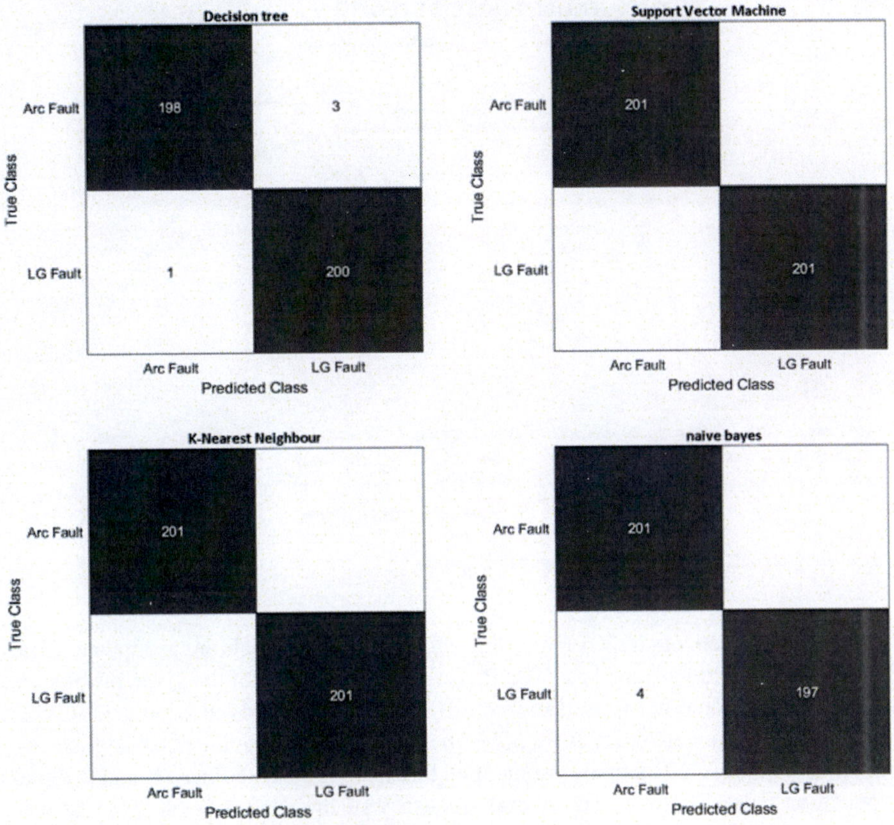

Fig. 5 Performance table: decision tree, SVM, KNN and Naive Bayes

3.2 *Detection of Fault Phase*

The arc fault is applied at bus number 5–7 as shown in Fig. 1. The fault is given to phase B of the microgrid test system. Voltage sensor is used to extract the signal at bus number 7. The Voltage data of Phase A, Phase B, Phase C is measured during arc fault. Discrete Wavelet transform is used for decomposing the signal up to 6 levels. The detailed values of voltages at three phases are measured, it is tabulated with chart, and it is shown in Fig. 6

In Fig. 5, it is concluded that Phase B is the faulty phase in the test microgrid system. This is done without human interfere by using various machine learning classifiers. The total 603 signals are used in this classification problem out of 201 are Phase A voltage signals, 201 are Phase B signals and 201 signals are Phase C Signals. The performance is evaluated through confusion matrix. This multiclass classification problem is solved, and the confusion matrix are given in Fig. 7.

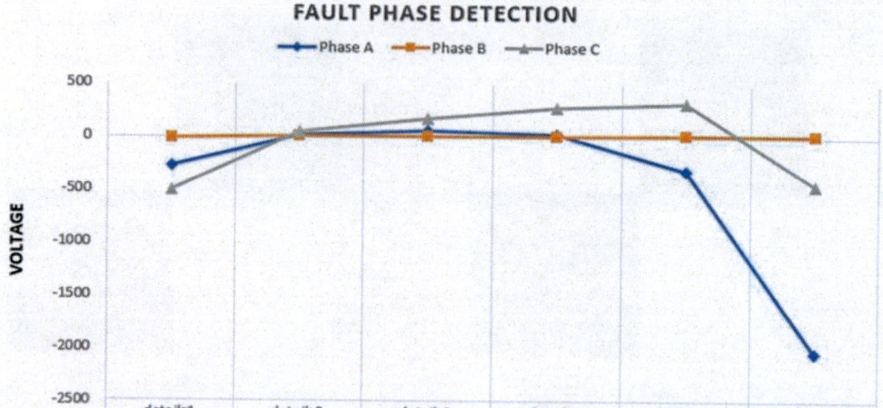

	details1	details2	details3	details4	details5	details6
Phase A	-279.2277837	19.51932927	60.68058245	24.80722843	-312.7756568	-2034.367244
Phase B	-11.5132296	1.244719035	4.263177464	7.950384347	13.87241019	13.01573663
Phase C	-501.3437305	49.58687663	167.0922622	273.3073818	321.4386107	-455.0362023

Fig. 6 Detail coefficients of phase A, phase B, phase C

In Decision Tree, out of 201 Fault Phase B signals 200 signals are grouped as True Phase B and 1 signal grouped as False Phase A. The accuracy of the given problem is almost 87.7% when DWT is used and overall efficiency is improved through DTCWT. The phase detection accuracy is 99.50%. In support vector machine, 402 signals are trained and in those 201 signals are mapped themselves as True Phase B and 0 signal grouped as False Phase A. The overall accuracy of the system is 68.8%, and it is corrected to 88.28%. The phase detection accuracy is 100.00%. In Naive Bayes, 402 signals are trained and out of those signals 201 signals are mapped themselves as True Phase B and 0 signal grouped as False Phase A. The overall accuracy of the classification problem is 89.55%. The phase detection accuracy is 100.00%.

4 Conclusions

The overall results are compared with dual-tree complex wavelet transform signal processing-based decomposition technique, and it is tabulated in Table. 2.

The above table concludes that DTCWT based proposed methodology outperformed the DWT based method. The overall fault phase detection accuracies are showing better results with DTCWT technique. The overall classification accuracy is improved, and the change in the microgrid is detected manually and with artificial intelligence in this paper. The proposed methodology is compared with existed microgrid change detection methods, and it is tabulated in Table 3.

The above table concludes that proposed methodology is outperformed the existed methods that available in literature.

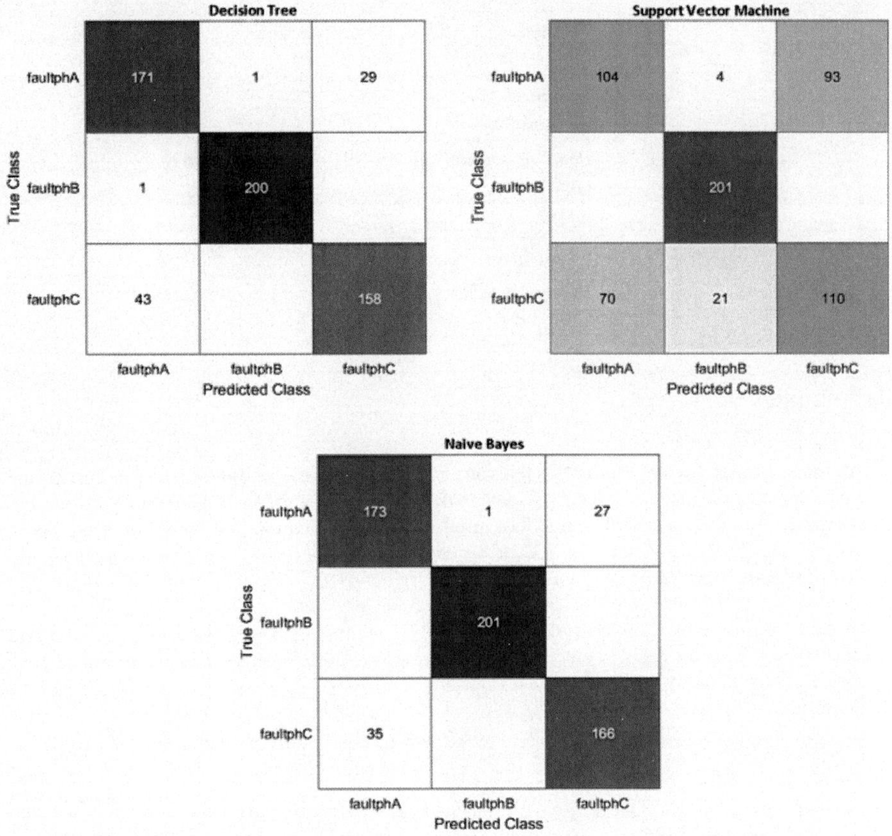

Fig. 7 Confusion matrix: fault phase—decision tree, SVM and Naïve Bayes

Table 2 Overall results comparison

Type of classifier	Type of fault accuracy		Phase of the fault accuracy	
	DWT (%)	DTCWT (%)	DWT (%)	DTCWT (%)
Decision tree	99	99.5	87.7	98.02
SVM	100	100	68.8	88.28
KNN	100	100	72.2	89.76
Naïve Bayes	99	99.5	89.55	98.56

Table 3 Comparison between existed methods

Method	Overall accuracy (%)	Response time (ms)
Proposed methodology	99.5	0.023
Pearson correlation coefficient	98.99	0.6
ANN	98.5	0.8
S transform	99	0.03
Hilbert transform and ANN	98.10	0.006

References

1. Ramana Kumar Joga S, Sinha P, Maharana MK (2019) Artificial intelligence in classifying high impedance faults in electrical power distribution system. In: Proceedings of international conference on recent trends in computing, communication and networking technologies (ICRTCCNT). Available at SSRN: https://ssrn.com/abstract=3430316 or https://doi.org/10.2139/ssrn.3430316
2. Joga S, Sinha P, Maharana MK (2021) Genetic algorithm and graph theory approach to select protection zone in distribution system. In: Zhou N, Hemamalini S (eds) Advances in smart grid technology. Lecture notes in electrical engineering, vol 688. Springer, Singapore. https://doi.org/10.1007/978-981-15-7241-8_13
3. Jarrahi MA, Samet H, Ghanbari T (2020) Novel change detection and fault classification scheme for AC microgrids. IEEE Syst J 14(3):3987–3998. https://doi.org/10.1109/JSYST.2020.2966686
4. Sharma NK, Samantaray SR (2020) PMU assisted Integrated impedance angle-based microgrid protection scheme. IEEE Trans Power Delivery 35(1):183–193. https://doi.org/10.1109/TPWRD.2019.2925887
5. Gong R, Ruan T (2020) A new convolutional network structure for power quality disturbance identification and classification in micro-grids. IEEE Access 8:88801–88814. https://doi.org/10.1109/ACCESS.2020.2993202
6. Yu JJQ, Hou Y, Lam AYS, Li VOK (2019) Intelligent fault detection scheme for microgrids with wavelet-based deep neural networks. IEEE Trans Smart Grid 10(2):1694–1703. https://doi.org/10.1109/TSG.2017.2776310
7. Dubey K, Jena P (2021) Impedance angle-based differential protection scheme for microgrid feeders. IEEE Syst J. https://doi.org/10.1109/JSYST.2020.3005645
8. Pradhan R, Verma P, Jena P (2020)Fault detection in Islanded microgrid based on positive power sequence component. In: IEEE international symposium on sustainable energy, signal processing and cyber security (iSSSC), Gunupur Odisha, India, pp 1–6.https://doi.org/10.1109/iSSSC50941.2020.9358875
9. Baghaee HR, Mlakić D, Nikolovski S, Dragicčvić T (2020) Anti-islanding protection of PV-based microgrids consisting of PHEVs using SVMs. IEEE Trans Smart Grid 11(1):483–500. https://doi.org/10.1109/TSG.2019.2924290
10. Haque M, Shaheed MN, Choi S (2018)Deep learning based micro-grid fault detection and classification in future smart vehicle. In: IEEE transportation electrification conference and expo (ITEC), Long Beach, CA, USA, pp 1082–1107.https://doi.org/10.1109/ITEC.2018.8450201
11. Sahoo A, Arunan A, Mahmud K, Ravishankar J, Nizami MSH, Hossain MJ (2019)Teager-Huang based fault detection in inverter-interfaced AC microgrid. In: IEEE international conference on environment and electrical engineering and 2019 IEEE industrial and commercial

power systems europe (EEEIC / I&CPS Europe), Genova, Italy, pp 1–5. https://doi.org/10.1109/EEEIC.2019.8783453
12. Swain PH, Hauska H (1977) The decision tree classifier: design and potential. IEEE Trans Geosci Electron 15(3):142–147. https://doi.org/10.1109/TGE.1977.6498972
13. Martinez-Arroyo M, Sucar LE (2006) Learning an optimal Naive Bayes classifier. In: 18th International conference on pattern recognition (ICPR'06), Hong Kong, China, pp 958–958. https://doi.org/10.1109/ICPR.2006.749.
14. Zhang Y (2012) Support vector machine classification algorithm and its application. In: Liu C, Wang L, Yang A (eds) Information computing and applications. ICICA 2012. Communications in computer and information science, vol 308. Springer, Berlin. https://doi.org/10.1007/978-3-642-34041-3_27
15. Guo G, Wang H, Bell D, Bi Y, Greer K (2003) KNN model-based approach in classification. In: Meersman R, Tari Z, Schmidt DC (eds) On the move to meaningful internet systems 2003: CoopIS, DOA, and ODBASE. OTM 2003. Lecture notes in computer science, vol 2888. Springer, Berlin. https://doi.org/10.1007/978-3-540-39964-3_62

Electroencephalogram Signal Processing with Independent Component Analysis and Cognitive Stress Classification Using Convolutional Neural Networks

Venkatakrishnan Sutharsan, Alagappan Swaminathan, Saisrinivasan Ramachandran, Madan Kumar Lakshmanan, and Balaji Mahadevan

Abstract Electroencephalogram (EEG) is the recording which is the result due to the activity of bio-electrical signals that is acquired from electrodes placed on the scalp. In electroencephalogram signal (EEG) recordings, the signals obtained are contaminated predominantly by the electrooculogram (EOG) signal. Since this artifact has higher magnitude compared to EEG signals, these noise signals have to be removed in order to have a better understanding regarding the functioning of a human brain for applications such as medical diagnosis. This paper proposes an idea of using independent component analysis (ICA) along with cross-correlation to denoise EEG signal. This is done by selecting the component based on the cross-correlation coefficient with a threshold value and reducing its effect instead of zeroing it out completely, thus reducing the information loss. The results of the recorded data show that this algorithm can eliminate the EOG signal artifact with little loss in EEG data. The denoising is verified by an increase in SNR value and the decrease in cross-correlation coefficient value. The denoised signals are used to train an artificial neural network (ANN) which would examine the features of the input EEG signal and predict the stress levels of the individual.

Keywords EEG · EOG · ICA · FastICA · Artifacts · CNN

V. Sutharsan · A. Swaminathan (✉) · S. Ramachandran · B. Mahadevan
Department of Electrical and Electronics Engineering, SSN College of Engineering, Chennai, India
e-mail: alagappan16011@eee.ssn.edu.in

V. Sutharsan
e-mail: venkatakrishnansutharsan16119@eee.ssn.edu.in

S. Ramachandran
e-mail: saisrinivasan16083@eee.ssn.edu.in

B. Mahadevan
e-mail: balajim@ssn.edu.in

M. K. Lakshmanan
CSIR—Central Electronics Engineering Research Institute, Pilani, Rajasthan, India
e-mail: mklakshmanan@ceeri.res.in

R. P. Mahapatra et al. (eds.), *Proceedings of International Conference on Recent Trends in Computing*, Lecture Notes in Networks and Systems 341,
https://doi.org/10.1007/978-981-16-7118-0_24

1 Introduction

An electroencephalogram (EEG) signal is that the recording of spontaneous electric pursuit of the mind recorded over a time period. The neurons of the human brain function work through ever-converting the movement of electrical currents across their membranes. These ever-converting currents generate electric and magnetic fields so as to record by using electrodes from the surface of the scalp. The differential potential of the electrodes is then amplified and recorded as the electroencephalogram signal. The main advantages of the method proposed by Radüntz et al. [1] is that it provides an automatic, reliable, real-time capable, and practical tool, which avoids the need for the time-consuming manual selection of ICs during artifact removal.

The recorded electroencephalogram signal is usually infected through spurious indicators from one-of-a-kind undesirable sources. This form of infection in scientific nomenclature is called as artifact. It is a crucial challenge to get rid of those artifacts from the electroencephalogram (EEG) sources for extra evaluation of EEG. The artifacts are hard to get rid of due to: (i) higher amplitude than the electroencephalogram signal, (ii) the huge frequency tiers of the elements, and (iii) because of their variable geographical distribution. Among those artifacts, the vast majority of them are electrooculogram (EOG) signals, which are because of the attention blinking or movement. Eye motion produces electrical interest (EOG signal); this is sufficient vast to be visible inside the electroencephalogram recording. The EOG is a sign that displays the charges distribution among the cornea and the retina which changes during eye motion. Since there is an overlapping of those artifacts over the specified interest data, there may be a massive loss of background electroencephalogram recording. An optimized path of action to accurately rectify this for an electroencephalogram mixed with eye movement is to first discover the EOG pattern and then to smooth the corresponding electroencephalogram signal component rather than cleansing the whole EEG recording. Usually, the reference electrodes are placed over the mastoid bone (which is that the bone in the back of the ear) of each the ears. Merits of this method are that it is reasonably priced and cost-efficient, however, suffers from the intense disadvantage of noise from environment. For this reason, this methodology is most popular for low-risk functions which includes BMI and many others. The algorithm by Mahajan & Morshed [2] achieves an average sensitivity of 90% and an average specificity of 98%. In the paper by Mammone et al. [3], a novel technique (Automatic Wavelet Independent Component Analysis, AWICA) for automatic EEG artifact removal is presented.

The International 10–20 electrode system is an internationally identified system to explain and observe the place of scalp electrodes for acquiring electroencephalogram recordings in non-invasive electroencephalogram. The device relies at the link among the placement of an electrode and the underlying area of the brain, in particular the cortex. The "10" and "20" confer with the truth that the actual distances among adjoining electrodes are either 10 or 20% of the entire front-lower back or right-left distance of the skull. The 10–20 international system of units has the structure as shown in Fig. 1.

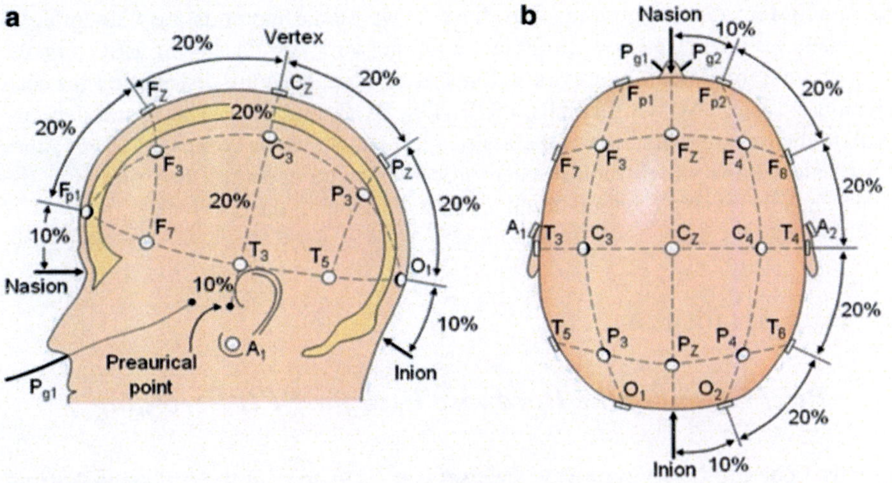

Fig. 1 10–20 international system electrode placement structure **a** Lateral view and **b** Frontal view

In the last few years, a wide range of methods have been carried out for the removal of artifacts in electroencephalogram. Among these artifacts, ocular artifacts are shown to have a huge deterioration of electroencephalogram signal. Many ways to dispose the ocular artifacts had been put forth for a long time, and Arjon Turnip et al. [4] have suggested the elimination of artifacts from electroencephalogram by the use of ICA and PCA. The authors have compared the two approaches for eliminating artifacts, i.e., ICA and PCA methodology, and from processing with ICA and PCA techniques, it turned that the ICA method is better than PCA in terms of the separation of the electroencephalogram components from blended signal. Anusha Zachariah et al. [5] have recommended the elimination of artifacts from EEG signal using critical electroencephalogram rhythms. Wavelet decomposition is used inside the preprocessing stage and has proven to increase redundancy and rejection of inappropriate wavelets. This algorithm has two-step identification of artifact content before and after ICA using kurtosis. Yuan Zou et al. [6] have planned the elimination of artifacts from electroencephalogram signal using hierarchical clustering to split artifacts. Jirayucharoensak et al. [7] have planned the artifact rejection from electroencephalogram using independent component analysis. He delivered an ahead technique to extract useful neural signals with the usage of lifting wavelet transform (LWT). Dwi Esti Kusumandari et al. (2015) have counseled the artifacts removal from electroencephalogram using ICA. The authors have as compared two strategies for removing artifacts, i.e., JADE and SOBI algorithms. From processing with JADE and SOBI methods, it is far discovered that SOBI technique is higher than the alternative technique in phrases of the separation of the electroencephalogram data from the mixed signal.

There is particular feature for each cognitive stress which is to be used as to classify the EEG signals. Those capabilities/features are power spectral density, wavelet based to name a few. On the grounds that those conventional methods are vulnerable to mistakes, we have used convolutional neural network for classifying EEG signals. Our convolutional neural network will research the underlying distribution for each signal kind like LAB, COLLEGE, FRIENDS, DOG, and CAT. The neural network will do the feature extraction and classification on its own as compared to any other set of algorithms which require each processes to be accomplished one after the other, thus increasing computational speed and reducing the complexity.

2 Data Acquisition

2.1 *Brief Layout of the Hardware Used for EEG Recording*

Emotiv Epoc Plus EEG brainwave headset is used to acquire the EEG signals from the scalp. A PC is used to coordinate data flow and also collect data from the sensors attached to it (hence, acting like a data acquisition system). A monitor views the waveform of the EEG and EOG signal to ensure proper recording of signals. A Holter is a device commonly used to record ECG signals. In this project, it has been used to acquire EOG signals. A pulse generation system produces impulses of high magnitude and then injects them into the body of the subject (this is used to synchronize the EEG and EOG signals). The Emotiv Epoc Plus EEG headset is placed on the head of the person. This headset is connected to the computer by Bluetooth. The EOG electrodes are connected to the Holter through a HDMI cable (Fig. 2). Pulse from a programmable pulse generator is given into the electrode. The purpose of giving the pulse is to create markings in the EEG signal waveform corresponding to each window of data.

The pulse generated from the external square waveform generator is injected into the ECG signal using two clamps placed on the hands of the subject. The main purpose of injecting a number of pulses into the ECG waveform is to mark the start time of recording and the end time of recording. This is also used for the time synchronizing of the EEG and EOG signals. Since EOG and ECG signals are from the same device, synchronizing ECG signal with the EEG signal inherently synchronizes EEG and EOG signal (Fig. 3).

2.2 *Procedure for Recording EEG Signals*

The Emotiv Epoc Plus EEG headset is placed on the head of the subject. The reference electrodes are placed over the mastoid bone (the bone behind the ear). In the Emotiv Pro software, the connectivity between the headset and the PC is checked. The headset is adjusted to get maximum connectivity. Now, the EOG electrodes are connected to the Holter. The recording is started for both the devices. After 2–3 min of recording

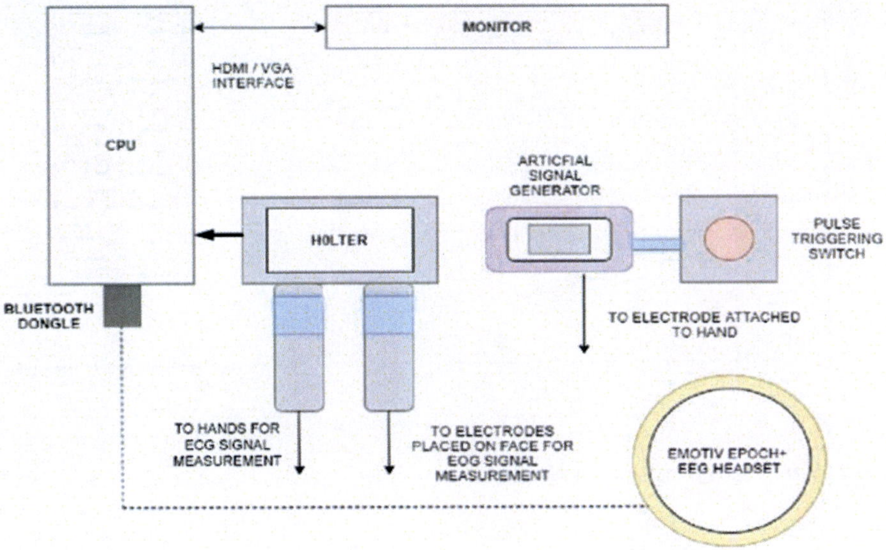

Fig. 2 Data acquisition setup

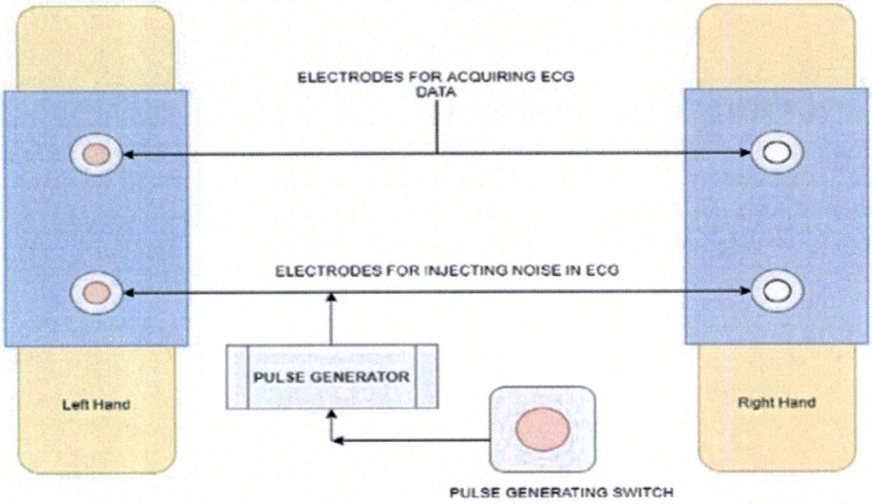

Fig. 3 Pulse injection setup

both EEG, ECG and EOG data, a train of pulses is injected for a duration of 30 s–1 min. This indicates the starting of EEG signal for processing. The patient is asked to perform certain activity based on the objective of the experiment (like word thinking). Data is recorded for a minimum of 10 min and a maximum of 1 h. Once sufficient data is recorded, a train of pulses with approximately the same duration is given.

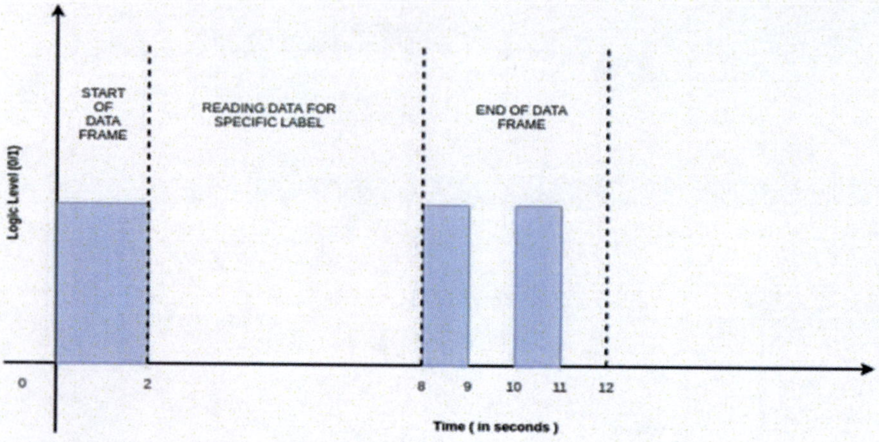

Fig. 4 Pulse waveform generation setup

This serves as an indication for the end time of taking readings from both Holter and headset. After the recording is terminated, data from the Holter is downloaded into local repository. Headset is removed from the subject's head, and the silica gel pads are removed.

2.3 Pulse Waveform Generation

The pulse waveform is generated in an endless loop. To indicate the start of a recording frame, a single pulse of width 2 s is given. Then, the output is low for a period of 6 s. Then, two pulses are generated with a 50% duty cycle and time period of 2 s. The pulse shown above is generated using an ATmega microcontroller (Fig. 4).

2.4 Collection of Data for Feature Classification

The subject under observation is asked to look on a monitor. Once a word is displayed on the screen, the subject should think about that in his mind for a certain period of time for 6 s. Then, the patient is given time to relax his mind and then process repeats till end time of experiment.

2.5 *Procedure for Classification of Cognitive Stress*

The main aim of the experiment is to predict what the person is thinking of. For that, we have to collect EEG data pertaining to some specific objects. Once all the equipment have been arranged, the PULSE START SWITCH of the interfacing circuit is pressed which starts the pulse generation from the microcontroller. At the same time, a program which would display the commands/instructions is executed. The program is as follows:

The word START appears on the screen, indicating the subject that the experiment is about to begin. This screen is displayed for 2 s. Then, a label is displayed for a period of 6 s. Then, the screen displays RELAX for 4 s and the loop starts over again.

3 Analysis Paradigms

As shown in Fig. 5, there are six steps in this denoising algorithm. Firstly, the EEG data and the EOG data recorded are preprocessed as per the preprocessing algorithm. After preprocessing the original signals, the second step is to decompose the EEG signal into multiple intermediate components (ICs) using a source separation

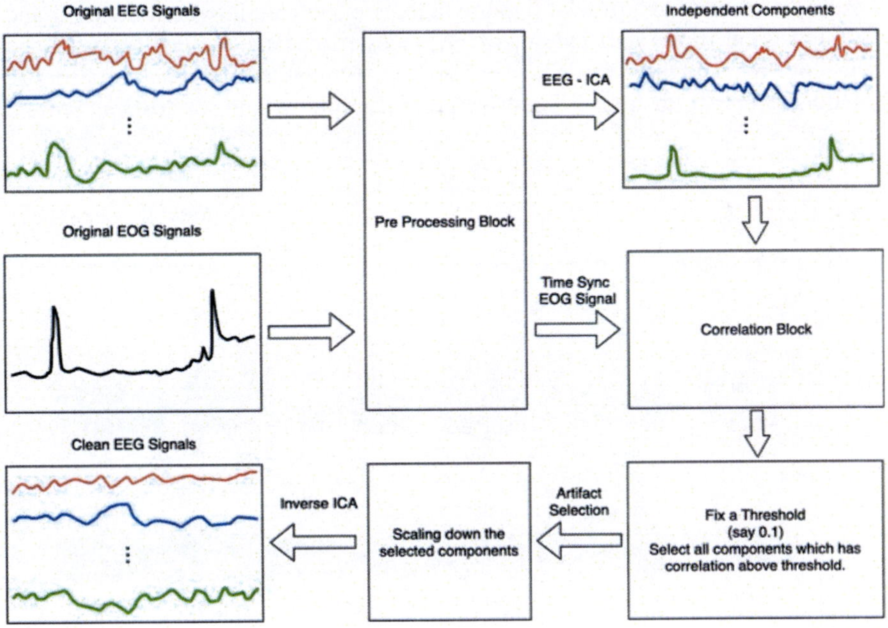

Fig. 5 Algorithm process diagram

technique called ICA. Thirdly, the correlation coefficient between all the EEG channels and the time synchronized EOG signal is calculated. Then, in the fourth step, a threshold is set and all the components having correlation value above the threshold are selected. In the fifth step, the selected components are scaled down rather than setting it to zero so that the information is not lost completely. Finally, after the scaling down of the components selected as artifact sources, the components are projected back into the group of noise-free components. Subsequently, electrooculogram-free signals are reconstructed with inverse ICA for artifact-free components. This noise-free signal is then sliced (640 samples each) into data and its corresponding labels (word associated with the data). This data-label pair is then used to train the artificial net to predict each word that the subject is thinking. According to the block diagram, this denoising algorithm can be described in detail as follows:

3.1 Preprocessing

The preprocessing step consists of two main processes which include time synchronization and band pass filtering. Devices which are used to capture the signals of electroencephalogram and electrooculogram have different starting time. To synchronize the time frame between both the device's signal, we use artificially generated pulses. A plot of the recorded signal is shown in Fig. 6.

Pulses used here are artificial impulse signals inoculated into both the EEG and EOG signals during the start and stop of the process. Since, the two devices cannot be started together, and therefore, a lag between two devices can cause inaccurate noise detection. Butterworth filter is used for preprocessing to remove artifacts present

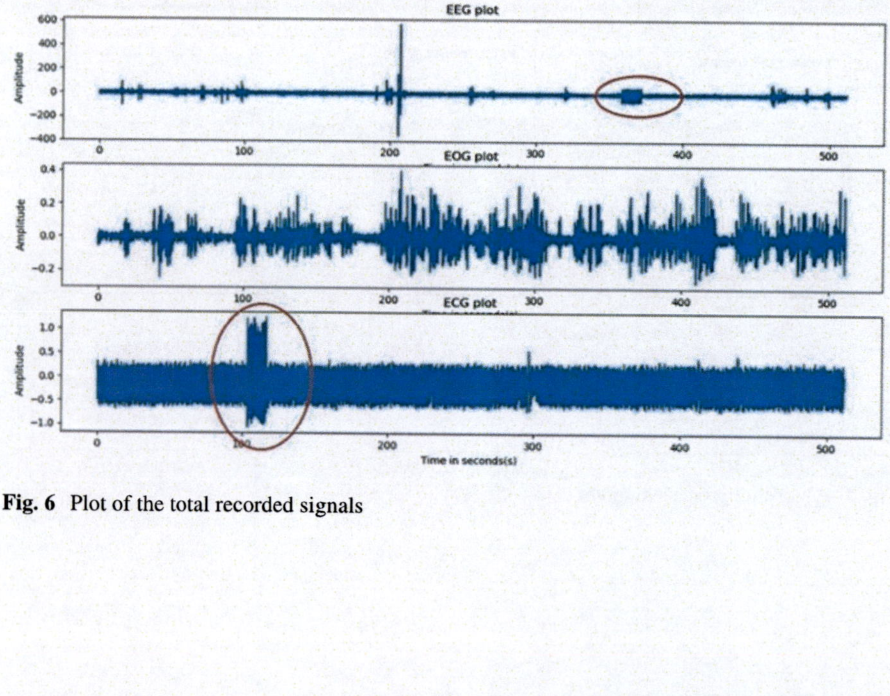

Fig. 6 Plot of the total recorded signals

above and below region of interest. It is a signal filtering mechanism designed to have a frequency reaction as even as possible in the bandwidth of the pass band region. This filter is also known as the maximally flat magnitude filter. It was defined originally by physicist Stephen Butterworth. The process is done for a sliding window of 20 s. The point of synchronization after the peaks are synchronized is shown in Fig. 7. A plot of the Fourier transform before and after preprocessing is illustrated in Fig. 8.

Program Algorithm: Data from the devices is imported into the program. Plot the whole recording for both the devices. Input the indexes of the synchronizing point of both the signals. Slice/crop the signals according to the indexes provided by the user. Filter the new data (time synchronized) using Butterworth filter (fifth

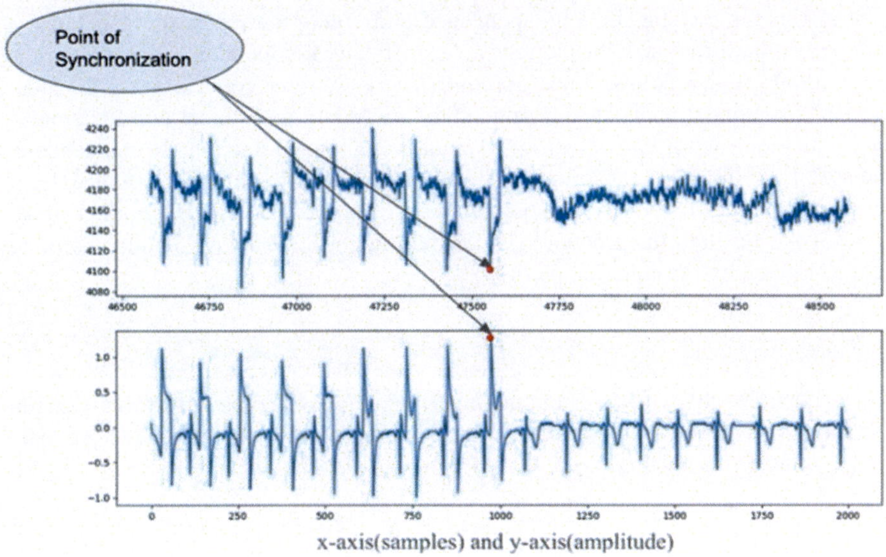

Fig. 7 Point of synchronization

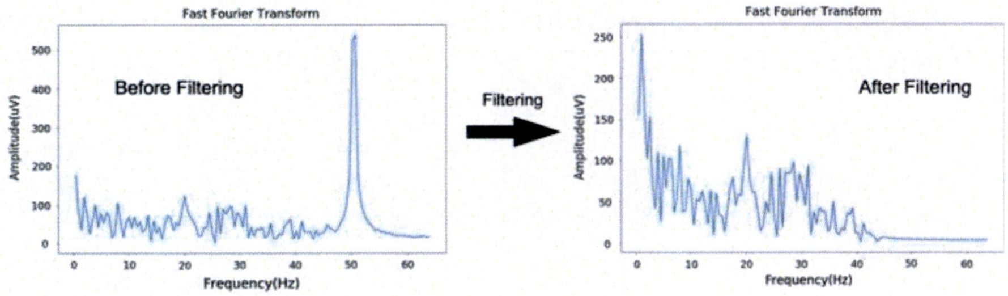

Fig. 8 Butterworth filter plot

order—bandpass (0.1–40 Hz). Write the time synchronized and filtered data into a CSV file.

3.2 *Independent Component Analysis*

ICA is a well-established technique on decomposing a signal into independent sources using kurtosis as the cost function. Independent component analysis is an effective technique to de-mix linear mixtures of signals into underlying independent sub-components. ICA utilizes higher order statistics like kurtosis to seek out the independent components. ICA is an extension of a statistical model called principal component analysis (PCA). Main supremacy in ICA is that it extracts the sources by investigating the statistical independence underlying the recorded data. Thus, it involves higher order statistics to recover statistically independent signals from the observations of an unknown linear mixture. There are three conditions for the independent component analysis algorithm. First, it should be a linear combination of the originating signals. Second, original signals ought to be statistically independent, and at last, the independent components should be non-Gaussian. For instance, let A be the mixing matrix, which is a square matrix of order two. Now, the mixed signal vector X is a result of the product between the source signal vector S and the mixing matrix A given by the equation below:

$$X = A * S \tag{1}$$

The core objective of ICA is to find the originating/source signals from the mixed signals. This is done by discovering the de-mixing matrix (given by W). The de-mixing equation is given as:

$$Y = A^{-1} * X \tag{2}$$

3.3 *Computation of Correlation Value*

The correlation value between two functions defines the trend between two variables on how strong the pair of variables is related. The correlation used here is Spearman correlation named after Charles Spearman. The Pearson correlation calculates linear relationships, whereas the Spearman correlation determines monotonic relationships. The value usually lies between $+1$ and -1 where the $+1$ shows positive trend; that is, both variables move in same direction, and -1 shows negative trend where the two variables move in opposite direction, while 0 shows no relation. The correlation coefficient was found between each channel of EEG signal and EOG data. The absolute estimate of the correlation is utilized in this research study as to avoid one particular limitation of ICA and the change in polarity of the electrodes. The value is calculated using the formula presented below:

$$\rho = 1 \frac{6 \sum d_i^2}{n * (n^2 - 1)} \tag{3}$$

where d is the difference between the rank of the value in the corresponding dataset.

3.4 Selection of Artifactual Components

The eye artifact is found by estimating the component which closely resembles the EOG data. The correlation values calculated using a statistical Spearman correlation are used to select the artifactual component by thresholding a value. Due to practical problems such as voltage bias, the correlation value can change from device to device. The thresholding value was found by experimentation to be 0.1. The independent components having correlation above the threshold are selected as EOG-related components. Additionally, optical examination is done to supervise the selection of all the components.

3.5 Reducing of Artifact Effect on Signal

The selected independent components are through a process of scaling down to reduce the effect on EEG signal. This scaling is done in such a way that the component having the most correlation with the EOG data is scaled down more than the component having less correlation. This can be explained by the two formulas defined below

$$\text{component} = \text{component} * (1 - 2 * \text{correlation}) \tag{4}$$

$$\text{component} = \text{component} * (1 - \text{correlation}) \tag{5}$$

Equation 4 is used when the correlation is between 0.1 and 0.5, while Eq. 5 is used when correlation is above 0.5, but the latter formula is applied as an exceptional case. This process of scaling the independent components helps in reducing the noise in the signal while retaining the feature of the EEG signal. Thus, the process of scaling down allows the denoising algorithm to remove the noise part and not affect the characteristics of the signal.

3.6 Signal Reconstruction

The artifact signals are selected as EOG-linked ICs, and their effect is reduced using the scaling algorithm. After the EOG-linked ICs are scaled down, the set of components are projected back on to the same space (inverse ICA) as the original components, therefore having EOG-free EEG signal.

3.7 Signal Learning Using Convolutional Neural Network

Apart from the denoising algorithm, in this section, we will be training a convolutional neural network to classify the input brain waves. The convolutional layer is the center of a convolutional neural network. The layer's specific parameters comprise of a bunch of teachable channels (or portions), which have a little open field, yet stretch out through the full profundity of the information volume. During the forward pass, each channel is convolved across the batch size, channel, and tests of the info volume, processing the scalar item between the sections of the channel and the information and creating a one-dimensional actuation guide of that across channels. Therefore, the organization learns channels that actuate once it recognizes some particular assortment of highlight at some spatial situation inside the info. The signs recorded are then parted with regards to the name and their relating signal for instructing. We have utilized 80% of the recorded information for preparing the neural organization and 20% to check the organizations exactness. Convolutional neural networks are normally utilized for pictures (two dimension), while this can be utilized in 1-D space (time-arrangement information) to perceive designs in mind waves. In the classifier, every one of the thick layers utilizes the ReLU (corrected straight unit) as the actuation work aside from the last yield layer (sigmoid). The reduction of the samples is by three stages of convolution (kernel-3, stride-1, padding-0 along with batch normalization and max pooling) with the ultimate layer giving three hundred channels. Dense layers have 23400 $\rightarrow$ 1024 (in the first hidden layer), 1024 $\rightarrow$ 512 (in the second hidden layer), 512 $\rightarrow$ 256 (in the third hidden layer), and 256 $\rightarrow$ 5 (in the fourth hidden layer) mapping to each of the five words.

4 Result

4.1 Quantitative Validation Measures

After processing the EEG signals, there are standard validation measures such as signal noise ratio and Spearman correlation. Signal noise ratio is one standard methodology to demonstrate the signal data over noisy data. The SNR of the signal can be found by

$$\mathrm{SNR} = 10 * \log(\mathrm{Pow}(S)/\mathrm{Pow}(N)) \tag{6}$$

where S relates to the signal and N happens to be the noise. The power can be calculated like the variance of the time signal by Hjorth activity. The increase in SNR value shows a strong increase in signal which is useful in applications such as sleep study for medical experimentation.

The correlation is also another useful parameter to measure the degree to which two variables move in relation to each other. This can be calculated by using the Spearman correlation using Eq. 3.

For neural network, the accuracy of the prediction is found out by dividing the count of images predicted correctly by the total number of images subjected to the network.

$$\text{Accuracy} = (\text{Total correctly predicted images})/(\text{Total images to the network}) \tag{7}$$

4.2 Experimental Data

In this study, the planned technique is used to get rid of EOG-related artifacts from electroencephalogram signals. A classic multiple channel of true electroencephalogram signals which are polluted with ocular blinks and with the original noise-free electroencephalogram signals once EOG-related signal removal is illustrated in Fig. 9. Especially, the channels close to the eye have shown massive amplitude of noise signals corresponding to the electrooculogram. The first signal bandwidth containing channel refers to the EOG signal.

We can use correlation values linking the electroencephalogram channels and electrooculogram signal to estimate whether the EOG-related signals were removed or not. The average correlation value of each electrode before and after denoising was calculated for the data and is illustrated in Table 1. It can be easily understood that the correlation value after denoising is less than the correlation value before denoising. The difference in correlation before and after denoising is dominant in the channels where the EOG-related signals have high influence. The channel having lower influence of eye artifact has less denoising effect, thus preserving the details of the signal.

Additionally, comparing the waveforms and the correlation value before and after denoising, it can be found that the denoising algorithm has effectively denoised the EEG signal without loss of too much information, thus preserving the data for further analysis.

As we do not know the noise and feature signals before denoising, we cannot calculate the SNR value before denoising. But after denoising since we know the denoised signal, the noise can be calculated using Formula 8.

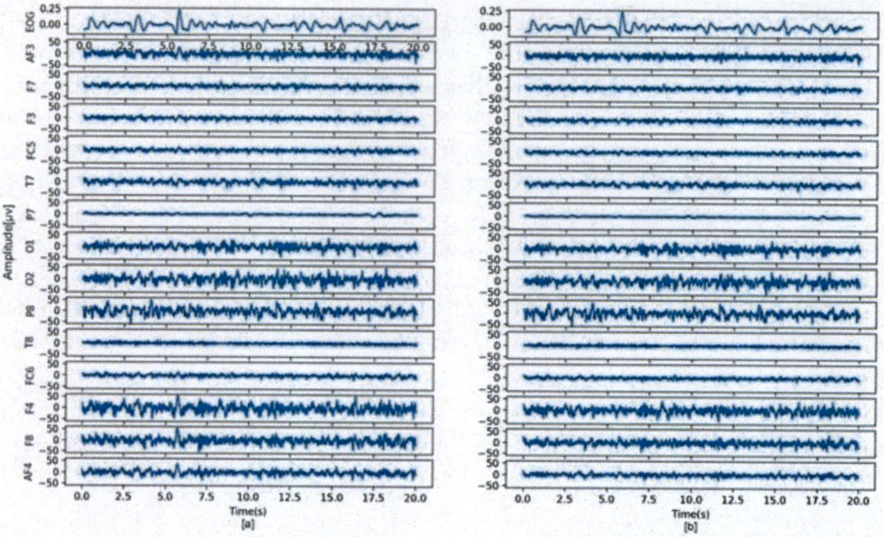

Fig. 9 **a** Typical example of a 14-channel electroencephalogram signal with effect of eye blinks, and the first channel is EOG reference signal. **b** 14-channel artifact-free EEG signal with first-channel EOG reference signal

$$\text{Noise} = \text{Input EEG Signal} - \text{Clean EEG Signal} \quad (8)$$

With the noise and the clean EEG signal, we can calculate the SNR value by using Formula 6. The average SNR value for all the subjects for the corresponding electrodes is shown in Table 2.

The SNR value shows that there is a significant increase in signal strength. The SNR is found to be high for electrodes which have a higher decrease in correlation between the corresponding electrode and the EOG signal.

Finally, we have successfully developed a convolutional neural network which can predict the state of mind (thinking of DOG, CAT, COLLEGE, LAB, and FRIENDS). After the neural network was trained, we subjected the test dataset to the network for label prediction and found to have an accuracy of 89.91%. This means that the neural net is able to precisely forecast one from the list of labels for 90 of every 100 images in the dataset. The learning loss of the neural network is shown in Fig. 10.

5 Conclusion

In this research paper, a technique proposed for cognitive stress classification was assessed by true and original electroencephalogram signals acquired from Emotiv Epoc Plus headset. The results indicate that the proposed method can eliminate EOG-related signals from acquired data with minimal loss of information and predict

Table 1 Value of correlation before and after denoising

Ch. name	AF3	F7	F3	FC5	T7	P7	O1	O2	P8	T8	FC6	F4	F8	AF4
Before	0.336	0.041	0.233	0.242	0.015	0.107	0.063	0.043	0.092	0.040	0.170	0.294	0.288	0.450
After	0.139	0.019	0.103	0.142	0.005	0.064	0.001	0.023	0.040	0.008	0.014	0.039	0.020	0.186

Table 2 SNR value for all the 14-electrodes after denoising

Ch. Name	AF3	F7	F3	FC5	T7	P7	O1	O2	P8	T8	FC6	F4	F8	AF4 SNR
Value	11.56	45.58	24.26	25.05	22.82	17.86	42.10	29.47	37.74	46.50	14.97	2.67	3.80	4.12

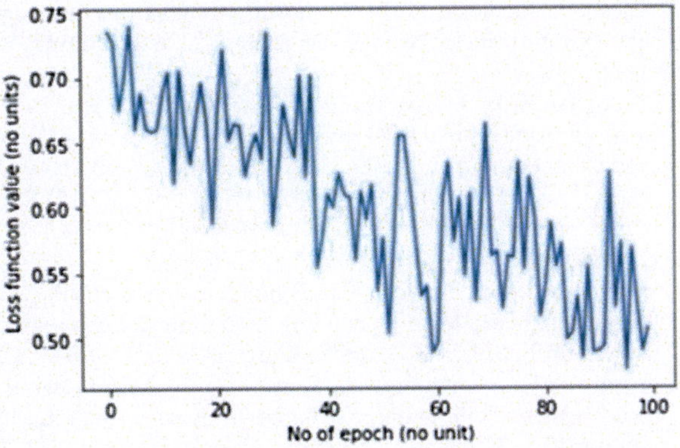

Fig. 10 Loss curve of the neural network

the cognitive stress. This can be seen by the increase in SNR value and also by the decrease in cross-correlation coefficient value. The method proves to remove dominant eye-related activity as the effect of EOG signal is higher compared to ECG and EMG signals. Also, the results from the convolutional neural network indicate a very high level of accuracy in classifying the cognitive stress.

Acknowledgements We would like to thank the Council of Scientific and Industrial Research—Central Electronic Engineering Research Institute, for supporting us with the laboratory and for providing us the necessary hardware equipment for data collection. We would also like to thank the three Homo sapiens who were in good physical health (gentlemen, average age 21, spanning 20–22 years old) who willingly took on to active participation in this study.

Declaration Authors' Contributions VKS finished the experimental research, participated within the evaluation of the signals, participated in the process of analyzing the various signals recorded, and drafted the manuscript. AS carried out the experimental studies, participated in the process of analyzing the various signals recorded, and drafted the manuscript. SR conducted literature survey and performed information acquisition. MKL coordinated the ethics application, came up with the direction for the evaluation of the conclusion, and confirmed the validity of the outcomes. BM coordinated the ethics utility, model recommendations for the evaluation of the terminal outcome and confirmed the validity of the effects. All authors read and accepted the very last manuscript. All authors read and permitted the very last manuscript.

References

1. Radu¨ntz T, Scouten J, Hochmuth O et al (2017) Automated EEG artifact elimination by applying machine learning algorithms to ICA-based features. J Neural Eng 14:8–15
2. Mahajan R, Morshed BI (2015) Unsupervised eye blink artifact denoising of EEG data with modified multiscale sample entropy, kurtosis, and wavelet-ICA. IEEE J Biomed Health Inform 19:158–165
3. Mammone N, La Foresta F, Morabito FC (2012) Automatic artifact rejection from multichannel scalp EEG by wavelet ICA. IEEE Sens J 12(3):533–541

4. Turnip A, Junaidi E (2014) Removal artifacts from EEG signal using independent component analysis and principal component analysis. 296–302. https://doi.org/10.1109/TIME-E.2014.7011635
5. Zachariah A, Jai J, Titus G (2013) Automatic EEG artifact removal by independent component analysis using critical EEG rhythms. 2013 International Conference on Control Communication and Computing (ICCC), 364–367. https://doi.org/10.1109/ICCC.2013.6731680
6. Yuan Qi, Zhou Weidong, Liu Yinxia, Wang Jiwen (2012) Epileptic seizure detection with linear and nonlinear features. Epilepsy & Behavior 24(4):415–421. https://doi.org/10.1016/j.yebeh.2012.05.009
7. Jirayucharoensak Suwicha, Pan-Ngum Setha, Israsena Pasin (2014) EEG-based emotion recognition using deep learning network with principal component based covariate shift adaptation. Sci World J 2014:1–10. https://doi.org/10.1155/2014/627892
8. Turnip Arjon, Soetraprawata Demi, Kusumandari Dwi E. (2013) A comparison of EEG processing methods to improve the performance of BCI. Int J Signal Process Syst 63–67. https://doi.org/10.12720/ijsps.1.1.63-67

Perceiving Machine Learning Algorithms to Analyze COVID-19 Radiographs

Shilpa Hari Prakash, Adithya Narayan K. A., Gouri S. Nair, and Sandhya Harikumar

Abstract Coronavirus disease, also referred to as COVID-19, is a contagious illness generated by a respiratory virus. There has been an exponential increase with the amount of patients affected with COVID-19 that has put an exceptional burden on the medical care frameworks across the world. Analysis of COVID-19 disease from the images of Chest X-ray may help isolate high-risk patients, while test results are anticipated upon. With most X-ray frameworks currently digitized, there is no transportation time required for the samples, hence making it easier for the health care workers to analyze it. In this work, we demonstrate the potential of ResNet, which is a CNN, to diagnose Chest X-ray images. These images can be classified into Normal, COVID, or Viral Pneumonia efficiently using ResNet. As a result, the probability of detecting patients with COVID-19 is maximized through higher accuracy. Empirical analysis exhibits that the proposed neural network strategy is better than Support Vector Machine, Naive Bayes algorithm, Logistic Regression, and k-NN.

Keywords Machine learning algorithms · Convolutional neural network · ResNet

1 Introduction

Novel Corona Virus epidemic (COVID-19) is an infectious illness that is primarily spread through droplets that are manifested when an infected patient sneezes, exhales or coughs [1]. This disease has caused a global pandemic with 176,693,988 confirmed cases and nearly 3,830,304 deaths [2]. Even though pre-cautionary measures such as lockdown and social distancing were imposed, the steep rise in the number of COVID cases could not be controlled. The computed tomography (CT) scans and

S. Hari Prakash (✉) · Adithya Narayan K. A. · G. S. Nair · S. Harikumar
Department of Computer Science and Engineering, Amrita Vishwa Vidyapeetham, Amritapuri, India

S. Harikumar
e-mail: sandhyaharikumar@am.amrita.edu

R. P. Mahapatra et al. (eds.), *Proceedings of International Conference on Recent Trends in Computing*, Lecture Notes in Networks and Systems 341,
https://doi.org/10.1007/978-981-16-7118-0_25

chest radio graphs (X-rays) of people are the common tests to check for the symptoms of this disease. Such scans and X-rays have been made available in the form of various datasets [3] for predictive analytics. These datasets are being used to produce insightful knowledge for early detection of COVID-19 [4]. Because of the unavailability of remedial treatment or antibodies for everyone suffering from COVID-19 [5], early detection is of utmost value to give the convenience of quick confinement of the COVID suspected individual and to diminish the scope of spread of the disease to the healthy population. One of the most prominent strategies to explore the concealed patterns for such diagnosis is machine learning.

Machine learning is a strong pillar of Artificial Intelligence, which learns and generates models from the available data to produce some generalized concepts. Supervised learning is a type of machine learning algorithm that creates model from a labeled dataset having output requirement explicitly mentioned [6]. It is used for predicting the output of a given input by creating a target function from the available dataset. The apt learning methodology for diagnosing COVID from chest X-ray PNGs, is supervised learning. Various algorithms and architectures are at everyone's disposal. Machine learning also has a subset namely Deep Learning with a multilayer neural network architecture suitable for learning very complex models from huge datasets [7, 8]. Convolution Neural Network, a class of Deep Learning Neural Network is commonly used to analyze image datasets. CNNs are based on the idea that a local understanding of an image is sufficient. Having a few parameters itself improves learning time and also reduces the amount of data required to train the model. Without using a fully connected network of weights from every pixel, CNN uses only sufficient weights to analyze a small patch of the image. Convolution layer and fully connected layers require certain parameter setting and are predominantly used to predict or classify the image datasets [9]. It is the most easiest CNN possible and often users choose to add these layers with predefined models to improve the working. The other pre-trained CNN used for image classification include ResNet50, VGG-16, Inceptionv3, EfficientNet.

The main contributions of this paper are as follows. Adapt Deep Learning models namely ResNet to classify the X-rays into Normal, COVID, or Viral Pneumonia. Compare ResNet with various Machine Learning (ML) algorithms such as Support Vector Machine, Logistic Regression, Naive Bayes algorithm, and k-NN and a 13 layer sequential neural network model. The experimental analysis reveals that ResNet achieves 10% increase in accuracy.

Related Works A lot of research has been occurring with respect to artificial intelligence and machine learning on how it can help in classifying images [7, 10–13]. ML algorithms like k-Nearest Neighbors, Decision tree, Logistic Regression, Random Forest, Naive Bayes and Back Propagation algorithm based on Neural Networks are applied on fruits' image dataset for classification [12]. It shows that LR resulted in the maximum accuracy and the highest precision score of 92% and 91%, respectively, among 6 algorithms used. In a study on binary image face recognition technique [11], Logistic Regression and Neural network concepts have been proposed. Logistic Regression resulted in an accuracy of 100% for both training as well as testing while

3-Layer Back Propagation Neural Network gave a training accuracy equal to 100% and a testing accuracy equal to 97%.

Subsequently, there have been several attempts at using these Machine Learning algorithms for helping the medical workers in predicting COVID-19. The following research papers show how this ability of training the machine to think and predict results by itself has proved to be a boon to the human society. In [14], they suggest a small evaluation of X-ray scans for patients which would aid in controlling the transmission of COVID among potentially infected patients. A Neural Network based model was tested on a minor dataset with the initial samples containing COVID infected, Normal, and Viral Pneumonia infected patients. It resulted in an accuracy of 90.5% with 100% sensitivity upon validation. In [15], a deep feature learning model containing SMOTE and various Machine Learning classifiers are used on the Chest X-ray image dataset. A total of seven different models were used. Random Forest obtained an accuracy of 97.3% and XGBoost predictive classifiers produced an accuracy of 97.7%. In [16], many supervised ML models such as Decision tree, Support Vector Machines, Naive Bayes, Logistic Regression, and Artificial Neural Network were applied on Epidemiology labeled dataset. The results produced an accuracy of 94.99% as the highest accuracy in Decision Tree, while the highest sensitivity of 93.34% was observed in Support Vector Machines, and the highest specificity of 94.30% was observed in Naive Bayes. In [17], many factors were taken into consideration to understand how the corona virus affected people. Machine Learning algorithms like SVM, k-NN, SVM with grid search, Logistic Regression, Decision Tree, non-linear and linear regression were utilized for classification and prediction. The study reports that there are 3 factors affecting the classification which were the PIH, the gender and the number of days. SVM with grid search resulted with a maximum accuracy of 95% and then Decision Tree model with a modest accuracy of 94%. In [18], a dataset with 5000 images was prepared by collecting data from different public datasets in order to detect the possibility of COVID-19 infected patients. 4 popular Convolutionary Neural Network models, including SqueezeNet, DenseNet-121, ResNet50 and ResNet18 were used to train 2000 radiograms using transfer learning. Upon evaluation on the 3000 images, it was found that most of them produced a sensitivity rate ranging near 98%, while obtaining a specificity rate of nearly 90%. All these papers conclude that machine learning algorithms are viable and efficient in detecting COVID-19.

The COVID-19 Radiography Dataset is a dynamic dataset that gets updated with new data over a period of time. Hence, this dataset was found to be more relevant compared to the previously used datasets. This paper provides a comparison of both traditional as well as Neural Network based algorithms, thus providing a better domain of comparison. This paper presents a comparison of different machine learning algorithms such as Logistic Regression, k-NN, 13 layer Sequential CNN, Naive Bayes, SVM and ResNet and suggests the best possible algorithm among the chosen traditional and Neural Network algorithms to detect COVID infected patients.

2 Methodology

In this work, Supervised Machine Learning models such as K-Nearest Neighbors, Support Vector Machine, Naive Bayes's Algorithm, Logistic Regression, 13 layer Sequential CNN and predefined ResNet was used to classify the images from chest X-ray dataset into Normal, COVID-19, or Pneumonia class.

The block diagram in Fig. 1 depicts the strategy used to implement various algorithms. The follow up phases include preprocessing and visualizing the data, splitting them into train and test dataset as per the requirement and finally implementing various models, training and validating them with the preprocessed dataset and testing them to learn their accuracies for comparison purpose.

Data Preprocessing The COVID-19 Radiography dataset which contains nearly 3800 images in total, were preprocessed and visualized as per the algorithm used. For algorithms such as k-NN, SVM, Naive Bayes and Logistic Regression, raw pixel data as array and feature vector array were created from the pixel values. The raw pixel array has 3829 rows and 3072 columns, whereas for feature vector, and its shape is 3829×512 because feature vector consist of only necessary feature values whereas the former contains all the values stored.

Convolutional Neural Network has a slightly different way of preprocessing the image data. Large image dataset was loaded from the directory and preprocessed by resizing, converting to RGB and shuffling. Image data generator allows augmenting images in actual time, while the model is still being trained, by which one make use of arbitrary transformations on individual training data when it is fed on to the model. This makes the model robust and also allows to save up on the overhead memory. Validation split of 0.2 was given so that there is enough validation data.

For ResNet-18, *torchvision's transform* method chained using *compose* function was used to preprocess the model. The images in the folders were resized, horizontally flipped and the normalized here. The function *ToTensor* within transform was used

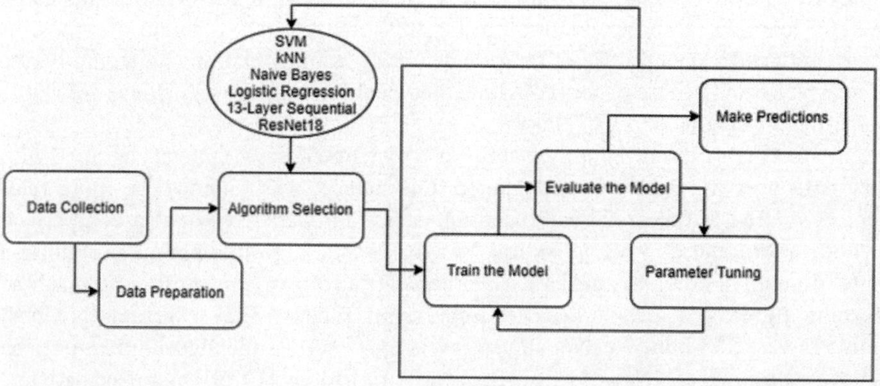

Fig. 1 Block diagram depicting the strategy adapted to implement various machine learning algorithms. This includes seven major steps from data collection to model's prediction accuracy

Fig. 2 Data visualization was performed after resizing and transforming the images for training the ResNet model. Image preprocessing included normalization and random horizontal flip. These images were resized to 300 × 300 pixel size so that the images has an even size throughout and can be better learned to further classify them.

here to convert values from 0 to 255 into a range from 0 to 1. The dataset was later randomly split for validation of the model. Upon data visualization as seen in Fig. 2, the images were augmented and transformed accordingly for both the ResNet-18 and Sequential model.

2.1 Conventional Machine Learning Algorithms

SVM is a discriminative supervised ML algorithm [6] which is used for classification or regression based problems. Here, the task is to classify linear or non-linear data into its corresponding classes by finding an appropriate hyper plane [19]. At first, the data items are copied into an m-dimensional feature space in which m stands for the number of rows. Then, a hyper plane is found out to perform classification which is used to classify the data with the help of a maximum margin.

Here, predefined SVM model was defined with the kernel parameter as linear and random state of 42, and its accuracy, recall, precision and f1 score were calculated. The same was done with raw images and with *k-fold cross validation.* Model performance for each of these methods were plotted for better understanding.

k-NN is an elementary supervised machine learning algorithms that is used for either classification or regression problems [19, 20]. The main motive of this algorithm is to measure and compare the distance of the testing data point with every other data. It then checks k-nearest members of that particular point in order to classify that point into a particular class.

The goal was to find the raw pixel accuracy of k-NN with 5 cross validation. A neighbor matrix with values 1, 3, 5, 10 and 25 was also defined. For these values, the model was fit with the training data which was already preprocessed and converted into arrays of image and label values. The raw pixel accuracy and recall values were plotted with and without *k-fold cross validation.*

Naive Bayes Algorithm is a generative supervised ML algorithm which is mostly used for classification problem and based on the concept of Bayes rule by Thomas Baye. Naive Bayes theorem is used for the calculation of the posterior probability from likelihood, class prior and predictor prior probabilities.

$$P(A|B) = (P(B|A)P(A))/P(B) \tag{1}$$

The term naive is used since it is assumed that the manifestation of a particular feature is autonomous of the manifestation of other features [21]. Naive Bayes classifier finds the probability of every features and then selects the outcome with highest probability among them.

The raw dataset was fit to the imported Gaussian Naive Bayes model with *k-fold cross validation* where *k* value is 5. The test dataset was further used to predict the model's accuracy and recall for each folds.

Logistic Regression is a parametric classification algorithm that frames a binary output model from a logistic function [19]. Hyper parameters such as regularization parameter and learning rate must be tuned properly to achieve high accuracy. Logistic Regression unlike SVM or k-NN can only handle linear solution and is also faster than k-NN [22].

Training raw dataset was fit to the predefined Logistic Regression function. Prepared test data was used to predict the accuracy and recall of the model.

2.2 13 Layer Sequential CNN Model

Sequential CNN Model is the easiest Convolutional Neural Network in *Keras* where layers are stacked one after the other. The function *add()* is used to add each different layer one after the other.

Sequential model was defined after importing Sequential function from *Keras*. The layers of the model consisted of *Conv2D*, *BatchNormalization*, *MaxPooling2D*, *Dropout*, *Flatten* and *Dense* layers making a total of 13 layers. *Batch Normalization* layer helps to train deep neural networks which standardizes its input into a layer all its mini-batch. Here for the first Convolutional layer, the number of filters are 32 with and then a *Max Pooling* layer that reduces the spatial dimensions of the output [23]. The activation functions used were *ReLu* and *Softmax* where *ReLu* is a simple to achieve, fast convergent speed achieving activation function that deactivates the neurons when the output of the linear transformation becomes less than 0 and *Softmax* on the other hand returns the probability that a data point is part of each individual class [24]. Various dropout values were used for each dropout function in this model and the output layer consist of 3 neurons. The augmented data with class names as 0 for COVID-19, 1 for Normal and 2 for Viral Pneumonia were loaded to the prepared model. Early stopping with patience equal to 5 was used to avoid overfitting, Adam optimizer was used here for optimization and *Sparse Categorical Cross entropy* to computes the loss. The model was fit with batch size = 32 and epochs = 20.

2.3 *ResNet18 Model*

ResNet18 is an 18 layer Deep Residual Convolutional Neural Network for image classification task [25]. ResNet-50 is acclaimed for its outstanding generalization performance with very less error rates on recognition work [26].

The pre-trained ResNet18 model contains *Conv2D* layers, *BatchNorm2D* layers, *ReLu* activation function layers, *AdaptiveAvgPool2d* layer with 512 input features and 3 output features. *CrossEntropyLoss* was used as loss function and *Adaptive Moment Estimation (Adam)* optimizer for optimization since it is a faster and reliably reaching a global minimum optimization algorithm than *stochastic gradient descent(SGD)*. The training of the model took place in 4 epochs and at various steps the accuracy and validation loss were printed. This was stopped when the performance condition was satisfied.

3 Experiment

3.1 *Experimental Setup*

For experimental analysis, Google Colab, a free notebook environment that runs in Cloud created by Google is adopted. It offers a robust GPU and provides fully configured runtime with AI libraries [27]. It also allows to import required dataset, train and finally evaluate the model. Thus, programmers can collaborate remotely to produce various solutions. Google Colab has a RAM size of 12 GB which is upgradable upto to 26.75 GB, runs in a cloud virtual machine and has a GPU Memory of 12 GB or 16 GB, respectively, for NVIDIA K80 or T4 GPU type. It has a disk space of 25 GB, and hence is the right choice when the task is memory consuming. After 12 h, the virtual machine and files are lost, and hence, the user needs to setup its run time from scratch. Even then, its reconfiguration is comparatively easy since the notebooks are saved in Google Drive, even though it takes some time [27].

COVID-19 Radiography Database is an X-ray database prepared by a team of researchers and medical doctors. It consists of chest X-ray PNGs of positive COVID-19 patients along with viral pneumonia and normal images. This dataset contains 1143 COVID positive PNGs, 1341 normal PNGs and 1345 viral pneumonia PNGs. The dataset is of resolution 1024-by-1024 pixel resolution. Here, we have used the dataset to prepare a machine learning model that rightly classifies a new test chest X-ray image into any of the above classes.

3.2 Experimental Results and Discussion

This project is an enhanced study of various Machine Learning and Deep Neural Network algorithms to compare and utilize the best one among the existing ones to classify if the X-ray image is of a patient suffering from Covid-19. Already existing papers completely focus on the comparison of either different Machine Learning algorithms or various Neural Network models on image datasets. Hence, a comparison among both Machine Learning algorithms and Neural Network models was necessary to understand that CNN is better at classifying X-ray images of the Covid affected patients. The proof of comparison is depicted in this section. Various models were trained and validated with the COVID-19 Radiography Database using various algorithms. Later, these models were tested with some unseen data to evaluate their performances. The accuracies of each model are specified here with respect to their training and test data.

The accuracy for SVM model using feature vectors and raw images without using k-fold cross validation were 72.26%, 91.59%, respectively. The accuracies for different k values during *k-fold cross validation* were well within the range of 91% with highest percent as 91.79% as depicted in Fig. 3.

The average accuracy and recall of k-NN Model are found out for each *k* value which is nearly in the range of 89.31% to 91.35% and 89.63% to 91.06%, respectively,

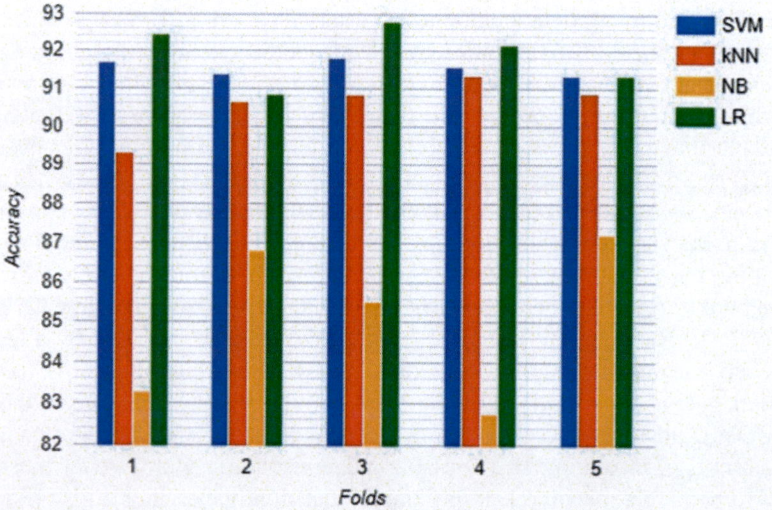

Fig. 3 Conventional machine learning algorithms like support vector machines, *k*-nearest neighbors, Naive Bayes and logistic regression's accuracies upon *k*-fold cross validation were plotted across the 5 folds. Each model was prepared using various parameters to increase its accuracies. SVM was prepared with a linear kernel and a random state as 42. k-NN on the other had *n* neighbor parameter set using a matrix with values 1, 3, 5, 10 and 25 and average parameter as macro. Logistic regression and Naive Bayes have the above mentioned average parameter as macro where Naive Bayes also has an array like shape with various fold values

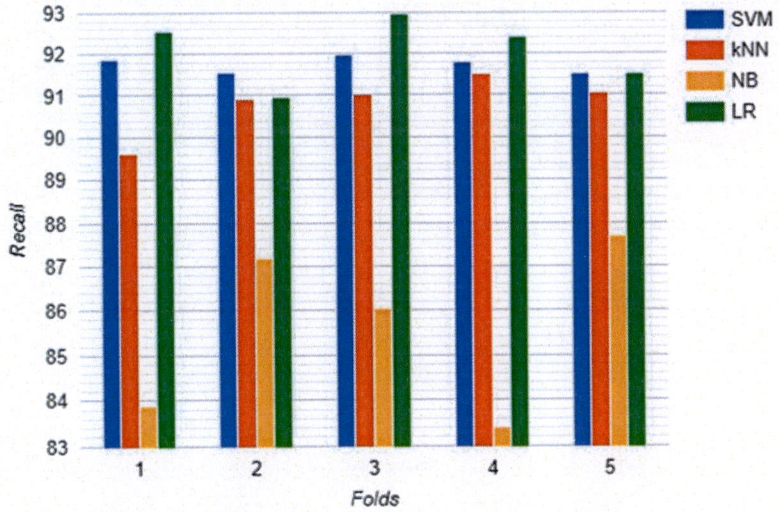

Fig. 4 Recall specifies how accurate the models were to specify pertinent data and hence serves an important role, while preparing models for health care purposes. Here, it is necessary to find how many people were really identified as COVID positive from the available dataset. Conventional machine learning algorithms like support vector machines, k-nearest neighbors, Naive Bayes and logistic regression's recalls upon *k*-fold cross validation were plotted across the 5 folds

as shown in Figs. 3 and 4. The same algorithm was also performed without fivefold cross validation where the neighbor matrix uses the same previous values which resulted in accuracies for these *k* values ranging from 90.40% to 93.32% Clearly, raw data accuracy without *k*-fold cross validation in k-NN yielded a better accuracy compared to the other.

In Naive Bayes model, the average of the accuracy and recall are 85.11% and 85.64%, respectively. A graph was prepared showing the accuracies and recalls for each fold to understand the performance in a better way as shown in Figs. 3 and 4.

For Logistic Regression, the average accuracy is 91.93% and average recall is 92.08% for raw pixel. Accuracy for each fold fall under the range of 90–93%. A graph was plotted showing the accuracies and recalls for each fold to understand the performance of the model better as demonstrated in Figs. 3 and 4.

The accuracy of the sequential CNN model when evaluated with test data was 93.725% at the end of the 12th epoch. Training and validation accuracies and losses were plotted to get a better understanding of the model.

The accuracy obtained from ResNet peaked to 98.83% in the 2nd epoch along with a validation loss of 0.0532, and hence produced a good result. The plots of both the epochs are plotted as shown Figs. 5 and 6.

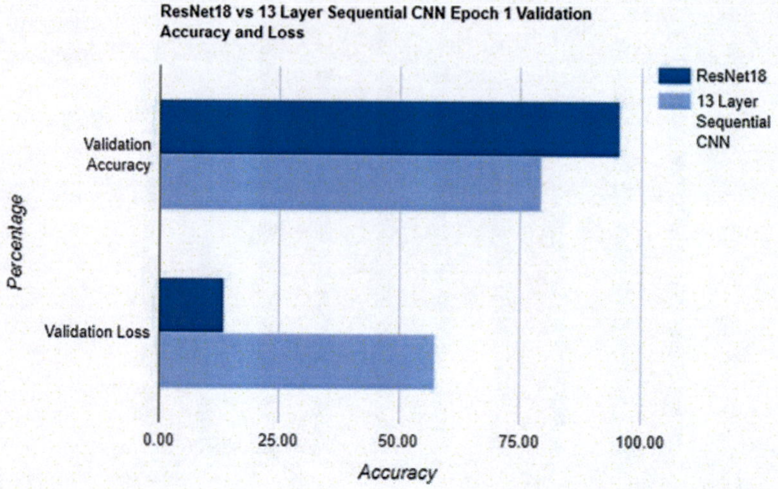

Fig. 5 Comparison of validation accuracy and loss of ResNet and sequential CNN algorithms at epoch 1 is plotted. A validation accuracy of 95.56% and loss of 13.58% was obtained for ResNet at the end of the 1st epoch which is comparatively better than the accuracy of the sequential CNN algorithm

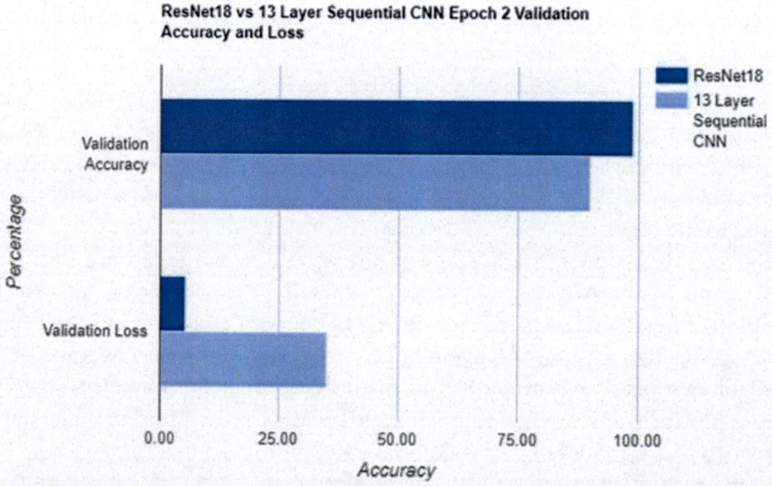

Fig. 6 Comparison of validation accuracy and loss of ResNet and sequential CNN algorithms at epoch 2 is plotted. A validation accuracy of 98.83% and loss of 5.32% was obtained for ResNet at the end of the 1st epoch which is comparatively better than the accuracy of the sequential CNN algorithm. ResNet has a ReLu activation function with 18 layers where input kernel is of size 3×3, with an adaptive average pooling, momentum equal to 0.01, CrossEntropyLoss, Adams optimizer and 2 epochs validation, where it stopped since an accuracy greater than 98% was obtained. 13 layer sequential CNN on the other hand has 13 layers, input layer kernel size of 3×3, ReLu and sigmoid activation functions, dropout values such as 0.3,0.2 and 0.15, sparse categorical CrossEntropyloss, Adams optimizer, batch size equal to 32 and 12 epochs validation after which early stopping took place

4 Conclusion

In this study, we compared various kinds of ML algorithms like Support Vector Machine, k-Nearest Neighbors, Naive Bayes's Algorithm, Logistic Regression, Sequential CNN and predefined ResNet model on COVID-19 Radiography database as shown in Fig. 7.

The goal of this comparative research was to find the most accurate machine learning model that can act as a tool for classifying chest X-ray as COVID, viral pneumonia or normal. Since more and more clinical data are being produced and documented for various researches on this infectious disease, a comparison of the best possible machine learning model is only possible when a common dataset is used on each model. Hence, COVID-19 Radiography Database, awarded as the victor of the COVID-19 Datasets by the Kaggle Community was used for this study. As per the prediction results shown in Fig. 7, ResNet which is a predefined CNN model has the highest accuracy of 98.83%. Hence, this model will be effective to detect COVID-19 infections through Chest X-ray images.

Acknowledgements The authors are extremely thankful to the faculties of the Department of CSE, Amrita Vishwa Vidyapeetham, Amritapuri Campus for their immense support and fruitful assistance in the fulfillment of this work.

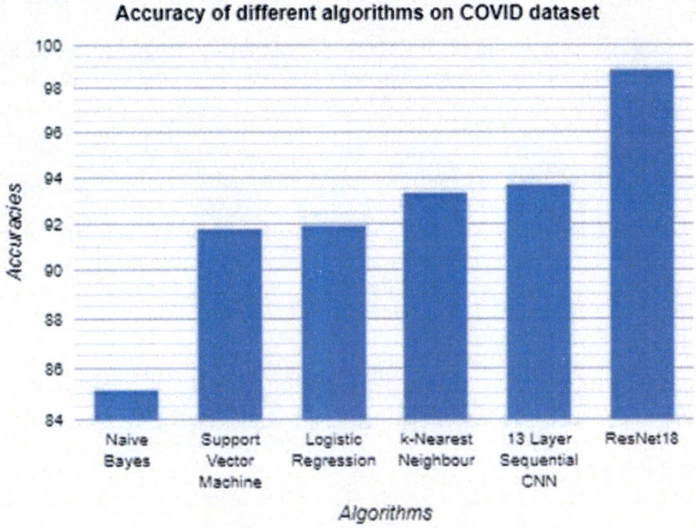

Fig. 7 Comparison of Naive Bayes algorithm, support vector machine, logistic regression, k-nearest neighbor, 13 layer sequential CNN and ResNet-18 models on COVID-19 radiography dataset. The accuracy of ResNet is the highest which is nearly 5% more than the 13 layer sequential CNN

References

1. Shereen MA, Khan S, Kazmi A, Bashir N, Siddique R (2020) Covid-19 infection: origin, transmission, and characteristics of human coronaviruses. J Adv Res 24:91–98. https://doi.org/10.1016/j.jare.2020.03.005, https://www.sciencedirect.com/science/article/pii/S2090123220300540
2. (2021) WHO coronavirus disease (COVID-19) dashboard. https://covid19.who.int/. Accessed 18 June 2021
3. Shi F, Wang J, Shi J, Wu Z, Wang Q, Tang Z, He K, Shi Y, Shen D (2021) Review of artificial intelligence techniques in imaging data acquisition, segmentation, and diagnosis for covid-19. IEEE Rev Biomed Eng 14
4. Chamola V, Hassija V, Gupta V, Guizani M (2020) A comprehensive review of the covid-19 pandemic and the role of IOT, drones, AI, blockchain, and 5g in managing its impact. IEEE Access 8:90225–90265. https://doi.org/10.1109/ACCESS.2020.2992341
5. Adithya J, Nair B, Aishwarya S, Nath L (2020) The plausible role of indian traditional medicine in combating corona virus (sars-cov 2): a mini-review. Curr Pharm Biotechnol 21. https://doi.org/10.2174/1389201021666200807111359
6. Joevivek V, Hemalatha T, Soman KP (2009) Determining an efficient supervised classification method for hyperspectral image. In: International conference on advances in recent technologies in communication and computing. pp 384–386. https://doi.org/10.1109/ARTCom.2009.174
7. Jia X (2017) Image recognition method based on deep learning. In: 29th Chinese control and decision conference (CCDC). pp 4730–4735. https://doi.org/10.1109/CCDC.2017.7979332
8. Ramachandran R, Rajeev D, Krishnan S, Palaniappan S (2015) Deep learning—an overview. Int J Appl Eng Res 10:25433–25448
9. Albawi S, Mohammed TA, Al-Zawi S (2017) Understanding of a convolutional neural network. In: International conference on engineering and technology (ICET). pp 1–6. https://doi.org/10.1109/ICEngTechnol.2017.8308186
10. Abdolmaleki P, Yarmohammadi M, Gity M (2004) Comparison of logistic regression and neural network models in predicting the outcome of biopsy in breast cancer from MRI findings. Int J Radiat Res 1(4). http://ijrr.com/article-1-33-en.html
11. Hruaia V, Kirani Y, Singh N (2017) Binary face image recognition using logistic regression and neural network. pp 3883–3888. https://doi.org/10.1109/ICECDS.2017.8390191
12. Vaishnav D, Rao BR (2018) Comparison of machine learning algorithms and fruit classification using orange data mining tool. In: 3rd International conference on inventive computation technologies (ICICT). pp 603–607. https://doi.org/10.1109/ICICT43934.2018.9034442
13. Zorgani MA, Ugail H (2018) Comparative study of image classification using machine learning algorithms. EasyChair Preprint no. 332 (EasyChair). https://doi.org/10.29007/4vbp
14. Mangal A, Kalia S, Rajgopal H, Rangarajan K, Namboodiri V, Banerjee S, Arora C (2020) Covidaid: covid-19 detection using chest X-ray
15. Kumar R, Arora R, Bansal V, Sahayasheela VJ, Buckchash H, Imran J, Narayanan N, Pandian GN, Raman B (2020) Accurate prediction of covid-19 using chest X-ray images through deep feature learning model with smote and machine learning classifiers. medRxiv. https://doi.org/10.1101/2020.04.13.20063461, https://www.medrxiv.org/content/early/2020/04/17/2020.04.13.20063461
16. Muhammad L, Algehyne EA, Usman SS, Ahmad A, Chakraborty C, Mohammed I (2021) Supervised machine learning models for prediction of covid-19 infec- tion using epidemiology dataset. SN Comput Sci 2(1):1–13
17. Dubey AK, Narang S, Kumar A, Sasubilli SM, Vicente GD (2021) Performance estimation of machine learning algorithms in the factor analysis of covid-19 dataset. Comput Mater Continua 66(2):1921–1936. https://doi.org/10.32604/cmc.2020.012151, http://www.techscience.com/cmc/v66n2/40629
18. Minaee S, Kafieh R, Sonka M, Yazdani S, Jamalipour Soufi G (2020) Deep-covid: predicting covid-19 from chest X-ray images using deep transfer learning. Med Image Anal 65:101794.

https://doi.org/10.1016/j.media.2020.101794, https://www.sciencedirect.com/science/article/pii/S1361841520301584
19. Uddin S, Khan A, Hossain ME, Moni M (2019) Comparing different supervised machine learning algorithms for disease prediction. BMC Med Inform Dec Making 19
20. Harikumar S, Akhil A, Kaimal R (2019) A depth-based nearest neighbor algorithm for high-dimensional data classification. Turk J Electr Eng Comput Sci 27(6):4082–4101
21. Kaviani P, Dhotre S (2017) Short survey on Naive Bayes algorithm. Int J Adv Res Comput Sci Manage 04
22. Isaac J, Harikumar S (2016) Logistic regression within DBMS. In: 2nd International conference on contemporary computing and informatics (IC3I). pp 661–666. https://doi.org/10.1109/IC3I.2016.7918045
23. O'Shea K, Nash R (2015) An introduction to convolutional neural networks. ArXiv e-prints
24. Lin G, Shen W (2018) Research on convolutional neural network based on improved relu piecewise activation function. Procedia Comput Sci 131:977–984. https://doi.org/10.1016/j.procs.2018.04.239, www.sciencedirect.com/science/article/pii/S1877050918306197. recent Advancement in Information and Communication Technology
25. Aydogdu MF, Celik V, Demirci MF (2017) Comparison of three different CNN architectures for age classification. pp 372–377. https://doi.org/10.1109/ICSC.2017.61
26. He K, Zhang X, Ren S, Sun J (2016) Deep residual learning for image recognition. In: IEEE conference on computer vision and pattern recognition (CVPR). pp 770–778. https://doi.org/10.1109/CVPR.2016.90
27. Carneiro T, Medeiros Da N´oBrega RV, Nepomuceno T, Bian G, De Albuquerque VHC, Filho PPR (2018) Performance analysis of google colaboratory as a tool for accelerating deep learning applications. IEEE Access 6:61677–61685 (2018). https://doi.org/10.1109/ACCESS.2018.2874767

Identification of Anomalies in Images Using CNN and Autoencoders Techniques

V. Navya and Kavita Avinash Patil

Abstract Identification of anomalies in a given arrangement of images is a responsibility of high practical significance for visual quality review, observation, or clinical picture examination. Autoencoder neural systems figure out how to reproduce typical pictures and, henceforth, can characterize those pictures as inconsistencies, where the remaking blunder exceeds some edge. Here, we examine an essential issue of this methodology when the preparation set is corrupted with a little portion of exceptions. We locate the reserved preparing of autoencoders unavoidably lessens the reproduction mistake of exceptions and henceforth corrupts the abnormality identification execution. So as to neutralize this impact, a compatible autoencoder design is adjusted, which forces earlier dissemination on the inactive portrayal, regularly putting irregularities into low-probability areas. Using the probability model, potential irregularities can be distinguished and dismissed previously during preparing, which bring about an abnormality indicator that is overall progressively strong to the nearness of anomalies during preparing. Ordered approaches are frequently hard to get a lot of named preparing information. In this paper, the proposed method is based on convolutional neural system to reconstruct the errors in pixel.

Keywords Anomaly identification · Autoencoders · Fault detection · Convolutional neural systems · Unsupervised learning · Image processing · Machine learning algorithms

1 Introduction

The objective of peculiarity discovery is to distinguish perceptions in a dataset that essentially stray from the rest of the perceptions [9]. Since abnormalities are

V. Navya (✉)
Department of Electronics and Communication, East Point College of Engineering and Technology, Virgonagar Post, Avalahalli, Bengaluru, Karnataka 560049, India

K. A. Patil
Department of Electronics and Communication, New Horizon College of Engineering, Marathahalli, Bengaluru, Karnataka 560103, India

R. P. Mahapatra et al. (eds.), *Proceedings of International Conference on Recent Trends in Computing*, Lecture Notes in Networks and Systems 341,
https://doi.org/10.1007/978-981-16-7118-0_26

uncommon and of different nature, it is not practical to acquire a named dataset delegate of every single imaginable inconsistency. A fruitful methodology for irregularity recognition [3] is to become familiar with a model of the ordinary class, under the suspicion that the preparation information comprises altogether of typical perceptions. In the event that a perception goes astray from that educated model, it is named an oddity [5]. Autoencoder neural systems have indicated unrivaled execution for distinguishing oddities on high-dimensional information, for example, pictures. Autoencoders comprise of an encoder organize, which performs nonlinear dimensionality decrease from the contribution to a lower-dimensional dormant portrayal, trailed by a decoder arrange, which recreates the first picture from this inert portrayal. Autoencoders do not require mark data since the information picture likewise speaks to the ideal yield. By figuring out how to separate highlights [4] and to reproduce the first pictures, the system yields a model that sums up to the remaking of pictures like those in the preparation set. Then again, pictures which show noteworthy deviations from those saw during preparing will prompt recreation blunders. The reproduction blunder of a picture would thus be able to be utilized as an inconsistency score. In spite of the fact that the autoencoder approach performs well on benchmark datasets [22], we recognize in this article a few significant inadequacies for true situations. To begin with, autoencoder strategies for oddity location depend on the suspicion that the preparation information comprises just of cases that were recently affirmed to be typical. Practically speaking, be that as it may, a clean dataset cannot generally be ensured, e.g., in view of comment mistakes, or in light of the fact that investigation of enormous datasets by area specialists is excessively costly or too tedious. It is along these lines alluring to get familiar with a model for irregularity identification from totally unlabeled information, in this manner taking a chance with that the preparation set is sullied with a little extent of peculiarities. Be that as it may, we find that autoencoder-based [7] inconsistency location strategies are touchy to try and slight infringement of the clean dataset suspicion. A modest number of inconsistencies tainting the preparation may result in the autoencoder figuring out how to recreate odd perceptions just as should be expected ones. We investigate the hidden foundations for this defenselessness of standard autoencoders and present a few key thoughts that make inconsistency recognition with autoencoders progressively powerful to preparing irregularities, along these lines improving the general peculiarity identification execution. In rundown, our commitments are: First, we use ill-disposed autoencoders [16], which permit to control the conveyance of dormant portrayals, consequently characterizing an earlier appropriation in the bottleneck layer. While (ill-disposed) autoencoders have been utilized for peculiarity identification before [15, 24], we here propose a novel rule for distinguishing inconsistencies comprising of both recreation blunder and probability in inactive space. Since inconsistencies are relied upon to have a low probability under the given earlier of the ordinary information, the blend of probability and recreation mistake yields an improved peculiarity score and in this manner better discovery execution. Second, we characterize a cycle refinement strategy for preparing test dismissal. Likely, irregularities in the preparation set are recognized in the lower-dimensional idle space by a variety of 1-class SVM [18], and by dismissing the least ordinary perceptions, we can expand power to sullied information. Third, we propose a retraining strategy to expand detachment in both dormant and picture space. We contrast our strategy with [10, 16], which just somewhat utilize the strategies joined in our methodology, and show that our proposed strategy brings about an altogether progressively strong peculiarity indicator against irregularities present during preparing.

In this work, a profound convolutional neural system is utilized for six consummation of surface pictures with the expect to distinguish stylish surface [12] deformities. Since the system is prepared only on ordinary information, it very well may be utilized to figure fault-free clones of the finished area. By taking away the relating inquiry district, pixel-wise irregular score is acquired, which may be used for identification absconds as shown in Fig. 1. Picture consummation errands ordinarily have the expect to finish missing areas of a picture in the most regular looking way. Other than being semantically important, the in-paint should likewise look as valid as could be expected under the circumstances. Therefore, feedforward DCNNs are regularly prepared together with an ill-disposed system, which was first done by Yu et al. in 2016. The ill-disposed system has the target to segregate among phony and genuine pictures. Conversely, the generative model [14] must build the blunder pace of the ill-disposed system by producing practical pictures. In spite of the fact that this extra ill-disposed misfortune in fact causes in-paints to look progressively sensible, it has no constructive outcome on pixel-wise coordinating the missing piece of the picture.

Preparing with the joint misfortune work even builds the pixel-wise remaking mistake, which is bothersome conduct for abnormality recognition. Hence, feedforward generation of DCNN is prepared using reproduction misfortune as it were. The reproduced flaw-free surface pictures of the contextual investigation are coordinated so intently that they are not really recognizable from the inquiry pictures. It is indicated that the recognition of well-unmistakable deformities on designed surfaces is conceivable with this methodology.

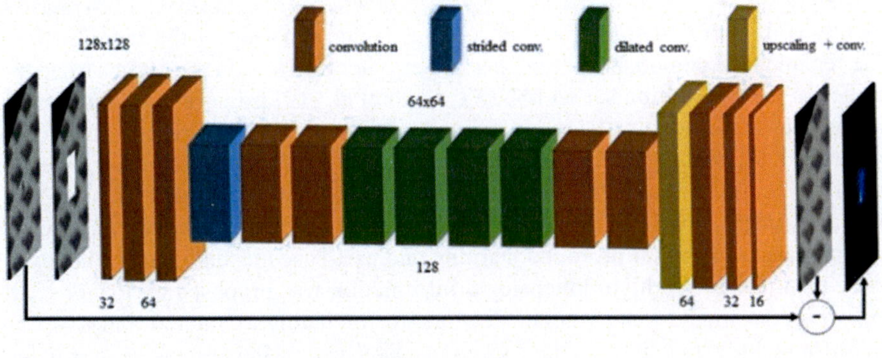

Fig. 1 Architecture of the proposed algorithm

2 Related Work

Autoencoders were initially planned for nonlinear dimensionality decrease and highlight extraction [10], yet it has been acknowledged from the get-go that their capacity to display the preparation information dispersion makes them reasonable for irregularity discovery [11]. Later work has proposed probabilistic translations [13] of profound autoencoders, which can straightforwardly show parts of the information producing process. Denoising autoencoders [20] take in recreation of pictures from clamor-ruined data sources. This type of regularization makes the inactive portrayal center around an information complex which encodes the most pertinent picture highlights. In [1, 2], it was demonstrated that regularized autoencoders certainly gage the information producing process and have set up joins between recreation blunder and the information creating thickness. Zhai et al. [24] applied these ideas to abnormality identification with profound organized vitality-based models, demonstrating that a measure dependent on a vitality score prompts preferable outcomes over the recreation blunder rule. Antagonistic autoencoders (AAE) [16] get familiar with a generative model of the information by joining the reproduction blunder with an ill-disposed preparing basis [8]. A discriminator arrange figures out how to recognize tests originating from the encoder and from an ideal discretionary earlier dissemination, which gives AAEs incredible adaptability to speak to presumptions about the information conveyance. AAEs for abnormality identification were first proposed in [15], utilizing a Gaussian blend model as earlier. It was discovered that a simply unaided methodology did not separate abnormalities and ordinary pictures into various groups, and it was proposed to either condition on class marks or train an express dismissal class with irregular pictures. Practically, all methodologies for peculiarity identification with autoencoders require the preparation information to comprise of ordinary models just; however, this by itself is no assurance for irregularities to have enormous reproduction blunders. Powerful profound autoencoders [25] address this issue by consolidating denoising autoencoders with strong PCA, subsequently confining commotion and exceptions from preparing of the reproduction. The technique accomplishes altogether better outcomes within the sight of irregularities in the preparation set on **MNIST**. Shah et al. [19] proposed utilizing a mix of vigorous misfortune work for autoencoder preparing along with semi-managed preparing of a classifier in inert space to beat the issue of adulterated preparing information. The technique accomplishes great recognition execution; in any case, their assessment shows that this expansion is primarily because of semi-administered [21] preparing. A mix of profound learning and part-based techniques for inconsistency identification in high-dimensional information was proposed by Erfani et al. [6], who consolidate a deep belief network for including extraction and a1-class SVM for oddity recognition in the packed inactive space. Their strategy can manage inconsistencies in the preparation information, yet it does not utilize this data to refine the preparation set. Conversely, [17] legitimately advanced the target capacity of a variety of 1-class SVM [23] in the yield space during system preparing. By doing as such, peculiarities can be distinguished promptly in the yield space, yet this data are not utilized during preparing for test dismissal. While considering recognition of potential antagonistic models, they have proposed thickness-based measures in a ConvNet to distinguish information focuses that lie outside the information complex. They increment the power of their technique by including a Bayesian vulnerability gage, which handles reciprocal circumstances.

2.1 Autoencoding

The autoencoder is prepared to limit the weighted recreation misfortune, where loads can change after some time. The equivalent-related example weight, wi, is utilized in the ill-disposed preparing methodology.

To iteratively refine the preparation set to make the model vigorous to inconsistencies present in preparing, the preparation technique is part into three stages:

Pretraining: The AAE is introduced via preparing on the total preparing set for a fixed number of ages where all loads are set to the indistinguishable worth, wi = 1. This is the beginning stage for peculiarity location in inert space with 1-class SVM in the resulting step.

Recognition and Refinement: A 1-class SVM is prepared on the idle portrayals with a steady expected peculiarity rate ν, yielding a lot of applicant oddities meant $A\hat{}$. All examples inside $A\hat{}$ are doled out another weight, wi = 0, along these lines expelling it from further preparing. The model is then prepared on the decreased preparing set $XnA\hat{}$ for a short number of ages. These two stages are rehashed d times where every redundancy expands the complete number of identified preparing irregularities. By iteratively barring up-and-comer peculiarities, the model of the ordinary class is refined.

Re-preparing: Subsequent to distinguishing inconsistencies in the preparation set and refining the model of the typical class, the model is re-prepared with the end goal that remaking mistakes on identified peculiarities increment. This can be accomplished by setting wi < 0; xi 2 $A\hat{}$, compelling a superior detachability of the two classes. The technique, in any case, made numerous bogus positive discoveries in the past refinement stage, which with this exacting reweighting, would wrongly done.

Architecture Design: A completely convolutional organize is utilized for picture culmination. By and large, the system comprises of 17 layers as shown in Fig. 1. Using third layer, goal of element maps is slitted by the strided convolution. Because of this expand open fields which give neurons, a progression of widening convolutions is utilized (layer of 6–10). Citified with back the information size at layers 14 is achieved by bilinear rescaled shadowed through via convolution. By the agreement, reflect cushioning is utilized using all convoluted layer. The exponential linear unit (ELU) actuation capacities are utilized.

Loss's function: The system stays prepared through L1 remaking misfortune. The 62×62 focus area, characterized through double veil J, is weighted uniquely in contrast to the rest of the locale. With Y being the picture fix to be investigated,

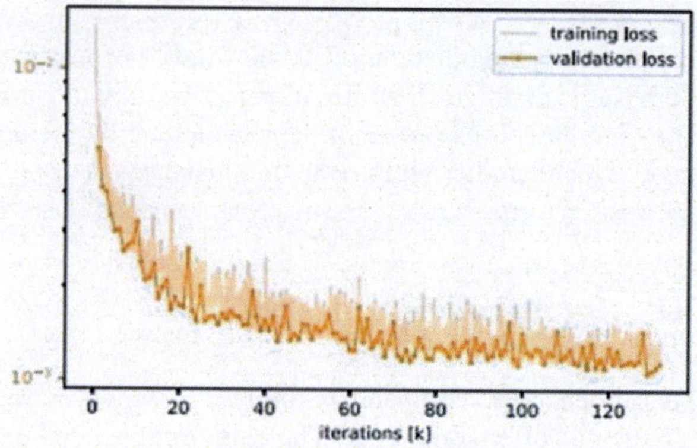

Fig. 2 Training of deep CNN for dataset

the system is prepared with the misfortune work:

$$L_i(Y) = \left(\lambda_1 \left|\left|JY - F\left(J^{-}Y\right)\right|\right| 1/S + (1 - \lambda_2)\right) \left|\left|(J^{-}Y - F\left(J^{-}Y\right)\right|\right| 1/S) \quad (1)$$

Where this indicates component insightful framework increase, J^{-} signifies the supplement veil of J, λ_1 is the weighting boundary between the inside and the rest of the area, S is the quantity of pixels of Y, and ||•||1 signifies the L1 network standard. The boundary λ_2 was picked to be 0.9 for the led tests. Since just the remaking mistake is important for peculiarity identification, an antagonistic misfortune is not included. Just, the pixel-wise recreation mistake is important for abnormality location. Thus, an ill-disposed misfortune is excluded, since an increasingly common looking picture does not diminish the recreation blunder.

Training Details: The picture finishing system was prepared without any preparation utilizing the ADAM streamlining agent [25] with hyperboundaries $\alpha = 0.0002$, $\beta 1 = 0.9$, $\beta 2 = 0.999$, = 10–9, a bunch size of 129. All loads remained introduced since a shortened Gaussian function dispersion as well as $\mu = 0$ and $\sigma = 1$. Predispositions are 0 announced. Individually, prototypical prepared on 200–400 k clumps inside 14 h on a GT 1090Ti is as seen in Fig. 2.

2.2 *Preprocessing*

The image finishing network is taken care of with picture area of size 138 × 138. The patches is degraded through veiling out of the focal region of size 64 × 64. This enormous proportion among known and obscure picture content furnishes the system with progressively semantic data to finish the inside district. Picture patches are separated progressively from high-goal surface pictures on randomized positions. Thusly, all potential patches are extricated from the crude information during preparing. Moreover, information enlargement comprising of randomized picture changes is performed. This, be that as it may, ordinarily prompts undesirable outskirt impacts.

So as to maintain a strategic distance from these impacts, patches bigger than the objective size are separated before applying these changes and focus trimmed to estimate 128 × 128 a while later.

After remaking of the undermined picture by the system, the pixel-wise outright distinction between the recreated picture and the inquiry picture is figured. For oddity recognition, just the 64 × 64 focus locale of this total distinction picture is utilized. Picture patches in which deformities show up near the outskirt of the cut out 22 × 22 focus district, and the neural system appears to create nearby continuations of the circumscribing surrenders. By considering just the 34 × 34 focus area, these undesirable continuations are generally barred. So as to guarantee that inconsistencies in the test information end up in this window in general at any rate once, a moving window with stride 16 is utilized.

2.3 Evaluation and Results

Evaluation

We train three distinct systems for each arrangement of boundaries, for example, number of units $h = 16/32$ just as with/without tests from the CIFAR informational index in the preparation information. For assessing imperfection discovery execution, we utilize both the general population and the opposition DAGM informational collections. Models are made by taking care of 500 non-faulty tests for each class into the system and taking the mean over the subsequent highlights. For every one of the remaining blemished and non-inadequate examples we at that point compute the Euclidean good ways from the model, yielding a separation framework of measurements 241 × 241. To survey how well non-defective and flawed examples are segregated, we figure the most extreme incentive for each separation lattice and compute the measure of genuine positives and bogus positives for changing limits. Genuine positives in this way indicate blemished examples accurately named flawed and bogus positives signify non-imperfect examples wrongly named damaged. The outcomes are communicated as recipient working trademark (ROC) bends. To evaluate the exhibition of each system in a solitary number, we ascertain the zone under the ROC bend (AUC). As rundown insights over the three systems prepared for each arrangement of boundaries, we utilized middle and standard deviations.

Results

Figure 3 shows imperfection discovery models for every one of the ten classes. Table 1 shows the region under the bend (AUC) scores for all classes and tests. It tends to be seen that exhibition differs a great deal contingent upon the surface sort. While the best outcomes for classes one, three, five, and six reach exhibitions that are valuable for a modern application, the strategy neglects to segregate blemished and non-deficient territories for classes two and four. It can additionally be seen that the preparation information has impressive impact on the deformity recognition execution. Adding tests from CIFAR-100 to the preparation information improves

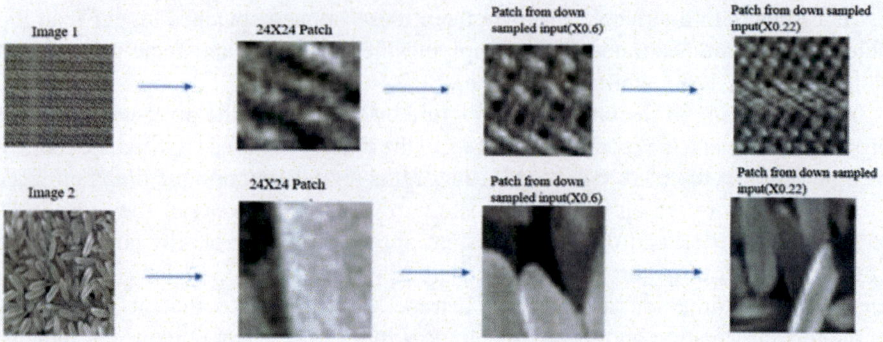

Fig. 3 Left to right: input texture and corresponding patches of 32 × 32 pixels for decreasing input sizes

Table 1 Region under the curve (AUC) scores for all classes as well as tests

Class	$H = 18$	$H = 18$, +Cifar	$H = 36$	$H = 36$, +Cifar
1	0.94 ± 0.02	0.94 ± 0.02	0.93 ± 0.02	0.93 ± 0.02
2	0.47 ± 0.01	0.42 ± 0.01	0.42 ± 0.0	0.49 ± 0.02
3	0.70 ± 0.03	1.0 ± 0.02	0.83 ± 0.02	1.0 ± 0.02
4	0.56 ± 0.01	0.39 ± 0.02	0.54 ± 0.02	0.54 ± 0.01
5	0.2 ± 0.03	2.0 ± 0.02	1.2 ± 0.02	0.43 ± 0.02
6	0.93 ± 0.01	0.94 ± 0.02	0.93 ± 0.02	1.0 ± 0.02
7	0.72 ± 0.08	0.93 ± 0.13	0.73 ± 0.02	0.62 ± 0.01
8	0.92 ± 0.01	2.0 ± 0.0	0.94 ± 0.02	2.90 ± 0.02
9	0.45 ± 0.02	0.92 ± 0.01	0.64 ± 0.02	0.86 ± 0.01
10	0.54 ± 0.02	0.53 ± 0.02	0.34 ± 0.02	0.64 ± 0.04
Mean	0.73	0.82	0.72	0.80

generally execution, particularly for the novel surface classes from the opposition DAGM informational collection. Regarding system boundaries, we locate no reasonable distinction between utilizing 16 or 32 units, for example, setting $h = 17$ or $h = 34$. For the most part, the strategy makes an interpretation of well to novel contributions as it can accomplish generally excellent segregation for classes eight and nine and great discriminative force for class seven.

3 Conclusion and Future Work

In this paper, the evaluation of image information is obtained by an intelligent convolutional neural system, trained to fix the images with the input textures and region of convergence. The system is prepared solely on issue-free information, and it finishes the picture patches with a deficiency-free form of the lost picture regions. Reconstruction of image gives a one pixel at a time anomaly score, which is along these lines utilized for deformity recognition. The system is prepared with L1 recreation misfortune with the missing focus area being weighted distinctively contrasted with the rest of the local regions. Regardless of not using an ill-disposed misfortune, the produced picture patches can scarcely be recognized from genuine ones. Particularly, differentiated irregularities can be identified, while decrepit, differentiated ones are regularly mistaken for remaining examples. On informational index A, the algorithm designed c accomplishes a pixel-wise AUROC (zone under region of convergence bend) of 0.94 and a one pixel at a time AUPRC (territory under PRC bend) of 0.44 and in this manner unmistakably outperforms the other tried strategies. Various machine learning training algorithms can be used to improve the accuracy of the result.

References

1. Alain G, Bengio Y (2014) What regularized auto-encoders learn from the datagenerating distribution. J Mach Learn Res 15(1):3563–3593
2. Bengio Y, Yao L, Alain G, Vincent P (2013) Generalized denoising auto-encoders as generative models. In: Advances in neural information processing systems. pp 899–907
3. Brodersen KH, Ong CS, Stephan KE, Buhmann JM (2010) The balanced accuracy and its posterior distribution. In: International conference on pattern recognition. IEEE, pp 3121–3124
4. Chan PP, Lin Z, Hu X, Tsang EC, Yeung DS (2017) Sensitivity based robust learning for stacked autoencoder against evasion attack. Neurocomputing 267:572–580
5. Chandola V, Banerjee A, Kumar V (2009) Anomaly detection: a survey. ACM Comput Surv (CSUR) 41(3):15
6. Erfani SM, Rajasegarar S, Karunasekera S, Leckie C (2016) High-dimensional and large-scale anomaly detection using a linear one-class SVM with deep learning. Pattern Recogn 58:121–134
7. Feinman R, Curtin RR, Shintre S, Gardner AB (2017) Detecting adversarial samples from artifacts. arXiv preprint arXiv:1703.00410
8. Goodfellow I, Pouget-Abadie J, Mirza M, Xu B, Warde-Farley D, Ozair S, Courville A, Bengio Y (2014) Generative adversarial nets. In: Advances in neural information processing systems. pp 2672–2680
9. Hawkins DM (1980) Identification of outliers, vol 11. Springer, Dordrecht
10. Hinton GE, Salakhutdinov RR (2006) Reducing the dimensionality of data with neural networks. Science 313(5786):504–507
11. Japkowicz N, Myers C, Gluck M et al (1995) A novelty detection approach to classification. In: Proceedings of the international joint conference on artificial intelligence, vol 1. pp 518–523
12. Kingma DP, Ba J (2015) Adam: a method for stochastic optimization. In: Proceedings of the international conference on learning representations. http://arxiv.org/abs/1412.6980
13. Kingma DP, Welling M (2014) Auto-encoding variational bayes. In: Proceedings of the international conference on learning representations
14. LeCun Y, Bottou L, Bengio Y, Haffner P (1998) Gradient-based learning applied to document recognition. Proc IEEE 86(11):2278–2324
15. Leveau V, Joly A (2017) Adversarial autoencoders for novelty detection. Technical report Inria—Sophia Antipolis. https://hal.inria.fr/hal-01636617
16. Makhzani A, Shlens J, Jaitly N, Goodfellow I, Frey B (2015) Adversarial autoencoders. arXiv preprint arXiv:1511.05644

17. Ruff L, Görnitz N, Deecke L, Siddiqui SA, Vandermeulen R, Binder A, Müller E, Kloft M (2018) Deep one-class classification. In: Proceedings of the international conference on machine learning. pp 390–4399
18. Schölkopf B, Platt JC, Shawe-Taylor J, Smola AJ, Williamson RC (2001) Estimating the support of a high-dimensional distribution. Neural Comput 13(7):1443–1471
19. Shah MP, Merchant S, Awate SP (2018) Abnormality detection using deep neural networks with robust quasi norm autoencoding and semi-supervised learning. In: Proceedings of the 15th international symposium on biomedical imaging. IEEE, pp 568–572
20. Vincent P, Larochelle H, Bengio Y, Manzagol P (2008) Extracting and composing robust features with denoising autoencoders. In: Proceedings of the international conference on machine learning. ACM, pp 1096–1103
21. Vincent P, Larochelle H, Lajoie I, Bengio Y, Manzagol P (2010) Stacked denoising autoencoders: learning useful representations in a deep network with a local denoising criterion. J Mach Learn Res 11:3371–3408
22. Williams G, Baxter R, He H, Hawkins S, Gu L (2002) A comparative study of RNN for outlier detection in data mining. In: Proceedings of the 2002 IEEE international conference on data mining. IEEE, pp 709–712
23. Xiao H, Rasul K, Vollgraf R (2017) Fashion-MNIST: a novel image dataset for benchmarking machine learning algorithms. arXiv preprint arXiv:1708.07747
24. Zhai S, Cheng Y, Lu W, Zhang Z (2016) Deep structured energy based models for anomaly detection. In: Proceedings of the international conference on machine learning. pp 1100–1109
25. Zhou C, Paffenroth RC (2017) Anomaly detection with robust deep autoencoders. In: Proceedings of the 23rd ACM SIGKDD international conference on knowledge discovery and data mining. ACM, pp 665–674

Big Data Privacy Preserved of END to END Delivery Solution with TRIZ Mechanism

Ashutosh Dixit, Nidhi Tyagi, and Rajeev Kumar

Abstract Data security is perhaps the most important issues of worry to deal with monstrous datasets in big data applications. The significant test of protection conservation over the distributed environment is dealing with the steady information on the grounds that the big data might be made and refreshed ceaselessly. Huge data began new issues related not exclusively to the volume or the assortment of the information, yet additionally to information security and protection. Techniques/statistical examination: Leading mechanical catchphrases, for example, big data. Security and ten significant standards, for example, segmentation, copying, feedback, homogeneity, beforehand cushioning, intermediary, self-service, merging, universality, and extraction among the 40 TRIZ standards are utilized in this investigation. This examination gathers the essential and auxiliary-related words with respect to significant watchwords. In the wake of gathering the related words utilizing text mining investigation, it finds again discovers words connecting significant catchphrases in innovations and standards in TRIZ.

Keywords Big data · TRIZ · Hadoop · Map reduce

1 Introduction

Today, the market is getting full grown step by step and everything in the market is getting digitalized. There can be numerous components for driving big data in the market as far as business where new and creative plans of action are coming. In technical terms, where persistent fizzling of customary arrangements in fulfilling the new market prerequisites and information is additionally expanding wherever that

A. Dixit
Bhagwant University, Ajmer, Rajasthan, India

N. Tyagi
MIET, Meerut, U.P, India

R. Kumar (✉)
Teerthanker Mahaveer University, Moradabad, India

R. P. Mahapatra et al. (eds.), *Proceedings of International Conference on Recent Trends in Computing*, Lecture Notes in Networks and Systems 341,
https://doi.org/10.1007/978-981-16-7118-0_27

too in different various configurations, information, we can see is filling dramatically on the lookout and last yet not the least money area is likewise the vital factor of driving big data in the market as the expense of keeping up the old and conventional information frameworks is ceaselessly developing to keep up the need of market, cost of equipment, and open-source programming likewise expanding [1–3].

1.1 Big Data

Big data is a clear as crystal word which itself implies an assortment of information that is huge in number. These informational collections are too perplexing to even consider being manage the conventional or existing information frameworks. These informational collections can be structure, unstructured, or a combination of both [4, 5].

The Three Versus of Big Data are Volume, Velocity, and Variety

1. **Volume**: The volume alludes to "the measure of information," which is expanding step by step and quick. The size of information produced by their connection on people, media, machines, and Web-based media is conspicuous.

Velocity: It alludes to the speed of the information that is gotten and followed up on. This is the speed which is utilized to fulfill the needs and difficulties that come in the way of advancement. Some Web-based items need constant assessment and activity [6–8].

Variety: It alludes to the sort and nature of the information accessible. Prior the customary and leaving information types were structure yet now daily's information comes in unstructured and unstructured arrangements like writings sound, and video that require extra preprocessing to infer meaning.

Alongside the over three significant versus of big data, two more versus have arisen in the course of recent years: Which is veracity and worth.

Veracity: It alludes to the ill-advised, untidy dubious, or uncertain information. As the information is coming fundamentally in unstructured structure, there is its more possibility containing a lot of unsure and loose information.

Value: Last yet not the least, indeed the main one as data have natural worth. It is the value of the information being extricated. Discovering an incentive in enormous information is not just about examining it [9–11].

Innovations and apparatuses to oversee big data

Apache Hadoop is a Java-based programming system that can store information in enormous amounts adequately cluster. MapReduce is another programming model for preparing huge informational indexes with an equal, appropriated calculation on a bunch. MapReduce when combined with HDFS can be utilized to deal with huge information.

1.2 Hadoop

Hadoop is an open-source structure that permits big data to be put away and prepared in the conveyed climate all through the utilization of straightforward programming models of PC gatherings. It has been made to scale a large number of single workers, offering every neighborhood estimation, and capacity. Living in a neighborhood record arrangement of the PC framework, in Hadoop, the information takes after the information lives in an appropriated document framework which is called as a Hadoop disseminated record framework [12].

TRIZ

TRIZ is a hypothesis for imaginative critical thinking created by Genrich Altshuller of Russia. It is the contraction of the Russian name, "Teoriya Resheniya Izobreatatelskikh Zadatch," and it is deciphered in English as "hypothesis of solving inventive problems" or "hypothesis of inventive problems solving (TIPS)." This is a strategy for getting creative thoughts or answers for tackle issues. Altshuller expressed that one of the qualities of TRIZ is that the revelation of groundbreaking thoughts is currently conquering opposing clashes. He additionally accentuated that on the off chance that we locate the opposing components in the issue and comprehend the marvels of shared impact between logical inconsistencies, we can discover new choices through this. He proposed the 40 standards of innovation, 76 standard arrangements, and critical thinking process (ARIZ) as TRIZ's imaginative standards (Table 1).

Table 1 A 40 inventive principles of TRIZ

1	Segmentation	15	Dynamics	28	Mechanics substitution
2	Taking out	16	Partial or excessive actions	29	Pneumatics and hydraulics
3	Local quality	17	Another dimension	30	Flexible shells and thin films
4	Asymmetry	18	Mechanical vibration	31	Porous materials
5	Merging	19	Periodic action	32	Color changes
6	University	20	Continuity of useful action	33	Homogeneity
7	Nested dot'	21	Skipping	34	Discarding and recovering
8	Anti-weight	22	Blessing in disguise	35	Parameter changes
9	Preliminary anti-action	23	Feedback	36	Phase transitions
10	Preliminary action	24	Intermediary	37	Thermal expansion
11	Beforehand cushioning	26	Self-service	38	Strong oxidants
12	Equipotentiality	26	Copying	39	Inert atmosphere
13	The other way around	27	Cheap short living	40	Composite material films
14	Spheroidality				

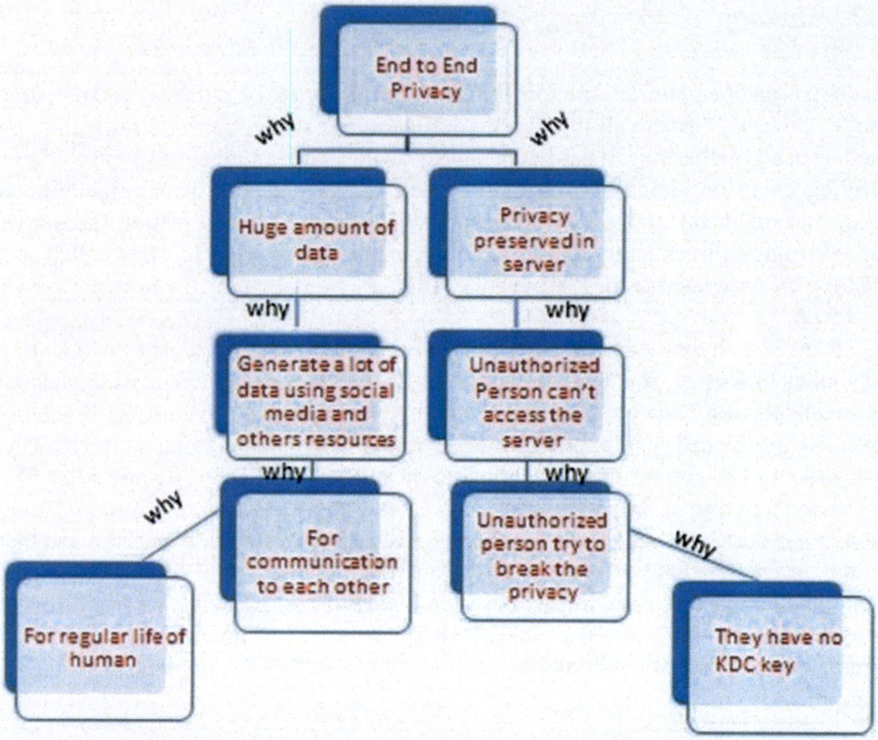

Fig. 1 Cause and effect chain analysis

1.3 Data Collection and Method

This study has following research procedure as shown in Fig. 1. Firstly, data are collected from portal size. Through the portal site, we search for the selected keyword and collect associated words with these keywords. We then select ten among the 40 TRIZ principles as keywords and collect associated words with them.

2 Analysis and Results

2.1 Problem Identified

In distributed environment security and privacy are happens with a few issues.

Appropriated Nodes

A disseminated hub is a structural issues and issue. The computation is done in any arrangement of hubs. It is exceptionally hard to guarantee the security of the entire calculation is finished.

(i) **Distributed Data**

Colossal measure of informational index can be put away in numerous pieces across numerous machines. In the event that a specific piece is ruined, the information can be recovered from its duplicates. In distributed environment, it is very hard to track down precisely where bits of record are put away since all connected information is absent in one spot and it changes.

(ii) **Internodes Communication**

The majority of Hadoop disseminations is use RPC over TCP/IP for client information/operational information move between hubs. Conceivable outcomes of altering the internodes correspondence for breaking into framework are high.

(iii) **Data Protection**

Most in HDFS, Hadoop stores the information for what it's worth without encryption to improve productivity. On the off chance that a programmer can get to a bunch of machines, it is highly unlikely to stop him. Hence, the likelihood of the programmer taking of basic information is high.

(iv) **Administrative Rights for Nodes**

A hub has managerial rights and access any information is risky as a malevolent hub can take or controlled basic client information.

3 Methodology

As a conveyance model for IT administrations, distributed computing can possibly upgrade business readiness and profitability while empowering more noteworthy efficiencies and diminishing expenses. The two advances keep on developing. Associations are moving past inquiries of what and how to store huge information to tending to how to infer significant examination that reacts to genuine business needs (Fig. 2).

3.1 Unlocking the Potential of Big Data in Clouds

Distributed computing models can help quicken the potential for adaptable investigation arrangements. Mists offer adaptability and efficiencies for getting to information, conveying bits of knowledge, and driving worth. Notwithstanding, cloud-based enormous information investigation is definitely not a one-size-fits-all arrangement.

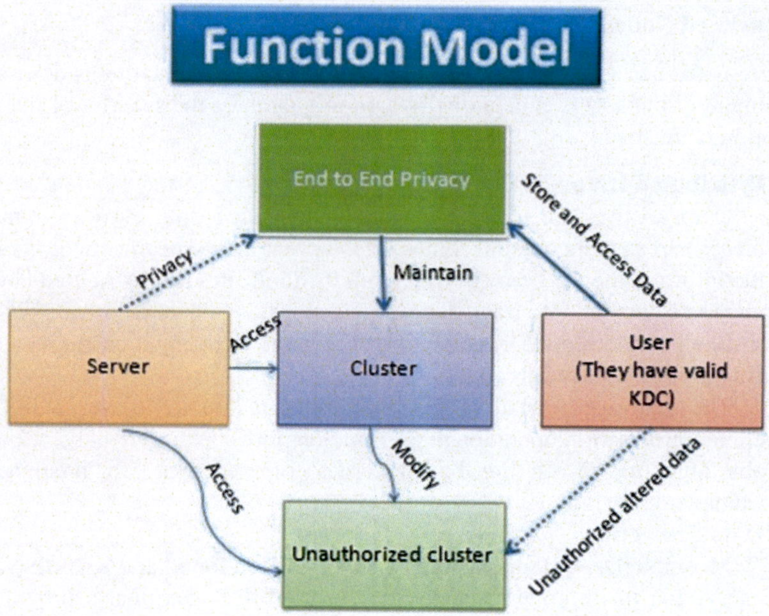

Fig. 2 Function model

Associations utilizing cloud framework to give an oh dear have numerous alternatives (Fig. 3).

Regardless of which cloud conveyance model bodes well, organizations with changing requirements, and financial plans can open the capability of large information in c uproarious conditions.

Fig. 3 Altshuller matrix result

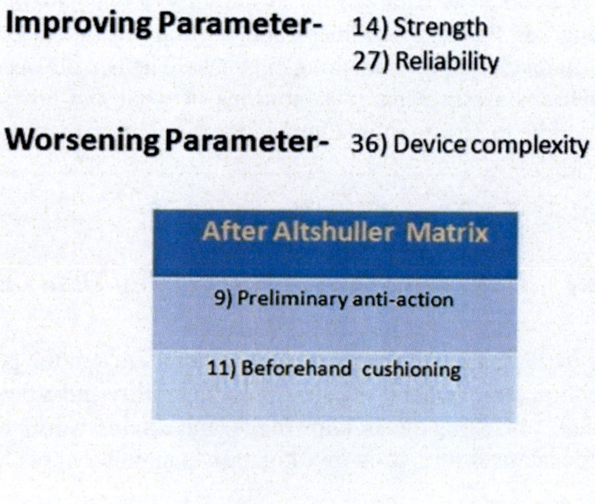

4 Conclusion

In this study, we suggest a text mining analysis to investigate the possible convergence among technologies using the principles of TRIZ. Based on the results from analysis, three cases of convergence were explained. For the analysis, each keyword was searched through the portal site and associated words with each keyword were collected. The associated word of a keyword is expressed as frequency.

We have enhanced the mechanism privacy preserved end to end safe delivery of the data in cloud as changes required. First of all, we should be assured in the distributed environment that the data are safe between the nodes and the data transfer is also safe. In the distributed environment, the machine serves the role of a third party publisher, which stores the sensitive data in cloud.

The limitations of this study are as follows. Firstly, there were some meaningful words but not excluded from the result because of the frequency difference of words related to keywords. Secondly, although there are 40 inventive principles of TRIZ, only ten populars were used in this study. Future studies will be able to obtain more diverse results by using more TRIZ principles.

References

1. Mallick PK (ed) (2015) Research advances in the integration of big data and smart computing. IGI Global
2. Risha T, Tyagi N (2017) Hadoop identity authentication using public private key concept. Int J Eng Trends Technol 45(9)
3. Mavi V, Tyagi N (2017) Data compression technique on Hadoop's next generation: yarn. Int J Sci Progress Res 92(33)
4. Mavi V, Tyagi N (2017) Hadoop's second generation—YARN. Int J Contemp Res Eng Technol 7(1). ISSN: 2250–0510
5. Dixit A, Dr. Kumar R, Tyagi N (2019) Big data in computer cyber security as an emergent infrastructure. Int J Anal Exp Mod Anal (IJAEMA) 11(12)
6. Saren N, Tyagi N (2016) Big data- a review for problem recognition. Int J Contemp Res Eng Technol 6(1). ISSN: 2250-0510
7. Dixit A, Tyagi N (2020) MapReduce: simplified data processing on clusters with privacy preserving by using anonymization techniques. Int J Recent Technol Eng (IJRTE) 8(6). ISSN: 2277–3878
8. Dixit A, Kumar R (2019) Enhancement of big data privacy preserved of map reduce by using data anonymization technique. Int J Anal Exp Mod Anal (IJAEMA) 11. ISSN: 0886–9367
9. Risha T, Tyagi N (2016) Issues and approaches for big data security. Int J Latest Technol Eng Manage Appl Sci 5(7)
10. Dixit A, Tyagi N (2019) Big data privacy for end to end delivery. Int J Recent Technol Eng (IJRTE) 8(2S8). ISSN:2277–3878
11. Mhamdi C, Al-Emran M, Salloum SA (2018) Text mining and analytics: a case study from news channels posts on Facebook. In: Intelligent natural language processing: trends and applications. Springer, pp 399–415
12. Amado A, Cortez P, Rita P, Moro S (2018) Research trends on big data in marketing: a text mining and topic modeling based literature analysis. Eur Res Manag Bus Econ 24(1):1–7

Heart Disease Prediction Using Various Machine Learning Algorithms

Arshdeep Singh Chudhey, Aditya Sharma, and Mrityunjay Singh

Abstract Cardiovascular disease is a major problem in modern society. The World Health Organization claims that cardiovascular diseases take the most lives when compared to any other disease taking an estimated 17.9 million lives each year. 4 in 5 deaths are due to strokes and heart attacks. Recently, data mining has proved to be an effective technique to analyze large amounts of data and provide valuable insights into it. When combined with machine learning techniques it can help to predict different heart diseases, therefore, helping in easy diagnosis and hence early treatment. In the course of this paper, various state-of-the-art machine learning models were used to achieve the best performance possible for predicting heart disease. The algorithms used are Random Forest classifier, Support Vector Machine classifier, Logistic regression classifier, and K nearest neighbor classifier. The best performance results were achieved by Logistic Regression when we obtained an accuracy of 91.80%. This approach can help to improve heart disease detection.

Keywords Heart disease detection · Machine learning · Feature selection

1 Introduction

Today, the leading cause of death is heart disease. Heart disease refers to various conditions of the heart, the most common among them is coronary artery disease. It affects the flow of blood towards the heart. The sedentary lifestyle of the modern world has led to an increase in heart problems. W.H.O. states heart disease causes an estimated 17.9 million deaths each year [1]. People with obesity and an unhealthy lifestyle are at a greater risk of heart diseases [2]. Unhealthy food has also turned out to be a major factor. Although heart diseases are a major threat to human life, they are very treatable today [3]. Therefore, detecting a heart disease at early stages can avoid it from becoming fatal.

A. S. Chudhey · A. Sharma (✉) · M. Singh
Department of Computer Science and Engineering, Jaypee University Of Information Technology, Waknaghat, India

R. P. Mahapatra et al. (eds.), *Proceedings of International Conference on Recent Trends in Computing*, Lecture Notes in Networks and Systems 341,
https://doi.org/10.1007/978-981-16-7118-0_28

A number of factors play their role in detecting heart disease [4]. The data available on the subject is very large and discrete, making it very complex to analyze. Data mining plays an important role for extracting important features from a given set of raw data. This helps to get hold of valuable data that can be used as training data for various machine learning algorithms. This helps to make our task easier by avoiding complex analysis of raw data. Machine learning algorithms can help to predict whether heart disease is present or not. There are a lot of machine learning algorithms present with respective pros and cons. This makes it necessary to find an algorithm that performs best for a specific application. In the proposed art, various machine learning algorithms are compared to predict whether a patient is susceptible to getting heart disease or not. The different accuracy of the algorithms helps to make the best choice for predicting the disease.

The paper follows a fixed set of proceedings: Section 2 gives the related work to our study in this area. Section 3 discusses the proposed approach; the experimental setup is discussed in Sect. 4. Section 5 sheds light on result and discussion and Sect. 6 gives a conclusion and feasible future work related to the proposed method.

2 Previous Works

A great deal of research work has been done to predict heart diseases. Apurb Rajdhan used Data mining techniques to obtain a patient's chances of having heart disease and his/her risk level [5]. They had the highest accuracy of 90.16% by using the Random Forest classifier. Singh [6] in their paper calculated the accuracy of machine learning algorithms in their task of predicting heart disease, and they achieved the highest accuracy of 87% for K-Nearest Neighbour. Dwivedi [7] in their work evaluated six machine learning models to predict cardiovascular disease and they achieved the best classification accuracy of 85% for logistic regression. Haq [8] used seven ML based algorithms along with 3 feature selection methods for the prediction of cardiovascular diseases and they achieved the highest accuracy of 89% using Logistic Regression. Milan Kumari[9] proposed a 10 fold cross-validation which was used to calculate the unbiased estimate of the predicted models and in their model they achieved the highest accuracy of 84.12% using SVM. Vinayaka [10] has proposed a classification method using Random Forest Classifier and has achieved an accuracy of 86.84%.

3 Proposed Framework

This section presents the proposed approach for Heart disease Prediction. This classification approach has adopted the use of the various state-of-the-art Classifiers like K Nearest Neighbor, and Random Forest Classifier. Our proposed approach works in the following phases: data processing phase, feature selection phase, model training phase, and validation phase. The detailed description of each phase is presented in the following subsections:

3.1 Data Pre-processing Phase

In this phase, we prepare our dataset using transformation and normalization then the missing values are checked in the dataset. Then, we decided to apply Normalization and Standardization on our dataset as they can simply increase the numerical stability of the model and can even help to speed up the training processes. We applied the Min-Max scaler for normalizing the categorical features of our dataset. After Normalization and Standardization, our dataset was later randomly distributed into 80% train and 20% test, and then feature selection was applied for the selection of the most relevant features.

3.2 Feature Selection Phase

In this phase, we applied feature extraction to choose the most relevant features given in the dataset, as these features help to make a more accurate prediction of heart disease. Feature importance can be used by the models to increase their performance as more features it's more likely that data will suffer overfitting. It is used for variable selection in which the variables that do not contribute significantly to model performance are removed. In the proposed model, three variables namely chest pain type, thalassemia, and st slope all were noticed to be categorical (i.e., they were not contributing significantly in the model) so we decided to turn them into dummy variables, and then we dropped those variables out of the dataset.

3.3 Model Training Phase

In our model, we have made use of the K-Nearest-Neighbor classifier, Support Vector Machine classifier, Random forest classifier and Logistic Regression to predict if the patient is suffering from heart disease. In this phase, we explain the detailed architecture of the machine learning algorithms used. Figure 1 denotes a detailed flowchart for the proposed architecture.

- *Logistic regression*—Logistic regression is an algorithm which is used for classification of data, and it consists of a cost function to make predictions and segregate them into different classes [11]. The cost is calculated using the sig- moid function after which optimization is performed. Equation (1) provides the value of the sigmoid function. The optimization is done using gradient descent and conjugate gradient.

$$S(x) = \frac{1}{1 + \exp^{-\text{val}}} \tag{1}$$

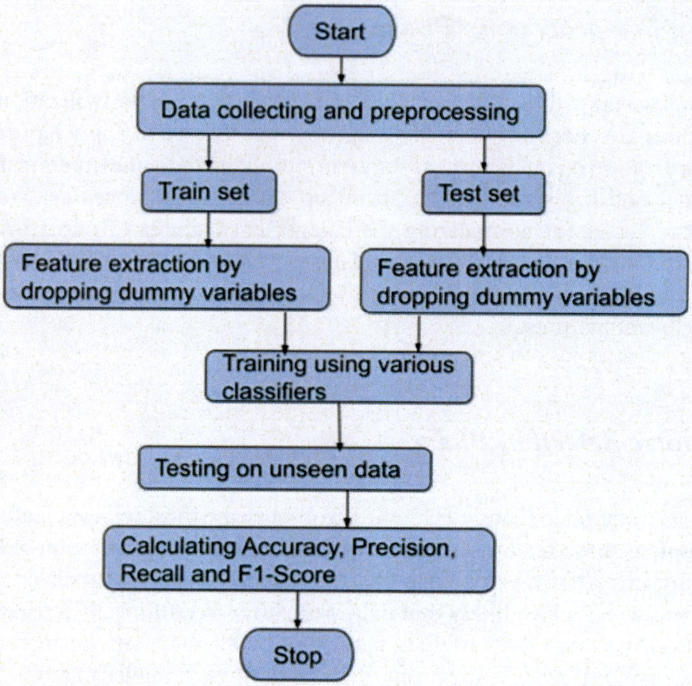

Fig. 1 The flowchart of the method proposed

- *K Nearest Neighbour*—KNN classifier is employed to segre- gate the input based on the features matching with previous data. It used Euclidean distance to allocate new classes based on the training data [12]. In this method, no assumption on underlying data is made. It provides a method to differentiate between the old cases and the new cases.

$$\text{distance}(x, y) = \sqrt{\sum_{i=1}^{n} (x_i - y_i)^2} \quad (2)$$

- *Support Vector machine*—Support Vector Machine (SVM) classifier learn- ing algorithm that uses supervised learning, and it assigns new inputs to different categories [13]. It makes use of a given set of training examples and uses margins to create a decision boundary pro- viding appropriate space between the objects (also known as Large margin classifier). The radial basis function kernel is used to implement this algorithm. The radial basis function is given by Eq. (3). A support vector machine is an alternative view of logistic regression and makes use of the "one-versus-one" approach for multi-class classification.

$$f(x, y) = \sum_{i}^{N} \alpha_i * y_i * k(x_i, x) + b \quad (3)$$

- *Random Forest classifier*—The random forest classifier is a supervised learning tech- nique that can be used as both regression and classification, here we are using it as a classifier [14]. It makes use of several decision trees. The decision trees make use of random samples independent of each other. The class which has been chosen by the maximum number of decision trees is selected as the final class. Random forests are in many cases better than decision trees.

3.4 Model Validation Phase

In this phase, we tested all 4 proposed machine learning models on the simple dataset, standardized dataset, and normalized dataset after feature selection and compared the performances of the model on various factors like precision, recall, accuracy, and F1-Score. Initially, for the random forest classifier, we achieved an accuracy of 86.88%, 0.8889 precision, 0.8276 recall, and 0.8572 F-1 score. Then, we applied Logistic regression and achieved an accuracy of 91.80%, 0.9259 preci- sion, 0.8929 recall, and 0.9091 F1-score. Then, we applied *K* Nearest Neighbour and we achieved an accuracy of 85.25%, 0.8621 precision,0.8334 recall, and, 0.8628 F1-score. Then, we applied Support Vector Machine and we got an accuracy of 86.89%, precision was 0.9655, recall was 0.8000, and, F1-Score was 0.8750.

4 Experimental Setup

All the experiments of our model were performed in a Google Colaboratory Server. During implementation, we have used python language and all its inbuilt libraries like Scikit, Numpy, Matplotlib and pandas. All the machine learning models were used on our initial dataset where the random state was taken as 0, the normalized dataset where the random state was taken as 27, and the standardized dataset where the random state was set taken as 42. The learning rate was sent to be 0.001 for all these models. Firstly for Random Forest Classiers numbers of trees were set to be 1000 and all other parameters were set to default. Then for Logistic Regression maximum iterations were taken as 1500 and all other parameters were set to default. Then for *K* nearest Neighbor numbers of the nearest neighbor was set to be 11 and all other parameters were set to their default. Then at last for Support Vector Machine regularization parameter was taken as 0.7 and gamma was taken as 0.1, and the remaining parameters were set to be the default.

4.1 Dataset

In our experiment, the dataset we have used is extracted from UCI Machine Learning [15] and it contains 303 instances out of which 138 people were healthy and the remaining 165 were suffering from heart disease. This dataset contains 14 different types of attributes like blood pressure, cholesterol, and maximum heart rate out of which 13 are predictors and 1 is a target class. In the target class, 1 represents patients who have heart disease and 0 represents patients who do not have a heart disease.

4.2 Performance Metrics

Various criteria are employed for performance evaluation of different machine learning models in our experiment are:

$$\text{Accuracy} = \frac{\text{TN} + \text{TP}}{\text{FP} + \text{FN} + \text{TP} + \text{TN}} \tag{4}$$

$$\text{Recall score} = \frac{\text{TP}}{\text{FN} + \text{TP}} \tag{5}$$

$$\text{Precesion} = \frac{\text{TP}}{\text{TP} + \text{FP}} \tag{6}$$

$$\text{F1-Score} = 2 * \frac{\text{Precision} * \text{Recall}}{\text{Precision} + \text{Recall}} \tag{7}$$

In Eqs. (4)–(7) FP, FN ,TP, TN, represents False Positive, False Negative, True Positive and True Negative, respectively. TP tells the proportion of people that were correctly labeled as suffering from heart disease. TN is the proportion of people that were correctly labeled as normal. FP is the proportion of people that were incorrectly labeled as suffering from heart disease. FN is the proportion of people that were incorrectly labeled as Normal.

5 Results and Discussions

In our study, we applied different machine learning algorithms for the predict- ing whether the patient has a heart disease or not. A number of machine learning models like Support Vector Machine, Logistic Regression, *K* Nearest Neighbor, and Random Forest Classifier have been trained to detect a heart disease. The highest accuracy has been shown by Logistic Regression when the data was standardized, followed by Random Forest and Support Vector Machine and the least accuracy was achieved by K nearest Neighbor. We noticed that Normalization and Standardization helped to increase the accuracy in all the Models. Although Standardization and Normalization had no impact on Random Forest Classifier. Figures 2 and 3 exhibit the confusion matrix for all the machine learning models. The random forest model predicts the 24 people correctly labeled as suffering from heart disease and 29 people correctly labeled as normal. The logistic regression predicts 25 people correctly labeled as

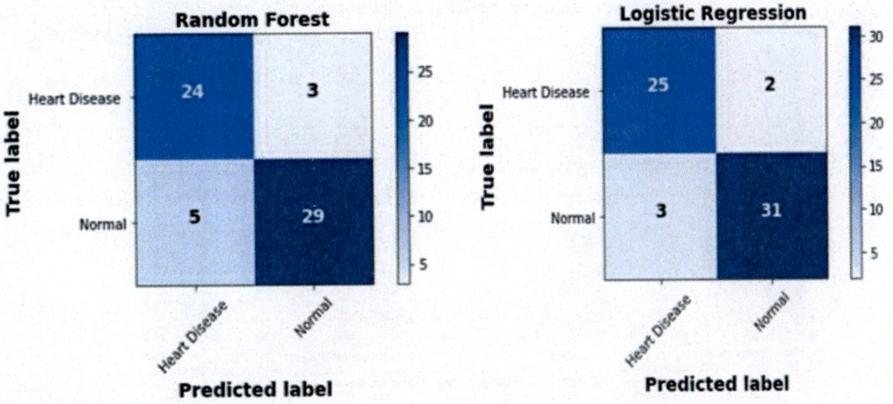

Fig. 2 Confusion matrix of random forest classifier and logistic regression classifier

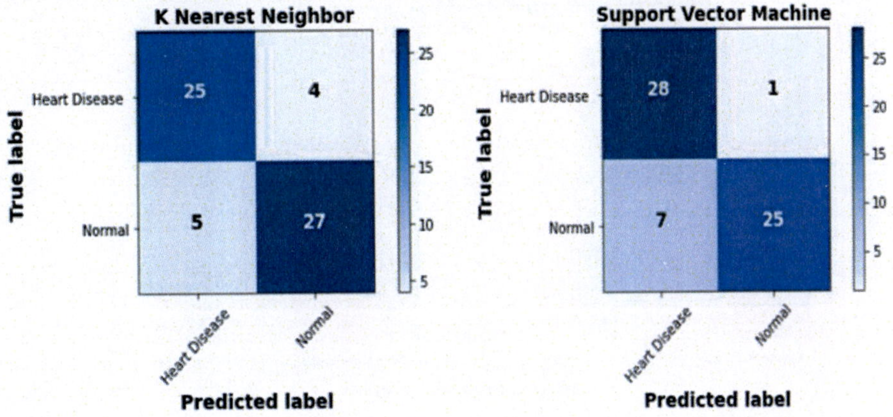

Fig. 3 Confusion matrix of K-nearest-neighbour classifier and support vector machine classifier

suffering from heart disease and 31 people correctly labeled as normal. The KNN classifier predicts 25 patients correctly labeled as suffering from heart disease and 27 patients correctly labeled as normal. At last, the SVM classifier predicts 28 patients correctly labeled as suffering from heart disease and 25 patients correctly labeled as Normal.

On the other hand, Figs. 4, 5, 6, and 7 exhibit the performance comparison of the various machine learning models used in this work in terms of various performance metrics. From the figures, it has been observed that the best performance obtained by our model was the accuracy of 91.80%, Recall of 0.8929, Precision of 0.9655, and F1-score of 0.9091. These figures represent a detailed comparison of various machine learning algorithm used in this work in form of accuracy, recall, precision and F1-Score for both cases with feature selection and without feature selection used during the preprocessing. Table 1 exhibits the accuracy comparison of our approach

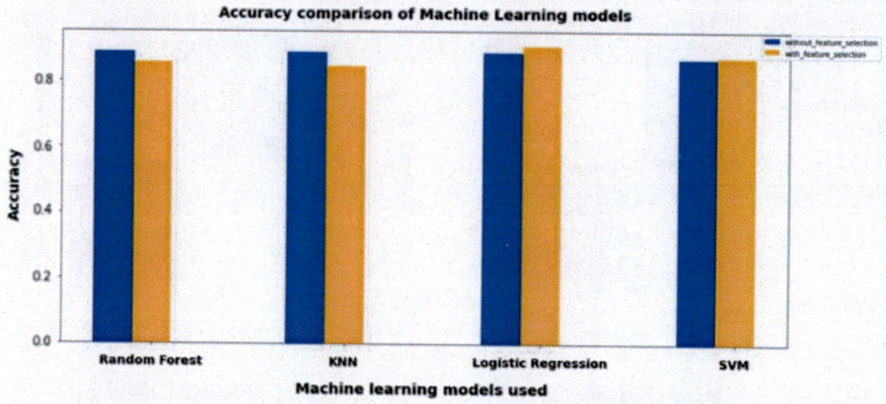

Fig. 4 Accuracy comparison of all the models

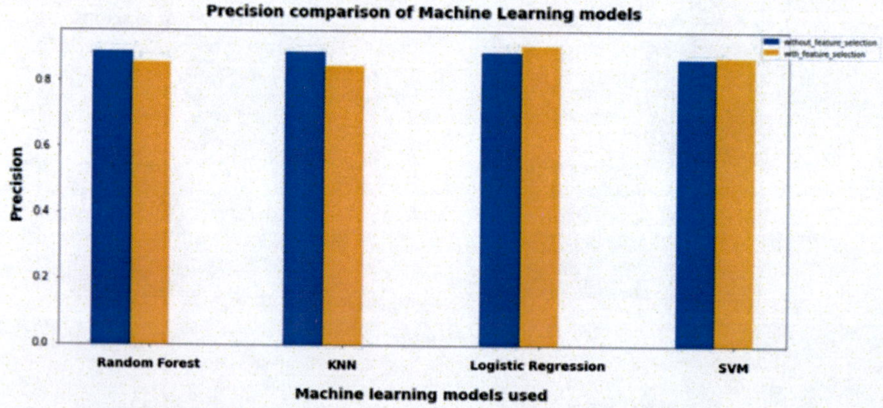

Fig. 5 Precision comparison of all the models

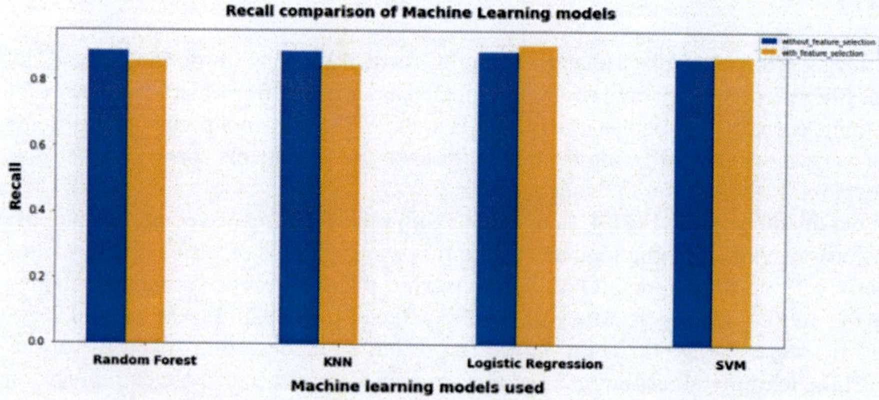

Fig. 6 Recall comparison of various models

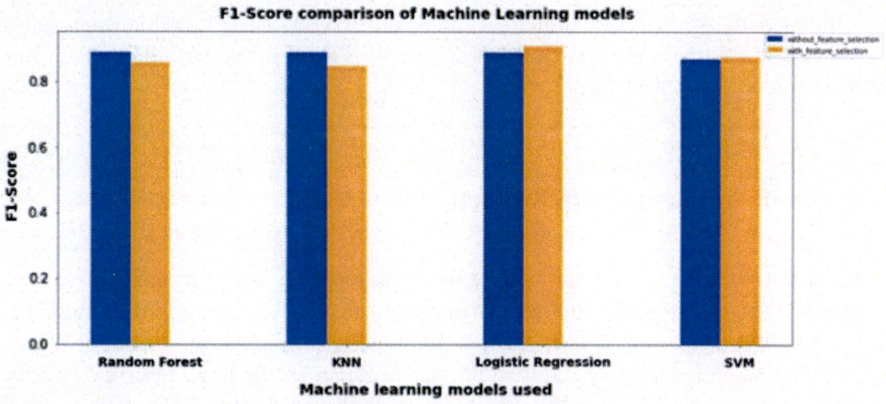

Fig. 7 F1-Score comparison of various models

Table 1 A comparison among the existing machine learning framework with proposed framework

Authors	Technique	Accuracy (%)
Rajdhan [5]	Decision tree	81.97
	Logistic regression	85.25
	Random forest	90.16
	Naïve Bayes	85.25
Singh [6]	Support vector machine	83
	Decision tree	79
	Linear regression	78
	K-nearest neighbor	87
Dwivedi [7]	Classification tree	77
	K-NN	80
	Naïve Bayes	83
	Logistic regression	85
	SVM	82
	ANN	84
Haq [8]	Logistic regression	89
	SVM (RBF)	88
Kumari [9]	RIPPER	81
	Decision tree	79
	ANN	80
	SVM	84
Vinayaka [10]	Random forest	86.84
Proposed framework	Random forest	86.89
	KNN	85.25
	Logistic regression	91.80
	SVM	86.89

with the existing work. From Table 1, it can be observed that the Logistic Regression classifier has achieved the higher accuracy of 91.8% with feature selection in our proposed in our proposed framework.

6 Conclusion and Future Works

In the proposed art, we have obtained a machine learning based framework to predict a patient's susceptibility for cardiovascular diseases. In the proposed framework, feature selection was done, followed by training of the state-of-the-art machine learning algorithms on the dataset and then the performance of the model was calculated using accuracy, precision, recall, and F1-Score, while predicting heart disease. The best performance was achieved by the Logistic Regression classifier that was 91.80% accuracy. The proposed method gives better results as the size of the dataset increases. This study can help in the early prediction of heart disease and prioritizing the treatment leading to precious time being saved for the patient. We may use some other existing machine learning models in the near future with the combination of existing feature selection techniques for better improvement in prediction tasks.

References

1. Organisation WH (2021) Cardiovascular diseases. https://www.who.int/health-topics/cardiovascular-diseases
2. Bohnert AS, Guy Jr GP, Losby JL (2018) Opioid prescribing in the united states before and after the centers for disease control and prevention's 2016 opioid guideline. Ann Intern Med 169(6):367–75
3. Cleaveland Clinic (2018) Heart disease. https://health.clevelandclinic.org/is-heart-disease-curable/
4. Virani (2020) Heart disease and stroke statistics-2020 update: a report from the American heart association. Circulation
5. Rajdhan A (2020) Heart disease prediction using machine learning. Int J Eng Res Technol
6. Singh A, Kumar R (2020) Heart disease prediction using machine learning algorithms. In: 2020 international conference on electrical and electronics engineering (ICE3), IEEE, pp 452–457
7. Dwivedi AK (2018) Performance evaluation of different machine learning techniques for prediction of heart disease. Neural Comput Appl 29(10):685–93
8. Haq AU (2018) A hybrid intelligent system framework for the prediction of heart disease using machine learning algorithms. Mob Inf Syst 2018
9. Kumari M, Godara S (2011) Comparative study of data mining classification methods in cardiovascular disease prediction 1. Int J Comput Sci Technol
10. Vinayaka S, Gupta P (2020) Heart disease prediction system using classification algorithms. In: International conference on advances in computing and data sciences. Springer, pp 395–404
11. Gasso G (2019) Logistic regression
12. Guo G, Wang H, Bell D, Bi Y, Greer K (2003) Knn model-based approach in classification. In: OTM confederated international conferences on the move to meaningful internet systems. Springer, pp 986–996
13. Cherkassky V, Ma Y (2004) Practical selection of SVM parameters and noise estimation for SVM regression. Neural Netw 17(1):113–126

14. Ani R, Augustine A, Akhil N, Deepa O (2016) Random forest ensemble classifier to predict the coronary heart disease using risk factors. In: Proceedings of the international conference on soft computing systems, Springer, pp 701–710
15. Aha DW (2020) Heart disease data set. https://archive.ics.uci.edu/ml/datasets/heart+disease

Home Security Using Facial Recognition System

Vibhu Dixit, Srishti Agarwal, and Nidhi Tyagi

Abstract Security is very important issue in today's time. Face recognition, or facial recognition, is one of the largest areas of research within computer vision. The development of facial recognition systems has also developed and has been implemented for home door locking systems and is an option that is quite simple and easy to use and is quite accurate in recognizing the face of homeowners. We can now use face recognition to unlock our mobile phones, verify identification at security gates, and in some countries, make purchases. Although lot of progress has been made in domain of face detection and recognition for security, identification and attendance purpose, but still there are issues hindering the progress to reach or surpass human level accuracy. This paper consists of face recognition and Machine Learning for detecting an intruder and recognition.

Keywords Machine learning · Security · Face recognition · Image processing · Python

1 Introduction

Nowadays, biometric-identification technology is being used almost by every active organization/firm. This technology is capable in identifying any individual via its image from any existing source. The Basic Working Principle behind this technology is comparing selected facial features from a database that contains a number of images. It requires well-designed and well-organized systems installed at different places including public places that can identify any person among the crowd using

V. Dixit · S. Agarwal
Meerut Institute of Engineering & Technology, Meerut, India
e-mail: vibhu.dixit.it.2017@miet.ac.in

S. Agarwal
e-mail: srishti.agarwal.it.2017@miet.ac.in

N. Tyagi (✉)
Shobhit Institute of Engineering & Technology, Meerut, India

R. P. Mahapatra et al. (eds.), *Proceedings of International Conference on Recent Trends in Computing*, Lecture Notes in Networks and Systems 341,
https://doi.org/10.1007/978-981-16-7118-0_29

their personal physiological characteristics and that too without letting them know about it. The result of every individual differs from others since it depends upon few factors like angle expression, facial structure, shape, size, etc., that means unique results every time. Quality measure plays a significant role in facial recognition as variation is also possible.

Few main objectives are:

- To uniquely identify a person.
- To represent data that the system can easily consume, interpret, and process.
- To ensure that this process is carried out in real-time.
- To Help Anti-terrorists Squad for recognizing terrorists.
- To reduce Hardware cost.

1.1 Present System

In our study, we concentrate on the various paradigms, which are used in machine learning. The existing literature also examines the comparative study of machine learning technique with suitable application area. The existing system consists of a camera which is connected to a screen and a database to store data, then this system can be monitored manually [1]. In the second paper, there are a large number of applications of image processing in diverse spectrum of human activities-from remotely sensed scene interpretation to biomedical image interpretation [2, 4]. Due to a number of images feature data, complex calculation, and the larger storage space and high processing capacity, currently, most face recognition are realized only by PCs with high performance; so portability and mobility in this process are restricted greatly. At present, the embedded system [3, 9] is widely used in the front-end of entrance guard system and attendance system in order to collect face images. The existing system consists of static recognition which increases the complexity and understanding; hence, it is difficult to use. It uses basic operations which consists of image capturing through camera. The most useful and unique features of the face image are extracted in the feature extraction phase [11].

2 Proposed System

The given system focuses on identifying the image of an object or a person. Though there exists different circumstance which makes it difficult for execution, there are some disturbing situations:

- Background with same color as of skin.
- Environment (Dark).
- Different objects at the same time.

It is very important that color recognizing algorithm should work properly. Our proposed system is able to work for different skin color and is able to function effectively in different lighting system. It can be combined with the current system easily. Computer vision is a rapidly growing field, partly as a result of both cheaper and more capable cameras, partly because of affordable processing power, and partly because vision algorithms are starting to mature. OpenCV itself has played a role in the growth of computer vision by enabling thousands of people to do more productive work in vision [10].

2.1 System Architecture

The architecture of the proposed security system is represented in Fig. 1, which reflects different functionality as process.

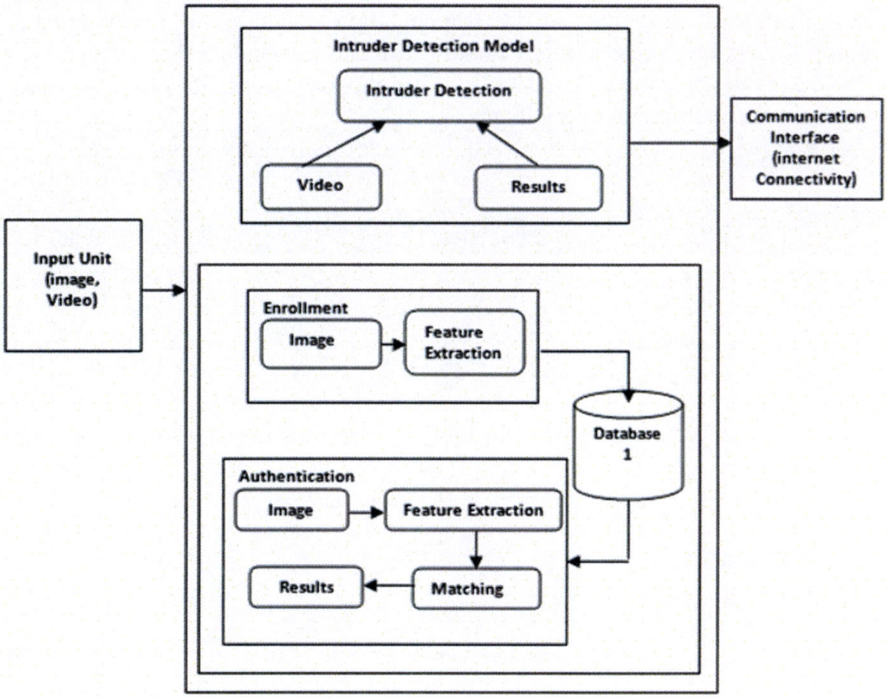

Fig. 1 Architecture of Security System

2.1.1 Different Units in the Systems

- **Input Unit**: Images captured by camera for detection (unknown person) recognition.
- **Processing Unit**: Data which is captured by the camera is fed into the processing unit in which the processing or calculations are performed on the proposed intruder detection and door lock system module, here the processing unit is "Raspberry pi" along with code scripts of implemented modules.
- **Communication Interface**: Communication interface includes wireless internet connecting devices which are connected to the unknown person detection module which is used to send an alert as E-mail through internet connectivity.

2.2 Algorithm

The system will detect image from stored database. Now with the help of Eigenface, we can start using them for facial recognition of new faces. This is done as follows. Known face images (those already seen and enrolled by the system, saved on the database) are saved as a collection of weights which describe how much each Eigenface contributes to that face image.

Let $A = \{a_1, a_2, \ldots, a_N\}$ be some random vector with observations $a_i \varepsilon R^d$.

$$\mu = {}^1/_{N=1} ai \tag{1}$$

$$S = ({}^1/_{N=1} i - \mu)(a_i - \mu)T \tag{2}$$

$$Sv_i = \lambda_i c_i, \quad i = 1, 2, \ldots, N \tag{3}$$

Now, the eigenvectors are ordered according to their Eigen value with descending.

$$b = W^T(a - \mu) \tag{4}$$

where $W = (c_1, c_2, \ldots, c_k)$.

Reconstruction value

$$a = Wb + \mu \tag{5}$$

Now, face recognition is performed by Eigen faces by forecasting the samples into PCA and training and query image (both forecasted) nearest neighbor is found. Figure 2 depicts the working /processing of the proposed algorithm.

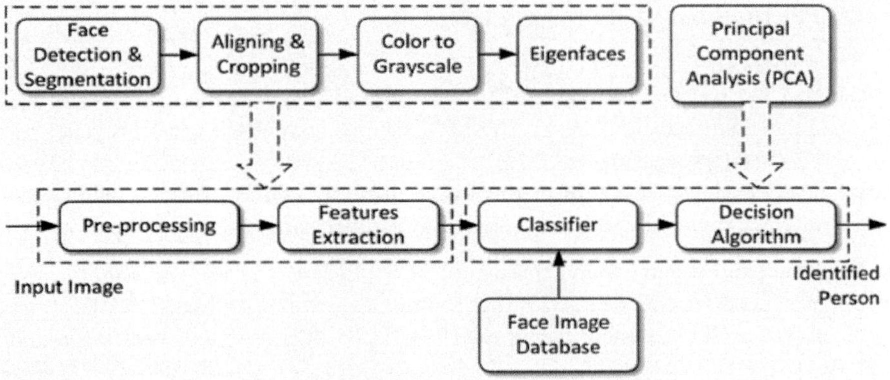

Fig. 2 Representation of the algorithm flow

3 How it is Performing

It has two segments:

(a) Profile Identification.
(b) Automatic message notification sending.

3.1 Execution of Profile Identification

Implementation takes place using an algorithm known as *K*-nearest algorithm and following python libraries are used.

- opencv
- numpy
- matplotlib
- cv2.
- **dataset_maker**: Input will be taken in the form of images. With the help of external document, we convert image input into black and white output image. Then these black & white images are converted into NUMPY array through EIGEN values.
- **Train**: For attaining higher precision, it trains the System when id is already present, otherwise new user_id is taken and data for a new id is created.
- **Facedetect**: Continuously intake the input(data) in real-time.
- **Detect**: Dataset_maker will be called and make the image data will be converted.

3.1.1 Implementation of Auto Alert Sending

Message alerts will be sent to authorized user G-mail_id. These libraries are used:

- **SMTPLIB** (Simple Mail Transfer Protocol):
  - o SMPTLIB data sets (TLIBs) are used as temporary storage for relative files that are loaded from SMPPTFIN during RECEIVE processing. They are deleted when the associated SYSMOD is deleted by REJECT, RESTORE, or ACCEPT processing.
  - o You can have SMP/E dynamically allocate the TLIB datasets, or you can allocate them yourself before RECEIVE processing.
- If you need to specify a unit that is not SYSALLDA, and the unit is not set by use of a STORCLAS or an ACS filter routine, then you must use a DDDEF entry instead of a DD statement to allocate SMPTLIB data sets and specify the unit value in the DDDEF entry.
- If you are using SMS to manage your data sets, you can set up the unit, volume, and space allocation through a STORCLAS or an ACS filter routine, instead of specifying them on a DD statement or DDDEF entry.
- If you are using SMS to manage your data sets, do not specify dummy volumes on a DD statement or in a DDDEF entry for the SMPTLIB allocation. Dummy volumes used as indicators to ACS routines for SMS class selection can cause operator mount messages to be issued for the non-existent dummy volumes. This is because SMP/E first attempts to allocate the SMPTLIB data sets as if they already exist on the volume, which can cause operator mount messages to be issued for the dummy volume.

Here is a simple syntax to create one SMTP object, which can later be used to send an e-mail—

import smtplib.
smtpObj = smtplib.SMTP([host [, port [, local_hostname]]]).

3.1.2 Mime

Other than ASCII, there is the inclusion or attachment of other types of information too. For example-pictures, audio, etc. The set of Non-ASCII char has Header info which is a part of Message bodies.

There we have to make import of the file name smtplib "MIMEMultipart" and "MIMEText":

import smtplib.
from email.mime.multipart import MIMEMultipart.
from email.mime.text import.
It permit to send files. (like a master key).

The system has been organized accordingly, now here comes the time for logging, which will have inclusion of ML. The working of input and output will be based on algorithm, including pictures. Profile identification and finding of unknown will be according to KNN and SMTPLIB.

4 Result

(i) Initialization of face detection (Fig. 3)

The above picture is showing the first step of working of the proposed system. This has refer as System Initialization.

Pictures format will be grayscale. (both clicking and storing) (Fig. 4).

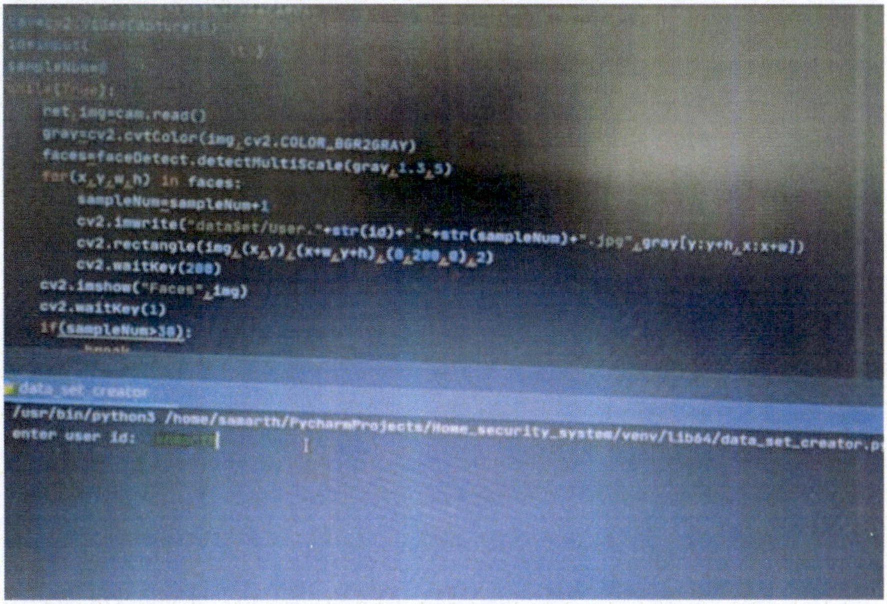

Fig. 3 System initialization

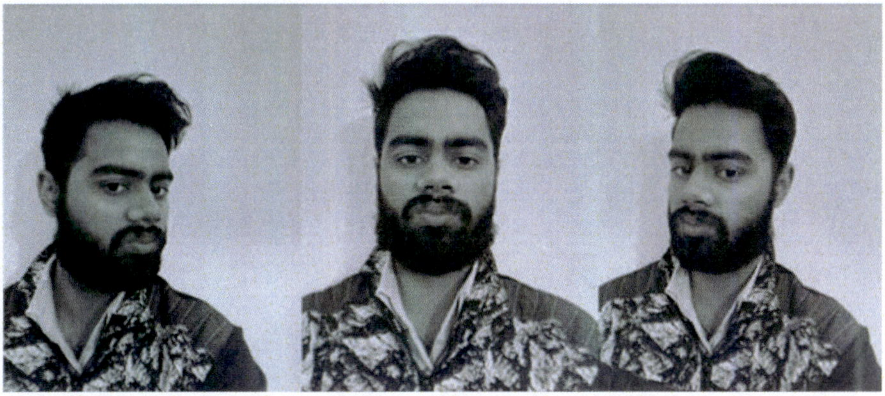

Fig. 4 Training images for algorithm

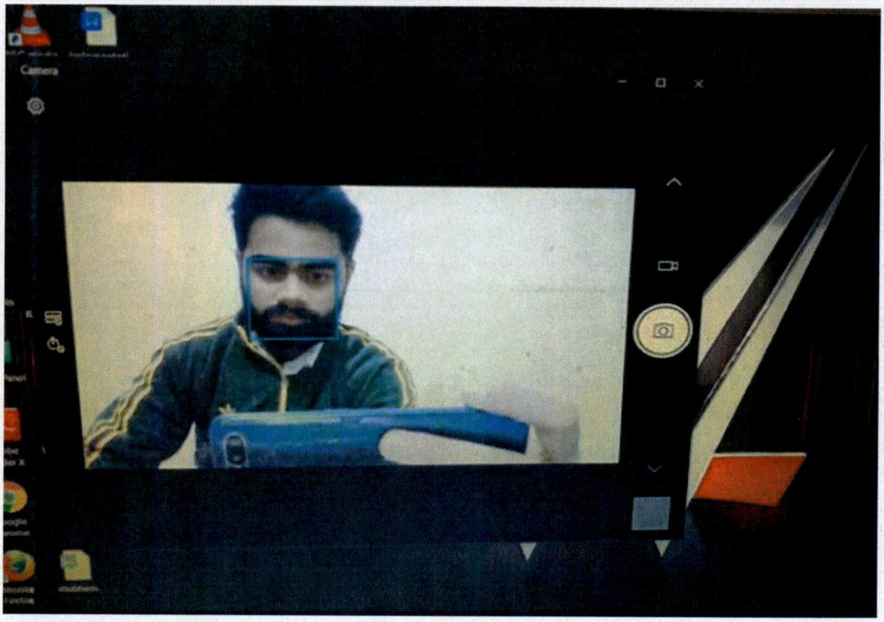

Fig. 5 Activating web camera

(iii) As per the structure of architecture, there will be activation of camera and its linkage with the system (Fig. 5).

(iv) Machine Learning algorithm detecting object (Fig. 6).

(v) Activities are recorded and stored. The permitted/allowed person will receive a notification alert on his/her E-mail address.

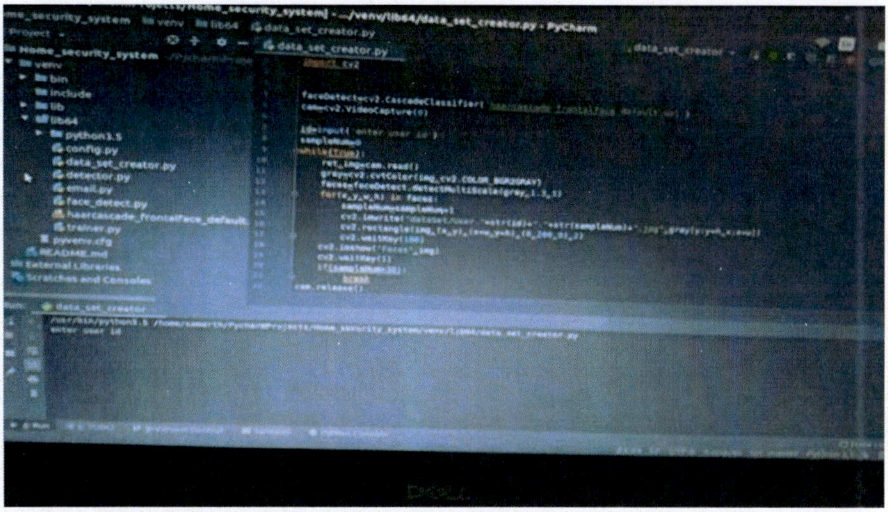

Fig. 6 Screenshot of object detection through algorithm

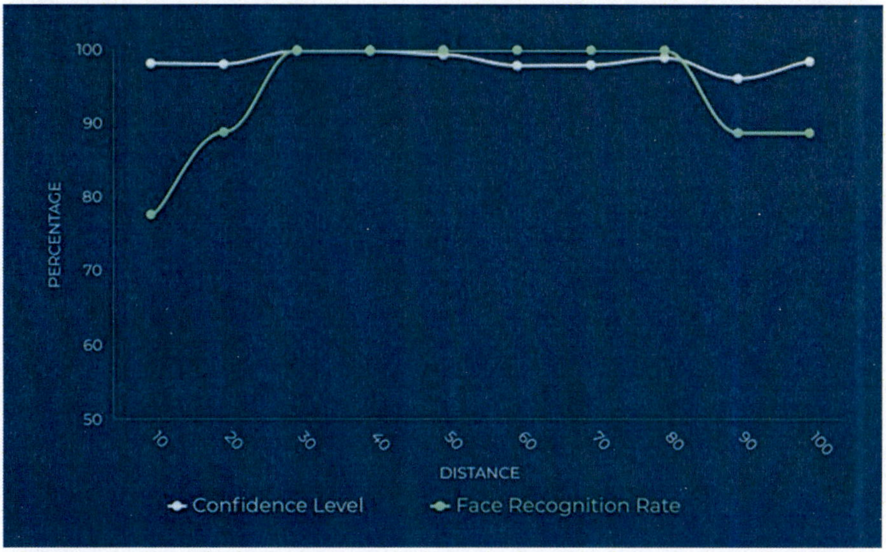

The above graphical representation is based on results of various distance vs percentage assessment.

5 Conclusion

The process of identifying the unknown has been designed in terms of the personal security or security for your place through the proposed home security system. This is very effective and adoptable and acceptable in regards with its price. After applying multiple test cases, every time quite accurate results were achieved when an unknown person makes his entry at home the owner of the house gets an alert message on his/her E-mail. More enhancements the system, this is going to have its application or implementation in some other different areas as well. For example—To track child and animals.

References

1. Singh Y, Bhatia PK, Sangwan O (2007) A review of studies on machine learning techniques. Int J Comput Sci Secur 1:70–84
2. Hegadi RS (2010) Image processing: research opportunities and challenges. Natl Sem Res Comput. Bharathiar University, Coimbatore, India
3. Ping W (2008) Research on the embedded system teaching. In: Proceedings of the international workshop on education technology and training and the international workshop on geoscience and remote sensing, vol 1. Shanghai, China, pp 19–21
4. Rathore MR, Chowdhary PS, Tyagi N (2014) Verification of data integrity using public auditability and data dynamics for storage security in cloud computing. Int J Adv Res Sci Eng 3(5):79–84
5. Gaikwad AT (2018) LBP and PCA based on face recognition system. Glob J Eng Sci Res 368–373
6. do Prado KS (2017) Face recognition using local binary patterns (LBP)
7. Hong P, Wen Z, Huang T (2001) IFace: a 3D synthetic talking face. Int J Image Graph 1:19–26
8. Vishnu Priya TS, Vinitha Sanchez G, Raajan NR (2018) Facial recognition system using local binary patterns (LBP). Int J Pure Appl Math 119(15):1895–1899
9. Zafaruddin GM, Fadewar DH (2014) Face recognition: a holistic approach review. In: International conference on contemporary computing and informatics. IEEE, Mysore, India, pp 175–178
10. Bradski G, Kaehler AO (2008) Learning OpenCV, O'REILLY Media, Sebastopol
11. Zbeda FG, Abdulaziz MH, Saleh AE (2016) PCA-HOG descriptors for face recognition in very small images. Int J Adv Res Comput Sci Softw Eng 6(9):449–451

Mechanism for Big Data Security Related to GMPLS/MPLS Networks

Rajani Singh and Ashutosh Dixit

Abstract The requirement for gainful ways to deal with the preparing of Big Data, set apart by huge amounts, various structures and quick rates, is basic and has as of late drawn the premium of many investigation associations. This is especially the situation as ordinary methodologies and capacities for information examination have demonstrated to be deficient around there. As highlighted in this article, another factor that is comparatively essential during the formation of Big Data is its security. Consequently, we suggest the treatment of big information at two separate speeds. The principal level portrays the information depending upon its development and whether security is required. In assessment, in light of length, arrangement, and speed factors, the resulting level examinations and cycles results. As the consequences of the reproduction uncovered that the use of characterization contribution as of the center MPLS/GMPLS switching network end up being basic in limiting information evaluation and handling time.

Keywords Big data · MPLS · GMPLS · Data encryption

1 Introduction

Big data is a cutting edge idea that applies not exclusively to enormous size information, yet in addition to data (for example video, sound, unstructured content, and interpersonal organization data) with unstructured trademark structures. Numerous administration arrangements, particularly the US organization in 2012 when it revealed the Big Data R&D drive [1], have as of late requested approaches to adapt to enormous information issues. The point of the undertaking is to explore pertinent and effective techniques for using large information to address difficulties and

R. Singh (✉)
IIMT College of Engineering Greater Noida, Noida, U.P, India

A. Dixit
IIMT College of Management Greater Noida, Noida, U.P, India

R. P. Mahapatra et al. (eds.), *Proceedings of International Conference on Recent Trends in Computing*, Lecture Notes in Networks and Systems 341,
https://doi.org/10.1007/978-981-16-7118-0_30

dangers confronting the nation, government, and industry. It can likewise be recollected that it could be helpful to assess large information in various territories like medical services, tutoring, account and public safety. The accompanying things are future obstructions to the administration of Big Data [2]:

(i) **Examination**: In solicitation to remove critical nuances, this system relies upon information catch, appraisal and diversion.
(ii) **Strategy and Change**: This methodology is used to assemble and join data got from different sources to ensure supportive information show, upkeep, and reuse.
(iii) **Searching**: In Big Data mining, this procedure is viewed as the most fundamental endeavor since it revolves around the most appropriate ways to deal with inspect inside information that is wide and not facilitated from one viewpoint, and on the situation and rightness of the glanced through information recuperated on the other hand.
(iv) **Storage**: This cycle includes best practices and techniques for the endeavor, depiction, and pressing factor of huge data, similarly as limit and yield chain of significance.
(v) **Modeling**: This methodology requires the reflection of enormous information and accordingly permits to communicate information basically and successfully.
(vi) **Security and sharing**: Notwithstanding reasonable and dependable methodologies for information trade, this stage centers around data security and encryption, just as ongoing handling of coded data.

The developing pattern in the use of data administrations and the approach in information examination methods likewise added to the development of the utilization of Big Data. Various issues rise out of the boundless use of Big Data, including information assortment, handling and interpretation, looking, stockpiling, recreation, assurance, and protection. safety measures and protection of huge amount of data are potential issues in the disseminated storage world as the expanding utilization of large information adds toward novel dangers to data, especially, while associating through private and significant records, for example, organization insider facts, individual and monetary subtleties. Each disappointment that may happen to this information will detrimentally affect the trust of the organization and could hurt its picture. Furthermore, moving Big Data inside numerous mists with shifting levels of affectability may open significant information to assaults.

In any case, when huge informational collections are utilized, the standard methodologies don't meet with huge information insurance models. To address data dangers and hazard control, current Big Data security and protection procedures are likewise required. Enormous information assurance and protection hazards and proposes a double way to deal with the characterization and precautions related to huge amount of data toward moderate information breaks and in the direction of uphold protection controls from end-to-end information distribution. As the matter of significant commitments, it examination in the direction of the insurance and security of Big Data are portrayed such as:

(i) The gathering of Big Data as per its structure that will in general limit the time it takes to consolidate information security methodology.
(ii) Multiprotocol Mark Swapping (MPLS) to arrive at elite broadcast communications networks using an information conveying procedure.
(iii) To forestall dynamic inquiries in a steering table, pass enormous information starting with one hub then onto the next zeroed in on short way marks instead of long organization addresses.
(iv) The use of images to separate between traffic subtleties beginning from different organizations.
(v) Networks Gate's examination and preparing of Big Data approaches to aid the heap conveyance of large information traffic and improve the effectiveness of huge information investigation and handling measures.

2 Related Work

Big Data investigation research began as of late (in 2012) after the White House dispatched the Big Data drive [1]. Up until this point, the writing on enormous information has focused on streamlining information dealing with and proficiency. Big Data insurance observing, on the opposite side, is now creating and has just started to draw the interest of many examination associations. In this segment, we introduce and think about the key exploration work pertinent to Big Data insurance that has been proposed up until now. Our work, in reality, is particular from a few.

The initial segment addresses the authenticity of the talks of safety experts in light of the data they seem, by all accounts, to be activating, while the subsequent part offers a bunch of scholarly trades around information, associations, and cycles to handle a portion of the impediments of existing Big Data security gathering tasks. Likewise, the dependability of constant Big Data in cloud administrations was respected in [3]. The work is centered around a complex insurance model that can get information at the accompanying security levels progressively: firewall and access control, character the executives, evasion of assault, and merged encryption. Another investigation that objectives constant material are introduced in [4], in which the mission for recordings in Big Data volumes is proposed through a semantic-based video putting together apparatus. A semantic progressive organization model is utilized by the proposed philosophy to mine and orchestrate video assets dependent on their associations, while the creators in [5] proposed a Dynamic Key Length Protection Mechanism (DLSeF) in view of a typical key emerging from synchronized indivisible numbers. To stop man in the middle attacks, the key is powerfully altered in a word stretches. The creators in [6], by examination, focused on the issue of Big Data media data inside a cloud climate. They proposed an imaginative answer for getting to safe monetary assets using Semantic-Based Access Control (SBAC) methods. Also, the examination in [7] focused on the issue of security and suggested a Dynamic Data Encryption Technique (D2ES) type of information encryption. The framework specifically written code of data in planning limitations utilizing techniques for protection order.

Likewise, a few scholarly reports have regularly thought about features of huge information security and arrangements [8–16].

3 Big Data Keys Term

Enormous data may join information of different sorts, similar to content, video, money related data and logs, similarly as information that is gotten or dishonest. The treatment of these different wellsprings of data can, in any case, not be something practically the same. Moreover, the proposed portrayal approach ought to consider the going with examinations [17].

Speed: The speed of conveying and taking apart data.

Volume: The size of the delivered data and sufficient additional room.

Variety: The data type and its credits.

The proposed portrayal estimation is connected with the consistent assessment of tremendous data. The going with thoughts should be considered in our computation, as the treatment of protected data is specific from plaintext information.

Confidentiality: The security viewpoint is associated with whether the subtleties can be scrambled. Huge information taking care of with ensured data is likewise not a simple interaction in light of secrecy and consequently needs separate consideration.

Verification: Authentication, for example information protection from adjustment, could be required for certain Big Data. Furthermore, confirmation works with client and affirmation authority (CA) verification.

The proposed calculation would likewise perceive other security contemplations, like Denial of Services (DoS) insurance and right to use control List.

The particulars of the recommended answer for dealing with large information security are tended to in the accompanying sections.

3.1 Mechanism for GMPLS/MPLS Networks

Big Data is prepared through two layers of the pecking order in the proposed technique. The errand of the primary level (Tier 1) is to recognize the fundamental information to be prepared. In different terms, this level initially decides whether the approaching traffic in large information is coordinated or unstructured. In at that point tests, the type of security administration that is added to the information, for example regardless of whether encryption is applied to the information being prepared, or whether security of the information being handled is upheld or required. Based on the results gathered from the main stage, and based on the investigation of speed, sum and assortment factors, the subsequent level (Stage 2) concurs lying on the right behavior of huge amount of data. Figure 1 shows a stream map for the overall plan of the proposed framework.

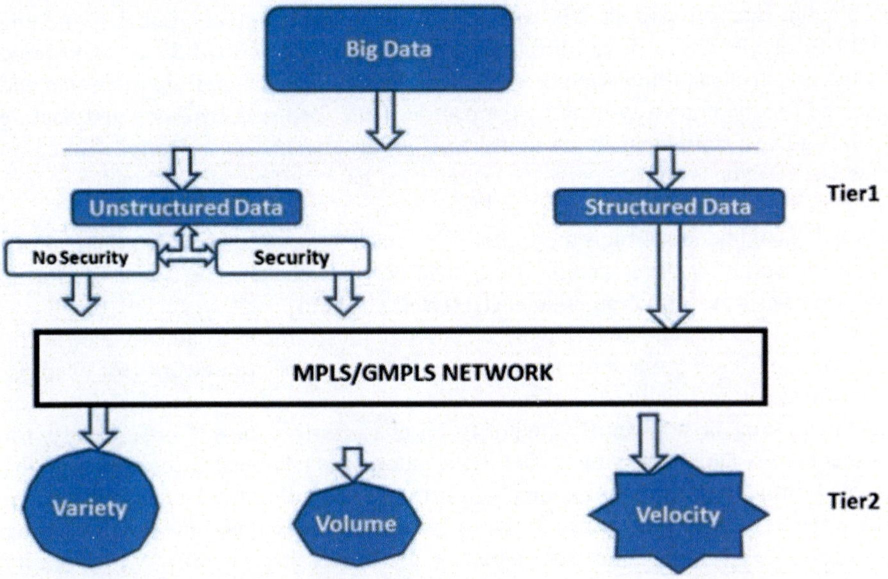

Fig. 1 The general architecture MPLS/GMPLS network mechanism

Stage1: Categorization of Data. The algorithm used for classifying large data knowledge (Tier 1) (i.e., whether data is organized or unstructured and whether or not encryption is applied) is presented in this chapter.

There ought to be a compelling technique prior to breaking down the huge information to recognize it on whether it is coordinated and afterward decide the wellbeing status of each gathering. The recommended calculation depends on numerous logical factors and is illustrated in this way:

Information Source and Destination (DSD): Originally, the data source and objective be able to assist with speculating the type of construction of the approaching information. This factor is utilized in this calculation as a rescanning point, however, it's anything but a main consideration. Consequently, without the need to do a careful survey of approaching outcomes, it permits to accelerate information order. The calculation, be that as it may, utilizes control contribution to redesign.

Information Header data (DH): Arriving information's have been accepted to be typified inside headers. Subsequently, in information arrangement, header subtleties may assume a significant part. For instance, there is a Type of Service (ToS) field in the IP network traffic header, which gives a sign of the sort of information (ongoing information, video sound, record information, and so on). Besides, the convention territory shows the upper layers, like UDP, TCP, ESP assurance, AH security, and so forth.

What's more, by using traffic labeling, the Stage 1 ranking component might be enhanced. Names (L) may, all in all, be utilized to separate or recognize approaching traffic information. We in this way assume that Multiprotocol Label Switching (MPLS) or Generalized Multiprotocol Label Switching (GMPLS) is empowered by the organization foundation heart, and marks can thus be helpfully presented and planned. At the entryway of the organization liable for breaking down and dealing with Big Data, traffic that shows up from different organizations is sorted. To recognize traffic data, a MPLS network center uses names. The MPLS header is four bytes in measurement lengthwise, and from network parcel subtleties, the marks are created. Insights about such the traffic (i.e., continuous, sound, video, and so on) might be borne by the imprints. It ought to be recollected so as to the label(s) is developed as of information open at (DH) & (DSD) [15].

The gateway is liable for arranging the approaching traffic into names named traffic (Lm) in the Tier 1 framework found in Fig. 2. The traffic framework (for example coordinated or unstructured) and sort (for example carried out or essential assurance administration, or no security) ought to be characterized now. Consequently, for further review and processing at Tier 2, the gateway is responsible for spreading the labeled traffic to the relevant node (Nk). If traffic has little or no safety criteria, the gateway can forward the traffic to the relevant node(s) assigned for the processing of traffic (i.e., certain nodes are liable for the processing of traffic with security service requirements, while other nodes are designated for the processing of traffic data without safety requirements). As seen later in this article, this method aids in the allocation of loads for large data traffic, and thus increases the efficiency of the analysis and processing phases.

Fig. 2 Stage1 structure

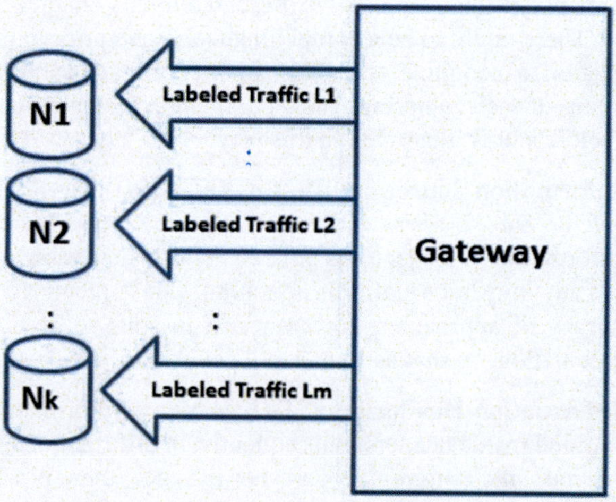

3.2 *Stage2: Variety of information's*

Level 2 assumes responsibility for the survey and arranging at this stage of the approaching marked large information traffic that Tier 1 has effectively screened. Furthermore, Tier 2 is liable for examining the approaching traffic as per the factors of speed, volume, and assortment. Every hub is likewise answerable for assessing and preparing the apportioned expansive information traffic reliant upon these factors.

(i) When dealing with large information traffic, per Tier 2 hub applies Algorithms 1 and 2. The primary calculation (Algorithm 1) settles on the volume cause-based insurance estimation and preparing, even as the successive calculation (Algorithm 2) manages the factors of speed and assortment.

(ii) You ought to depict Algorithms 1 and 2 as must be followed:

(iii) The two-level technique is utilized in two stages previous to any additional examination to deal with approaching outcomes.

(iv) By deciding if it is organized or non-organized, Tier 1 is answerable for separating approaching information. It is additionally more likely that weakness survey would be reached out to coordinated information or somewhere else relying upon choice.

(v) Tier 2 is liable for dealing with and assessing huge information traffic zeroed in on factors like volume, speed, and assortment. In the proposed calculations, the focal idea depends lying on the utilization of imprints to arrange and classes the huge information passage handled (Fig. 3).

Algorithm 1 Big Data Security Manipulated by Volume

Stage 1: Obtain (label Big Data passage, Gateway numeral, Factor Vol, Security Service);

Capacity for receiving Big Data traffic by Name hub somewhere:

Marked Big Data Traffic → traffic name (TL), Data (D),

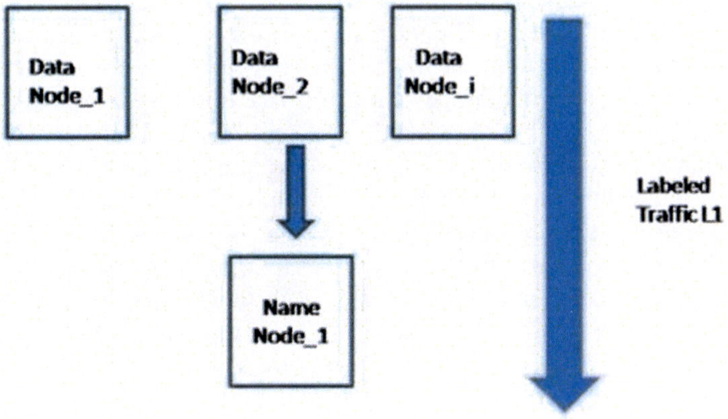

Fig. 3 Node architecture

Entryway quantity → Gateway advantage for presenting named Big Data traffic (GN) Factor Vol → Form of volume factor (V):

Security Service → The name importance of each traffic is dissected by classification (C), verification (Auth).

Stage 2: frontward (Data Node, Security Service).

Strategy used for appropriating marked passage to insurance administration task intended for predefined information node(s):

Stage 3: examination of information hubs (bundle headers, names).

Execute trying to header and imprint data:

→ result check does mark tests.

→ Header looking at conveys header check.

Presumptions: Secured content, like an ESP header, accompanies extra header size.

Step 4: Return to step 1.

To assist level clusters to settle on structure and classification of prepared information, the organization center names are utilized. Subsequently, the utilization of MPLS names diminishes the tension on level node(s) to implement the characterization assignment and along these lines builds the productivity of this technique. In the opposite side, if hubs don't acknowledge MPLS usefulness, it will need additional timeslot and additional transfer speed to order of specific organization directing conventions.

3.2.1 Organization Architecture with Big Data Node Module

The node model utilized in the accumulation and grouping with huge information's have been introduced up until now. In any case, to describe the prepared information, the recommended arrangement regularly includes contributions from the organization. The presence of a fundamental organization community that works with data checking is our assumption here. As expressed in the past portion, our favored option is MPLS, as most Internet Service Providers (ISPs) have now accepted it. Concerning execution, design, and execution, the MPLS header and marking conveyance conventions permit the arrangement of huge amount of data at the preparing clusters more successful.

The planning flanked by the organization community, which here is dared to be a Generalized Multiprotocol Label Switching (GMPLS) or MPLS Organization. Notwithstanding parcel exchanging, the GMPLS grows the MPLS engineering by empowering switches for frequency, space, and time exchanging. The focal organization is comprised of vendor switches named P switches here and numbered A, B, and so on Complex supplier Edge switches named here in these work entryways end the GMPLS/MPLS organization. It is the obligation of the Gateways to finish and keep up the planning between the hubs answerable for dealing with the huge amount of data passage upcoming as of the central organization. Furthermore, the key component utilized for information arranging utilized for Stage 1 and Stage

2 segment is the active named passage entryways. The primary columns utilized for planning between the organization community and the enormous information handling clusters are calculations 1 and 2.

Algorithm 2 Big Data Security Manipulated by Velocity and Variety

Stage 1: Obtain (Labeled huge amount of data stream, Variety of Factor Velocity, Number of Gateway, DSD prob);

Named huge amount of data Traffic → traffic mark (TL), Data (D),

Door token → Assessment of Gateway carriage named Big Data traffic (GN) Aspect Value → Aspect type (Variety || Velocity).

Stage 2: Frontward (Cluster, Security tune-up).

Technique for circulating the named traffic with assurance administration task for the predetermined information node(s):

The choice of the protection supplier administration is taken ward on the DSD risk assessment(s)?

Valid, at that point go to the information node(s) apportioned for security administrations checking and handling.

Stage 3: return stage 1.

3.2.2 Security Key Constraints

The planned work encourages protection features present at birth from the structural design of the switching techniques, which are given below:

Traffic Separation

From first to last using fundamental Private Network (VPN) stamping and the stacking fragment (S) region that is empowered by the GMPLS/MPLS headers, the utilization of the switching network center organization exertion permits traffic separation. To clarify auxiliary, separation of passage is a vital assurance include important. For example in the event that a similar ISP is utilized by two unique firms, it is truly significant not to blend and communicate traffic between the contending parties. All things considered, by executing verification encryption strategies, traffic separation can be refined, yet this would clearly impact the organization's effectiveness because of the overhead impacts of expanded idleness and dormancy. This overhead doesn't happen in the proposed GMPLS/MPLS execution since traffic confinement is refined naturally by using the MPLS VPN capacity, and our methodology hence doing enhanced in such a manner.

Hiding Network Internal Structure

One major part of the engineering and design of the GMPLS/MPLS network is that attention to taking interest switches inside the central organization isn't required for approaching or active traffic. Utilizing names just (for example not utilizing IP header data), the traffic is straightforwardly sent/exchanged inside. Measures, for example, IP parodying and Denial of Service (DoS) may likewise be kept away from successfully.

Reliability and Availability
Significant security guidelines are known to give secure information move, execution, and fast recuperation from disappointments and in this manner improve wellbeing. Utilizing a GMPLS/MPLS design based hidden organization center permits recuperation from hub or association disappointments quick and adequately. Numerous recuperation approaches in the writing have exhibited that GMPLS/MPLS center organizations can altogether support security and accessibility.

Big Data Encryption through Credential
GMPLS/MPLS isn't intended to advance techniques for encryption and confirmation as this will minimize the organization's effectiveness. Assurance execution for enormous information is then carried out at the limits of the organization (for example network entryways and nodes for the transmission of enormous information). Be that as it may, on account of the utilization of the GMPLS/MPLS foundation, Virtual Private Networks (VPNs) capacities might be empowered. Traffic disengagement, however, with no encryption, is the VPN include that might benefit from some intervention in the present circumstance. In the event that encryption is essential, it will be empowered utilizing reasonable encryption procedures on nodes.

4 Results

The vital upgrade of our proposed work is the utilization of the great speed organizing convention (i.e., GMPLS/MPLS) as a hidden framework that can be utilized to arrange big data traffic through handling nodes at network edges. To be sure, how traffic labeling is utilized to distinguish traffic has been tended to beforehand.

Presently, our motivation in this section is to assess and look at through reproductions the impact of utilizing the marking strategy on improving Big Data order and subsequently improving insurance. Along these lines, reenactment considers have been completed in this fragment to evaluate the effect of marking on outcomes. Transfer speed overhead, stacking time, and data arrangement distinguishing proof achievement is the yield factors considered in the reproductions. The NS2 reproduction apparatus (NS-2.35) was utilized to play out the reenactments.

5 Conclusion and Future Work

For Big Data, another insurance taking care of technique was proposed in this article. The methodology recommended is centered around the division of large information into two layers (i.e., Tier 1 and Tier 2). An organization foundation that upholds

GMPLS/MPLS abilities is required for the order. The GMPLS/MPLS work on the arrangement by providing the gathered Big Data traffic with checking tasks. The discoveries acquired delineate the arrangement's effectiveness. Improvements when testing models like ID, handling time, and overhead. The introduction of huge information will be finished by likely examinations on the theoretical strategy to incorporate dynamic grouping investigation. More security observing standards, like genuineness and constant examination of Big Data, can likewise be researched.

References

1. Gai K, Qiu M, Zhao H (2016) Security-aware efficient mass distributed storage approach for cloud systems in big data. In: 2016 IEEE 2nd international conference on big data security on cloud (bigdatasecurity), New York, NY, Usa, pp 140–145
2. Claudia A, Blanke T (2015) The (Big) Data-security assemblage: knowledge and critique. Big Data Secur 2:12
3. Chang V, Kuo Y-H, Ramachandran M (2016) Cloud computing adoption framework: a security framework for business clouds. Futur Gener Comput Syst 57:24–41
4. Liang H and Gai K (2015) Internet-based anti-counterfeiting pattern with using big data in China. In: The IEEE international symposium on big data security on cloud, pp 1387–1392
5. Yan Z, Ding W, Yu X, Zhu H, Deng RH (2016) Deduplication on encrypted big data in cloud. IEEE Trans Big Data 2:138–150
6. Gholami A, Laure E (2016) Big data security and privacy issues in the coud. Int J Netw Secur Its Appl (IJNSA) 8(1)
7. Li Y, Gai K, Qiu L, Qiu M, Zhao H (2017) Intelligent cryptography approach for secure distributed big data storage in cloud computing. Inf Sci 387:103–115
8. Narayanan A, Huey J, Felten EW (2016) A precautionary approach to big data privacy. In: Data protection on the move, vol. 24 of law, Governance and technology series, Springer Netherlands, Dordrecht, pp 357–385
9. Kang S, Veeravalli B, Aung KMM (2016) A security-aware data placement mechanism for big data cloud storage systems. In: Proceedings of the 2nd IEEE international conference on big data security on cloud, New York, NY, USA, April 2016, pp 327–332
10. Domingo-Ferrer J, Soria-Comas J (2016) Anonymization in the time of big data. In: Privacy in statistical databases, vol 9867 of lecture notes in computer science. Springer, pp 57–68
11. Jeong Y-S, Shin S-S (2016) An efficient authentication scheme to protect user privacy in seamless big data services. Wirel Pers Commun 86(1):7–19
12. Babiceanu RF, Seker R (2016) Big Data and virtualization for manufacturing cyber-physical systems: a survey of the current status and future outlook. Comput Ind 81:128–137
13. Xu Z, Wu Z, Li Z et al (2016) High fidelity data reduction for big data security dependency analyses. In: The 2016 ACM SIGSAC conference. New York, NY, USA, October 2016, pp 504–516
14. Chang V, Ramachandran M (2016) Towards achieving data security with the cloud computing adoption framework. IEEE Trans Serv Comput 9(1):138–151
15. Puthal D, Nepal S, Ranjan R, Chen J (2015) A dynamic key length based approach for real-time security verification of big sensing data stream. In: WISE 2015: 16th international conference, Miami, FL, USA, November 1–3, 2015, pp 93–108
16. Li Y, Gai K, Ming Z, Zhao H, Qiu M (2016) Intercrossed access controls for secure financial services on multimedia big data in cloud systems. ACM Trans Multim Comput Commun Appl (TOMM) 12(4s). (Article no. 67)
17. Xu Z, Liu Y, Mei L, Hu C, Chen L (2015) Semantic based representing and organizing surveillance big data using video structural description technology. J Syst Softw 102:217–225

18. Liu C, Yang C, Zhang X, Chen J (2015) External integrity verification for outsourced big data in cloud and IoT: a big picture. Futur Gener Comput Syst 49:58–67
19. Gai K, Qiu M, Zhao H, Xiong J (2016) Privacy-aware adaptive data encryption strategy of big data in cloud computing. In: Proceedings of the 3rd IEEE international conference on cyber security, China, June 2016, pp 273–278

Deep Learning Algorithms based Vehicle Mobility Prediction from Satellite Imagery During Pandemic

M. Deva Priya, A. Sahaya Gebin, S. Selva Kumar, and R. G. Vipin

Abstract The COVID-19 epidemic has made governments around the world to enforce lockdowns and isolations to stop the spread of virus. Both human and financial activities are affected throughout the globe. It takes time to recover from these losses. Financial actions influence social activities which incorporate signatures in satellite images that can be perceived and categorized. Satellite imagery aids in making decisions of predictors and decision makers by offering diverse types of perceptibility in the relating financial changes. In this paper, deep learning methods including Fast Region-based Convolutional Network (Fast R-CNN) and You Only Look Once (YOLO) are employed to identify the detailed elements in satellite images that can be used to find the financial indicators based on it. The proposed system uses Histogram Equalizer (HE) for enhancing the satellite pictures to provide accurate analysis about human movements. The system shows results on genuine instances of various destinations when COVID-19 flares up to delineate extraordinary quantifiable markers. The area is partitioned into different sections and the human and economic activities are identified. Mobility of people shows the spreading of COVID-19. YOLO offers the best performance in object (vehicle) identification from which the presence of economic downfall is predicted.

Keywords Fast R-CNN · YOLO · Histogram equalization · COVID-19 · Satellite imagery

M. Deva Priya (✉) · A. Sahaya Gebin · S. Selva Kumar · R. G. Vipin
Department of Computer Science and Engineering, Sri Krishna College of Technology, Coimbatore, Tamilnadu, India
e-mail: m.devapriya@skct.edu.in

A. Sahaya Gebin
e-mail: 17tucs201@skct.edu.in

S. Selva Kumar
e-mail: 17tucs210@skct.edu.in

R. G. Vipin
e-mail: 17tucs251@skct.edu.in

R. P. Mahapatra et al. (eds.), *Proceedings of International Conference on Recent Trends in Computing*, Lecture Notes in Networks and Systems 341,
https://doi.org/10.1007/978-981-16-7118-0_31

1 Introduction

COVID-19 has changed the world dramatically. Lockdowns and isolations are executed throughout the world. There is a momentous reduction in nitrogen dioxide emissions in metropolitan cities as a result of financial breakdown. The probable use of distant sensing has increased. Satellite imagery connected with the pandemic can be considered. These images can be used to regulate traffic, know about medical infrastructure, economically improve resources and implement social distancing.

Images with high resolution as delivered by classy satellites like WorldView-3 [1] gather images with a Ground Sample Distance (GSD) of 0.3 m. It can be used to find the influences of COVID-19.

Traveling is highly reduced during this pandemic which has a significant influence on vehicle motion. Information about traffic and social distancing can be obtained by identifying and categorizing vehicles, staying updated with the details of medical infrastructure, etc., Fig. 1 shows an example scenario.

Satellite images can be observed to support new pointers and areas of attention with less effort as they are in the same format. These systems can be scaled to a global level by adopting the satellite coverage and computational power. This reduces the difficulties related to data gathering from varied sources.

To analyze satellite imagery, automated Machine Learning (ML) based algorithms are needed to mine information without the need for increased manual labor. In this paper, Fast Region-based Convolutional Network (Fast R-CNN) and You Only Look

Fig. 1 Highway in Wuhan, China on Oct 17, 2019 (left) and February 25, 2020 (Right)

Once (YOLO) are applied to identify and classify objects in aerial images. The xView dataset deals with the COVID-19 problem as it uses WorldView-3 as the source for many images with improved resolution around an area of 1400 km^2.

2 Related Work

Data collected using satellite sensors confirm the growth of primary cautioning systems for identifying diseases like malaria which takes the life of around 1–2 million people every year. Rogers et al. [2] have used tools to examine factors that deal with the dynamics of vector populaces and malaria pest communication. As the mosquito populace methods and incubation stages in the form of vectors differ with temperature and moisture on the ground, distantly detected images of seasonal climate are the controlling forecasters of mosquito dissemination patterns and mean levels of communication of parasites. Forms of infection may have extrinsic or intrinsic effects. The stability of these factors is determined based on the levels of malaria spread in every place and changes over time with resistance to deal with parasites and vectors. Early cautioning systems are in want of models that include intrinsic and extrinsic factors.

It is obvious that the vulnerability of huge populations to infectious illnesses leads to the change of natural environments. Multidrug-resistant tuberculosis was discovered in 2007 and made the United States and global public health infrastructure focus on it. The health experts comprehended that today's affluence of air travel can expose many to a fatal disease associated with a transferrable infectious disease agent. As satellite imaging plays a vital role in predicting aquatic and vector-borne sickness, Ford et al. [3] have used remote sensing for disease prediction, listed the lessons learnt and provided ideas for modeling.

Elvidge et al. [4] have focused on a category of poverty map that advances over time by including reference data for normalization of poverty estimations and enhancements made based on the human activities associated with the economic activity and technology. Location-based disaggregated global map of poverty numbers are obtained from satellite data and different sensors. The map is generated based on count of Landscan population and DMSP night time lights. It is assumed that areas with increased population in growing countries will be less lit and hence have better proportions of underprivileged people and vice versa. By considering the disaggregated data, the national and sub-national estimations of poverty stages for a huge portion of the globe is produced.

Krizhevsky et al. [5] have trained a huge Deep Convolutional Neural Network (DCNN) to categorize images with high resolution in the ImageNet LSVRC-2010 contest into different classes. The NN includes millions of parameters and neurons with 5 convolutional layers, in which some are followed by max-pooling layers and 3 fully connected layers with a 1000-way softmax. To support faster training, non-saturating neurons and a very effectual GPU of convolution are used. To decrease overfitting in the fully connected layers, regularization is done. The network includes many new and unfamiliar features that offer better performance and lessen the training time.

Super-resolution is an image handling technique that finds images of high resolution from sole or consecutive less-resolution images. Deep CNNs can be used to handle super-resolution. Lei et al. [6] have proposed a sole image super-resolution algorithm called Local–Global Combined networks (LGCNet) for distant sensing using deep CNNs. The scheme is designed with "multifork" structure to study multilevel illustrations of distant sensing images with both local and global environments. It is seen that the mechanism offers better results on UC Merced dataset in terms of accuracy and visual features.

Jean et al. [7] have provided a precise, cheap and scalable technique for calculating consumption from satellite imagery with high resolution. They have used CNN to detect image characteristics for dealing with the variation in local level economic consequences. The poverty in emerging countries is dealt with. It shows that ML algorithms can be applied on limited training data, signifying extensive possible application across scientific areas. The prevailing high-resolution satellite imagery during day time is used to make precise predictions about spatial dissemination of economic welfare across 5 African countries. This model is capable of dealing with the timing and cluster location in the training data.

An innovative and extensive dataset is introduced by Lam et al. [8] for improving object detection and overcoming the challenges faced in object detection. Four keys are involved in computer vision frontier which is enabled by using the satellite imaginary dataset. For Geospatial category detection and bounding box, a novel method is employed in the three phases of quality control. Data is composed from the worldview-3 satellites at 0.3 m ground sample which provides high resolution when compared to the public satellite imaginary dataset. There are more than one million objects in xView object detection datasets. It supports diverse wide ranging applications which unify dataset for land-use objects including buildings, roads with large and small vehicles and mini scenes. Other than overheads, xView holds a variety of classes and visual backgrounds.

Gupta et al. [9] have dealt with xBD on disaster recovery with a wide scale dataset for the progression of change recognition and damage of building for humanitarian support. It is challenging to face the after effects of disasters, logistics, resource preparation and damage estimation. The analysis on satellite imagery is done by a passive method to assess the damage resulting in less manpower, thus lowering the risk factor. The xBD extends multi-band satellite imagery for pre-event and post-event of various calamities which involves building structures, kind of damage caused by events, labels of damage level and metadata involved in the satellite capture. The data which involves water bodies, smoke and lava environmental labels are part of the dataset with bounding boxes. Satellite imagery is inefficient in addressing the building damage. xBD covers the complex and dynamic data like nature of the damage, large quantities and type of damage. The data set is curated with responses obtained from diverse places around the world that expertises in diverse types of disaster, creating a clear view and pointing out the real-world damage status. To provide a detailed report with quality control, a verifiable annotation process is executed repeatedly and rigorously.

The recovery of distant sensing images plays a dominant role in physical data mining. Much focus is on extraction, metric similarity and significance response. Tong et al. [10] have used three essential aspects of distant sensing image retrieval. The behavior of systematic scrutiny on assessing correlated issues may affect the performance of deep features. The results provide a guiding role for analyzing content-based distant sensing image retrieval. Visual aspects are focused and 3 levels of feature extraction based on low, mid and high-level features are used. Low-level features are chosen based on the foundation of production skills and area expertise. Varied low-level features have been exploited for distant sensing retrieval, mostly with spectral, texture and shape factors. In divergence, the mid-level can signify discriminated data by encrypting original features using Bag-of-Words (BoW), Vector Locally Added Descriptors (VLAD) or Fisher vector (FV). Nevertheless, owing to the variations of picture scale, alignment and brightness, changes are applicable to the ground objects that have fairly dissimilar appearance. The described handmade structures drop their efficiency of refining the high-resolution distant sensing images.

The deep learning training models predict the survey-oriented estimations of prosperity approximately more than 20,000 African communities from multispectral satellite imagery. The variation of the model can be viewed from the village wealth in countries which use untrained model to outperform the earlier standards from well-refined resolution imagery. The evaluation with self-determining wealth sizes from surveys that causes errors in satellite approximations are similar to the errors in ground data. The study of the imagery can be used to precisely calculate the local welfare of space and time as demonstrated in the coarse during night. The light imagery can be used to evaluate the country's economic performance, and higher resolution imagery from private subdivision providers can be used to evaluate the spatial difference in the local financial results in some of the developing countries. Yeh et al. [11] have emphasized on using numerous sources of spatially coarser community imagery to conclude the spatial and temporal variances in local level financial comfort throughout the sub-saharan Africa including countries, where trusted survey data do not occur and survey based on interpolation approaches may fail to generate precise values.

3 Proposed Framework

The proposed system includes image enhancement using Histogram Equalization (HE) and detection of objects in the image using Convolutional Network (Fast R-CNN) and You Only Look Once (YOLO).

3.1 Image Enhancement Using Histogram Equalization

Pre-processing of images deals with least level of abstraction. The image related information is not improved if the entropy is an information measure. The goal of pre-processing is to enhance the image data that subdues undesired distortions or improves the image characteristics related to handling and analysis.

There are different types of image pre-processing schemes.

- Pixel intensity alterations/ Brightness improvement
- Geometric alteration
- Image filtering and segmentation
- Fourier transform and image refurbishment

Histogram equalization is useful for enhancing contrast. It is a pixel brightness conversion technique. Contrast is the variance in intensity among items in an image. This contrast enhancement technique offers better performance on all kinds of image. Contrast may not constantly increase. There are some cases where histogram equalization does not offer better results. In that case, the contrast is to be decreased. It is an illustration of frequency dissemination. It forms the basis for many spatial domain handling schemes. In case the contrast is reduced, it is impossible to differentiate objects, and it is found as a distinct object. It is an enhancement mechanism in image handling as it is extremely effectual and easy. It is the finest scheme for varying the range and contrast by modifying the image such that the intensity histogram is of the anticipated shape. It can be categorized into two types based on the transformation function.

Global Histogram Equalization (GHE):
Global Histogram Equalization (GHE) is very fast, but the efficiency of contrast improvement is less. The histogram of the complete image is used to calculate the transformation function. The active range of image histogram is made flat and stretched. The complete contrast is enhanced.

Local Histogram Equalization (LHE):
Local Histogram Equalization (LHE) is used to improve the complete contrast in a better way. The disadvantages of HE are that the average brightness of an image decreases considerably as a significance of flattening, and this is not the desired property as conserving the original brightness of an image is essential. Bi-HE overcomes this issue.

The inputs are given and the histogram is generated. The local minimum is found followed by dividing it depending on the local minima. Specific gray levels of every partition are seen and HE is applied on each partition.

The histogram of pixel values of the image is computed. The pixel $f_n[x, y]$ is considered as 'H' uniformly spaced buckets $B[i]$.

$$B[i] = \sum_{i=1}^{n}\sum_{j=1}^{m}\begin{cases} 1, \text{ if} & f_n[x, y] = k \\ 0, \text{ Otherwise} & \end{cases} \tag{1}$$

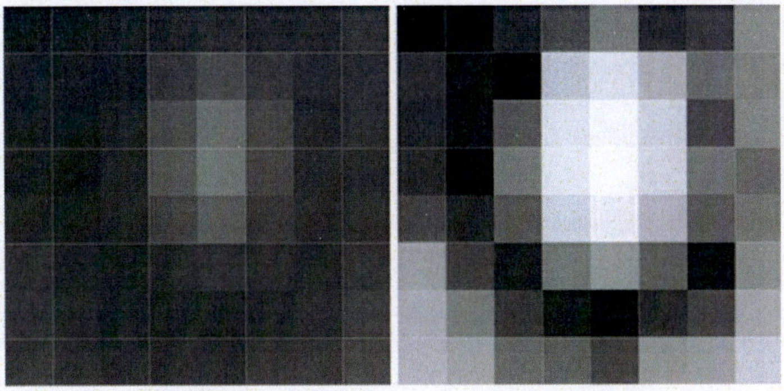

Fig. 2 Sample original and equalized image

where, $H = 2^8$ and the image dimension is $m \times n$.

Cumulative Distribution Function (CDF) is given by,

$$\text{CDF}[l] = \sum_{a=1}^{b} B[i] \tag{2}$$

The image is scaled using CDF to generate the output.

$$h[x, y] = \frac{\text{CDF}[f_n[x, y] - \text{CDF}_{\text{min}}}{(n \times m) - CDF_{\text{min}}} \times (H - 1) \tag{3}$$

where, 'CDF_{min}' is the least non-zero value of the CDF. Figure 1 shows a sampled and equalized image. Figure 2 shows how an image is equalized and Fig. 3 shows an example.

3.2 Fast Region-Based Convolutional Neural Network (Fast R-CNN)

For an image, Fast Region-based Convolutional Neural Network (Fast R-CNN) involves selective search to mine Regions of Interest (RoI), where every RoI is a rectangle that represents the borderline of an item in an image. Based on the situation, the number of RoIs may vary. Each RoI is fed into a Neural Network (NN) to generate output features. For each output feature, a pool of Support Vector Machine (SVM) classifiers is employed to find the object in the RoI. Following challenges are faced with R-CNN.

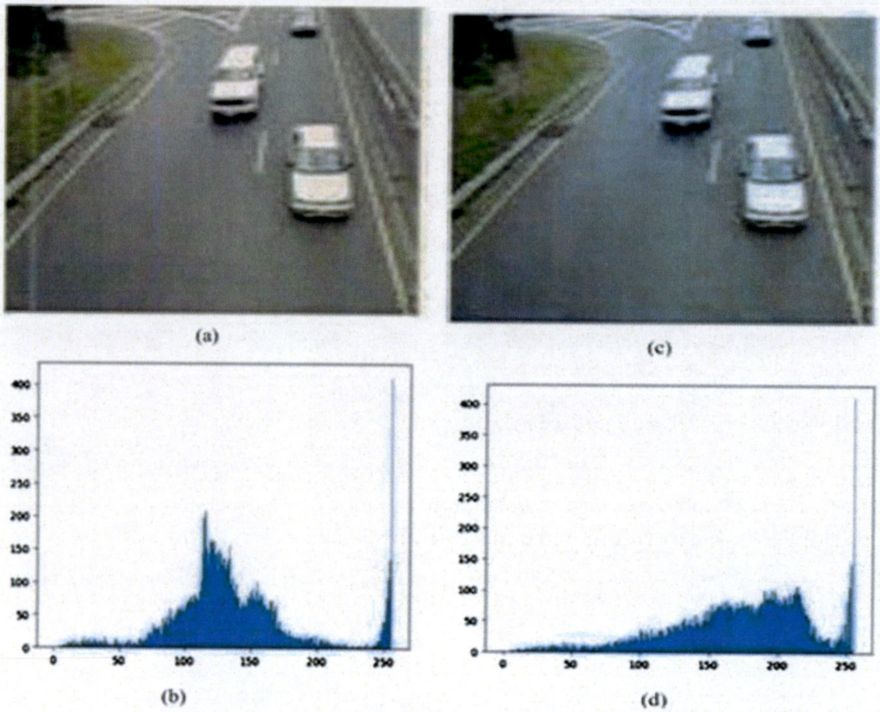

Fig. 3 Before and after histogram equalization

- More time is involved in training the network as 2000 regions per image are classified
- Real-time implementation is difficult as it takes about 47 s for every test image
- The selective search algorithm is fixed and hence no learning is done at this level. This might cause production of irrelevant candidate regions.

To overcome these demerits, Fast R-CNN was developed [1]. In case of R-CNN, it individually computes the NN features on nearly 2000 as RoI. Fast R-CNN runs the NN on the complete image. Finally, RoI Pooling is done to slice out RoI from the network's output tensor, redesign it, and categorize it. Fast R-CNN also employs selective search to produce region proposals.

It performs object discovery faster, wherein the input image is fed into the CNN instead of region proposals and a convolutional feature map is generated. From the map, the RoIs are identified and distorted into squares by using a pooling layer. They are reshaped into stable size so that it can be added to a fully connected layer. From the feature vector, by using a softmax layer, the class of the propounded region along with the offset values for the bounding box can be found.

Fast R-CNN is slowed down at testing time, as incorporating region proposals decelerate the algorithm considerably. Hence, region proposals are blockages leading to reduced performance.

3.3 *You Only Look Once (YOLO) Algorithm*

You Only Look Once (YOLO) is the most commanding real-time object identification algorithm. In case of former algorithms like R-CNN and Faster R-CNN, the image or video are to be sent into the network once. Several regions that identify the objects are observed. YOLO considers the whole image. It uses an exclusive NN based on the features of the whole image to forecast several boxes, each containing a specific object.

The image is split into regions of uniform size and the region containing the center of the object is accountable for identifying the object. The cells in this grid are accountable for envisaging boxes which include an object, and a score demonstrating the level of sureness for the entity existing in the box. In case there are no objects in the cell, the score is zero. Else, in case an object is present in a cell, the score will be equal to the Intersection over Union (IoU) amid the expected box and the image's ground truth.

The class-definite confidence values of every box are obtained using a CNN that is based on the GoogLeNet network. These algorithms will give an image or video as output, which serves as input with objects being localized and class added to it.

YOLO examines the entire picture rather than consecutively investigating many regions. This leads to predominant increase in the identification speed but leads to a reduction in the accuracy of object recognition in contrast to other mechanisms. It is a powerful algorithm used in object detection used to identify autonomous vehicles, poker fraud recognition, and etc., YOLO offers better performance and is based on regression. Instead of choosing the RoI, it envisages classes and bounding boxes for the entire image in a run of the algorithm. It is essential to know what is essentially being forecast. Eventually, the class of the object and the bounding box identifying the object position can be found. The bounding boxes are labeled using the following descriptors namely, Box's Center, Width, Height and class of the object. The probability of the presence of an object in the bounding box is also found. The algorithm does not search for RoIs that include an object, but splits the image into cells, naturally 19×19 grid. Every cell is accountable for envisaging 'K' bounding boxes. An object is said to be in a particular cell only when the co-ordinates of the anchor box's center is found in the cell. These co-ordinates are computed comparative to the cell, while the height and width are found relative to the complete image size. In a pass of forward propagation, the algorithm finds the probability of an object present in the cell. The class that has the maximum probability is found and allocated to the specific grid cell. Comparable processes are carried out for all the grid cells in the image (Fig. 4).

Once the class probabilities are predicted, non-max suppression is done to remove needless anchor boxes. The bounding boxes that are found to be very close are removed by preforming IoU with the one having the maximum probability. IoU for every bounding box based on the one with the maximum class probability is found, and the bounding boxes with IoU greater than a threshold are rejected. It is understood that the bounding boxes cover the same object, but the one with low probability is

Fig. 4 Sample outputs of fast R-CNN (left) and YOLO (right)

eliminated. All the bounding boxes are different. Finally, the essential vector that gives the information of the bounding box of the particular class is generated as output.

Figures 5 and 6 show how Fast R-CNN and YOLO can be used to identify objects in images. It is seen that YOLO offers the best results. YOLO accurately identifies the objects (vehicles). It is evident that the number of vehicles is very much reduced due to COVID-19 which shows an economic breakdown.

Fig. 5 Output of fast R-CNN

Fig. 6 Output of YOLO

4 Conclusion

In this paper, Histogram Equalizer (HE) is used to improve the quality of the image. Fast R-CNN and YOLO are applied to identify objects especially vehicles in the image. The main aim is to analyze whether there is an economic downfall due to COVID-19 pandemic. As mobility is seen to be reduced, the movement of commodities, people, etc., are very much affected. This has led to the collapse of the normal functioning of human life. YOLO offers improved performance as it deals with the image as a whole. It offers 95.5% accuracy by detecting all the vehicles which are missed by Fast R-NN.

References

1. Girshick R (2015) Fast r-CNN. In: Proceedings of the IEEE international conference on computer vision, pp 1440–1448
2. Rogers DJ, Randolph SE, Snow RW, Hay SI (2002) Satellite imagery in the study and forecast of malaria. Nature 415(6872):710–715
3. Ford TE, Colwell RR, Rose JB, Morse SS, Rogers DJ, Yates TL (2009) Using satellite images of environmental changes to predict infectious disease outbreaks. Emerg Infect Dis 15(9):1341
4. Elvidge CD, Sutton PC, Ghosh T, Tuttle BT, Baugh KE, Bhaduri B, Bright E (2009) A global poverty map derived from satellite data. Comput Geosci 35(8):1652–1660
5. Krizhevsky A, Sutskever I, Hinton GE (2012) Imagenet classification with deep convolutional neural networks. Adv Neural Inf Process Syst 25:1097–1105
6. Lei S, Shi Z, Zou Z (2017) Super-resolution for remote sensing images via local—global combined network. IEEE Geosci Remote Sens Lett 14(8):1243–1247
7. Jean N, Burke M, Xie M, Davis WM, Lobell DB, Ermon S (2016) Combining satellite imagery and machine learning to predict poverty. Science 353(6301):790–794
8. Lam D, Kuzma R, McGee K, Dooley S, Laielli M, Klaric M, Bulatov Y, McCord B (2018) xview: Objects in context in overhead imagery. arXiv preprint arXiv:1802.07856
9. Gupta R, Goodman B, Patel N, Hosfelt R, Sajeev S, Heim E, Doshi J, Lucas L, Choset H, Gaston M (2019) Creating xBD: A dataset for assessing building damage from satellite imagery. In: Proceedings of the IEEE/CVF conference on computer vision and pattern recognition workshops, pp 10–17
10. Tong XY, Xia GS, Hu F, Zhong Y, Datcu M, Zhang L (2019) Exploiting deep features for remote sensing image retrieval: a systematic investigation. IEEE Trans Big Data 6(3):507–521
11. Yeh C, Perez A, Driscoll A, Azzari G, Tang Z, Lobell D, Ermon S, Burke M (2020) Using publicly available satellite imagery and deep learning to understand economic well-being in Africa. Nat Commun 11(1):1–11

Textual Posts Analysis Using Voting Classifier

Naman Goel, Yash Rastogi, Yash Verma, and Rajesh Kumar Singh

Abstract Web-based media is a tremendous virtual space to communicate and impart singular insight. Despite the fact that there are numerous positive parts of web-based media, there are additionally chances. Strict and standing predisposition, religious and caste bias, sexual provocation, cyberbullying and sexting are at present netizen's propensity via online media. This affects the humans round the globe. Bias identification is a perplexing cycle that involves a few measurements to be considered and it is interwoven with social, political and monetary issues. This necessitates the requirement of a model which is capable enough to segregate the posts in their initial phase and curb those posts which are threat or can lead to civil disobedience and mass agitation. The proposed model presented in this paper can recognize the biasedness of client's post and characterize it into positive or negative. This characterization will help in blocking their further propagation. In the experimental study, the textual posts are gathered from different sources. The proposed model scored highest accuracy of 80.30%.

Keywords Machine learning · Natural language processing · Polarity analysis · Voting classifier

N. Goel · Y. Rastogi (✉) · Y. Verma · R. K. Singh
Department of Computer Science Engineering, Meerut Institute of Engineering and Technology (MIET), Meerut, U.P, India
e-mail: yash.rastogi.cse.2017@miet.ac.in

N. Goel
e-mail: naman.r.goel.cse.2017@miet.ac.in

Y. Verma
e-mail: yash.verma.cse.2017@miet.ac.in

R. K. Singh
e-mail: rajesh.singh@miet.ac.in

R. P. Mahapatra et al. (eds.), *Proceedings of International Conference on Recent Trends in Computing*, Lecture Notes in Networks and Systems 341,
https://doi.org/10.1007/978-981-16-7118-0_32

1 Introduction

Web-based media assumes an essential function in our everyday lives. Currently, increasing popularity of online social media networks has accumulated billion of users. According to Statisca.com, the social media users in 2020 totalling 3.6 billion are expected to increase sharply to 4.41 billion in 2025. With widespread access to internet as well as social platforms, users can post more easily whatever they want which can hurt sentiments of a person or a group, leading to differences in the society. Following are the obscurity of social media:

1. Political Bias: Political bias is a predisposition including the inclining and adjusting of data to cause a political candidate or political position to appear to be more alluring. Social media posts made at the time of general elections or at the time of other political activities are biased towards a specific political party in a multiparty system. They tend to favour or oppose those in power. These posts tend to create an illusion among the masses and hamper their political temperament.
2. Hate Mongering: Social media platforms are accessible to all. They have become the people's platform for spreading hate against each other in the name of religion, caste, etc. Since social media platforms do not impose aggressive checks and balances on each post put on their platform, the hate content is easily spread which leads to violence at times.
3. Communal Riots: Various social media posts are intended at sparking the communal riots. Be it in the form of criticizing the other person or mainly using abusive language for others religion. They have come up as the root cause of various communal riots like the one observed in Bengaluru recently.

This research paper looks into the post's sentiment, as represented in huge collections of posts collected from different sources like Twitter, Facebook, YouTube and classify the user's post against the dataset using various algorithms. This can be used to predict the biasedness of posts.

Our research in this paper is to tackle the above mentioned problems by characterizing the posts as positive and negative on the basis of the combination of words used in the textual posts. The classification algorithms help in classifying them accordingly.

This paper is organized as follows: Section 2 describes the dataset creation and preprocessing. Section 3 gives brief description of the classification algorithms applied. Section 4 shows experiment and the analysis of the results obtained. Section 5 concludes our findings and chances for future works.

Sr No.	Textual Posts	Feature
1	brahminism is the root of all that is evil in india	0
2	the quran is a deeply progressive text and modern day hinduism is deeply regressive	0
3	hinduism is against all form of violence	1
4	modi is an awesome friend of the president of USA	1

Fig. 1 Sample dataset

2 Methodology

2.1 Dataset Creation

In order to cover the wide spectrum of textual posts, it was essential to have a dataset which covers the same. We have created a dataset which comprises of posts of all possible thoughts. The dataset comprises of about 1500 textual posts. The dataset contains two labels one textual posts and the other feature. The textual posts label contains the social media posts, and the feature label contains the numerical feature of them 0.The feature can have only two values either 0 or 1.

As in the table shown below, the textual post is 'Brahmanism is the root of all that is evil in India' and '0' represents its feature or polarity. As per the conventions used by us, '0' indicates negative and '1' represents positive textual posts, respectively (Fig. 1).

Full dataset is available at www.kaggle.com/dataset/18a3e6d67d78b6c9e967a64b3d03eb5447dd459db7f8370e30c9470d046ff655

2.2 Data Preprocessing

Preprocessing of the textual posts is done as follows:

1. Allow only the string of characters to be used for analysis. The string of characters is case insensitive, i.e. it can contain both lowercase and uppercase characters.
2. The second step removes punctuation marks like semicolon, period, colon, question mark, exclamation point, apostrophe, etc., as they are of no use for the analysis.
3. Convert the series of characters to lowercase for simpler working with the dataset.
4. Each textual post is converted to a list of words and each post is subjected to stemming. Stemming is a process which produces the morphological variants

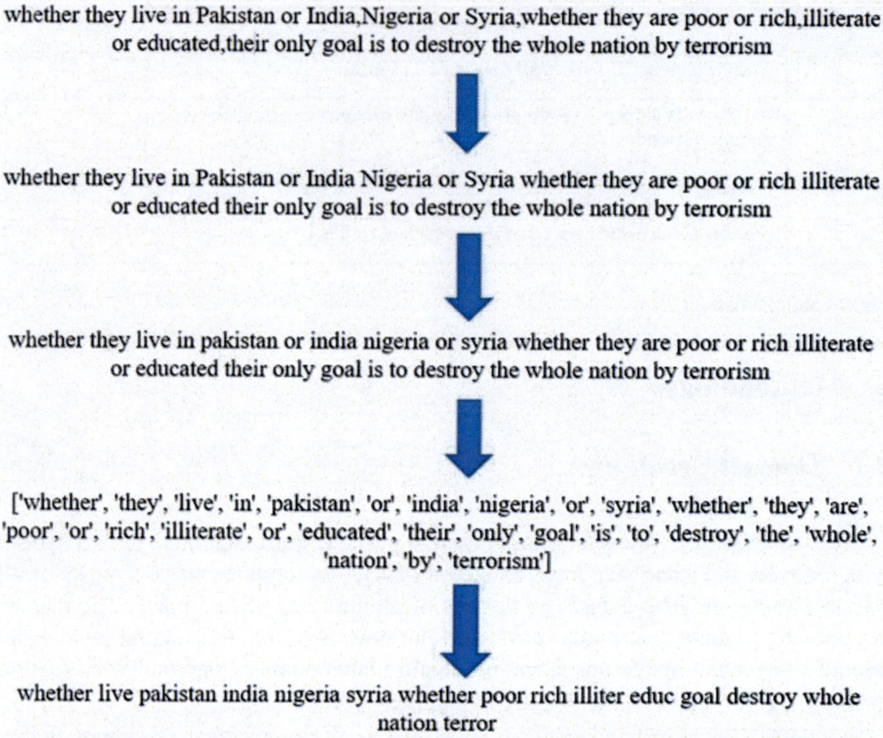

Fig. 2 Data preprocessing

of a root word. Eg: The word 'retrieve' has morphologically similar words like 'retrieval' and 'retrieves'.

5. The last step comprised of removing the noise (stopwords in this case) from the textual posts. Stopwords comprise of commonly used words like a, an, the. Their removal is done to save space in our database and also to save valuable processing time (Fig. 2).

2.3 Tokenization and Sparse Matrix Generation

The textual posts have been tokenized into a bag of words where each word in a post represents a specific column. Each column will contain a binary outcome 0 or 1, representing absence and presence of a specific word in a post, respectively. The combinations of these binary outcomes form a correlation with the feature of the particular post. The binary outcomes of each word in the post form a correlation with the feature assigned to it which helps in easy classification of the sample post as positive or negative.

3 Classification Algorithms

A lot of classification algorithms is available to classify the textual posts data. After applying all the available classification algorithms to a random dataset, we found Naïve Bayes, logistic regression and random forest best suited for our dataset. The final outcome of the model is the combination of these three algorithms only implemented with the help of another classifier, i.e. voting classifier.

3.1 *Naive Bayes*

Naive Bayes is a simple Bayesian classifier which is based on Bayes theorem from probability theory. The Naive Bayes classifier [1] finds its application mainly in categorization of documents and filtering out spam email. It expects that the presence of a specific element in a class is unrelated with the presence of some other element. The rule is described as follows:

$$P(a|b) = \frac{P(b|a) * P(a)}{P(b)}$$

where

$P(a|b)$	posterior probability
$P(b|a)$	likelihood probability
$P(a)$	prior probability
$P(b)$	evidence probability.

3.2 *Logistic Regression*

It is a supervised classification algorithm. The classification algorithm finds application in problems where there are more than one independent variable which determine the outcome. The outcome is binary, i.e. can have values 0 or 1. The logistic regression classifier [2] models the dataset using sigmoid function

$$f(x) = 1/(1 + 1/e^{(x)})$$

where '$f(x)$' is sigmoid function and 'e' represents Euler's number.

3.3 Random Forest

Random forest [3, 4], a classification algorithm based on supervised learning, is used for both classification and regression. Random forest classifier creates decision trees on the data sample and gets prediction from each of them and finally selects best solution through voting. The two key concepts of random forest are

1. Sampling data points randomly when making trees.
2. Considering random subsets of features whilst splitting nodes.

3.4 Voting Classifier

It trains on an ensemble of number of classification techniques and then selects the class having the highest probability. It combines the outcome of each classifier and predicts the output based on the majority of voting [5]. The voting classification technique is categorized into two categories:

1. Hard
2. Soft (Fig. 3).

4 Experiment

1. The dataset is first preprocessed which mainly includes removing stopwords and the process of stemming. The preprocessed data are then split into training

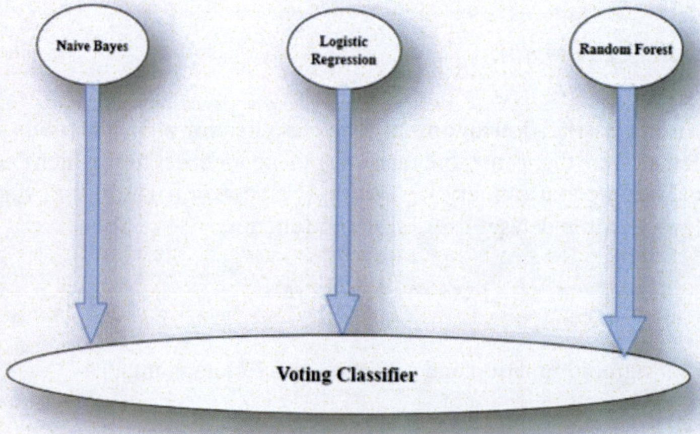

Fig. 3 Voting classifier

```
#Voting Classifier
estimator=[]
estimator.append(('LR',LogisticRegression()))
estimator.append(('NB',GaussianNB()))
estimator.append(('RF',RandomForestClassifier()))
vot_hard = VotingClassifier(estimators = estimator, voting ='hard')
vot_hard.fit(x_train, y_train)
```

Fig. 4 Code snippet of voting classifier

dataset and testing dataset. Training dataset is used to train the different classification algorithms used and testing dataset helps in finding the accuracy of each classification algorithm. 90% of the data is used for training whilst the remaining is used for testing the classification algorithm.

2. Each classifier is fitted into the training set. Three classification algorithms have been used mainly: Naive Bayes, logistic regression and random forest.
3. All three classification algorithms produced different accuracies and different confusion matrices. In order to arrive at a final result, which reflects upon the result all three classification algorithms, voting classifier was used. A voting classifier takes in input of all the three classification algorithms output and then predict the final output as per the voting type applied (voting is categorized in two types—soft voting and hard voting).In this research, hard voting has been used to arrive at the results (Fig. 4).

5 Result Analysis

5.1 Confusion Matrix

Confusion matrix is a table which is used to analyze how effectively a classification algorithm works. In a two class problem, it helps to find the correctly predicted values and the incorrectly predicted values on the testing dataset. In the confusion matrices described above, **'0'** represents **negative** and **'1'** represents **positive**. The four count values, which explain the confusion matrix, are as depicted in Table 1 as follows:

True Negative: Model correctly predicted the post to be negative, i.e. model predicted that the textual post is negative and it actually was positive.

Table 1 Confusion matrix

	Predicted		
Actual		**0**	**1**
	0	True negative	False positive
	1	False negative	True positive

True Positive: Model correctly predicted the post to be positive, i.e. model predicted that the textual post is positive and it actually was negative.

False Negative: Model incorrectly predicted the post to be negative, i.e. model predicted that the textual post is negative whilst actually it was positive.

False Positive: Model incorrectly predicted the post to be positive, i.e. model predicted that the textual post is positive whilst actually it was negative.

Out of the total textual posts considered for testing the trained model, different classification algorithms showed different results. The results obtained have been depicted in the form of confusion matrix.

From Fig. 5, it is clear that out of 75(47 + 28) actual negative posts classifier predicted that 47 of them are negative posts and 28 are positive posts. Out of 57(14 + 43) actual positive posts classifier predicted that 14 of them are negative posts and 43 are positive posts.

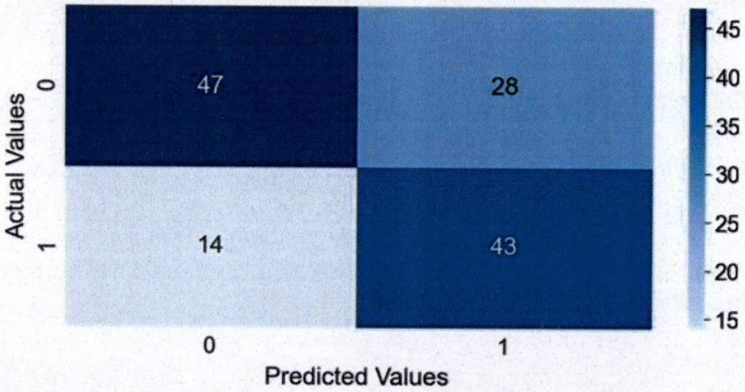

Fig. 5 Naive Bayes—confusion matrix

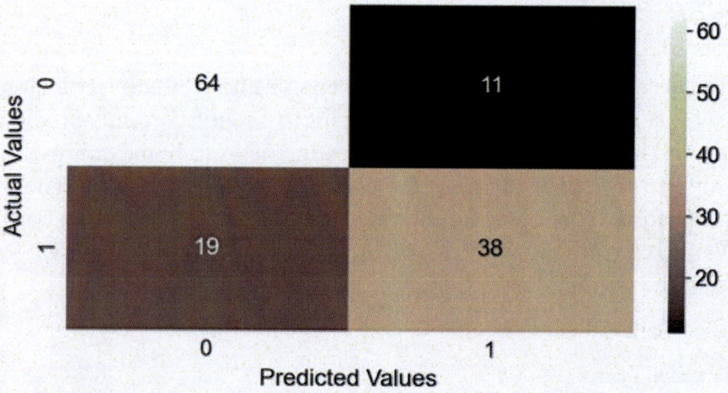

Fig. 6 Logistic regression—confusion matrix

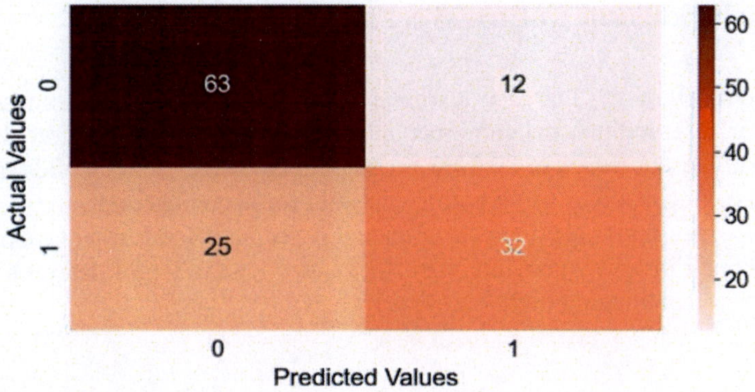

Fig. 7 Random forest—confusion matrix

In Fig. 6, out of 75(64 + 11) actual negative posts classifier predicted that 64 of them are negative posts and 11 are positive posts. Out of 57(19 + 38) actual positive posts classifier predicted that 19 of them are negative posts and 38 are positive posts.

In Fig. 7, out of 75(63 + 12) actual negative posts classifier predicted that 63 of them are negative posts and 12 are positive posts. Out of 57(25 + 32) actual positive posts classifier predicted that 25 of them are negative posts and 32 are positive posts.

5.2 Accuracy Results

Table 2 represents the accuracies obtained using different algorithms. Out of the three classification algorithms used mainly, Naïve Bayes gives the lowest accuracy of about 68.18% and logistic regression gives the highest accuracy of 77.27%.On combining the three classifiers result through the voting classifier, it was found that it gave the highest final accuracy of 80.30% which was highest among the previously obtained results. It was seen that the accuracy increased on combining these three algorithms (Naïve Bayes, logistic regression, random forest).

Table 2 Accuracy table

S. No	Algorithm	Accuracy (%)
1	Naive Bayes	68.18
2	Logistic regression	77.27
3	Random forest	71.96
4	Voting classifier	80.30

6 Conclusion and Future Works

This paper presented a machine learning model to train the textual posts dataset. It has been observed that the combined result of the three classification algorithms produced highest accuracy of 80.30% as compared to using them individually which leads to the conclusion of using voting classifier for accurate results.

The paper includes only analysis of textual posts but it can further be extended to posts including image, video and other. It can also lead to larger dataset and hence can also help in achieving higher accuracy.

References

1. Xue Z, Wei J, Guo W (2020) A real-time naive bayes classifier accelerator on FPGA. IEEE Access 8:40755–40766. https://doi.org/10.1109/ACCESS.2020.2976879
2. Wright RE (1995) Logistic regression. In: Grimm LG, Yarnold PR (eds) Reading and understanding multivariate statistics. American Psychological Association, pp 217–244
3. Ali J, Khan R, Ahmad N, Maqsood I (2012) Random forests and decision trees. Int J Comput Sci Iss (IJCSI) 9
4. AIP Conference Proceedings 1820, 080020 (2017). https://doi.org/10.1063/1.4977376
5. Ruta D, Gabrys B (2005) Classifier selection for majority voting. Inf Fusion 6:63–81
6. Dorle S, Pise N (2018) Political sentiment analysis through social media. In: 2018 2nd international conference on computing methodologies and communication (ICCMC), Erode, pp 869–873. https://doi.org/10.1109/ICCMC.2018.8487879

People Monitoring and Social Distancing Using Computer Vision

Nidhi Tyagi, Riya Upadhyay, Vatsal Gupta, and Shekhar Chauhan

Abstract The goal of this studies is to help minimize COVID-19 cases. Social distancing practices are considered to forestall unwellness transmission by reducing the contact rate of infected and uninfected people. One of the easiest ways for people to lower their risk of infection during an epidemic is for them to reduce their rate of contact with infectious individuals (Timothy in, PLoS Comput Biol 6(5):e1000793, (2010) [1]). As the pandemic situation increases day by day, so there is a need to understand the situation and take possible actions against them. Social distancing comes under the major precaution against pandemic. Apart from these, there is a need to monitor people whether they take the situation seriously about social distancing or not. So, the main objective of this research is to provide a possible solution of how to monitor people using CCTV/drone. This research provides an idea about an application that can be used to monitor people for prevention against social distancing.

Keywords Computer vision · Social distancing · YOLO

1 Introduction

Computer vision is the field computer science that is used to visualize objects from images and videos. The concept of computer vision is used to visualize the living and nonliving objects in the computer for their better understanding. Computer vision has real-life applications to improve the lifestyle of people. Some real-life applications of

N. Tyagi (✉)
Professor, Shobhit Institute of Engineering & Technology, Meerut, India

R. Upadhyay · V. Gupta · S. Chauhan
Student, Meerut Institute of Engineering & Technology, Meerut, India
e-mail: riya.upadhyay.it.2017@miet.ac.in

S. Chauhan
e-mail: shekhar.chauhan.it.2017@miet.ac.in

R. P. Mahapatra et al. (eds.), *Proceedings of International Conference on Recent Trends in Computing*, Lecture Notes in Networks and Systems 341,
https://doi.org/10.1007/978-981-16-7118-0_33

computer vision are face detection, people counting, object recognition, image classification, color detection, object tracking, and so on. In the fight against coronavirus, social distancing has proven to be very effective. In the absence of any pharmaceutical intervention, the only strategy against pandemic is to reduce the mixing of susceptible and infectious people through early ascertainment of cases or reduction of contact [2]. The Government sets some strict rules and regulations to fight against this pandemic. One of the rules is to follow social distancing among people which is accomplished by a restricted limited number of people at a specific place and time according to the requirement and situation and specific distance between everyone. However, some of us do not take this pandemic situation seriously and defy the rules, which makes the situation worse. So, to prevent disobedient behavior of peoples, people monitoring can be used which help to detect the insubordinate behaviors so that the higher authorities can take strict actions against them. An application can be made to monitor people using Python for social distancing. In the applications, two considerable concepts of computer vision taken—people counting and object distancing are merged. People counting in computer vision is the technique used to count or estimate the number of people present in an image or video. This idea of people counting in people monitoring can be used for crowd detection. Estimating the quantity of persons during a public place provides helpful info for video-based police investigation and observance applications [3]. Object distancing in computer vision is a technique used to estimate the distance between individuals. The idea of object distancing in people monitoring can be used to detect the required distance between individuals which fulfilled the criteria of social distancing. Analysis of data from large repository need to be done using big data [4, 5].

1.1 COVID-19

Today's circumstances of coronavirus are very vital. World Health Organization initial learned of the virus on December 31, 2019, following a report of a cluster of cases of respiratory disorder in urban areas, People's Republic of China. On March 11, 2020, the rapid increases in the number of cases outside China led the WHO director-general to announce that the irruption might be characterized as a deadly disease. The temporal order of vaccines is a contentious subject around the world. It is an extremely infectious virus that causes varied clinical manifestations with a wide range of severity. Every young person has been known to become severely ill or die, though not as often as the elderly. The level of viral load a person is exposed to the immunity status or pre-existing health conditions may be factors that influence this severity. As cases increase day by day, the Government has enforced "social distancing" measures to prevent the valuable life of the peoples [6] (Fig. 1).

The rate of disease spreads from a single infected person within 6 weeks and is reduced by more than 60%, that's why social distancing is considered as the major factor to stand against the pandemic in this world.

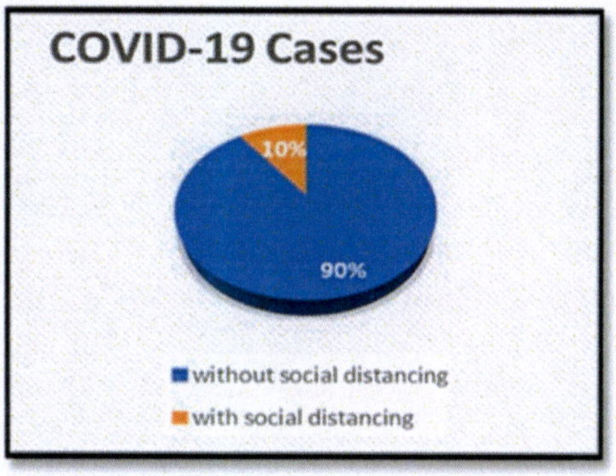

Fig. 1 Reflects COVID-19 cases

1.2 Scenario of COVID-19

There are rapid increases in coronavirus cases and deaths almost all over the world [7]. There are almost 105,422,292 confirmed corona cases and 2,296,036 confirmed deaths covered under almost 222 countries, areas, or territories. In recent pandemics, isolation, and quarantine (more extreme forms of social distancing) have precipitated depression and anxiety [8]. Managing a certain meter distance (prescribed by WHO) between everyone helps to reduce the disease transmission from infected to uninfected persons and minimizes infected patients. India has the largest number of confirmed cases in Asia. From March so far, there were total of 105,469,801 cases reported and 77,171,418 people recovered (Fig. 2).

2 Background Study

Object distancing and people counting are two major concerns in recent research. The object distancing in people monitoring calculates the safe distance between everyone recorded and showing the details in the computer where the frame is shared by the CCTV/drone. The object distancing calculates every human in the frame record and displays that social distancing violation occurred. People counting can be loosely understood as the instantaneous estimation of the number of persons present in a scene [9].People counting calculates the total number of people in the frame, number of people in a safe distance, number of people in the alert zone, and number of people who are very close to each other. People counting estimates the crowd shown in the frame and if the crowd detected to exceed the criteria, then the alert will be displayed in the system. The alert generated by the CCTV/drone can be represented to the higher authorities by producing a pop-up alarm so that the authorities would be able to take the action against these. The application used in CCTV/drone by applying the concept of computer vision can be made by using Python (Fig. 3).

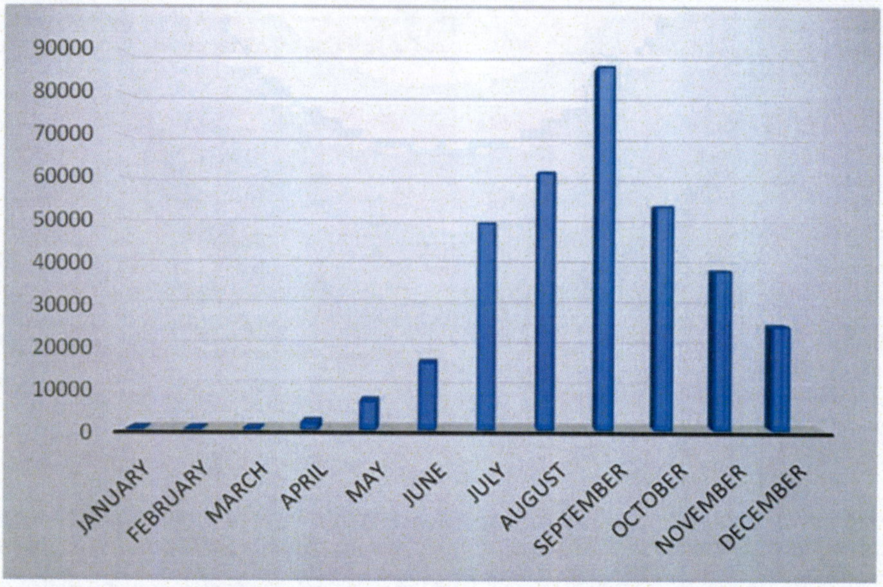

Fig. 2 Reflect COVID-19 cases in India

Fig. 3 Shows social distancing using RGB boxes

3 Proposed Work

The idea of the application can be used in CCTV/drone by recording a video in it. So, the recorded video can be processed by using the code which helps to monitor peoples. After processed video, the required output to check generated to check whether the cities around the area under the monitoring follow the social distancing rules or not. Hence, the main work of the application is to record a video and processed the recorded video using the code, and at last, generates the desired output. The processing of video can be done locally or by using the cloud. An application can also be made which sends the recorded video directly to the cloud so that with the cloud processed the recorded video and generates the output.

3.1 Flowchart

CCTV/drone can be used to record video around the locality for people to monitor. After the video recorded, the size of the video will be checked as

- If the size of the video is less than 1 then the system exits with a message "PLEASE GIVE CORRECT INPUT."
- Else input processing will be done in our application for people counting and social distancing. In this process, two conditions will be checked.

As the main precaution for social distancing is to maintain the distance between everyone. Figure 4 represents the diagrammatic representation of the flow of activities to monitor the people. This application will check that the distance between each person must be greater than 10 m.

- If the distance between everyone is greater than 10 m, then the system returns a message "SOCIAL DISTANCING NOT MAINTAINED" with RED LINE ALERT.
- Else the system returns a message "SOCIAL DISTANCE MAINTAINED" with GREEN LINE ALERT.

The second main precaution for social distancing is to maintain the crowd of people. As there is limited number of people that can be present in a particular area, these applications will check whether the number of peoples covered under the video at the time will not exceed 100 people.

If the number of people covered under the video at a certain time is less than 100, then the system will return a message "PEOPLE LESS THAN 100" with GREEN LINE ALERT.

Else the system will return a message "PEOPLE GREATER THAN 100" with RED LINE ALERT.

The output of both the conditions distance between everyone and the people count will be processed together in a process video output. After all the processing is done

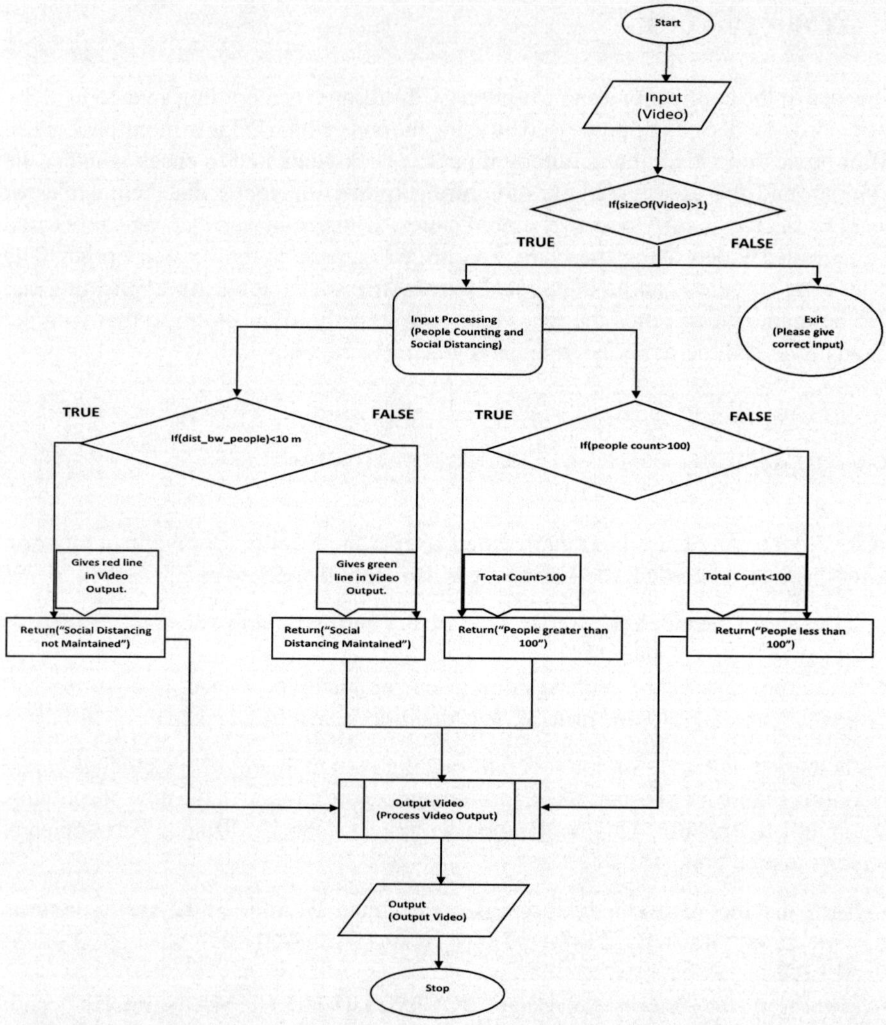

Fig. 4 Flowchart represents people monitoring

by the application, the output of the processing in the application will be displayed in the system.

According to the output returned by the system using the application, the higher authorities will be able to act against the situation. So, this application can be used to prevent the peoples against the worst condition of today's scenario.

3.2 *Algorithm*

In the creation of the application, the YOLO algorithm can be applied. It stands for you only look once. Joseph Redmon developed in 2016 which is the cornerstone of object recognition research and it makes the computer vision algorithm better and faster. Real-time object detection is used for multiple image recognition in a single frame. It is fast as compared to other classification algorithms. It makes localization errors but predicts fewer false positives in the background [10]. In the prototype of the application, YOLOv3 can be used. As the 4th generation named YOLOv4 released in April 2020, which is developed by three developers Alexey Bochkovskiy, Chien-Yao Wang, and Hong-Yuan Mark Liao can also be used for people monitoring application. It can be used in the application when people are detected, the application draws a bounding box around them and the centroid of the boxes can be used to measure the distances between them. Red and green colored boxes are used to represent the unsafe and safe zone in the application. It will also count the total number of people because crowed also results in social distancing disobey behavior.

3.2.1 YOLOv3 Versus YOLOv4

YOLOv2	YOLOv3
Second version (v2) of YOLO	Third version (v3) of YOLO
More accurate and faster than predecessor	More accurate and faster than version 2
Predicts output in the last layer	Predicts output in three different scales
Framework is Darknet-19	Framework is Darknet-53
Uses anchor boxes	Uses residual blocks

3.2.2 YOLOv3

YOLOv3, the third variant of the popular object detection algorithm, is extremely fast and accurate compared from the previous algorithm variants. Its accuracy is same as the single shot multi-box (SSD). It is better than version 2 for the detection of small images. It uses residual boxes to predict their output at three different scales. The algorithm is applied to a single neural network to the full image. Firstly, the image is divided into a 13×13 grid of cells. Depending on the size of the input, the size of these 169 cells varies. The network divides the image into regions and predicts bounding boxes and probabilities for each region. By using YOLOv3, human can be detected in the frame and then distance between each human will be calculated who is detected in the frame which will shows the number of peoples who are at low, high, or no risk [11].

3.3 Distance Calculation Using Cosine

After finding the distance between every people in the frame using camera, cosine law can be used for finding the distance between people. Subset of two people will be taken at a time in the frame. Distance between the two people is already taken from the camera but there is a need of having the horizontal distance between them. By using the coordinates from the bounding boxes, we can average the x values in the midpoints to get the pixel distance. And by using similar triangles, we can calculate the real-life horizontal distance. We are considering three points C, A, and B as HUMAN 1, HUMAN 2, and CAMERA, respectively. Therefore, AB is the distance between the HUMAN 2 and CAMERA, BC is the distance between HUMAN 1 and CAMERA, CD is the horizontal distance, and AC is the distance between the humans. Since, it is two-dimensional. BC is approximately equaling to BD as the humans are parallel in the frame. Subtracting the distance from object 2 to the camera and BD so that we get line BD.

$$\mathrm{Cos}A = \left(b^2 + c^2 - a^2\right)/2bc \tag{1}$$

$$\mathrm{Cos}B = \left(c^2 + a^2 - b^2\right)/2ca \tag{2}$$

$$\mathrm{Cos}\,C = \left(a^2 + b^2 - c^2\right)/2ab \tag{3}$$

The inverse cosine of $(b^2 + c^2 - d^2)/2bc$ can be found by using the above formulas as we know all the sides of the triangle CDB.

$$< \mathrm{CDA} = 180- < \mathrm{CDB} \tag{4}$$

Again, we can plug the values of the lines CD, AD, and the angle CDA to get the line AC. Line AC represents the distance between the two people.

OpenCV can be used to draw lines from the center of people's bounding boxes to represent the distance. Green lines represent people follow social distancing while red lines represent people do not follow social distancing. These lines are drawn from every person to the other in every frame [12].

4 Conclusion

Physical interaction or closeness between infectious and non-infectious persons maximizes the rate of spread of the virus among people. If these situations will not take them seriously then they will take a shape of huge uncontrollable conditions. The Government sets some controlled measures about social distancing which helps to control the spreading behavior of COVID-19. But some peoples still defy the rules of social distancing, so to help the higher authority for acting against them which minimize the spread rate of pandemic, it provides a possible solution of an application that can be used to monitor those peoples. It suggests an idea about the application which can be made to monitor the crowded gathering and the distance between everyone. The suggestion of the application in this paper can be used in CCTV/drone. If the virus spreading condition becomes worse or the people may have to a faced situation like that in future also like a new virus STRAIN is rumored as infection in future, then this application will be taken as a possible solution for people monitoring so the condition will be taken under controlled. Hence, the application generates number of peoples at risk on a certain area recognized by drone/CCTV as a result.

References

1. Reluga TC (2010) Game theory of social distancing in response to an epidemic. PLoS Comput Biol 6(5):e1000793
2. Lewnard J, Lo N (2020) The scientific and ethical basis for social-distancing interventions against COVID-19. Lancet Infect Dis 20(6):631–633
3. Celik H, Hanjalic A, Hendriks EA (2006) Towards a robust solution to people counting. In: International conference on image processing, Atlanta, GA, pp 2401–2404
4. Dixit A, Tyagi N (2020) MapReduce: simplified data processing on clusters with privacy preserving by using anonymization techniques. Int J Recent Technol Eng (IJRTE) 8(6). ISSN: 2277-3878
5. Dixit A, Tyagi N (2019) Big data privacy for end-to-end delivery. Int J Recent Technol Eng (IJRTE) 8(2S8). ISSN: 2277–3878
6. Hawryluck L, Gold W, Robinson S, Pogorski S, Galea S, Styra R (2004) SARS control and psychological effects of quarantine, Toronto, Canada. Emerg Infect Dis 10(7):1206–1212
7. Mahmoudi M, Baleanu D, Mansor Z, Tuan B, Pho K (2020) Fuzzy clustering method to compare the spread rate of Covid-19 in the high risks countries. Chaos Solitons Fractals 140:110230
8. Venkatesh A, Edirappuli S (2020) Social distancing in covid-19: what are the mental health implications. BMJ, p.m1379
9. Bondi E, Seidenari L, Bagdanov AD, Del Bimbo A (2014) Real-time people counting from depth imagery of crowded environments. In; IEEE international conference on advanced video and signal based surveillance (AVSS), Seoul, pp 337–342
10. Geethapriya S, Duraimurugan N, Chokkalingam SP (2019) Real-time object detection with Yolo. Int J Eng Adv Technol (IJEAT) 8(3S). ISSN: 2249-8958.
11. Vinitha, Velantina (2020) Social distancing detection system with artificial intelligence using computer vision and deep learning. Int Res J Eng Technol (IRJET) 07 (08). ISSN: 2395-0056
12. Chaudhary K (2020) Maintaining social distancing using artificial intelligence. Int J Sci Res (IJSR) 9(9). ISSN: 2319-7064

Radiological Image Synthesis Using Cycle-Consistent Generative Adversarial Network

Rahul Nehra, Abhisikta Pal, and B. Baranidharan

Abstract Radiology is the branch of science that deals with the study of energetic radiations and their use in generating medical images. MRI (Magnetic Resonance imaging) and CT (Computed tomography) are the two widely used modalities in radiology. CT comes with the disadvantage of high radiation risk which may have side effects. Thus, medical image from MRI-only radiation which is much safer than CT can be used to synthesize CT images using Deep Learning techniques. In this paper, we propose to build an architecture of fully convolutional neural network (FCN) along with a cyclic Generative Adversarial network (GAN). Our model has successfully generated CT images from the given MRI images from an unpaired ADNI image dataset.

Keywords Deep learning · Radiology · cGAN · FCN · CT generation · ADNI dataset · MRI

1 Introduction

Radiology is the branch of science that deals with the study of energetic radiations and their use in generating medical images. Radiology plays a huge role in disease management by providing detailed structural information about disease-related changes. Amongst the various modalities in this field, CT (Computed tomography) and MRI (Magnetic resonance imaging) are the most widely used ones. In this paper, we derive deeper into the opportunities that deep learning structures have in applications in the field of radiology. Computed tomography (CT) comes with

R. Nehra · A. Pal · B. Baranidharan (✉)
SRM Institute of Science and Technology, Kattankulathur, Chennai, India
e-mail: baranidb@srmist.edu.in

R. Nehra
e-mail: rv7822@srmist.edu.in

A. Pal
e-mail: ap9785@srmist.edu.in

R. P. Mahapatra et al. (eds.), *Proceedings of International Conference on Recent Trends in Computing*, Lecture Notes in Networks and Systems 341,
https://doi.org/10.1007/978-981-16-7118-0_34

the disadvantage of high radiation risk which may have side effects. 8 mSv dosage used in one abdominal tomography image can be approximated to be 0.05% risk in a lifetime, or a chance of developing cancer by one in 2000 cases. Alternative methods to generate CT should be devised, i.e., synthesis of images of different modalities from a single modality. In situations where a tomography scan is not available but is essential for diagnosis such as radiotherapy, this process of generating images will prove vital. This is important as it:

- Reduces the risk of radiation from CT
- Reduces the overall time and cost of diagnosis
- Doesn't depend on the patient to return for getting images of all modalities

We have studied the inherent meaning of radiology and the basic deep learning algorithms that are widely used in the present world. Deep learning has contributed to a series of advances in tackling image recognition and visual understanding problems. Deep learning based methods have gained pace recently in variety of medical field applications. Thus, our paper aims at

- Applying deep learning applications to produce synthetic CT images from MRI images.
- Comparing other state of art methods to get better results.

In deep learning, the extensive field of neural networks has been increasing the capabilities of models exponentially. It has proved at par with trained medical representatives and physicians in visually assessing medical images for the prediction, detection, diagnosis, characterization and monitoring of diseases. ML (Machine Learning) algorithms have also demonstrated remarkable progress in image recognition tasks.

Our work aims at using the FCN (Fully Convolutional Network) [1] and cycle consistent GAN (Generative Adversarial Network) [2] algorithms for synthesis of CT image from MRI. With the usage of a GAN framework, we can excel at improving assessments of radiographic characteristics. Medical imaging of different modalities together cost higher and radiation dose is more. MRI-only radiotherapy can drastically decrease time, cost and risk of diagnosis. Generative Adversarial Networks can be applied to increase accuracy of the model.

2 Literature Review

Segmentation is a popular approach which partitions medical images into different meaningful segments. Medical images can be represented using one or more Atlases. An Atlas is a model generated by learning parameters on training a data set containing a number of specific images. Atlas based segmentation [3] is the process of segmenting unseen images based on manually labelled images which forms the Atlas. This paper [4] aims at comparing 6 MRI based synthetic CT image generation methods. This paper [4] also indulged in organ auto contouring based on the methods evaluated. Amongst the 6 methods 4 were Atlas based methods—A-Median, A-VW, A-Bone, A-Iter. A voxel represents the average of itself and its neighbours when the brain images are considered. Voxel wise comparison gives us the local concentration difference amount in the brain tissues. The 5th method used deep learning convolutional neural networks and the 6th one was based on tissue segmentation based on two classes—water and air. This paper considered medical images of patients containing image-image linked MRI and CT images of the areas such as bladder, rectum and bone. This paper [4] found out that DL CNN resulted in a better performance giving dice matrices of 0.93, 0.90, 0.93, respectively.

This paper [5] used four different MRI protocols to generate synthetic CT images. The 4 protocols included T1- or T2-weighted MRI, Dixon MRI, UTE MRI (RESOLUTE), and ZTE MRI. The study [5] found that Dixon MRI was more efficient in producing accurate CT images, whilst ZTE MRI provided deeper understanding of the bowel air distribution. They used a data set of 26 patients with 10 of them for training and 16 of them for validation. MRI images for pelvic region was considered. They used a N4 bias correction method for pre-processing and a deep-convolutional neural network (DCNN) for synthetic image generation. The study suggested further work on optimization of MRI inputs features such as spatial resolution and degree of contrast. They also suggested evaluation on ZTE MRI of brain region that includes distinguishing bone and air in the sinuses for evaluation.

This paper [6] explained the advantages of CT scans over rRt-PCR. rRT-PCR is a variation of PCR (Polymerase Chain Reaction) used in COVID-19 testing purposes. rRt-PCR is a resource intensive method and relatively less sensitive for COVID-19 testing. They used a data set containing 829 lung CT slices from 9 patients. This paper pointed out how synthetic CT images can save the medical staff from the high risk of communicating COVID-19 from the infected patients. They applied a deep learning algorithm based on conditional generative adversarial network that successfully generated realistic COVID-19 CT images based on a segmentation map provided. The work [6] was further improved with the proposed multi resolution discriminator by following the patch GAN method.

The goal of this paper [7] was to synthesize arterial spin labelling (ASL) image which is an important fMRI module. ASL images are useful in diagnosis of diseases such as dementia. The paper designed a deep discriminant learning based model along with a ResNet substructure. The model consisted of 3 branches having a number of ResNet substructures. This research also considered remodelling of the substructure by using 3 blocks: Input block, Basic block and Bottleneck block. They [7] successfully synthesized ASL images which passed the voxel based PVE (partial volume effects) correction tests. They suggested future work on synthesizing ASL from images in ADNI data set.

This study [8] aimed at reducing the blurriness in target images that are generated. They proposed a fully convolutional network (FCN) to synthesize CT images from source MRI images and then use adversarial learning strategies along with an auto context model (ACM) to increase the quality of the artificially generated images. The usage of the difference in image gradient as the loss function helped in reducing the blurriness of the target images. They successfully generated CT images from MRI and 7T MRI images from 3T MRI images. The study [9] also successfully used a similar approach of FCN combining a U-net to generate sCT images from given MRI dataset.

The aim of this paper [10] was to synthesize CT images from MR images by using a residual learning based U-shaped deep neural network (RUN). In residual learning, the skip connections in a residual block between layers add output from previous layers to output of stack layers thus training deeper networks. This paper handles the degradation problem in neural networks when it gives higher training error as the deepness of the model increases suitably. The run network in this paper had 16 residual units. It applied the algorithm on a data set of 35 patients that had more than 5000 T1/T2 weighted MRI and CT image pairs they achieve the mean absolute error of 66.12 ± 5.95 HU. This paper [10] pointed out the limitation of having data set related to only the head region of the body.

In this research [11], the author evaluated a data set of the lumbar spine from the SpineWeb library data set having a total of 18 patients with both MRI and CT images. A fully unsupervised model was trained. Normalization and segmentation of vertical bodies and pedicles were done by first designing a fully convolutional network and then cascading a FC-ResNet to it. They successfully applied a pseudo 3D cycle GAN architecture along with a cyclic loss function for the generation of CT images from MRI images. They [11] achieved an average mean absolute error of 125.65.

The aim of this paper [12] was to apply state-of-art Atlas based methods to produce pseudo-CT. This study pointed out the limitation of lack of work on regions other than the head and neck region data sets. They also noted there are a smaller number of datasets where CT and MRI images are paired. Thus, they used a discrete optimization registration framelet with a self-similarity-based metric to generate the CT atlas from the respective patient anatomy. The model was tested on dataset of 3D-CT and 3D-MRI whole body scans of the Visceral Concept Extraction Challenge in Radiotherapy. The study compared and made improvement on all previously tested atlas based models but did not compare atlas based methods to machine learning approaches.

In this research [13], the author made an alternative improvement to the existing generative adversarial network model, GANai, where new images are composed from source images keeping a high-level attainable goal that leads to the synthesis of images having similar standard to the original ones. This resulted in technical improvements such as phase specific loss function, phase specific training data and usage of ensemble learning. This study also pointed out the disadvantage of analyzing large scale CT images as the protocols used to produces then are not standardized which disrupt the radiomic features. They used a data set containing 2448 chest CT images of lung cancer patients that was collected at University of Kentucky medical centre. The stability of the proposed model was evaluated using the metrices of cumulative sum control chart (CUSUM) where a value of 0.13 was found for GANai model.

In this research [14], the author successfully used the DECNN model to synthesize CT images from given T1 weighted MRI image dataset giving a MAE (mean absolute error) of 42.5 and PSNR (Peak signal-to-noise ratio) of 33.5.

On reviewing the literature, we have observed the following defects.

- There are lesser number of paired MRI and CT images datasets.
- Deep neural networks seemed to get more training errors as they got deeper.
- Number of whole body scan datasets are lesser.
- Data is available in raw format and thus, paired datasets are less in number.

Many theories have been proposed to explain the various methods that can benefit the task of image recognition and generation in an efficient way. Although the above-mentioned researches cover a wide variety of fundamental applications and methods, our work will focus on two major algorithms. These elements are:

- FCN
- Cycle consistent GAN.

3 Proposed Model

GAN starts with producing random noise and goes on to produce high quality CT images. Deep neural networks have been previously used for supervised learning; however, GAN uses neural networks for the purpose of generative modelling. In this, images which closely resemble images from the original data set are generated by discovering the regularities in the source data images. We can observe the working of a GAN in the following Fig. 1. Data is processed and divided into two classes MRI and CT image datasets. CT images are normalized and fed to the discriminator. MRI images are sent to the generator which generates fake CT images. A binary classification model is used to judge between real and fake and the losses are backpropagated to the discriminator and the generator model is updated.

We will be designing a FCN model to execute the job of generating the target image given a source image. To incorporate a nonlinear mapping on the given data of unpaired images of CT and MRI scans, we have used the cycle consistent GAN architecture. Cyclic GAN is an unsupervised architecture that does image-to-image translation on unpaired datasets, i.e., where target and source images are not linked to each other.

3.1 Dataset

We have used the Alzheimer's disease Neuroimaging Initiative (ADNI) dataset. It consisted of unpaired datasets of CT and MRI scans of 15 patients. A total of 315 images were used. The data was in DICOM file format which was processed using pydicom package.

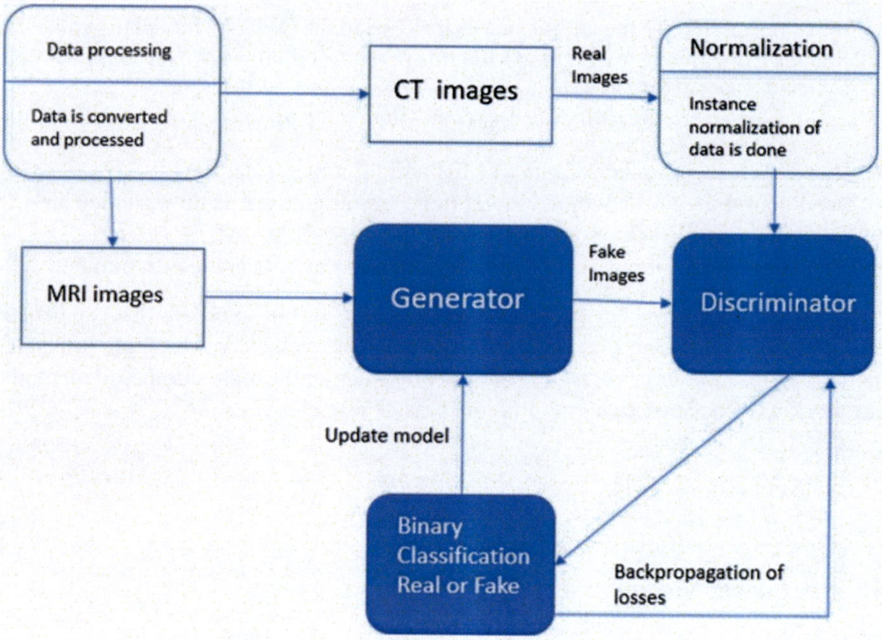

Fig. 1 Data flow diagram for GAN

3.2 Method

A generative adversarial network has two different neural networks a generator network and a discriminator network. The input to a generator model is a random vector containing noise that generates an image. The discriminator network distinguishes between the real images and the generated images from the generator.

1. *Pre-processor*: Image_folder class from torch vision has been used to load the data. The image has been resized to a dimension of 64 × 64 px and batch size is set to 128. The pixel values are then normalized with the following values each channel and a range of (−1, 1) is set.
    - Mean = 0.5
    - Standard deviation = 0.5
2. *Helper functions*: These were used to de-normalize image. Torch vision utility function make_grid was used to display images in the batch. A function denorm was used to de-normalize the values back to a range of (0,1). In the same module, we display the grid of images to visualize the working of the model.
3. *Parameters*: The parameters used for training the model is noted in Table 1.
4. *Discriminator network*: The discriminator's work is to take a batch of images as the input parameter and the try to distinguish the discriminated image and

Table 1 Training parameters

Parameters	Description
lambda_ABA, lambda_BAB	Sets the cycle consistency loses
learning_rate_D, learning_rate_G	Learning rates for generator and discriminator
generator_iterations, discriminator_iterations	Times every batch of image is trained
synthetic_pool_size	Sets size of the image pool
beta_1 and beta_2	Parameters for Adam optimizer
batch_size	Number of images for each update
epochs	Number of training epochs

the real image. We are using a fully convolutional neural network (FCNN) to output a number for each image. This number indicates the probability that the image was derived from the real data set. Conv2d layers were used with a kernel size of 4 and stride of 2. After which a Batch normalization was done by BatchNorm2d function. Following this a LeakyReLU activation function was applied. The final output is flattened from a $1 \times 1 \times 1$ tensor to a single vector and finally sigmoid function is applied.

5. *Activation function*: ReLU is a function that simply ignores negative values. Leaky ReLU is a variation of ReLU where it allows a slight amount of negatively valued gradient descent signal to pass (Eq. (1)). It is varied by the alpha variable a. It overcomes the limitation of ReLU where a lot of output from the generator is lost.

$$\text{Parametric ReLU: } y = a * x \tag{1}$$

In Eq. (1) value of 0.2 is used for a.

6. *Generator network*: The input to a generator network is a matrix of random numbers. Generator network performed the conversion of the latent tensor into an image tensor having shapes (128, 1, 1) and (3, 28, 28), respectively. ConvTranspose2d layer available in PyTorch is used. Latent size of 128 is used. Kernel size of 4 and stride of 1 was used. After that Batch normalization was done and ReLU activation function was used. In addiction a Tanh function was used. This function reduces values in range of $(-1, 1)$.
7. *Layers*: The individual layers used to train the model is shown in the above Fig. 2.

8. *Discriminator training*: Discriminator indicates whether images real or not. Thus, it is a binary classification model having classes real image or fake so we use the binary-cross-entropy [15] loss function (Eq. 2). Discriminator function takes real images and opt_d argument as input. Opt_d optimizes the parameters from the discriminator. We generate fake images using the generator and pass

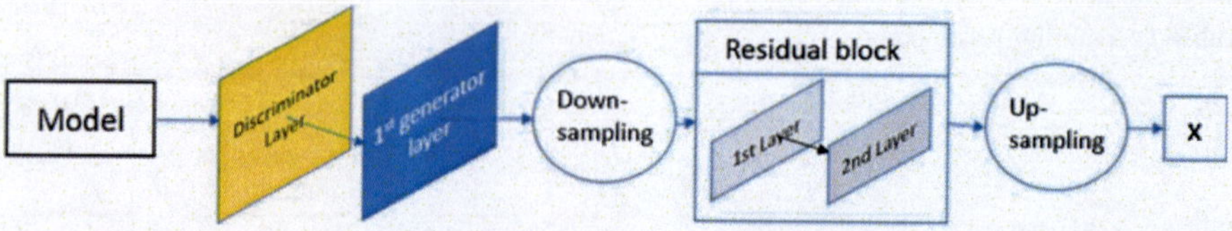

Fig. 2 Describes the various layers of implementation in the model. Discriminator layer uses Conv2D and gives an output which is used by generator to learn to produce more realistic images. Residual blocks are used to do reflection padding and Convolutional layers are applied. During up-sampling resize convolution is used

those through discriminator and get the fake loss. On passing real images to the discriminator, we get the real loss. The overall loss for the discriminator is the summation of both losses. Finally, we perform optimization of the discriminator by adjusting the weight to so that, the next batch of generated images is slightly better.

$$H_p(q) = -\frac{1}{N}\sum_{i=1}^{N} y_i \cdot \log(p(y_i)) + (1 - y_i) \cdot \log(1 - p(y_i)) \quad (2)$$

In Eq. (2), entropy H is calculated. y is the label and $p(y)$ is the probability of the point belonging to one class for all N points.

9. *Generator training*: Loss function consists of the output from discriminator used to train the generator. Firstly, we generate fake images and then the target labels are set to real valued output of 1 which means real image class output satisfying our primary aim. Gradient descent is performed based on the loss that was out after the binary classification and it was backpropagated to the generator. This resulted in change of input weights to the generator so that it gets better at "fooling" the discriminator. Thus, it starts giving us real life images.
10. *Loss functions*: MAE (Mean Absolute Error) and MSE (Mean Squared Error) are considered to calculate cycle consistency losses. Finally, we calculated the PSNR (Peak Signal Noise Ratio) and SSIM (Structural Similarity Index) for to compare the final synthetically generated image and the original image.

$$\text{MAE} = \frac{\sum_{i=1}^{n} |y_i - x_i|}{n} \quad (3)$$

In Eq. (3), MAE is calculated by taking the average of the distance between predicted values (y) and real values (x) over n data points

$$\text{MSE} = \frac{1}{n}\sum_{i=1}^{n}(y_i - \hat{y}_i)^2 \quad (4)$$

In Eq. (4), MSE is calculated by taking the average of the square of difference between the observed value (Y) and the predicted value (Y')

$$\mathrm{SSIM}(x, y) = \frac{(2\mu_x\mu_y + c_1)(2\sigma_{xy} + c_2)}{(\mu_x^2 + \mu_y^2 + c_1)(\sigma_x^2 + \sigma_y^2 + c_2)} \tag{5}$$

In Eq. (5), SSIM is calculated between the real and fake imafe of same window size of 64 px × 64 px

$$\mathrm{PSNR} = 10\log_{10}\left(\frac{(L-1)^2}{\mathrm{MSE}}\right) \tag{6}$$

In Eq. (6), PSNR has been calculated using the MSE found.

4 Experimental Results

On implementation of the designed adversarial deep learning techniques, fully convolutional neural network (FCNN) along with a cycle consistent generative adversarial neural network, we have successfully achieved our aim of generating Computed Tomography images synthetically from the given MRI images. A comparison between the real and fake synthetically generated images are shown in Fig. 3.

Initially, the generator loss can be seen as 15.43, on proceeding the value decreases and finally we get a value close to 2.029 after training for up to 200 epochs. We have visualized the effect of changing loss over time. We can observe that the generator loss has come quite close to the discriminator loss. The output loss values have been tabulated in Table 2.

We have visualized in Fig. 4. The working of our model with a line graph below. The x label indicates the ephocs and the y label indicates the losses. On observing the graph, we can see that the generator loss line (G_losses) has come close to the discriminator loss line (D_losses). This indicates the success of the designed model.

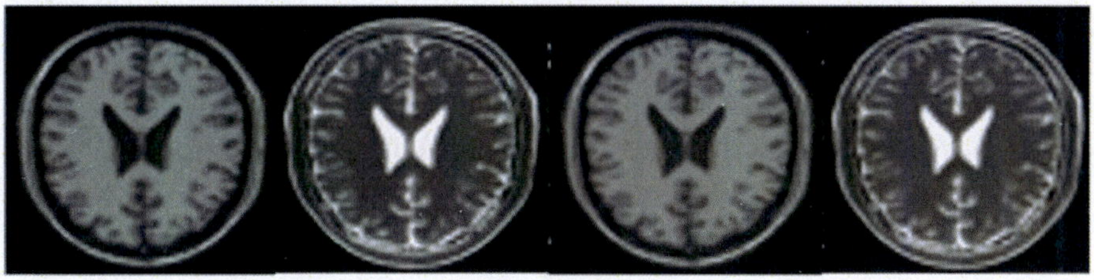

Fig. 3 The 1st and 2nd images represent real MRI and CT scans, respectively, and the 3rd and 4th images represent the synthetically generated images

Table 2 Output losses

Output losses	Values	Description
ABA_reconstruction_losses	0.001313	Class A to B reconstruction loss
BAB_reconstruction_losses	002047	Class B to A reconstruction loss
DA_losses	7.12E-06	Discriminator loss for class A
DB_losses	3.99E-05	Discriminator loss for class B
D_losses	4.70E-05	Total Discriminator loss
G_AB_adversarial_losses	0.998336	Generator adversarial loss for class A to B
G_BA_adversarial_losses	0.997466	Generator adversarial loss for class B to A
G_losses	2.029403	Total Generator loss
reconstruction_losses	0.00336	Final reconstruction loss

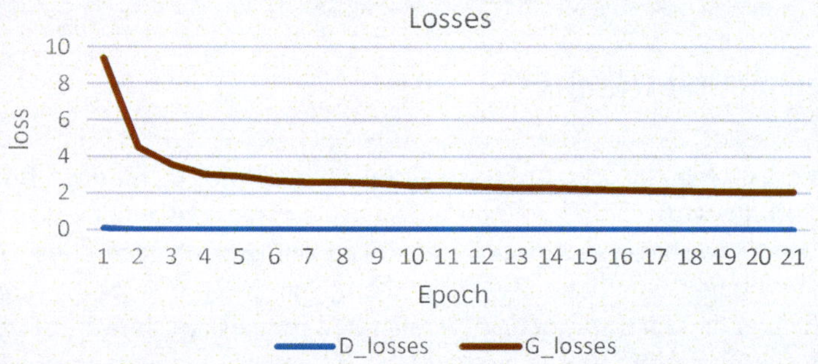

Fig. 4 Graph to indicate loss over time

Finally, we computed the values of PSNR and SSIM for the final synthetically generated CT image and the original image and got the following values.

PSNR = 32.335870 dB.

SSIM = 0.95191956.

In the research [8], the authors have used a FCN with auto context model and got a PSNR value of 27.6 for brain image dataset. In research [10], the authors have applied a residual unit based U-shaped network (RUN) and got a PSNR value of 28.52 and SSIM score of 97.6. Comparing with previous works [8, 10], we can observe that our proposed model has given a higher PSNR and appropriate SSIM score. Our model also solved the problem of Unpaired images datasets by using a cycle consistent GAN thus widening the scope of application on various other datasets.

5 Conclusion

Radiation risk in certain radiological modalities can result into adverse effects on the patient undergoing diagnosis. It is important to device methods to artificially generate medical images of other modalities without actually scanning based on scans of lower radiation risk. We intend to help the doctors in diagonising under lower cost and lesser time. Experimental results show that Deep Learning architectures FCNN combined with cycle consistent GAN has been successful in generating Synthetic CT images from a source dataset of MRI images. This helps the medical professionals in MRI-only Radiotherapy. Thus, this reduces various constraints of time, cost and the availability of the patients for getting the CT scans. It results in faster diagnosis with the patient being exposed to much lesser risk of radiation. For the model, it was observed that larger datasets take a longer time span to train and the complexity increases. The ADNI dataset consisted of head scans only. The synthetically generated CT images can be utitlized to predict Alzheimer's disease and compared with results from prediction using real images as a extension to the future work on this paper. Modification on GAN network can also be experimented such as using a Wasserstein GAN can be researched. Whole body scans can be tested on with the proposed model. The N4 bias structure can also be used at the input to further normalize it.

References

1. Liu A, Yang Y, Sun Q, Xu Q (2018) A deep fully convolution neural network for semantic segmentation based on adaptive feature fusion. In: 2018 5th International Conference on Information Science and Control Engineering (ICISCE), Zhengzhou, China, pp 16–20. https://doi.org/10.1109/ICISCE.2018.00013
2. Zhu J, Park T, Isola P, Efros AA (2017) Unpaired image-to-image translation using cycle-consistent adversarial networks. In: 2017 IEEE International Conference on Computer Vision (ICCV), Venice, Italy, pp 2242–2251. https://doi.org/10.1109/ICCV.2017.244
3. Lauritzen AD, Papademetris X, Turovets S, Onofrey JA (2019) Evaluation of CT image synthesis methods: from atlas-based registration to deep learning. arXiv:1906.04467 [physics.med-ph]
4. Arabi H et al (2018) Comparison of synthetic CT generation algorithms for MRI-only radiation planning in the pelvic region. In: 2018 IEEE Nuclear Science Symposium and Medical Imaging Conference Proceedings (NSS/MIC), Sydney, NSW, Australia, pp 1–3. https://doi.org/10.1109/NSSMIC.2018.8824321
5. Leynes AP, Larson PEZ (2018) Synthetic CT generation using MRI with Deep Learning: how does the selection of input images affect the resulting synthetic CT? In: 2018 IEEE International Conference on Acoustics, Speech and Signal Processing (ICASSP), Calgary, AB, Canada, pp 6692–6696. https://doi.org/10.1109/ICASSP.2018.8462419
6. Jiang Y, Chen H, Loew M, Ko H (2021) COVID-19 CT image synthesis with a conditional generative adversarial network. IEEE J Biomed Health Inform 25(2):441–452. https://doi.org/10.1109/JBHI.2020.3042523
7. Huang W et al (2019) Arterial spin labeling images synthesis from sMRI using unbalanced deep discriminant learning. IEEE Trans Med Imaging 38(10):2338–2351. https://doi.org/10.1109/TMI.2019.2906677

8. Nie D et al (2018) Medical image synthesis with deep convolutional adversarial networks. IEEE Trans Biomed Eng 65(12):2720–2730. https://doi.org/10.1109/TBME.2018.2814538
9. Li G, Bai L, Zhu C, Wu E, Ma R (2018) A novel method of synthetic CT generation from MR images based on convolutional neural networks. In: 2018 11th international Congress on Image and Signal Processing, BioMedical Engineering and Informatics (CISP-BMEI), Beijing, China, pp 1–5. https://doi.org/10.1109/CISP-BMEI.2018.8633142
10. Li Y, Li W, He P, Xiong J, Xia J, Xie Y (2019) CT synthesis from MRI images based on deep learning methods for MRI-only radiotherapy. In: 2019 International Conference on Medical Imaging Physics and Engineering (ICMIPE), Shenzhen, China, pp 1–6. https://doi.org/10.1109/ICMIPE47306.2019.9098190
11. Oulbacha R, Kadoury S (2020) MRI to CT synthesis of the lumbar spine from a Pseudo-3D cycle GAN. In: 2020 IEEE 17th International Symposium on Biomedical Imaging (ISBI), Iowa City, IA, USA, pp 1784–1787. https://doi.org/10.1109/ISBI45749.2020.9098421
12. Degen J, Heinrich MP (2016) Multi-atlas based Pseudo-CT synthesis using multimodal image registration and local atlas fusion strategies. In: 2016 IEEE Conference on Computer Vision and Pattern Recognition Workshops (CVPRW), Las Vegas, NV, USA, pp 600–608. https://doi.org/10.1109/CVPRW.2016.81
13. Liang G, Fouladvand S, Zhang J, Brooks MA, Jacobs N, Chen J (2019) GANai: standardizing CT images using generative adversarial network with alternative improvement. In: 2019 IEEE International Conference on Healthcare Informatics (ICHI), Xi'an, China, pp 1–11. https://doi.org/10.1109/ICHI.2019.8904763
14. Xiang L, Wang Q, Nie D, Zhang L, Jin X, Qiao Y, Shen D (2018) Deep embedding convolutional neural network for synthesizing CT image from T1-Weighted MR image. Med Image Anal 47:31–44
15. Liu L, Qi H (2017) Learning effective binary descriptors via cross entropy. In: 2017 IEEE Winter Conference on Applications of Computer Vision (WACV), Santa Rosa, CA, pp 1251–1258. https://doi.org/10.1109/WACV.2017.144

Mouse Assistance for Motor-Disabled People Using Computer Vision

P. Anantha Prabha, K. Srinivash, S. Vigneshwar, and E. Viswa

Abstract Technological innovations draw the attention of all people, but not for those with motor disabilities as the contact with devices and such users draw on a line of frustration. One of the major problems faced by them is that they are unable to have full mouse access, the device which plays a major role in human–computer interaction. Several solutions have been made to address this issue, but they have limitations like using external devices like sensors which may not be affordable to all and require high-end computing because of the processing of data generated by external devices. To overcome this, a system is proposed for hands-free mouse control using facial gesture recognition techniques that can benefit people with motor disabilities. The proposed system intends to eliminate the use of external equipment and also simplify the interface for the disabled which was the major drawback in most of the existing systems. It makes use of face recognition and eye gestures by using only a webcam and uses this for mouse operations which makes the system much more cost effective and simpler. The proposed system uses a face detection algorithm using Histogram of Oriented Gradients (HOG) and was trained using the iBUG 300-W dataset which made the system a suitable solution for all lighting and environmental conditions. The evaluation of experimental performance indicates that the proposed system has extensive performance and never compromises in terms of stability and sensitivity.

Keywords Human–computer interaction · Computer vision · Face landmark detection · Virtual mouse

1 Introduction

"For most people, technology makes things easier. For people with disabilities, however, technology makes things possible. In some cases, especially in the workplace, technology becomes the great equalizer and provides the person with a

P. Anantha Prabha (✉) · K. Srinivash · S. Vigneshwar · E. Viswa
Sri Krishna College of Technology, Coimbatore, Tamil Nadu, India

R. P. Mahapatra et al. (eds.), *Proceedings of International Conference on Recent Trends in Computing*, Lecture Notes in Networks and Systems 341,
https://doi.org/10.1007/978-981-16-7118-0_35

disability a level playing field on which to compete", said Mary Pat Radabaugh, Former Director of IBM National Support Center for Persons with Disabilities [1]. Even though people with disabilities may benefit greatly from computers, they typically find them inaccessible. Many existing computers are not intended for individuals with physical disabilities and enable them to take measures beyond their capacity. Human–computer interaction (HCI) is a multidisciplinary area that focuses on the design and user interface of computers. HCI has grown from concerning generic and individual activities to social and organizational computing, assisting with the aged, cognitive, and physical disabilities, and all kinds of human experience and activities [2].

Motor disorders are characterized as a partial or complete loss of control of the leg, hand, or other part of the body. People with physical/motor disabilities can have difficulty interacting with computers the same way as the public does. For example, people with motor disabilities typically struggle with navigation while using computers or other devices [3]. Since people with physical disabilities do not have the precision that general users of technology do when using a mouse, devices should be designed to be as accurate as possible for users. For people with motor disabilities, buttons with small hit radius that can be difficult to reach, buttons should have larger hit radius, interactions/actions should not be packed close to each other on any device, as disabled users can take the wrong action. Even though these modifications would be much better, a better alternative can be sought for people with motor impairment.

The rest of the paper is organised as follows. Related works done by various authors are described in Sect. 2, a detailed description of the proposed system design is given in Sect. 3. Implementation and experimental results of the proposed system are described in Sect. 4. Finally, in Sect. 5, the paper is concluded.

2 Related Works

The existing systems developed for differently abled people, make use of some external devices like sensors [4] and also some other similar technologies [5] which may not be affordable to all and also require extensive computing power to process the data generated by the external devices.

Reddy et al. (2020) and Reza et al. (2015) suggest using fingers tapped with different colors and using hand gestures captured over the camera to perform mouse operations [6, 7], but this solution is not applicable for motor-disabled people as it requires using their hand gestures for mouse operations. Luo et al. (2020) suggest using acoustic sound signals captured over a microphone that enables the user to navigate the mouse cursor by processing those signals and translating them into mouse cursor operations [8], but this solution requires high-quality noise cancellation microphones which is not affordable to all and also less responsive to users because of signal translations to mouse operations. Buragohain et al. (2019) suggest a solution that requires the use of head-connected sensors and provides mouse control by receiving neuron signals from a human brain and passing through several electrodes for amplifying the signals [9], but the main downside is that the neuron signals need to be amplified and preprocessed often, resulting in high latency and high external hardware costs.

Jennifer M. Vojtech et al. (2018) suggest a solution using surface electromyography technology that requires the use of EMG sensors in the user's face to perform mouse control operations [10], but this system requires the use of external sensors which is not affordable and also produces high latency in mouse operations. Meena et al. (2020) and Fahim et al. (2020) suggest a solution using eye-gaze tracking that requires the use of real-time images captured through a webcam to navigate the cursor on the screen [11, 12], but as this system navigate the cursor using eye-tracking, it is highly difficult for the system to detect and process the image in low-lighting conditions. But all of these solutions lack certain features which, in turn, lead to the proposed solution that ticks almost all perfection boxes with minimal requirements.

The proposed system includes the use of computer vision to detect facial orientation and movements to activate sufficient mouse functionality. To overcome the disadvantages of the existing systems, the proposed system eliminates the need of using external sensors and also uses simple techniques to provide hands-free mouse control operations. Computer vision-based head motion detection relies on image processing and/or pattern recognition techniques [13] to estimate head pose from user's real-time images, and the program is configured to analyze the users face gestures [14] to determine the type of action to be taken by matching the previously stored expressions to identify the procedure to be performed [15] and, if necessary, to take the parameter when the users head moves, the change in the relative orientation of the central location tells the direction to be moved.

This system also supports macro gestures that allow the user to map a face gesture to commonly use applications (like GUI shortcuts) without having to navigate across the entire screen and perform mouse click operations to invoke the application. The proposed method does not require any special hardware, and only a webcam is required.

3 System Design

The overall architecture of the proposed system is depicted in Fig. 1. The preliminary stage of the system focuses on obtaining real-time images from the webcam. Then, the obtained image is passed through a face detection algorithm based on the histogram of oriented gradients [16] which gives the bounding box of the face and, this bounding box is marked in the image and passed through a landmark detection algorithm [13] which gives the coordinates of the key points of the user's face in the obtained image. The next stage of the system focuses on providing stability and sensitivity by introducing a threshold checker for eye aspect ratio and a contiguous frame counter. The purpose of this is to detect the eye blink [17] of the user and toggling between the two modes of the system, cursor enabled mode, and cursor-disabled mode. Then,

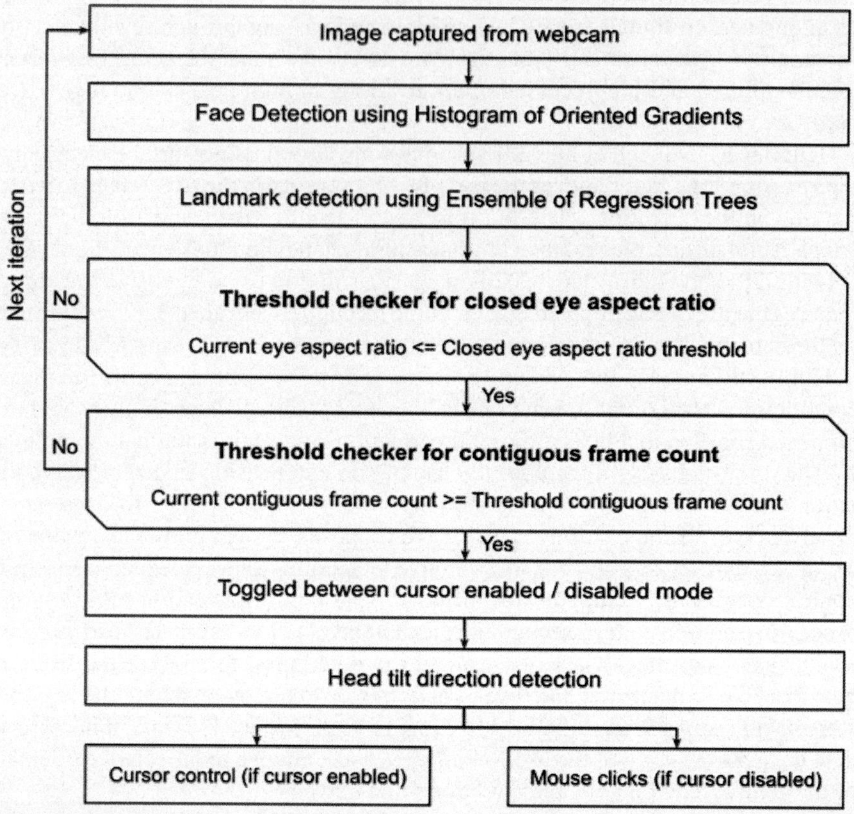

Fig. 1 Overall architecture of the proposed system

the final stage of the system detects the angle of inclination of the user's head on the image, and based on the mode, it performs either a mouse cursor movement or mouse click.

3.1 Facial Landmark Detection and Mouse Cursor Control

Facial landmarks are used to mark key regions in the face, such as eyes, eyebrows, nose, mouth, chin, and more. Detection of facial landmarks is a subproblem in the broad range of problems in shape detection and prediction techniques. Given the input image with the region of interest (ROI), the shape predictor localizes the key points of interest along with the shape which is then used to control the cursor as depicted in Fig. 2.

In the context of predicting facial landmarks, the system aims to detect important facial points on the face using the method of shape prediction. For this, the system

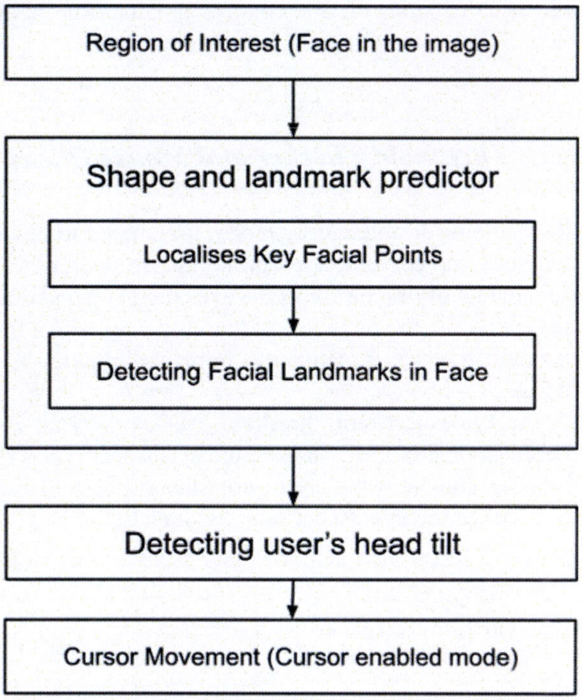

Fig. 2 Shape predictor for predicting facial landmarks in the region of interest

first locates the face in the picture and then detects the key facial regions in the region of interest (ROI). For this, the system uses a face detection algorithm using the histogram of oriented gradients (HOG) which internally uses a linear classifier, an image pyramid, and a sliding window detection scheme and has been trained on the iBUG 300-W face landmark dataset.

Then, the next step is to detect landmarks on the user's face. The dlib's shape predictor is used for this purpose and was trained on the iBUG 300-W face data collection. The dataset itself consists of 68 pairs of integer values, which are the (x, y) coordinates that describe key points on the face such as points 36–42 and 42–48 specifies the coordinates of the left and right eye, respectively [18]. This algorithm processes an array of pixel intensities to estimate key points and transfers these points to the ensemble regression trees (ERT) to improve the precision through retrograde cascades which generates a new estimate each time when it is passed through, which in turn improves calculated coordinate points.

Using these predicted landmarks of the face, the system is designed to detect certain actions like the user's head tilts and the mouse can be programmed to navigate in real-time according to the direction of the user's head tilt, and similarly, when the user is in cursor-disabled mode, the user may tilt their head to their needs and widen their eyes to perform the mouse click process. The wide opening of the eyes is detected by the same definition as the eye blink detection, the eye aspect ratio of the current eye landmarks is obtained and passed through the threshold checker to check whether the current frame count is greater than the threshold frame count and the mouse click operation is allowed based on that.

3.2 Threshold Checker and Mouse Clicks

To enable or disable cursor mode, users need to close their eyes for a second, and then they can perform appropriate mouse operations. However, this technique has trouble identifying blinks as the eyes travel up and down rapidly. To take account of this condition, the proposed system is set up with a threshold checker, so that it can distinguish between regular eye blink and blinking for mouse operation as depicted in Fig. 3.

The eye aspect ratio threshold checker uses the landmark points obtained from the shape predictor to calculate the eye aspect ratio [19]. Here, the proposed system needs to compress the set of coordinate points it determined to a flattened value to calculate the eye aspect ratio as depicted in Fig. 3. But, the problem is that the

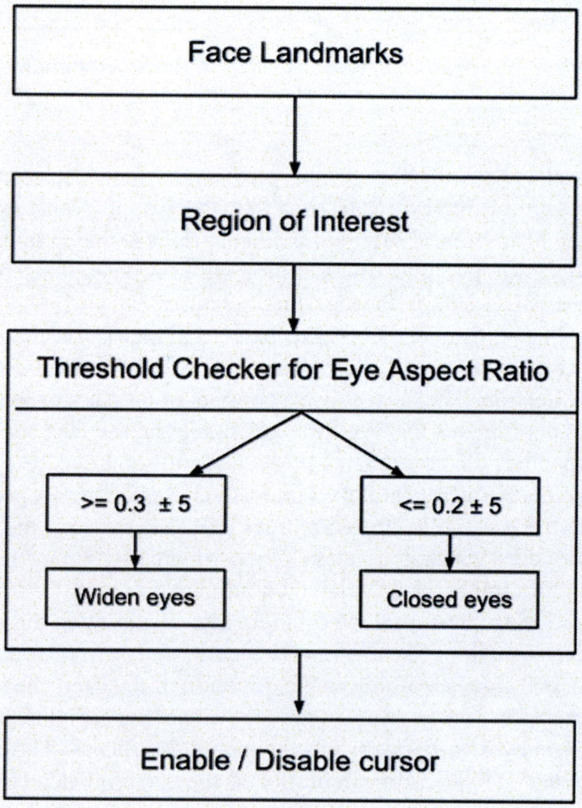

Fig. 3 Eye aspect ratio—threshold checker for enabling or disabling mouse cursor

user might just naturally blink their eyes. To account for this problem, the system uses another threshold checker, Consecutive Frame Counter (CFC) which counts the number of consecutive counts the user has kept their eyes opened or closed and check if it is longer than the CFC threshold frame count to determine whether it is just a normal eye blink or a blink for a mouse operation.

Once the system is passed through the eye aspect ratio threshold checker, it is processed further and sent to the consecutive frame counter. The system is set to either mouse cursor enabled or disabled mode, and these inputs are further sent to the head tilt detector to find the direction of head inclination of the user to perform appropriate mouse operations like navigating cursor on the screen or performing mouse clicks.

4 System Implementation and Experimental Results

The proposed system's implementation begins by detecting faces in a real-time webcam image. The system is developed using the Python programming language, and since the system depends on a C++ library, to make the dlib's library accessible to Python, the system uses a binding called CMake. This function internally uses an object that is a method for running a fixed-size sliding window classifier over an image pyramid. Using this method, the bounding box of the face is obtained.

The next step toward facial recognition is face landmark prediction, for which the proposed system uses the dlib library's landmark predictor, which was trained on the iBUG-300 W dataset. The model training phase is controlled by a set of parameters and these parameters affect the size, accuracy, and speed of the model produced. So by fine-tuning the training options properly, the proposed system met the constraints of the proposed system like faster execution speed, less memory footprint, and overall accuracy and robustness.

Now, the system has all the preliminary components to begin the actual processing. To perform mouse click operations and to enable or disable the mouse cursor, the system needs to find whether the user has opened or closed their eyes. For this reason, the proposed system finds the ratio of two eyes to the eye aspect. Once the eye coordinates are obtained from the face landmark detection algorithm, the eye aspect ratio (EAR) threshold checker is called to check if the user has opened or closed their eyes as depicted in Fig. 4.

To calculate the eye aspect ratio, the system uses the L2 norm which takes the sum of the squared values, taking the square root at the end as depicted in Eq. (1). One of the use cases of the L2 norm is to find the shortest path from A to B and here, the proposed system uses it to calculate the distance between the upper and lower eyelids.

$$||x||_2 = \left(\lim_{i=1 \to N} \sum |x_i|^2\right)^{1/2} = \sqrt{x1^2 + x2^2 + x3^2 + \ldots + xn^2} \tag{1}$$

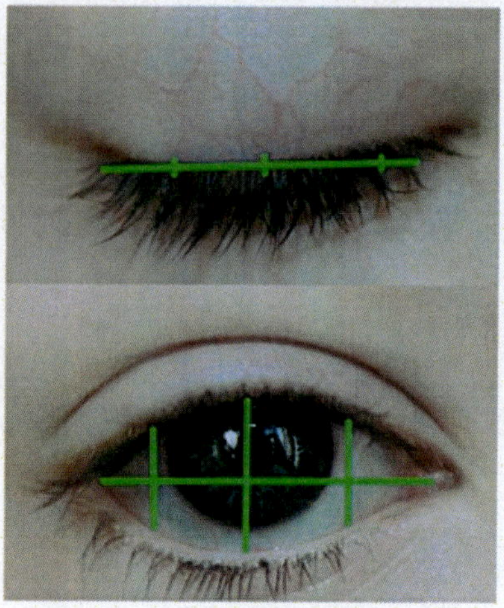

Fig. 4 Eye aspect ratio differences when the user's eyes are opened and closed

(1) Equation for L2 norm to calculate the distance between the top and bottom eyelids.

To detect the direction of head tilt, the proposed system uses the nose reference points as the nose is a central region of the user's face. The anchor point for the current session is generated by obtaining the current coordinate of the nose point on the screen once the device is switched to cursor-disabled mode as depicted in Fig. 5.

To move the cursor on the screen and to perform mouse clicks, the proposed system uses a library called PyAutoGUI. The positions on the computer screen are shown in PyAutoGUI by X and Y Cartesian coordinates. The X coordinate starts at 0 on the left and increases to the right. Compared to arithmetic, the Y coordinate starts at 0 at the top and rises at the bottom.

The facial detection component is then evaluated in both good and bad resolution and lighting environments at first and has shown promising results as depicted in Fig. 6, and the system has correctly recognized the faces in the tests which were

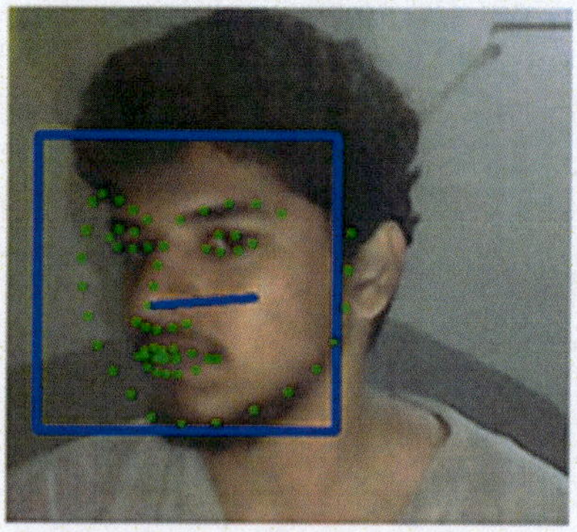

Fig. 5 User tilts head toward right from the anchor point to move the cursor

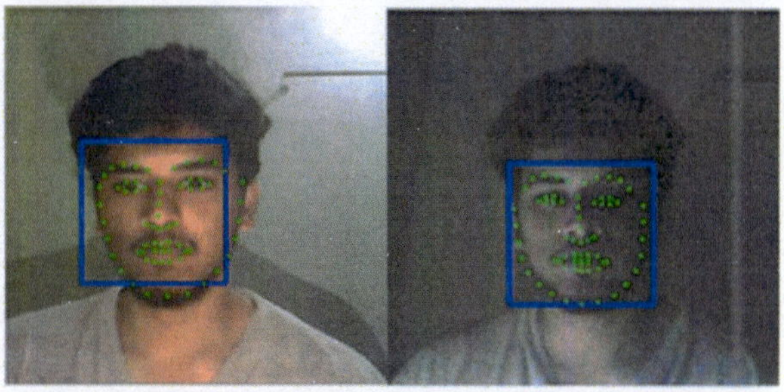

Fig. 6 Face landmark detection with the bounding box in both good and bad light conditions

conducted in three kinds of lighting environments, i) excessive lighting, ii) adequate lighting, and iii) insufficient lighting as described in Chart 1.

The experiment is conducted in a system with quad-core CPU, 4GiB DDR4 memory module, Ubuntu 20.04 machine with Linux kernel version 5.44.0–66 generic. From the experiment, it is found that the system utilizes only an ample amount of CPU and memory (around 50–60% of CPU and 4 to 6% of memory) for processing. The resource usage of the system is tracked using the htop tool which is available in most Linux systems, and the results are depicted in Chart 2.

The system further classifies face gestures and activates mouse functions that have been explicitly checked to verify pointer speed and face gestures. Mouse movements and eye motions are too sensitive, and so threshold values have been introduced through recursive testing of the device with fine-tuned values, so that movements and gestures are regulated in terms of sensitivity. The threshold conditions showed accurate results which, in turn, led to positive feedback from the proposed system. Eye winks were first used to view mouse clicks, such as the left-click and the right-click. But that was not the best solution for mouse clicking, as successive eye winks could cause eye strain on users. So mouse clicks have been applied in a much smoother

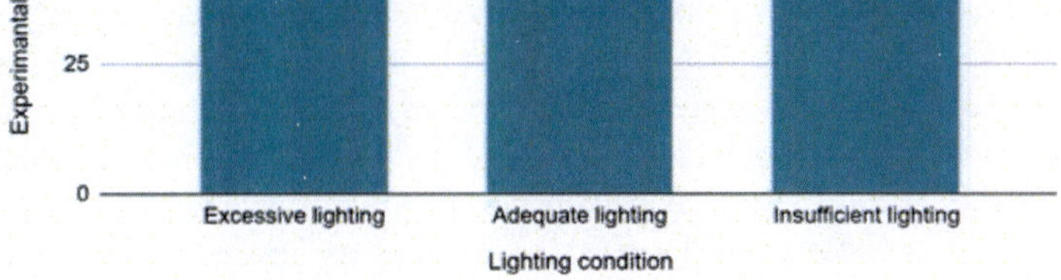

Chart 1 Statistics of the experiments conducted on the proposed system

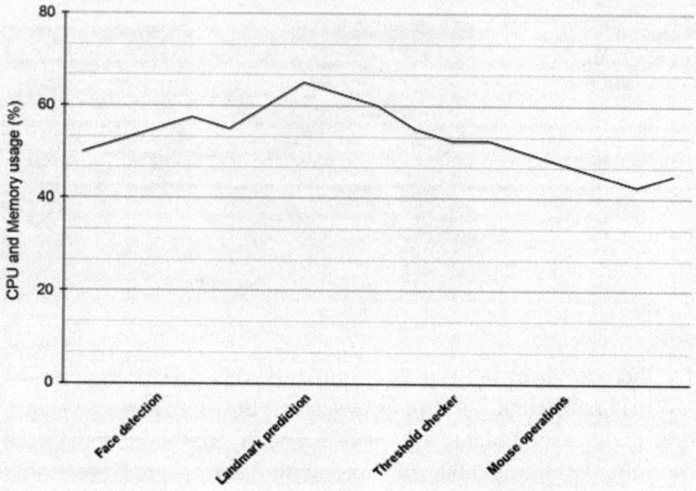

Chart 2 Statistics of the resource usage by the proposed system in experiments

and safer way by changing the user interaction designs. Finally, the proposed system is made to be a system with a very good user interface, precision, and speed.

5 Conclusion

The mouse assistance for motor-disabled people using computer vision has been successfully implemented. It has an easy mouse cursor movement using head tilt movements and mouse clicks using facial and eye gestures. The proposed system helps the user to navigate the cursor easily and also aids motor-disabled people with a virtual mouse. The proposed system overcomes the trade-off problem in the existing systems between speed and accuracy by removing the need for external sensors like inertial measurement units which will make the system slow, and it also improves the accuracy in low-light and low-resolution conditions by using matured computer vision algorithms. This system has a broad range of applications other than being a virtual mouse for motor-disabled people such as the eye aspect ratio checker feature can be used to detect drowsiness of drivers, while they are driving and also the head tilt recognition system can be used to analyze students' behavior while they are attending online classes. The experimental results showed that the proposed system is far better than the existing systems and promises to perform better over time. The future work of this research is to enhance the proposed system with more mouse functionality like scrolling and also incorporating natural language processing to provide a virtual keyboard interface that will serve as keyboard assistance to the motor disabled.

References

1. Bouck EC (2010) Technology and students with disabilities: does it solve all the problems. In: Current issues and trends in special education: research, technology, and teacher preparation. Emerald Group Publishing Limited
2. Li Y, Kumar R, Lasecki WS, Hilliges O (2020) Artificial intelligence for HCI: a modern approach. In: Extended Abstracts of the 2020 CHI conference on human factors in computing systems, pp 1–8
3. Ghosh P, Chingtham TS, Ghose MK, Dutta R (2013) A review of modern HCI challenges for betterment of differently abled people. IJCIR 9(2):79–88
4. Kim S, Park M, Anumas S, Yoo J (2010) Head mouse system based on gyro-and Opto-sensors. In: 2010 3rd international conference on biomedical engineering and informatics, vol 4. IEEE, pp 1503–1506
5. Pereira CAM, Bolliger Neto R, Reynaldo AC, Luzo MCDM, Oliveira RP (2009) Development and evaluation of a head-controlled human-computer interface with mouse-like functions for physically disabled users. Clinics 64(10):975–981
6. Reddy VV, Dhyanchand T, Krishna GV, Maheshwaram S (2020) Virtual mouse control using colored fingertips and hand gesture recognition. In: 2020 IEEE-HYDCON. IEEE, pp 1–5
7. Reza MN, Hossain MS, Ahmad M (2015) Real-time mouse cursor control based on bare finger movement using a webcam to improve HCI. In: 2015 international conference on Electrical Engineering and Information Communication Technology (ICEEICT). IEEE, pp 1–5
8. Luo G, Yang P, Chen M, Li P (2020) HCI on the table: robust gesture recognition using acoustic sensing in your hand. IEEE Access 8:31481–31498
9. Buragohain S, Kashyap K (2019) Mouse cursor control using EEG signal and machine learning. Int J Comput Sci Eng 7(5):1623–1627
10. Vojtech JM, Cler GJ, Stepp CE (2018) Prediction of optimal facial electromyographic sensor configurations for human–machine interface control. IEEE Trans Neural Syst Rehabil Eng 26(8):1566–1576
11. Meena K, Kumar M, Jangra M (2020) Controlling mouse motions using eye tracking using computer vision. In: 2020 4th international conference on Intelligent Computing and Control Systems (ICICCS). IEEE, pp 1001–1005
12. Fahim SR, Datta D, Sheikh MDRI, Dey S, Sarker Y, Sarker SK, Badal FR, Das SK (2020) A visual analytic in deep learning approach to eye movement for human-machine interaction based on inertia measurement. IEEE Access 8:45924–45937
13. Kazemi V, Sullivan J (2014) One millisecond face alignment with an ensemble of regression trees. In: Proceedings of the IEEE conference on computer vision and pattern recognition, pp 1867–1874
14. Shah HM, Dinesh A, Sharmila TS (2019) Analysis of facial landmark features to determine the best subset for finding face orientation. In: 2019 International Conference on Computational Intelligence in Data Science (ICCIDS). IEEE, pp 1–4
15. Zarkasi A, Nurmaini S, Stiawan D, Ubaya H, Sanjaya Y, Kunang YN (2018) Face movement detection using template matching. In: 2018 International Conference on Electrical Engineering and Computer Science (ICECOS). IEEE, pp 333–338
16. Déniz O, Bueno G, Salido J, De la Torre F (2011) Face recognition using histograms of oriented gradients. Pattern Recogn Lett 32(12):1598–1603
17. Panning A, Al-Hamadi A, Michaelis B (2011) A color-based approach for eye blink detection in image sequences. In: 2011 IEEE International Conference on Signal and Image Processing Applications (ICSIPA). IEEE, pp 40–45
18. Trejo K, Angulo C (2016) Single-camera automatic landmarking for people recognition with an ensemble of regression trees. Computación y Sistemas 20(1):19–28
19. Mohanty S, Hegde SV, Prasad S, Manikandan J (2019) Design of real-time drowsiness detection system using Dlib. In: 2019 IEEE international WIE Conference on Electrical and Computer Engineering (WIECON-ECE). IEEE, pp 1–4

Two-Way Economical Smart Device Control and Power Consumption Prediction System

P. Anantha Prabha, N. Arjun, J. Gogul, and S. Divya Prasanth

Abstract Due to the availability of high-speed broadband networks such as 3G and Long-Term Evolution (LTE), as well as more affordable and open smart phones, the mobile industry has seen exponential growth in terms of offering numerous services and apps at citizens' fingertips. The Internet of Things (IoT) is a promising technology that can be used to link, manipulate, and manage intelligent objects. But a large portion of the population were not able to adapt to IoT due to a lack of knowledge about IoT technology and usage, unfriendly user experience, high cost, and low-range wireless transmission. This paper discusses a smart economical device control system based on IoT in which, devices in home are connected with low-cost two-way smart switches and in turn which are connected with mobile application. The idea behind the proposed system is to integrate the mobile app with cloud networking using wireless communication. The mobile application provides users with electricity usage and also helps in predicting the bill beforehand. Thus, users can have control over the usage of electricity and can adjust it accordingly. The system sustains out to be an economical, flexible, and scalable one.

Keywords Arduino UNO · Device control · ESP32 · Home automation · Internet of Things

1 Introduction

Automation systems can assist in the day-to-day life and are helpful in making our surroundings smart, be it our home, workplace, institution, etc. Mobile technologies made it easy to control and monitor IoT devices, and with the advent of cloud technology, things started to click via the Internet. The survey from Deloitte states Internet of Things along with AI, big data/analytics, and cloud technology has the potential to support organizations to connect, generate, and provide intelligent insights from data to carry data-driven decision making and operations. According to Microsoft's

P. Anantha Prabha (✉) · N. Arjun · J. Gogul · S. Divya Prasanth
Sri Krishna College of Technology, Coimbatore, Tamil Nadu, India

R. P. Mahapatra et al. (eds.), *Proceedings of International Conference on Recent Trends in Computing*, Lecture Notes in Networks and Systems 341,
https://doi.org/10.1007/978-981-16-7118-0_36

IoT signals report of 2019, many organizations fail to estimate the budget before the start of the project as it usually takes around mid- to long-range and 29% of the organizations reported that they lacked the resources. So, these reports clearly show that having a cost-efficient and budget-friendly IoT automation system is still a far cry. The data from N.I.T.I Aayog give an idea of the common household appliances that contribute to the majority of the electricity usage in the country. The data from the year 2016 show that 90% of the current consumption comes from appliances like AC, TV, lights, fans, air coolers, water heaters, and refrigerators. It showed that lights and bulbs are the major contributors to household power consumption. The rapid growth of IoT, mobile, cloud technology made the interaction between all three easy.

The above two reports clearly state that the need of the hour is to develop an IoT system that can control the common household appliances like light and fan that must be economical, scalable, and controlled via an application. The resource and budget estimation will be relatively easy for a small IoT project with a concise hardware setup, ease of access, and control of the mobile application along with the cloud technology.

Data have shown that the cloud-based IoT systems with real-time data synchronization and remote control of the appliances in the home or workplace from anywhere have proven to be a great choice. Wireless communications enable data transmission to be carried out in a seamless way without any interruptions. With the growing trends of BaaS architecture, it becomes easy for the user to concentrate on the application side without worrying about the backend and cloud connectivity.

The main objective is to forecast the future electricity bill with the current usage statistics in hand. This hardware setup makes the system a centralized one; i.e., there is no need to make each device smart. By doing so, the cost of the entire system is reduced.

2 Literature Survey

Survey was carried out to know about the existing methodologies available in the market. It helped in adding further enhancements in the proposed model.

The system proposed by Kodali et al. [1] made use of GSM, MQTT, and Wi-Fi module for controlling devices at the remote level. Estimated battery usage was less than 0.06%, but the system suffered a major disadvantage of user-friendliness and methods of data visualization.

The system proposed by Pujaria et al. [2] had made use of an Arduino microcontroller to estimate temperature, air quality, humidity, and other parameters to forecast bad weather. It exclusively uses ESP8266 Node MCU for network interaction, but its disadvantage is the Bluetooth network that could spare only a short distance.

The system proposed by Mandula et al. [3] uses both Bluetooth and Wi-Fi as communication mediums in the Arduino UNO-based home automation system.

This IoT-based system controls appliances based on instructions from an android application.

The system proposed by Gill et al. [4] had made use of ZigBee technology with Arduino UNO to coordinate all the functions of electronic appliances like a bulb, fan, air conditioner, etc. Though the connection setup was successful, the system failed to communicate with one another even though it was a one-to-one operation.

Jiang et al. [5] had made use of current sensors to predict electricity bills and displayed usage statistics via a Web application. The disadvantages associated with the system include data storage and system delay. The disadvantages of other network communication protocols over Wi-Fi can be understood from this paper.

The system proposed by Pavithra et al. [6] makes use of Raspberry Pi as the master controller. Its main disadvantage is its cost. Studies show that high cost and beforehand resource estimation make it tough to adopt IoT everywhere.

The system proposed by Chang et al. [7] suggests the need for an efficient and easy way to control and monitor home appliances anywhere from the house without compromising security. He visualizes the need and comfort to switch on/off home devices like television from anywhere in the home and also notifying the status to the user.

Mowad et al. [8] explain that there are various subsystems available within the smart home system where each subsystem controls and monitors a certain portion like fire detection, surveillance, and intruder detection and control.

In the system proposed by Yashiro et al. [9] raises major issues associated with designing a home automation system. The need for interactive user-friendly interfaces and cost efficiency is the key factors in IoT applications. The use of cloud networks manages network connectivity both remotely and locally.

In the system proposed by Anusha et al. [10], the use of control systems in the field of production of goods and services shows us the upper-hand IoT along with information technology can provide to reduce human work. It speaks volumes of how automation can ease production and help in uplifting the world economy.

In this proposed system by the author Shuhaiber et al. [11] makes his study based on structural equation modeling-partial least squares (SEM-PLS). Risks are the major concerns, and the study emphasizes educating dwellers to know about automation systems and to increase awareness to increase the quality of living.

In the proposed system by Monteriù et al. puts forward the advantages of using low-cost sensors in developing a domestic smart home architecture. Similarly, the robustness and stability provided by commercial cloud platforms make it a better fit for low-cost smart home architectures [12].

Dickey et al. [13] state that the extension from the Internet is addressed as the Internet of Things, where devices or things are connected, communicated, and internetworked to exchange information. IoT will be adopted in each and every field and is considered to take a huge hop forward.

The concern about the environment and tackling its issues is always challenging. Adopting IoT is nature-friendly and features like turning on the light when it becomes dark, taking care of children when no one around helps humans in a significant way [14].

This study by ASUS Smarthome [15] provides an example with ASUS SmartHome gateway. It accumulates independent and small unit sensors and stacks one above the other for connectivity, and the whole system acts as a centralized one, but it popped out to be a costly setup.

Taking all these into consideration, a system is proposed that interacts using Wi-Fi technology. The system consists of Arduino UNO and ESP32 to support the hardware connectivity and to maintain the ease of interaction with the added appliances. Cloud is used to store real-time continuous data and helps in electricity bill estimation along with usage statistics. This information is made available to the user via a cross-platform mobile application that makes it available in both iOS and android operating systems.

3 Proposed Model

Figure 1 shows the proposed system consisting of NodeMCU that has Bluetooth low energy, Wi-Fi, and dual-core. ESP32 is additionally supported by an Arduino microcontroller to support all the hardware functionalities. Relay switches is shown in Fig. 1, and the architecture diagram is connected with the devices and helps the entire system to act as two-way controllable system. The integration of IoT hardware to the circuit makes the system a centralized one.

The system is connected with the cloud for maintaining and storing real-time data. Cloud is crucial to any IoT system, and it acts as an interface between hardware and mobile application. The interaction happens only between the cloud and hardware. There is no direct communication between the mobile application and hardware. The data in the cloud are used to maintain the status of the appliances. The status is

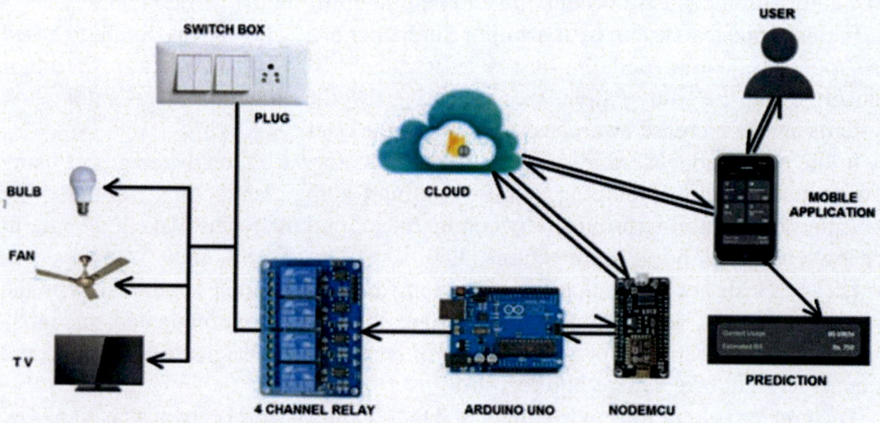

Fig. 1 Architecture of the proposed IoT system

updated whenever the mobile application or the manual switch is toggled and there it makes a change in the system.

The mobile application part of the proposed system consists of buttons widgets each denoting the status of the device and a toggle button available to control the devices. The button turns green when the status is active. The interface is provided with the option where the user can add a new device from the available set of devices, or if not available, an option is provided to add a new device manually. This newly added device gets added to the cloud database, and it is provided with a button to control its status. The user interface is also provided with electricity consumption usage and also predicts month-end power estimation.

3.1 Hardware Setup

NodeMCU is used to establish the connectivity between the database and the hardware. The functional code is dumped into the NodeMCU. The NodeMCU is connected with the Arduino UNO which acts as an interface between the NodeMCU and relay. Relays act as electronic switches that can be controlled through a mobile application. The manual switch and relay act as a two-way switch. The functionality of the code is such that whenever the button in the mobile application has touched the value of the relay in the database gets toggled. This changes the status of the circuit. If the state of the device was ON, it will be turned OFF and vice versa. Feedback is obtained from the circuit so that the status of the circuit is known. This feedback current is stepped down and converted using a step-down transformer. This stepped-down current is given as an input to ESP32 pins. Using this input feedback, the status of the circuit is known, and the signal is sent to the ESP32 module and the ESP32 updates the changes in the database if any.

The D0 and D1 pins act as OUTPUT pins in ESP32. The ESP32 reads the relay value from the database and sends the signal to Arduino UNO. The Arduino UNO acts as an interface between ESP32 and relay. The Arduino UNO pins 12 and 13 are input pins, and 4 and 5 are output pins.

This input is processed, and if the input for pin 12 is high, the output in pin 4 is high; similarly, if the input in the pin 12 is low, the output in pin 4 is low. Similarly, for pin 13, the input gets reflected as output in pin 5. Pins 4 and 5 are connected to a relay each as shown in Fig. 2 in such a way that the relays are controlled. An alternating power source is connected with the center port of the three ports in the relay. The top and bottom ports of the relay are connected with the two-way switch's top and bottom port, and the center port is connected to an electrical device and a step-down transformer. The step-down transformer converts 220–5 V. This 5 V is given as an input to the D5 and D6 of ESP32. This pin reads and updates the status of the device to the database whether the device is ON/OFF. Figure 3 highlights the hardware setup.

Arduino UNO: Arduino UNO developed by Arduino.cc. is based on Microchip ATmega328P microcontroller. The accepted voltage ranges are between 7 and 20 V.

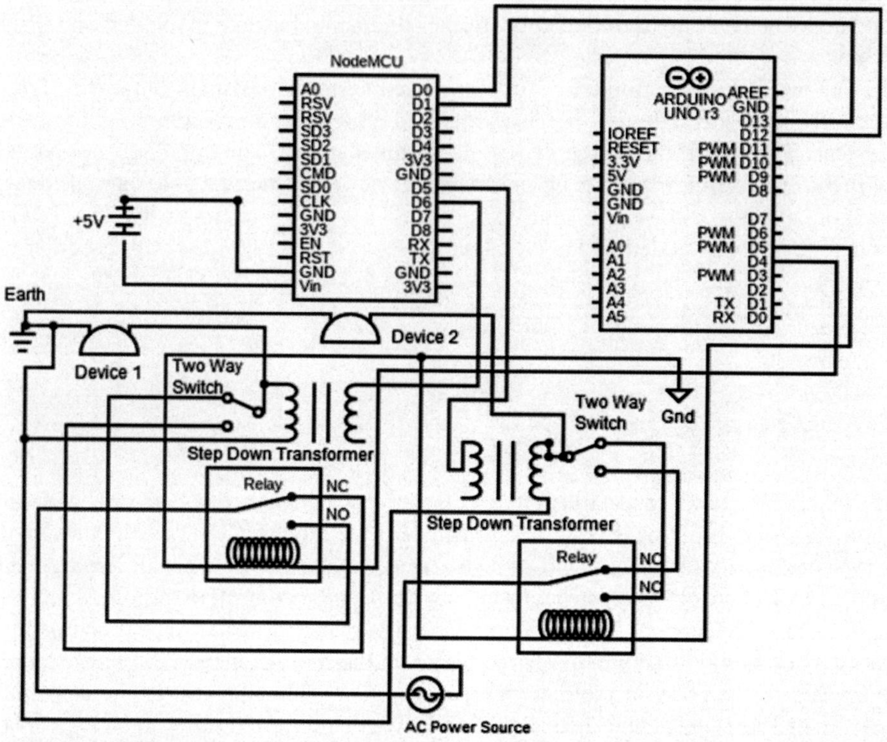

Fig. 2 Circuit diagram

UNO can be powered by a 9 V battery or by USB cable. Arduino IDE helps in controlling and monitoring Arduino code and helps in device interaction connected to its pin seamlessly. The Arduino UNO acts as an interface between the relay and ESP32. The Arduino reads the signal for the control of the relay from the ESP32. It sends the signal to the relay. The Arduino UNO also reads the status of each electrical device connected with the circuit.

ESP32: ESP32 consumes low power and has inbuilt Wi-Fi and dual-mode Bluetooth in it. It has enough capability to stand alone as a hardware component and is the successor of ESP8266. The ESP32 continuously reads the data from the database, and if there is a change in the database, it gets reflected in the hardware, like when the value of the relay in the database is toggled through the button from the mobile application. The ESP32 sends the signal to the Arduino UNO.

Step-down Transformer: A step-down transformer is used to step down high-voltage current to low-voltage current. The input voltage for ESP32 analog pins is 3.3 V. The transformer reduces the voltage and is given as an input for ESP32 analog pins.

Relay: Relay is nothing but an electrically operated switch. It makes the interaction with an electrical or electronic device possible way. It can have any number of contact

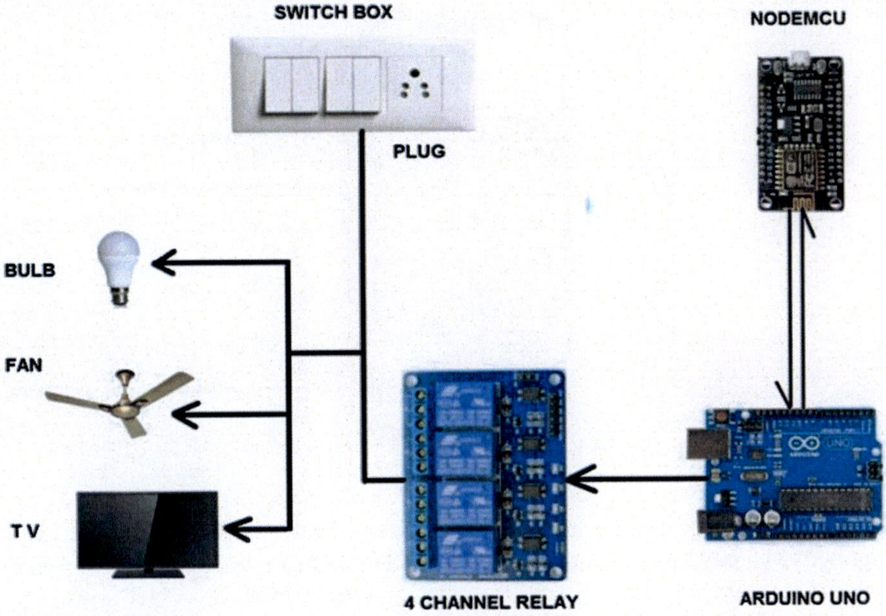

Fig. 3 Hardware setup

forms, like break contacts, make contacts, or a combination of both. The input for the relay is from the Arduino. The signal for the relay is sent when the button in the mobile application is toggled, and it changes in the database. The ESP32 reads and sends the signal to the Arduino UNO.

3.2 *Frontend Development—User Interface*

The main feature of a user interface should be user-friendliness and providing a smooth user experience. It increases the responsiveness of the application, and when things are put neatly inside the layout, it increases the readability of the page.

The user interface consists of several buttons; each button controls the power supply for an electrical device connected to it. Near the button, the status of the device is shown, i.e., whether the power supply to the device is ON/OFF.

The power usage of each device is shown. The future power consumption prediction using the current usage is featured as shown in Fig. 4.

Flutter and Dart: Flutter is used as it is a popular open-source cross-platform UI tool kit. It has everything in the form of widgets. Widgets enable the user to build UI and rebuilds state when needed. Flutter framework makes use of Dart programming language.

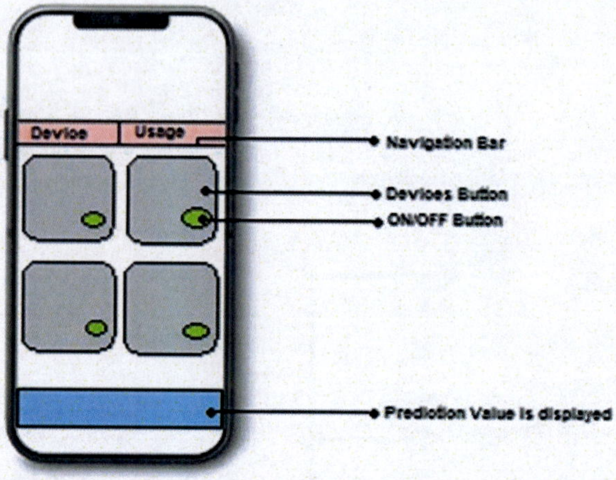

Fig. 4 Mobile application layout

Main.dart is the starting point for the application. It has the Material.dart package imported. MaterialApp helps to provide the title and leads to HomeScreen. Some basic widgets include Row, Stack, Column, and Text. Inkwell widget is used to produce an effect when the user touches on the device switches. TabController used in the application makes the transition between HomeScreen and Usage screen possible.

Since the aim was to develop a cross-platform mobile application, flutter was considered the best option available in the market. It is also open source. It is easy to build a cross-platform application as it functions using just-in-time and hot reload feature. Flutter has many inbuilt library function that ease interaction with Firebase cloud.

3.3 Backend Service—Cloud Storage

Instead of relying on traditional storage like servers and hard drives, cloud-enabled businesses to store and compute large amounts of data with less maintenance on the user side. Storing their software on the servers without proper security can lead to many consequences often making way for intruders. Another problem with the traditional storage methods is scalability, where multiple users at the same time can result in system crashes. Cloud provides the option to store our data at multiple locations which provides data backup. The option to access our data from anywhere with the help of the cloud is an added advantage. In this proposed system, cloud is used for storage and computing purposes. Cloud provides various services like Software as a Service (SaaS), Platform as a Service (PaaS), Infrastructure as a Service (IaaS), and Backend as a Service (BaaS).

Firebase: Firebase is a Backend as a Service (BaaS) that provides all the functionalities ranging from database with real-time data synchronization, cloud functionalities to crash reports, and analytics. It provides a real-time database (JSON tree) and Cloud Firestore (collections and documents) to store and retrieve NoSQL data. The interaction between the Wi-Fi module and cloud database happens through the inbuilt Wi-Fi connection of ESP32 NodeMCU. Here, Firebase acts as a central storage depot between hardware and mobile application. Whenever a user connects a device and makes an entry in the mobile application, Firebase adds a new entity to store all the data associated with it. These values can be further used for bill calculation and prediction.

3.4 Electricity Prediction

The main enhancement on the user interface side is the information the user gets in terms of the current bill prediction of the upcoming month. Each time when the status of any device that is connected to the relay gets changed from ON to OFF, the timestamp gets saved in the database. Using this, time interval is calculated. For every ON and OFF interval taken for a particular device, the power consumption duration is known along with the status of the currently active relays. Hence, the total power consumption prediction is found. With the power consumed data, the EB bill is predicted using the EB bill calculation algorithm.

4 Implementation and Results

The implemented system is two-way controllable, both via manual switch and via mobile application. The user interface side consists of ways to add new devices either from the available set of devices or can also manually add a device. The user can also view the usage statistics in the second window and inform the user about the upcoming electricity charges given the current usage scenario.

The home screen of the application contains all the devices that are connected through the relay as shown in Fig. 5. When a particular device is clicked, the device screen of that particular device opens up. The device screen shows all the related details of the device such as its status, name of the device, power consumption details in watts, etc. The second 'Usage' tab gives a pictorial representation of the chart to display the electricity bill contribution of different devices. Usage value for the last month is available as with the electricity bill amount. Predicted month-end consumption and month-end electricity bill amount are also displayed at the bottom of the screen as shown in Fig. 6.

The live data from the devices are sent via the Wi-Fi module to the cloud storage; i.e., if the user switches on light bulb of relay1 in the mobile application, the Firebase data reflect this change and gets updated to '1' accordingly. As shown in Fig. 7, the

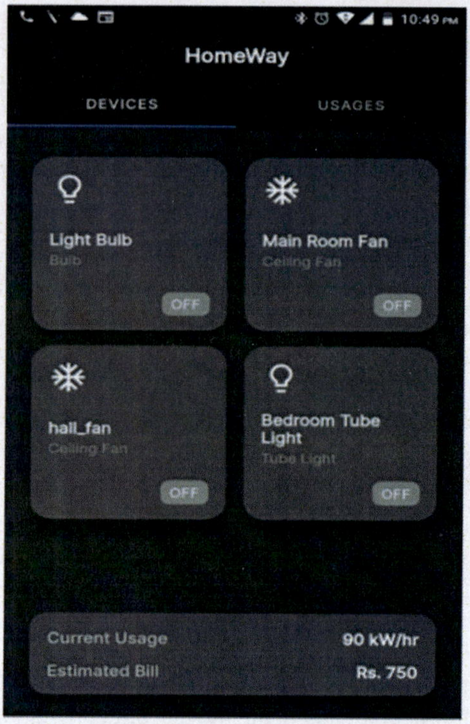

Fig. 5 Home Screen of Mobile Application

real-time database consists of various relay fields. Relay fields have a history, relay, flag, name, number, type, and watt under each. Figure 7 has four relay switches representing four relay switches in chronological numbering. Each relay gets connected to a real-world device and stores its corresponding status in 'Status' column in the firebase. ESP32 (NodeMCU) is used to establish the connectivity between the database and the hardware. The functional code is dumped into the ESP32. The ESP32 continuously reads the database. If there is any change in the database, it gets reflected in the hardware. The ESP32 reads the relay record from the database, and if there is any change in the value of the relay, it gets reflected in the hardware relay. If the button in the mobile application is clicked, the flag in the database is set 1 and the relay value is toggled. Whenever the flag is set to 1, the ESP32 reads the relay value from the database and sends the signal to the Arduino from there to the relay. A feedback current is got from the circuit and is stepped down using a transformer. The stepped-down current is sent to the ESP32 analog pin.

The analog pin reads the current if high set the status in the database as '1' else if low sets the status as '0'. The application reads the status from the database and if 1 it displays the ON/OFF button in green color and ON, else if the status is '0' it sets the ON/OFF button in gray color and OFF. There are two switches; one is a relay and the other manual. So, if the relay value in the database is '1' the status of the device not necessarily is ON as shown in Fig. 8. It depends on the status of the

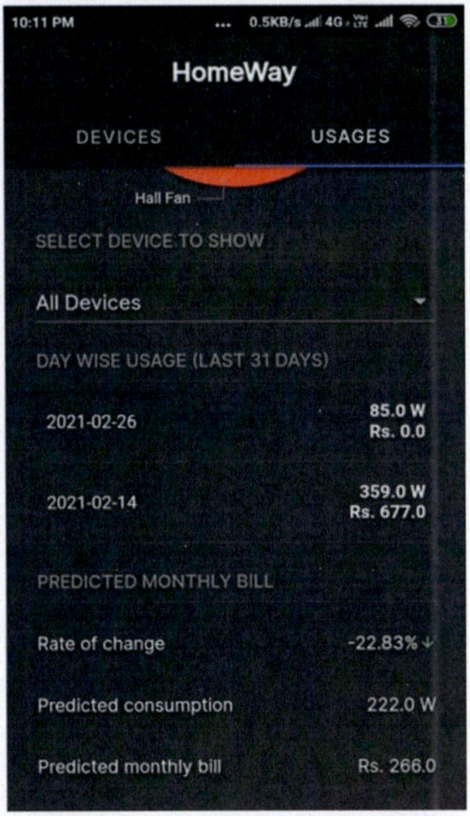

Fig. 6 Usage Screen of Mobile Application

manual switch since the manual switch and relays act as a two-way switch. So, the feed is necessary from the circuit to know the status of the device.

There is always the asynchronous flow of data between ESP32, mobile applications, and Firebase. The results from the literature survey clearly state that NodeMCU ESP32 is the better choice compared to Bluetooth, ZigBee, LAN, or Wi-Fi. Its low-power consumption with different operating modes makes it favorable to other devices. This setup is ideally suited for IoT-based automation systems, and this approach without much load on the backend will increase the performance and system accuracy.

The operation and correctness of the system are checked with a 9 W bulb connected with the hardware. Whenever the switch is turned ON/OFF, the time value gets stored in the database.

The raw data stored in the database consist of the date and time of the change in status in the circuit. Using this data, the time duration for which the switch is ON is calculated.

The values thus recorded and stored are used to calculate the average periodic growth rate. The average periodic (monthly) growth rate from the value at time 1 to

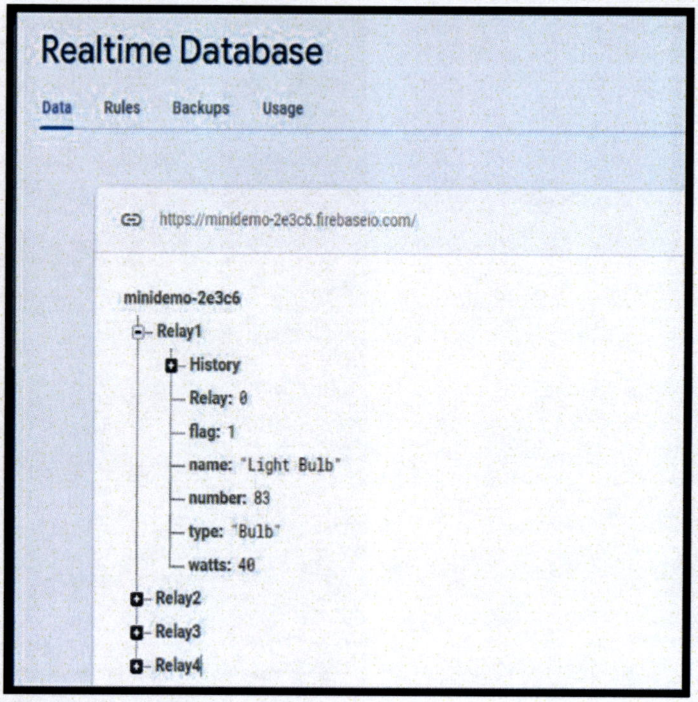

Fig. 7 Firebase real-time database

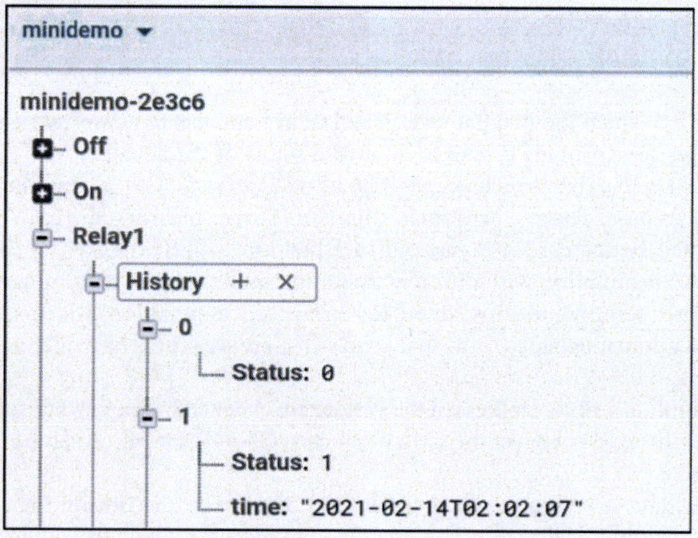

Fig. 8 Firebase real-time database history

value at time 2 is given by

$$r = (V_2/V_1)^{1/n} - 1 \quad (1)$$

where r = decimal rate
V_1 = value at time 1
V_2 = value at time 2
n = number of periods.

After obtaining the rate value, the electricity prediction for upcoming month is calculated. It is the time 2 value. Value at time 2 with constant or average periodic (monthly) growth rate from time 1:

$$V_2 = V_1(1 + r)^n \quad (2)$$

where r is growth rate
n = number of periods (months)
V_1 = value at time 1
V_2 = value at time 2.

For example, consider the above bar chart in Fig. 9, total power consumed during the months May to November is available. Growth rate value (1) is first found out with the variation in the data between the months available. Then, the electricity bill for the upcoming month, i.e., December, is calculated using (2). A similar pattern is followed to calculate electricity predictions for upcoming months.

The hardware responds to whatever change in the database which is controlled by the mobile application. Any automation can be easy to be implemented in the mobile application and is reflected in the system. There is a scheduling feature such that like setting an alarm the ON/OFF for any electrical. For example, the AC can be turned OFF after 3 am and lights should be turned ON in sync with the alarm.

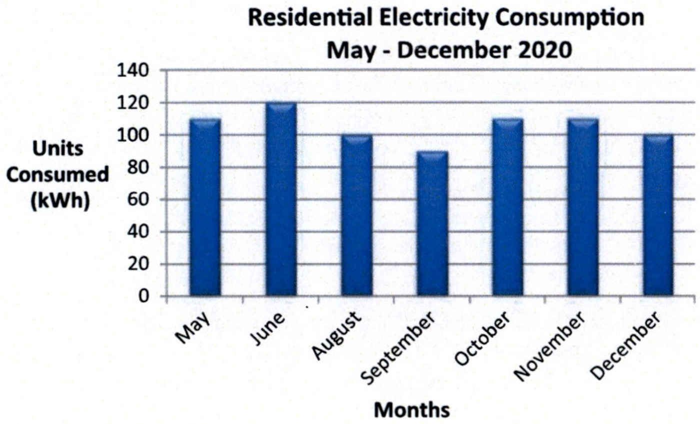

Fig. 9 Residential electricity consumption May–December 2020

5 Conclusion

This project is implemented with the aim of developing a low-cost IoT system that can be controlled using a mobile application. The system makes use of ESP32 NodeMCU and Arduino to control the appliances from the hardware side. The cloud support from Firebase remains a better alternative as all the backend services are taken care of single-handedly. The proposed system is capable of detecting new devices as it gets added, and the system is also scalable. The system as a whole can be managed from anywhere and shows power usage and future power predictions too. The future scope of this project includes enhancing it further and building it as a complete product. The fear of not knowing whether a particular electrical or an electronic device's power supply is turned off when not near is eradicated. The report of the power consumption and power usage in the future is predicted. The proposed system is scalable where one server can manage many hardware interface modules as long as it exists on Wi-Fi network coverage.

References

1. Kodali, Ravi Kishore, SreeRamya Soratkal (2016) MQTT based home automation system using ESP8266. In: 2016 IEEE region 10 humanitarian technology conference (R10-HTC IEEE, pp 1–5
2. Pujaria, Uma, Patil, Bahadure, Asnodkar M (2020) Internet of Things based integrated smart home automation system. Available at SSRN 3645458
3. Mandula, Kumar, Ramu Parupalli CH, Murty AS, Magesh E, Lunagariya R (2015) Mobile based home automation using Internet of Things (IoT). In: 2015 International Conference on Control, Instrumentation, Communication, and Computational Technologies (ICCICCT) IEEE, pp 340–343
4. Gill K, Yang S-H, Yao F, Lu X (2009) A Zigbee-based home automation system. IEEE Trans Consumer Electronics 55(2):422–430
5. Jiang H, Cai C, Ma X, Yang Y, Liu J (2018) Smart home based on WiFi sensing: a survey. IEEE Access 6:13317–13325
6. Pavithra D, Balakrishnan R (2015) IoT based monitoring and control system for home automation. In: 2015 Global Conference on Communication Technologies (GCCT) IEEE, pp 169–173
7. Chang C-Y, Kuo C-H, Chen J-C, Wang T-C (2015) Design and implementation of an IoT access point for smart home. Appl Sci 5(4):1882–1903
8. Mowad MAE, Fathy A, Hafez A (2014) Smart home automated control system using android application and microcontroller. Int J Sci Eng Res 5(5):935–939
9. Yashiro T, Kobayashi S, Koshizuka N, Sakamura K (2013) An Internet of Things (IoT) architecture for embedded appliances. In: 2013 IEEE region 10 humanitarian technology conference, IEEE, pp 314–319
10. Anusha S, Madhavi M, Hemalatha R (2015) Home automation using atmega328 microcontroller and android application. Int Res J Eng Technol (IRJET) 2(06):865–868
11. Shuhaiber A, Mashal I (2019) Understanding users' acceptance of smart homes. Technol Society 58:101110

12. Monteriù A, Prist MR, Frontoni E, Longhi S, Pietroni F, Casaccia S, Scalise L et al (2018) A smart sensing architecture for domestic monitoring: methodological approach and experimental validation. Sensors 18(7):2310
13. Dickey N, Banks D, Sukittanon S (2012) Home automation using cloud network and mobile devices. In: 2012 proceedings of IEEE Southeast. IEEE, pp 1–4
14. Amudha A (2017) Home automation using IoT. Int J Electronics Eng Res 9(6):939–944
15. ASUS SmartHome (Online). Available: https://www.asus.com

Image Colorization Using Deep Convolution Autoencoder

Vaibhav Singh, Deepak Arora, and Puneet Sharma

Abstract Authors present a deep convolution autoencoder based image colorization system that colorize the black and white images without any direct human interaction. It takes input of grayscale images and return the colored images of inputted image. By using these kind of deep learning model and artificial intelligence, the human effort and mistakes can be reduced on a large-scale works. For implementing this approach, various kind of deep neural network architecture have been explored like VGG16, ImageNet and capsule network. Different suitable Image data have been chosen to train the image colorization model. The concept of autoencoder is used in this approach, because it was very efficient in learning the abstract representation of any king of data, and the decoder can built the approximately same representation of data by using the data encoded by the encoder. CIELAB color space is used to get L (The lightness of an image which is similar to the grayscale image) and ab channels (a varies from red to green and b varies from blue to yellow) of image. Mean squared error (MSE) loss function is used to calculate the loss which is calculated by comparing actual ab channel and outputted ab channel and gradient descent is used to reduce the loss. The whole model is trained on the Flickr dataset and some custom-made dataset. Trained model is tested on various sized images and results are very satisfying. In this paper, results have been compared with the ground truth of the image.

Keywords Image colorization · VGG16 · Autoencoder · Automatic coloring · Deep learning autoencoder · Convolution neural network

1 Introduction

Colorization of grayscale images is a hot research topic in field of artificial intelligence and computer vision. By diving in deep learning, we can make such tasks easy image enhancement and generate duplicate faces and many more. Here, authors take

V. Singh (✉) · D. Arora · P. Sharma
Department of Computer Science and Engineering, Amity University, Lucknow, India
e-mail: vaibhav.singh18@s.amity.edu

R. P. Mahapatra et al. (eds.), *Proceedings of International Conference on Recent Trends in Computing*, Lecture Notes in Networks and Systems 341,
https://doi.org/10.1007/978-981-16-7118-0_37

a deep learning approach to solve this particular problem. They design an autoencoder that accepts grayscale images and generate a colored image. In this era, we have lot of historic pictures that is in black and white. To colorize these images, we need a professional artist who have great knowledge of colors of different things, lot of time and lot of patience. In recent years, convolution neural network have emerged to solve classification problems, pattern recognition at rate of less than 5% error. CNN performs very well to detect objects, recognize patterns and shapes in an image. Authors believe that CNN can colorize the black and white images by having these kinds of ability to perform tasks. Adding colors to any black and white image can make it realistic and get a lot of attention in the computer vision community. Autoencoders are designed for machine learning tasks to learn some kind of data like images and videos. Use of autoencoders is a traditional approach to enhance the image, reconstruct the image, upscale the image and now this is used for the automatic image colorization of black and white images. This is very simple approach to colorize the image because it requires only 3 steps of preprocessing. Colorization of black and white images is a very difficult task for a human but not for computers. That is why there are lot of methods has been developed to fill appropriate colors in the grayscale images [1]. Motivation behind this paper is many world class movies are in black and white color and our human eye is used to see colored movies and videos nowadays. So that I have decided to develop fast and efficient system that convert these videos and photos of black and white color to a colored image. There are so many proposed methods like automatic image colorization by using support vector regression and image colorization using Capsule Network. The composition of encoder and decoder is called autoencoder. If we have a well-trained decoder, then decoder can generate identical image of inputted image. In this paper, we have used VGG16 as our encoder part and 11 layers of convolution neural network as decoder. Authors have used VGG16 model architecture for the encoder part. VGG16 is a very famous neural network architecture consisting of 19 layers with including 1 fully connected layer. Generative and feature extraction capabilities of pretrained VGG16 model is inherited and used them as weights encoder. Encoder is the most important part in autoencoder, because it recognizes the feature or segmented part of the image. Decoder is widely used in deep learning or machine learning to generate the same image or data from the given set of representation. Decoder is able to generate those data also by using set of representation which is never seen by the decoder. Each images are converted from RGB to CIELAB color space, so that L and ab channels can be separated for that input and output of autoencoder. L channel of represents the lightness of an image where ab channels vary from -128 to $+127$. Figure 1 shows an example of inputted black and white image and generated colorful image.

Fig. 1 Inputted image (left) output image (right)

2 Related Work

Many related research works is in progress computer vision and artificial intelligence community. These papers are based on different approaches of colorizing a black and image or video. This paper is inspired from the published paper on arXive named "Image colorization using Capsule Network" and authors of this paper present a better approach to fill colors in black and white images by using Capsule Network [2]. Authors have inherited the generative and segmentation capabilities of capsule network and updated the weights and biases of their network using pretrained Vgg16 model. The other published paper named "Image Colorization with deep convolution neural network" is based on the baseline regression model [3]. In this paper the approach is classification approach. The range of AB channel is [−100,100] and they are divided in 50 bins [0, 49] using bin() function. And these are classified using the baseline regression approach. There is one more published paper named "Very Deep Convolution Neural Network for large-scale image recognition" in which the classification of images based or recognition of objects in an image is proposed [4]. In a published research paper, named "Image colorization using similar images," the proposed method is about the colorization of black and white image by training with one approximately similar Image [5]. This method can work for these conditions when there are two similar image in which one is colored and other one is black and white. By learning features from colored images algorithms can fill colors in that black and white image as well. A research paper named "colorization of black and white images using deep convolution network" is about automatic colorization of cartoon images that is obtained from video sequencing [6–10]. There are two deep convolution networks is used in this paper, plane convolution network and residual network. In terms of results of each research papers, it is very promising but the approach is very complex. There would be very difficulties during implementation of these approaches [11–16].

3 Methodology

In recent years, there is huge impact of deep learning in artificial intelligence community. It requires a huge amount of training data for the training and testing of any model. Because by increasing the training data we can increase the accuracy of the model. Convolution neural network is very famous and important neural structure for the classification of images and object detection. Method that is used for this paper is based on the autoencoding of images. It means inputted black and white image pass through autoencoder and encoder parts learns the features of image like color intensity, blurring area, scaling, and many more and decoder generated the images based on the learned features of images by encoder. Actually, decoder builds a new ab channel for images based on learning of featured. The generated ab channel is matched with the actual ab channel. After that loss is calculated using mean squared error (MSE).

3.1 Preprocessing

Preprocessing is done before feeding training data to the network. Each images are read in pixel dimension of 224 × 224 in the RGB format and convert it into CIELAB color space. Flickr and ImageNet colored images dataset is used to create training data. Total amount of training images is approx. hundred thousand. It is such a huge amount of data for the training and it requires more hardware specification like minimum 8 GB RAM and NVIDIA GeForce 1650 graphics card. Following samples are from Flickr and custom-created datasets (Fig. 2).

Lab color space is designed by International Commission on illumination. The 'L' channel is named 'lightness' and ranged from black (0) to white (100). 'A' channel is used to code from green (−) to red (+). Similarly, 'B' channel is used to code from blue (−) to yellow (+). For the A/B channels, it ranges from −128 to +128. VGG16 architecture accepts the images with size (224, 224). So that each images has to be resized consist in the flickr30k image dataset to 224 × 224. We read all images into

Fig. 2 Sample colored Images from dataset

RGB (Red, Green, Blue) mode and after that we have to convert all images into LAB color space using following method. Formula of convert an image from XYZ to LAB is as follows. After performing all these preprocessing steps, it has to be separated L and ab channels from given Lab images. The 'L' channel is fed to the network as input and 'ab' channels are extracted for the Target Value.

3.2 Approach

The approach is to feed a set of preprocessed data of colored images to the deep neural network of autoencoder and that will be only L channel of each images. One composition of autoencoder that is encoder will extract the features from images and will gain pass that information to the other composition that is decoder. Decoder will construct ab channel for that L channel. Ten ab channel is compared with the actual ab channel of the images. The loss function that is used fort this paper will calculate loss from the generated ab channel and actual ab channel. We build a pipeline for the learning of model. At the time of training our program read all the images consist in the dataset into pixel dimension of 224 × 224 and 3 channels. And these channels are Red, green and Blue. VGG16 model is already trained on the ImageNet model. So that we can transfer the weights and bias of this model to our autoencoder. In each convolution and dense layer, we have applied rectified linear unit and in the final convolution layer "tanh" activation function is applied. Following is the formula to compute the "tanh" activation function:

$$f(x) = \tanh(x) = \frac{2}{1 + e^{-2x}} - 1$$

3.3 Network Design

Pretrained VGG16 and ImageNet model are used for the classification and as discussed above, VGG16 model architecture is also used as the Encoder and several layers of convolution and Up sampling is used to construct the decoder. In both cases, classification and encoder the architecture of VGG16 will be same except the fully connected layer in the encoder (Fig. 3).

where

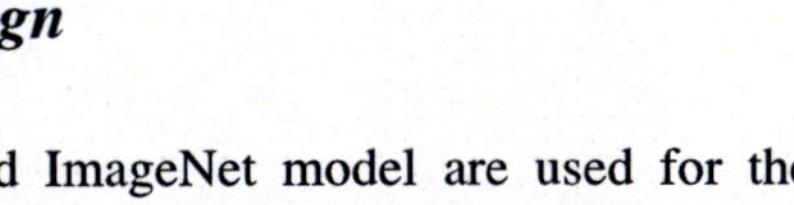

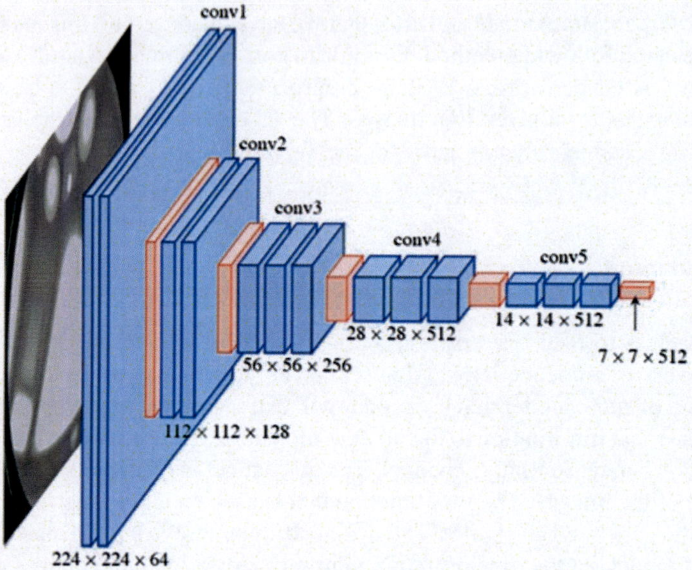

Fig. 3 VGG16 architecture (excluded fully connected layer)

Fig. 4 Architecture of decoder

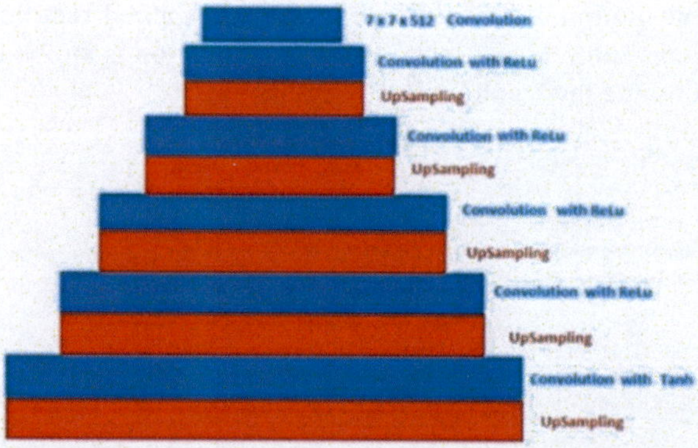

In Fig. 4, it is shown the architecture of decoder. By composing both architecture, we can make an autoencoder to perform the proposed task. All these architectures can be programmed by the Google's TensorFlow or Facebook's PyTorch.

3.4 Parameters

To develop any system testing and experiments is very important things because by doing testing we'll be able to know that if there are any faults in our system or not.

But by experimenting, we would be able to know that by changing any parameter my model is performing better than previous or worse than previous. In our project, we have experimented these following parameters, and it gives a better result than previous results but on some parameters our model was not performing well than previous results.

3.4.1 Loss Function

Loss function is also known as cost function. In deep learning specially in convolution neural network loss functions play an important role because balancing the weights and biases are totally dependent on the loss function. By minimizing the error, approximately target value can be achieved. In this paper, mean squared error is used as a cost function or loss function, and it gives very satisfying result to balance the weights and biases of neurons. Formula of mean squared error is as follow.

$$\textbf{MSE} = \frac{1}{n}\sum_{i=1}^{n}(y - \widetilde{y})^2$$

Here "y" is the actual value which is "ab" in our project and "$\widetilde{y}$" is the observed value which is generated by the autoencoder. By taking average of the differences of actual value and observed value, we can get the mean squared error.

3.4.2 Optimizers

In deep learning, optimizers play an important role in the reduction of loss. Optimizers are used to change the weights and biases of the neurons. It is very important to know that what optimization method should be used for particular problem. There are many optimizers are developed in the neural network. Adam Optimizer is used in this paper. **Adam optimizer** is names as Adaptive Moment Estimation. It was presented by "DiederikKingma" from OpenAI and Jimmy Ba from the University of Toronto in 2015. The very basic advantage of using this Optimizer is that it is very straight forward to implement and requires very less amount of memory. It works with the first and second order momentum.

3.4.3 Epochs

Epochs works as an iterator. It iterates of whole process of by number of epochs set by developer. In this project, model is trained on different epochs and is seen such a huge difference in loss and accuracy of the model. The system is experimented on different epochs from 50 to 100 to 200 and 300. And by the completion of 300 epochs, accuracy was nearly 83% measured by accuracy matrices.

4 Experimental Setup

The complete model is trained on the Google's Collaboratory and AWS lambda. Google's Collaboratory provides 3 type of runtime. 1. Only CPU with 12.75 GB RAM and 67 GB of storage. 2. GPU with 25 GB RAM and 67 GB Storage. 3. TPU with 35 GB RAM and 107 GB of Storage. Each model is trained on the GPU runtime of 18 h a day. And it is a great cloud computing platform for free. Accuracy of any model can be increased by adding more related training data. It should be observed one thing that in the training image dataset there should not be any black and white images, because it might create some noise in the result or maximize the error. Whole model is trained on approximately 50 K images in batch-size of 100 images.

5 Result

Figure 5 presents the accuracy and the loss representation. The accuracy is increasing, and the loss is decreasing on increasing epochs. Table 1 presents the summary of accuracy and loss in complete training time.

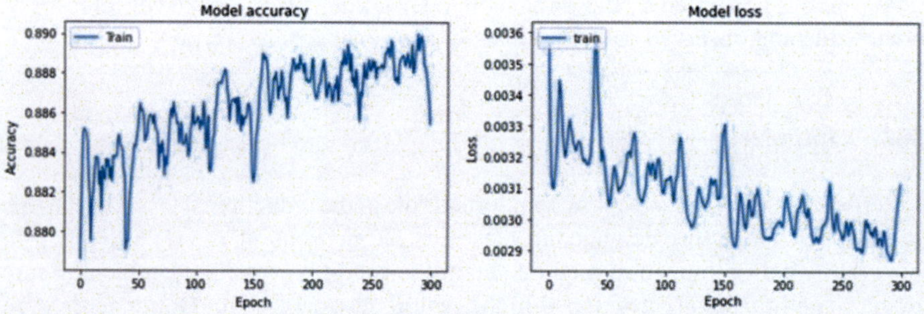

Fig. 5 Accuracy (left) and loss (right) graph

Table 1 Accuracy and loss on each 50 epochs

Epoch	Accuracy	Loss
0	0.885	0.0031
50	0.873	0.0031
100	0.884	0.0030
150	0.883	0.0029
200	0.887	0.0030
250	0.886	0.0030
300	0.888	0.0029

Table 2 Result obtained from proposed method

S.N.	Inputted Image	Resulted Image	Ground Truth
1			
2			
3			
4			
5			
6			
7			

(continued)

Table 2 (continued)

8			
9			
10			

And Table 2 presents the many colorization results obtained from the proposed method with the ground truth and inputted image. Generated ab channel after running whole epochs is concatenated with the L channel of inputted image, and the resulted lab image is the colored imaged colorized by the model.

6 Conclusions and Future Scope

Since 1970s, progress of image processing is at much heigher rate but colorization of any black and white image was a mannual intensive and time-consuming process. But in this paper, a method is proposed to accomplish such tasks with zero deirect human interaction. Through the experiments, it is demonstrated the potential and efficiency of the using autoencoders to colorize the black and white image. Using right loss function, optimizers and suitable color space, it is shown that the proposed method is capable of producing accuracte colored images of any grayscale image. In this paper, LAB color representation is used in which L represents the lightness of an image, "a" varies from red to green and "b" varies from blue to yellow. And autoencoder is used to learn the representation of "ab" channels and decode or generate the approximately similar kind of data. In this framework, users have to just upload a black and white image and trained colorization model will generate an output image. Obtained result is shown along with the ground truth of that image in Table 2. Accuracy and epochs are also shown above in Table 1 with respect to the number of epochs.

In furure work, classification and image segmentation technique will be used along with this proposed method. Additonally, we plan to explore more optimized method so that a low scale black and white image can also be filled with appropriate color.

There are lot of historical images which are very low contrasted and low scaled. We plan to make image suitable contrasted and high scaled before colorizing the image. It may give better result if there is any low scaled image.

References

1. Lata K, Khan S (2019) Experimental analysis of machine learning algorithms based on agricultural dataset for ımproving crop yield prediction. Int J Eng Adv Technol (IJEAT) ISSN: 2249–8958, pp 123–134, vol 9, issue 1
2. Özbulak G (2019) Image colorization by capsule networks. In: 2019 IEEE/CVF conference on Computer Vision and Pattern Recognition Workshops (CVPRW), Long Beach, CA, USA, pp 2150–2158
3. Hwang J, Zhou Y (2016) Image colorization with deep convolutional neural networks. Stanford University, Tech Rep
4. Simonyan K, Zisserman A (2014) Very deep convolutional networks for large-scale image recognition. arXiv preprint arXiv:1409.1556
5. Gupta RK, Yong-Sang Chia A, Rajan D, Ng ES, Zhiyong H (2012) Image colorization using similar images. In: Proceedings of the 20th ACM international conference on multimedia, pp 369–378
6. Futschik D (2018) Colorization of black-and-white images using deep neural networks
7. Levin A, Lischinski D, Weiss Y (2004) Colorization using optimization. In: ACM SIGGRAPH 2004 Papers, pp 689–694
8. Varga DI, Szirányi T (2017) Convolutional Neural Networks for automatic image colorization, pp 1–15
9. Nielsen MA (2015) Neural networks and deep learning, vol 2018. Determination press, San Francisco
10. Szegedy C, Ioffe S, Vanhouche V, Alemi A (2016) Inception-v4, inception-resnet and the impact of residual connections on learning. Arxiv, 1602.07261
11. Hinton GE, Srivastava N, Krizhevsky A, Sutskever I, Salakhutdinov R (2012) Improving neural networks by preventing co-adaptation of feature detectors. CoRR, abs/1207.0580
12. Jordan MI (1990) Artificial Neural Network. IEEE Press, pp 112–127
13. Krizhevsky A, Sutskever I, Hinton GE (2012) Imagenet classification with deep convolutional neural networks. Advances in neural information processing systems, pp 1097–1105
14. LeCun Y, Bottou L, Bengio Y, Haffner P (1998) Gradient-based learning applied to document recognition. IEEE Commun Mag 27(11):41–46
15. LeCun Y, Jackel L, Boser B, Denker J, Graf H, Guyon I, Henderson D, Howard R, Hubbard W (1998) Handwritten digit recognition: applications of neural networks chipsand automatic learning. Proc IEEE 86(11):2278–2324
16. Glorot X, Bengio Y (2010) Understanding the difficulty of training deep feedforward neural networks. In: International conference on artificial intelligence and statistics, pp 249–256

Binary Tree-Based Asymmetric Moderate Algorithm for Secured DNS

Puneet Sharma, Deepak Arora, and Anil Kumar

Abstract With the advent of recent advancements in different Web technologies, domain name system (DNS) has become vital for the entire network infrastructure. Still, it lacks robust security mechanism implementation for ensuring integrity and authentication of information. The extensions to the DNS for implementing these services are represented with the help of cryptanalytic digital signatures. Keys associated with DNS names will be retrieved to support other protocols also. In addition to that, security extensions provide authentication of DNS protocol transactions also. In this research work, the authors have proposed a moderate algorithm, a stream cipher algorithm, which converts plaintext into cipher text with stringent levels of time complexity in order to produce security by combining the concept of each the digital signature and asymmetric key (public key) cryptography. Here, the public key is used instead of the private key with a combination of the SHA-512 algorithm to compress the message (text file) and pseudo-random number generator (PRNG) algorithm for generating further public and private keys. The proposed encryption algorithm encrypts the data using ASCII values of the characters and properties of binary trees. The authors have evaluated the complexity of the proposed moderate method and found it satisfactory with the $O(n)$ nature of time complexity theoretically. Authors have shown the performance of both encryption and decryption algorithms with the help of figures and graphs also.

Keywords DNS · SHA-512 · RC4 · Message security

1 Introduction

An encryption algorithm converts some plaintext into some cipher text, i.e. from some readable form to a non-readable form. In this paper, author shave introduced moderate algorithm, which converts an array of N elements. This array contains the plaintext to an unreadable form or the cipher text. The proposed algorithm encrypts

P. Sharma (✉) · D. Arora · A. Kumar
Department of Computer Science and Engineering, Amity University, Uttar Pradesh, India

R. P. Mahapatra et al. (eds.), *Proceedings of International Conference on Recent Trends in Computing*, Lecture Notes in Networks and Systems 341,
https://doi.org/10.1007/978-981-16-7118-0_38

the data by making use of the ASCII values of the characters of the plaintext, it then defines a key based on the length of the plaintext, then constructs a binary tree by taking the elements of the array as the leaf nodes and obtaining the parent nodes by taking the average of the two adjacent leaf nodes. This process is redone till the last node is obtained. When the binary tree is completely constructed, all its right child nodes are replaced by zeros, now the key is added to all the nodes, this array of decimal values are converted into hexadecimal which is the final encrypted text or the decrypted text [1]. In the decryption process, the received encrypted text which is in hexadecimal is first converted into decimal (base 10), a binary tree is constructed with these values, now the right nodes of the tree are computed by the formula 2*PN – LN, where PN denotes the parent node and LN denotes the left child node. An array is constructed with the leaf nodes of this binary tree, and now, the key is computed by counting the number of the leaf nodes and converting the obtained value to its corresponding ASCII code. Now, the key is subtracted from all the elements of this array to obtain a new array.

The final decrypted text is obtained by converting these ASCII values to their corresponding characters. The message combines with the private key to create a signature using DSA algorithmic rule, and it is encrypted using RC4 algorithm, now the encrypted message, signature is sent in conjunction with the public key [2]. The receiver uses the public key and DSA algorithm to create a signature. If this signature matches with the signature of the message received, the message is decrypted and read else discarded [3]. The time complexity of the algorithm has been examined theoretically and found of $O(n)$ in nature.

2 Proposed Methodology

2.1 Encryption

Step 1: Replace the elements of the array by their corresponding ASCII values.
Step 2: Compute the key.
Step 3: Using these ASCII values and their averages construct a binary tree.
Step 4: Replace all the right child nodes by zero.
Step 5: Add the key to all the nodes.
Step 6: Change these decimal values into its respective hexadecimal.

2.2 Pseudocode for Encryption

Here, original message is 'M', length is 'L'; position of array is 'Ap'; encrypted message is 'EM'; temporary maximum array index is 'I'; type of array is 'T'; encryption recursion method is 'ER' and ASCII message is 'AM'.

```
Start Encryption of M
        Initialise L to M.L
        If L less than 9
            Set Ap to L
        Else
            Set Ap to L modulus 9
        EndIf
    Set Ap to Ap plus 48
     For i is 0, i is less than L, i increments by 1
         Set i of AM to sum of (integer) i of M and Ap
     EndFor
     Initialise I to 2*L minus 1
     Initialise Ap to 0
     If L modulus 2 is equal to 0
         For i is L minus 1, i is greater than or equal to 0, i decrements by 1
             Set Ap of EM to position i in AM
             Ap++;
         EndFor
         Call method ER with EM, Ap, 0, Ap divide 2 ,Ap, I
     Else
         For i is L minus 2, i is greater than or equal to 0, i decrements by 1
             Set Ap of EM to position i in AM
             Ap++;
         EndFor
         Set Ap of EM to position in AM
         Ap++;
         Call method ER with EM, Ap, 0, Ap divide 2 ,Ap, I
     EndIf
     For i is 0, i is less than I minus 1, i is equal to i sum 2
         Set i of EM to Ap
     EndFor
     For i is 0, i is less than high, i increments by 1
         Set i of EncryptedStringArray to Double.toHexString with position i in EM
     EndFor
     Return EncryptedStringArray
End Encryption

Start ER of EM, Ap, baseValue, tempValue ,T, I
     Initialise templ to T
     If tempValue is greater than 0
         Set Ap of EM to ( position base in EM plus position in EM ) divide 2
         tempValue--;
         Ap++;
         Call method ER with EM, Ap, base plus 2, tempValue, T, I
     Else if T modulus 2 is not equal to 0 and Ap is less than I
         Set Ap of EM to ( position in EM plus position templ in EM ) divide 2
         If Ap plus 1 is less than I
             Call method ER with EM, Ap plus 1, templ plus 1, ( Ap minus templ ) divide 2, Ap plus 1, I
         EndIf
     Else if T modulus 2 is equal to 0
         Call method ER with EM, Ap, templ, ( Ap minus templ ) divide 2, Ap , I
     EndIf
End ER
```

2.3 Decryption

Step 1: Convert the hexadecimal cipher text into decimal.
Step 2: Construct a binary tree with these values.
Step 3: Compute the values of the right child nodes by the formula—2*PN – LN, PN is parent node and LN is the left child node.
Step 4: Compute the key.
Step 5: Subtract the key from the leaf nodes.
Step 6: Convert this array of ASCII values into its corresponding characters.

2.4 Pseudocode for Decryption

Here, length is 'L'; original message length is 'OML'; encrypted message is 'EM' and decrypted message is 'DM'.

```
Start Decryption of Message
1:    For i is 0, i is less than StringArray.L, i
increments by 1
2:        Set i of EM to Double.parseDouble with
position i in StringArray
3:    EndFor
4:    Initialise L to EM.L
5:    OML = (L+1)/2;
6:    If OML is less than 10
7:        Set key to OML
8:    Else
9:       Set key to OML modulus 9
10:   EndIf
11:   Set key to key plus 48
12:   Call method RecurOfDecryption with cc, L minus
1, L minus 3
13:   If OML modulus 2 is equal to 0
14:       Initialise temp to OML minus 1
15:       For i is 0, i is less than OML, i
increments by 1
16:            Set i of DM to position temp in EM
minus key
17:       EndFor
18:   Else
19:       Initialise temp to OML minus 2
20:       For i is 0, i is less than OML minus 1, i
increments by 1
21:           Set i of DM to position temp in EM
minus key
22:       EndFor
23:       Set OML minus 1 of DM to position in EM
minus key
24:   EndIf
End Decryption

Start RecurOfDecryption of EM, max, temp
25:    If temp is greater than or equal to 0
26:        Set temp of EM to ( position max in EM
multiplied by 2 )
27:        Set temp to temp minus 2
28:        Set max to max minus 1
29:        Call method RecurOfDecryption with EM,
max, temp
30:    EndIf
End RecurOfDecryption
```

3 Theoretical Evaluation

3.1 Encryption

Finding the average and constructing the tree are the major part of this algorithm. Dependency on the auto-generated key makes this algorithm independent, and it only requires user input (i.e. plaintext) for encryption [4, 5]. Let plaintext be 'creative' as an input as shown in Fig. 1. The key is auto-generated from the message length. Now key = 8, so, no modulo operation is required. After that, key is converted to ASCII value, i.e. 56, and plaintext is converted to ASCII value character by character. So, now the plaintext is converted to 99, 114, 101, 97,116,105,118 and 101. In step 2 of Fig. 1, the binary tree is generated using plaintext (i.e. ASCII value) as input. The average operation is performed on the each step to generate new node. Like the

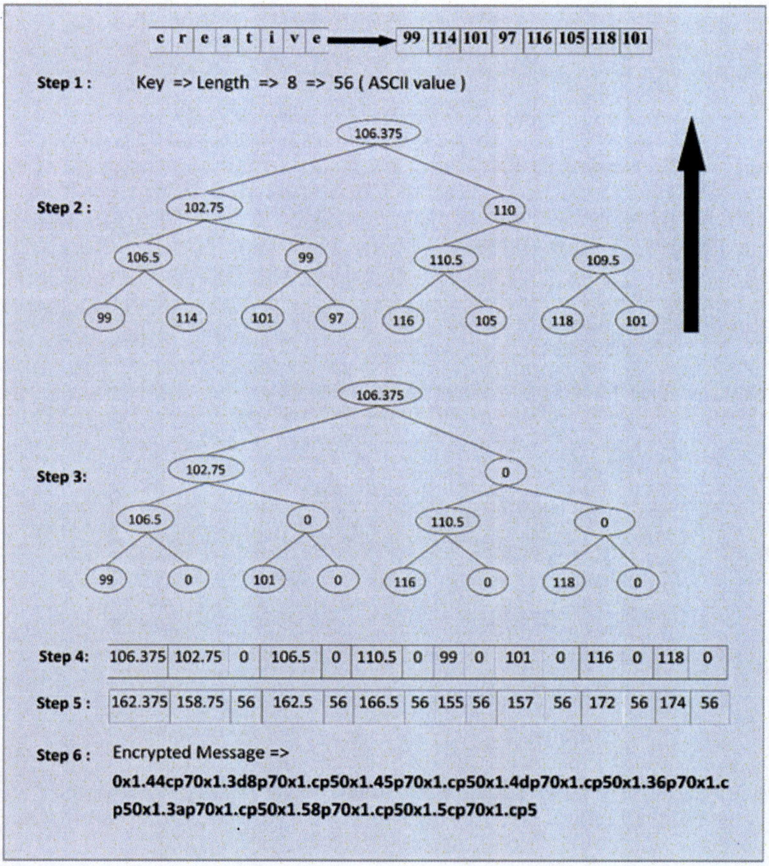

Fig. 1 Encrypting method applied

average of 99 and 114 is 106.5, the average of 101 and 97 is 99; the average of 106.5 and 99 is 102.75. Similarly, in the RHS of this tree, the average of 116 and 105 is 110.5; the average of 118 and 101 is 109.5, and the average of 110.5 and 109.5 is 110. Finally, the average of 102.75 and 110 is 106.375, and the required tree is generated. In steps 3, 4 and 5 of Fig. 1, data from the tree are added with the key, after setting all right side node to zero. The array of decimal values is converted into hexadecimal which is the final encrypted text or the cipher text.

Let $T(n) = \#$, for investigating the complexity of the encryption algorithm of array n size (i.e. length of the plaintext). This algorithm requires $n - 1$ operation on n size array. Here, $(n - 1)$ is the total number of non-leaf node in the tree. Cn is the complexity of other sideway operation in this algorithm.

$$T(n) = T(n - 1) + Cn \tag{1}$$

At message, length = 1 takes constant time

$$T(1) = K(\text{constant}) \tag{2}$$

Author has used substitution technique to yield (8) from (1) and (2).

$$T(n - 1) = T(n - 2) + Cn \tag{3}$$

$$T(n - 2) = T(n - 3) + Cn \tag{4}$$

Getting general equation

$$T(n) = T(n - k) + Cn \tag{5}$$

From (2) and (5) together, we get

$$n{-}k = k(\text{where } k \text{ is constant}) \tag{6}$$

$$n = 2k(\text{i.e. equal to } k) \tag{7}$$

Substituting the value of k in (1), we get

$$T(n) = k + Cn \tag{8}$$

Thus $T(n) = O(n)$.

3.2 Decryption

Similarly, in Fig. 2 decryption takes place. The encrypted hexadecimal text obtained is converted into decimal (base 10). In step 3, the tree is generated from step 2 decimal array. The right child of the tree is calculated using formula 2 * PN−LN, where PN denotes the parent node and LN denotes the left child node. All the leaf nodes in the tree are always the message characters in encrypted form. The leaf nodes obtained from step 4 are stored in an array serially (i.e. left → right). Here, in Fig. 2 step 5, the obtained array is 155, 170, 157, 153, 172, 161, 174 and 154.

Key = (total number of nodes in the tree + 1)/2 is then converted to its corresponding ASCII value (i.e. 56). At last, key = 56 is subtracted from each array value to obtain 99, 114, 101, 97, 116, 105, 118 and 101. The obtained value in step 6 is the ASCII value, which is then converted to its corresponding character.

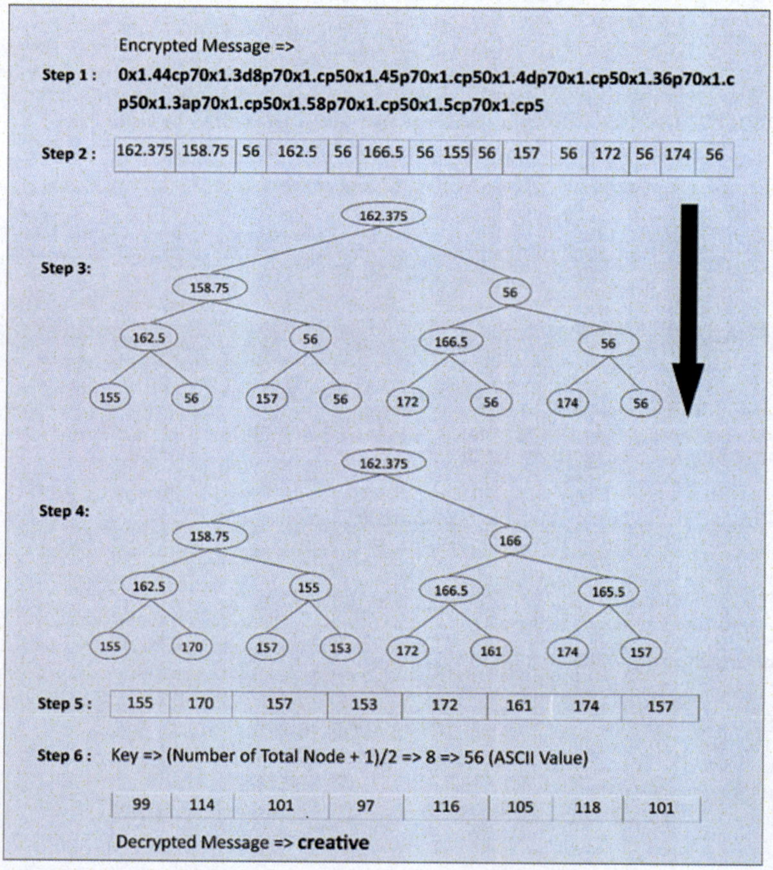

Fig. 2 Decrypting method applied

Finally, the decrypted message (i.e. plaintext) obtained is creative. Decryption input is proportional to encryption input thus, their complexity is equal to $O(n)$.

4 Result Analysis and Discussion

The time complexity of this algorithm is obtained using system clock time. The tests were executed on PC running Windows 10 and with following specifications: Intel® Core™ i3-2328 M CPU at 2.20 GHz with 2 GB of RAM. Algorithms are run in Java.

Table 1 is a quantitative approach to analyses the encryption time (i.e. in millisecond) of the plaintext length as an input.

Figure 3 shows a graph that is plotted using Table 1 values. This figure shows that the encryption algorithm's time complexity is linear. Considering N (i.e. length of original message) along x-axis and Time in ms (i.e. millisecond) along y-axis. Encrypted time in millisecond obtained is the output average of 50 values observed for each length.

Table 2 has three columns showing the dependency of the decryption time ms (i.e. millisecond) on the plaintext and the encrypted text message length.

Figure 4 shows a similar linear graph like in Fig. 3 of the encryption algorithm. Here, to verify data are fetched from Table 2. This figure shows length of encrypted message along x-axis and time in ms (i.e. millisecond) along the y-axis. Here, also, the average of 50 random values is considered for complexity computation.

Table 1 Encryption time computation with message

Length (N)	0
0	0
1	0.0127822
5	0.0155078
10	0.0157678
50	0.0264509
100	0.32469
500	0.1799207
1000	0.3758204
1500	0.5248701
2000	0.5830907
2500	0.7863954
3000	0.8902405
3500	0.9851751
4000	1.11735
4500	1.3525982
5000	1.502068

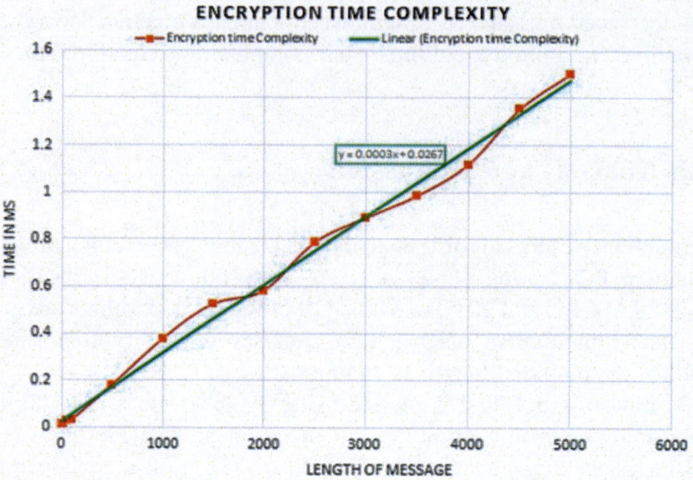

Fig. 3 Time duration for encryption

Table 2 Decryption time computation with message

Length of encrypted message ((2*N)–1)	Length of original message (N)	Decryption time complexity (ms)
0	0	0
1	1	0.0163276
5	9	0.0167944
10	19	0.0174006
50	99	0.0208063
100	199	0.0284103
500	999	0.1170845
1000	1999	0.1943946
1500	2999	0.2786464
2000	3999	0.3430248
2500	4999	0.4706617
3000	5999	0.5072828
3500	6999	0.5804781
4000	7999	0.6908076
4500	8999	0.7478625
5000	9999	0.846622

5 Conclusion

In this paper, authors have proposed moderate algorithm, which is an encryption algorithm. The proposed moderate algorithm uses ASCII values of characters and

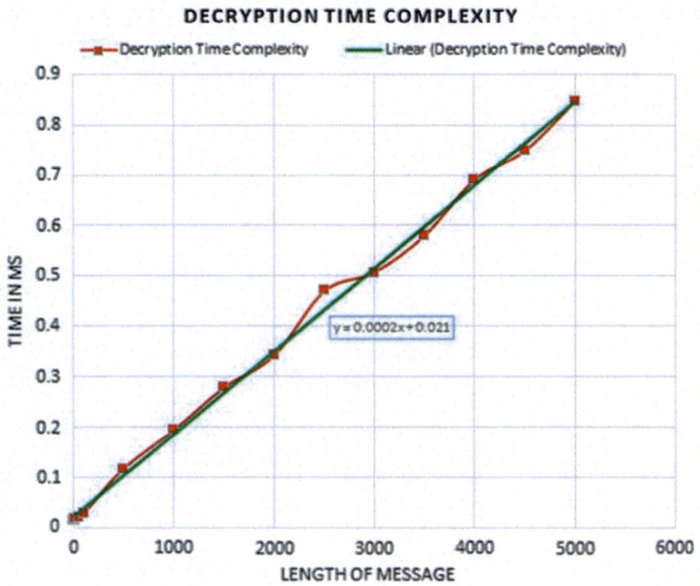

Fig. 4 Time duration for decryption

properties of a binary tree to produce encrypted data. The theoretical analysis of the proposed algorithm shows that moderate algorithm is having a time complexity of *O*(*n*), which is found to be satisfactory specifically to produce more secured DNS system design.

References

1. IETF DNSSEC WG (1994) DNS Security (dnssec) Charter, IETF
2. Albitz P, Liu C (1997) DNS and Bind, 2nd edn. O'Reilly & Associates, Sebastopol, CA, pp 1–9
3. Ateniese G, Mangard S (2001) A new approach to DNS security (DNSSEC). In: Proceedings of the 8th ACM conference on computer and communications security, pp 86–95
4. Bhujade VV, Yadav JR, Mandal BB, Raut SV (2013) Deploying cryptography in domain name system: an overview. Int J Appl Innov Eng Manage 2(4)
5. Schildt H (2003) The complete reference JAVA 2. Tata McGraw Hill Publications

Multimodal Analysis of TED Talks Toward Designing a Recommendation System

M. N. Renuka Devi and Gowri Srinivasa

Abstract The efficacy of any oratory performance is primarily determined by the content of talk (text/vocabulary), the tone and modulation of an orator's voice (audio) and nonverbal cues: facial expressions or body language (video). In particular, body language in this context may be considered to include hand gestures and pose (stance taken by a speaker) in addition to facial expressions. In this paper, we present an analysis and interpretation of audio, text and visual features of a video recording of a speech to delineate features of the speech. We use these features to design a recommendation system that profiles features from the foregoing modalities for a novice speaker and suggests improvements. The video dataset considered in the study is taken from TED talks. Through an analysis of these videos, we have evolved three coarse categories and a total of 27 fine categories to describe the ability of a speaker. The 27 categories describe the strengths and weaknesses of a speaker in each modality—audio, visual and textual content and determines the placement of a speaker based on the yardstick evolved from the reference league of speakers. These analyses and recommendations based on experimental results are envisaged to help the speaker understand their strengths and weakness better and suggest some direction for improvement. This, in turn, would help the speaker work strategically on specific areas to be a better orator in the future.

Keywords Facial expression · Hand gestures · Audio features · Visual cues · Ext features · Recommendation system

M. N. Renuka Devi (✉)
Dayanada Sagar University, Kudlu Gate Campus, Hosur Road, Bengaluru, Karnataka, India

G. Srinivasa
Department of Computer Science and Engineering, PES University, Bengaluru, Karnataka, India
e-mail: gsrinivasa@pes.edu

R. P. Mahapatra et al. (eds.), *Proceedings of International Conference on Recent Trends in Computing*, Lecture Notes in Networks and Systems 341,
https://doi.org/10.1007/978-981-16-7118-0_39

1 Introduction

The art of speaking is an important feature of human interaction. A good speaker is expressive, maintains a good posture, connects with the audience and can create an impact. In general, public speaking provides one an opportunity to share ideas with a large audience in an impactful manner. Hence, it is important to craft and deliver an effective speech, ensuring all parameters are in the right proportion. It is proven after critical analysis that duration of a speech is an important aspect and evidently, many of the popular speeches through the years have been short ones (a maximum duration of 15–20 min). Given this understanding on input, it is acknowledged that to communicate any content, particularly one that an audience is not familiar with, and more so, to a large audience in a short duration is a nontrivial challenge. There have been quite a few studies done on effective communication by researchers and different recommendation systems have been published for novice speakers to become more effective in communicating with audience.

According to one of the surveys, there are eight scientifically proven characteristics of effective public speaking [1]. These are: passion, confidence, tone of the voice, articulation, appropriate body language, no filler words and connecting with the audience.

In this paper, given an input snippet of a speech, we analyze the behavior and ability of a speaker against the pre-analyzed patterns of speeches of orators sampled from the TED database. We then present a brief description of the key features of the input speech and provide recommendations on improving various facets of the speech to be an effective orator. We have found that it is not only the visual appearance including facial gestures, emotions and pose of the speaker which makes a speaker more effective, but also audio quality, content of the speech, types of words used, vocabulary of the speaker etc., are critical. Thus, the recommendation is a holistic one focusing on all these aspects.

2 Literature Review

"Sharma, Tanya and Gaurav" [2] have analyzed visual behavior in public speaking using multichannel attention network. This study analyzes human-centered visual cues related to physical and facial appearance and pose. The study was conducted by constructing database of 1800 videos from TED talk using convolutional neural network (CNN) connected to LSTM (long short-term memory) to predict the video popularity.

The above framework may be improved by including audio features. The accuracy of analysis may be further improved significantly if verbal cues including speech delivery, pitch contour and style of speaking are included in the framework.

Zeng and Xingbo [3–5] performed a visual analysis of emotion coherence in presentation videos. In this work, authors introduced a collaborative visual analytics system to enable effective investigation of sentiment consistency across audio, face and text modalities in presentation videos. The authors proposed a system that comprises five views, which allows users to perform detailed explorations of emotions in video, sentence and text levels. It incorporates well-documented visualization methods and innovative plans to support visual analysis of videos. It involved TED talk videos and interviews. However, the system fails to address the study of additional modalities like hand gestures and detailed analysis of emotion detection algorithms.

The difference between previous work and the work we propose is that our algorithms not only provide a system to detect and characterize behavioral patterns through public speaking as a case study, but also a computationally viable and useful application through an automated recommendation system. The results are promising and open up avenues for further research on the automated analysis of public speaking videos and recommendation systems for honing behavioral patterns and aid a novice speaker to be more effective at communication.

3 Datasets

We have sampled some of the most inspirational TED talks [6] for this study. TED talks is a not-for-profit organization that offers achievers a stage to express themselves. The format of presentation is scientifically designed to transmit information to an audience effectively; they are story-driven and typically last for 10–18 min depending on the topic. Some of the TED talks have become very popular (i.e., garnered a large number of views) within a short period of time after they were posted in the public domain. Hence, TED talks is a viable database for this study.

Features considered for effective public speaking.

Factors that we have considered for effective public speaking are: audio, visual and text.

Table 1 presents the aspects of an input video we elicit features from: audio, visual and text with the statistical description of the features.

The first audio feature is referred to as 'Inconsistent Factor,' calculated through hierarchical clustering of audio features such as zero crossing rate, standard deviation of magnitude and average magnitude for short time intervals. Next, visual features are computed through counting the number of frames in a TED video that are classified as happy, sad and miscellaneous based on the facial expressions and also through counting the number of video frames in which speakers have used their hand gestures (gesticulating) while speaking.

Visual features also include the pose of a speaker, computed as video frames. These are classified as 'front,' 'side' and 'back' poses of the speaker, indicating they are not stiff and possibly shifting to face different sections of the audience. Text analysis includes bi-modal features (i.e., textual content in conjunction with audio features) at the tone-level to decide whether the speech is analytical, confident or

Table 1 Features considered for effective public speaking

Feature type	Feature	Statistical description
1. Audio-based features	1. Inconsistent-factors	ZCR (zero crossing rate) Joule/meter square
	2. Energy	SD (standard deviation) Pascal Magnitude (decibels) STE (short-term energy) Mean Power (decibels)
2. Vision-based features	1. Face	Happy Sad Miscellaneous
	2. Hand gesture	Expressive through hand gestures No hand gestures
	3. Pose	Front, side and back facing
3. Text: Bi-modal features (Text + Audio)	1. Tone-level analysis	Analytical, confident and tentative
	2. Language-level analysis	Positive, neutral and negative

tentative. Further, sentiment analysis is performed on the text to decide the polarity of the text as positive, neutral or negative.

TED talks research cell (HBC works on the study for TED talks) publishes best speakers regularly based on their popularity on the media along with study results. The research is primarily conducted to understand why some talks become viral on the media, which may be classified as 'popular category' and some do not become so famous, which may be classified under 'not so popular category'.

We have computed a set of over 200 TED talks in our study. Based on the results of research elaborated, we have classified TED talks broadly into following three categories [7, 8].

Class A Speakers	Class B Speakers	Class C Speakers
Audio Feature: high energy (1 or > 1), standard deviation of magnitude (0.1 and > 0.1), inconsistency coefficient = 0 and ZCR (0 to 3 kHz) (A1) **Video Features**: 65–75% frames with happy expressions and hand gestures (V1) **Text features**: 65–75% of the words are positive and analytical (T1)	**Audio feature**: mid-level energy (0.55–0.90), standard deviation of magnitude (0.09–0.05), 0 < inconsistency coefficient < 1 **Video Features**: 50–65% of the frames with happy expressions and hand gestures **Text features**: 50–65% of the words are positive and 50–65% are tentative	**Audio features:** low to mid-level energy (< 0.55), standard deviation and magnitude (<0.05), inconsistency coefficient > 1 **Video Features**: 40–50% of the frames are with happy expressions and hand gestures **Text features:** 40–50% of words are positive and 40–50% tentative

TED Class	ATV Class	Number of Speakers
Popular	A	43
Popular	B	136
Not-so popular	C	38

Sl.No	Speaker Name	ATV Class (fine-grained)
1.	Sir Ken Robinson	A_1 T_1 V_1
2.	Beth Malone	A_3 T_3 V_2
3.	Shameem	A_2 T_2 V_2
4.	Anthony Veneziale	A_2 T_2 V_2
6.	Howard Steveson	A_2 T_1 V_2
7.	Ivonne Roman	A_2 T_2 V_2
8.	Sonaar Luthra	A_2 T_1 V_1
9.	Kavitha Ramdas	A_2 T_2 V_1
10.	Meera Vijayann	A_3 T_2 V_2

Fig. 1 Distribution of TED speakers in various classes and ten sample speakers

---

Of the 217 videos processed for the study, 43 speakers are in class 'A,' 136 speakers in class 'B' and 38 speakers in class 'C' as presented in Fig. 1.

3.1 Automated Talk-quality Valuation (ATV): Multimodal analysis of speech towards designing a recommendation system

Based on audio, visual and text features, we divided each class in to finer categories. This helped us understand how little the difference could be between the speeches of two speakers belonging to two subdivisions, and how big the impact of specific features could be on the output of the speaker's performance. This is what we call an automated talk-quality valuation or the ATV system. It serves as a precursor to an automated system that presents a detailed analysis of a speech and specific areas, if any, for improvement.

Figure 2 illustrates the 27 subdivisions of speakers on the basis of the gradation of their speech's audio, visual and text content.

The speeches of 217 speakers used as training data for the study were mapped to one of the 27 classes based on the analysis of the respective features. These, in turn, were tallied with the TED categorization. Based on our analysis, we observe that $A_1T_1V_1$ and $A_1T_2V_1$ belong to class A as these classes have good audio, visual expression and text quality in the speech, we have totally 43 speakers in Class A. Class B consists of $A_1T_3V_1$, $A_1T_1V_2$, $A_1T_3V_2$, $A_2T_1V_1$, $A_2T_2V_1$, $A_2T_3V_1$, $A_2T_1V_2$, $A_2T_2V_2$, $A_2T_3V_2$, $A_3T_1V_1$, $A_3T_2V_1$ and $A_3T_3V_1$ subdivisions with 136 speakers from the training set. These speakers have mid-level visual features, audio and content in the text. Class C comprises subdivisions $A_3T_1V_2$, $A_3T_2V_2$, $A_2T_3V_3$ and A3T3V2

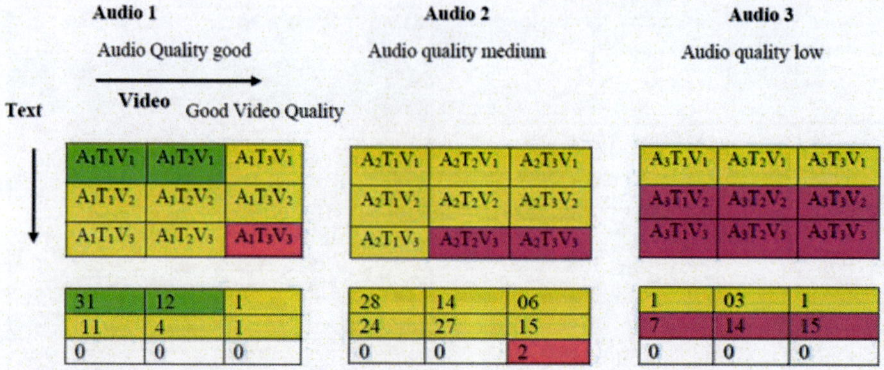

Fig. 2 Automated Text-Quality Valuation (ATV) categories for the analysis of speech quality: the first row depicts the subdivisions based on the gradation of content (in terms of the audio, text and visual features); depicted in green are the top speakers ('Class A'), depicted in yellow are the speakers who are good orators but have some room for improvement ('Class B') and finally, in red are those whose speech has much room for improvement ('Class C'). The second row of the table presents the number of training samples in the TED dataset in each of the categories

consisting of 38 speakers. These speakers have fewer hand gestures and happy face expressions, lower voice modulations and medium level of text content in the speech. Out of 217 speakers we have classified, the following categories $A_1T_1V_3$, $A_1T_2V_3$, $A_1T_3V_3$, $A_2T_2V_3$, $A_3T_1V_3$, $A_3T_2V_3$ and $A_3T_3V_3$ do not have any speakers in them. We understand that speakers of TED talk are mostly experts and achievers in their fields and high impact orators, thus exhibiting high quality in most factors identified. Hence, there are no speakers in the last few categories of classification. However, having a provision for these would allow us to analyze the features of any input speech and make suitable recommendations.

Our model categorizes speakers with the speech ratings $A_1T_1V_1$ and $A_1T_2V_1$ as 'Class A' with the speakers demonstrating excellent audio (volume, range of pitch, etc.), visual expressions and text quality in their performance. We have classified 43 speakers in total in 'Class A.' In 'Class B,' we have speakers with ratings $A_1T_3V_1$, $A_1T_1V_2$, $A_1T_3V_2$, $A_2T_1V_1$, $A_2T_2V_1$, $A_2T_3V_1$, $A_2T_1V_2$, $A_2T_2V_2$, $A_2T_3V_2$, $A_3T_1V_1$, $A_3T_2V_1$ and $A_3T_3V_1$. A total of 136 speakers from the training set are classified as belonging to this category. In 'Class C,' we have speakers with ratings $A_3T_1V_2$, $A_3T_2V_2$, $A_2T_3V_3$ and $A_3T_3V_2$. There are 38 speakers who come under this division. This class of speakers show relatively fewer hand gestures and a smaller percentage of positive facial expressions, inconsistent voice modulations and a medium level of text content in their speech with audio ratings of low-mid-level energy.

Out of 217 speakers we have classified, no speaker comes under the following categories: $A_1T_1V_3$, $A_1T_2V_3$, $A_1T_3V_3$, $A_2T_2V_3$, $A_3T_1V_3$, $A_3T_2V_3$ and $A_3T_3V_3$. Since this is a rule-based classifier, most of the rules for 'Class C' have been designed as a default class (i.e., 'not Class A and not Class B'). Further, it is noted that speakers invited to deliver a TED talk are mostly experts and achievers in their fields and, being passionate about their work and mostly strong communicators, present high quality in most factors identified. Hence, there are no speakers in the training set under the last classification as per TED's classification system.

TED talks have not published in detail the yardstick for their classification of speakers. However, they have revealed, at a high-level, parameters considered for research and the basis for classification of speakers. As per these guidelines, the speech of good speakers comprises a high quality of the following features: visual features/body language (nonverbal communications), audio features/voice modulation and text quality/vocabulary etc. An effective, popular speaker would mostly amalgamate these factors in the right proportion, the mix of features may be interpreted as approximately 70:20:10, respectively, for better understanding [9].

There is no separate classification other than the popular category as per TED cell survey, but the rest may be classified as not-so-popular category [10]. Their videos hold a large number of views on the website next to the popular set of speakers. All good speakers as per the TED cell whose videos have gone viral on the media are included in 'Class A' in our study. The not-so-popular category of speakers as per TED cell survey are classified into one of the 27 sub-categories along with popular speakers in our study based on the gradation of the various features extracted. It is evident that speakers having smile on their face through the length of talk, using their hands effectively showing complimenting gestures and with good vocabulary/voice modulation are classified as good speakers those have become popular as well. This defines class 'A.'

4 Experimental Results of the Combined Audio, Text and Video (ATV) Content Analysis

We analyzed speeches of 217 speakers from TED talks to arrive at a partitioning of the combined feature space of audio, text and visual features. This section presents an experimental validation of the approach through running test video clips of speakers of TED talks not previously used to derive the rules. We validate the broad category against the 'popular' or 'not-so-popular' tag from the TED cell and proceed to present a detailed analysis of various aspects of the speech, followed by a recommendation, if there is room for improvement.

Test Speech 1: "Your body language may shape who you are."

Speaker—Amy Cuddy (Social psychologist); the talk was delivered in September 2012.

Some talks are analyzed in detail by a human behavior consultancy group called 'Science of people' to understand and analyze what drives the popularity of TED talk speakers. According to their results, which is mostly based on the views garnered, Ms. Amy Cuddy scores high and finds her a place in their top league. She receives an astonishing 55,420,269 views and belongs to their 'Most popular ted talks of all-time' list. Let us understand where Amy stands according to our research. Tables 2

Table 2 Audio parameters for test speech 1 by Amy Cuddy

Speaker 1	STE Joule/meter square	SD Pascal	ZCR kHz	Mean Power DB	Magnitude	Inconsistency coefficient
Ms. Amy Cuddy	0.92	0.10	36	70	0.06	0

and 3 are derived from a graph of amplitude plotted against time which elaborates with numerical results against each parameter.

Video Frames analysis.

No-HG → No-Hand gestures H-G → Hand Gestures (Fig. 3).

Table 4 presents an analysis of video frames for facial expressions, hand gestures and speaker pose classification using CNN classifier. The entries indicate number of frames classified for each emotion, hand gestures and poses and percentages are indicated for the most significant entries.

Table 3 Text parameters for test speech 1 by Amy Cuddy

Analytical	Confident	Tentative
0.76	0.73	0.33
Positive	Neutral	Negative
17.7	70.7	11.6

Fig. 3 Visual expressions of an accomplished TED Speaker captured by video frames

Table 4 Video frame analysis

Facial expressions			Hand gestures		Pose classification				Class
Happy/smiling	Sad	Miscellaneous	Expressive/Hand Gestures	No-Hand Gesture	Front	Side	Back	Total frames	A
268 (~67%)	50	87	287 (~70%)	118	187	213	5	405	$A_1T_1V_1$

Table 5 Overall analysis for audio, video and text features

Audio A			Text T						Video V				
STE	SD	Po	P	Neu	N	A	C	T	Happy	Sad	Misce	HG	Recommended class
0.92	0.10	69.3	70.7	17.7	11.6	0.76	0.73	0.33	66%	12.3%	21.4%	70%	A1 T1 V1

4.1 Combined ATV Analysis

Table 5 presents the results of the overall analysis for audio, text and video features. Following are definitions of abbreviations used in the table.

IC-Inconsistency coefficient, STE-Short-Term Energy, SD-Standard Deviation, Po-Mean Power, P-Positive, Neu-Neutral, N-Negative, A-Analytical, C-Confidence, T-Tentative, Facial Expressions—happy, sad and miscellaneous and HG-Hand Gesture.

Ms. Amy Cuddy is categorized as Class-A based on the experimental results. From the above Table, it is clear that audio features for Ms. Amy Cuddy including energy, standard deviation and power are at higher levels and at text level, she is having higher positive tone. Speaker having 66% of happy frames throughout the speech probably is a necessary criterion to be a class-A Speaker. The values in the table indicate that the speaker is consistent throughout the speech, so no further recommendation for improvement is required. The audio, text and video classification results are $A_1 T_1 V_1$, respectively, which means audio, text and video features considered are at the highest level as defined in Table 5, So the experimental results are in line with the audience survey and research conducted by Human behavior consultancy as well. As explained earlier, following are the values defined for class-A speakers.

4.2 Automated Talk-Quality Valuation (ATV) Based Recommendation

Our research is based on the analysis of the TED talk performances with regards to three aspects of a speaker's presentation: audio, text and video. The combined analysis of speakers of all three classes A, B and C showcase the values for their performances under audio, text and video parameters. With the above evidence, it is clear that though the topic and other few factors have important roles to play from audience perspective, voice modulation with good control on the volume, good vocabulary with positive phrases complimenting with positive body language share an equal percentage, in making any performance a popular one.

Hence, any speaker in class-B or Class-C considering their score in the fields of performance as scientifically projected by our program can work on further to improve their ability on various aspects to become one of the better presenter in the field.

Sample recommendations for various components of the ATV Classes and structure of the recommendations for various categories that are strung together from the components are presented in Table 6.

Table 6 Structure of recommendation system

ATV Label	Recommendation
A_1	The audio is audible, with good modulation and vocal consistency
T_1	The language and tone indicate confidence and provoke the audience to think with analytical content
V_1	The video content presents a pleasant expression for __% of the frames, i.e., most of the frames with plenty of hand gestures and change of pose to connect with the audience
Class A	Congratulations on a high quality speech! __________ As such, you are already an excellent speaker whose speech is likely to engage the audience
$A_1T_1V_1$ (Class A)	Congratulations on a high quality speech! An analysis of the speech reveals (a) the audio is audible, with good modulation and vocal consistency (b) the video content presents a pleasant expression for ___% of the frames, i.e., most of the frames with plenty of hand gestures and change of pose to connect with the audience and (c) the language and tone indicate confidence and provoke the audience to think with analytical content. As such, you are already an excellent speaker whose speech is likely to engage the audience
A_2	The audio could be marginally for louder volume and greater vocal consistency. Playing back a recorded version of the speech may help identify specific portions that need work
T_2	The use of a tentative tone and words such as 'perhaps,' 'like' and 'maybe' could be mitigated
V_2	The video content is good. It has been found that a pleasant expression helps to win over the audience. Try to smile more often or feel free to emote. Practicing before a mirror and visualization techniques may help you be more relaxed on stage
Class B	Well done on the solid talk! ____________ With a little more work, you would be among the best speakers!
A_3	The audio content needs a little bit of work; it has been found speakers who are audible and whose speech is intelligible win the audience's attention. You could try to record and playback your speech to identify portions that need a little more effort. Vocal exercises would also help enhance the volume and improve the range and consistency
T_3	The tone seems tentative and use of certain words such as 'perhaps,' 'like' and 'maybe' could be mitigated. Being confident in tone and causing the audiences to think would engage your listeners
V_3	Body language plays an important role in communicating with the audience. Do make an effort to smile more and use your hands to gesticulate for emphasis or help paint a visual picture. Practicing before a mirror and visualization exercises may help to relax and be more expressive on stage
Class C	Kudos on the solid effort! It takes courage to speak before an audience! Here is some analysis and suggestions for improvement: __________ With deliberate planning and a little more practice, you will get there!

5 Conclusion

In summary, our system takes as input a snippet of a speech, we analyze the behavior and ability of a speaker against the pre-analyzed patterns of speeches of orators from the TED database. The system analyzes feature in three modalities: (1) visual content, (2) text and (3) audio. We have found that it is not only the visual appearance including facial gestures, emotions, pose of the speaker which makes a speaker more effective, but also audio quality and content of the speech, types of words used, vocabulary of the speaker etc. Our study is based on analysis provided by human psychology expert as well as factors we have come up with it plots each modality in to category and combination adjudged to belong either one of classes A, B or C and provide suitable recommendations on improving various facets of the speech to be a more effective orator. Thus, the recommendation is a holistic one focusing on all these aspects.

References

1. Khoury P (2018) The top 9 characteristics of effective public speakers
2. Sharma R, Guha T, Sharma G (2018) Multichannel attention network for analyzing visual behavior in public speaking. In: Proceedings—2018 IEEE Winter Conf Appl Comput Vision, WACV 2018, vol. 2018-January, pp 476–484
3. Iftekhar Tanveer M, Liu J, Ehsan Hoque M (2015) Unsupervised extraction of human-interpretable nonverbal behavioral cues in a public speaking scenario, MM 2015 - Proc. 2015 ACM Multimed Conf, pp 863–866
4. Zeng H et al. EmoCo: visual analysis of emotion coherence in presentation videos
5. Yuan L, Chen Y, Fu S, Wu A, Qu H (2019) SpeechLens: a visual analytics approach for exploring speech strategies with textural and acoustic features. In: 2019 IEEE International Conference on Big Data and Smart Computing, BigComp 2019 - Proceedings
6. Schwemmer C, Jungkunz S (2019) Whose ideas are worth spreading? The representation of women and ethnic groups in TED talks. Polit Res Exch 1(1):1646102
7. Alisonprato (2015) TED Blog. Stories for 'Science of people'
8. Khajornphaiboon P, Vungthong S (1984) Analyzing the introduction of TED talks : a corpus—based analysis of discourse organization, pp 52–70
9. Lapakko D, Lapakko D (2007) Communication is 93 % Nonverbal: an urban legend proliferates, vol 34, no. January, pp 7–19
10. Hassan K, Gildea D, Hoque ME (2018) Predicting TED talk ratings from language and prosody

Effective Resource Allocation in Network Slicing Using Artificial Intelligence

Anushka Svara and Ramkumar Jayaraman

Abstract Research in the fifth generation (5G) of the mobile communication system has been gaining momentum in order to reinforce mobile systems and, hence, gratify the growing consumer and business requirements of 2021 and beyond. In the era of air interface technology and smart devices, mobile Internet user density and traffic volume is surging at an exponential number. Hence, the burden of rapidly expanding capital expenditures and operating expense besets the mobile operators. There is a dire need to remodel the resource allocation techniques of network slicing. Artificial intelligence methodologies can be a game-changing team player in achieving this. This paper presents those different methodologies and the upcoming challenges that can be addressed.

Keywords 5G · Radio access network · Quality of service (QoS) · Network slicing

1 Introduction

In past 40 years, mobile communication networks have advanced from the first to the fifth generation, revolutionising various different features of human communication and transaction taking place in our society. The fifth-generation standard for mobile networks is being explored to gratify the ever-growing needs of the consumer as well as the vertical industries. Beyond 2020, the 5G communication system is expected to fulfil and satisfy the enhanced demands of business, consumers and services. It is anticipated to substantiate a plethora of services, applications and use cases and yield a superior quality of network connection. With the advent of an unusually huge number of new forms of services and newly linked devices, the 5G communication networks are increasingly becoming more complex and cumbersome [1]. Support is to be provided by 5G communication network to a variety of other areas as well, such as automotive, construction, energy, environment, healthcare, logistic,

A. Svara · R. Jayaraman (✉)
Department of Computing Technologies, SRM Institute of Science and Technology, Kattankulathur, Chennai, India

R. P. Mahapatra et al. (eds.), *Proceedings of International Conference on Recent Trends in Computing*, Lecture Notes in Networks and Systems 341,
https://doi.org/10.1007/978-981-16-7118-0_40

manufacturing, apart from huge user equipment, enhanced traffic in the network and increased QoS needs of voice/data services.

Earlier communication system fifth-generation (5G) networks were made in such a fashion so as to attain 1000 times catered to all types of services from one common network, whether needed by different user cases, application type, data requirements/speed/error rates/efficiency, etc. The design of enhanced capacity of existing, five times reduced latency and ten times enhanced battery life. This development is mainly due to enhancement of user-requests from mobile applications, and the daily increased traffic conspired worldwide through billions of mobile wireless devices [2]. One common network architecture may not be efficient enough to harness and make optimum use of 5G and beyond communication systems.

The virtualisation of the network provides a suitable and enduring solution called 'network slicing', which supports a diversity of service and offers distinct resources to each individual service [3]. Network slicing permits the operator to split the network over a shared physical infrastructure, in an organised, flexible, expandable and automated fashion. With the advent of network slicing, the existing infrastructure could be virtually bifurcated into various different network slices, to suit performance, bandwidth, reliability, and latency requirements for optimum resource booking and enhanced incorporation into a multiplex signal. The next-generation mobile network alliance (NGMN) gave the definition of network slicing as multiple autonomous logical networks running on a common physical infrastructure as independent business operations [4]. A separate and virtual end-to-end network service, as requested by each individual network, allows network operators to execute multiple different architectures parallel. The 5G communication network should be capable of satisfying various and different services requirements of a fully connected smart society, viz. enhanced mobile broadband, massive machine-type communication and ultra-reliable low-latency communication [1]. Table 1 demonstrates these use cases, characteristic examples, and the significant diversity of their associated requirements.

The network slicing shares the network resources most efficiently, in order to handle complex and various different multiple service requirements, and it is also vital that this is achieved in an economical and cost-effective manner. New entrants in the mobile network environment are accommodated by the network slicing, confirming that infrastructure providers and network slice tenants have a separate entity. Network slicing can be configured to back individual mobile services, thereby giving it more

Table 1 Major 5G use cases

Case	Application	Requirements
Ultra-reliable communication	Autonomous vehicles, traffic navigation, industrial automation	1 ms latency, 10–9 error rate
Massive machine-type communication	Smart wearables, trackers and sensors	1 million machine/km^2, high energy efficiency
Enhanced mobile broadband	Broadband connection in dense areas, 4 K/8 K video streaming	Peak speed 50 Gbps, edge area 100 Mbps

resilience than radio access network (RAN) sharing techniques in 4G communication networks. This survey work can be referred to know about the use of artificial intelligence, using machine learning (ML) for automation of slicing function, in a 5G communication network. The paper has been structured as per the sequence mentioned here: Sect. 2 describes the network architecture based on network slicing in 5G. Section 3 illustrates the process of network slicing. Section 4 highlights the importance of network slicing in 5G communication system, Sect. 5 gives essential prerequisite for slice-based resource allocation, Sect. 6 gives a brief overview of design and structure of virtualised networking environment, Sect. 7 gives effective and efficient way of distributing the transportation and operational burden of a network slice, and Sect. 8 concludes the paper, mentioning future communication network technologies after 5G, the roadmap to next-generation wireless network (NGWN)—cable-free.

2 Network Slicing

2.1 Overview

Network slicing concept has been put forward to support diverse requirements of the vertical industries and booming traffic demands from the consumers. The basic notion behind network slicing is the virtualisation of the traditional physical network into several logical end-to-end networks. Each network slice comprises mainly three components, namely network functions, portion of core network resources and RAN resources. The owner of the network slice, a service vendor, is known as a tenant. According to the network slicing concept, every individual tenant is provided in Fig. 1. Network slicing for 5G with a localised control unit is proficient enough for administering the network slice which is reliable and secure in compliance with service of a particular quality of service (QoS) and type standards comprising latency, data rate. The network slicing comprises of the following salient features:

Flexible, adaptable, expandable and extendable: to assist in fulfilment of a large variety of requirements for different types of services and quality of services (QoS) and to help in expanding and extending of slice management after the placement of slices; As on request, network slices can be created, amended or cancelled as and when required, so network slicing basically enhances the expandability and extendibility in network management [5].

Adjustable and Automated: to assist in automatic resource allocation to cloud network and RAN and adjustment depending upon network performance and data traffic and to help in automatic creation, adaptation and performance monitoring of network slice.

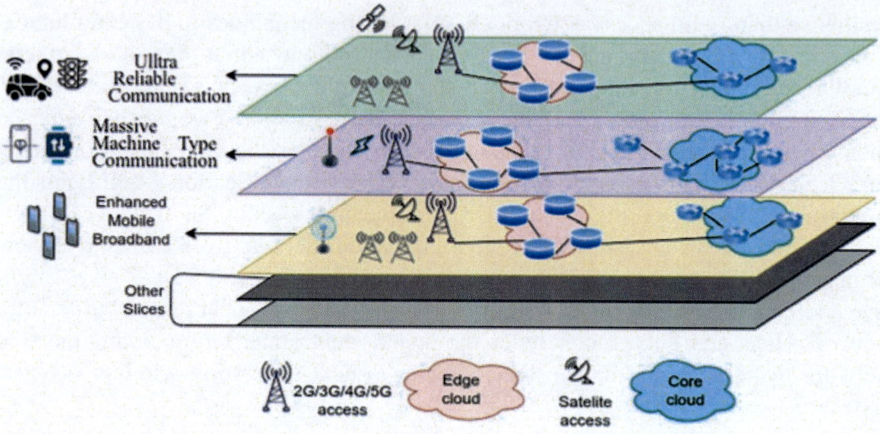

Fig. 1 Network slicing for 5G

Open and modularity: to assist in defining or operation of custom-made network slices by virtual network operator (VNO) and opening of some network management functions to third parties.

Multi-occupancy: T-network slicing, by way of using multiplexing of virtual networks, helps in sharing of the same physical network infrastructure by multiple virtual network operators, known as multi-occupancy [6], and thereby reducing capital expenditure (CAPEX) in deployment and operation of the network the same physical network infrastructure [6], and thereby reducing capital expenditure (CAPEX) in deployment and operation of the network.

Customising: network slicing offers the capability to create custom-made network slices for various service types having different quality of service (QoS) requirements, and for each type of service, can obtain differentiated service and secured service-level agreement (SLA).

Speedy, certain and reliable resource allocation is quite tough and difficult task, basically due to the certain aspects. Precise functionalities are required by individual network slices, possibly performed in a specific order. Network slice functioning includes and needs close coordination of many active players like cloud provider, slice owners and network controller. The bandwidth and processing power are the two types of resources required by an individual network slice. And between these two elements, the resource allocation must be balanced. At the same time, the idea of fairness and efficiency becomes much more complex. Network slices can be considered to be a group of virtual networks all interconnected. Essential prerequisite for slice-based resource allocation:

Configuration: one of the key objectives is to allow lodgers to customise functionality and resource allocation of individual slices to meet the customer's requirements.

Clearly defined interfaces must be given so as to allow lodgers to actively adjust their network slices allocations to fulfil space–time varying demands of the customer.

Intricacy: the cost of enforcement, intricacy and execution should be kept as less as possible. The excessive cost may arise due to too much signalling related with active customisation of network slices, their arrangement and demolition.

Productivity: in order to be economical, the communications through the network and resources used for calculation should be highly utilised. This will result in reduced expenses on the operational cost as well as the capital deployed. And, it is the outcome of versatile sharing.

Segregation and Protection: the main attributes of network slicing rule are that each individual network slice should be considered as 'a virtual network' separated from all other network slices and having guaranteed network resources available, and which are isolated from other network slices.

Secrecy and Confidentiality: it is vital that during sharing of infrastructure resources, there is no leakage of important information between individual slices. As secrecy is linked with isolation, so increased privacy may arrive at enhanced cost and/or loss of efficiency.

Cost Anticipation: a particular resource allocation model is chosen only if its cost can be foreseen beforehand. With reference to mobile services, expectations are for more authentic and foreseeable cost models.

3 Network Architecture Based on 5G Network Slicing

The network slicing can be divided into two steps. Creation of network slice—different exaggerated services having unique quality of service (QoS) requirements send requests for creation of network slices to ensure independent service. Subsequent to receiving a request for slice creation, the network controller accepts or rejects the request, depending on network resources availability. Once accepted by the network controller, a new slice will be created based on cluster templates and network activity conditions. Arrangement of network resources—based on the service-level agreements (SLAs), the network resources shall be allocated to network slices. Since advent of real-time mobile services, such as AR, VR and autonomous driving, can consume multitasking resources (computer, communications and cache), a fragment will be shared with multiple network resources. A resource mapping algorithm shall be used for mapping of virtualised network resources on to the physical network infrastructure.

Network slicing means the placement of virtual network functions (VNFs) onto individual slices, limiting to the restrictions imposed by the physical infrastructure and conditions for quality of service (QoS), the establishment of VNFs logical arrangement in each isolated network slice, and then, from the VNFs, mapping to the inherent physical network infrastructure.

There are numerous driving functions for network slicing [7]. One of them is software-defined networking (SDN). Both SDN and NFV are used to flexibly set up the virtual network resources, which include network bandwidth, network element processing capability, server processing capability, to construct the base network slices for particular service requirements.

SDN is premised on the key concept of the physical segregation of data handling in the forwarding plane from the routing process in the control plane. This provides an easy and efficient network configuration which enables flexibility. SDN influences the cloud computing model in the orchestration of networks, such a fashion so as to obtain an optimised performance from the network having a centralised supervisor for dynamically steering and managing the flow of traffic and orchestrating distribution of network resource allocation [8]. For creating a slice network, the SDN controller offers the outlined resources and control logic, and that isolated network slice may be considered as an SDN client [9]. Based on the service requirements and characteristics, the SDN helps in the establishment of slice schematics as well as the generation of on-demand slice instances.

NFV is based on the concept of segregation of network functions from the physical network apparatus, and virtualisation of these network functions into that is linked together to create a particular kind of communication service. All network functions such as address translation, firewall, load balancing are implemented by network function virtualisation (NFV), as software cases, called as virtual network functions (VNFs), which run on virtual machines over standard servers (NFV nodes) without the need for any special hardware [10, 11]. Network service in network function virtualisation (NFV) can be considered part of a slice, whereas one or more VNFs are contained in a network slice [12]. In the deployment of any network slicing, the NFV supplements the SDN, as the control plane functions get established through SDN, which enable network slicing, while provisions for services are made by NFV, which also control the life cycle of any network slice and organises slice resources through realising VNFs [13].

3.1 *Major Stakeholders in Virtualised Networking Environment*

In the 5G communication network, the unique design concept introduced is network slicing. The virtualised networking environment framework is depicted by the following three stakeholders [14]:

1. Virtual network operator
2. Infrastructure provider
3. Application service provider

The virtualised resources are quite flexible and can be dynamically amended (i.e. increased or decreased). Unlike conventional networks, the virtualised network slices can harness a great deal of computation and storage resources from the network nodes also. Besides the usual data forwarding functions, the functions such as media transcoding, caching, directory service in network data processing and traffic engineering are hosted by the virtualised network nodes (VNNs). Depending on the demand for the scaling of network functions, fresh nodes are introduced or abolished or their resources enhanced/reduced dynamically.

3.1.1 Virtual Network Operator (VNO)

The VNO analyses the requirements of the application's QoS and specifies the protocols, functions and capabilities of the supporting network. QoS application requirements are interpreted according to network performance metrics such as network bandwidth, network latency and packet loss rate by VNO. A virtual topology gets mapped by the requirements, containing the needed amount of networking resources and computing. For providing the needed network resources and leasing the required amount of virtualised resources, the request is sent by VNO to the infrastructure provider.

Installation of necessary network functions, protocols, customer registry and software platform on the leased virtualised resources is done by the virtual network operator (VNO). Virtual network operator (VNO) regularly monitors the performance after deploying the virtual network for the application service and checks if currently allocated resources are in appropriate amounts to execute the necessary network functions in order to process the given workload.

3.1.2 Infrastructure Provider

The infrastructure provider maps the topology of the logical network onto the virtualised resources of the base network, also referred to as virtual network embedding (VNE). The infrastructure provider allocates the virtualised resources demanded and also provides the corresponding management interfaces and resource control to virtual network operator (VNO). The infrastructure provider utilises tools for node and network virtualisation (e.g. OpenFlow and OpenStack) for monitoring and adjusting resources. The foremost open-source SDN controller for building next-generation SDN/NFV solutions is Open Network Operating System (ONOS), which supports both real-time network configuration and control, eliminates the need to run routes and change control protocols within the network fabric.

3.1.3 Application Service Provider

Using virtual network slices that is managed and operated by virtual network operator (VNO), the application service provider offers application services such as automated driving and smart metering, to customers. For this purpose, it provides the virtual network operator (VNO), the service requirements such as number and types of devices to be connected, its application type and reliability needed. Effective and efficient way of distributing the transportation and operational burden of a network slice:

1. To auto-scale and make provision for network slices in real time.
2. A utility function to reconfigure based on the cloud's processing power capacities and network's bandwidth and clearly express the issue of resource allocation in a network slice. The designer, taking clue from the solution, could drive the system in any way, swapping between computing-fairness and traffic-fairness.
3. Authenticating the iterative algorithm's (ADMM) speedy convergence, in quasi-stationary and dynamic settings, for automatic scaling of slices based on variable workloads or re-engineering fresh new slices on-demand.
4. The iterative algorithm, ADMM, is proposed to solve complex resource allocation problem in network slicing. The ADMM assimilates the basic roles of stakeholders, viz.:

   (1) *the cloud controller* (to allocate processor resources for serving incoming network flows).
   (2) *the network controller* (to allocate bandwidth to network slices and make provision end-to-end paths interconnecting cloud locations); and
   (3) *the slice owners* (requires scaling of the slice service to admit user sessions);

Alternating direction method of multipliers (ADMM)-based solution schemes endeavour to break a convex optimisation problem into smaller ones, which then become simple to solve. It accurately enables the intermediate level of centralisation, thus providing significant performance benefits during the merger.

4 AI Support in Network Slicing for Resource Allocation

The three main techniques in AI, which are coming out in wireless communication and recently having good successes in complex decision--making, resource optimisation, detailed knowledge discovery for complicated wireless networking environments and wireless network management, are machine learning (ML), deep reinforcement learning (DRL) and deep learning (DL) [15].

A lot of research and development work are going on at a faster pace in machine learning, considered one of the most potent AI tools, for deployment in self-driven autonomous vehicles, voice recognition and image processing. Speedy developments in ML, further enhanced by advancements in technology of hardware, paves the way for implementing AI for NGWN [16]. The key benefit of using ML is its capability to solve complex problems, which makes ML a strong tool for incorporating salient features of NGWNs, viz. decentralised, heterogeneous and dynamic characteristics. Obtaining faster convergence and improved performance in network performance optimisation and management automation in large scale systems is some of the possible advantages of implementing ML. The network's computing resources, communication, and caching or combined problem of content delivery and caching [17], can be studied for allocating them jointly for network resource management, using ML-based techniques and methods [18].

4.1 AI-Based Network Slicing

For precise and accurate forecasts for service-oriented traffic, AI-based methods can be aptly utilised. In coming future, the adjustments in network service requirements for resource allocation shall be effectively managed through RAN slicing, for such precisely forecasted service-oriented traffic only. Recent studies have shown that methods based on AI, for network slicing, are quite adapt at the precise prediction of service-oriented traffic burden, viz. long short-term memory and deep neural network; for example a DNN can be used in cellular networks for forecasting aggregated data traffic based on historical service demands. On the basis of a modified LSTM network for type fine-grained service-specific traffic, a forecasting model is provided to precisely forecast the average traffic burden, whereas for peak service traffic forecasting, a deep learning structure is suggested to assist in reducing over-provisioning of resources and violations of service-level agreements (SLAs).

Two-tier controllers are employed in AI-assisted network architecture, viz.

1. Rational centralised SDN controller located at the central cloud.
2. Native SDN controllers at the specific RANs.

As evidenced by the extensive research works lately, some of the complicated resource management problem areas, in wireless networks, where AI-based approaches can be widely used in solutions are:

1. Space air integrated networks computing work offloading [19].
2. Cellular networks SBS on/off scheduling [20].
3. Cloud radio access networks (CRANs) resource block allocation [21].
4. Interference management power allocation [22].

The resource allocation problem [23] and [24], as a whole [25], is developed as a Markov decision-making technique [26] and [27] for making online decisions, and for the Markov decision process (MDP) problem, a reinforcement learning (RL) framework is developed. RL-based methods can be used with the objective of optimising network performance under the restrictions imposed for satisfying quality of service (QoS) requirements, as the RAN slicing can be considered a problem of efficiency.

4.2 Different AI Methodologies Being Used in Network Slicing

By providing accurate predictions of service traffic, AI-based methods can be used. In future, RAN slicing can adequately assist in the allocation of network resources to address service needs, with such speculation of a particular service only. Recent studies have shown that AI-based network cutting techniques are able to accurately predict service-related traffic loads, such as long short-term memory and deep neural network. For example DNN can be used to predict aggregated data traffic on mobile networks based on historical service requests [20].

A forecasting model based on an amended LSTM network is proposed to precisely forecast the average traffic load, for fine-grained service-specific traffic, while a deep learning framework is suggested for the peak service traffic forecasts, which can facilitate in reducing resource over-provisioning and SLA violations. There is software that sets up power and controls a piece of network. The 5G network networks have a natural ability to create desired network fragments with good resources, utilising software-defined networking technology and network function virtualization [1] (Table 2).

5 Conclusion

The unique challenges facing RAN-based AI are such as obtaining solid QoS certification within the RL framework. How to achieve QoS limitations in the RL framework requires creative and innovative solutions to the use of RAN slicing. The QoS requirements usually are merged into the reward function, due to constraints of the Q-value limits modelling, by some predefined weights [20]. Especially, where QoS requirements are of varying scope, it is difficult to find the right tools for QoS requirements, so in this way, ensuring specific QoS requirements can be challenging or difficult. Most current solutions can only satisfy the soft requirements of QoS. Further research is needed to develop an effective RL-based slicing methodology, to meet the strict QoS requirements.

In a network operation, challenges are being posed for processing information of very huge traffic volume, sustaining against all potential security threats and yielding to adjust and accommodate the workload which is changing with time, efficient resource allocation to help facilitate the needs of virtual network slices created for providing optimum services to various different types of users and applications. It can be very challenging to monitor and control the SLA of the network piece,

Table 2 Network functions and relevant machine learning techniques

Network functions	Machine learning techniques	Purposes
Design and planning	Gradient boosting decision tree, support vector machine, spectral clustering, reinforcement learning	Classification of service requirements Forecasting trend, user behaviour Configuration of parameters
Working and management	K-mean clustering, DNN	Clustering cells, users, devices Reinforcement learning Executing decisions for dynamic resource control, policy formulation Reconfiguration of parameters Routing, forwarding, traffic control
Monitoring	Spectral clustering, K-mean clustering, SVM, DNN	Clustering of syslog data Classification of operation modes Forecasting resource utilisation trend
Exposing defect	Principal component analysis, independent component analysis, logistic regression, Bayesian networks	Classification of operation data Detection of network anomaly Predicting unusual behaviour
Security	DNN Principal component analysis	Clustering users and devices Detecting malicious behaviour Intrusion detection

and the acceptance of the network piece depending on the variability of network traffic and flexibility, as resource allocation between the slice, is based on slice segmentation. The challenges mentioned above can be solved by following standard model-based methods, which generally involve making perfect decisions in a continuously changing environment with unidentified information. Therefore, additional developments and innovations are needed in the network architecture to address the above challenges.

References

1. Kafle VP, Fukushima Y, Martinez-Julia P, Miyazawa T (2018) Consideration on automation of 5G network slicing with machine learning. In: 2018 ITU Kaleidoscope: machine learning for a 5G future (ITU K), Santa Fe, Argentina, pp 1–8. https://doi.org/10.23919/ITU-WT.2018.8597639
2. Ji H, Park S, Yeo J, Kim Y, Lee J, Shim B (2018) Ultra-reliable and low-latency communications in 5G downlink: physical layer aspects. IEEE Wireless Commun 25(3):124–130. https://doi.org/10.1109/MWC.2018.1700294
3. Abbas K, Afaq M, Ahmed Khan T, Rafiq A, Song W-C (2020) Slicing the core network and radio access network domains through intent-based networking for 5G networks. Electronics 9(10):1710. https://doi.org/10.3390/electronics9101710

4. Hedman P (2016) Description of network slicing concept. Next Gener Mob Netw Alliance. https://www.ngmn.org/publications/description-of-network-slicing-concept.html
5. Guo T, Suárez A (2019) Enabling 5G RAN slicing with EDF slice scheduling. IEEE Trans Veh Technol 68(3):2865–2877. https://doi.org/10.1109/TVT.2019.2894695
6. Samdanis K, Costa-Perez X, Sciancalepore V (2016) From network sharing to multi-tenancy: the 5G network slice broker. IEEE Commun Mag 54(7):32–39. https://doi.org/10.1109/MCOM.2016.7514161
7. Leconte M, Paschos GS, Mertikopoulos P, Kozat UC (2018) A resource allocation framework for network slicing. In: IEEE INFOCOM 2018—IEEE conference on computer communications, Honolulu, HI, USA, pp 2177–2185. https://doi.org/10.1109/INFOCOM.2018.8486303
8. Yousaf FZ, Bredel M, Schaller S, Schneider F (2017) NFV and SDN—key technology enablers for 5G networks. IEEE J Sel Areas Commun 35(11):2468–2478. https://doi.org/10.1109/JSAC.2017.2760418
9. 5G network slicing using SDN and NFV: a survey of taxonomy, architectures and future challenges. Comput Netw 167, 2020. https://doi.org/10.1016/j.comnet.2019.106984
10. Ye Q, Zhuang W, Zhang S, Jin A, Shen X, Li X (2018) Dynamic radio resource slicing for a two-tier heterogeneous wireless network. IEEE Trans Veh Technol 67(10):9896–9910. https://doi.org/10.1109/TVT.2018.2859740
11. Zhuang W, Ye Q, Lyu F, Cheng N, Ren J (2020) SDN/NFV-empowered future IoV with enhanced communication, computing, and caching. Proc IEEE 108(2):274–291. https://doi.org/10.1109/JPROC.2019.2951169
12. Network Functions Virtualization (NFV) Release 3; Evolution and Ecosystem; Report on Network Slicing Support with ETSI NFV Architecture Framework
13. Ordonez-Lucena J, Ameigeiras P, Lopez D, Ramos-Munoz JJ, Lorca J, Folgueira J (2017) Network slicing for 5G with SDN/NFV: concepts, architectures, and challenges. IEEE Commun Mag 55(5):80–87. https://doi.org/10.1109/MCOM.2017.1600935
14. Miyazawa T, Jibiki M, Kafle VP, Harai H (2018) Autonomic resource arbitration and service-continuable network function migration along service function chains. In: NOMS 2018—2018 IEEE/IFIP network operations and management symposium, Taipei, Taiwan, 2018, pp 1–9. https://doi.org/10.1109/NOMS.2018.8406235
15. Mohammadi M, Al-Fuqaha A (2018) Enabling cognitive smart cities using big data and machine learning: approaches and challenges. IEEE Commun Mag 56(2):94–101. https://doi.org/10.1109/MCOM.2018.1700298
16. SDN/NFV empowered future IoV with enhanced communication, computing, and caching
17. Somuyiwa SO, György A, Gündüz D (2018) A reinforcement-learning approach to proactive caching in wireless networks. IEEE J Sel Areas Commun 36(6):1331–1344. https://doi.org/10.1109/JSAC.2018.2844985
18. He Y, Zhao N, Yin H (2018) Integrated networking, caching, and computing for connected vehicles: a deep reinforcement learning approach. IEEE Trans Veh Technol 67(1):44–55. https://doi.org/10.1109/TVT.2017.2760281
19. Cheng N et al (2019) Space/aerial-assisted computing offloading for IoT applications: a learning-based approach. IEEE J Sel Areas Commun 37(5):1117–1129. https://doi.org/10.1109/JSAC.2019.2906789
20. Ye J, Zhang YJA (2020) DRAG: deep reinforcement learning based base station activation in heterogeneous networks. IEEE Trans Mobile Comput 19(9):2076–2087. https://doi.org/10.1109/TMC.2019.2922602
21. Alqerm I, Shihada B (2018) Sophisticated online learning scheme for green resource allocation in 5G heterogeneous cloud radio access networks. IEEE Trans Mobile Comput 1–1. https://doi.org/10.1109/TMC.2018.2797166
22. Sun H, Chen X, Shi Q, Hong M, Fu X, Sidiropoulos ND (2018) Learning to optimize: training deep neural networks for interference management. IEEE Trans Signal Process 66(20):5438–5453. https://doi.org/10.1109/TSP.2018.2866382

23. de Veciana BG, Sciancalepore V, Costa-Perez X (2020) Resource allocation for network slicing in mobile networks. IEEE Access 8:214696–214706. https://doi.org/10.1109/ACCESS.2020.3040949
24. Habibi MA, Han B, Schotten H (2017) Network slicing in 5G mobile communication architecture, profit modeling, and challenges. ArXiv, abs/1707.00852
25. Pereira RS, Lieira DD, Silva MACD et al (2020) RELIABLE: resource allocation mechanism for 5G network using mobile edge computing. Sensors (Basel) 20(19):5449. Published 2020 Sep 23. https://doi.org/10.3390/s20195449
26. Rost P, Mannweiler C, Michalopoulos D, Sartori C, Sciancalepore V, Sastry N, Holland O, Tayade S, Han B, Bega D, Aziz D, Bakker H (2017) Network slicing to enable scalability and flexibility in 5G mobile networks. IEEE Commun Mag 55. https://doi.org/10.1109/MCOM.2017.1600920
27. Shen X et al (2020) AI-assisted network-slicing based next-generation wireless networks. IEEE Open J Vehicular Technol 1:45–66. https://doi.org/10.1109/OJVT.2020.2965100

Pulmonary Image Analysis with Computer Vision Using Binary Classification Method for COVID-19 Detection

S. V. N. S. Vaibhav Ram and Samiappan Dhanalakshmi

Abstract The advent of COVID-19 raised a terrorizing situation across the globe. The virus is spreading at an exponential rate since its beginning which was first identified in the Wuhan province, China, in the year 2019 in the month of December. The virus belongs to a family similar to that of the severe acute respiratory syndrome (SARS) which was identified in the year 2002. It has a complex structure which makes it difficult for scientists to get the exact action and cure for the disease. The spread of the virus takes place only by body fluids (saliva, mucosa, etc.). Various research organizations, pharmaceuticals, and institutes are working on the production of testing kits and vaccinations, though some of them are already being produced in the market. The testing kits produced are less in numbers due to the lack of resources and knowledge gathered to detect and fight the virus properly because of which not everyone is getting the chance to get themselves checked. Computer tomography and X-rays of the pulmonary region along with various acquaintance and methods of AI deep learning give an effective alternative that can be employed. This stratagem can be applied using a dataset consisting of images of various X-rays and CT-scans which are of patients who are COVID-19 positive and also of healthy people. This diagnosing tool uses the binomial classification method. The accuracy and the working of the tool primarily depend on accessible information and data for better processing. Post testing of the tool shows us that it is flexible and accurate to use.

Keywords Computer vision · COVID-19 classification · CT-scan analysis · X-ray analysis · Deep learning · Classification

S. V. N. S. Vaibhav Ram · S. Dhanalakshmi (✉)
Department of Electronics and Communications Engineering, Faculty of Engineering and Technology, SRM Institute of Science and Technology, College of Engineering and Technology, SRM Nagar, Chengalpattu Dt., Kancheepuram, Tamil Nadu 603203, India
e-mail: dhanalas@srmist.edu.in

S. V. N. S. Vaibhav Ram
e-mail: vs7961@srmist.edu.in

R. P. Mahapatra et al. (eds.), *Proceedings of International Conference on Recent Trends in Computing*, Lecture Notes in Networks and Systems 341,
https://doi.org/10.1007/978-981-16-7118-0_41

1 Introduction

In the year 2019, one of the most unexpected storm had started its journey and went on throughout the year 2020 and is continuing to threaten human existence. It is the COVID-19 (Corona Virus Infectious Disease—2019). WHO (World health organization) had identified the symptoms of the COVID-19 ranges from most common symptoms to serious symptoms. The most common symptoms include fever, dry cough, and tiredness; the least common one includes various symptoms like loss of taste, sore throat, and diarrhea; and the serious symptoms are loss of speech, chest pain, and pneumonia [1, 2].

This disease spreads via droplets that come out when people sneeze or cough. These droplets when consumed or inhaled causes COVID-19. The eyes, nose, and mouth are the primary gateways for the virus to get into the body. Once the virus gets inside the body, the host attacks the pulmonary system (lungs, mucosa membranes, the pulmonary tract, etc.), and in a period of 7–14 days, it starts showing the symptoms and rapidly multiplies the number of COVID cells inside the body [2, 3]. These patients exhibit lymphopenia, lymphocyte activation and dysfunction, granulocyte and monocyte abnormalities, high cytokine levels, and an increase in immunoglobulin G (IgG) and total antibodies [4]. This disease is more likely to go deeper than other viruses as the lower airways have more of the ACE2 receptors [5]. The COVID-19 is continuously evolving to this day. There are new strains identified in the Asian, European, and African continents which are not yet completely identified or analyzed.

The disease has got no particular cure but various treatment methods help to reduce the effects and also remove the virus completely out of the body. People with mild symptoms are usually quarantined in their homes for treatment with some prescribed medicines, whereas people with less common and serious symptoms are admitted into the hospitals for treatment using ventilators, multiple medicine doses, etc.

COVID-19 as of now has made a severe impact during its first wave, and the world has started going through the second wave of this disease. The analysis of the COVID-19 published by the WHO as of 20/3/2021 shows that there are a total of 123,859,209 coronavirus cases with a record of 2,727,680 deaths occurred and about 99,784,365 recovered cases across the globe. A total of 21,347,164 cases are still active and are getting treated. Daily a total minimum of 200,000 new cases are reported, and about a minimum of 4000 cases are closed due to deaths. The countries which are in an extremely alarming condition are the United States of America with a total of 30,000,000 + cases, Brazil and India with a total of 11,500,000 + cases, and various other countries like Russia, France, and Italy are also in same state [6]. These figures clearly show that we need to accelerate our process for helping us have a non-disruptive existence.

The number of people who are getting affected is increasing, and the number of test kits is reducing which gets us to a situation where we are not able to identify and treat all the people across the globe which brings in a factor of threat that there might be even more COVID-19 cases which might be unidentified and treated apart from the numbers mentioned by WHO. The hospitals are also unable to do the testing for every patient in a full-fledged manner to date.

Many institutes are still developing rapid testing kits for the hospitals to use but are also facing various constraints with regards to the effect the kits can show. Various other factors also make the currently supplied kits questionable such as the accuracy, the time taken to process and analyze, and the transparency regarding the result of the testing. A lot of country's economy is getting drained due to the amount put in R&D for new and better test kits. The world has started its supply of vaccinations, but still, the effect is not reducing as the vaccinations are taking time to reach globally and as mentioned earlier the factor of identification is also affecting.

Hence, the organizations across the globe are considering different approaches to get the testing done, one of them is COVID-19 detection using AI (Deep learning). Currently, deep learning is being applied in the healthcare sector in large amounts to solve various problem statements [7] scenarios such as for drug discovery [8], personalized medical treatment [9], and to maintain smart health records [10].

Here, we are using deep learning for disease detection. This can be termed as "pulmonary imaging for COVID-19 detection" using various image modalities such as X-rays (radiographs), magnetic resonance imaging (MRI), computed tomography (CT), and ultrasound (see Fig. 1).

Using various base networks, i.e., 3D CNN, CNN, U-Net and CNN, U-Net and ResNet-50, etc. [11], this will help us to detect COVID-19 with already existing resources as our primary inputs, the detection time is also reduced, and the accuracy

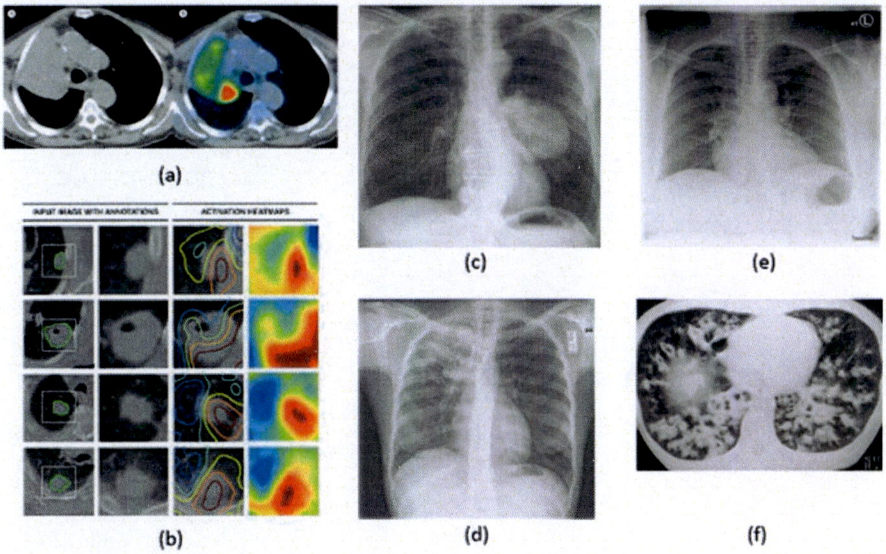

Fig. 1 Pulmonary imaging using various modalities. **a** CT image of pulmonary path way. **b** Deep learning analysis on CT-scan. **c** Image of lungs with cancer. **d** Images of lungs with COVID-19. **e** Image of lungs with tuberculosis. **f** Image of lungs with pneumonia

is determined which helps us to know how well the prediction is done for that particular image input.

2 Anticipated Classification Method

Here in, we are proposing an approach for COVID-19 detection using deep learning with a different method and process which was researched and analyzed by us using the binary classification for detecting COVID-19. The image modalities used are CT-scan and X-ray of the pulmonary system (i.e., lungs, pulmonary canal, etc.) and can be extended for other modalities too by making some small iterations in the program. Table 1 shows the methods which scientists have done this research and got numerous results for different modalities used.

Table 1 Deep learning application for medical image analysis to detect COVID-19

Authors	Image modality	Base network	Method
Butt et al. [12]	CT	3D CNN	It uses a 3D CNN for segmentation from CT scanning after preprocess. Uses noisy or Bayesian function
Wang et al. [13]	CXR	CNN	Using CNN to extract feature map with classification result, regression result, and the needed mask
Li et al. [14]	CT	U-Net and ResNet-50	CT-scan is preprocessed and lungs are extracted using ROI UNET AUC Recorded 0.967
Hasan et al. [15]	CT	CNN	Framework that uses histogram thresholding to isolate the background of CT-scan and then feature extraction
Kassani et al. [16]	CXR, CT	DenseNet-121, ResNet-50, Mobile Net, Xception, Inception-V3, Inception-ResNet-V2, VGG and NASNe	The extracted features were then fed into several machine learning classifiers to classify subjects as either a case of COVID-19 or a control. This approach avoided task-specific data preprocessing methods to support a better generalization ability for unseen data

The dataset used is a custom dataset that was made using Bing web crawler provided by Azure web services and Kaggle repository. The dataset contains five different sets which are combined COVID, normal, test, and sample. The combined set consists of images that have both COVID-19 and normal (chest X-ray and CT-scan images), the test set consists of the images which are going to be used as the testing data, and the sample set consists of the images which are the training data all these sets together consist of 3465 images.

The sample set is loaded into the program by using OpenCV and OS, respectively, and accessing them using the list directory command. Once the images are loaded, they are sorted based on the format whether they are jpg or png, and once sorted it is appended into a list for further process. Later, the dataset is formatted into a data frame, categorizing the images based on the labels whether it is COVID-19 or not by using 1 and 0 (COVID-19 -0 and normal-1). The CNN architecture consists of conv2D, batch normalization, Max pooling2D, and dropout layers with a percentage of 25% with relu and softmax as activation functions for removing the negative values from the data considering it is binary classification.

The optimizer is RMS prop, the loss is categorical cross-entropy, and the metrics used is accuracy. We are using the image data generator for the training process in the program because it helps to do image augmentation in real time even when the data is undergoing training. In the image data generator, we made it into a train, validate, and test data gen for augmentation and for the flow of data frame into these generators. The data trained is pushed into the predicted set. After all this process, the test data is finally visualized, and we get to know what the image is. Post-visualization, the part comes up which helps in detecting whether a scan or a modality of the lungs is COVID-19 positive or normal. We have extended the program to help detect the state of images taken from any XYZ source outside the datasheet we created. When an image is fed to the cell of the program, it gives the result in the form of 0/1 that is if it is COVID then it gives 0 and if it is normal then it gives 1 as the result. The flow of program (see Fig. 2). This final step is done by importing the PIL library and using the expand dimension for the image data augmentation as we do not know the dimensions of the image taken to our set dimensions according to flow (see Fig. 3).

3 Results and Discussion

The program was iterated multiple times and we got results that are good for the purpose to be fulfilled. As mentioned above, the metrics we used was accuracy, and the loss was categorical cross-entropy. The accuracy that we got with 3465 images was 0.88–1 that is 88–100%, and this can be changed if we use an even bigger dataset that has more than 5000 images. In computer vision, "larger the dataset more the training and more the training better the accuracy." The various trends of accuracy attained by various scientists' algorithms are given in Table 2. Ucar and Korkmaz have a better validation accuracy than us, but the modality used is X-ray, whereas CT-scans provide an additional advantage as they give the imaging of the bones, the

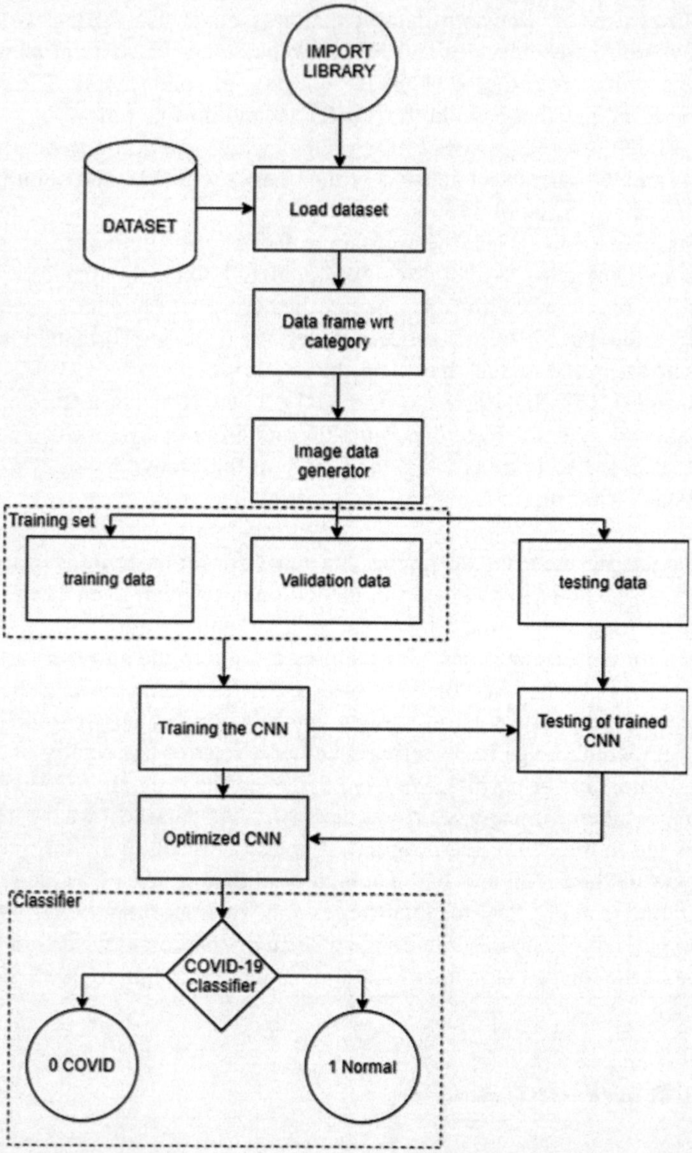

Fig. 2 Program flow process

tissues, and the vessels which is very important for COVID-19 detection and our method can be used for both X-ray and CT which gives an added advantage other the other methods. Post-visualization result shows the predicted output (see Fig. 4).

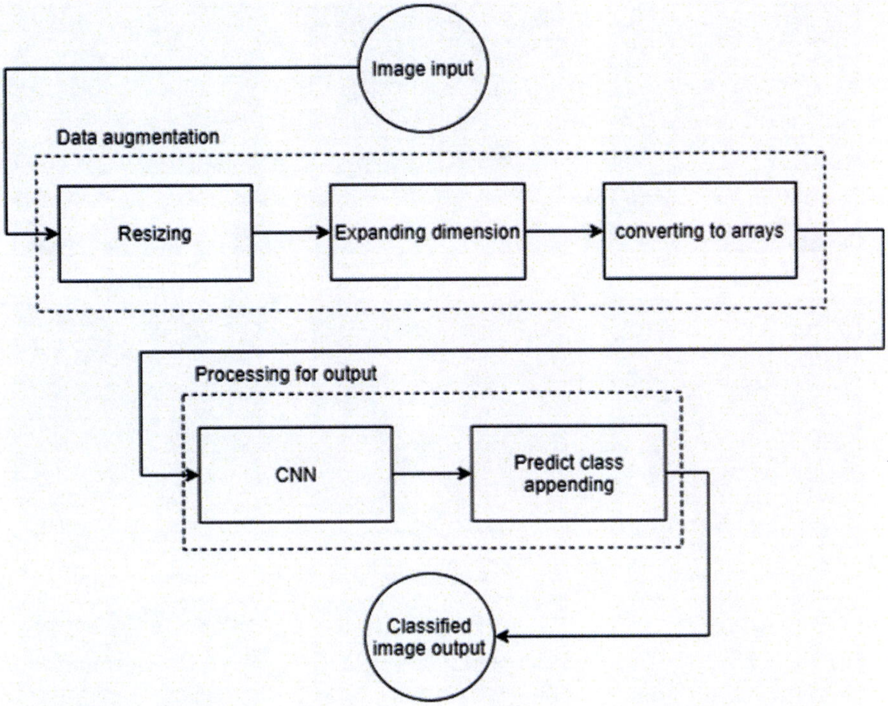

Fig. 3 Classification process

Table 2 Competitive metric analysis

Name of researchers	Modality used and architecture	Validation accuracy (%)	Test accuracy (%)
Wu et al. [17]	CT, ResNet 50	70	76
Ucar and Korkmaz [18]	X-ray, Squeeze Net	98	100
Butt et al. [12]	CT, ResNet 23	95	97.8
Our proposed classification	CT, CNN binary	88	100

4 Conclusion

The program framework for detecting COVID in X-rays and CT-scans is dependent on the information accessible for loading in. All these methods used are a fusion of various algorithms and various layers and activation functions which help in making an effective approach toward detecting COVID-19 without using the test kits but by using the already available resources such as the X-ray and CT-scan machines. The program not only shows good accuracy but also shows that it is flexible enough

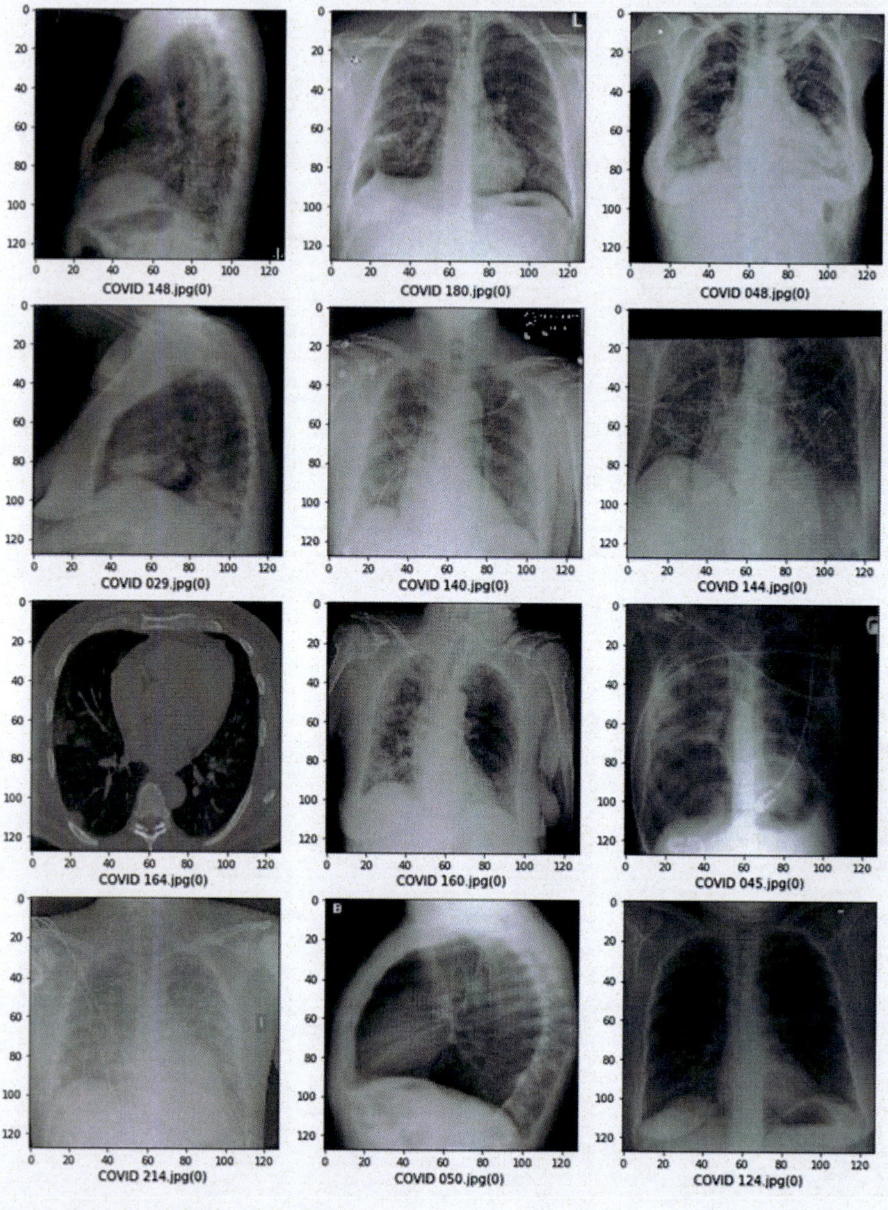

Fig. 4 Visualized COVID predicted data

to get altered for further advancements as the field of research and development is continuously evolving which might help this tool to be used for various other applications too.

References

1. WHO. (n.d.) (2021) Coronavirus disease data page. https://www.google.com/search?q=covid+19+cases+worldwide&rlz=1C1SQJL_enIN861IN861&oq=COVID+19+cases+world&aqs=chrome.0.0i131i433j69i57j0l8.9992j0j15&sourceid=chrome&ie=UTF-8
2. Cao X (2020) COVID-19: immunopathology and its implications for therapy. Nat Rev Immunol 20:269–270
3. Anthony S, Fauci MH (2020) COVID-19-navigating the uncharted. N Engl J Med 382(13):1268–1269
4. Yang L, Liu S, Liu J, Zhang Z, Wan X, Huang B, Chen Y, Zhang Y (2020) COVID-19: immunopathogenesis and immunotherapeutics. Sig Transduct Target Ther. 5:128
5. Bhargava HD (2021) Coronavirus: what happens when you get infected?. Lungs diseasse and respiratory health. https://www.webmd.com/lung/coronavirus-covid-19-affects-body#2
6. COVID-19 (2021) Corona virua pandemic. https://www.worldometers.info/coronavirus/
7. Gibson E, Li W, Sudre C, Fidon L, Shakir DI, Wang G, Eaton-Rosen Z, Gray R, Doel T, Hu Y (2018) NiftyNet: a deep-learning platform for medical imaging. Comput Meth Prog Biomed 158:113–122
8. Gawehn E, Hiss JA, Schneider G (2016) Deep learning in drug discovery. Mol Inf 35(1):3–14
9. Katzman JL, Shaham U, Cloninger A, Bates J, Jiang T, Kluger Y (2018) DeepSurv: personalized treatment recommender system using a Cox proportional hazards deep neural network. BMC Med Res Method 18(24):1–12
10. Shweta, Ekbal A, Saha S, Bhattacharyya P (2016) Deep learning architecture for patient data de-identification in clinical. In: Proceedings of the clinical natural language processing workshop. Osaka, Japan, pp 32–41
11. Farhat H, Sakr GE, Kilany R (2019) Deep learning applications in pulmonary medical imaging: recent updates and insights on COVID-19. Nat Public Health Emerg Collect 31(6):1–42
12. Butt C, Gill J, Chun D, Babu BA (2019) Deep learning system to screen coronavirus disease 2019 pneumonia. Appl Intell 1573–7497
13. Wang BW (2019) Deep convolutional neural network with segmentation techniques for chest x-ray analysis. In: 14th IEEE conference on industrial electronics and applications, pp 1212–1216
14. Li L, Qin L, Zeguo X, Yin Y, Wang X, Kong B, Bai J (2020) Using artificial intelligence to detect COVID-19 and community-acquired pneumonia based on pulmonary CT: evaluation of the diagnostic accuracy. Radiology 296:E65–E71
15. Hasan AM, Al-Jawad MM, Jalab HA, Shaiba H, Ibrahim RW, Al Shamasneh AR (2020) Classification of Covid-19 coronavirus. Pneumonia Entropy 22(5):517
16. Kassani SH, Kassasni PH, Wesolowski MJ, Schneider KA, Deters R (2020) Automatic detection of coronavirus disease (covid-19) in x-ray and ct images: a machine learning-based approach. arXiv preprint arXiv:2004.10641
17. Xiangjun W, Hui H, Niu M, Li L, Wang L, He B, Yang X, Li L, Li H, Tian J, Zha Y (2020) Deep learning-based multi-view fusion model for screening novel coronavirus pneumonia: a multicentre study. Euro J Radiol 128:10941
18. Ucar F, Korkmaz D (2020) COVIDiagnosis-Net: Deep Bayes-SqueezeNet based diagnosis of the coronavirus disease 2019 (COVID-19) from X-ray images. Med Hypotheses 140:109761

Bone Fracture Detection Using Image Processing Methods

M. Uma Devi, Sweta Nayak, and Aman Gupta

Abstract Automation is a profoundly arising subject of social, technical, and monetary significance. Automation will progressively affect the universe of work during the following not many years. Automation is changing medical services today, preparing for considerably more progressions to come. In medical care, automation could seem as though a day by day cluster of instant messages that help patients to remember their arrangements the following day, or on the other hand, computerization could conjure a dataset that distinguishes and cautions clinicians of known contraindications and medication associations. The bone is an essential piece of the body framing. Bone enables the body to move. The bone cracks are extremely ordinary ailment which emerges due to a mishap or bone sickness. The specialists utilize the X-shaft pictures to investigate the broken bone. The manual break disclosure system is monotonous, and furthermore, the mistake likelihood chance is high. In this manner, an automated structure needs to make to investigate the cracked bone. In this paper, we have attempted to plan a mechanized framework which decides if the bone is broken or not, utilizing the X-ray pictures. A X-ray picture is gone through the discrete wavelet transform (DWT) channel to decrease the commotion from the pictures, and the features are isolated from the photographs utilizing factual technique for inspecting surface that considers the spatial relationship of pixels called gray-level co-occurrence matrix (GLCM). Surface highlights determined utilizing GLCM are contrast, correlation, entropy, energy, and homogeneity. Classification of the X-beam images is finished by the backpropagation neural network (BPNN), which can precisely arrange if the bone is cracked or not. This automated framework will lessen the time and increment the proficiency to distinguish the cracked bone in clinical science.

M. Uma Devi · S. Nayak · A. Gupta (✉)
Department of Computer Science and Engineering, SRM Institute of Science and Technology, Kattankulathur, Chennai, Tamil Nadu 603203, India
e-mail: as4944@srmist.edu.in

M. Uma Devi
e-mail: umadevim@srmist.edu.in

S. Nayak
e-mail: sn4183@srmist.edu.in

R. P. Mahapatra et al. (eds.), *Proceedings of International Conference on Recent Trends in Computing*, Lecture Notes in Networks and Systems 341,
https://doi.org/10.1007/978-981-16-7118-0_42

Keywords Discrete wavelet transform (DWT) · Gray-level co-occurrence matrix (GLCM) · X-ray · Backpropagation neural network (BPNN).

1 Introduction

Various observations demonstrate that the majority of bone cracks happen because of mishaps. Different major and minor bone cracks happen because of a mishap or bone infections. Minor cracks are hard to recognize in the X-ray pictures by unaided eyes. The manual break discovery procedure is tedious, and furthermore, the mistake likelihood chance is high. Consequently, a computerized framework needs to be created to analyze the broken bone X-ray pictures. Programmed absconds location in X-beam pictures is vital in numerous indicative and helpful applications. During recent years, bone break division has become a developing examination region in the field of clinical imaging framework. Precise identification of size and area of bone break assumes an essential part in the analysis of break. The conclusion technique comprises of three phases, pre-preparing of X-beam pictures, highlight extraction, and classification. A picture is an exhibit or a grid of square pixels (picture components) organized in segments and rows. Image handling is a technique to work out some methodology on a picture, to get an improved photograph or to isolate some important information from it. It is a sort of sign planning where the data is an image and yield may be an image or characteristics/features related with that picture. After gathering the helpful data from radiographs utilizing picture handling strategies, classification methods are utilized to arrange the strange, cracked, or intermittence in the bone. The primary inspiration of this paper is to recognize breaks, injury, and unusual bones utilizing high exactness in less time. This structure can assist specialists with distinguishing the cracks and anomalies in bone radiographs with high precision in less measure of time additionally help to identify the minor cracks and irregularities which cannot be distinguished by the exposed eyes.

2 Literature Survey

2.1 Related Work

Article [1] proposed a backpropagation neural network (BPNN) along with Canny edge segmentation strategy and preservation sleeking procedure to identify the bone cracks. But, the proposed philosophy gives less precision, while distinguishing breaks in the bended bones additionally the blend of calculation expands the intricacy of the structure.

Article [2] gives the correlation concentrate between different classical picture division strategies like thresholding, edge-based division strategies. Also, pattern-acknowledge-based clustering or group examination.

Article [3] proposed another strategy for X-beam bone picture division in which the converse of the first X-beam picture applies to a limited item administrator and fluffy regulator to control the difference of the backward picture. Versatile thresholding of inclination extent of the picture is accomplished to recognize the edges in the X-beam pictures.

Article [4] proposed the deep learning model to group the crack and solid bone and uses the information enlargement method to defeat the over-fitting issue. Yet, the frame work does not get approved on the bigger dataset.

Article [5] proposed the two-line-based crack identification plot, adaptive differential parameter optimized and standard line-based crack recognition plans. Characterization is finished utilizing artificial neural organizations. The impediment of their work lies on the two focuses given to depict the lines.

Article [6] proposed the discrete wavelet transform (DWT) along with the thresholding methods to denoise the X-ray picture utilizing DWT-TM calculation.

Article [7] utilizes the gray-level co-event matrix for surface element extraction and for characterization utilizing logistic regression, support vector machine, and the decision support tool called decision tree. Proposed system can be improved more with various capabilities.

Article [8] proposed a summed up bone break discovery strategy that is material to various bone breaks types. Utilizations the highlights separated from up-and-comer patches in X-ray pictures in Stacked Random Forest Feature Fusion. The result of this strategy is various break bouncing boxes positioned from the well on the way to the least extent liable to contain a crack. Yet, the location precision could be also improved by combining more kinds of close by features.

Article [9] proposed a strategy to mandibular crack (break of lower jaw in mandibular bone) through orthopantomogram (OPG). They utilized gray-level co-occurrence matrix (GLCM) for including extraction, and crack identification is finished by dissecting picture surface utilizing a channel cover.

3 Proposed Methodology

The projected strategy comprises of three significant parts, in particular.

(1) Image preprocessing,
(2) Highlight extraction,
(3) Classification.

The square design of suggested philosophy is portrayed under in Fig. 1.

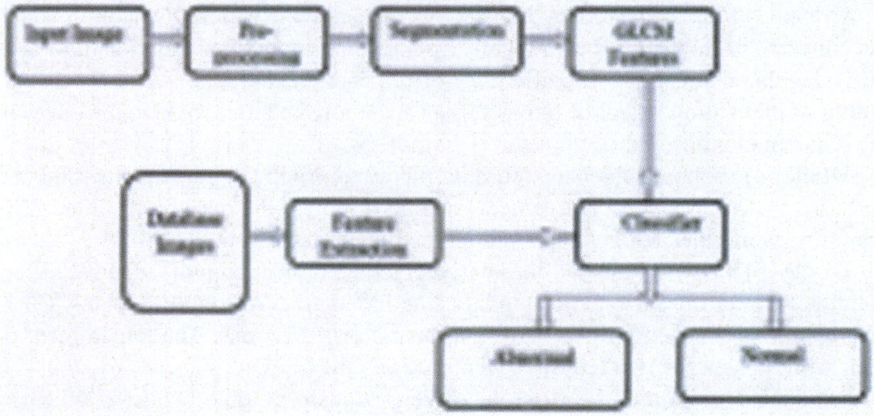

Fig. 1 Square layout of projected methodology

3.1 Picture Pre-treatment Picture

Pre-treatment is an average name for assignments with pictures at the most diminished level of deliberation. Its data and yield are power pictures. The mark of pre-planning is the development of the image data that smother unwanted mutilations or updates some image features huge for extra readiness. The point is to diminish the commotion in X-beam pictures to get precise highlights from the X-beam picture. To decrease the commotion in X-ray pictures, we utilized a novel denoising and clamor limitation methodology by utilizing discrete wavelet change (DWT). The low X-ray portion picture was deteriorated using the DWT. Subsequently, we used edge to take out low-power energy coefficients, and we switched DWT which reproduced the denoised X-ray picture.

3.2 Feature Extraction

The conspicuous confirmation of articles in an image would doubtlessly start with picture taking care of systems, for instance, upheaval removal, followed by (low level) incorporation to discover lines, areas and maybe zones with explicit surfaces. A picture is a cluster or a grid of square pixels (picture components) organized in sections and rows. Image preparation is a technique to apply some procedure on an image, to get an overhauled photograph or to extricate some significant information from it. Highlight extraction is such a dimensionality decline where endless pixels of the image are profitably tended to so that intriguing bits of the image are obtained sufficiently. For inclusion extraction, we utilized factual technique for inspecting the surface. The gray-level co-occurrence matrix (GLCM) uncovers certain characteristics about the spatial appointment of the faint levels in the surface photograph. For example, if by far most of the areas in the GLCM are concentrated along the corner to corner, the surface is coarse concerning the foreordained equilibrium.

3.3 Classification Model

Classification is characterized as the interaction of acknowledgment, comprehension, and gathering of articles and thoughts into preset classes. Classification calculations use input preparing information to foresee the probability or likelihood that the information that follows can be categorized as one of the foreordained classifications. For classification, we utilized backpropagation neural network (BPNN). Back-spread is the quintessence of a neural net getting ready. It is the strategy for tweaking the heaps of a neural net subject to the mix-up rate gained in the past age. Genuine tuning of the heaps allows you to decrease goof rates and to make the model strong by extending its theory. Back-inducing is a short construction for "in turn around expansion of errors." It is a standard method for getting ready phony neural associations. This method helps with registering the point of an adversity fill in concerning all of the heaps in the association. Works on the organization structure by components weighted connections that have minimal impact on the prepared organization.

4 Implementation

4.1 Discrete Wavelet Transform (DWT)

Beginning with the picture denoising, we utilize discrete wavelet transform (DWT). Exactly, when a sign is conveyed over to some distance, it is routinely contaminated by racket. The most untroublesome model for getting of upheaval by sign is added substance noise that has the design (Fig. 2).

$$q'(x, y) = q(x, y) + z(x, y)$$

where

$q'(x, y)$: polluted sign.

$q\ (x, y)$: unique signal.

$z\ (x,y)$: noise signal.

Picture $f(k)$ is deteriorated utilizing DWT into different wavelet coefficients (generally approximation coefficients and details coefficients). These details coefficients are thresholded by thresholding functions like universal edge (VisuShrink), Bayes Shrink, minmaxi, and afterward, we get thresholded details coefficients, and we take inverse transform to get a denoised picture (Fig. 3).

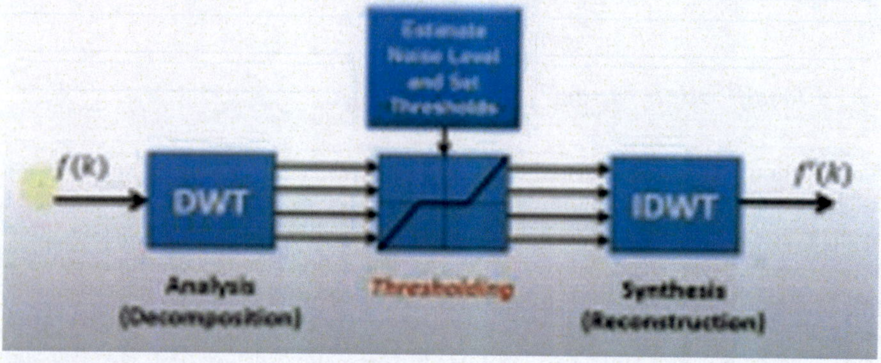

Fig. 2 Denoising of X-ray image using DWT

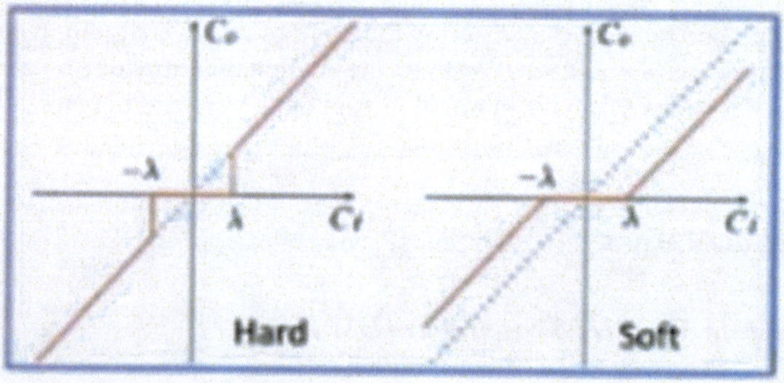

Fig. 3 Shrinked wavelet details coefficient

4.2 *Gray-Level Co-Occurrence Matrix (GLCM)*

Subsequent to denoising the X-beam picture, we separated the surface highlights from the denoised X-beam picture utilizing factual technique for inspecting surface, i.e., gray-level co-occurrence matrix (GLCM). A picture is made out of pixels each with an intensity (a particular gray level), and the GLCM is a classification of how often various blends of gray levels co-occur in a picture or picture segment. GLCM utilizes second request insights for highlights that can be utilized to surmise the correlation between the sets of pixels. For instance, taking two sets of pixels, one of every intensely surface territory and other in a perfect zone, and afterward, it tells the level of correlation between them. It utilizes the sets of pixels where the client can characterize the distance and angles between the pixels. Suppose, distance = 1 (it derives the correlation between the adjoining pixels), and if the distance is any longer, it gives the correlation at a more drawn out scale (Fig. 4).

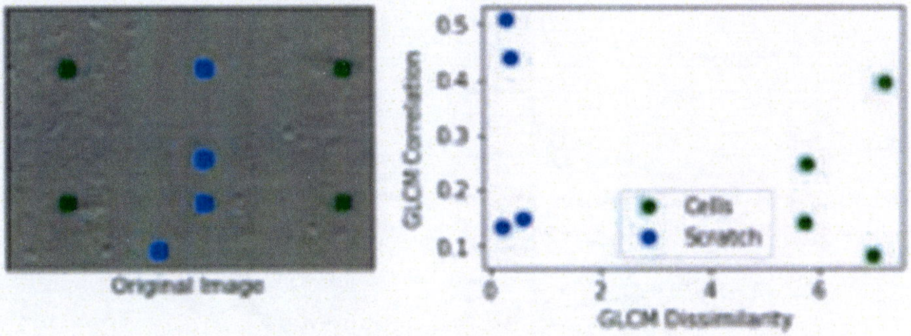

Fig. 4 Dissimilarity verses correlation plot

For extracting the correlation and texture features, we used haralick features method that uses the gray-level co-occurrence matrix and calculates 14 textural features. Those 14 different types of textural features are calculated based on some statistical theory. Normally, the feature vector is calculated in 13 dimensions only because computing the fourteenth dimension is very complex which increases the computational time.

4.3 *Support Vector Machine*

Support vector machine (SVM) is a very popular, supervised machine learning algorithm which is used for classification as well as regression analysis. Here, we used the support vector machine algorithm for the classification of normal and fractured bone images. It is basically used as a binary classifier which assumes that the data in the question contain only two possible target values. In the SVM calculation, we plot every information thing as a point in n-dimensional space with the estimation of each element being the estimation of a specific coordinate. At that point, we perform characterization by tracking down the hyperplane that separates the two classes well over all four neural networks. Neural networks are a bunch of algorithms, shown openly after the human brain, that are proposed to see plans. Neural networks help us group and order. They provide help to total unlabeled data according to resemblances among the model information sources, and they describe data when they have a checked dataset to plan on. To order the bone break X-ray images into cracked and normal, we utilized the backpropagation neural network (BPNN) (Fig. 5).

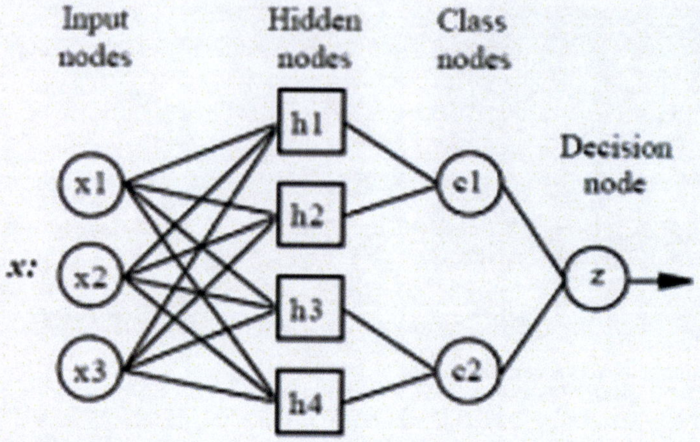

Fig. 5 Design of neural network

4.4 Backpropagation Neural Network (BPNN)

Back-spread is the quintessence of neural net planning. It is the procedure for changing the heaps of a neural net subject to the slip-up rate procured in the past age. Fitting tuning of the heaps allows you to diminish botch rates and to make the model trustworthy by growing its hypothesis. Back-engendering is a short construction for "in reverse proliferation of mistakes." This method helps with calculating the tendency of an adversity work concerning every one of the loads in the organization. Improves on the organization structure by components weighted connections that have minimal impact on the prepared organization. You need to examine a gathering of information and initiation esteems to build up the connection between the information and covered up unit layers. It assists with evaluating the effect that a given info variable has on an organization yield. The information acquired from this examination ought to be addressed in rules. Backpropagation is particularly helpful for profound neural organizations dealing with mistake inclined tasks, like picture or discourse acknowledgment. Backpropagation exploits the chain, and force rules permit backpropagation to work with quite a few yields (Fig. 6).

5 Results Discussion

Using the various algorithms for image preprocessing along with the GLCM feature extraction method, we applied various classification algorithms to get the comparative study of the model accuracy. We applied a support vector machine which is a supervised machine learning algorithm used for classification provides the results with the accuracy of 70 to 72% based on the dataset. After that, we build the neural network

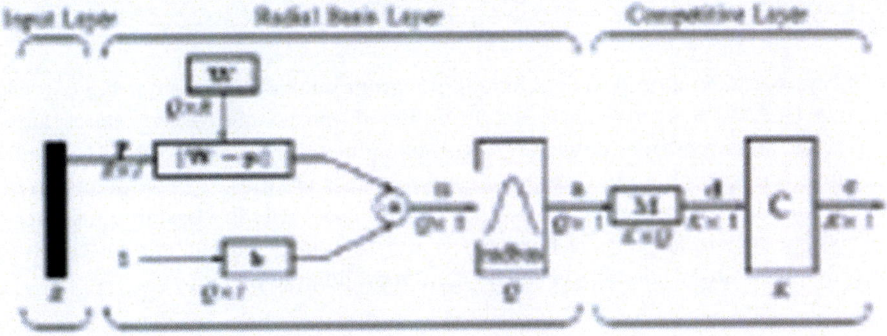

Fig. 6 Proposed neural network

model based on convolutional neural network and backpropagation neural network which provides the results out of which convolutional neural network provides the best accuracy.

Algorithms	Accuracy (%)
Support vector machine (SVM)	70–72
Convolutional neural network (CNN)	85–88
Back-propagation neural network	75–80

Through this machine learning application, the adaptability alongside the heartiness of the programmed insight framework has been illustrated. The distinguishing proof of the pictures yielded 82% where 57 information pictures out of the dataset that had 70 bone pictures gave the ideal yield. Our investigation shows that GLCM had the option to deliver sensible outcomes for the vast majority of the pictures. For every region, there are actually four qualities determined in four ways which prevalently are [1, 0], [0, 1], [1, 1], [−1, −1]. The normal of the four qualities will be taken, and subsequently, every picture will have four qualities by and large. The qualities given by energy demonstrate the surface of the region. Energy gives the amount of squared components. Subsequently, if the territory has a worth 1, it implies the surface is near uniform and there will be no break and the other way around. Neighborhood varieties are estimated by the difference. Correlation assists with figuring joint likelihood events of any pre-defined pair of pixels. Homogeneity checks how close the dissemination of parts in the GLCM is to the GLCM corner to corner. Subsequent to figuring, every one of the four qualities to be specific energy, relationship, homogeneity, contrast, the actual calculation will itself consequently compute and think about and check if there exists any crack in the ideal territory or not.

6 Conclusion

We have expressed a proposed technique for programmed detection of bone crack utilizing GLCM mechanized systems. From this proposed strategy, we can arrange the cracked and sound bone dependent on the boundary obtained from GLCM esteem. The limit esteem lining the crack and sound bone is set to 0.95. The precision of the calculation is at present covered in any event 85% which possibly guarantees a proficient system to perceive bone crack in a computerized climate. Discoveries show that the framework can give steady and reproducible outcomes.

References

1. Karimunnisa S, Raj R, Madupu RK, Basha Z, Neelakanteshwara P (2020) Detection of bone fracture automatically with enhanced performance with better combination of filtering and neural network. In: Proceedings of the second international conference on inventive research in computing applications (IC IRCA-2020) IEEE Xplore Part Number: CFP20N67-ART; ISBN: 978-1-7281-5374-22
2. Mall PK, Singh PK, Yadav D (2019) GLCM Based element ex traction and medical X-Ray picture order utilizing machine learning techniques. In: 2019 IEEE conference on information and communication technology (CICT)
3. Stolojescu-Crisan C, Holban S (2013) A Comparison of X-ray image segmentation technique. Adv Electric Comput Eng 13(3)
4. Naik A, Tikhe S, Bhide S, Saravanan T (2016) Algorithm to detect fracture from OPG image using texture analysis. In: 2016 IEEE 6th international conference on advanced computing
5. Cao Y, Wang H, Moradi M, Prasanna P, Syeda Mahmood TF, Crack detection in X-Ray images through stacked random forests features fusion. IBM Research—Almaden. San Jose, CA, USA
6. Naouel G, Olfa MC, Mokhtar M, Jerome M (2018) Evaluation of DWT denoise method on X-Ray images acquired using flat detector. 2018 IEEE 4th middle east conference on biomedical engineering (MECBME)
7. Nashaat M, Hassan H, Exceptionally proficient methods for programmed division of X-Ray Bone pictures dependent on fluffy rationale and edge location procedure. Electronics and Communication Engineering Department, Fayoum University, Fayoum 63514, Egypt
8. Yadav DP, Rathor S (2020) Bone fracture detection and classification utilizing deep learning approach. In: 2020 international conference on power electronics & IoT applications in renewable energy and its control (PARC) GLA University, Mathura, UP, India
9. Yang A, Cheng L, Shimponda M, Zhu H-Y (2019) Long bone break discovery utilizing Artificial neural organization dependent on line highlights of X-Ray Images. In: 2019 IEEE symposium series on computational intelligence (SSCI). Xiamen, China

An Inventory Model with Pareto Distribution Deterioration with Weibull Demand Rate and Time-Dependent Shortages

S. Sindhuja and P. Arathi

Abstract The objective of this paper is to develop an inventory model with Pareto distribution, which is considered as the deterioration. The Weibull demand rate and shortage of costs are also taken into consideration. The objective of the paper is to focus on the possible effects of minimizing the total cost. The proposed model determines the best operating plan and overall cost. A numerical example and sensitivity analysis are discussed to illustrate the inventory model. The sensitivity analysis is carried out to examine the effects of changing different inventory parameters one at a time. The changes in parameters are carried out through the graphical representations that validate the proposed model. Complex equations are solved using Matlab2013a.

Keywords Inventory · Pareto distribution · Deterioration · Weibull demand · Shortages

1 Introduction

In the manufacturing sector, inventory plays a major role, where demand and deteriorating of items are the most important factors. Inventory models are noted to be constructed under the influence of several other factors, including demand rate, shortage cost, and item deterioration. The impact of these various factors has been discussed by many researchers since 1915.

Demand is categorized based on a variety of factors, including the nature of the product, how it is used, the number of customers who buy it, and the number of suppliers who sell it. In different cases, the demand for a particular product will be different. Constant demand, time-dependent demand, price-dependent demand, and time- and price-dependent demand are some of the types of demand that have

S. Sindhuja (✉) · P. Arathi
Department of Mathematics and Actuarial Science, B.S. Abdur Rahman Crescent Institute of Science and Technology, Chennai, India

P. Arathi
e-mail: arathi@crescent.education

R. P. Mahapatra et al. (eds.), *Proceedings of International Conference on Recent Trends in Computing*, Lecture Notes in Networks and Systems 341,
https://doi.org/10.1007/978-981-16-7118-0_43

been studied so far. The above demands do not accurately depict the demand for newly introduced fashion items (e.g., cosmetics, garments, etc.) for which demand increases over time as they are introduced to the market and then stabilizes. The ramp-type demand is the name for this kind of demand.

Some modelers felt compelled to take deterioration into account. Deterioration is characterized as spoilage, obsolescence, loss of utility (or) loss of marginal value of a commodity, all of which decrease the utility of the commodity compared to its original state. Blood, fish, fruits, and vegetables, as well as alcohol, gasoline, radioactive compounds, and medications, are all examples. As a result, the deterioration of physical goods in stock is an important feature to consider.

In this paper, an inventory model for deteriorating items with Pareto distribution deterioration with a time-dependent shortage and with the Weibull demand rate is presented, as well. Tripathi and Uniyal [1] suggest an inventory model that reduces total cost by treating demand as a linear demand. This existing model is compared with the newly proposed model, which results in a higher reduction in total cost.

The rest of the paper is structured as follows: A thorough literature review is given in Sect. 2. Section 3 consists of notations and assumptions. The mathematical model is derived in Sect. 4. Section 5 provides a numerical example and a comparison of a linear demand model and the suggested Weibull demand with the Pareto distribution deterioration model. In Sect. 6, a detailed sensitivity analysis is presented. Conclusion and the scope of future work are given in Sect. 7.

2 Literature Survey

In the year 1915, F.W. Harris proposed the classic EOQ formula, also known as Wilson's square root formula. The inventory model was developed by Ghare and Schrader [2], who discussed about a constant rate of production, a constant rate of depletion, and no shortages. Another class of inventory model was developed by Dave and Patel [3] for the deterioration of products with time-proportional demand and no shortages.

An inventory model for deteriorating objects was created by Chakrabarti and Chaudhri [4]. Su et al. [5] proposed an inventory model for declining goods in the year 1996. Chang and Dye [6] developed an inventory model in which the proportion of clients willing to accept backlog is the reciprocal of a linear waiting time function. Teng and Yang [7] developed a fraction of unmet demand with a decreasing waiting time function before the next replenishment. According to this model, the demand rate of any commodity in the real market is always in a dynamic situation.

Different types of demand can be considered in various functional forms depending on time. Teng et al. [8] added a partial backlog to the inventory EOQ model proposed by Ghosh and Chaudhari [9]. A quadratic demand rate EOQ model for non-instantaneous declining products of the backlog on waiting time was proposed Begum et al. [10]. The rate of deterioration is time proportional in this model. Darwish et al. [11] used a stochastic inventory model with finite supply and partial backorders and found that the demand-to-production ratio has a subtended effect on the main parameters of inventory models, such as ordering quantity, ordering time, and overall cost. In the case of deterioration, Tripathi and Uniyal [1] developed an EOQ model with linear demand.

The production inventory model has recently been the subject of a lot of studies. For declining goods with variable demand rates, Khurana et al. [12] proposed an Economic Production Quantity Model (EPQ) in 2018, permitting shortages. Khara et al. [13] created a quantity model for economic development. This model deals with products of both perfect and imperfect nature. In this model, demand is determined by the purchase price, product reliability, and advertising.

A finite horizon is needed for the creation of an EOQ model. In 2016, Palanivel et al. [14] created an EOQ model for non-instantaneous deteriorating objects that is called a probabilistic function. This model was developed with inflation and the time value of money in mind over a finite planning horizon. Saha et al. [15] suggest an inventory model for declining goods with time and price-based demand.

A Weibull time-varying demand, according to the literature, signifies a uniform variance in the item demand rate per unit time, which is unusual in the real world. In this paper, a Pareto distribution deterioration inventory model for deteriorating goods is addressed, as well as a time-dependent shortage with the rate of demand as Weibull is also considered.

3 Notations and Assumptions

The symbols that are being used in the model are as follows:

A: Cost of placing an order	a, b: Pareto distribution parameters for deterioration
α, β: Demand parameters	
h: Cost of holding the stock	D: Demand rate $D = \alpha\beta\gamma^{(\beta-1)}$
d: Purchase cost of one unit	D_n: Number of deteriorated units
Q_i: Initial stock	d_r: Rate of deterioration at any time
T_1: Time of positive stock	q_1: Maximum stock level
T: Cycle time	q_2: Maximum stock shortage level
T_1^*: Optimal time of positive inventory	
T^*: Optimal time cycle	s: stock-out cost of shortage
z: Total cost per cycle	$z*$: Optimal total cost per cycle

4 The Proposed Mathematical Model

The aim of this work is to extend the EOQ model proposed by Tripathi and Uniyal (2015) to the case of the deterioration and demand in an EOQ model. Considering the effect of the Weibull demand and the Pareto deterioration rate, the inventory level at any point of time t, $I(t)$ can be expressed by the following differential equation:

$$\frac{\mathrm{d}I(t)}{\mathrm{d}t} + \frac{1}{a - bt} I(t) = -\alpha\beta\gamma^{(\beta-1)} \quad 0 \le t \le T_1 \tag{1}$$

With the initial condition $I(0) = q_1$ and boundary condition $I(T_1) = 0$, the inventory level can be expressed as:

$$\frac{\mathrm{d}I(t)}{\mathrm{d}t} = -\alpha\beta\gamma^{(\beta-1)} \quad T_1 \le t \le T \tag{2}$$

The solution of Eq. (1) with condition $I(T_1)$ can be obtained as:

$$I(t) = \frac{-\alpha\beta\gamma^{-1+\beta}(-\alpha + \gamma\beta)}{(-1+\beta)} + (-\alpha + \gamma\beta)^{\frac{1}{\beta}} \tag{3}$$

Using the initial condition $I(0) = q_1$, the value of q_1 is obtained.

$$q_1 = (-\alpha)^{\frac{1}{\beta}} \tag{4}$$

The solution of Eq. (2), with the condition $I(T_1) = 0$, is given by

$$I(t) = -T_1\alpha\beta(T_1)^{-1+\beta} + t\alpha\beta T \tag{5}$$

Maximum storage quantity q_2 is given by

$$q_2 = a(T - T_1) + \frac{b}{2}(T^2 - T_1^2) \tag{6}$$

Initial order quantity is $Q_i = q_1 + q_2$

$$Q_i = \frac{\alpha[b\alpha + ba(-1+2\beta)]}{(-1+\beta)(-1+2\beta)} + (-\alpha)^{\frac{1}{\beta}} + a(T - T_1) + \frac{b}{2}(T^2 - T_1^2) \tag{7}$$

Inventory carrying cost in the system during the time intervals $(0, T_1)$ is given by

$$\begin{aligned} C_{ih} &= h\int_0^{T_1} I(t)\mathrm{d}t \\ &= h\left[-\frac{(\gamma\beta - a)T_1(2b\alpha + a(-2+4\beta) + b(-1+\beta)T_1)}{2 - 6\beta + 4\beta^2} + (\gamma\beta - \alpha)^{\frac{1}{\beta}} T_1\right] \end{aligned} \tag{8}$$

The stock out cost between the time intervals (T_1, T) is given by:

$$C_s = s\int_{T_1}^{T} -I(t)\mathrm{d}t = s\Bigg[(\gamma\beta-\alpha)^{\frac{1}{\beta}}(T-T_1)$$
$$-\left[\frac{(\gamma\beta-\alpha)(T-T_1)(2b\alpha+bT(-1+\beta)+a(-2+4\beta)+b(-1+\beta)T_1)}{2-6\beta+4\beta^2}\right]\Bigg] \tag{9}$$

No. of units purchased in the beginning is $Q_i = q_1 + q_2$.
The number of units deteriorated is $Q = T_1\left(\alpha + \beta\frac{T_1}{2}\right)$.
Hence, the deterioration cost is $D_r = d\left\{q_1 - T_1\left(\alpha + \beta\frac{T_1}{2}\right)\right\}$.
By substituting the value of q_1, the following equation is obtained.

$$D_r = d\left\{\frac{\alpha[b\alpha + ba(-1+2\beta)]}{(-1+\beta)(-1+2\beta)} + (-\alpha)^{\frac{1}{\beta}} - T_1\left(\alpha+\beta\frac{T_1}{2}\right)\right\} \tag{10}$$

The total cost per cycle z is defined as $z = \frac{1}{T}[A + C_{ih} + C_s + D_r]$, and substituting the value of Eqs. (8), (9) and (10), it is obtained as:

$$z = \frac{1}{T}\left\{\begin{array}{l} A + h\left[-\frac{(\gamma\beta-a)T_1(2b\alpha+a(-2+4\beta)+b(-1+\beta)T_1)}{2-6\beta+4\beta^2} + (\gamma\beta-\alpha)^{\frac{1}{\beta}}T_1\right] \\ +s\Bigg[(\gamma\beta-\alpha)^{\frac{1}{\beta}}(T-T_1) \\ \quad -\left[\frac{(\gamma\beta-\alpha)(T-T_1)(2b\alpha+bT(-1+\beta)+a(-2+4\beta)+b(-1+\beta)T_1)}{2-6\beta+4\beta^2}\right]\Bigg] \\ +d\left\{\frac{\alpha[b\alpha+ba(-1+2\beta)]}{(-1+\beta)(-1+2\beta)} + (-\alpha)^{\frac{1}{\beta}} - T_1\left(\alpha+\beta\frac{T_1}{2}\right)\right\} \end{array}\right. \tag{11}$$

Taking the first and the second order partial differentiation for Eq. (11) with respect to T_1 and T, Eqs. (12) and (13) are obtained:

$$\frac{\partial z}{\partial T_1} = \frac{1}{T}\left\{\begin{array}{l} h\left[\begin{array}{l}\frac{-\mu_1(3(\alpha a+ab(-1+2b))+4\beta(-1+b)bT_1)}{6(-1+b)(-1+2b)} - \\ \frac{6a(\alpha-\beta a-2\alpha b)+3(\beta a+\alpha b(-1+2b))T_1+2\beta(-1+b)bT_1^2}{6(-1+\beta)(-1+2\beta)} \\ -\frac{a}{b}+T_1\end{array}\right] \\ -s(T-T_1)(\alpha+\beta T_1) \\ +d(-\alpha-\beta T_1)\end{array}\right\} \tag{12}$$

$$\frac{\partial z}{\partial T} = -\frac{1}{T^2}\left\{\begin{array}{l} A+h\left[\begin{array}{l}\frac{-T_1\left(6a(\alpha-\beta a-2\alpha b)+3(\beta a+b\beta(-1+2b))T_1+2d(-1+b)bT_1^2\right)}{6(-1+\beta)(-1+2\beta)} \\ -\frac{\alpha T_1}{\beta}+\frac{T_1^2}{2}\end{array}\right] \\ +\frac{s(T-T_1)^2}{6}\{3\alpha+\beta(T+2T_1\} \\ +d\left\{\frac{\alpha[d\alpha+dc(-1+2\beta)]}{(-1+\beta)(-1+2\beta)} + (-a)^{\frac{1}{\beta}} - T_1\left(\alpha+\beta\frac{T_1}{2}\right)\right\}\end{array}\right.$$
$$+\frac{1}{T}\left\{\frac{s(T-T_1)}{2}(2\alpha+\beta T-\beta T_1)\right\} \tag{13}$$

The optimal values of $T_1 = T_1^*$ and $T = T^*$ are obtained by solving simultaneously for $\frac{\partial z}{\partial T_1} = 0$ and $\frac{\partial z}{\partial T} = 0$,
Hence, the following equations are obtained below:

$$\frac{\partial z}{\partial T_1} = 0$$

$$\begin{aligned}
& h\begin{bmatrix} 3\alpha\beta^2 T_1 - 6\alpha\beta^3 T_1 - 6\alpha\beta c - 6d\alpha^2\beta + 12\alpha c\beta^2 - 3c\beta^2 T \\ 1 + 6c\beta^3 T_1 - 6\alpha + 18\alpha\beta - 12\beta^2 + 6\beta T_1 - 18\beta^2 T_1 + 12\beta^3 T_1 \end{bmatrix} \\
& + s\begin{bmatrix} 6cT\beta + 6d\beta T T_1 - 6a\beta T_1 - 6\alpha\beta b T_1^2 - 18\alpha T b^2 - 18\beta b^2 T T \\ 1 + 18\alpha b^2 T_1 + 18\alpha\beta b^2 T_1^2 + 12b^3\alpha T + 12\beta b^3 T T_1 - 12cb^3 T_1 - 12\beta\beta b^3 \mu_1^2 \end{bmatrix} \\
& + pc\left[-6\alpha b - 6\beta b T_1 + 18\alpha b^2 + 18\beta b^2 T_1 - 12c\beta^3 - 12d\beta^3 T_1\right] = 0
\end{aligned} \tag{14}$$

$$\frac{\partial z}{\partial T} = 0$$

$$\begin{aligned}
& 6A\left(b - 3b^2 + 2b^3\right) \\
& + h\begin{bmatrix} -6\alpha a T_1 b + 6\beta a^2 T_1 b + 12a\alpha b^2 T_1 - 3\beta a b T_1 + 3b^2 T_1^2 - 6\beta b^3 T_1^2{}_1 \\ +2d\beta^2 T_1^3 - 6aT + 18abT_1 - 12ab^2 T_1 + 3bT_1^2 - 9b^2 T_1^2 + 6b^3 T_1^2 \end{bmatrix} \\
& + s\begin{bmatrix} 3\alpha T^2 + \beta b T^3 + 2\alpha T_1 b - 9\alpha T^2 b - 3\alpha b^2 T^3 - 6b^2\alpha T^2 T_1 + 6\alpha b^2 T^2 \\ +2\beta b^3 T^3 + 4\beta T^2 T_1 b^3 - 6\alpha T T_1 - 2\beta b T^2 T_1 - 4\beta T T_1^2 b + 18\alpha b T T_1 \\ +6\beta b^2 T^2 T_1 + 12\alpha b^2 \mu\mu_1^2 - 12\alpha b^2 \mu\mu_1 - 4\beta b^3 T^2 T_1^2 - 8\beta b^3 T T_1^2 + 3\alpha T_1^2 \\ +\beta b T_1 T_1^2 + 2\beta T_1^3 b - 9\alpha b T_1^2 - 3\beta b^2 T T_1^2 - 6b^2\beta T_1^3 + 6\alpha b^2 T_1^2 + 2\beta b^3 T T_1^2 + 4\beta T_1^3 b^3 \end{bmatrix} \\
& + pc\begin{bmatrix} 6a(\beta a - \beta\alpha + 2\alpha\beta b) + (-\alpha)^{\frac{1}{\beta}}\left(6\beta - 18\beta^2 + 12\beta^3\right) \\ -T_1(2c + dT_1)\left(3\beta - 9\beta^2 + 6\beta^3\right) \end{bmatrix} \\
& + 3s(T - T_1)(2\alpha + bT - T_1)\left(b - 3b^2 + 2b^3\right) = 0
\end{aligned} \tag{15}$$

Solving Eqs. (14) and (15) simultaneously, the optimal (minimum) values of $T_1 = T_1^*$ and $T = T^*$ are obtained.

Note: MATLAB R2013a is used for finding optimal solution of $T_1 = T_1^*$, $T = T^*$, $Q_i = Q_i^*$, $z = z^*$.

5 Numerical Example

The numerical example presented in this section demonstrates the efficacy of the proposed analytical model. Let's consider the system using the following information:

A—300 units per annually
α —150 units per annually
β—10 units per annually
γ—15 units per annually
s—20 per annually
h—10 per annually
d—2
a—0.001
b—0

Table 1 Comparison of the results

Parameters	τ_1^*	τ^*	Q_i^*	z^*
Tripathi and Uniyal [1]	0.5747	0.605	93.78	965.19
Proposed model	0.535	0.905	103.93	746.13

The values of $T_1 = T_1^* = 0.53, T = T^* = 0.905$, $Q_i = Q_i^* = 89.9327$ units and $z = z^* = 746.1315$ are obtained from Eq. (11).

The optimal solutions of the existing model and the proposed model are compared in Table 1. The initial stock requirement and optimal time cycle are now higher than in the existing model. The proposed model is an improvement over Tripathi and Uniyal's (2015) model since it demonstrates that the optimal total cost per cycle is lower than the existing model.

6 Sensitivity Analysis on A, α, β, γ, d, a, b, h and s

In the above example, the parameters A, a, b, d, h, s, γ and α, β are varied to examine an illustration of the proposed model. The table below shows how variance in a parameter affects the inventory system's decision variable (Table 2).

6.1 *Sensitivity Analysis with Graphical Representation*

See Figs. 1, 2, 3, 4, 5, 6, 7, and 8.

7 Conclusion

A model for deteriorating goods with a Pareto distribution deterioration and Weibull demand under shortages has been discussed in this paper. A Weibull time-dependent demand rate is found to be a more realistic demand approach among different time-varying demands in EOQ models, as it reflects both accelerated and retarded demand growth. The Weibull demand parameters (β, α, γ) and Pareto distribution deterioration parameters (a and b) lead to significant effects on reducing the total cost. As compared to the existing model, the proposed model has a higher initial stock requirement Q_i and a near-optimal time length. When compared to the existing model, the total cost of the proposed model is slightly lower, i.e., the total cost of an existing model is 965.19 , while the total cost of the proposed model is 746.13. A large amount of money is spent on inventory control and management in the company system. Deterioration and demand trends are two critical aspects of successful business growth.

Table 2 Sensitivity analysis on A, α, β, γ, d, a, b, h and s

Parameter		$T = T1^*$	$T = T^*$	$Qi = Qi^*$	$Z = Z^*$
β	11	0.53	0.9	109.06	752.93
	12	0.52	0.89	108.19	759.41
	13	0.51	0.89	107.47	765.6
	14	0.49	0.88	106.74	771.55
	15	0.49	0.88	105.97	777.4
	16	0.49	0.77	105.29	782.83
	17	0.48	0.77	104.72	788.2
	18	0.48	0.77	104.14	793.41
	19	0.47	0.76	103.63	798.48
	20	0.47	0.76	103.08	803.4
A	310	0.54	0.82	101.97	750.2
	320	0.55	0.82	103.8	764.04
	330	0.56	0.84	105.62	777.67
	340	0.56	0.85	107.42	791.1
	350	0.57	0.86	109.2	801.34
	360	0.58	0.87	110.95	817.39
	370	0.58	0.88	112.67	820.27
	380	0.59	0.89	114.37	822.97
	390	0.6	0.9	116.05	845.51
	400	0.6	0.91	117.71	867.89
h	11	0.53	0.8	109.98	747.43
	12	0.53	0.8	109.89	748.72
	13	0.53	0.8	109.79	750
	14	0.53	0.8	109.71	751.27
	15	0.53	0.8	109.62	752.54
	16	0.53	0.7	109.54	753.79
	17	0.52	0.7	109.44	755.05
	18	0.52	0.7	109.36	756.29
	19	0.52	0.7	109.29	757.52
	20	0.52	0.69	109.2	758.75
d	2.1	0.42	0.65	100.22	745.67
	2.2	0.42	0.65	100.37	745.21
	2.3	0.42	0.65	100.51	744.74
	2.4	0.42	0.65	100.66	744.27
	2.5	0.42	0.65	100.82	744.01
	2.6	0.42	0.65	100.97	743.33

(continued)

Table 2 (continued)

Parameter		$T = T1*$	$T = T*$	$Qi = Qi*$	$Z = Z*$
	2.7	0.42	0.65	101.11	742.85
	2.8	0.42	0.65	101.27	742.37
	2.9	0.42	0.65	101.42	741.89
γ	3	0.43	0.65	101.58	741.4
	16	0.44	0.71	109.8	748.11
	17	0.44	0.71	109.53	750.08
	18	0.44	0.71	109.2	752.03
	19	0.44	0.71	108.92	753.97
	20	0.43	0.71	108.71	755.89
	21	0.43	0.7	108.45	757.8
	22	0.43	0.7	108.19	759.69
	23	0.43	0.7	107.93	761.56
	24	0.42	0.7	107.68	763.42
	25	0.42	0.7	107.43	765.26
s	25	0.46	0.67	104.96	780.35
	30	0.45	0.66	103.28	805.55
	35	0.44	0.64	100.32	824.89
	40	0.43	0.63	98.83	840.28
	45	0.42	0.63	97.68	852.56
	50	0.49	0.62	96.75	862.78
	55	0.49	0.61	96.01	871.35
	60	0.5	0.61	96.3	878.63-
	65	0.51	0.61	95.7	884.88
a	0.002	0.45	0.68	104.96	780.35
	0.003	0.44	0.67	103.28	783.55
	0.004	0.43	0.66	100.32	782.89
	0.005	0.41	0.65	98.83	780.28
	0.006	0.41	0.64	97.68	780.12
	0.007	0.39	0.63	96.75	772.78
	0.008	0.39	0.62	96.01	771.35
	0.009	0.34	0.6	96.3	770.63
	0.1	0.33	0.6	95.7	764.88
b	0.2	0.44	0.68	104.96	770.35
	0.3	0.44	0.67	103.28	769.55
	0.4	0.43	0.66	100.32	762.89
	0.5	0.41	0.65	98.83	760.28

(continued)

Table 2 (continued)

Parameter		$T = T1*$	$T = T*$	$Qi = Qi*$	$Z = Z*$
	0.6	0.41	0.64	97.68	760.12
	0.7	0.39	0.63	96.75	759.78
	0.8	0.39	0.62	96.01	751.35
	0.9	0.34	0.6	96.3	750.63
	0.91	0.33	0.6	95.7	744.88

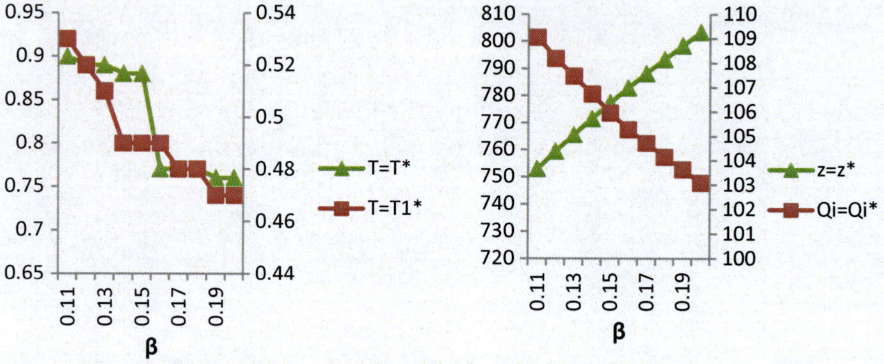

Fig. 1 Sensitivity analysis on β

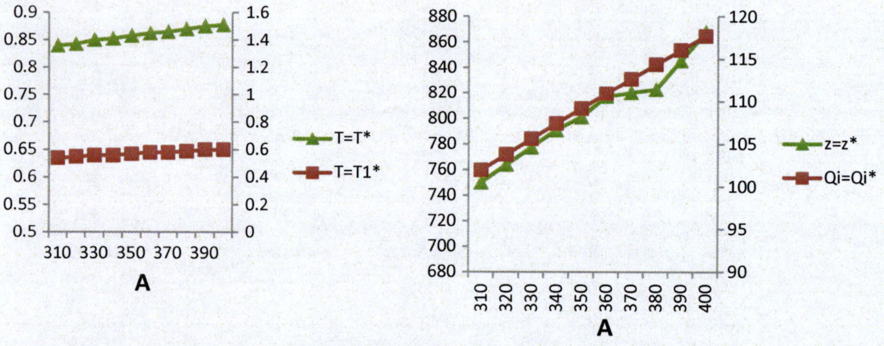

Fig. 2 Sensitivity analysis on ordering cost A

As a consequence, a deterministic inventory model is built in this paper, which is more efficient than linear demand. Partial backlogging and the advance payment may be used to further investigate the proposed inventory model.

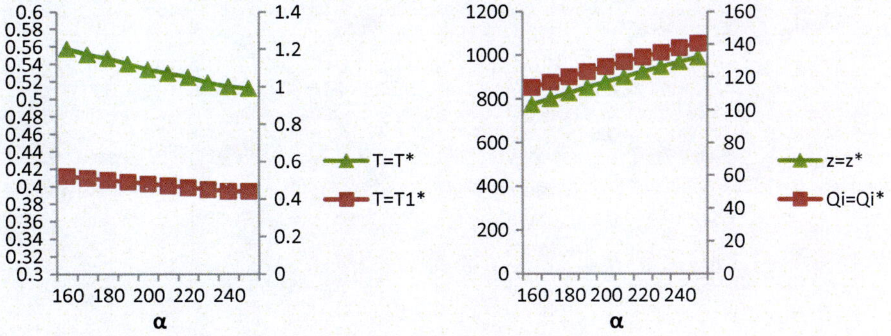

Fig. 3 Sensitivity analysis on α

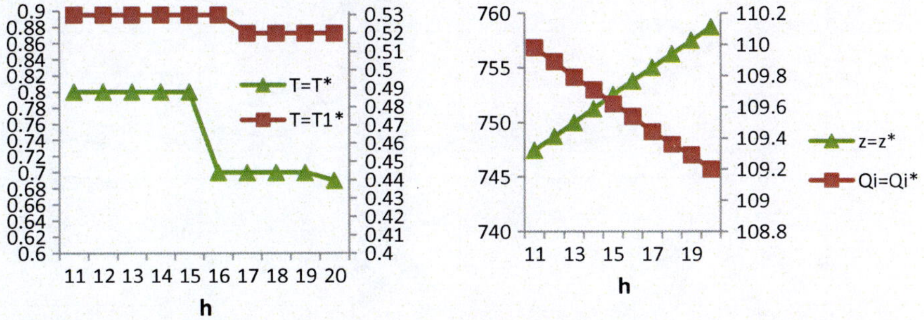

Fig. 4 Sensitivity analysis on h

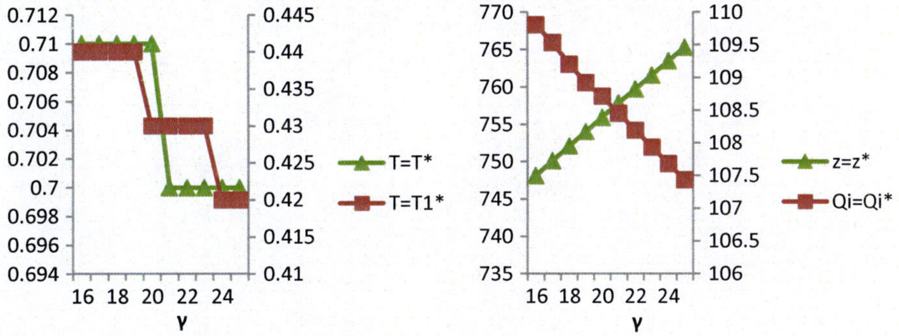

Fig. 5 Sensitivity analysis on γ

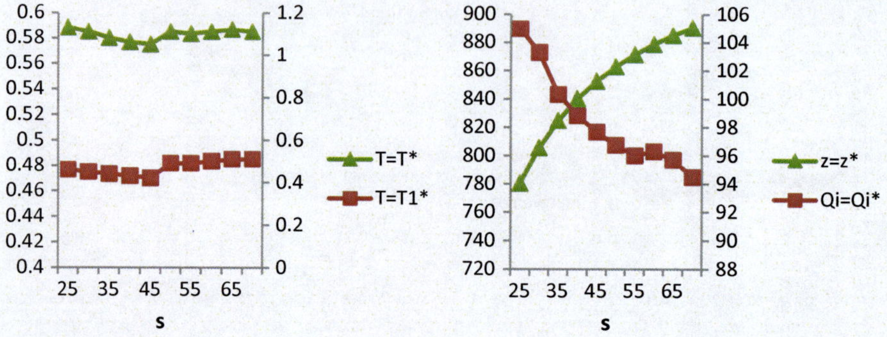

Fig. 6 Sensitivity analysis on *s*

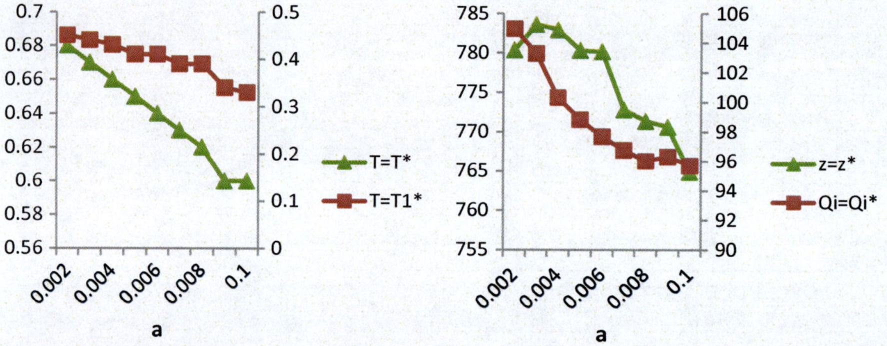

Fig. 7 Sensitivity analysis on *a*

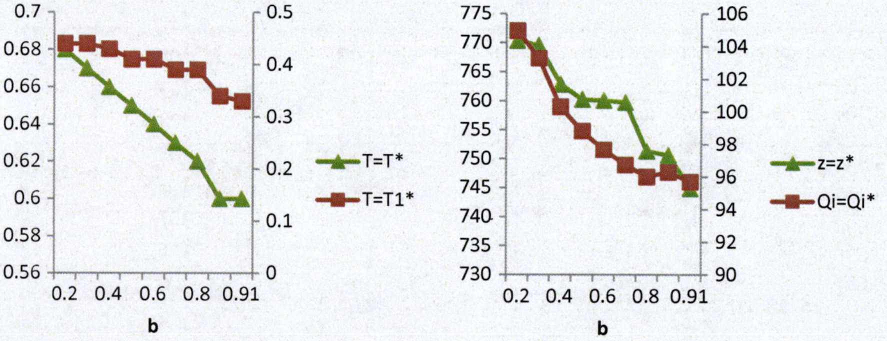

Fig. 8 Sensitivity analysis on *b*

References

1. Tripathi RP, Uniyal AK (2015) EOQ model for deteriorating items with time dependent demand rate under time varying shortages. Int J Math Oper Res 7(6):706–720
2. Ghare PN, Schrader GF (1963) A model for exponentially decaying inventories. J Ind Eng 15(5):238–243
3. Dave U, Patel LK (1981) (T, Si) policy inventory model for deteriorating items with time proportional demand. J Oper Res Soc 32(2):137–142
4. Chakrabarti T, Chaudhuri KS (1997) An EOQ model for deteriorating items with a linear trend in demand and shortages in all cycles. Int J Prod Econ 49(3):205–213
5. Su CT, Tong LI, Liao HC (1996) An inventory model under inflation for stock-dependent consumption rate and exponential decay. Opsearch 33(2):72–82
6. Chang HJ, Dye CY (1999) An EOQ model for deteriorating items with time varying demand and partial backlogging. J Oper Res Soc 50(11):1176–1182
7. Teng JT, Yang HL (2004) Deterministic economic order quantity models with partial backlogging when demand and cost are fluctuating with time. J Oper Res Soc 55(6):495–503
8. Teng J-T, Yang H-L, Chern M-S (2011) EOQ models for deteriorating items and partial back logging when demand is quadratic in time. Euro J Ind Eng 5(2):198–214
9. Ghosh SK, Chaudhuri KS (2006) An EOQ model with a quadratic demand, time-proportional deterioration and shortages in all cycles. Int J Syst Sci 37(10):663–672
10. Begum R, Sahu SK, Sahoo RR (2012) An inventory model for deteriorating items with quadratic demand and artial back logging. Brit J Appl Sci Technol 2(2):112–131
11. Darwish MA, Goyal SK, Alenezi AR (2014) Stochastic inventory model with finite production rate and partial back orders. Int J Logistics Syst Manag 17(3):401–434
12. Khurana D, Tayal S, Singh SR (2018) An EPQ model for deteriorating items with variable demand rate and allowable shortages. Int J Math Oper Res 12(1):117–128
13. Khara B, Dey JK, Mondal SK (2018) Effects of product reliability dependent demand in an EPQ model considering partially imperfect production. Int J Math Oper Res 15(2):242–264
14. Palanivel M, Uthayakumar R (2018) Finite horizon EOQ model for non-instantaneous deteriorating items with probabilistic deterioration and partial backlogging under inflation. Int J Math Oper Res 8(4):449–476
15. Saha S, Sen N (2018) An inventory model for deteriorating items with time and price dependent demand and shortages under the effect of inflation. Int J Math Oper Res 14(3):377–388

Performance Evaluation of Full Adder Using Magnetic Tunnel Junction

Jyoti Garg and Subodh Wairya

Abstract As technology nodes scale down, many challenges emerge, such as increased leakage current causes high power dissipation, which has become a major issue in low-power circuit design due to transistor size reduction and lower voltage supply. To address these issues, nanotechnology devices are combined with low-power design methods. Spintronics is a new emerging technology that has a high speed, almost no leakage current, and high endurance. This paper proposes a hybrid MTJ-based circuit -XOR/XNOR logic gate and full adder. The hybrid logic gate and hybrid full adder surpass earlier designs in terms of device count, propagation latency, and power consumption. Transient simulations were used to simulate all of the designs using the HSPICE tool 32 nm technology.

Keywords Magnetic tunnel junction · Logic design · Adder · Non-volatile logic design

1 Introduction

We are surrounded by low-power, portable, and high-speed devices in today's world. All of these devices are built with CMOS technology (complementary metal–oxide-semiconductor). Due to scaling of CMOS devices, leakage current in circuits has increased in recent decades, affecting overall power dissipation in circuits [1, 2]. To achieve high-speed devices, the clock frequency is increased, resulting in increased dynamic power dissipation. Due to that, the overall power dissipation (sum of dynamic and static power) of the device is enlarged. As a result, there is a trade-off between power and speed. But now, static power is playing a significant role in the power dissipation of circuits. There are numerous techniques for improving

J. Garg (✉)
Department of Electronics Engineering, Dr. A.P.J. Abdul Kalam Technical University, Uttar Pradesh, Lucknow, India

S. Wairya
Department of Electronics and Communication Engineering, IET, Uttar Pradesh, Lucknow, India

R. P. Mahapatra et al. (eds.), *Proceedings of International Conference on Recent Trends in Computing*, Lecture Notes in Networks and Systems 341,
https://doi.org/10.1007/978-981-16-7118-0_44

VLSI circuit performance, but they all increase power dissipation. It exemplifies the trade-off between power, area, and performance. There are numerous other techniques available to address these issues, such as CNFET, QCA, and nanomagnetic devices [3–5]. Spintronics is a new field that has the potential to solve the issue of increased power dissipation in VLSI circuits [6].

There are numerous spintronics devices, but the magnetic tunnel junction based on total magnetoresistance (TMR) is the most promising [7]. The location of a transistor in spintronics may be reformed by switching the spin of electrons. It has no standby power, a fast CPU, and strong CMOS compatibility as a result of this. The STT mechanism has a high density and indefinite durability, and it is non-volatile. It has an important scalability feature because only deep sub-micron CMOS scaling causes a leakage current to flow. For decades, CMOS technology has been used in logic gates designs. Unlike CMOS, MTJ has the property of non-volatility, and it can be seen in application circuits [8, 9]. This paper proposes a hybrid logic gate and full adder based on MTJ. Existing MTJ designs save power and time, but they have a slew of additional flaws [10–13]. While it has a tiny area, this research indicates that as the number of transistors rises, the write energy increases, which is a considerable drawback. The installation of additional circuitry in this article is necessitated by the increase in area, power consumption, and delay. Low-power circuits are also designed using the adiabatic method with MTJ [14]. Our main contributions in this paper are summarized as follows: (i) design and simulation of Hybrid MTJ based XOR gate and full adder; (ii) in terms of device count, static power, dynamic power, and propagation latency, the proposed hybrid circuits compare favorably to existing designs. The following is how the rest of the paper is organized: Magnetic Tunnel Junction: Background, Hybrid Logic Gate, Proposed Hybrid Circuit Design, Analysis, Results, and Conclusions.

2 Background

2.1 Magnetic Tunnel Junction

In MTJ, 1-Bit STT-MRAM cell comprises a transistor and an MTJ, as depicted in Fig. 1. The magnetic tunnel junction comprises of a fixed layer (magnetic orientation fixed) and a free layer (variable magnetic orientation). A tunneling oxide separates the magnetic layers. The relative magnets of the pinned and fixed layers determine how resistant the magnetic tunnel junction is. The two resistance states Rp and R_{AP} encode that logic '0' represents for parallel orientation, and logic '1' is for antiparallel orientation. The resistance difference between these two states is called total magneto resistance (TMR) and is expressed as in Eq. (1).

$$\mathrm{TMR} = \frac{R_{\mathrm{AP}} - R_{\mathrm{P}}}{R_{\mathrm{P}}} \tag{1}$$

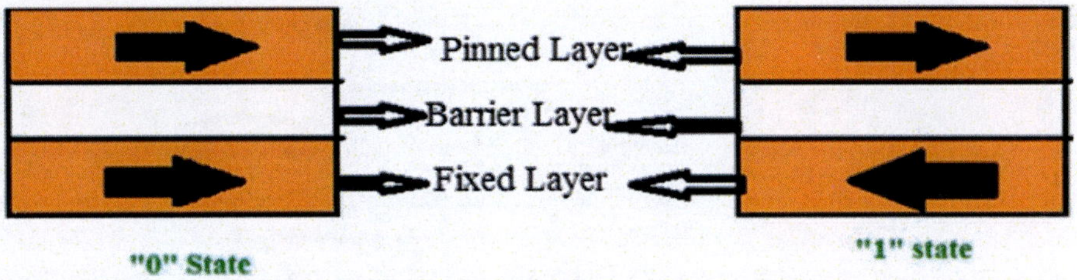

Fig. 1 MTJ state '0' (means low resistance state, magnetization is in same direction) '1' (high resistance state, magnetization is in opposite direction)

A current greater than the MTJ switch's critical current is sent through a bit cell to write it. The written value depends on the current direction shown in Fig. 1, and it can be changed by flipping the direction. Magnitude and duration of write current determine write rate in writing operations [15].

2.2 Hybrid Logic Gates

The proposed generic STT-based magnetic circuit format is depicted in Fig. 2. In Fig. 2, there are three sections: a pre-charge sensing amplifier circuit that analysis outputs result, CMOS cell structure to a write logic block, and a logic control data block. Logic gates are composed of PCSA cells; volatile data is written in CMOS logic while non-volatile data is written in STT-MRAM part. For two reasons, we are using the PCSA circuit in the hybrid MTJ-based circuit designs: firstly, the dynamical sensing enables the amplification of ultra-low-power from analog to digital data; secondly, there can be a significant reduction in reading perturbation induced by sensing activities [16]. This is significant for embedded STT-MRAM because it is

Fig. 2 An MTJ-based hybrid logic circuit's structure

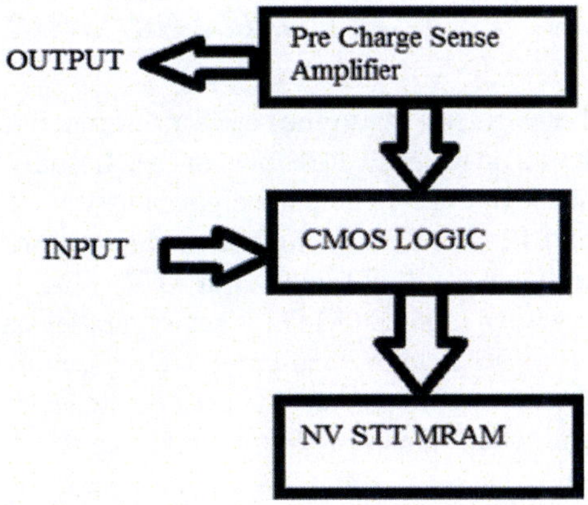

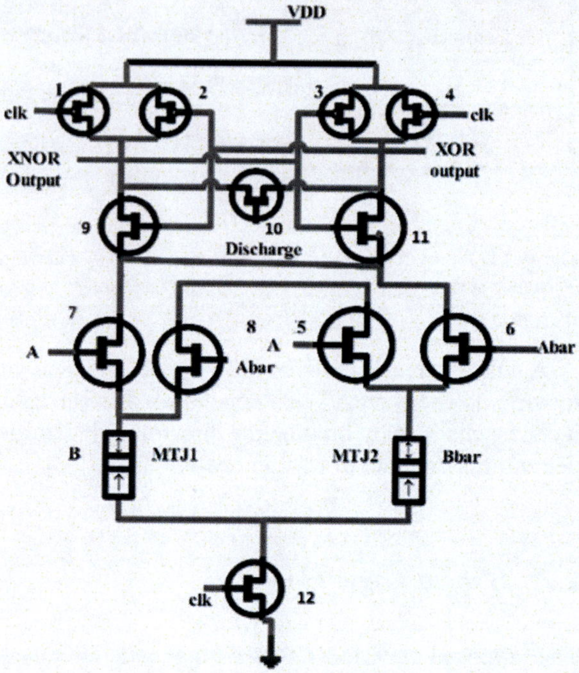

Fig. 3 Design of hybrid XOR/XNOR logic gate

difficult to modify the logic circuit when a complicated ERC is prevented to ensure rapid computation of the system (Fig. 3).

3 Design of Hybrid Magnetic Full Adder Based on STT-MRAM

3.1 Proposed Hybrid XOR/XNOR Gate

Logic gates serve as the base for combinational circuits. The XOR/XNOR logic gates based on the MTJ are shown in Fig. 4. Both inputs have taken on their complimentary shape in Fig. 4. There are eight transistors in the sense amplifier: 1, 4, 2, 3, 9, 10, 11, and 12. The CMOS logic is made up of transistors 5, 6, 7, and 8. Input A is given in transistor 7 and input B at MTJ. When MTJ1 is in antiparallel state then ‘B’ is regarded high and MTJ2 is set to parallel state, or conversely. If ‘A’ is set to low and ‘B’ is set to high, then input ‘01’ is used to illustrate how the system works. When the input at transistor 7 is set low signal, transistors 1, 2, 3, and 4 are in ON state and when transistor 7 is set high signal, then 1, 2, 3, and 4 are in the OFF state. The transistors 5 and 7 are both turned off. MTJ2 is running in low state, whereas MTJ1 is running in high state. The circuit’s left side component is disabled, and the

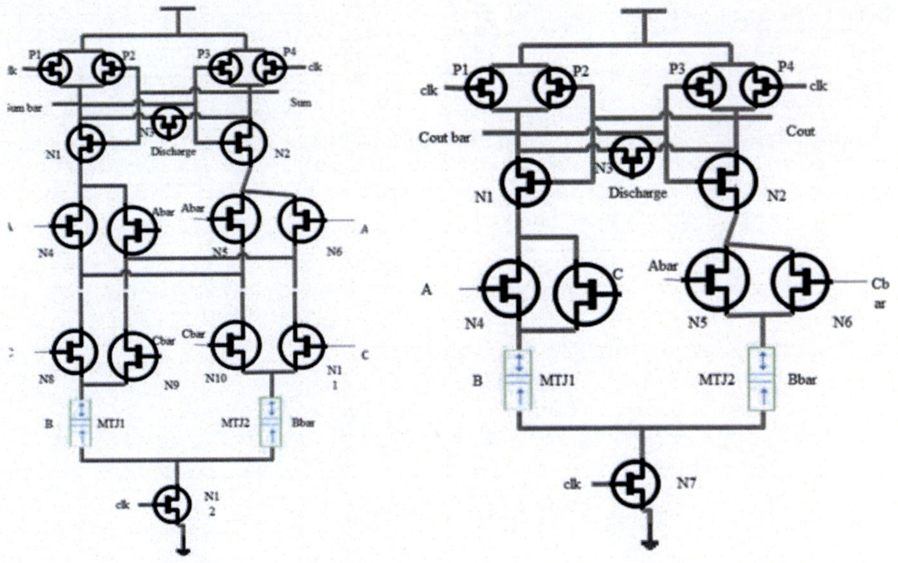

(a) (b)

Fig. 4 **a** Sum. **b** Cout of proposed hybrid full adder

XNOR logic output is low. Meanwhile, the maximum supply voltage is applied to the XOR gate logic. During the recharge and evaluation phases, a signal (that has to be discharged) is provided to transistor 10, which have a varied voltage. Since magnetic tunnel junction 1 is in a state of '1,' the XOR gate arm has a more resistance and causes delayed discharge. The XNOR gate branch drains rapidly, therefore it is a good idea to use it. Transistor 2 surpasses its threshold level, causing transistor 3 to switch on. In addition, transistor 10 approaches the limit of its supply. The branch of the XNOR tends to discharge and exceeds a threshold voltage of the transistor 9, which reduces the node voltage to a minimum amount, boosting the node voltage to a top level. At the XNOR gate logic, this yields 0 while at the XOR gate logic, it yields 1.

3.2 *Proposed Hybrid Full Adder Design*

Figure 4a, b depicts a 1-bit MTJ-based magnetic full adder circuit with the inputs 'A,' 'B,' and 'C,' and the outputs 'SUM' and 'Cout.' In Eqs. 2 through 5, the sum and c_{out} are presented.

$$\text{Sum} = A\bar{B}\bar{C} + \bar{A}B\bar{C} + \bar{A}\bar{B}C + ABC \qquad (2)$$

Table 1 Truth table of sum and cout

A	B	C	Sum	Cout
0	0	0	0	0
0	0	1	1	0
0	1	0	1	0
0	1	1	0	1
1	0	0	1	0
1	0	1	0	1
1	1	0	0	1
1	1	1	1	1

$$\text{Sumbar} = AB\bar{C} + A\bar{B}C + \bar{A}BC + \bar{A}\bar{B}\bar{C} \tag{3}$$

$$\text{Cout} = AB + BC + CA \tag{4}$$

$$\text{Coutbar} = \bar{A}\bar{B} + \bar{B}\bar{C} + \bar{C}\bar{A} \tag{5}$$

For the verification of functionality of full adder, truth table is shown in Table 1. The proposed hybrid full adder works in the same way as the proposed hybrid XOR gate circuit. The input 'B' refers to STT-MRAM non-volatile storage among all the inputs. While retaining high-speed performance, the precharge sensing amplifier circuit (PCSA) delivers outstanding sensing reliability and power efficiency. The discharge currents in both branches (left and right) change based on the logic tree MOS state and the STT-MRAM element state, as well as 'Sum,' 'Sumbar' and 'Cout,' and 'Coutbar' in that order.

4 Validation and Analysis

For performance assessment, the logic circuits XOR gate and full adder simulation using MTJ were used. The designs were simulated in the HSPICE tool using 32 nm CMOS technology [17] and the MTJ model [18], and the power usage was compared to a conventional CMOS design. For the simulation of the full adder, Table 1 has been verified. The transient simulation results of proposed XOR/XNOR logic gate and full adder using MTJ are shown in Figs. 5 and 6, respectively. For transient emulation of the XOR gate, input B stored in the MTJ logic is supposed to be highly logical. From Fig. 6, the functionality of the full adder can be verified. Table 2 describes all the parameters used in STT-MRAM Model [11]. In terms of latency and power consumption, hybrid MTJ/CMOS-based basic logic gates and full adder are

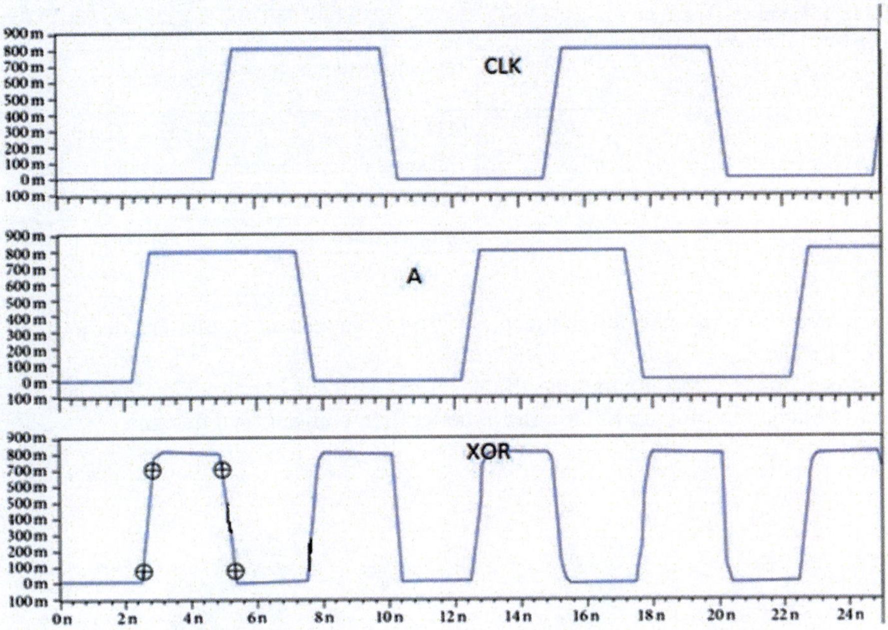

Fig. 5 Simulation result of hybrid XOR gate

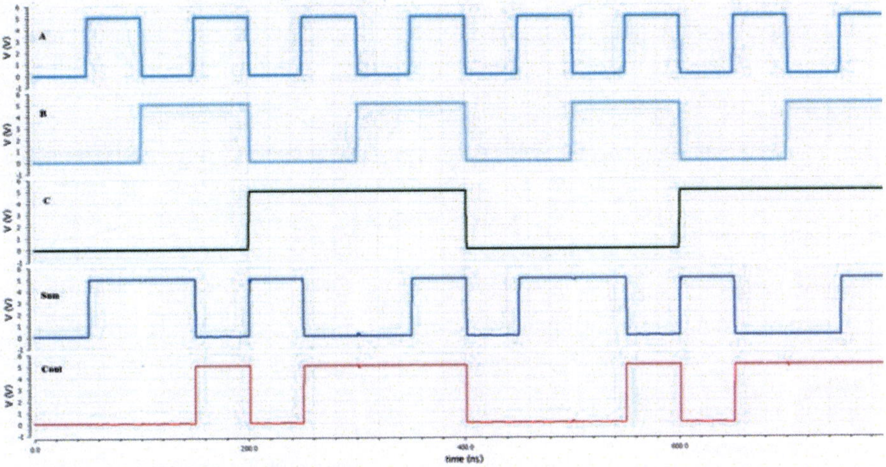

Fig. 6 Simulation result of hybrid full adder

Table 2 Parameters used in STT-MRAM model

Parameters	Description	Value
TMR (0)	Total magneto resistance at 0 bias	120%
Area	MTJ surface	32 × 32 nm
Tox	Thickness of oxide layer	1.4 nm
Jco	Critical current density	5.7×10^5 A/cm^2
V	Volume of free layer	Surface × 1.1 nm

compared to traditional designs (Fig. 7). The Comparison results are depicted in Tables 3 and 4.

Based on the foregoing findings (Tables 3 and 4), it can be stated that logic circuits and full adder employing MTJ perform better than conventional designs.

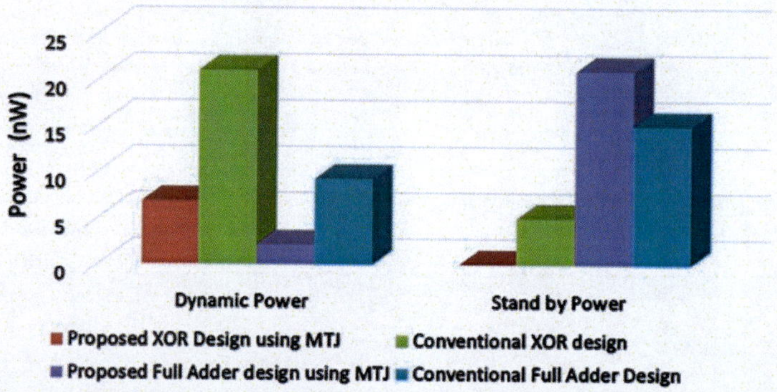

Fig. 7 Power consumption of hybrid XOR gate and hybrid full adder with traditional CMOS design

Table 3 Performance of an XOR gate utilizing MTJ versus traditional design

Parameters	MTJ design	Conventional design
Dynamic power (nw)	6.8	21
Propagation delay (ns)	42	37
Stand by power (nW)	0.01	5

Table 4 Performance evaluation of full adder utilizing MTJ verses traditional design

Parameters	MTJ design	Conventional design
Dynamic power (μw)	2.1	9.25
Propagation delay (ps)	31	39.27
Stand by power (pW)	21	15

5 Conclusion

The goal of this study was to show that hybrid logic circuits using MTJ outperform traditional designs. Magnetic tunnel junctions' impact on power and propagation delay has also been investigated. In this study, the XOR/XNOR logic gate and full adder circuits were simulated. It can be concluded that the proposed hybrid designs result in a significant reduction in power consumption of more than 35%. This demonstrates a significant improvement in power and propagation delay.

References

1. Augustine C, Panagopoulos G, Behin-Aein B, Srinivasan S, Sarkar A, Roy K (2011) Low-power functionality enhanced computation architecture using spin-based devices. In: Proceedings of the IEEE/ACM international symposium on nanoscale architectures. pp 129–136
2. Kim NS, Austin T, Baauw D, Mudge T, Flautner K, Hu JS (2003) Leakage current: Moore's law meets static power. Computer 36:68–75
3. Goswami M, Kumar B, Tibrewal H, Mazumdar S (2014) Efficient realization of digital logic circuit using QCA multiplexer. In: 2nd international conference on business and information management (ICBIM). Durgapur, India, pp 165–170
4. Sahoo R, Sahoo SK, Sankisa KC (2015) Design of an efficient CNTFET using optimum number of CNT in channel region for logic gate implementation. In: International conference on VLSI systems, architecture, technology and applications (VLSI-SATA). Bangalore, India, pp 1-4
5. Rajasekaran S, Sundari G (2017) Design and analysis of logic gates using single electron nanodevices. In: International conference on advances in electrical technology for green energy (ICAETGT). Coimbatore, India, pp 64–68
6. Joshi VK (2016) Spintronics: a contemporary review of emerging electronics devices. Eng Sci Technol 19(3):1503–1513
7. Matsunaga S, et al. (2009) MTJ-based nonvolatile logic-in-memory circuit, future prospects and issues. In: 2009 Design, Automation & Test in Europe Conference & Exhibition, Nice, France, pp 433–435
8. Barla P, Joshi VK, Bhat S (2021) Spintronic devices: a promising alternative to CMOS devices. J Comput Electron 20:805–837. https://doi.org/10.1007/s10825-020-01648-6
9. Zhang Y, Zhao W, Lakys Y, Klein JO, Kim JV, Ravelosona D, ChappertC (2012) Compact modeling of perpendicular-anisotropy CoFeB/MgO magnetic tunnel junction. IEEE Trans Electron Devices
10. Jaiswal A, Andrawis R, Roy K (2018) Area-efficient nonvolatile flip-flop based on spin hall effect. IEEE Magn Lett 9:1–4
11. Zhang D, Zeng L, Gao T, Gong F, Qin X, Kang W, Zhang Y, Klein JO, Zhao W (2017) Reliability-enhanced separated pre-charge sensing amplifier for hybrid CMOS/MTJ logic circuits. IEEE Trans Magn 53(9):1–5
12. Trinh HP, Zhao W, Klein JO, Zhang Y, Ravelsona D, Chappert C (2013) Magnetic adder based on racetrack memory. IEEE Trans Circuits Syst I 60(6):1469–1477
13. Gupta MK, Hasan M (2016) A low-power robust easily cascaded pentaMTJ-based combinational and sequential circuits. IEEE Trans Very Large Scale Integr (VLSI) Syst 24(1):218–222
14. Sharifi F, Saifullah ZM, Badawy A-H (2017) Design of adiabatic MTJ-CMOS hybrid circuits. https://doi.org/10.1109/MWSCAS.2017.8053023
15. Zhao WS, Zhang Y, Devolder T, Klein JO, Ravelosona D, Chappert C, Mazoyer P (2012) Failure and reliability analysis of STT-MRAM. Microelectr Reliab

16. Barla P, Shet D, Joshi VK, Bhat S (2020) Design and analysis of LIM hybrid MTJ/CMOS logic gates. In: 5th international conference on devices, circuits and systems (ICDCS. Coimbatore, India, pp 41–45
17. Zhao W, Cao Y (2006) New generation of Predictive Technology Model for sub-45nm early design exploration. IEEE Trans Electron Devices 53(11):2816–2823
18. Kim J, Chen A, Behin-Aein A, Kumar S, Wang JP, Kim CH (2015) A technology-agnostic MTJ SPICE model with user-defined dimensions for STT-MRAM scalability studies. In: Custom integrated circuits conference (CICC). IEEE, pp 1–4

Intrusion Detection System and Vulnerability Identification Using Various Machine Learning Algorithms

Kishor P. Jadhav and Mohit Gangwar

Abstract In the current data transmission environment, network and device protection is of extreme significance. Hackers and intruders will make several good attempts at triggering an illegitimate intrusion to collapse networks and software platforms. In conjunction with secure device growth, new threats and related solutions arise to avoid these attacks. Several intrusion detection opportunities have been suggested before. Still, none shows acceptable results, so systems are investigating for a better outcome in this region. It has been categorized into network-based IDS and host-based IDS. Intrusions and abuse are constantly threatening to comprehensive internet service use. Therefore, intrusion detection is the most important component of the machine and its network security. Intrusion detection system (IDS) is an algorithm-focused computer network surveillance system that detects malevolent interference in the network. The IDS system has been recognized for maintaining high safety standards, meaning that information is exchanged with confidence and security among dissimilar organizations. This paper proposed a new approach for intrusion detection system (IDS) using a deep learning-based neural network (RNN) classification algorithm. The research suggested even takes a description of different kinds of malicious activity detection techniques for IDS. The system can able to detect network intrusion as well as host base intrusion simultaneously. This research additionally researches these extraordinary methodologies, their exactness, and also false-positive proportions.

Keywords Intrusion detection system · Soft computing · Classification techniques · NIDS · HIDS · IPS · IRS

K. P. Jadhav
Department of Computer Science and Engineering, Bhabha University, Bhopal, Madhya Pradesh, India

M. Gangwar (✉)
Engineering, Bhabha University, Bhopal, Madhya Pradesh, India

R. P. Mahapatra et al. (eds.), *Proceedings of International Conference on Recent Trends in Computing*, Lecture Notes in Networks and Systems 341,
https://doi.org/10.1007/978-981-16-7118-0_45

1 Introduction

The massive rise in the use of computer systems in the general public today, particularly the sudden increase in e-business hugeness toward the wealthy community, has rendered computer asylum a global goal. Since creating a system without any defects is not feasible, an important field of study has been noted for interference. The three most commonly applied defense innovations are safety at night, network-based security, and host-based support. Host-based security is difficult to implement, given the reality that the majority utilize it often. The hard part is to force it on every single entity. While this is quite convenient in a small network of the same machines as the network expands and becomes heterogeneous, it becomes a big headache from an administrative point of view.

They infringe upon the trustworthiness, secrecy, and quality of the system resources of the program. The lost or unfinished data cannot be responded to in the system. Consequently, intrusion detection systems (IDSs) are needed to reduce the serious impact of such assaults. The intrusion detection program is defined as the device or software method to detect unauthorized access to a network or a computer system. IDS can detect all types of attacks, such as disruptive, hazardous attacks, vulnerabilities, data-driven attacks, and host attacks, such as authority breaches and sensitive file access. The firewall works to prevent the interference of internal remote traffic network, and this system is able to detect and prevent the network IDS and host IDS, respectively.

2 Review of Literature

This section is focused on the study of various existing systems, and the work has done by various existing research in IDS.

2.1 Related Work

IDS have been categorized into network and host IDS, which defines as special characterization and execution into different environments. It has also done the work with the signature base and rule base anomaly detection [1]. Various researches have done work on network as well as host-based intrusion detection systems. Data mining algorithms and machine learning algorithms introduce accepted level accuracy on synthetic dataset like KDDCUP99 [2, 3]. But still, some issues have left to detect the network environment malicious activities. KDDCUP99 consists of 41 attributes with 23 sub-attacks that hold master attacks like DOS, PROBE, U2R, and R2L. Another data-sheet was introduced NSLKDD in 2010; it is quite similar to KDDCUP99. Still, it contains 38 attacks that provide drastic supervision and attack detection in both environments.

2.2 Existing Methodologies

According to Vijayanand et al. [4], they utilize deep learning techniques, a new attack detection technique to detect attempts by analyzing smart meter traffic. The proposed model has several complex multilayer methods, organized to efficiently see the attackers in a hierarchical structure. By comparing the integrative solution learning algorithm and deep relational features with classifiers using the standard synthetic dataset, the proposed system's result is analyzed. A significant security measure that aims to monitor attacks through the analysis of network traffic was its exploit to prevent the training database from using the feature selection technique. The simulation framework results demonstrate that the designed DL-based IDS reliably identify and conduct attacks at a quick rate. The equations are ordered based on the importance of the attacks in a chronological fashion. Each classifier is compositional and is intended to detect the special attack available data throughout the network activity. In the IDS, the structure for classifiers is based on the price of the type of attack. When the classification model identifies the connection contains malicious packets, the data is sent to the user. The information is enabled for further analysis.

According to study, Otomo et al. [5] describe the use of computer and machine learning technologies in wireless sensor environments for IDS systems. It introduces deep Boltzmann computer-distributed IDS (DBCD-IDS). A possible need plenty of IDS technique for WSNs to track sensitive systems to execute that RBC-IDS performance is studied and compared to the previously suggested efficient machine teaching IDS: the adaptively monitored and grouped hybrid IDS. This system experiments indicate that duplicate detection and accuracy rates are obtained by RBC-IDS and ASCH-IDS, while RBC-IDS provides better detection over the ASCH-IDS. Security vulnerabilities, including intrusions into network systems and sink nodes, may occur in either a cyber or digital portfolio. As an essential solution for computer security, remote monitoring has been implemented to deal with external aggression in communication systems and immediately identify a different intrusion. The restricted Boltzmann machine (RBM) is a two-layer neural, energized network: visible (V) and hidden (V) (H).

A classifier is investigated in Vinayakumar et al. [1] deep Naive Bayes to build a scalable and efficient IDS to detect and identify unexpected cyber-attacks. The constant change in network architecture and rapid attack development makes it essential to evaluate different datasets created by simple and complex strategies over the years. This form of research promotes discovering the best algorithm that can work effectively to identify potential cyber-attacks. On different publicly accessible benchmark minicom, a thorough assessment of observations on datasets and other robust machine learning SVM classifiers seen. Using super configuration selection methods with the KDDCup 99 dataset, optimum system configurations and networking protocols for DNNs have been picked. All DNNs experiments have performed up to 1000 iterations with a learning rate varying in the [0.01–0.5] range. The DNN template that worked well on KDDCup 99 added to the test's performance on specific datasets like NSLKDD, UNSW-NB15, Kyoto, WSN-DS, and CICIDS 2017. Bypassing then into several hidden layers, this DNN model can learn about the complicated but high detailed image classification of the IDS data. It was verified by rigorous experimental research that DNNs performs well compared to classical classifiers. A category for ANN is defined as a directed graph to transfer different display adapter along edges through one node to that without generating a cycle recurrent neural system.

According to Sheu et al. [6], the malicious activity identification using a structural steel part identifying formulation and construction provided by IDS-DLA framework. There is a massive volume of forms but found helpful, provided the composite materials. Using mathematical and CNN triplet filters, IDS-DLA performs high precision and intrusion recognition from the point cloud database. The IDS-DLA also tests the Hu moments ranking to pick its top-five ranking forecasts as final results obtained. Undoubtedly, it was found from the experiments that high precision is achieved compared to the previous benchmarks. Using hybrid feature extraction approach and multi-filtering approach concludes that it achieves greater efficiency. The main factors will be the coordination between humans, materials, and the corresponding machines with correct recipes to improve or increase the production lines' overall efficiency.

In combination with Abou et al. [2], the durability of IDS focused on computer vision against intrusion detection. This refers to the min–max method in the UNSW-NB 15 dataset to train intrusion detection systems against crown prosecution samples. On the other hand, this uses the current min strategy as a security strategy to optimize the detection mechanism that minimizes the loss during training data of the integrated adversarial samples. This research tests and calculates the efficacy of the techniques of malicious attacks and the resistance of the training images against such attacks. This system used a methods of remote attack that have been built in binary environments which should be used in trustworthy environment and generate network attacks. Finally, this shows that removing the primary component analysis function will improve the robustness of the sensor network using a learning algorithm. Each column has 49 characteristics, including the class name, in the dataset. You pick the best 28 features known as feature ranking (or score). Regularization is also used to select several features with a broader composition to enhance the classifier's effectiveness and reliability.

Self-taught learning (STL) with an inter SVM has used and according to Arthur [3] IDS to preserve the high highest accuracy of the IDS, except in unknown waters. The neural learning, a deep learning-based algorithm of vehicle routing modeling to promote the continued survival home, uses a self-healing approach in the IDS recovery phase. Simulation results indicate the efficacy of precision, precision, and accuracy of the proposed IDS against cyber-security attacks on UAVs. This system able to prevent potential cyber-attacks such as GPS hacking and bumping, a drone's contact patterns lost, compromised, or hijacked.

Framework against cyber-security threats, even in missing signals or unforeseen attacks or uses a learning algorithm, ensures every unscrewed aerial vehicle. Return back to home or to the nearest safe zone and design software lightweight, responsive and efficient detection mechanism using a deep learning model with a multi-class support vector machine for massive and small attacks on surveillance drones (UAV).

The method, datasets, recent work on Re-ID, challenges, approaches, and techniques that have been applied using computer vision systems are briefly discussed in Jaiswal and Dinesh [7]. Some of the most widely used re-Id applications include video monitoring, pedestrian counting, multiple camera tracking, and multiple camera activity detection and analysis. Its purpose is to classify individuals.

In the photograph, there is only one picture, no physical examination, or effects of treatment. Its purpose is to provide period, image location, moving goals, and nearly people in a crowded scene. The challenge here, unlike detection, is time. It recognized an individual in a group of individuals in a camera network at various locations that do not have a similar field of view. The state-of-the-art outcomes are obtained through CNN-based techniques on the individual re-ID method. The necessary process of recognizing individuals object called as re-ID of individuals. Concerning deep learning methods and the recent detection methods has proposed for this review. This has also addressed the ongoing studies, tasks, and difficulties of individual re-identification. It turns out to be a complicated task as applications start to develop in a wide variety of varieties. Individual re-identification is catching eyes not only in academia but also because of its diverse applications in industries.

According to Atefi et al. [8], to optimize the detection rate and reduce the error, a detailed experimental analysis based on different binaries is presented. Various studies were also carried out on intrusion detection systems with the old dataset, such as the Kddcup'99 dataset. Most of them did not accurately detect intrusion with the new intrusion because the old dataset does not protect the current attacks. The basic objective of this system is to detect zero-day attacks using deep learning (DL) and binary algorithm (BA) hybrid anomaly classification of IDS. It also demonstrates detection accuracy on numerous synthetic dataset called KDDCUP99, NSLKKD, etc. Besides, this study performs the deep neural network (DNN) anomaly classification of IDS as the deep learning (DL) platform and binary algorithms (BA) in terms of conditional bat algorithm, binary evolutionary algorithms (BGA), and binary magnetic simulated annealing as an optimization method to increase detection speeds. Some of the results for DNN and the hybrid version that were regarded and reached are accuracy, retrieval, correctness, classification error, tolerance, specificity, but cost error. The study researcher will use visualization for the results. This can also be used for laboratory experiments and procedures using MATLAB. Furthermore, researchers produce graphs and diagrams to be used as a calculation stage and visualize the experiments' effects.

Using a deep learning model that can identify different types of attacks without human-generated rules or signature mapping, an intrusion detection technique is used in Chockwanich et al. [9]. This uses RNN, stacked RNN, and CNN supervised deep learning technologies to identify five common types of attacks by using Keras at TensorFlow's top. This technique only requires knowledge about the packet header and does not require any payload from the user. This uses MAWI datasets, which are Winpcap files, to check the performance and compare them with Snort IDS. The results show that Snort could not detect the network scan attack via ICMP and UDP due to the lack of user payloads. This system also indicates that RNN and CNN can be used to identify port scan attacks, ICMP network scans, UDP network scans, TCP network scans, and provides high detection accuracy for DoS and DDoS attacks. Using deep learning with TensorFlow, it is able to detect five types of common attacks, which are associated with DoS, port scan, network scan via UDP, network scan via TCP, and network scan via ICMP, RNN, and provides the highest accuracy for all network and host-based attacks. It will analyze and compare the proposed system according to the output matrix.

A compressive and comparative study of the NSLKDD and CIDDS-001 benchmark network audit datasets with machine learning classification has been done in Rashid et al. [10]. This used hybrid feature selection and ranking methods before applying self-learning (machine or deep learning) classification algorithmic approaches to achieve optimal results, SVM, Naïve Bayes, k-NN, neural networks, DNN, and DAE, for example. Via some prominent performance indicator metrics such as precision, this has evaluated the performance of IDS.

The system can manually as well as dynamically monitor the network- and host-based traffic for vulnerable and malicious threats. It also generates false-positive warnings when any new attack generated on terminal. Anomaly-based cyber-security refers to the problem of identification in supervised learning in which a model of intrusion is created. Based on the training dataset, legitimate activities, any deviation occurring in the standard model, is known as an attack or anomaly.

3 Proposed Works

3.1 Problem Statement

In this research work, the system aims to design and develop an approach for intrusion Detection for fast learning-based neural network and a machine learning approach to evaluate the proposed system evaluation on different network datasets that will produce the classification accuracy of the system.

3.2 System Architecture

The proposed anomaly and signature-based network intrusion detection and prevention system aims to cultivate the detection accuracy and minimize the false-positive rate, detector generation time and two-phase in the proposed system. The system has NSLKDD dataset for system training as well as testing purposes. The system deals

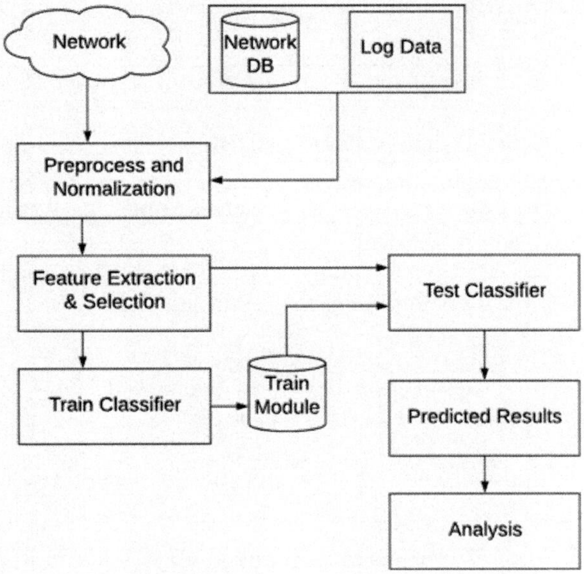

Fig. 1 Proposed system architecture

with deep learning base fine-tuning heterogeneous untrusted network configuration. When two or more combinations form a new model commonly called an ensemble model, this model incorporates input from a deep learning classifier. This conceptual structure consists of numbers for the classifiers. First, the system receives data from various sources, both online and offline. Once the system collects the data, it will be applied in proposed classification approaches to identify the packet as malicious or normal (Fig. 1).

The system initially collects the input packet from various sources numerous synthetic datasets and real-time network data packets. The entire execution holds three different phases which are listed below.

Module 1: This module consists of data collection, data preprocessing, data normalization, feature extraction, and selection with classification algorithm for training and testing, respectively, called IDS.

Module 2: Intrusion prevention system (IPS) works to prevent known attacks that have already been generated from remote sources. The various machine learning classification algorithms have been used to prevent the malicious activities of remote user. The system works like reinforcement learning when new attack has generated, and it stores automatically in rule repository dynamically.

Module 3: Intrusion response system (IRS) carried out security from various unknown attacks and malicious behaviors. The system executes the ensemble modules with the help of multiple machine learning classifiers for detecting malicious activity.

3.3 Algorithm Design

3.3.1 Customized Recurrent Neural Network (CRNN)

Input: Training dataset TestDBLits [], Train dataset TrainDBLits[] and Threshold Th.
Output: Resulset < class_name, Similarity_Weight > all set which weight is greater than Th.
Step 1: For each testing records as given below equation, it works in convolutional layer for both training as well as testing

$$\text{testFeature}(k) = \sum_{m=1}^{n}(.\ \text{featureSet}[A[i] \ldots A[n] \leftarrow \ \text{TestDBLits}\ )$$

Step 2: Create feature vector from testFeature(m) using below function.

$$\text{Extracted\_FeatureSet\_}x[t \ldots n] = \sum_{x=1}^{n}(t) \leftarrow \text{testFeature}(k)$$

Extracted_FeatureSet_$x[t]$ is the outcome of each pooling layer that is extracted from each convolutional layer and forward to net convolutional layer. This layer holds the extracted feature of each instance for testing dataset.
Step 3: For each train instances as using below function,

$$\text{trainFeature}(l) = \sum_{m=1}^{n}(.\ \text{featureSet}[A[i] \ldots A[n] \leftarrow \ \text{TrainDBList}\ )$$

Step 4: Generate new feature vector from trainFeature(m) using below function.

$$\text{Extracted\_FeatureSet\_}Y[t \ldots n] = \sum_{x=1}^{n}(t) \leftarrow \text{Train Feature}(l)$$

Extracted_FeatureSet_$Y[t]$ is the outcome of each pooling layer that is extracted from each convolutional layer and forward to net convolutional layer. This layer holds the extracted feature of each instance for training dataset.
Step 5: Now evaluate each test records with entire training dataset, in dense layer

$$\text{weight} = \text{calcSim}\left(\text{FeatureSetx} \ || \sum_{i=1}^{n} \text{FeatureSety}[y]\right)$$

Step 6: Return Weight.

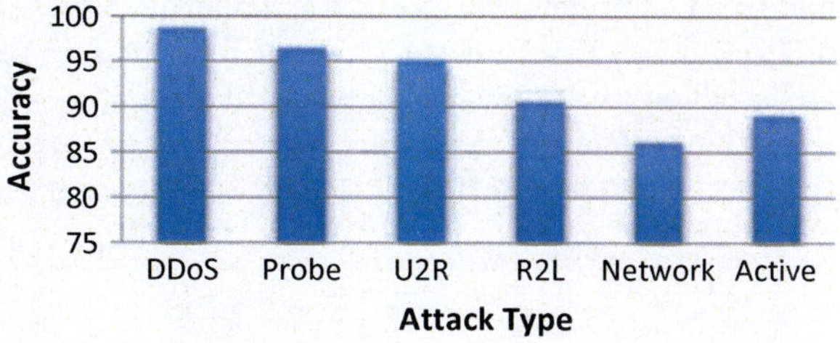

Fig. 2 Accuracy of the proposed system for various attacks using machine learning algorithms

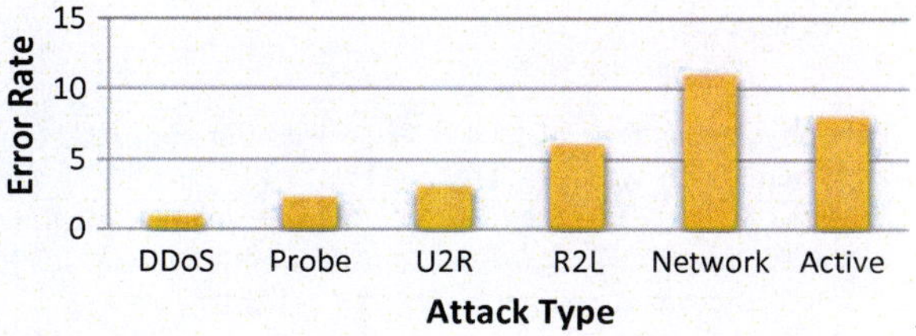

Fig. 3 Error rate of the system for various attacks for proposed algorithms

4 Results and Discussions

The proposed research focuses on the grid computing approach and classification-based recognition, and fundamentally this approach is consuming the respectable exposure rate, nevertheless generating sometimes more false-positive ratio. Some systems are also not applicable in real-time environments, and some may not focus on misclassified anomalies. As noted, the mark is still missing in most applications because no program currently provides a discovery rate of 100% and the sky is the limit (Fig. 2).

The above figure shows the detection accuracy of proposed for various network attacks using synthetic real-time and network datasets. We have tested KDDCUP99, NSLKDD, BOTNET, and ISCX datasets for multiple experimental analyses. The error rate of each attack is also low for all attacks. Figure 3 demonstrates in detail the error rate of the system.

5 Conclusions

The system suggested an ensemble method for network traffic anomaly detection in this research work. Our approach focused on building the model of anomaly detection of normal traffic profile. Through experiments, the system also showed some features of the various KDD dataset and ISCX dataset with the normal profile are efficient. With input training data, system uses RNN algorithm for classification. The experiments showed that our approach works well, even with a small training sample, including precise recognition. The system will use an open-source IDS to build and play with the proposed model in our plan's existing network. Several different ideas have emerged to confront this problem since about ten years ago; the intrusion detection concept began to gain momentum in the security community. Detection systems for intrusion vary in the approaches used to collect the data and the specific techniques used to analyze them. Evaluate the system with some combination of network and synthetic dataset and generate the dynamic rules for strongly unknown attack detection in a vulnerable environment.

References

1. Vinayakumar R et al (2019) Deep learning approach for the intelligent intrusion detection system. IEEE Access 7:41525–41550
2. Abou K, Rana M, Omair S, Ashraf M (2020) Investigating resistance of deep learning-based ids against adversaries using min-max optimization. In: ICC 2020–2020 IEEE international conference on communications (ICC). IEEE
3. Arthur MP (2019) Detecting signal spoofing and jamming attacks in UAV networks using a lightweight IDS. In: 2019 international conference on computer, information, and telecommunication systems (CITS). IEEE
4. Vijayanand R, Devaraj D, Kannapiran B (2019) A novel deep learning-based intrusion detection system for smart meter communication network. In: 2019 IEEE international conference on intelligent techniques in control, optimization, and signal processing (INCOS). IEEE
5. Otomo S, Kantarci B, Mouftah HT (2019) On the feasibility of deep learning in sensor network intrusion detection. IEEE Networking Lett 1(2):68–71
6. Sheu RK et al (2020) IDS-DLA: sheet metal part identification system for process automation using deep learning algorithms. IEEE Access 8:127329–127342
7. Jaiswal S, Dinesh KV (2019) State-of-the-arts person re-identification using deep learning. In: 2019 6th international conference on signal processing and integrated networks (SPIN). IEEE
8. Atefi K, Habibah H, Touraj K (2020) A hybrid anomaly classification with deep learning (DL) and binary algorithms (BA) as optimizer in the intrusion detection system (IDS). In: 2020 16th IEEE international colloquium on signal processing & its applications (CSPA). IEEE
9. Chockwanich N, Vasaka V (2019) Intrusion detection by deep learning with tensor flow. In: 2019 21st international conference on advanced communication technology (ICACT). IEEE
10. Rashid A, Muhammad JS, Shahid MA (2020) Machine and deep learning based comparative analysis using hybrid approaches for intrusion detection system. In: 2020 3rd international conference on advancements in computational sciences (ICACS). IEEE

Multi-entity Topic Modeling and Aspect-Based Sentiment Classification Using Machine Learning Approach

Ganesh N. Jorvekar and Mohit Gangwar

Abstract Online reviews are essential for measuring the quality of business using sentiment analysis, opinion mining, and sentiment classification, which has achieved much popularity in recent years. The product quality has been evaluated based on a large number of user's comments on social media. To evaluate such a kind of large data is a very tedious ask task in data mining. Choosing any product or service after checking existing users' opinion is the general thing for new users. Still, it is too hard to evaluate such kind of large in single features. Such techniques have already been proposed by existing analysis, but numerous systems are facing negation handing, accuracy-related issues, and eliminating those problems. We proposed sentiment classification on restaurant review dataset using naive Bayes classification. This work also carried out data preprocessing using natural language processing. Various feature extraction technique has been used to collect unique and important features itself. Different cross-validation techniques are also used for experimental investigation. The dataset consists of approximately 3000 records. Each form is being analyzed utilizing the hybrid naive Bayes procedure, using a succession of NLP and a WordNet dictionary; this is a probabilistic method. Eventually, the result is a concern in the form of a matrix. The correctness of the intended approach is yet calculating with the help of the confusion matrix. The system provides better classification accuracy according to given aspect for real time and synthetic dataset.

Keywords Sentiment analysis · Opinion mining · Supervised learning · Classification · Naïve Bayes · NLP · Features extraction · Features selection · Machine learning

G. N. Jorvekar
Department of Computer Science and Engineering, Bhabha University, Bhopal, Madhya Pradesh, India

M. Gangwar (✉)
Engineering, Bhabha University, Bhopal, Madhya Pradesh, India

R. P. Mahapatra et al. (eds.), *Proceedings of International Conference on Recent Trends in Computing*, Lecture Notes in Networks and Systems 341,
https://doi.org/10.1007/978-981-16-7118-0_46

1 Introduction

The sentiment classification faces many problems nowadays due to unstructured data statement, negation handling, and some other technical data errors [1]. The first is a word of display view as a positive and similar word in another statement as negative. A second problem is people do not also have to share views. Most conventional text processing is focused on the assumption that there are minor variations between these two bits of text that do not so much change the context. However, in sentiment analysis, "the food was great" is "the food was not great" really specific [2]. Throughout their comments, individuals may be inconsistent. Most comments are set to have both constructive and negative feedbacks. They can be addressed more by evaluating sentences one at a time. The text review is also more essential rather than rating views, and the large text sometime demonstrates positive and negative views, respectively. In such neutral classification, problem has occurred, similar to multi class classification problems. Even, we have trouble knowing what someone means based on a brief piece of text in the meaning is missing.

The high utilization of the Internet has changed the way people opine social media user's perceptions. Nowadays, it has done using different social blog posts, online forum discussions, product review web applications, etc. To a large extent, users usually trust the content created by the end-user when it comes to carrying out the required action. When users need to obtain a product online, they must first search the ratings on that particular product review page online before making a purchasing decision. Several analyses have to be done on all such comments. The final result shows that the company product is good or not to purchase. There are different techniques for sentiment analysis available with many applications for different domains, like in business to get users' comments on products. Knowledge-based recommendation and machine learning techniques are two methods that are used mainly for the study of emotional identification as well as sentiment classification. The knowledge-based approach includes a broad collection of predefined emotions and usable representation of knowledge to classify sentiment class comment. In the machine learning approach, no predefined collection of emotions is required. This allows using a training feature to generate the training rules that runtime classifies the test data using supervised classification algorithms. Numerous machine learning techniques are used to classify a large amount of unstructured data using classifiers like naïve Bayes classifier, SVM, DT, RF, ANN [3, 4].

In the NB classification process, users eventually planned the number of classes for the data based on various application programs [5]. We then retrieved characteristics from the fabricated data, names, lengths, styles, sizes, and parents' folders. On a Bayesian classifier, the features extracted were designed to predict the system file activity of different functional classes. The framework defines then used to classify the new activities of the system files. The block diagram of the overall procedure is shown in Fig. 1.

The analysis is a technique used to implement a variety of various sets of data into other types. These classification procedures have been apportioned into two sections,

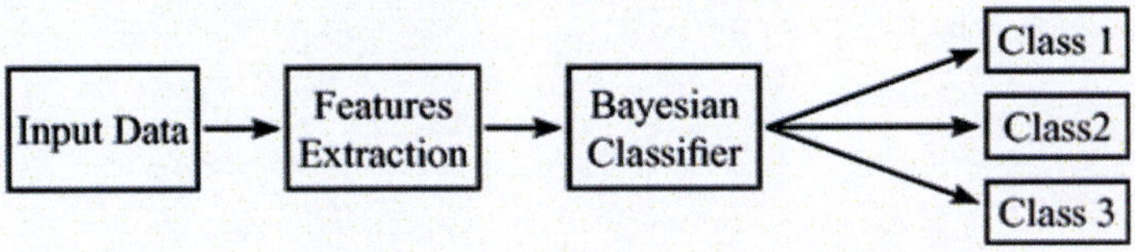

Fig. 1 Block diagram of the classification process using naïve Bayes

supervised and unsupervised. We concentrate on one specific activity of categorizing text, interpreting emotions, extracting sentiments, the positive or negative inclination a user communicates toward any topic. An opinion about the movie, book, or product on the web described the reviewer's sentiment for the service. In contrast, an editorial or political text expresses sentiment for a party or government action. Extracting customer or public opinion is also important in fields ranging from ads to government. In this work, we proposed sentiment classification using machine learning techniques. The naïve Bayes classification algorithm has been used to train and test the entire dataset and grabs the system result. Various feature extraction and selection techniques have been carried out to achieve the implementation of a system and finally tested the system on multiple user reviews dataset.

2 Review of Literature

This section focused on the study of numerous existing systems, and the work has done by various existing research in IDS.

In this segment, we present the naive Bayes classifier, popular as the be-naive Bayes classifier, since it is a Bayesian classifier that simplifies (naive) inference about how the features communicate. In Fig. 2, we depict a text document as though it was a bag-of-words, i.e., an unordered collection of terms ignoring their location and maintaining their frequency in the document. In the example "I love this movie" and "I would recommend it," we simply note that the word I occurred five times in the whole statement, so such words are removed by stop word removal and lemitization algorithms that generated over-fitting problem. Customer reviews are a significant tool for consumers who want to buy a product, watch a video, or go to a restaurant, as well as companies that monitor customer input. Exponential growth in knowledge created by social networks online needs efficient recommendation mechanisms to produce valuable outcomes. The fusion of social contextual factors that are derived is important and challenging from improved performance to social behavior to suggestions.

Ma et al. [1] improved the LSTM network with a hierarchical attention mechanism. The main influence of this work is that common sense knowledge of sentiment-related concepts is integrated into an attentive LSMT. They completed both task, i.e., aspect extraction and sentiment classification.

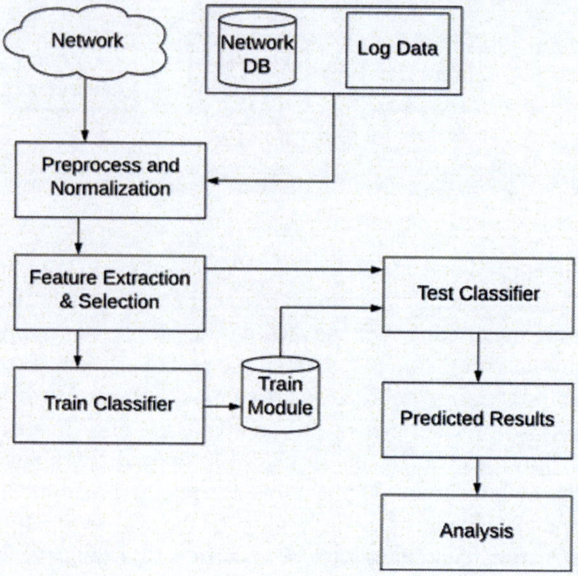

Fig. 2 Proposed system architecture

Zeng et al. [6] used a convolutional neural network (CNN) with linguistic resources and gating mechanism for aspect-based topic modeling. Al-Smadi et al. [2] proposed a supervised learning approach for the ABSA. In this research, aspect extraction and sentiment classification are implemented. The classifiers are trained on lexical, morphological, syntactic, and semantic features. By using SVM with six different pattern classes, Baas et al. [7] proposed a new method. This includes lexical, syntactic, semantic, hybrid, and surface feelings. They used lexico-semantic patterns in customer reviews to find aspect topic. Synset bigram, negator POS bigram, and POS bigram are used to enhance the extraction of aspects based on feelings.

Latent Dirichlet allocation topic model is applied by Amplayo et al. [8]. This work is an extension of the aspect and sentiment unification model (ASUM). It considers seller sentiments for aspect extraction and sentiment classification. Seller-aided aspect-based sentiment model (SA-ASM) and seller-aided product-based sentiment model (SA-PBM) are proposed. SA-ASM provides improved results for sentiment classification and SA-PBM for aspect extraction. Xu et al. [3] proposed a new implicit aspect identification approach which depends on non-negative matrix factorization (NMF). This method clusters aspects of a product by merging co-occurrence data with intra-relations of aspects and sentiments.

De Kok et al. [9] presented ontology-centered approach for review-level aspect-based sentiment analysis. In this work, they mainly focus on ontology-enhanced methods that complement a standard ML algorithm. For these two different algorithms are used, a review-based algorithm and a sentence aggregation algorithm. Their work contains feature generators and feature adaptors. Several feature generators are independent of the ontology which are aspect, sentence count, lemma and several are dependent on an ontology which are ontology concepts, sentiment count.

Also, they used several feature adaptors which are ontology concept score, negation handling, synonyms, weight, word window, etc.

In the current research, several approaches are applied. The bulk of the actual work involves machine learning algorithms [4], including SVM, random forest, KNN. The current research provides a basis for the process suggested. Various studies are performed in various forms of emotional research, and each approach requires different stages of the study. In several instances, the consequence would be corresponding in all works that identify the text as beneficial, destructive. Some scholars will describe the decision in the form of charts and some in records, etc. On analysis, it is found that KNN has the deepest level of correctness in classification. Most of the past work involves a probabilistic way of research.

Sayali et al. [10] generated a Twitter API list and compiled all the blue whale game tweets. Their key goal is to carry out nostalgic tweet research. They used naïve Bayes, help vector machines, ensemble classifier, and maximum entropy. Classifiers SVM and naive Bayes are introduced utilizing the built-in MATLAB functions. A full classifier for entropy is introduced using maximum entropy. Naïve Bayes has higher consistency and marginally poorer recall and consistency based on comparable tests, i.e., 89%. Other classifiers have similar accuracy rates, i.e., 90%. The outcome reveals a pie chart with percentages of good, negative, and neutral hashtags.

Jain and Dandannava et al. [5] recommended a framework for conducting a sentiment study on the Apple, bank, and BSNL datasets. They have 75% of the dataset used for preparation and the remaining 25% for processing. They used two different classification techniques, i.e., the classifier for multinomial naïve Bayes and the classifier for decision tree. Preprocessing of tweets is carried out using the extraction of the feature. They used the portable Apache Spark platform, which provides quicker, reliable results.

In this work [11], two separate approaches have been used, supervised and unsupervised learning. The co-occurrence of the terms is known, and the association of certain terms is evaluated. The major demerit of the proposed approach is that some of the parameters required for interpretation must be established initially, which is very complicated. The authors have defined the heuristics to establish certain parameters. All strategies have a level ranking of 74 and 78%, respectively. As potential research, the writers intended to use even more information to produce better performance and use even more machine learning methods to boost precision.

The proposed work [12] involves developing a Hindi multidomain sentiment conscious dictionary that contains a dataset from various domains and is analyzed using supervised learning techniques. The list is comprised of numerous consumer feedbacks. This research analyzes the interaction between terms originating from comments in various domains. For classification purposes, the authors choose the unlabeled results. The outcome received contains approximately 200 ratings, consisting of approximately 100 favorable feedback and 100 negatives. This dictionary assists in certain cases that require the classification details called.

For this work [13], the authors suggested a wording sentiment knowledge-based model using Gibb's sampling process. Here, the authors suggested a weakly guided learning method that would that the issue of term sparsity. The system utilizes the laptop analysis dataset for classification purposes of aspects. The latent Dirichlet allocation (LDA) paradigm is being implemented to generate the clusters of each term sentiment and finally generate similar cluster groups of reviews like positive or negative, respectively.

In the traditional recommender systems [14], the phenomenon of overloading the knowledge continues. Therefore, mining the text is a lot needed to resolve the issue. Thus, the authors have described a rating prediction framework that involves user feedback or review comments such as user purchasing history, geographic place, product type, emotional, relational feedback of various users profile parameters. The user's sentimental similarity and reputation of the item are also considered for better prediction of rating. For prediction, the Yelp dataset is being considered. The results show that the user sentiment highly helps characterize the user's preferences and improves the recommendation performance.

The sentimental classification method proposed here [15] implies a support vector machine's strategy to locate favorable and derogatory feedback. The analysis is being done of the term frequency, precision, recall of technique. The dataset used for review is the data collection for the film. The method's precision is measured, and the recommended approach is shown to be outperforming in all situations. The dataset is preprocessed where the unacceptable values are deleted and only considered for classification. The scientists have agreed to incorporate more machine learning methods as potential research and have a comparative review.

According to García-Díaz et al. [16], it has been developed to enhance the accuracy of probabilistic classifiers by utilizing decision support methods for classifying sentiments. This method operated on basic probabilistic classifiers focused on machine learning, testing a naive Bayes classifier, the foundation of one of the widely popular soft computing techniques. This framework helps boost consumer emotions against a film relative to what will happen with a traditional approach.

According to .Shahana and Omman [17], the sentiment analysis's key problems are (i) appropriate preprocessing for the datasets is not completed. (ii) The overlap between the two terms shall not be recognized. (iii) The treatment of the document through negation is a very challenging job in certain theoretical cases.

3 Proposed Works

The proposed anomaly and signature-based network intrusion detection and prevention system aims to cultivate the detection accuracy, minimize the false positive rate and detector generation time and two-phase in the proposed system. The system has NSLKDD dataset for system training as well as testing purposes. The system deals with deep learning-based fine-tuning heterogeneous untrusted network configuration. When two or more combinations form a new model commonly called an ensemble model, this model incorporates input from a deep learning classifier. This conceptual structure consists of numbers for the classifiers. First, the system receives data from various sources, both online and offline. Once the system collects the data, it will be applied in proposed classification approaches to identify the packet as malicious or normal.

The system initially collects the input packet from various sources numerous synthetic datasets and real-time network data packets. The entire execution holds three different phases which are listed below.

Module 1: This module consists of data collection, data preprocessing, data normalization, feature extraction, and selection with classification algorithm for training and testing, respectively, called IDS.
Module 2: Intrusion prevention system (IPS): This works to prevent known attacks that have already been generated from remote sources. The various machine learning classification algorithms have been used to prevent the malicious activities of remote user. The system work like reinforcement learning when new attack has generated, and it stores automatically in rule repository dynamically.
Module 3: Intrusion response system (IRS) carried out security from various unknown attacks and malicious behaviors. The system executes the ensemble modules with the help of multiple machine learning classifiers for detecting malicious activity.

3.1 Algorithm Design

Algorithm 1: Training

Input: Training dataset TrainData[], various activation functions[], threshold Th.
Output: Extracted features Feature_set[] for completed trained module.
Step 1: Set input block of data d[], activation function, epoch size,
Step 2: Features.pkl ← ExtractFeatures(d[]).
Step 3: Feature_set[] ← optimized(Features.pkl).
Step 4: Return Feature_set[].

Algorithm 2: Testing

Input: Training dataset TestDBLits [], train dataset TrainDBLits[], and threshold Th.
Output: Result set < class_name, Similarity_Weight > all set which weight is greater than Th.
Step 1: For each testing records as given below equation

$$\text{testFeature}(k) = \sum_{m=1}^{n} (.\text{featureSet}[A[i] \ldots A[n]\text{TestDBLits}\ )$$

Step 2: Create a feature vector from testFeature(m) using the below function.

$$\text{Extracted\_Feature Set\_}x[t\ldots n] = \sum_{x=1}^{n}(t) \leftarrow \text{testFeature}(k)$$

Extracted_FeatureSet_$x[t]$ holds the extracted feature of each instance for the testing dataset.

Step 3: For each train instances as using the below function

$$\text{trainFeature}(l) = \sum_{m=1}^{n}(.\ \text{featureSet}[A[i]\ldots A[n]\text{TrainDBList})$$

Step 4: Generate new feature vector from trainFeature(m) using below function.

$$\text{Extracted\_FeatureSet\_}Y[t\ldots n] = \sum_{x=1}^{n}(t) \leftarrow \text{TrainFeature}(l)$$

Extracted_FeatureSet_$Y[t]$ holds the extracted feature of each instance for the training dataset.

Step 5: Now, evaluate each test records with the entire training dataset

$$\text{weight} = \text{calcSim}\left(\text{FeatureSetx}\ ||\sum_{i=1}^{n}\text{FeatureSety}[y]\right)$$

Step 6: Return weight

4 Results and Discussions

Fig. 3 shows the train and test dataset distributions of the entire dataset. Each section shows a number of occurrences of the dataset in percentage. The 66–34 data splitting standard approach has been used for different (e.g., 5, 10, 15) cross-fold data validations. We have given around 2706 instances for training the system and 1394 instances for testing the dataset.

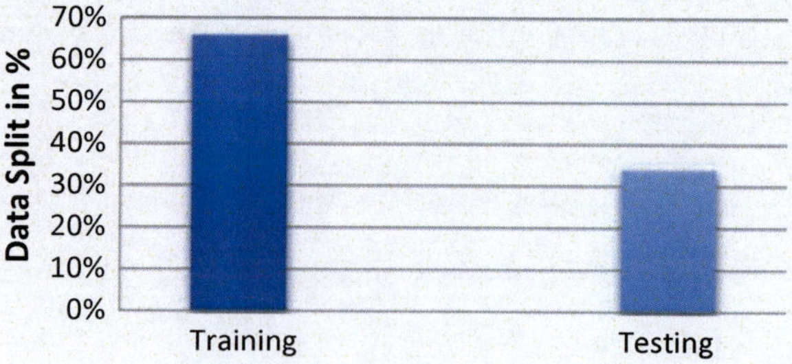

Fig. 3 Data splitting for training and testing, respectively, with different cross-validation

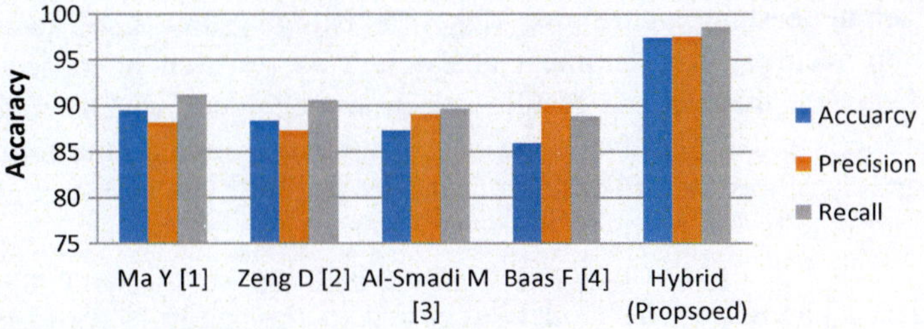

Fig. 4 Performance evaluation of the proposed system

Table 1 Comparison of classification

Prediction	Class	Proposed method	Naive Bayes		
			10-fold	15-fold	20-fold
True	Positive	85.99	94.43	55.25	87.34
	Negative	93.96	15.43	74.35	91.15
False	Positive	14.01	5.57	44.75	12.66
	Negative	6.04	84.60	23.65	8.85

To determine the efficiency of such classification algorithms, we used another dataset as test instances created by various social medial data sources. The performance analysis has been done with the various existing systems that have developed similar techniques with different machine learning algorithms (Fig. 4).

Another experiment has done with different data samples and evaluated the classification accuracy accordingly. The outcome of first experiment demonstrated in Table 1 that clearly showed proposed method provides better classification accuracy over the traditional naïve Bayes classifier. We have done three type of cross-validation of data and showed each splitting result in table also. Standard classification of naïve Bayes uses categorized text or sentences as a training class. The consequence of using different positive subset and the pessimistic dataset is in Table 1.

In comparison, the system has evaluated with traditional naïve Bayes classification with our proposed NLP and SentiWordNet hybrid naïve Bayes algorithm.

5 Conclusions

In this research, the extraction of sentiment knowledge from feedback by online users suggests a recommendations model. We combine consumer opinion, the behavioral influence of emotion, and a comparison of credibility into a single matrix factorization process to accomplish ranking prediction. This research also proposed sentiment classification according to the user's perspective on social media reviews. Various existing systems illustrated similar kinds of work and sentiment classification with several machine learning algorithms. Most of existing system is still facing proper classification accuracy for sentiment analysis. This research is carried out with different feature extraction and feature selection techniques which reduce redundant feature selection and cultivates the classification accuracy for the entire system. This research explores the utilization of various feature extraction techniques and classifiers for classifying business text reviews on a large restaurant review dataset. We built a model that performs using machine learning to determine emotions on large dataset technology. The model suggested in this study is constructed with natural language processing (NLP) technique on the dataset, with tweets. Concept bag-of-words is included, which contains separate words, both positive as well as negative. We opted for an appropriate social media option, but dataset that increases classification's efficacy and accuracy.

This system generates that class imbalance problem data are unstructured, and most of NLP features generate over-fitting problem that reduces the accuracy. To implement some deep learning classifiers on heterogeneous dataset will be interesting research in the future direction.

References

1. Ma Y, Peng H, Cambria E (2018) Targeted aspect-based sentiment analysis via embedding commonsense knowledge into an attentive LSTM. In: Thirty-second AAAI conference on artificial intelligence. pp 5876–5883
2. Al-Smadi M, Al-Ayyoub M, Jararweh Y, Qawasmeh O (2019) Enhancing aspect-based sentiment analysis of Arabic hotels' reviews using morphological, syntactic and semantic features. Inf Proc Manag 56:308–319. https://doi.org/10.1016/j.ipm.2018.01.006
3. Xu Q, Zhu L, Dai T, Guo L, Cao S (2019) Non-negative matrix factorization for implicit aspect identification. J Ambient Intell Humanized Comput
4. Neethu MS, Rajasree R (2013) Sentiment analysis in Twitter using machine learning techniques. In: 2013 fourth international conference on computing, communications and networking technologies (ICCCNT)
5. Jain AP, Dandannavar P (2016) Application of machine learning techniques to sentiment analysis. In: 2016 2nd international conference on applied and theoretical computing and communication technology (iCATccT) pp 628–632
6. Zeng D, Dai Y, Li F, Wang J, Sangaiah AK (2019) Aspect based sentiment analysis by a linguistically regularized CNN with gated mechanism. J Intell Fuzzy Syst 36:3971–3980. https://doi.org/10.3233/JIFS-169958
7. Baas F, Bus O, Osinga A, van de Ven N, Van Loenhout S, Vrolijk L, Schouten K, Frasincar F (2019) Exploring lexico-semantic patterns for aspect-based sentiment analysis. In: Proceedings of the 34th ACM/SIGAPP symposium on applied computing. pp 984–992. https://doi.org/10.1145/3297280.3297377
8. Amplayo RK, Lee S, Song M (2018) Incorporating product description to sentiment topic models for improved aspect-based sentiment analysis. Inf Sci 454–455:200–215. https://doi.org/10.1016/j.ins.2018.04.079
9. de Kok S, Punt L, van den Puttelaar R, Ranta K, Schouten K, Frasincar F (2018) Review-aggregated aspect-based sentiment analysis with ontology features. Prog Artif Intell 7:295–306. https://doi.org/10.1007/s13748-018-0163-7
10. Nazare SP, Nar PS, Phate AS, Ingle DR (2018) Sentiment analysis in Twitter. Int Res J Eng Technol (IRJET) 05

11. Schouten K, van der Weijde O, Frasincar F, Dekker R (2018) Supervised and unsupervised aspect category detection for sentiment analysis with co-occurrence data. IEEE Trans Cyber 48(4)
12. Jha V, Savitha R, Deepa Shenoy P, Sangaiah AK, Venugopal KR (2018) A novel sentiment aware dictionary for multi-domain sentiment classification. J Comput Electric Eng 69:585–597
13. ShufengXiong KuiyiWang, Donghong JB (2018) A short text sentiment topic model for product reviews. J Neurocomput 297(5):94–102
14. Lei X, Qian X, Zhao G (2017) Rating prediction based on social sentiment from textual reviews. IEEE Trans Multimedia 40(4)
15. Tripathy A, Agrawal A, Rath SK (2015) Classification of sentimental reviews using machine learning techniques. Procedia Comput Sci 57:821–829
16. García-Díaz V, Crespo RG, Bustelo BCPG, Lovelle JMC, Espada JP (2018) An approach to improve the accuracy of probabilistic classifiers for decision support systems in sentiment analysis. Appl Soft Comput 67:822–833
17. Shahana PH, Omman B (2015) Evaluation of features on sentimental analysis. Procedia Comput Sci 46:1585–1592

Demographic Filtering for Movie Recommendation System Using Machine Learning

Sanket Rastogi, Divyanshu Agarwal, Jatin Jain, and K. P. Arjun

Abstract In e-commerce, a film advisory scheme aims to tailor the user interface by showing things that the user will most likely be interested in based on related items that the user was interested in or items similar to users. Two facets and approaches of online recommendation systems, using the consumer and collective filtering frameworks, are the primary subject of our project. To complete the assessment, a film advisory system is developed and intended for a forecast user-backed ranking from users of related films, then evaluate and compare various algorithms and prediction models to construct the recommendation system. The film rating system often produces predicted user-based outcomes. To determine the accuracy, we take advantage of the knowledge from Movie Lens (ML) Web sites. Following research, the item-based root medium square derivation (RMSD) filtering was observed as opposed to the user-based assessment of filtering efficiency determined by an average bug between the rating and the rating expected. Our work finds the possibility for small-to-medium (SME) e-commerce and information filtering company to refine and scan through online businesses for new goods, so that demands can be satisfied and more sales can be created. Building on an algorithm to evaluate user liking, discouragement, and related customer tastes in order to produce the best outcomes with their goods or services is one of the most common methods for companies to help consumers communicate online or find objects. Including content, games, and collaborative, there are several types of suggestions framework models. By dynamic means, we mean a model that runs on the database's Web business and offers ratings or estimates at a future period that helps and maintains the interaction of consumers with the pages and applications. The explicit data input is used to form a model of choice for the buyers. Such considerations such as purchasing experience and search history should be considered. However, the five-star computer ranking datasets will be used for review of the recommendation framework in our work. The principal problems are enhancing filtering scalability, sparsity, and efficiency. In our research, we solve these problems by using algorithms based on user and material for the recommendation systems for different approaches, as we explored with users or item-based ranking estimation on ML datasets using filter users and item-based techniques.

S. Rastogi (✉) · D. Agarwal (✉) · J. Jain (✉) · K. P. Arjun (✉)
Computer Science and Engineering, Galgotias University Greater, Noida, Uttar Pradesh, India

R. P. Mahapatra et al. (eds.), *Proceedings of International Conference on Recent Trends in Computing*, Lecture Notes in Networks and Systems 341,
https://doi.org/10.1007/978-981-16-7118-0_47

Keywords Movie recommendation · E-commerce · Collaborative filtering · RMSD · Machine learning

1 Introduction

The movie recommendation method uses an algorithm to forecast unknown information with available user expertise and relevant objects [1, 2]. A film suggestion framework in e-commerce aims to personalize the user experience by showing the things that the user is most likely to be interested in, based on similar items that the user has been interested in or items that similar users have been interested in. The key goal of our research is to concentrate on two dimensions and approaches of online recommendation systems, using user-based and point-based collective filtering models [3].

Content filtering—Objects are recommended in content-related filters based on contrasts of object profile and user profile [4]. The user profile is the content considered to be important to the user in the form of keywords (or features). An object's item profile includes a series of keywords (or features). Consider for example a situation in which a person purchases his favorite 'X' cake for a cakes. Cake 'X' was sadly sold out and, because of that, it is suggested that the individual purchase cake 'Y' which contains cake-like ingredients. Figure 1a show the content-based filtering case.

Collaborative filtering-based systems—The shared screening framework proposes products based on consumer and/or object similarity tests [5]. The scheme suggests products that related consumers prefer. This is focused on the case in which

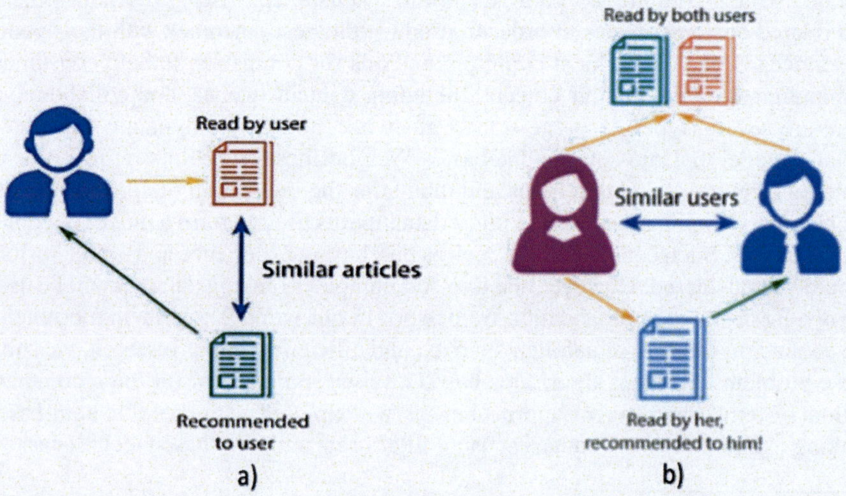

Fig. 1 **a** Content-based filtering. **b** Collaborative filtering

a person asks his friends, who have common preferences, to suggest such movies to him. Figure 1b shows the abstract view of collaborative filtering.

2 Related Works

MOVREC is a recommendation framework for movies presented by Yadav et al. [6] centered on joint activities and approach to filtering. Collaborative filtering makes use of user-provided information. The information is information that the consumers who are arranged first with the film with the highest ranking are evaluated and a film is recommended. The framework also has a provision for the user to pick attributes from which to recommend the film. Two conventional recommendation mechanisms were studied by Capos et al. [7], i.e., content-based filtering and filtering. Filtering collaboratively. He suggested a modern method, which is a new system, because all of them have their own disadvantages, Bayesian network mix and shared routing. The device proposed is tailored for the provided problems and offers distributions of probabilities to make useful inferences. Kaur et al. [8], proposed a hybrid framework. The framework uses a combination of content and an algorithm for shared filtering. The context of the films is often taken into account when suggesting them. The customer relationship plays a part in the suggestion as well as the user–object relationship. They clubbed user-specific information or item-specific information to create a cluster. This is an important methodology for the suggested method based on hierarchical clustering. The ranking of an object voting scheme is used to forecast it. The system suggested has lower errors and higher efficiency. Clustering of things that are identical. Clustering as a way to work with suggestion structures was suggested by Kużelewska et al. [9]. Two strategies for representatives of the computing clusters were presented and analyzed. Solution based on centroid and memory-based collaborative methods of filtering have been used as a framework for comparing the efficiency of the two methods suggested. The result was a considerable improvement in the accuracy of the recommendations produced as compared to only method-based centroid. Chiru et al. [10] proposed Movie Recommender, a software that uses proven knowledge to have film reviews for the consumer. This scheme aims to solve the particular dilemma of recommendations that come from ignoring user-specific results. The personality profile of the users, their viewing history, and data are gathered from other sources that contain movie ratings. They are built on calculation of aggregate similarities. The framework is a hybrid model that uses both filtering based on content and filtering collaboratively. LIn et al. [11] also proposed a tool called content boosted to estimate the complexity level of each situation for each trainee. The algorithm is broken into two steps, the first being the content-based algorithm. Filtering that increases current case ratings data for trainees, and the second is collective filtering that the final predictions are given. The CBCF algorithm incorporates the benefits of both CBF and CF, while the drawbacks of both CBF and CF are overcoming both of their drawbacks at the same time.

3 Existing Movie Recommendation Models

Here, we explained about different existing models of movie recommendation models.

3.1 Cosine Similarity Method

$$u.v = \sum_{i=1}^{n} u_i v_i \tag{1}$$

$$\text{Similarity} = \cos(\theta) = \frac{u.v}{|u||v|} = \frac{\sum_{(i=1)}^{n} u_i v_i}{\sqrt{\sum_{(i=1)}^{n} u_i^2 \sqrt{\sum_{(i=1)}^{n} v_i^2}}} \tag{2}$$

Equations 1 and 2 represents the Cosine similarity equations, where *u*, *v* are the two vectors.

3.2 Collaborative Filtering Recommendation System

Figure 1b shows the working of collaborative filtering recommendation system.

3.3 K-Nearest Neighbor (KNN)

The KNN algorithm works on the principle that if the majority of a sample's *k* most close neighbors in the feature space belong to the same category, then the sample belongs to that category. As seen in Fig. 2, the majority of w's closest neighbors are in the *x* group, and w is in the *X* category as well. KNN Collaborative Filtering Algorithm-Based Movie Recommendation System Design and Implementation.

We use rating provided by user and predict outcome in Eq. 3

$$r_{ij=\frac{\sum_k \text{similaries}(u_i u_k) r_{kj}}{\text{number of ratings}}} \tag{3}$$

and for users who do not provide rating in Eq. 4.

$$r_{ij} = \underline{r}_i + \frac{\sum_k \text{similaries}(u_{i,} u_k)(r_{kj} - \underline{r}_k)}{\text{number of ratings}} \tag{4}$$

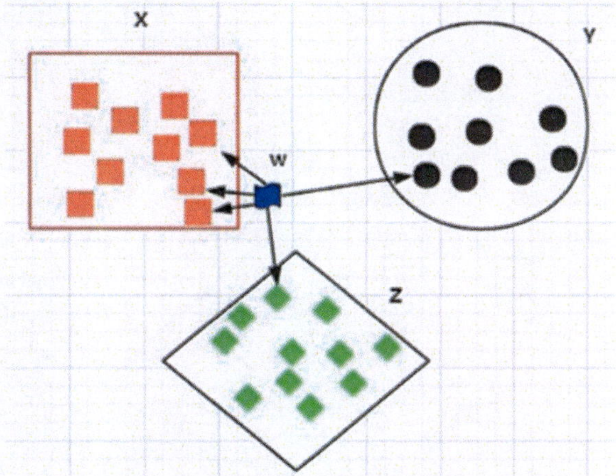

Fig. 2 K-nearest neighbor method

r is rating, and *u* is factors. Figure 2 shows the KNN similarity of different features.

3.4 Matrix Factorization

To develop collective filtering recommendation systems, scarcity could be a major problem. Our method produces matrices in which rows are special in our setting, thus columns represent various films, and the values are the ratings given to films by different users. But the fact that not all films are scored by everyone is very clear. Therefore, the sparsity that must be solved faces this matrix of ours. We use matrix factorization for that reason. In this procedure, we decompose the first sparse matrix with latent characteristics to low-dimensional matrices. The matrix factorization thus demonstrates how much a person is associated with a group of latent functionality and how much a film is aligned with this set of latent functionalities. Figure 3 shows the method of matrix factorization method.

4 Proposed Work

We studied various models and came up with a hybrid recommendation model. It has basically three types of filtering methods.

- Demographic filtering algorithm
- Content-based filtering—Cosine algorithm
- Collaborative filtering—Hybrid algorithm.

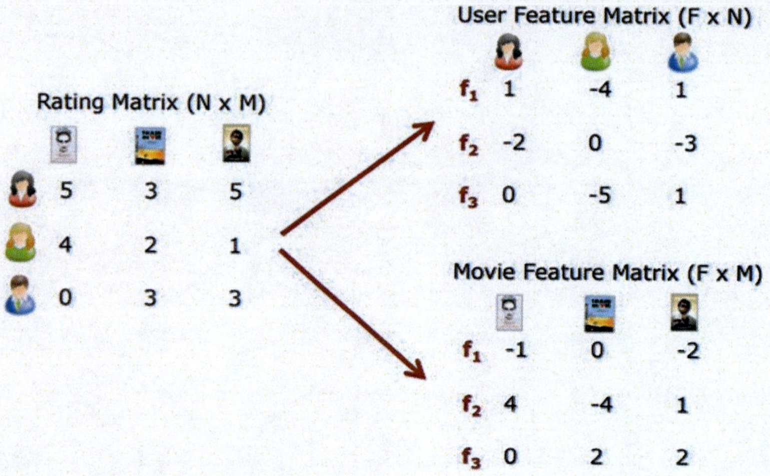

Fig. 3 Matrix factorization method

4.1 Demographic Filtering Algorithm

We use our dataset and filter out the movies on a certain criterion and recommend the top results to users. This filtering also helps the model to resolve cold start problem as the model will recommend the trending movies. The weighted rating equation represent in Eq. 5.

$$\text{Weighted Rating (WR)} = \left(\frac{v}{v+m}.R\right) + \left(\frac{m}{v+m}.C\right) \tag{5}$$

- *V* is the film's number of votes.
- M must be registered in the chart with minimum votes.
- *R* is the film's average.
- *C* means voting in the whole article.

The movies that qualify the parameter will be recommended.

4.2 Content-Based Filtering—Cosine Algorithm

The model predicts the next movie based on the watch history of user; the next movie recommended is based in the last movie's genre. Here, we are using the hybrid method (Cosine similarity method + linear kernel). Figure 1a shows the working of content-based filtering recommendation system.

Algorithm of Hybrid method

Step 1. Get the film index with its title.

Step 2. Get the list for that film of all films with cosine similarity ratings. Turn it into a tuple list, in which the first element is its location, and the second element is its similitude stage.

Step 3. Order the above list of tuples according to the parallel values; the second part, that is.

Step 4. Get the top ten list elements. Ignoring the first element of the self.

Step 5. Return the titles corresponding to the indices of the top elements.

4.3 Collaborative Filtering—Hybrid Algorithm

Content-based filtering is limited to predict movies only based on the watch history of the user, whereas collaborative filtering also takes movies watched by different users into account. The model uses Pearson correlation to recommend the movie. It takes rating provided by different users and apply Pearson correlation to find the similarities between the users

We have to recommend movies to user *E*. User *A*−*E* rate different movies, and we calculate Pearson correlation between all users and user *E*, and based on the calculation, we recommend the movies to *E*. Since user *A* and *F* have no common rating, so Pearson correlation cannot be calculated, while for rest of the users it is calculated in Table 1.

The dataset that we used for our model is from Kaggle. It is a labeled dataset. We are provided with the dataset which is used to test our model as it works on certain formula and algorithms. We have to upload dataset on Jupyter notebook whenever we have to test our model. We have used 5 K movie lens dataset for better performance. Dataset is divided in main 9 files credits with 3 columns and 45,505 rows, keywords with 2 columns and 46,420 rows, links with 3 columns and 45,844 rows, links_small, movies_metadata with 24 columns and 45,467 rows, ratings with 4 columns and 1,048,576 rows, ratings_small, tmdb_5000_credits with 4 columns (movie id, title, cast, crew), and 4814 rows, tmdb_5000_movies with 21 columns and 4804 rows. Therefore, we have a total dataset with 1,241,430 rows to test and train our model.

Table 1 Similarity index

	The avenger	Sherlock	Transformer	Matrix	Titanic	Me before you	Similarity between (*i*, *E*)
A	2		2	4	5		NA
B	5		4			1	0.87
C			5		2		1
D		1		5		4	−1
E			4			2	1

Comparison between previous and our model is that after comparing previous and our model we can clearly see the result between the previous and our model. After seeing the result, we get to know that our model has predicted much similar movies than the previous model; hence, we can say that our model is much accurate and efficient.

5 Conclusion

We illustrated how to build a scalable collaborative filtering and content filtering-based hybrid movie recommendation system. Our model has taken a small step against the cold problem too by introducing demographic filtering in which we filter out the most trending movies of time and recommend them to users. We used learning kernel instead of similarity cosine for our content filtering and Pearson correlation for collaborative filtering. We used movie lens 100 k dataset for better results and can use movie lens 1 m dataset for even better results.

Acknowledgements We want to thank our guide Mr. Arjun K P for giving us the chance to do this project in terms of the film recommendation system, which has also allowed us to carry out a lot of testing, and we find many new things that we really are grateful for it.

References

1. Pazzani MJ, Billsus D (2007) 'Content-based recommendation systems. In: The adaptive web. Springer, Berlin, Germany, pp 325–341
2. Zhang R, Mao Y (2019) Movie recommendation via markovian factorization of matrix processes. IEEE 7:13189–13199
3. Su X, Khoshgoftaar TM (2009) A survey of collaborative filtering techniques. Adv Artif Intell 2009:421425
4. Arjun KP et al (2020) Emerging IoT-big data platform oriented technologies. The Internet of Things and Big Data Analytics, Auerbach Publications, 1st Edition, ISBN 9781003036739.
5. Arjun KP, Achuthshankar A, Soumya MK, Sreenarayanan NM, Priya VV, Faby KA (2016) PROvacy: Protecting image privacy in social networking sites using reversible data hiding. In: 2016 10th International conference on intelligent systems and control (ISCO), pp 1–4
6. Manoj Kumar DKY, Singh A, Gupta VKr (2015) Article: a movie recommender system: MOVREC. Int J Comput Appl 124(3):7–11. Published by Foundation of Computer Science (FCS), NY, USA
7. de Campos LM, Fernández-Luna JM, Huete JF, Rueda-Morales MA (2010) Combining content-based and collaborative recommendations: a hybrid approach based on Bayesian networks. Int J Approx Reason Rev
8. Virk HK, Maninder Singh Er (2015) Analysis and design of hybrid online movie recommender system. Int J Innovat Eng Technol (IJIET) 5(2)
9. Kużelewska U (2014) Clustering algorithms in hybrid recommender system on movielens data. Studies in Logic, Grammarand Rhetoric 37(50)
10. Chiru C-G, Dinu V-N, Preda C, Macri M (2015) Movie recommender system using the user's psychological profile. In: IEEE international conference on ICCP

11. Lin H, Yang X, Wang W (2014) A Content-boosted collaborative filtering algorithm for personalized training in interpretation of radiological imaging. J Digit Imaging 27(4):449–456
12. Satish M, Srinivasa Rao P, Ramakrishna Murty M (2019) Identification of natural disaster affected area using twitter. In: International conference and publish the proceedings in AISC Springer ICETC-2019, Vol.3, pp 792–801 at Osmania University, Hyderabad, pp 792–801
13. Tarakeswara Rao B, Ramakrishna Murty M et al (2020) A comparative study on effective approaches for unsupervised statistical machine translation. In: International conference and published the proceedings in AISC Springer conference, vol 1076, pp 895–905

Kubernetes Continuous Development

Sakshi Sharma, Ayush Shankar Pathak, Harshit Malhotra, Soumalya Ghosh, and Garima Pandey

Abstract The scope of this research paper is to develop a continuous integration/continuous development pipeline for microservices application, decide which options will include in our continuous integration phase, utilization of Jenkins to execute blue/green or moving sending, utilizing Ansible or integration/continuous deployment, is the foundation to assemble our "framework"; i.e., Kubernetes Clusters and to make our undertaking stick out, we have actualized checks, for example, security examining, execution testing, joining testing, and so forth! Henceforth, AWS CloudFormation licenses you to exhibit your entire establishment and application resources with either a book record or programming vernaculars. The AWS CloudFormation Registry and CLI simplify it to manage untouchable resources with CloudFormation. This gives a lone wellspring of truth for all of your resources and urges you to standardize system parts used across your affiliation, engaging arrangement consistence, and faster examining. CloudFormation manages choosing the right errands to perform while managing your stack, orchestrating them in the most capable way, and moves back changes normally if botches are recognized. On the consummation of continuous integration, we have set up the continuous deployment which incorporates pushing the constructed Docker container(s) to the Docker storehouse and deploying these Docker container(s) to a little Kubernetes bunch. For our Kubernetes cluster, we have used AWS Kubernetes as a service; to deploy our Kubernetes cluster, we used cloud formation tool on AWS. The languages used for the deployment are PHP and JSON. Preferably, we have been able to run these from within Jenkins as an independent pipeline.

Keywords Amazon web services · AWS CloudFormation · CI/CD · Kubernetes · Docker · Jenkins pipeline

---

S. Sharma (✉) · A. S. Pathak · H. Malhotra
Bachelor of Computer Science and Engineering, Galgotias University, Greater Noida, India

S. Ghosh · G. Pandey
School of Computer Science and Engineering, Galgotias University, Greater Noida, India
e-mail: garima.pandey@galgotiasuniversity.edu.in

R. P. Mahapatra et al. (eds.), *Proceedings of International Conference on Recent Trends in Computing*, Lecture Notes in Networks and Systems 341,
https://doi.org/10.1007/978-981-16-7118-0_48

1 Introduction

A CI/CD pipeline usage, or continuous integration/continuous deployment, is the foundation of the advanced DevOps environment. It overcomes any barrier among improvement and activities groups via mechanizing the structure, testing, and organization of utilizations.

- With CI, each adjustment in code triggers a computerized fabricate and test arrangement for the given venture, giving criticism to the developer(s) who made the change [1]. The whole CI criticism circle should run in under 10 min.
- Continuous delivery incorporates framework provisioning and sending, which might be manual and comprise of various stages. What is significant is that every one of these cycles are completely mechanized, with each run completely logged and noticeable to the whole group.

Microservices are an acknowledgment of the administration arranged structural style for creating programming made out of little administrations that can be sent and scaled autonomously by completely mechanized organization apparatus, with least incorporated administration [2]. Microservices function on business discrete functions. Several microservices operate on its own period and use APIs regularly via lightweight components. The drawbacks of heavy applications are solved by microservices. They are tiny and can reset faster at the moment of recovery from redesign or disappointment. Microservices are roughly paired and irritated.

Two main steps for pipeline are: -

Step 1: A square that contains a progression of steps. A phase square can be named anything; it is utilized to imagine the pipeline cycle.

Step 2: An errand that says what to do. Steps are characterized inside a phase block.

Kubernetes is an open-source compartment organization stage made by Google. You can utilize Kubernetes for on-premises, cloud, or edge deployments [3]. At the point when utilized in blend with AWS, you use Kubernetes to oversee groups of Amazon Elastic Compute Cloud (EC2) cases that have your holders. When conveying Kubernetes in AWS as shown in Fig. 1, you can arrange and deal with your sending

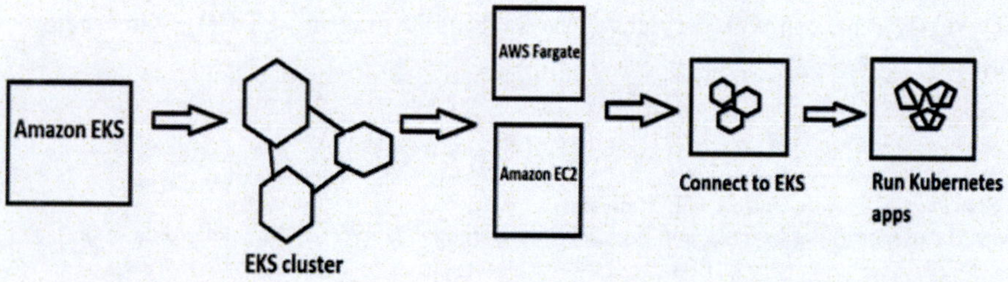

Fig. 1 Working of Kubernetes

without anyone else for full adaptability and control. You additionally have the alternative of utilizing either AWS-offered types of assistance or outsider administrations to deal with your execution.

2 Literature Review

Cerner Corporation conveys medical services innovation internationally to 3 million medical services experts and enhances to make a consistent and associated world in which everybody flourishes. The organization has gone through the most recent forty years digitizing medical care information and freeing clinician's workplaces of manila envelopes and file organizers. Presently, the cooperation of Cerner and AWS will convey information that is more open and significant and utilizes AWS AI and AI advancements to anticipate and possibly forestall medical conditions [4]. Likewise, as a feature of its work to modernize how it conveys arrangements and improves quiet results, Cerner has been relocating its secretly facilitated stages to AWS. One of the organization's objectives is to carry more bliss to the act of medication—keeping that in mind, Cerner is trying its Virtual Scribe innovation utilizing discourse acknowledgment and Amazon Transcribe Medical to drastically diminish manual information passage and give specialists more opportunity to go through with patients.

GE Healthcare is known for its clinical imaging gear and suggestive imaging trained professionals, yet has—all through the latest a long time—continued in its high level change. "Reliably, clinical benefits data travels through huge number of clinical devices, remembering for abundance of 500,000 GE Healthcare clinical imaging devices all around the world." "We need to build client respect from contraption use and information by connecting with supplemental effect of cloud register, amassing, and access." GE Healthcare intends to grow its utilization of AWS associations as it fosters the GE Health Cloud. For example, the affiliation is manhandling Amazon SageMaker, a managed association for building, arranging, and sending AI models. "We are all in on Amazon SageMaker for our basic learning limits proceeding," said by Andre Sublett. GE Healthcare will comparatively keep inclining toward AWS to help keep up the connection's significant level. "Our high level framework is related with improving receptiveness, flexibility, or perhaps usage of applied appraisal," said by Jackson. "We need to set up a structure at a general scale to accomplish these objectives, and we can do that by running our GE Health Cloud on AWS." Kellogg uses Amazon CloudWatch for checking, which helps the association with assigning costs to each office reliant on their individual system use [5]. "CloudWatch helps our family make better decisions around the cutoff they need, so they can keep an essential separation from waste," McIlwain says. We were reliably inadequate to do that with our on-premises foundation. AWS isolates use and cost to an especially granular level that we can recognize which costs come from which division like a cost model."

Philips HSDP investigates and stores 15 PB of patient information assembled from 390 million imaging considers, clinical records, and patient contributions to give medical care suppliers significant information, which they can use to straightforwardly affect quiet consideration. Running on AWS gives the steadfastness, execution, and flexibility that Philips needs to help guarantee open-minded data as its overall progressed stage creates at the speed of one petabyte for every month. It is like way utilizes information to drive choices, and, when its on-premises instructive file plans could not deal with the extent of information in 37 million records, it went to AWS. Philips set up Attunity CloudBeam, an AWS Marketplace thing for Amazon Redshift, in less than one second to revise, stimulate, and motorize data which moves to the AWS Cloud. Before AWS, the association's snappiest data moves were 434 records for each second [6]. Using AWS, the association moved 37 million records in 90 min and can upgrade tremendous instructive files inside two hours.

3 Feasibility Analysis

Amazon Web Services (AWS) is associate adjuvant of Amazon giving on-request distributed computing phases and APIs to individuals, organizations, and ministry, on determined pay— a lot of solely as prices arise basis [7]. The particular distributed computing Internet administrations provide associate assortment of essential abstract specialized framework and sent process architecting squares and apparatuses. One in every of these administrations is Amazon Elastic reason Cloud (EC2) that permits shoppers to possess accessible to them a virtual cluster of PCs, accessible perpetually, through the net.

AWS's variant of virtual computers imitates the larger vicinity of the proprietary of a true PC, as well as instrumentality focal getting ready units (CPUs) and illustrations handling entity (GPUs) preparing nearby/RAM memory, solid-circle/SSD stockpiling, an alternative of operating frameworks, standardizing, and pre-mutilated application computation, as an example, net staff, info bases, and shopper relationship the executives (CRM) [8]. A CD pipeline might be an automated explanation of your cycle for getting programming from interpretation control straightforwardly through to your customers and clients. Each change to your item (submitted in source control) encounters a momentous measure on its way to deal with being conveyed. This cycle incorporates building the product during a strong and repeatable way, even as propelling the manufactured programming through various periods of testing and association. Jenkins is, on a fundamental level, an automation engine which supports different computerization plans. Pipeline adds a weighty arrangement of automation contraptions onto Jenkins, supporting use cases that reach from clear reliable joining to broad CD pipelines. By showing a movement of related tasks, customers can abuse the various features of pipeline. A ceaseless conveyance (CD) pipeline is a mechanized articulation of your interaction for getting programming from variant control directly through to your clients and clients. Each change to your product (submitted in source control) goes through an unpredictable interaction while heading to being delivered. This interaction includes building the product in a solid and repeatable way, just as advancing the assembled programming through numerous phases of testing and organization. Jenkins is, essentially, a mechanization motor which upholds various computerization designs. Pipeline adds an incredible arrangement of robotization instruments onto Jenkins, supporting use cases that range from basic persistent incorporation to exhaustive CD pipelines. By displaying a progression of related undertakings, clients can exploit the numerous highlights of pipeline. A Jenkins file may be composed utilizing two forms of linguistic structure—Declarative and scripted. Revelatory and scripted pipelines are developed in a very general sense in an unexpected way. Definitive pipeline may be a later element of Jenkins pipeline which provides more extravagant linguistic highlights over scripted pipeline grammar, for example, Jenkins file (declarative and scripted pipeline):

Execute this pipeline or any of its stages, on any accessible specialist, then at that point characterize the "Form" stage, after that play out certain means identified with the "Form" stage, next characterize the "Test" stage, play out certain means identified with the "Test" stage, characterize the "Convey" stage, and finally play out certain means identified with the "Send" stage.

4 Proposed System

Security hub gives us an exhaustive perspective on your security state in AWS and encourages you check your current circumstance against security industry guidelines and best practices [9]. Security hub gathers security information from across AWS records, benefits, and upheld outsider accomplice items and encourages you examine your security drifts and recognize the most elevated need security issues. The following is the syntax required to create a Security Group: Type: AWS::EC2::SecurityGroup Properties:

GroupDescription:

String GroupName:

String SecurityGroupEgress:

Egress SecurityGroupIngress: Ingress Tags: Tag VpcId: String.

The Security Group Egress and Security Group Ingress property rules are the most critical to the security group as it defines where the traffic will go [10]. While Security Group Egress defines outbound traffic, Security Group Ingress defines the inbound traffic. Ingress rules restrict or allow traffic trying to reach our resources on specific ports. Egress rules restrict or allow traffic originating from our server—typically we are ok allowing all outbound traffic without restrictions as this does not pose a risk for a security breach. In cloud, traffic is completely blocked, so you have to explicitly open ports to allow traffic in and out [11].

Aqua is a Jenkins plugin designed for testing Docker containers. It is one of many security vulnerability scanners available, testing against common vulnerabilities and exposures (CVE). Linting is the process of running a program that checks the pipeline code for potential syntax errors. Performance testing is done by setting up a smaller scale environment as compared to production, with Jenkins and running simulated host calls into that environment to determine how the new environment performs under a particular workload. It is a two-stage process:

Stage 1—Run Apache JMeter—JMeter is a testing tool used for estimating the performance of the newly created Jenkins environment, and Stage 2—Capture Reports—Jenkins gives performance module to catch reports from well-known testing instruments, for example, JMeter, Selenium, and numerous others in the XML and CSV design.

The diagram represents the AWS account, and all resources it can access.

When planning creation frameworks, one should consider various availabilities zones in all circumstances. An availability zone is characterized as a bunch of at least one server farms. Virtual private cloud is a pool of arranged cloud assets. It can traverse more than one accessibility zone. What could be compared to this would be a server farm as portrayed in Fig. 2. A subnet is a subset of the overall VPC association, and it simply exists in a lone availability zone, not in any manner like its parent association, the VPC. Web gateway is an asset that empowers inbound and outbound traffic from the Web to your virtual private cloud. Organization address translation passage gives outbound—just Web door to private administrations to get to the Web as displayed in Fig. 2.

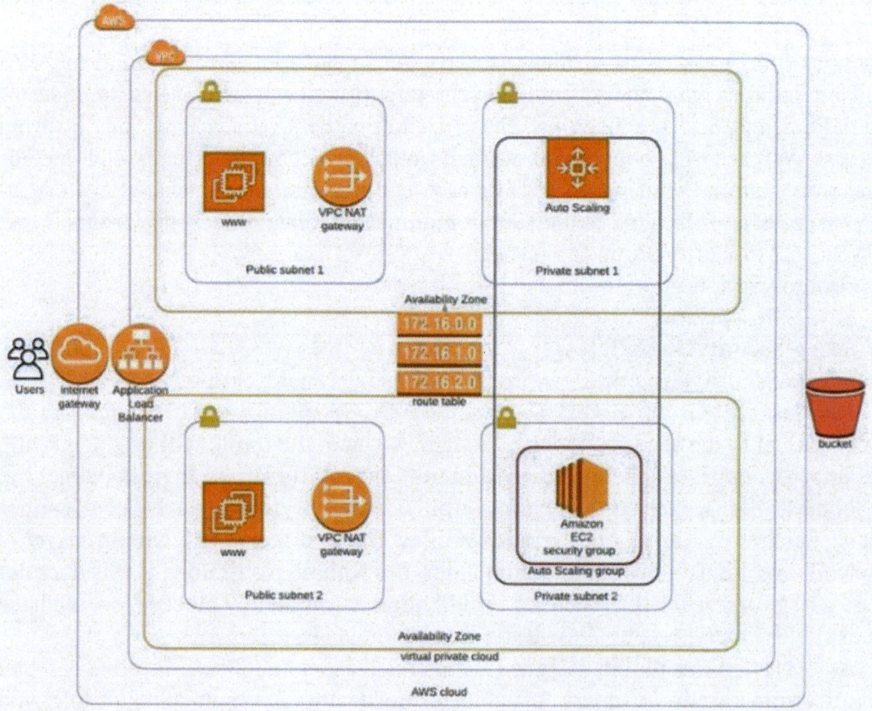

Fig. 2 AWS and its resources

In this research, we have been able to specify firewall rules using security groups, create autoscaling groups for the elasticity of servers, code the launch configuration for the Web application, and add the target groups and listeners. Avoiding hard-coding ever-changing values into your server infrastructure is by using parameters. We created a launch configuration for our application servers in order to deploy four servers, two located in each of your private subnets. The system (OS) to be used is Ubuntu18. So, we chose an instance size and machine image (AMI) that best fits this space. We downloaded an application archive from an S3 Bucket, for which we needed to create an IAM role that allows your instances to use the S3 service [12, 13]. Concerning outbound, the workers will require unlimited Web admittance to have the option to download and refresh its product. Outbound, it might be utilizing port 80 to arrive at the inside workers. The application needs to be deployed into private subnets with a load balancer located in a public subnet. We deployed our servers with an SSH key into public subnets while creating the script. This helps with troubleshooting [14–16]. Once done, move them to a private subnet and remove the SSH key from our launch configuration. It also helps to test directly without the load balancer.

5 Result Analysis

In this paper, we have deployed a static Web site through AWS using S3, CloudFront, and IAM. The files included are: -index.html—The Index document for the Web site. /css- CSS files for the Web site. /img—The background image file for the Web site. /vendor—Bootssrap CSS framework, Font, and JavaScript libraries needed for the Web site to function. At the point when CI/CD pipelines moved to the overall cloud, the management of executives did not advance with them. In case you are during the present circumstance, you will have seen assortment of strategic workarounds to remain Jenkins constructs address the administrations they rely on. The workarounds range from awful coding to simply agonizing, but their common feature is that they have a tendency to form copies of secrets beyond the reach of automation [17]. This increases the attack surface, makes routine key rotation impractical, and makes remediation difficult after a breach. AWS Secrets Manager may be a comprehensive solution for secure secret storage. You define a secret just one occasion for your whole AWS account and then you give your consumers permission to use the secrets. Making and imagining CD pipelines are a few things significant for a few Jenkins clients, and this is regularly shown inside the modules that the Jenkins people group has made to fulfill their necessities; this implies a prerequisite to return to how Jenkins right now communicates these ideas and consider conveyance pipelines as a focal subject to the Jenkins client experience. It is difficult CD ideas yet the devices that engineers utilize a day—Bitbucket, Github, Docker, or Slack; it is about very Jenkins—it is the designer work process which encompasses Jenkins that traverses different instruments [18].

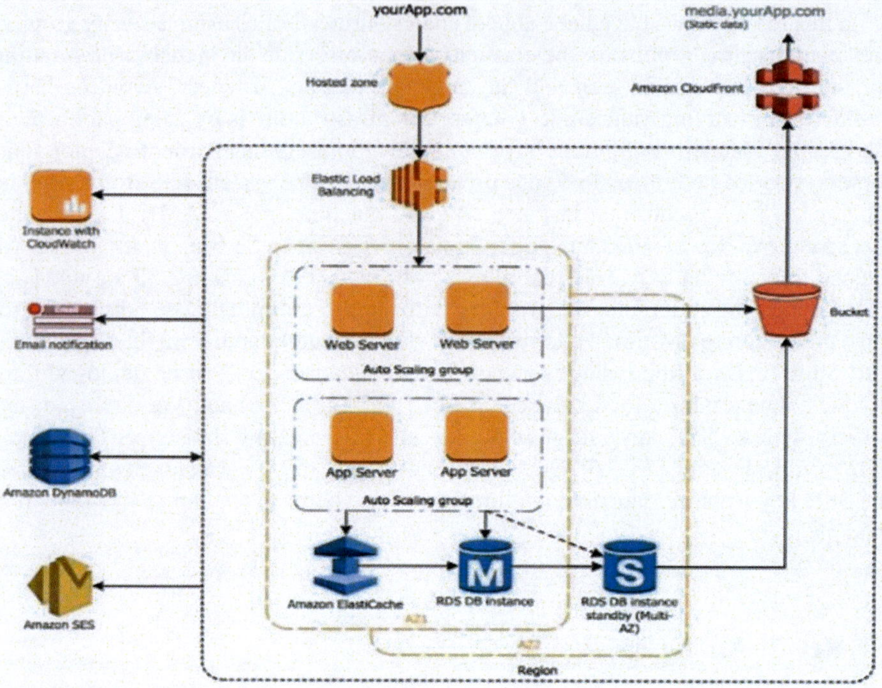

Fig. 3 AWS web hosting architecture

Load balancing along with elastic load balancing permits you to spread weight across various availability zones and Amazon EC2 Autoscaling bundles for reiteration and decoupling of organizations. Storing with Amazon ElastiCache gives holding organizations with Redis or to take out load from the application and informational index, and lower inactivity for standard requesting. Overseen database with Amazon RDS makes an especially available, Multi-AZ database plan with six potential DB engines as displayed in Fig. 3. Edge caching with Amazon CloudFront—Edge stores high-volume substance to decrease the inactivity to customers. Edge security for Amazon CloudFront with AWS WAF channels harmful traffic, including XSS and SQL mixture through customer described standards. Static storage and backups with Amazon S3 empowers fundamental HTTP-based article accumulating for fortifications and static assets like pictures and video.

6 Conclusion and Future Scope

In this research paper, we have applied the skills and knowledge developed through Cloud DevOps which includes working in AWS, using Jenkins to implement continuous integration and continuous deployment, building pipelines, working with

Ansible and CloudFormation to convey groups, building Kubernetes bunches, and building Docker holders in pipelines. In more open-ended way, we have been able to make some of our own choices in this capstone such as for the type of deployment we implemented, which services we will use, and the nature of the application we developed. A specialist is by and large a machine, or holder, which associates with executes commitments while coordinated with the guide of utilizing the regulator.

Plugins are not always terrible things. In fact, while they are used properly (because of this that extending capability past the center functions required of a software pro- gram platform), they are first-rate assets. They deliver customers the selection of including more functions to the equipment they use, without requiring them to devote assets to the ones functions in the event that they do not desire to apply them. But for Jenkins, plugins do now no longer offer admission to optionally available capability extending past the center functions required to apply the platform. Instead, Jenkins calls for groups to apply plugins to obtain obligations that, in lots of cases, are without a doubt pretty basic.

References

1. Deelman E, Singh G, Livny M, Berriman B, Good J (2008) The cost of doing science on the cloud: the montage example. In: Proceedings of the 2008 ACM/IEEE conference on Supercomputing, pp 1–12
2. https://docs.aws.amazon.com/index.html?nc2=h_ql_doc AWS Management Console, Login, Documentation under the scaling Group EC2I. Jacobs S, Bean CP (1963) Fine particles, thin films and exchange anisotropy. In: Magnetism, vol III, Rado GT, Suhl H (eds) Academic, New York, pp 271–350
3. Amazon (2014d) AWS Reference Architectures: Fault Tollerance & High Aavailability Online. http://media.amazonwebservices.com/architecturecenter/AWS_ac_ra_ftha_04.pdf. Accessed 30 July 2015
4. Amazon EC2 Service Level Agreement Online. http://aws.amazon.com/ec2/sla/; 2013 Accessed 30 July 2015
5. NGINX Ingress Controller for Kubernetes. Contribute to kubernetes/ingress- nginx development by creating an account on GitHub. Kubernetes, 2018.
6. Vayghan LA, Saied MA, Toeroe M, Khendek F (2018) Deploying microservice based applications with kubernetes: experiments and lessons learned. In: 2018 IEEE 11th international conference on cloud computing (CLOUD), pp 970–973
7. Mcluckie C (2014) Containers VMs Kubernetes and VMware. https://googlecloudplatform.blogspot.com/2014/08/containers-vms-kubernetes-and-vmware.html
8. “Container and Microservice Driven Design for Cloud Infrastructure DevOps - IEEE ConferencePublication.” [Online]. Available: https://ieeexplore.ieee.org/abstract/document/7484185. Accessed: 12 Oct 2018.
9. Thakrar U (2013) Introducing RightScale Cloud Appliance for vSphere”, Dec. 2013, [online] Available: www.rightscale.com/blog/enterprise-cloud-strategies/introducing-rightscale-cloud-appliance-vsphere.
10. Roy N, Dubey A, Gokhale A (2011) Efficient autoscaling in the cloud using predictive models for workload forecasting. In: Proceedings of the IEEE 4th international conference on cloud computing. IEEE, pp 500–507. https://doi.org/10.1109/CLOUD.2011.42
11. Khazaei H, Barna C, Beigi-Mohammadi N, Litoiu M (2016) Efficiency analysis of provisioning microservices. In: 2016 IEEE International conference on cloud computing technology and science (CloudCom), pp 261–268

12. Netto HV, Lung LC, Correia M, Luiz AF, Sá de Souza LM (2017) State mac replication in containers managed by Kubernetes. J Syst Architect 73:53–59
13. "Microservices," martinfowler.com. [Online]. Available: https://martinfowler.com/articles/microservices.html. Accessed: 01 Oct 2018
14. Arjun KP et al (2020) Emerging IoT-big data platform oriented technologies. The Internet of Things and Big Data Analytics, Auerbach Publications, 1st Edition, 2020, ISBN 9781003036739
15. Gupta J, Singh I, Arjun KP (2021) Artificial Intelligence for Blockchain I. Blockchain, Internet of Things, and Artificial Intelligence, CRC Press, vol 6
16. "Container and Microservice Driven Design for Cloud Infrastructure DevOps - IEEE Conference Publication." [Online]. Available: https://ieeexplore.ieee.org/abstract/document/7484185. Accessed: 12 Oct 2018
17. "Integrating Open SAF High Availability Solution with Open Stack - IEEE Conference Publication." [Online]. Available: https://ieeexplore.ieee.org/abstract/document/7196529. Accessed: 12 Oct 2018
18. Arvindhan M, Anand A (2019) Scheming an proficient auto scaling technique for minimizing response time in load balancing on Amazon AWS Cloud. International Conference on Advances in Engineering

Movie Recommendation and Mood Prediction System—Leisures

Poornima Dave, Shashi Bhushan Singh, Divyansh Chaurasia, and Avneesh Kumar

Abstract A recommendation system is a machine learning model to help the users to find items of their choice and prevent themselves from data overloading. It predicts the choices of a particular user and makes the recommendation as per the user's choice and requirements. The recommendation systems might adopt different techniques based on personal rating and interest and even sometimes based on the wish list of some other user having the same interest. So in order to make one's leisure good, we present our movie recommendation system and mood predictor, Leisures. Leisures is a platform where one can find everything they want to do in their free time with a built-in suggestion module. So, it analyzes what mood an individual is in and makes suggestions on what he/she should do in order to make their free time/leisure good. For example, if someone is gloomy and he opens Leisures, it will first analyze the person's temper, and after figuring out his distress, it would suggest to either listen to sad songs or watch movies and will fetch respective data for it as well. The user can then watch movies as per his mood, and when he returns again, he can log in to his account, and he can either choose from recommended movies as per his last choice or may use the mood prediction module again. The introduction of the chatbot was the latest and advanced idea to make recommendations for the user. Moreover, apart from this newest mood predictor chatbot, the recommendation system works efficiently as well, a content-based recommendation model is being used in the project (Raval and Khedkar in Int J Sci Tech Res 8(12) 2019 [1])

Keywords Movie recommendation system · Content-based filtering · Data analysis · Mood prediction

P. Dave (✉) · S. Bhushan Singh (✉) · D. Chaurasia (✉) · A. Kumar (✉)
Computer Science and Engineering, Galgotias University, Uttar Pradesh, Greater Noida, India
e-mail: shashibhushan.singh01_2017@galgotiasuniversity.edu.in

A. Kumar
e-mail: avneesh.kumar@galgotiasuniversity.edu.in

R. P. Mahapatra et al. (eds.), *Proceedings of International Conference on Recent Trends in Computing*, Lecture Notes in Networks and Systems 341,
https://doi.org/10.1007/978-981-16-7118-0_49

1 Introduction

In recent times, specifically, when we decided to design this project, the world was under the attack of pandemic COVID-19. Therefore, in such scenarios, it was necessary to fill their lives with something with which they can pass their time. They can make their free time fruitful, and they can make their lockdowns amazing. They can make their Leisures good. Thus, this obnoxiousness leads to the development of something that can relieve people from stress and can lighten their mood. Overall it can be used to create a healthy environment. The best part about Leisures, our project, is that it helps us in passing the time in the way we want. This project acted as the psychological treatment for the people who suffered from loneliness [2].

2 Literature Review

The project mood prediction and movie recommendation system is not new. Even before we had tools and applications that helped in spending the free time with fun, they also used various recommendation techniques that helped in making the Leisure time good for people. The application like Raspberry API had its existence in the market which used the method of converting the text into modules which are saved as record and are treated as data which is further used for making recommendations. The applications like Netflix, Amazon Prime, and many other Web sites collect such data and use them as feed for preparing their suggestion module. But all of them lack the perspective of understanding the user's state of mind. They used the technique of creating a user's favorite list based on the ratings the user had given. Many applications were even made for determining a user's mood by attaching a camera, so by reading and analyzing the user's face, they predicted the user's mood. But they faced major failure. Till now, we have a recommendation system and a mood predictor as the two separate applications. So we need one such application which makes movie recommendations and also the prediction of in what mood the user is and what he actually wants to do. Thus, the combination of prediction and recommendation is still to arrive [3, 4].

3 Methodology

For designing the efficient recommendation system, the foremost step is to have an in-depth study about the algorithm. Data collection is the most important step as the entire model is going to be designed and worked upon by the respective dataset. The data collected is then moved to the data analysis part wherein data is analyzed, examined, interpreted, and sometimes visualized, if needed, in order to remove the noises and find them insightful and required information that is to be used in the

```
head(rating_data)
```

```
##   userId movieId rating  timestamp
## 1      1      16    4.0 1217897793
## 2      1      24    1.5 1217895807
## 3      1      32    4.0 1217896246
## 4      1      47    4.0 1217896556
## 5      1      50    4.0 1217896523
## 6      1     110    4.0 1217896150
```

Fig. 1 Dataset having user id, rating, movie id, and timestamp

algorithm. Therefore, in order to have correct and veracious data, a large dataset is always recommended. Computational mathematics is generally used in data analysis [5] (Fig. 1).

4 Recommendation System

For such recommendations, we have three methods.

1. Collaborative Recommendation: This is the method wherein the recommended list is formed based on past events. It analyzes all movies the user has watched previously and prepares the list as per the past records.

Example: If two users A and B prefer the same television, then it might be a chance that other preferences of the users match. Thus, the wish list of user A is recommended to user B (reviewer point).

2. Content-Based Recommendation: In this method, the system works upon finding such a list that is based on the user's interest. It may use the set of keywords that are composed in some other movie of his choice. It may match the wish list of two such users of similar interest and recommend the same to the other user [6]

Example: If user A prefers some kind of television of a brand, then other television of the same brand or same type will be recommended to the user along with the list of similar previous items purchased by the user in the past.

3. Hybrid Recommendation: The hybrid method is more popular nowadays wherein both the collaborative filtering mechanism and content-based filtering mechanism are combined, this proves to be a more effective way of recommendation because both the techniques are used to provide better results and satisfaction.

4.1 Content-Based Recommendation System

The content-based recommendation is the technique or the filtering based on the choices or the mindset of the user. In this project, a content-based recommendation system is used to enable the user to create a watch list of theory choice. If a person watches Avatar and gives the movie a rating of more than 4, this means that he extremely liked the movie and is keenly interested in the movies like Avatar, whose genre are fantasy, adventure, and space. So for the next time when he visits the application, movies will similarly be reflected on his suggestion list. As the user keeps watching the movies, the list will get updated, the more the data, the better will be the suggestion.

4.2 Formulation and Computation

The content-based recommendation system follows the approach of the vector space modeling system; thus, term frequency is calculated on the basis of keywords. If we assume $j(i)$ for the keyword positioned at the ith place, in the content $d(x)$, wix will be the weight of ji for the content, and now the content $d(x)$ will be calculated as:

$$d(x) = \{w1x, w2x, \ldots \text{ so on}\}$$

The content-based recommendation system suggests the movies as per the user's past choices and ratings; thus, a preference vector is created as per the individual users. So let N be the movies he preferred the most, and 'a' as his preference, so the total movie's user preferred in past will be N(a), so the possible preference vector will be formulated as

$$a = (1/|N(a)|) * \text{Content}$$

Now we have the content and the vector, and in order to get the final list of suggestions in accordance with the user's choice, we need to draw some relationship between the two quantities. Hence, we define it by

$$S(a, d) = \text{similarly(preference vector, content)[7].}$$

After the evaluation for similarity, term frequency and inverse document frequency are calculated.

$$\text{TF(tf)} = \text{(Number of times the term appears in a document)}$$
$$\text{/(Total number of terms in the document)}$$
$$\text{IDF(idf)} = \text{log\_e(Total number of documents}$$
$$\text{/Number of documents with term t in it)[7].}$$

Since we have already formulated the similarity relation but for the successful and accurate model, we need to have a cosine similarity, which can be formulated as:

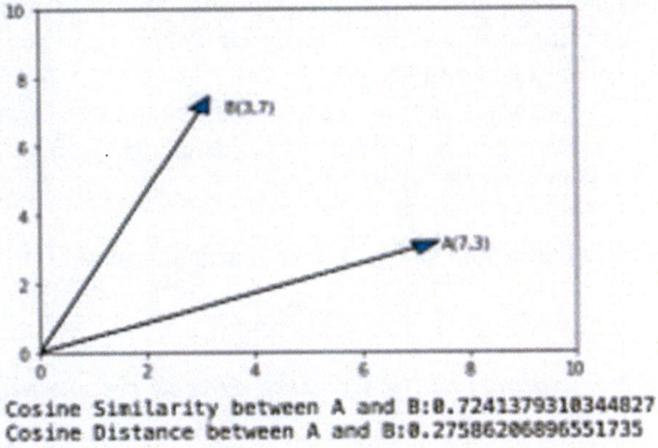

Fig. 2 Graph of cosine similarity between two vectors

$$\text{sim}(A, B) = \text{Cos(theta)} = A \,.\, B \,/\, ||A|| * ||B||$$

where A and B represent the data for the two movie id which represents the two vectors and cosine similarity is calculated as per the vector space model. For example, let vector $A = (3,7)$ and vector $B = (7,3)$, then (Fig. 2)

5 Design and Interface

5.1 User Interface

The application being discussed is a Web application, which is a better version of a native Android application. The user can sign into the application and create his own account, so that he might not have the trouble of starting the application from the very beginning and the recommendation system works properly. After signing in to the platform, the user gets his own account. Thus, he can avail the benefit of a built-in recommendation model. So for every time while logging into the application if a user enters his credentials, he can easily avail all the benefits the application offers. Now the user is again directed to the dashboard wherein he gets two options. The first is to use the integrated chatbot, and the second is to explore the movie and prepare his own watch list using the smart recommendation system [8].

5.2 Mood Prediction

The project not only makes recommendations but also predicts the mood of the user. The integrated chatbot enables the application to analyze in what mood a person is by recognizing the state of mind by using a chatbot. The integrated chatbot takes the user input in such a way that it interacts with the user and asks few questions and on quantifying the questions, it takes care of the mood an individual is in and makes the suggestions accordingly [9].

The chatbot is simple to use, the user can interact with the chatbot about how he feels and on getting certain keywords chatbot gains more data by putting a query more than what the user wants to watch, then on getting the set of the keyword the dataset fed into it, afterward it generates the response to the user and suggests some movies that would suit him in the particular mood [10].

5.3 Implementing Content-Based Recommendation

The option available for the user is to first explore and then rate his movie experience. The explore menu would also provide the facility to search for the movie the user wants to watch.

We use the concepts of term frequency (denoted as tf) and inverse document frequency (denoted as idf) to formulate the concepts of content-based recommendation systems. The preferences vector and similarly indexes that had been formulated earlier enables us to have a content ($d(x) = \{w1x, w2x, \ldots$ so on$\}$) for each word, but we need something concrete to calculate the total repetitions [11].

Term frequency as the word suggests is the number of times any word is repeated or used. Inverse document frequency or idf is defined as the inverse or opposite of term frequency. Both tf and idf play a much important role in determining the recommendation dataset and making suitable recommendations. However, when calculating TF-IDF, the stem is used to reduce the effect of very high names. Example: TF = 3 versus TF = 4 is very different from TF = 10 versus TF = 1000. In other words, the coherence of the word in the document cannot be measured as a simple calculation. The formula for weighted term frequency is given below:(reviewer) [12] (Fig. 3).

The above formula is used to find the weighted term frequency, so that it enables finding the most appropriate keywords which would satisfy the relation of similarly ($S(a, d) =$ similarly(preference vector, content)). After having a similarity relation, cosine similarity is calculated for the vectors in consideration according to the vector space model which is being adopted in the project.

$$\text{sim}(A, B) = \cos(\text{theta}) = A \,.\, B \,/\, ||A|| * ||B||[13]$$

Thus, the model is trained based on this formula majorly. The model thus helps to first analyze the data and on successful analysis of the data, the model has

Fig. 3 Formula for weighted term

$$w_{t,d} = \begin{cases} 1 + \log_{10} \text{tf}_{t,d}, & \text{if tf}_{t,d} > 0 \\ 0, & \text{otherwise} \end{cases}$$

trained accordingly. The recommendation thus requires a larger amount of data to be processed in order to make the perfect prediction [14].

5.4 *Merits of Proposed System*

The proposed application serves as a recommendation system and the mood prediction model. The following are the merits of the proposed system.

1. Analyzing a person's frame of mood and making suggestions accordingly
2. Fetching the respective data from various sources.
3. Integrated chatbot.
4. All sorts of entertainment sources on one platform [15].

6 Analysis

In order to make the data efficient and accurate, we split the dataset 70% as training data and 30% as testing data, and now we evaluate *tf* and *idf* for the algorithm using the similarity preference and the formula. We finally get the table suggesting the movie id and count for the user's choice [16] (Fig. 4).

6.1 *Feasibility Analysis*

The project **Leisures** is a progressive Web application that is supported on the browser as well as on phone, just like a native application. It is a Web-based project with the rest Web API from different sources, for fetching data of different formats. The ML prediction model with integrated chatbot is pretty feasible for a Web application. The

```
def recommend(item_id, num):
    print("Recommending " + str(num) + " products similar to " + item(item_id) + "...")
    print("-------")
    recs = results[item_id][:num]
    for rec in recs:
        print("Recommended: " + item(rec[1]) + " (score:" + str(rec[0]) + ")")

recommend(item_id=3, num=5)
```

```
Recommending 5 products similar to Active sport briefs...
-------
Recommended: Active sport boxer briefs (score:0.41816639921615745)
Recommended: Active boy shorts (score:0.1140184812203873)
Recommended: Active briefs (score:0.11053729446572887)
Recommended: Active briefs (score:0.10917640016582862)
Recommended: Active mesh bra (score:0.10172320448715227)
```

Fig. 4 Analysis for the data obtained

convenient storage will provide a smooth and good experience to the user. Thus, the project is feasible and will run on all the platforms [17].

7 Conclusion and Future Scope

Recommendation modules are exclusive and new technology integrated with every movie broadcasting or entertainment site. They use different techniques to include these modules in order to make the user experience smoother. In this paper, a new technique of movie recommendation system is discussed which involves mood predictions as well to transform the user experience to what it is called excellent. As the pace of growth of artificial intelligence and machine learning is increasing constantly, very soon new algorithms and new techniques for such recommendation systems are going to get introduced. The algorithms will be designed which not only make recommendations based on the user's rating but will also consider subtitle, color, and many such features for more accurate and precise recommendations. The data analysis technique would be made efficient enough to figure out the movies disliked by the user and made a separate list and take care not to reflect such movies holding the similarity to recommend user accurately, not missing those movies the user might like to watch [10, 18].

References

1. Raval N, Khedkar V (2019) A review paper on collaborative filtering based movie recommendation system. Int. J. Sci. Technol. Res. 8(12)
2. Davidson J, Liebald B, Liu J, Nandy P, Vleet TV, Gargi U, Gupta S, He Y, Lambert M, Livingston B, Sampath D (2010) The youtube video recommendation system. In: Proceedings of the 2010 ACM conference on recommender systems, RecSys 2010, Barcelona, Spain, September 26–30, pp 293–296. [Online]. Available: https://doi.org/10.1145/1864708.1864770
3. Zhou R, Khemmarat S, Gao L (2010) The impact of youtube recommendation system on video views. In: Proceedings of the 10th ACM SIGCOMM conference on internet measurement 2010, Melbourne, Australia - November 1–3, 2010, 2010, pp 404–410. [Online]. Available: https://doi.org/10.1145/1879141.1879193
4. Amazon, "Amazon instant video," http://www.amazon.com/Instant-Video/b?node=2858778011. Last access: 29 Mar 2015
5. Crackle, "Crackle home page," http://www.crackle.com, last access: 29 Mar 2015. [Online]. Available: http://www.crackle.com
6. Agarwal A, Srinivasan S (2020) Movie recommendation system. Int Res J Eng Technol (IRJET) 07(07)
7. Balu B Content-based recommendation system description and implementation with python. https://medium.com/@bindhubalu/content-based-recommender-system-4db1b3de03e7/[Online]
8. Master's Thesis at VionLabs Supervisor: Chang Gao, "Content-based Recommender System for Movie Website"https://www.diva-portal.org/smash/get/diva2:935353/FULLTEXT02.pdf/Draft-2016

9. Reddy SRS (2019) Sravani Nalluri, Subramanyam Kunisetti, S. Ashok and B. Venkatesh," Content-Based Movie RecommendationSystem Using Genre Correlation" © Springer Nature Singapore Pte Ltd. 2019S. Satapathy C et al (eds) Smart intelligent computing and applications, Smart Innovation, Systems and Technologies **105**, https://doi.org/10.1007/978-981-13-1927-3_42
10. Ricci F, Rokach L, Shapira B, Kantor P (2011) Recommender systems handbook. Springer Publishing Company, Incorporated, New York Dordrecht Heidelberg London
11. Cami BR, Hassanpour H, Mashayekhi H (2017) A content-based movie recommender system based on temporal user preferences. In: 2017 3rd Iranian conference on intelligent systems and signal processing (ICSPIS), Shahrood, Iran, pp 121–125. https://doi.org/10.1109/ICSPIS.2017.8311601
12. Masthoff J (2004) Group modelling: Selecting a sequence of television items to suit a group of viewers. User Modeling and User-Adapted Interaction 14(1):37–85. [Online]. Available: https://doi.org/10.1023/B3AUSER.0000010138.79319.fd
13. Praneel ASV, Srinivasa Rao T, Ramakrishna Murty M (2019) A survey on accelerating the classifier training using various boosting schemes within casecades of bossted ensembles. International Conference with Springer SIST Series, 169:809–825
14. Tarakeswara Rao B, Ramakrishna Murty M et al (2020) A comparative study on effective approaches for unsupervised statistical machine translation. In: International Conference and Published the Proceedings in AISC Springer conference, vol 1076, pp 895–905
15. O'Connor M, Cosley D, Konstan JA, Riedl J (2001) Polylens: a recommender system for groups of users. In: Proceedings of the seventh conference on european conference on computer supported cooperative work, ser. ECSCW'01. Norwell, MA, USA: Kluwer Academic Publishers, 2001, pp. 199–218. [Online]. Available: http://dl.acm.org/citation.cfm?id=1241867. 1241878
16. Bernier C, Brun A, Aghasaryan A, Bouzid M, Picault J, Senate C (2010) Topology of communities for the collaborative recommendations to groups. In: Information systems and economic intelligence—SITE 2010, Sousse, Tunisie, 6p. [Online]. Available: http://hal.archives-ouvertes.fr/hal-00546932
17. "Deezer—Discover music you'll love," http://www.deezer.com/. Last access: 29 Mar 2015. [Online]. Available: http://www.deezer.com/
18. Napster (2015) Rhapsody home page. http://www.napster.com/, Napster has joined Rhapsody! Millions of songs to play as much as you want, Last access: 29 Mar 2015. [Online]. Available: http://www.rhapsody.com/
19. Masthoff J (2011) Group recommender systems: combining individual models. In: Ricci F, Rokach L, Shapira B, Kantor PB (eds) Recommender Systems Handbook. Springer, US, pp 677–702

Brain Tumor Classification Using Convolution Neural Network

Preeti Sharma and Anand Prakash Shukla

Abstract Computer-aided diagnosis system is playing a significant role in brain tumor classification. Misdiagnosis of brain tumor types will prevent patients from responding effectively to medical intervention and will reduce their chances of survival. In recent years, classification of brain tumor types like meningioma, glioma, and pituitary tumor is done with the help of computer-aided diagnosis. There is only a single conventional method that differentiates magnetic resonance images (MRI) by manually inspecting them by radiologists. But this strategy becomes very tedious as well as inclined to errors by radiologists. Brain magnetic resonance images scans are used here to identify tumor in this paper. Deep learning is expediently confirming to be a great technique and accomplishing upgraded performance in medical field. The intension of this research is to distinguish the brain tumor into prevalent categories such as glioma, meningioma, pituitary tumors, and no tumor with the help of convolutional neural network. Through this research, 96.45% training accuracy and 93.38% validation accuracy are achieved.

Keywords MRI · Convolutional neural network · Brain tumor · Classification

1 Introduction

The term brain tumor refers to a collection of abnormal cells in some brain tissues. It is noticed that the most lethal type of cancer disease is brain tumor. As a result, it is critical to look for ways of identifying or alert people about the possibility of a brain tumor earlier because human brain's most neural motor is highly affected part, where even minor flaws will cost a lot. Saving the life of people and treatment of cancer can be increased by the identification of tumor at an early stage. Approximately 2.5 lakhs people are affected by brain tumor, but among those only 2% are cured. So computer-aided diagnosis (CAD) is playing a crucial role in this area. It is split into two categories: primary tumors and secondary tumors. The origin of primary brain

P. Sharma (✉) · A. P. Shukla
Department of Computer Science and Engineering, KIET Group of Institutions, Ghaziabad, U.P., India

R. P. Mahapatra et al. (eds.), *Proceedings of International Conference on Recent Trends in Computing*, Lecture Notes in Networks and Systems 341,
https://doi.org/10.1007/978-981-16-7118-0_50

tumor is in the brain, while secondary tumors originate in the body from other areas. A brain tumor may be malignant (cancerous) or benign (noncancerous). Meningioma, glioma, and pituitary adenoma are the three brain tumors categories determined by the position of the tumor in the brain. Brain tumors, or other different kind of tumors, are usually subdivided into two categories, i.e., benign tumor or non-cancerous tumor and malignant neoplasm, which is highly harmful and cancerous. The expansion of these two types of tumors within the skull puts pressure on the brain, which can be lethal for the life of an individual person. Description of brain tumor categories is as follows, and Fig. 1 is depicting brain tumor categories.

Glioma is a critical and dangerous form of brain tumor which is mostly targeting adults. Glial cells are the cause of glioma and the most affected component. There is less than 2 years time to survive if the person is suffering from high-grade gliomas. Some gliomas are particularly dangerous because they develop rapidly, but they can be healed if we can detect them on time. Glioblastoma multiforme, also known as GBM, is a type of glioma, i.e., grade IV which is typically fatal. As a result, patients should receive the most effective and appropriate care in order to protect and save their lives. To assess the tumor's location, tumor tissues must be compared to neighboring areas, and this is very time-consuming and costly process. Meningiomas grow in the thin membranes of the spinal cord and brain. If the origin of tumor in pituitary gland, we can call it as pituitary tumors. These tumors are round in shape and have an inherent nature that allows them to grow in any part of the brain.

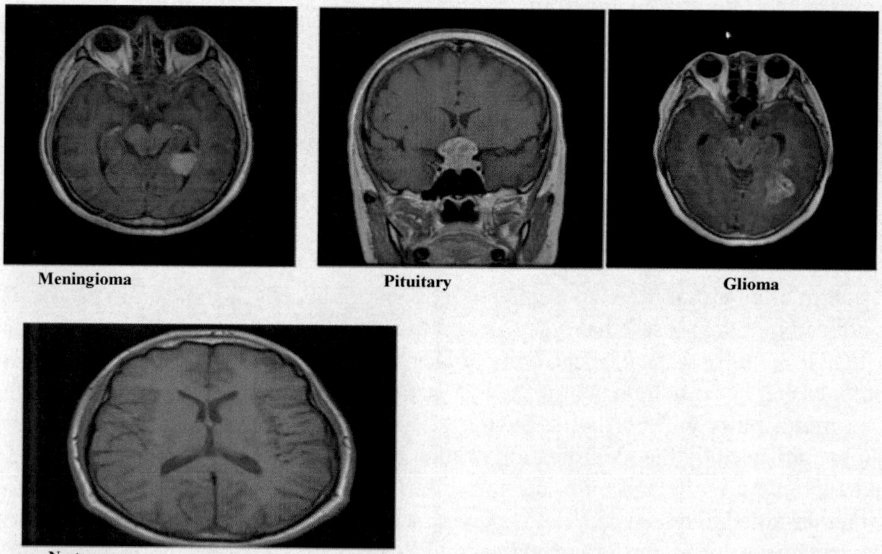

Fig. 1 Brain tumor categories

2 Literature Review

In [1], authors developed an adaptive tumors and no tumor cell classification system using an algorithm called as support vector machine. This algorithm is equipped via four parameters no tumor, tumor, non-enhancing, and enhancing.

In [2], to distinguish the part of the brain, i.e., non-tumorous as well as tumorous with the help of the fuzzy C-means segmentation, greater precision recognition as well as classification can potentially be achieved by incorporating DNN, i.e.—deep neural network. Multilevel discrete wavelet transform (DWT) was used to extract the wavelet function. Linear discriminant analysis (LDA), SMO, i.e., sequential minimal optimization, and K-nearest neighbors along with latter procedure are all measured together. The DNN-based brain tumor classification achieved a 96.97 percent accuracy score. However, because of the high degree of complexity, the execution was low.

In [3], authors proposed an empirical landmark-based feature representation for diagnosis of diseases. The proposed method boosts diagnostics accuracy, according to experimental results. To measure the severity of tumors, segment the tumor area from the MRI and then describe it as malignant or benign.

In [4], ReLU layer and a max-pool layer are containing 64 hidden neurons in architecture. On the training and validation packages, they performed 98.51% and 84.19%, respectively. To classify tumor in brain, a probabilistic neural network (PNN) is implemented.

In [5], Gabor wavelet features have been used in various studies, but they are limited to capturing only local MRI image structures such as regularity, position, and direction. Similarly, several writers called for the use of first-order, co-occurrence statistical features, gray-level matrix features, as well as gray-level run length matrix features.

In [6], in preprocessing, image filtering was introduced to improve contrast, sharpening, resizing, as well as GLCM features which has been extracted. The highest accuracy of 83.33% is attained. Statistical features are extracted using DWT as well as Gabor filter techniques. They used a back propagation multilayer perceptron neural network to classify the brain tumors and got the highest accuracy of 91.9%.

In [7], authors suggested a technique that employs watershed segmentation to detect cancerous nodules in lung CT scan images and a support vector machine (SVM) to classify nodules. There are six levels in this work that are: PCA is incorporated for reduction in features, to distinguishing and evaluation they introduced SVM, segmentation on images that have already been preprocessed, and the last preprocessing of images. In the end, they summarized and found that system was 86.6% accurate.

In [8], authors constructed a double stream layered network, where output feature maps linked to the second network simultaneously to obtain more contextual features from the input patches.

In [9], authors introduced a new model called capsule networks (CapsNets) to classify brain tumor. In order to improve accuracy, they modified the function maps in CapsNet's convolution layer. Using 64 function maps and one CapsNet convolution layer, they achieved the highest accuracy of 86.56%. In five different CNN architectures, found architecture 2 to be the most accurate in [10].

In [11], authors developed a scheme to extract features, then data is analyzed using texture-based features in ultrasound scans to distinguish breast lesions. Based on radiology and pathology findings, some patients were diagnosed with benign and malignant tumor that are 46 and 32 out of 78.

In [12], authors used deep neural network architecture. They generally categorized the images into two groups: tumorous and non-tumorous. For both CT and MR images, classification accuracy was greater than 97%. For the same reason, the frameworks AlexNet and ZFNet are compared.

In [13], GLCM was computed as well as infused in convolution neural network. Features that have been intensified are consolidated with GLCM gain sufficient correctness by 20%. Using this approach, they were able to obtain the highest accuracy of 82%.

3 Background Study

3.1 Deep Learning

In concrete language, DL is a subfield of ML that emphasizes the learning of successive layers of increasingly vivid manifestations as a substitute for learning representations from data. The term "deep" in the context of deep learning does not correspond to any particular understanding and knowledge reached by the method; instead, it refers to the concept of subsequent stages of interpretations. The strength of a model is defined as the percentage of layers which provide details to it. Proper names for the area, on the other hand, could include hierarchical interpretations learning.

3.2 Magnetic Resonance Imaging

There is a different method for obtaining a brain image that uses frequency signals it is known as magnetic resonance imaging. This image extraction technique is what we are concentrating on. This procedure is frequently used and assists in the identification and treatment plan of brain tumors in a number of neurological disorders. It generates one-of-a-kind images that are accompanied pixel by pixel. Using of a brain MRI image is limited to tumor treatment and prevention advancement. Among the most common approaches for diagnosing brain tumors is magnetic resonance imaging (MRI). MRIs are secure, non-invasive, and represent true proportions in comparison with other procedures for diagnosis, such as ultrasound, CT scans, and so on. As a result, they are commonly used in brain imaging.

4 Proposed Work

4.1 Dataset

The dataset used in this paper is taken from a publically available dataset at Kaggle. It contains images acquired from patients that are meningioma, glioma, pituitary tumor, and no tumor. It contains total 3264 images.

4.2 Convolution Neural Network

CNNs are a type of neural network in which the first layer extracts features, and the second layer uses the extracted features to perform classification. While CNN training is computationally intensive, the use of a graphics processing unit (GPU) can speed up the process. However, 3D kernel convolutions are more practically infeasible than 2D kernel convolutions, reducing the number of layers that can be added.

4.3 Adam Optimizer

Adam, which is originated from adaptive moment estimation, is an optimization technique that helps achieving learning rate decay by fixing the issues of vanishing gradients. It employs both the first and second moments. Adam is usually thought to be reasonably resistant to hyper parameter changes, though the learning rate might need to be adjusted from the recommended norm.

4.4 Convolution

In real-world applications, convolutional networks are highly useful. The term "convolutional neural network" refers to a mathematical procedure known as convolution, which the network conducts. This might be a highly accurate and sequential procedure. There will always be a layer which is going to include traditional calculation of matrix in place of convolution in CNN technique. Images are interpreted by computers as pixels. A group of learnable filters are utilized in the convolution layer. The training and testing stages of CNN architecture' classification are two categories. Using groups or labels such as regular and abnormal brain images, these images are divided into separate types. Figure 2 depicts basic architecture of CNN. CNN is a type of artificial neural network which contains a typical architecture as shown in Fig. 2 [14]. Generally, RGB images (3 channels) or gray scale images

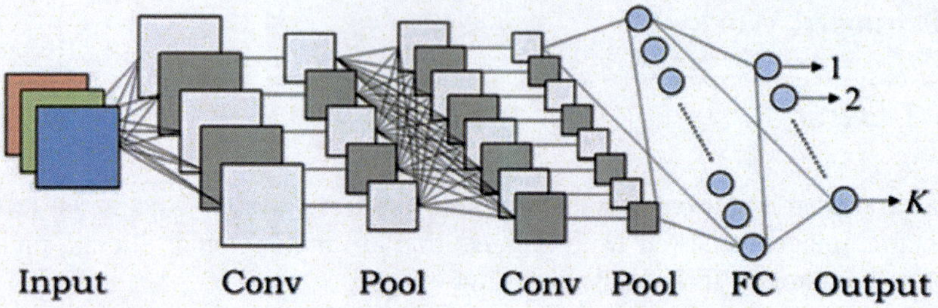

Fig. 2 Basic architecture of CNN [14]

(1 channel) are input of CNN. Input layer is followed by various convolution and pooling layers. FC (fully connected) layers are used for classification. Output layer is used for prediction values for objects of K types where the input is going to be classified.

4.5 The Pooling Layer

Pooling feature is chosen for output of the layer to be switched. The max-pooling process, for example, records the maximum output inside in an oblong neighborhood. In CNN architecture, it is frequently observed between the convolution layers. It aids in the elimination of the network's computation and parameter count. The pooling processing is done solely for the purpose of minimizing the image's spatial size. Since each depth dimension is completed separately, the image's depth stays consistent. The max-pooling layer is the most popular form of pooling layer.

4.6 Fully Connected Layer

Feed forward neural networks are what the fully connected layer is all about. Fully connected layers are the network's final layers. We gain some output by pooling or convolution layer, and this is flattened and finally given to the fully connected layer.

4.7 Tools and Parameters

On the TensorFlow platform, the CNN-based system is demonstrated in Python. We calculated and evaluated performance of our model with the help of accuracy curve and loss curve. Adam optimizer is used and ReLu activation function is employed. Pooling layer is max-pooling. Epoch and batch size are 50 and 40, respectively.

4.8 Proposed Architecture

In this proposed architecture, brain MRI images are used from Kaggle publically available dataset. In this dataset, total 3264 images are used. And this dataset is split into 2870 training data and 394 test data. In this architecture, image size is 150 and 5 convolution layers, and 2 fully connected layers are used. All the convolution layers use 5 × 5, 3 × 3, and 2 × 2 kernel size, and stride 2 is used by max-pool layers. Filter size is 64, 128, and 256. Padding "same" is used. Figure 3 depicts the basic overview of proposed work. Dropout is 0.25 and 0.3. Total epochs are 50, and batch size is 40. Images are extracted from source, and then they are preprocessed. Then, we performed labels loading. Data is partitioned into training and validation set. Then, we train our model and use Adam optimizer and ReLu activation function. Dropout layer is added to control overfitting, and softmax is also used. After training the model, we calculate the performance of the model. Steps that are followed:-

- We firstly used convolution neural network as a first layer.
- With the help of activation layer, we transform signals from one layer to the subsequent layers.

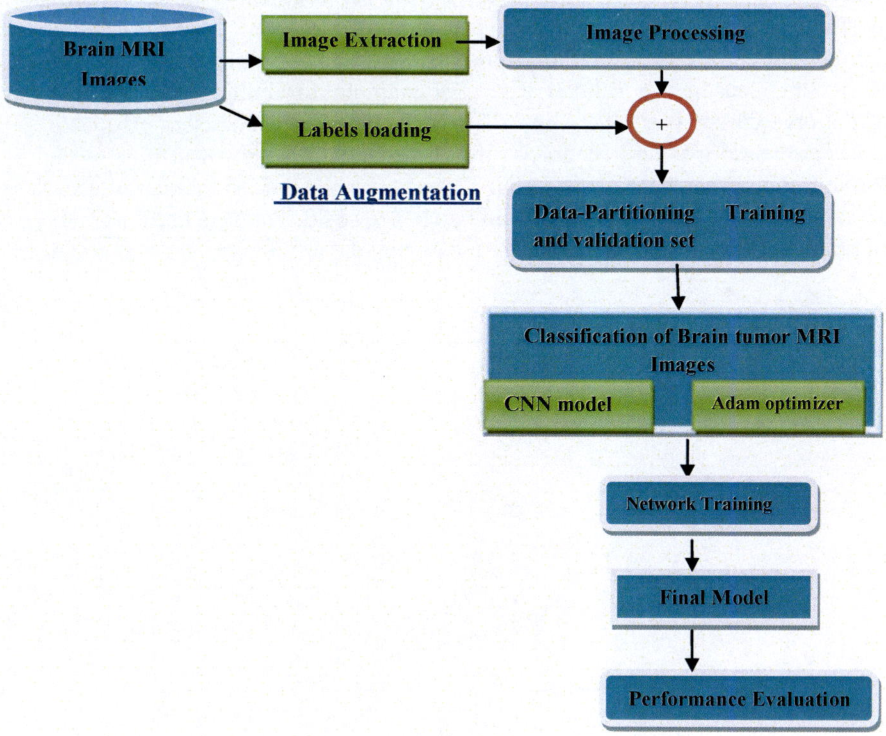

Fig. 3 Overview of proposed work

- Training time is fasten with the help of rectified linear unit (ReLu). It is written as $F(x) = \max(0, x)$ that means return 0 for negative input and return the value as it as when the input is positive.
- Finally, we add a fully connected layer at the end to get the classified output.

5 Experimental Results

5.1 *Loss Plot*

At the time of training, for debugging, a neural network loss plot is one of the most commonly used curves. It provides an overview of the training process as well as the network's learning trajectory.

5.2 *Accuracy Plot*

With the help of accuracy plot, we can explain the advancement of the neural network. Accuracy plot is a crucial part in deep learning.

Loss and acc are measured on training set, and on the other hand, val_loss and val_acc are measured on validation set. For example if training accuracy is $x\%$ and validation accuracy is $y\%$, then new data model will give accuracy of $y\%$.

Experimental loss and accuracy plot of our model is shown in Figs. 4 and 5. Through this research, 96.45% training accuracy and 93.38% validation accuracy are achieved with epoch size of 50 and batch size of 40. We have used dropout just for removing over-fitting. Here, we are measuring the performance of our model and

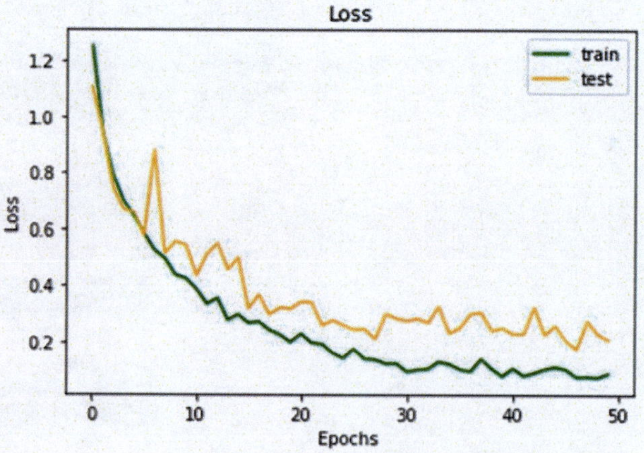

Fig. 4 Loss plot

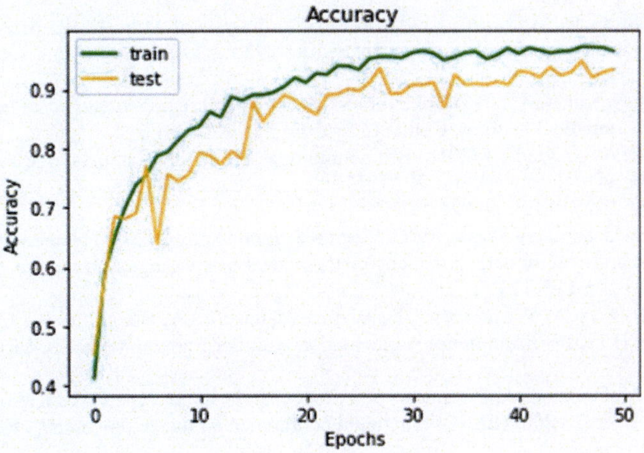

Fig. 5 Accuracy plot

predicting the model which is the last step in proposed work with the help of loss and accuracy plot.

6 Conclusion

With the help of CNN, we developed this model to classify brain MRI images into glioma, meningioma, pituitary, and no tumor types. The implementation is done on publically available Kaggle data. In this approach, highest accuracy is achieved, i.e., 96.45% training accuracy and 93.38% validation accuracy. As mentioned, brain tumor is classified into four types. We can define the severity of a disease and region, i.e., part of a body in which it is occurring with the help of its grading. Further, brain tumor can be classified using a different CNN architecture in future work with a better accuracy and results. We can also measure the shape and size of tumor using CAD and deep learning techniques.

References

1. Blumenthal D, Artzi M, Liberman G, Bokstein F, Aizenstein O, Ben Bashat D (2017) Classification of high-grade glioma into tumor and nontumor components using support vector machine. Am J Neuroradiol 38:908–914
2. Mohsen H, El-Dahshan ESA, El-Horbaty ESM, Salem ABM (2018) Classification using deep learning neural networks for brain tumors. Fut Comput Inf J 3(1):68–71
3. Liu M, Zhang J, Nie D, Yap PT, Shen D (2018) Anatomical landmark based deep feature representation for MR images in brain disease diagnosis. IEEE J Biomed Health Inform 22(5):1476–1485

4. Abir TA, Siraji JA, Ahmed E, Khulna B (2018) Analysis of a novel MRI based brain tumor classification using probabilistic neural network (PNN). Int J Sci Res Sci Eng Technol 4(8):65–79
5. Nabizadeh N, Kubat M (2015) Brain tumors detection and segmentation in MR images: Gabor wavelet vs. statistical features. Comput Electr Eng 45:286–301
6. Ismael MR, Abdel-Qader I (2018) Brain tumor classification via statistical features and back-propagation neural network. In: 2018 IEEE international conference on electro/information technology (EIT). IEEE, pp 0252–0257
7. Devkota B, Alsadoon A, Prasad PWC, Singh AK, Elchouemi A (2018) Image segmentation for early stage brain tumor detection using mathematical morphological reconstruction. Procedia Comput Sci 125:115–123
8. Havaei M, Davy A, Warde-Farley D, Biard A, Courville A, Bengio Y, Pal C, Jodoin PM, Larochelle H (2017) Brain tumor segmentation with deep neural networks. Med Image Anal 35:18–31
9. Afshar P, Mohammadi A, Plataniotis KN (2018) Brain tumor type classification via capsule networks. In: 2018 25th IEEE international conference on image processing (ICIP) pp 3129–3133. IEEE
10. Abiwinanda N, Hanif M, Hesaputra ST, Handayani A, Mengko TR (2019) Brain tumor classification using convolutional neural network. In: World congress on medical physics and biomedical engineering 2018. Springer, Singapore, pp 183–189
11. Sadeghi-Naini A, Suraweera H, Tran WT, Hadizad F, Bruni G, Rastegar RF, Curpen B, Czarnota GJ (2017) Breast-lesion characterization using textural features of quantitative ultrasound parametric maps. Sci Rep 7(1):1–10
12. Makde V, Bhavsar J, Jain S, Sharma P (2017) Deep neural network based classification of tumourous and non-tumorous medical images. In: International conference on information and communication technology for intelligent systems. Springer, Cham, pp 199–206
13. Widhiarso W, Yohannes Y, Prakarsah C (2018) Brain tumor classification using gray level co-occurrence matrix and convolutional neural network. IJEIS (Indonesian J Electron Inst Syst) 8(2):179–190
14. Hidaka A, Kurita T (2017) Consecutive dimensionality reduction by canonical correlation analysis for visualization of convolutional neural networks. In: Proceedings of the ISCIE international symposium on stochastic systems theory and its applications, pp 160–167. https://doi.org/10.5687/sss.2017.160

Perceptual Energy Weight Matrix Based Adaptive Block Compressed Sensing for Marine Image Compression

R. Monika, Samiappan Dhanalakshmi, and R. Kumar

Abstract The compressed sensing (CS) achieves compression at sub-nyquist sampling rate, and hence, it is widely applied for image compression applications. CS dictates random projection of samples which results in poor reconstruction. To overcome this issue, block level CS (BCS) can be used, as they capture the geometry of the blocks perfectly. BCS selects fixed samples from each block without considering block's nature. Adaptive block compressed sensing (ABCS) overcomes this by selecting different count of samples from various blocks. To orient the signal components more precisely, the human perceptual properties are utilized to construct an energy weight matrix. This weight matrix is applied on the signal vector to select samples which are more attractive to human eye. To improve the reconstruction quality of underwater images, we propose a combination of energy matrix and ABCS (PEM-ABCS) in this article. Experimental findings show that the proposed scheme has led to significant improvement in the reconstruction quality when compared to other non-weighted adaptive schemes in the literature.

Keywords Energy matrix · Block compressed sensing · Adaptive block compressed sensing · Sparse binary random matrix · Orthogonal matching pursuit

1 Introduction

CS [3] has been more widely used in the recent years as they can accomplish the compression process at very less sampling rate. It considers whole image for picking samples which results in poor reconstruction quality. A sub category of CS is BCS where division of the image into blocks is performed and each blocks are processed separately. But they share some common disadvantages such as how to pick the

R. Monika · S. Dhanalakshmi (✉) · R. Kumar
Department of ECE, College of Engineering and Technology, Faculty of Engineering and Technology, SRM Institute of Science and Technology, SRM Nagar, Kattankulathur, Chengalpattu Dt. 603203, Tamil Nadu, India
e-mail: dhanalas@srmist.edu.in

R. P. Mahapatra et al. (eds.), *Proceedings of International Conference on Recent Trends in Computing*, Lecture Notes in Networks and Systems 341,
https://doi.org/10.1007/978-981-16-7118-0_51

samples and how many samples to choose. The samples can be picked by using measurement matrix. However, the number of samples to be chosen gave way to a variant of BCS, i.e.; ABCS. In BCS only fixed measurements are chosen, whereas in ABCS samples are chosen differently according to block's nature. Entropy is used to decide on the number of samples that are to be fixed to each block. Higher the entropy of the block, more will be the information content it holds and more will be the samples that are to be allotted to that particular block and vice versa.

Improving the accuracy of the recovered image has recently become a significant concern. To achieve this, weighting strategy utilizing the weighting factors for the construction of weight matrix, can be incorporated into the ABCS. This process has to be carried out during the sampling process to achieve better sampling performance. We utilize the second order moment, i.e.; energy to construct the weight matrix. Energy can more precisely orient the signal components than the first order moments. Energy values of the various frequency components of the image are calculated and is placed on the leading diagonal of the matrix with zero elsewhere.

This weight matrix with energy values selects signal components with high energy. Further refinement is done using ABCS process which decides on the number of samples to be chosen from each block.

1.1 Literature Review

The iterative reweighted algorithm used in [1] is used to recover the image signals efficiently. A reweighted strategy is used to calculate the weighting factors. But these weighting factors are to be updated iteratively which makes the recovery more difficult process. To avoid this, reweighting should be done during the sampling process rather than the reconstruction phase. Weighting matrix is constructed using first order moments in [12] and seconds order moments in [5, 16]. In all these schemes, fixed sample count is assigned to all the blocks. But various image regions consists of various information contents and uniform measurement allocation affects the sampling efficiency, limiting the reconstructed image quality. Therefore, we focus on adaptive sampling with perceptual weighting in this article. In adaptive sampling, we assign varying measurements to various image blocks in non-uniform manner based on image block's characteristics. Various adaptive sampling/ABCS strategies are available in the literature to assign varying number of measurements. Standard deviation [13], error [7], energy [8], saliency [14], entropy [6], gradient scale [15], texture [10], edge information [2] were used for adaptive measurement allocation. However, they resulted in blocking artifacts. To overcome that we propose entropy based ABCS integrated with the perceptual weight matrix.

1.2 Contribution and Organization of This Work

The proposed method PEM-ABCS takes the following advantages

1. Blocking artifacts/ Block distortions have been completely eliminated.
2. Entropy calculation procedure is simple and easily implementable in real time.
3. Significant improvement in performance measuring parameters like PSNR, SSIM.

The content of the article is laid out as follows. We discuss the overview of BCS and ABCS in part 2. In part 3, proposed PEM-ABCS is discussed. Part 4 presents the findings and discussions. The paper is concluded in part 5.

2 Overview of BCS and ABCS

In BCS, a fixed count of measurements is selected from all blocks, while in ABCS, a variable count of measurements is selected from all blocks.

2.1 BCS [4]

In conventional CS, the measurement matrix size is same as that of the size of the image. So measurements are taken from the whole image at once without considering the information content at various regions of the image, leading to poor reconstruction quality. The image is split into $B \times B$ blocks in BCS and the measurement matrix is applied on the blocks separately and simultaneously. Each block is transformed to 1-D signal and fixed count of samples are chosen from all of the blocks to form sampled vector y_i

$$y_i = \Phi^{(B)} x_i \tag{1}$$

where $\Phi^{(B)}$ is measurement matrix of size $n^{(B)} \times B^2$. $n^{(B)}$ is fixed for BCS.

2.2 ABCS

In ABCS also, the image is divided into block. But different measurements are chosen form different blocks. The sampled vector y_i is represented in ABCS as

$$y_i = \Phi_a^{(B)} x_i \tag{2}$$

where $\Phi_a^{(B)}$ is the adaptive measurement matrix of size $n^{(B)} \times B^2$. $n^{(B)}$ is varying for ABCS.

3 Proposed Perceptual Energy Weight Matrix Based Adaptive Block Compressed Sensing (PEM-ABCS)

In the proposed method, the perceptual energy weight matrix is designed and applied on the signal vector to scale and select only important signal components. Then, ABCS is performed on all the blocks to fix adaptive sampling rate for each and every block. Combination of both of these strategies will improve the sampling efficiency.

3.1 *Perceptual Energy Weight Matrix Design*

Let $\{E_1, E_2, \ldots E_{B\times B}\}$ be the energy weighting coefficients calculated for all the frequency components of the image block. The energy weighting coefficient for any particular frequency component is calculated by using the formula,

$$E_j = \sum_{j=1}^{B^2} |C_i^j|^2 \tag{3}$$

where C_i^j is the coefficient obtained after applying transform for sparsification. The energy coefficients are placed along the leading diagonal of the matrix with zeroes elsewhere to form energy weight matrix. This matrix is multiplied with the signal vector to select only information contributing components from the vector.

3.2 *Adaptive Block Compressed Sensing Using Entropy*

Entropy is the information contained in the value of the image pixel. More the value of entropy of the block, less sparse will be the coefficients and more will be the information content of that block. Therefore extracting the entropy content of the image block provides convincing sample allocation to the blocks of the image. The sampling scheme includes the following steps

1. The entropy is calculated by using the formula

$$H_i = \sum_{j=0}^{255} P_{ij} \log_2 P_{ij} \tag{4}$$

2. The probability distribution of the entropy for each block is calculated by using

$$W_i = \frac{H_i}{\sum_{i=1}^{K} H_i} \tag{5}$$

where 'k' is the total block count in the image.

3. Let us assign 'M' to be the total count of measurements chosen from the entire image which is calculated by using the equation

$$M = \sum_{i=1}^{k} m_i \tag{6}$$

where m_i is the measurement count of each block
4. The total measurements set to each block is given by

$$m_i = \text{round}[W_i(M - km_0) + m_0] \tag{7}$$

where m_0 is the initial count of measurements fixed for each block.
5. The initial sampling rate for the block is given by the equation

$$R_i = \frac{m_i}{B \times B} \tag{8}$$

6. m_i rows are chosen from random measurement matrix, "sparse binary random matrix" to construct adaptive measurement matrix $\phi_a^{(B)}$
7. ABCS is then performed using Eq. (2).
8. Orthogonal matching pursuit (OMP) algorithm [9] is chosen for reconstruction.

4 Results and Discussion

Matlab R2019b is used for simulation of the various compression schemes. The images used are standard test images and underwater images available in the database [11]. The results are compared with other non-weighted adaptive algorithms [6, 13] and conventional BCS [4]. Performance measuring parameters such as PSNR and SSIM were used for comparison whose formulas are given as

$$\text{PSNR} = 20 \log_{10} \frac{\text{MAX}_I^2}{\text{MSE}} \tag{9}$$

where MSE is the acronym for mean square error which is given by the formula,

$$\text{MSE} = \frac{1}{\text{MN}} \sum_{m=1}^{M} \sum_{n=1}^{N} (O(m,n) - R(m,n))^2 \tag{10}$$

SSIM is given by,

$$SSIM = \frac{(2\mu_x\mu_y + S_1)(2\sigma_{xy} + S_2)}{(\mu_x^2 + \mu_y^2 + S_1)(\sigma_x^2 + \sigma_y^2 + S_2)} \quad (11)$$

where the stabilization factors $S_1 = (k_1 L)^2$, $S_2 = (k_2 L)^2$, $k_1 = 0.01$, $k_2 = 0.03$ and L is the dynamic range of pixel values

4.1 Standard Test Images

Table 1 shows the PSNR computation results for PEM-ABCS and other ABCS techniques from the literature.

Table 1 shows that for different sampling rates, the proposed PEM-ABCS has obtained higher PSNR values than other non-weighted ABCS schemes. SSIM comparison for various ABCS techniques and BCS is tabulated in Table 2.

Table 2 shows that the proposed PEM-ABCS scheme has a higher SSIM value than other schemes. SSIM values closer to 1 indicates closeness of the reconstructed image with the original image. The subjective results for the standard test images are shown in Fig. 1. Figure 1 shows that the proposed PEM-ABCS has recovered the image as closely as possible to the original image. Other ABCS methods have distortions in them. Standard test images give good results for most of the algorithm. To check the consistency of the proposed method in reconstructing images with good quality, few underwater images were chosen and tested. The results are given below.

Table 1 PSNR variations for different compression techniques

Image: Lena (512 × 512), $B \times B = 8 \times 8$			
Method	PSNR (dB)		
	$R = 0.1$	$R = 0.3$	$R = 0.5$
BCS [4]	30.68	32.68	34.43
ENT-ABCS [6]	31.95	35.27	36.97
STD-ABCS [13]	32.59	34.59	35.46
PEM-ABCS	34.96	36.72	37.98

Table 2 SSIM variations for different compression techniques techniques

Image: Lena (512 × 512), $B \times B = 8 \times 8$			
Technique	SSIM		
	$R = 0.1$	$R = 0.3$	$R = 0.5$
BCS [4]	0.6937	0.7067	0.7949
ENT-ABCS [6]	0.7950	0.8004	0.8245
STD-ABCS [13]	0.7479	0.8677	0.9002
PEM-ABCS	0.9002	0.9130	0.9895

Fig. 1 Subjective result comparison **a** Source image, **b–e** reconstructed using BCS, STD-ABCS, ENT-ABCS, Proposed PEM-ABCS respectively

Table 3 PSNR variations for different underwater images

$R = 0.1, B \times B = 8 \times 8$				
Image	Techniques			
	BCS [4]	ENT-ABCS [6]	STD-ABCS [13]	PEM-ABCS
Coral	26.83	27.92	27.54	30.22
Jelly fish	30.84	31.85	31.94	34.00
Fish	28.23	29.34	30.23	32.03
Sea horse	25.95	27.23	28.84	30.44
Crab	28.92	29.04	32.94	33.49

4.2 Underwater Images

Underwater images are chosen from the database available in [11]. Table 3 shows the PSNR comparison for various underwater images.

Table 3 shows that, as opposed to other strategies in the literature, the proposed PEM-ABCS has reached higher PSNR values.

Figure 2 shows that even at a low sampling rate of about $R = 0.1$, the proposed approach produced better results. Other now-weighted ABCS methods have more distortions in them.

5 Conclusions and Future Direction

In this research article, perceptual energy weight matrix based adaptive block compressed sensing is proposed for underwater image compression. It is found that the proposed method has produced better subjective and objective results for both standard test images and underwater images. There is a significant improvement in the performance measuring characteristics like PSNR and SSIM. As opposed to other approaches, the visual accuracy of the images reconstructed using the proposed approach has relatively few block distortions. The proposed system can be extended for monitoring underwater videos as a future work.

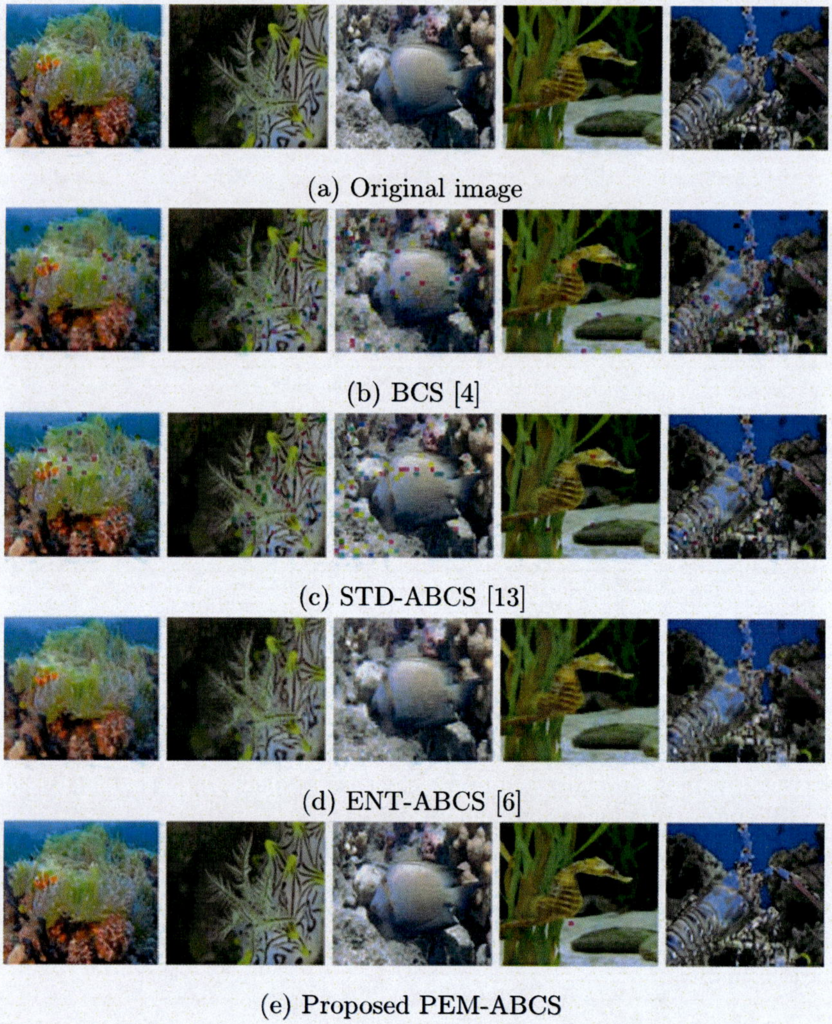

Fig. 2 Visual quality comparison of underwater images at $R = 0.1$ **a** Original image, **b–e** reconstructed using BCS, STD-ABCS, ENT-ABCS, Proposed PEM-ABCS respectively

References

1. Candes EJ, Wakin MB, Boyd SP (2008) Enhancing sparsity by reweighted l 1 minimization. J Four Anal Appl 14(5–6):877–905
2. Canh TN, Dinh KQ, Jeon B (2014) Edge-preserving nonlocal weighting scheme for total variation based compressive sensing recovery. In: 2014 IEEE international conference on multimedia and expo (ICME). IEEE, pp 1–5
3. Donoho DL (2006) Compressed sensing. IEEE Trans Inf Theo 52(4):1289–1306

4. Gan L (2007) Block compressed sensing of natural images. In: 2007 15th International conference on digital signal processing. IEEE, pp 403–406
5. Gao Z, Xiong C, Ding L, Zhou C (2013) Image representation using block compressive sensing for compression applications. J Vis Commun Image Representation 24(7):885–894
6. Li R, Duan X, Guo X, He W, Lv Y (2017) Adaptive compressive sensing of images using spatial entropy. Comput Intell Neurosci
7. Li R, Duan X, Lv Y (2018) Adaptive compressive sensing of images using error between blocks. Int J Distrib Sens Netw 14(6):1550147718781751
8. Monika R, Samiappan D, Kumar R (2020) Underwater image compression using energy based adaptive block compressive sensing for iout applications. Vis Comput 1–17
9. Shen Y, Li S (2015) Sparse signals recovery from noisy measurements by orthogonal matching pursuit. Inverse Prob Imag 9(1):231
10. Sun F, Xiao D, He W, Li R (2017) Adaptive image compressive sensing using texture contrast. Int J Dig Multimedia Broadcast **2017** (2017)
11. xahidbuffon: Underwater-Datasets. https://github.com/xahidbuffon/Underwater-Datasets/commits?author=xahidbuffon (2019) [Online; accessed 5-Apr-2021]
12. Yang Y, Au OC, Fang L, Wen X, Tang W (2009) Reweighted compressive sampling for image compression. In: 2009 Picture Coding Symposium. IEEE, pp 1–4
13. Zhang J, Xiang Q, Yin Y, Chen C, Luo X (2017) Adaptive compressed sensing for wireless image sensor networks. Multimedia Tools Appl 76(3):4227–4242
14. Zhang Z, Bi H, Kong X, Li N, Lu D (2020) Adaptive compressed sensing of color images based on salient region detection. Multimedia Tools Appl 79(21):14777–14791
15. Zhao HH, Rosin PL, Lai YK, Zheng JH, Wang YN (2020) Adaptive gradient-based block compressive sensing with sparsity for noisy images. Multimedia Tools Appl 79(21):14825–14847
16. Zhu S, Zeng B, Gabbouj M (2014) Adaptive reweighted compressed sensing for image compression. In: 2014 IEEE international symposium on circuits and systems (ISCAS). IEEE, pp 1–4

Self-Adaptive Cuckoo Search-Based Cluster Head Selection for Maximizing Network Lifetime in Wireless Sensor Networks

G . Rajeswarappa and S. Vasundra

Abstract Energy stabilization is considered as the core factor of concern in wireless sensor networks (WSNs) as they are responsible for lifetime longevity of sensor nodes leading to maximized network lifetime. In this context, novel metaheuristic algorithm is highly ideal for facilitating clustering process and subsequent cluster head selection process. In this paper, a self-adaptive cuckoo search-based cluster head selection (SACS-CHS) scheme is proposed for maximizing network lifetime with sustained energy stability of sensor nodes. This SACS-CHS scheme is proposed with adaptive parameters that attributes toward better cluster head selection without the need of tuning the utilized parameters. It included the reduced population proportion concept based on the fitness evaluated based on the previous and current best solution. This CH selection is attained through the fitness function formulated based on residual energy, intra-cluster distance, and inter-cluster distance. It also incorporated Gaussian sampling mechanism that improved the tendencies of exploitation and exploration. In addition, Weibull distributed probability switching is used for increasing the trade-off between exploitation and exploration. The simulation results confirmed better energy stability of 12.38% and improved network lifetime of 14.21%, excellent to the baseline schemes considered for investigation.

Keywords Self-Adaptive cuckoo search · Weibull distributed probability · Gaussian sampling mechanism · Network lifetime · Cluster head

1 Introduction

The rapid advancement of wireless communication and microelectromechanical systems (MEMSs) has significantly increased the mass production of low-cost sensor

G. . Rajeswarappa (✉)
Department of CSE, JNTUA, Ananthapuramu, India

S. Vasundra
Department of CSE, JNTUACEA, Ananthapuramu, India
e-mail: vasundras.cse@jntua.ac.in

R. P. Mahapatra et al. (eds.), *Proceedings of International Conference on Recent Trends in Computing*, Lecture Notes in Networks and Systems 341,
https://doi.org/10.1007/978-981-16-7118-0_52

nodes, thereby introducing the possibility of establishing wireless sensor networks (WSNs) with minimized cost [1]. The integration of sensor network terminals with the Internet's backbone has opened the way for most significant technologies such as Internet of things (IoT), big data analytics, and the cloud [2]. This WSNs are self-organized, infrastructure-less networks with a large number of sensor nodes capable of organizing themselves in an ad hoc manner [3]. WSNs are used in diversified environments, such as military operations, weather forecasting, health monitoring, and habitat observation, based on their appropriateness and applicability [4]. WSNs include hundreds and thousands of sensor nodes for performing sensing and data accumulation operations during their deployment in suitable applications [5]. During deployment, the sensor nodes need to be organized into clusters for handling the challenges of limited energy, memory, bandwidth, computing power, and sensing capability inherent with them [6]. This clustering process in WSNs can significantly affect the quality of service (QoS) parameters like throughput, lifetime of the network, and latency. The network lifespan of WSNs is wholly based on the availability of resources present in the network, and these resources cannot be replenished easily. The network performance can be improved by utilizing resources efficiently through inclusion of appropriate load balancing and clustering algorithms during the routing process. At this juncture, clustering schemes organize the closely located sensor nodes into groups to attain efficient energy management in the network. The clustering mechanism is considered to be essential in resource-constrained environments such as WSNs. This clustering process is heavily influenced by the selection mechanism of cluster head (CH) nodes.

Several clustering algorithms were proposed to balance energy consumption, by resolving issues, and challenges of WSNs from the past decades [7]. These clustering algorithms also focus on the ways of energy consumption in low-power devices, routing techniques, data aggregation, clock synchronization, and medium access control (MAC) in WSNs. These clustering algorithms are categorized into *classical* and *computational intelligence*-based approaches. The classical and computational intelligence (CI) algorithms focus mostly on the network lifetime. However, the classical clustering algorithms are identified to be less potent, as they incorporate iterative, probabilistic, and simple optimization methods during cluster head selection process. These classical clustering algorithms are found to be better in the organization of nodes in a network, but lacking in load balancing and energy efficiency [8]. On the other hand, computational intelligence algorithms are quite promising in facilitating prolonged network lifetime and energy efficiency. They incur low computing overhead for estimating optimal solutions as compared to the classical and iterative strategies. Therefore, CI-based metaheuristic approaches need to be explored for modeling and constructing energy efficient solutions in WSNs. However, most of the standalone metaheuristic algorithms are not capable of balancing the trade-off between the rate of exploitation and rate of exploration in the search process. Therefore, there is a need for hybrid metaheuristic computational intelligence-based algorithms to handle the exploitation and exploration limitations mutually.

In this paper, self-adaptive cuckoo search-based cluster head selection (SACS-CHS) scheme is proposed for maximizing network lifetime with sustained energy stability of sensor nodes. This proposed SACS-CHS scheme established better balance between exploitation and exploration by preventing the issue of premature convergence. It is proposed with an extensive self-adaptive knowledge model that derives experience based on individual history and population knowledge. It included the merits of threshold statistics learning approach that exploits the individual and population learning process that results in selecting only the superior fitness sensor nodes as cluster head. It also used historical knowledge for adjusting the position of the individual solution based on the communication achieved with other interacting individuals of the population during the process of optimization. Simulation experiments of the proposed SACS-CHS scheme are conducted for understanding its predominance over the baseline schemes with respect to network lifetime, energy consumptions, and throughput with scalable increase in the number of sensor nodes of the network.

2 Related Work

Time-to-live ant colony optimization (T-ANT) algorithm was proposed as clustered data distribution approach based on ant swarm behavior for attaining uniform dissemination of CHs [9]. This algorithm, during its implementation, considerably achieved significant energy conservation. In this T-ANT approach, the clusters are constructed based on the foraging and food source categorization process. A node checks to know whether it possesses an ant when the timer perishes. A node with an ant becomes a CH and publicizes its neighbors by disseminating its node ID. In the set-up phase, the CHs are elected using a swarm of ants, wherein an ant relates to a control message. The ant travels to the sink limited by its time-to-live (TTL). When an ant reaches a node, the subsequent node is arbitrarily selected. Ant colony optimization for clustering (ACO-C), an energy-based protocol, is propounded [10]. Objects are allocated to "K" clusters depending on the distance between objects and the cluster center. The BS chooses nodes with remaining energy above the threshold as CHs and uses ACO algorithm for finding the most acceptable solution based on the cost function. Once the route is found, the BS notifies the nodes about the elected CHs. Each CH coordinates data forwarding in their conforming clusters using TDMA. The CH aggregates the data received from the members and forwards it to the BS.

Genetic algorithm-based energy efficient clustering (GABEEC) [11] algorithm, similar to LEACH, includes setup and steady-state phases. The nodes join the cluster depending on their proximity with the CH. Data are forwarded to the CH using TDMA. The clusters remain the same, but the CHs may be changed based on the quantity of remaining energy. The initial population is randomly produced, and the nodes are evaluated for fitness. Another GA-based hierarchical clustering approach is propounded to focus on the transmission involving long distances between the nodes and the BS [12]. GA is combined with hierarchical clustering for lessening the distance, which in turn reduces the amount of energy consumed to the maximized level. In this scheme, the position of the BS, size of population, and number of runs are determined by applying GA is applied on the network. The CHs that are in proximity to the BS collects the data from the CHs and forwards it to the sink.

Bee colony optimization (BCO) schemes are stimulated based on honeybee foraging. Insects are proactive with self-organizing capacities [13]. Honeybees are grouped into colonies and live in a hive and are capable of auto-solving. Scoutbees are involved in exploring the hive so as to identify likely sources of food. When a source is identified, the bee returns to the hive to appoint forager bees through waggle dance. ABC-SD [14] is a cluster-based routing scheme that uses the ABC algorithm for effective searching. It is a semi-distributed scheme, wherein clustering is performed at the BS based on the remaining energy and location of nodes and their neighbors. Distributed data routing is performed. The BS builds clusters and chooses CHs by constructing a set of candidate CHs depending on the position and remaining energy. ABC algorithm is involved in allocating nodes to a CH and build clusters. Multi-hop TDMA based on B-MAC is employed wherein a sleep/wake-up scheme is designed to convey the collected data to the conforming CH, either in a single hop or multi-hop based on the cluster density and position of the CH. Each member commences route discovery to build a path to the CH.

3 Proposed Algorithm

The cuckoo search algorithm is a nature-inspired metaheuristic algorithm proposed based on the inspiration derived from the obligate brood parasitism features of some cuckoo species that generally lay their eggs in the nests of other host birds [15]. This obligate brood parasitism features of cuckoo species are modeled based on the Levy flight principles, which refer to a category of random walk with heavy tail. This cuckoo search algorithm is modeled based on three suitable rules such as (i) it randomly selects nest and lays only one egg at a time (the solutions are randomly selected and only one solution is explored at a time), (ii) the eggs of high quality present in the best nests are selected and they are carried over the subsequent generations (the solutions of better fitness value will be carried out from one generation to the other for cluster head selection process), and (iii) the host bird has the capability of discovering an alien egg from the fixed number of available host nests depending on the probability $p_S \in [0, 1]$. Similarly, the cuckoo search agents can identify the best and worst solutions from the search space to select them as cluster head and cluster member nodes, respectively. In this context, the d-dimensional vector $N_{\text{Pos(id)}} = \left(N_{i1}, N_{i2}, \ldots\ldots, N_{(iD)}\right)$ with $1 \leq i \leq n$ and $1 \leq d \leq D$ representing the nest position (search space where solutions can be explored for determining the optimal solutions from the compared solutions (offspring solution)). At this juncture, the new solution is determined based on Levey flight (random walk) as represented in Eq. (1)

$$S_i^{t+1} = S_i^t + \text{Info}_{SS} * \text{Levy}(\tau) \quad (1)$$

where S_i^{t+1} and S_i^t refer to the current and existing randomly selected solutions that are compared based on the evaluation of fitness function. Further, step size information used for controlling the random searching range is determined based on Eq. (2)

$$\text{Info}_{SS} = \text{Info}_{SS(0)} * \left(S_j^t - S_i^t\right) \tag{2}$$

Moreover, the Levy flight representing the random walk can be represented through the formula for simple power formula specified in Eq. (3).

$$\text{Levy}(\alpha) \sim \mu = t^{-1-\alpha} \text{ with } 1 \leq \alpha \leq 2 \tag{3}$$

$$\text{Levy}(\alpha) \sim \frac{\delta \times \mu}{|v|^{\frac{1}{\alpha}}} \tag{4}$$

$$\delta = \left(\frac{\sin\left(\frac{\pi\alpha}{2}\right) \times \gamma(1+\alpha)}{\alpha \times 2^{\left(\frac{\alpha-1}{2}\right)} \times \gamma\left(\frac{1+\alpha}{2}\right)}\right)^{\frac{1}{\alpha}} \tag{5}$$

where in Eq. (4) μ and v are normally distributed random numbers with the mean and standard deviation value of 0 and 1, respectively. In this context, the value is 1.5. In Eq. (5) δ was calculated. Then, the new solution S_i representing the random walk component of Levy flight is determined based on Eq. (6)

$$S_{G+1,j} = S_{G,j} + \beta_0 \times \frac{\delta \times \mu}{|v|^{\frac{1}{\beta}}} \tag{6}$$

where β_0 is the factor of scaling with the value set to 0.01.

3.1 Self-Adaptive Knowledge Learning Improved Cuckoo Searching-Based Cluster Head Selection

In this cluster head selection approach, the cuckoo search algorithm utilized the merits of learning model that derives the merits of individual and population knowledge. In this context, knowledge learning is the process through which the individual search agents interact with one another to exchange their knowledge and experience. In specific, the individual knowledge concentrates on studying the historical experience of every individual at each and every generation in order to determine the potential sensor node as cluster heads. The main merit of individual knowledge-based learning model targets not only on the individuals' historical optimal solution, but also utilized the individuals' complete historical solutions. This individual learning model aids in adjusting the weights-based knowledge potential energy and knowledge learning rate. The degree of knowledge acquired by historical learning aids in better exploration and guides in the evolution of the newly generated individual solutions. The knowledge potential energy possessed by the individual solution is impacted by the cooperation learning. At this juncture, the rate of knowledge learning ($K_{LR(i)}$) is presented in Eq. (7).

$$K_{\mathrm{LR}(i)} = \frac{\mathrm{SC}^t_{\mathrm{en}(i)}}{\mathrm{SC}^t_{\mathrm{en}(1)} + \mathrm{SC}^t_{\mathrm{en}(2)} + \cdots + \mathrm{SC}^t_{\mathrm{en}(m)}} \quad (7)$$

where $SC^t_{\mathrm{en}(i)}$ is the knowledge potential energy in individual knowledge learning model determined based on Eq. (8)

$$\mathrm{SC}^t_{\mathrm{en}(i)=} = \begin{cases} 1 & \mathrm{If}(\mathrm{Fit}_{\mathrm{Best}} = \mathrm{Fit}_{\mathrm{Worst}}) \\ \frac{\left(\mathrm{Fit}^t_{\mathrm{Worst}(i)} - \mathrm{Fit}^t_{(j)}\right)}{\left(\mathrm{Fit}^t_{\mathrm{Worst}} - \mathrm{Fit}^t_{\mathrm{Best}}\right)} & \mathrm{Otherwise} \end{cases} \quad (8)$$

It is identified that higher knowledge potential energy maximized rate of knowledge learning associated with the solutions that could be possibly explored in the search space. This algorithm concentrates on the optimal position of individuals and completes ignores the history of common individuals, which could not provide potential information for decision making process. This individual learning approach considers both the historical optimal solution and self-adaptively balances their weights based on the knowledge potential energy and knowledge learning weight. Moreover, the individual learning strategy is used for guiding the generated individuals that are better in exploring new areas based on the process of updating the Levy flights as specified in Eq. (9)

$$S_{G+1.j} = S_{G,j} + \beta_0 \times \frac{\delta \times \mu}{|v|^{\frac{1}{\beta}}}\left(S_{G,i} - S_{G,\mathrm{Best}}\right).\mathrm{rnd}(1) + \left(S_{G,i} - P_i^G\right).\mathrm{rnd}(2) \quad (9)$$

where rnd(1) and rnd(2) represent random numbers that vary between 0 and 1, which is primarily used for attaining randomization during the process of searching.

On the other hand, each individual pertains to the historical optimal position of the population and prevents interaction between individual solutions present in the population. For ensuring maximized interaction between individuals, population knowledge-based learning model is included into the primitive cuckoo search algorithm as specified in Eq. (10)

$$S_{G+1.j} = S_{G,j} + \beta_0 \times \frac{\delta \times \mu}{|v|^{\frac{1}{\beta}}}\left(S_{G,i} - S_{G,\mathrm{Best}}\right).\mathrm{rnd}(1) + \left(S_{G,i} - P^*\right).\mathrm{rnd}(2) \quad (10)$$

where P^* is the determined new position of the individual solution derived based on the population knowledge. This included population knowledge-based learning model is mainly used in the primary phases of the cuckoo algorithm for the purpose of

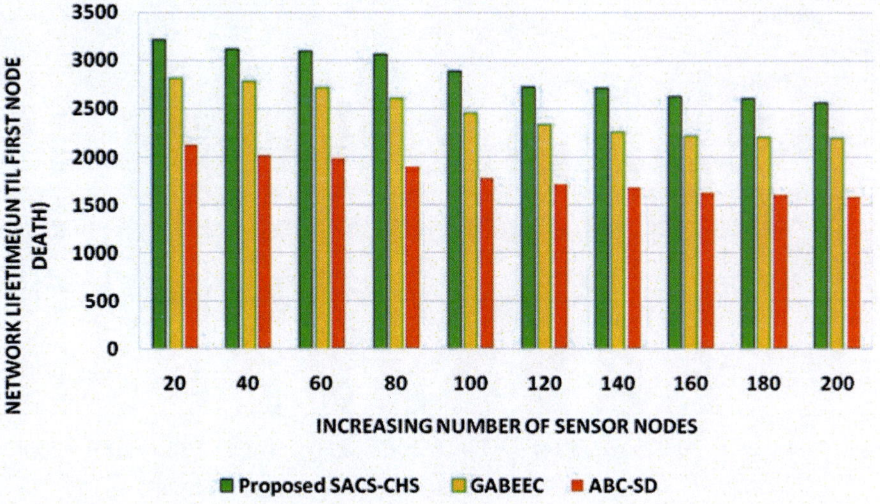

Fig. 1 Proposed SACS-CHS-network lifetime (until first node death)

improving the algorithmic diversity and expanding the scope of search to the maximized level. It included population knowledge-based learning model for utilizing individual knowledge information and population knowledge information for potentially escaping from the minimal local point of optimality. Hence, the inclusion of individual knowledge learning model and population knowledge model aided in better balance between the rate of exploitation and exploration. This capability of self-adaptive knowledge learning improved cuckoo searching aided in better cluster head selection process that prevented worst solution from being selected as cluster head and prevented frequent selection of cluster heads in the network.

4 Results and Discussion

The simulation experiments of the proposed SACS-CHS and the benchmarked GABEEC and ABC-SD approaches are conducted using MATLAB R2016a. The number of sensor nodes considered for implementation is 200. The network area considered for implementation is 1500 × 15,000 square meters with the base station location at the position of (250, 50) in the network [16]. The energy model considered for implementation is the Heinzelmann energy model.

Initially, network lifetime with respect to first, half, and last sensor nodes' death achieved by the proposed SACS-CHS and the benchmarked GABEEC and ABC-SD schemes are explored based on scalable rounds of implementation. Figures 1 and 2 present the network lifetime evaluated corresponding to first and half sensor nodes death. The first senor node death in the network during the implementation of SACS-CHS is visualized to be outsmarting compared to GABEEC and ABC-SD

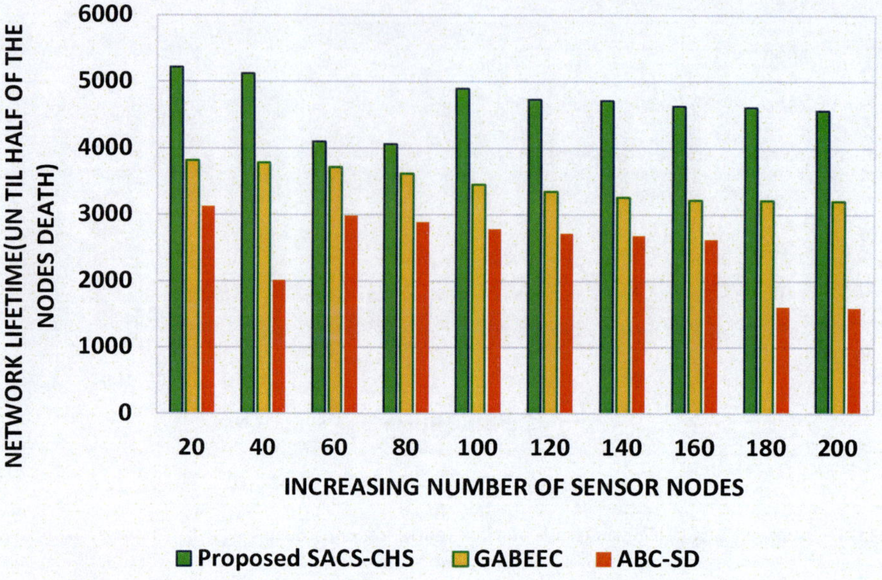

Fig. 2 Proposed SACS-CHS-network lifetime (until half node death)

schemes, since the inclusion of new searching strategy prevented the unnecessary depletion of node energy. Further, the lifetime until the death of half count of sensor nodes guaranteed by the proposed SACS-CHS is also confirmed to be potent, since it adopted mutation and crossover operation that always attained high energy and reliable sensor nodes as cluster heads. From Fig. 3, it is transparent that the proposed SACS-CHS is reliable enough in sustaining the network lifetime evaluated with respect to last sensor node death. With respect to first sensor node death, the network lifetime is improved by 7.29% and 8.52%, excellent to the compared GABEEC and ABC-SD approaches. With respect to half count of sensor node death, the proposed SACS-CHS is confirmed to get improved by 7.45 and 8.84%, better than the compared GABEEC and ABC-SD schemes. With respect to last sensor node death, the network lifetime attained by the proposed SACS-CHS is enhanced by 6.98 and 7.79%, better than the compared GABEEC and ABC-SD schemes.

Further, Figs. 4, 5, and 6 demonstrate the predominance of SADSS-IABCA-CHS over the compared approaches investigated using network lifetime, throughput, and energy consumptions. This comparative analysis identified that the network lifetime is considerably increased by the proposed SACS-CHS scheme at an average rate of 8.74 and 9.43%, better than the baseline proposed GABEEC and ABC-SD schemes used for investigation. The throughput guaranteed by the proposed SACS-CH Sunder scalable increase in sensor node density is enhanced in an average by 8.32 and 9.64%, better than the compared GABEEC and ABC-SD schemes used for investigation. The energy consumptions incurred by the proposed SACS-CHS scheme under a different number of sensor nodes was determined to be improved at an average by

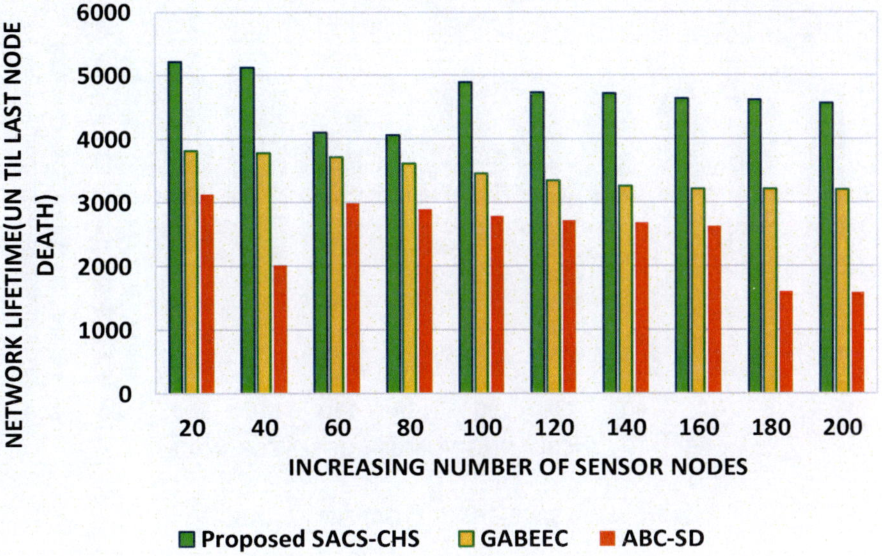

Fig. 3 Proposed SACS-CHS-network lifetime (until last node death)

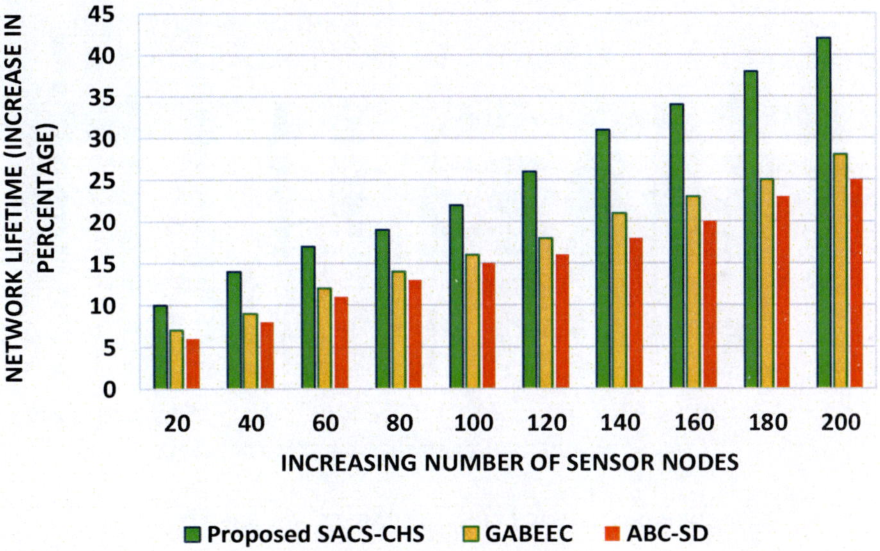

Fig. 4 Proposed SACS-CHS improvement in network lifetime

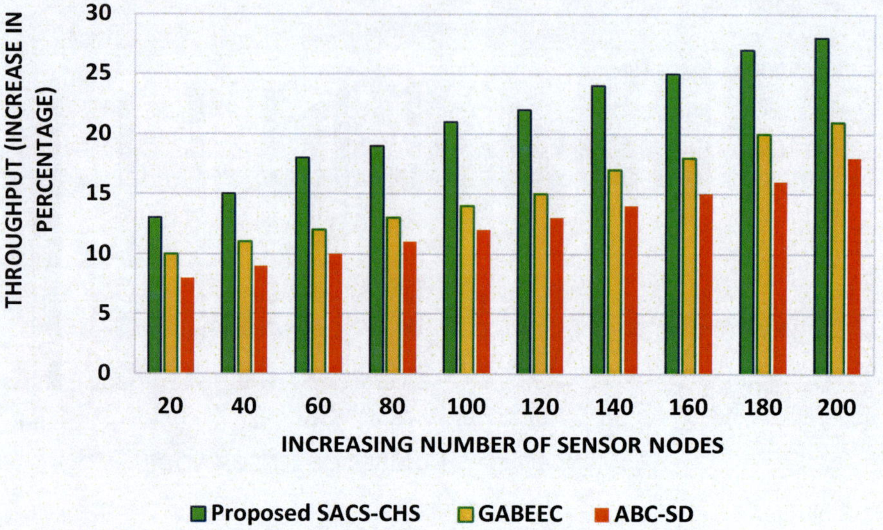

Fig. 5 Proposed SACS-CHS improvement in throughput

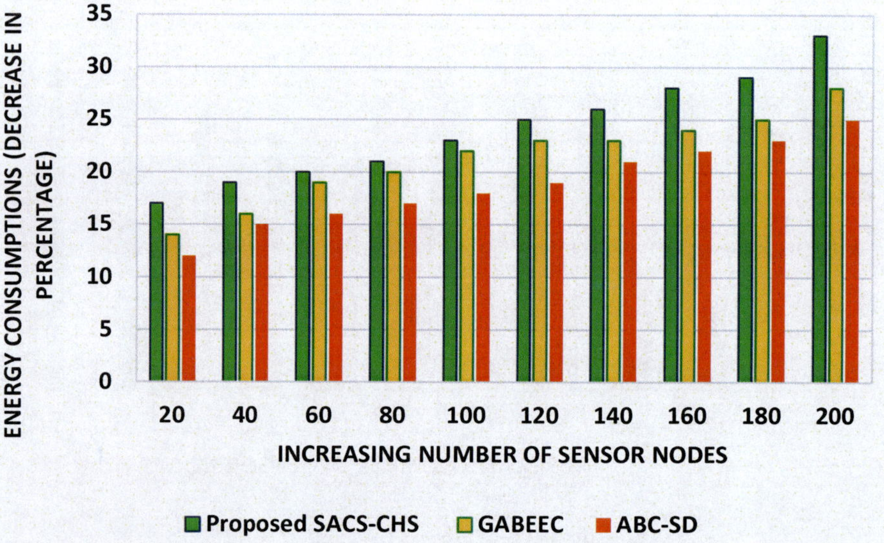

Fig. 6 Proposed SACS-CHS minimization in energy consumptions

7.56 and 8.54%, better than the compared GABEEC and ABC-SD approaches used for investigation.

In addition, Fig. 7 glorifies the performance of the proposed SACS-CHS using mean throughput under varying densities of sensor nodes. The mean throughput under

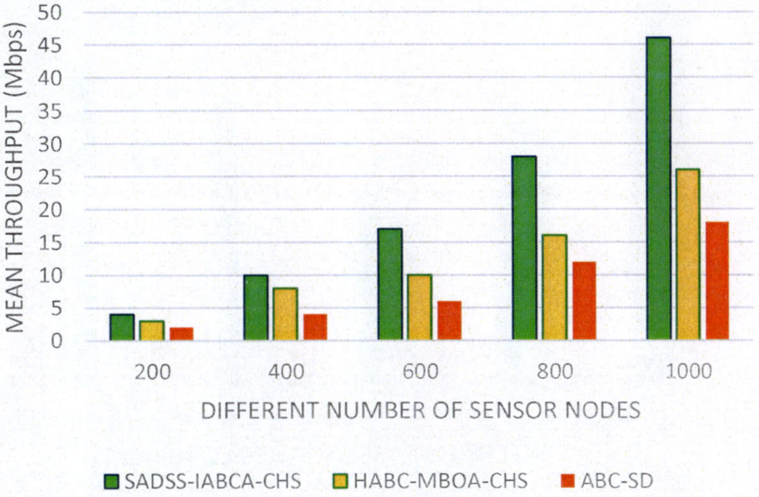

Fig. 7 Proposed SACS-CHS mean throughput with scalable increase in sensor nodes

a different number of sensor nodes facilitated by the proposed SACS-CHS scheme was determined to be significant on par with GABEEC and ABC-SD schemes, independent to the systematic increase in sensor nodes as it explores all possible criteria for selecting CH. The mean throughput enabled by the proposed SACS-CHS was determined to be improved at an average by 5.98% and 6.84%, better to the compared GABEEC and ABC-SD frameworks used for investigation.

Figure 8 demonstrates the comparative analysis of SADSS-IABCA, GABEEC, and ABC-SD schemes evaluated based on mean residual. The mean residual energy under a different number of sensor nodes in the network facilitated the proposed SACS-CHS scheme was determined to be predominant over the proposed GABEEC and ABC-SD schemes, independent to the systematic increase in the number of sensor nodes in the network. This improvement in performance attained by the proposed SACS-CHS is mainly due to benefits of individual and population knowledge used in quantifying the influence of each criteria over the selection of appropriate cluster head selection strategy. The mean residual energy ensured by the proposed SACS-CHS scheme is determined to be improved at an average by 6.24 and 7.18%, better than the comparable GABEEC and ABC-SD schemes used for investigation.

5 Conclusions

In this paper, SACS-CHS scheme was proposed for achieving better energy stability that results in maximized network lifetime of sensor nodes in the network. It achieved superior balance between exploitation and exploration by preventing premature convergence problem. It adopted significant self-adaptive knowledge model

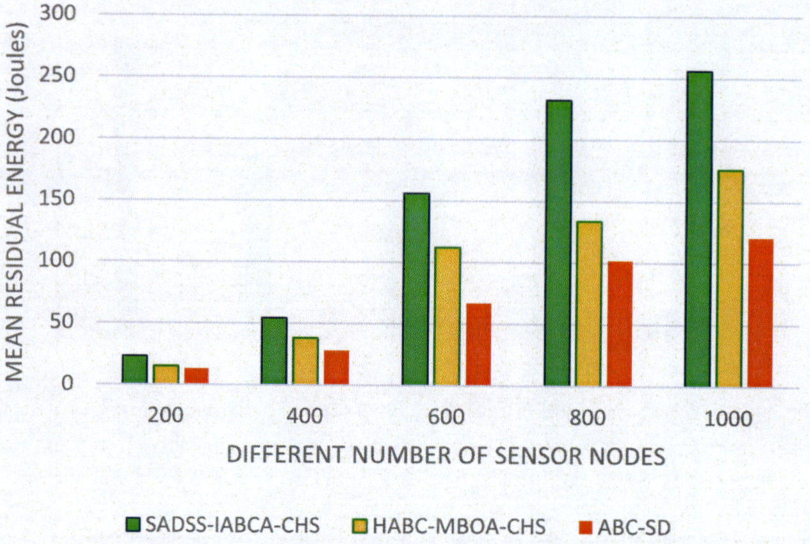

Fig. 8 Proposed SACS-CHS mean residual energy with scalable increase in sensor nodes

for extracting experience through individual history and population knowledge. It also derived the advantages of threshold statistics learning approach and exploited individual and population learning process for preventing frequent selection of cluster heads. The simulation results of the proposed SACS-CHS scheme confirmed improved throughput of 8.32 and 9.64%, better than the compared GABEEC and ABC-SD schemes used for investigation. The results proved that the energy consumptions incurred by the proposed SACS-CHS scheme under a different number of sensor nodes was determined to be improved at an average by 7.56% and 8.54% better than the benchmarked approaches. Moreover, the mean residual energy ensured by the proposed SACS-CHS scheme is determined to be improved at an average by 6.24 and 7.18%, better than the comparable GABEEC and ABC-SD schemes used for investigation.

References

1. Sarkar A, Murugan TS (2021) Analysis on dual algorithms for optimal cluster head selection in wireless sensor network. Evol Intell 2(1)
2. Janakiraman S (2018) A hybrid ant colony and artificial bee colony optimization algorithm-based cluster head selection for IoT. Proc Comput Sci 143(2):360–366
3. Senthil Murugan T, Sarkar A (2018) Optimal cluster head selection by hybridisation of firefly and grey wolf optimisation. Int J Wireless Mob Comput 14(3):296
4. Janakiraman S, Priya M, Devi S, Sandhya G, Nivedhitha G, Padmavathi S (2018) A markov process-based opportunistic trust factor estimation mechanism for efficient cluster head selection and extending the lifetime of wireless sensor networks. EAI Endorsed Transactions on

Energy Web 2(1):168093
5. Pour SE, Javidan R (2021) A new energy aware cluster head selection for LEACH in wireless sensor networks. IET Wireless Sens Syst 11(1):45–53
6. Qiang Y, Pei B, Wei W, Li Y (2015) An efficient cluster head selection approach for collaborative data processing in wireless sensor networks. Int J Distrib Sens Netw 11(6):794518
7. Subramanian P, Sahayaraj JM, Senthilkumar S, Alex DS (2020) A hybrid grey wolf and crow search optimization algorithm-based optimal cluster head selection scheme for wireless sensor networks. Wireless Pers Commun 113(2):905–925
8. Janakiraman S, MDP (2020) An energy-proficient clustering-inspired routing protocol using improved bkd-tree for enhanced node stability and network lifetime in wireless sensor networks. Int J Commun Syst 2(1):e4575
9. Kim J-Y, Sharma T, Kumar B, Tomar GS, Berry K, Lee W-H (2014) Intercluster ant colony optimization algorithm for wireless sensor network in dense environment. Int J Distrib Sens Netw
10. Rakhee, Srinivas M (2016) Cluster based energy efficient routing protocol using ANT colony optimization and breadth first search. Proc Comput Sci 89(2):124–133
11. Karimi M, Naji HR, Golestani S (2012) Optimizing cluster-head selection in wireless sensor networks using genetic algorithm and harmony search algorithm. In: 20th Iranian conference on electrical engineering (ICEE2012) 1(2):34–42
12. Pal V, Yogita SG, Yadav R (2015) Cluster head selection optimization based on genetic algorithm to prolong lifetime of wireless sensor networks. Procedia Comput Sci 57(3):1417–1423
13. Khushboo K, Daniel AK (2015) Section based hybrid routing protocol for WSN using artificial bee colony. In: 2015 International conference on advances in computer engineering and applications, vol 3, no 1, pp 23–3
14. Yang Y, Fu G (2015) Clustering routing algorithm in wireless sensor networks based on artificial bee colony and assistant cluster heads. MATEC Web of Conf 22:01021
15. Li J, Li Y, Tian S, Xia J (2019) An improved cuckoo search algorithm with self-adaptive knowledge learning. Neural Comput Appl 32(16):11967–11997
16. Vasundra S, Venkatesh D (2018) Performance evaluation of routing protocols for voice and video traffics AJCST- Asian J Comput Sci Technol ISSN: 2249-0701, **7**(3)

State-Wise Analysis and Prediction of Covid-19 in India

Kritika Malhotra and Amit Prakash Singh

Abstract Machine learning (ML) is an application of artificial intelligence (AI) and through ML, the system gets the capability of learning by refining from experience. Whereas deep learning, machine learning's subsection is based on artificial neural network. Its customs multiple layers to excerpt higher level features from raw input. In the paper machine learning algorithm—polynomial regression and deep learning model—convolutional neural network is used for state-wise analysis and prediction of Covid-19 in India. Covid-19 started from Wuhan from December 2019. From that day till now, coronavirus has increased exponentially and has affected large number of human populations throughout the world. By March 2019, it was avowed as pandemic by World Health Organization (WHO). Clinical Doctors and Scholars have been working 24 h, day and night since then to test the coronavirus infected patients and to develop vaccine to cure the disease. This paper attempts to build up a model that can help in prediction of total active cases in the states of India so that doctors can take some preventive measures to save more human lives and also so that vaccine could be provided as early as possible to the state that is most infected and will be most infected in future.

Keywords Machine learning · Deep learning · Polynomial regression · Convolutional neural network · Covid-19 · Coronavirus

1 Introduction

Coronavirus or Covid-19 that created pandemic in year 2020 started to spread from Wuhan (China) in year December 2019. According to World Health Organization (WHO), coronavirus is a class of virus ranging from common cold to severe acute respiratory syndrome (SARS). Coronavirus causes diseases in mammals and birds. It can transmit from animals to humans leading to respiratory tract infections in humans.

K. Malhotra (✉) · A. P. Singh
University School of Information, Communication and Technology, Guru Gobind Singh Indraprastha University, Delhi, India

R. P. Mahapatra et al. (eds.), *Proceedings of International Conference on Recent Trends in Computing*, Lecture Notes in Networks and Systems 341,
https://doi.org/10.1007/978-981-16-7118-0_53

Coronavirus was declared as global pandemic by WHO on March 11, 2020. It quickly got spread to more than 200 countries or territories or areas infecting more than nine million people globally. In India, the disease started spreading from Kerala from a student who returned from Wuhan tested positive on January 30, 2020. At present (September 1, 2020), there are total 3.69 million cases with 65,288 deaths in India with these numbers increasing exponentially daily.

Till now, there are no vaccines or antiviral drugs present to treat this disease, but various countries are doing research to get the vaccine, and in fact, some countries are under trial phase of vaccine, so sooner or later there will be a vaccine available to treat the disease.

In the paper, machine learning model—polynomial regression and deep learning model—convolutional neural network are used and compared for state-wise analysis of Covid-19 in India. The number of active cases in the next five days for the top five states having the highest number of active cases are being predicted so that preventive measures for that state can be decided accordingly to save more number of humans from getting infected and also so that vaccine if developed can be first provided to that states that will have higher spread rate of the virus in future to cure people and save human lives.

The paper is divided into following sections as follows—Sect. 2 includes the literature review that has been done; Sect. 3 explains about regression and polynomial regression; Sect. 4 explains the neural networks and convolutional neural networks. Section 5 discusses about the methodology used, Sect. 6 includes the experimental results obtained. Finally, the paper ends with Sect. 7 explaining the conclusion that is obtained, and then, last section contains the references.

2 Literature Review

Investigation on Covid-19 analysis and prediction have been done extensively in the past few months. Various related works that have been done in this field in past few months are explained in this section.

Ghosh et al. in [3] have developed three models to predict the Covid-19 infected people in various states of India in next 30 days. In the paper, they have also discussed the importance of state-wise analysis. The three models that have been developed by them are—exponential model, logistic model, and susceptible infectious susceptible (SIS) model. Various plots have been used to talk about the predicted number of cases in next 30 days that resulted from the three models.

L. Jia [4] et al. adopted three models—logistic model, Bertalanffy model, and Gompertz model to predict and analyze the coronavirus disease. The epidemic trends of Covid were first fitted and analyzed in order to prove the validity of the existing mathematical models. The results were then used to fit and analyze the situation of Covid-19. The prediction results of three different mathematical models are different for different parameters and in different regions. Regression coefficient was used to evaluate the fitting ability of the three models.

In study [11], Bandyopadhyay determines the feasibility of the usage of machine learning method in evaluation of prediction results. Prediction of confirmed cases, negative cases, released, and deceased cases of Covid-19 coronavirus was done. For this deep neural network—recurrent neural network method is used. Three models long short-term memory (LSTM), gated recurrent unit (GRU), and combined model of LSTM + GRU were proposed. The dataset was fed to all the three models, and based on accuracy and root mean square error, best model out of the three models is chosen. Experimental results showed that the combined approach LSTM-GRU-RNN provides quite better result over LSTM-RNN, GRU-RNN in terms of accuracy, RMSE metrics.

In [13], Huang et al. used the concept of deep learning to forecast the confirmed number of Covid-19 cases in upcoming days in China based on previous five days data. They have applied convolutional neural network (CNN), multilayer perceptron (MLP), long short-term memory (LSTM) neural network, and gate recurrent unit (GRU) on small dataset. Root mean squared error (RMSE), mean absolute error MAE, and various plots have been used to compare the performance of all the four models.

3 Regression and Polynomial Linear Regression

Regression is a statistical technique used for predicting a continuous value [1]. In regression, there are two types of variables used—one dependent variable (Y) and one or more independent variables (X) [2]. The dependent variable is the variable that we predict. The regression model relates Y, dependent variable to a function of X, independent variable. In regression, the independent variable can be categorical or continuous, but dependent variable should be continuous only, it cannot be categorical. Regression can be used for prediction of both linear and non-linear dataset (the linearity of the dataset depends upon the relationship between the variables) [5]. But, in the paper, regression for non-linear dataset is used and that is polynomial linear regression.

Polynomial regression is an extension of multiple linear regression (predict the value of Y based on multiple value of X where Y and X have linear relationship) in which the relationship between the independent variable and dependent variable is expressed as nth degree polynomial [7]. It is used when there is non-linear relationship between independent and dependent variable [9]. Though polynomial regression is used for non-linear relationship [6], but still it is called as linear because the coefficients (b_0, b_1, b_2...... b_n) of the polynomial regression equation are linear as explained by the equation:

$$Y = b_0 + b_1X + b_2X^2 + b_3X^3 + \ldots\ldots.. + b_nX^n \quad (1)$$

In the equation, n is the degree of the polynomial. To implement polynomial regression two steps are followed. Firstly, the feature attribute is transformed into polynomial features. And then, simple linear regression is used to fit these features and target attribute to the model [8].

4 Neural Network and Convolutional Neural Network

Artificial neural network (ANN) (sometimes called as neural networks) belongs to one of the family of machine learning that is deep learning. ANNs are information providing models inspired by human brain. Like human brain consists of billions of neurons that communicate to each other using electrical and chemical signals that enable humans to see, feel, and make decisions, ANNs also work by mathematically mimicking the human brain and connecting multiple artificial neurons in a multilayered manner [12]. It is similar to brain in following two ways [14]: -

1. It acquires knowledge from its environment by learning procedure.
2. Synaptic weights are used to store the acquired knowledge.

4.1 How Neural Networks Learn?

Most neural network architectures have following layers:-

1. Input layer
2. Output layer
3. Hidden layer

Suppose some input values have been applied to the network through input layer, then the activation function is applied at the hidden layer to learn complex features of the data by transforming the data into a slightly more abstract and composite representation, and finally, an output is obtained at output layer as shown in Fig. 1.

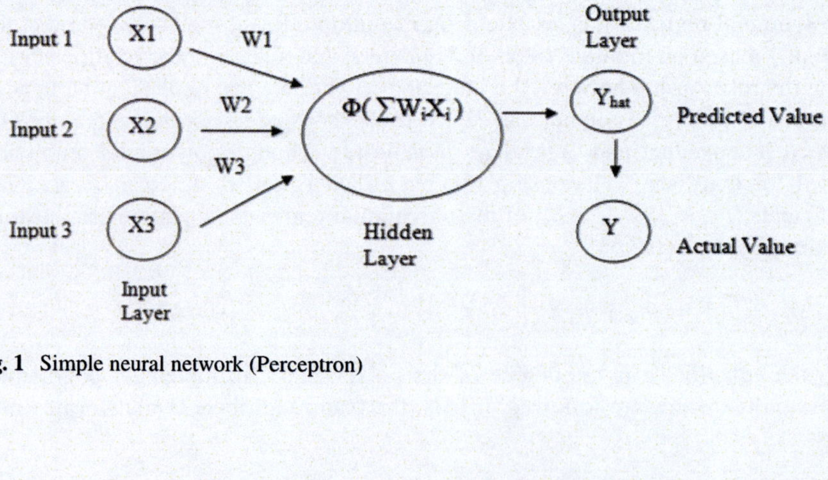

Fig. 1 Simple neural network (Perceptron)

In order to learn, predicted output values are compared with actual value. To calculate the difference between both the values, a cost function most commonly used is given as:

$$C = 1/2\left(Y_{\text{hat}}^2 - Y^2\right) \quad (2)$$

where
Y_{hat} = predicted values.
Y = actual values.

This cost function tells about the error in the predicted values and the actual values. The goal is to minimize the cost function because lower the cost function the closer the Y_{hat} is to Y. The cost function calculated is fed back to the network, it goes to the weights, and the weights get updated. This whole process followed during learning is called as back propagation [15].

4.2 *Convolutional Neural Network*

Convolutional neural network is a type of neural network that uses convolution mathematical operation to analyze the data given to it. It is a version of multilayer perceptron where each neuron in previous layer is connected to all the neurons in the next layer [16].

A convolutional neural network has an input layer, an output layer, and multiple hidden layers [17]. The hidden layers of the CNN consist of the convolutional layer, RELU layer, and pooling layer, and fully connected layers [18]. The following figure, Fig. 2 describes the architecture of convolutional neural Network.

1. Convolutional layer—The layer derived its name because of the convolutional operation of mathematics that is done on this layer. The main purpose of this

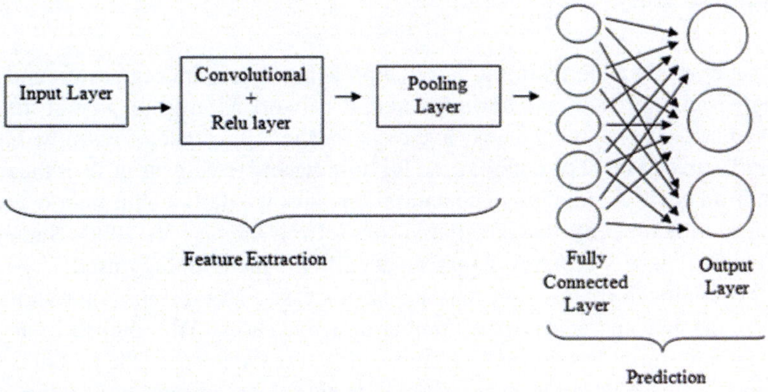

Fig. 2 Architecture of convolutional neural network

layer is to extract features from the input fed to it by performing a linear operation—convolution operation. Convolution operation is used for extraction of features by applying a small matrix of numbers called as kernel or filter [19]. This filter is slided over the input, and element-wise dot product of input and filter is computed thus forming an output called as feature map. The procedure is repeatedly done on the same input by applying multiple kernels having different values. This results in different feature maps for the same input. The value of these filters is learnt by CNN itself during the training process [19, 20].

2. ReLU—The output of the convolution operation—feature map is fed to ReLU. Rectified linear unit (ReLU) is a non-linear activation function that replaces all the negative values in the feature map by zero. It adds non-linearity in the network. There are other activation functions such as sigmoid and hyperbolic tangent that can be used in place of ReLU, but ReLU is most commonly used [19].
3. Pooling layer—This layer reduces the dimensionality of the feature map by retaining the important information pooling is done to:

- To reduce the number of computations and parameters of the network
- To make the network unchangeable to small transformations, translations, and distortions.
- The most commonly used pooling operations are max pooling and global average pooling [19].

4. Fully connected layers—The feature map obtained after applying pooling operation is converted into a one-dimensional (1-D) matrix and is made connected to one or more fully connected layers known as dense layers where each neuron in a layer is connected to every neuron in next layer by a learnable weight [19]. The purpose of the fully connected layers is to do the task of prediction or classification by using the feature map obtained.

5 Methodology Used

In this paper, a machine learning-based approach polynomial regression and deep learning-based approach convolutional neural network is used to predict the total number of active cases for the next five days for the top five states with the highest number of active cases. The dataset used for training and testing purpose of the model is taken from covid19india.org. The dataset contains the daily confirmed, deceased, and recovered number of cases for Indian states from January 30, 2020– September 6, 2020 [10]. Figure 3 explains diagrammatically the methodology used.

After obtaining the dataset, the data are preprocessed, and various graphs are made to analyze the data and analyze the number of active cases. After preprocessing and analyzing, the dataset is split into training and testing data. Then, the model is build and trained. Then, the model is applied on testing data, and predictions are made to get the total number of active cases in the state in the next five days. Then, root

Fig. 3 Workflow of methodology used

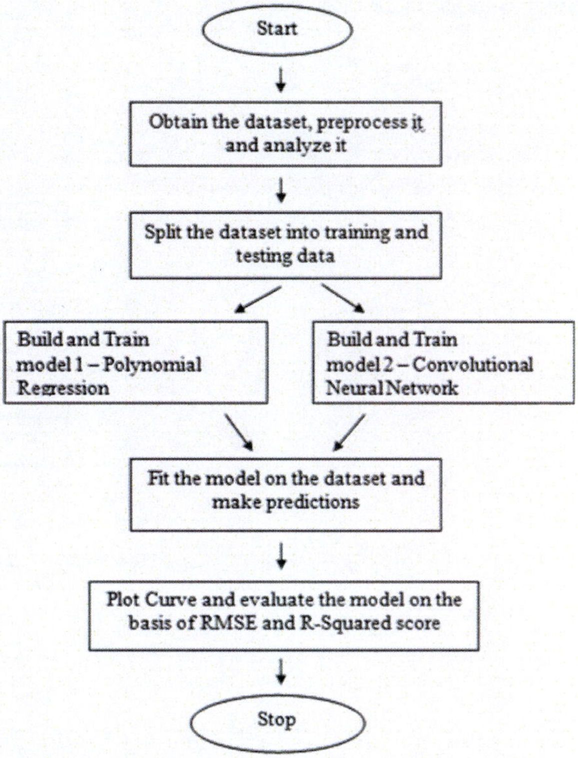

means squared error and *R*-squared score is used to evaluate the performance of the model, and curve is plotted to evaluate the results.

5.1 Structure of Convolutional Neural Network Used

2-Dimensional (2-D) convolutional model is build, trained, and used for prediction of total number of active cases in states of India in the next five days. The preprocessed, splitted data are fed to this model and performance is measured. Figure 4 describes the structure of the model used.

The input is given to the input layer of the model. The input is passed through a sequence of convolution layers and pooling layers that act as feature extractors. And finally through fully connected layer to interpret the features and output layer with a linear activation function for prediction.

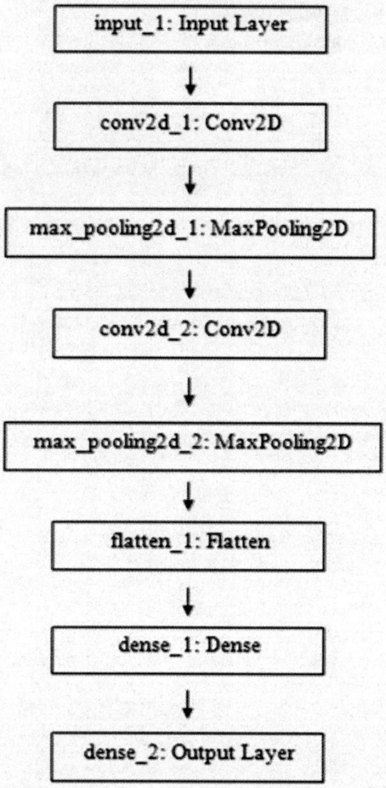

Fig. 4 Structure of the CNN model

5.2 Performance Metrics Used

To measure the performance of the model trained and used for prediction, root mean squared error (RMSE) and *R*-squared score are used.

- RMSE is a metric used to measure the difference between the predicted values by the model and observed values.

$$\text{RMSE} = \sqrt{\sum (Z_{\text{ia}} - Z_{\text{ip}})^2 / N} \quad (3)$$

  where:
  Z_{ip} = Predicted values.
  Z_{ia} = Actual values.
  N = Number of non-missing data points.
  $i = 1$ to N.

- *R*-squared score (R^2 score) is defined as percentage of the variance in dependent variable that is predictable from independent variable. R^2 score can be negative indicating poor model performance, but normally it lies between 0 and 1. The closer it is to 1, the better the model is. Adding more variable to model will never make R^2 score decrease, it will only either increase or remain same.

$$\mathrm{SS}_{\mathrm{res}} = \sum (Z_{\mathrm{ia}} - Z_{\mathrm{ip}})^2 \quad (4)$$

$$\mathrm{SS}_{\mathrm{total}} = \sum (Z_{\mathrm{ia}} - Z_{\mathrm{avg}})^2 \quad (5)$$

$$R^2\mathrm{score} = 1 - (\mathrm{SS}_{\mathrm{res}}/\mathrm{SS}_{\mathrm{total}}) \quad (6)$$

where:
Z_{ia} = Actual values.
Z_{ip} = Predicted values.
$i = 1$ to N.
N = Number of non-missing data points.
$\mathrm{SS}_{\mathrm{res}}$ = Sum of squares of residual.
$\mathrm{SS}_{\mathrm{total}}$ = Total sum of squares.

6 Experimental Results

This section is divided into three parts—First part has the plots that were obtained on analysis of the dataset. Second part has the plots showing the comparison of the actual and predicted values obtained for the top five states having the highest number of active cases for both the models. And the third part has a table comparing the RMSE and R-Squared values for both the models for the five states.

6.1 Plots Obtained on Analyzing the Dataset

The graph shown in Fig. 5 explains about the trend of total number of active cases starting from January 30, 2020–September 6, 2020. The *y*-axis represents the total number of active cases with the scale taken as 0–800 K with an interval of 200 K, while the *x*-axis represents the dates. The plot shows that the number of active cases almost keeps on increasing throughout the months. By the last date that is September 6 the total number of active cases crossed 800 K and is about 900 K.

From the plot obtained after analysis as shown in Fig. 6, it can be observed that the top five states with the highest number of total active cases is Kerala, Maharashtra, Andhra Pradesh, Karnataka, and Uttar Pradesh. Among all, Kerala is having the highest number of active cases with more than 800,000 (800 K).

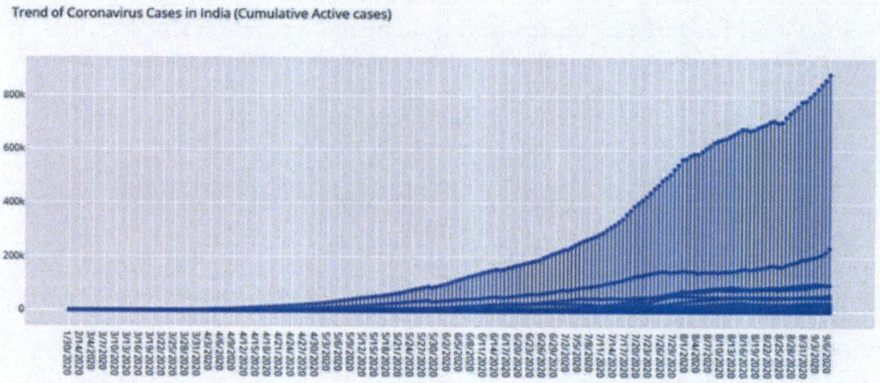

Fig. 5 Trend of active cases of Covid-19 in India

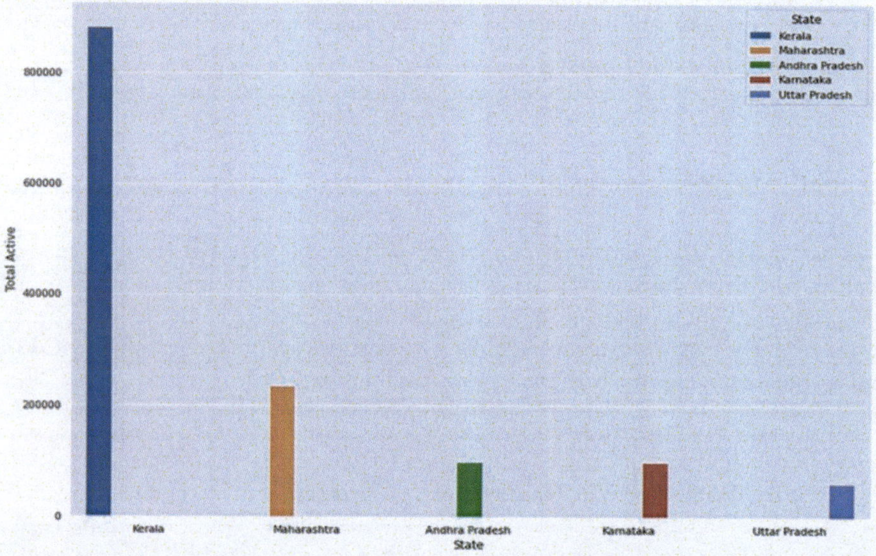

Fig. 6 Top five states with the highest number of active cases of Covid-19 in India

6.2 Plots Obtained for Actual versus Predicted Values

All the plots below are for the actual versus predicted values of the total active cases obtained on applying polynomial regression and convolutional neural network. The red line represents the predicted values, while the blue line represents the actual values, and the Y-axis represents the number of active cases whose values have been scaled down so that they can be handled easily.

1 For Maharashtra

(See Fig. 7)

2 For Andhra Pradesh
(See Fig. 8)
3 For Karnataka
(See Fig. 9)
4 For Uttar Pradesh
(See Fig. 10)
5 For Kerala
(See Fig. 11)

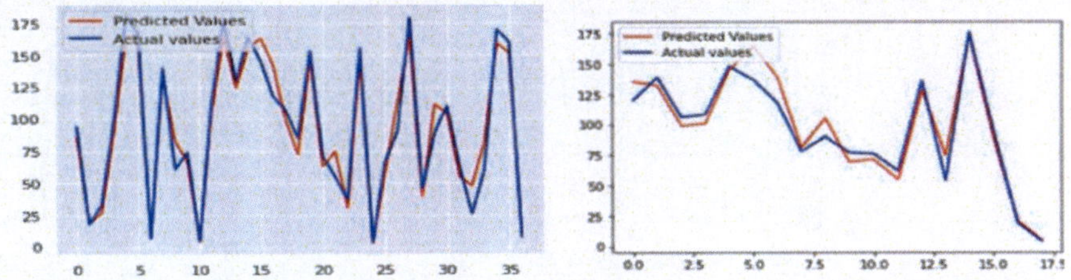

Fig. 7 Actual versus predicted values on applying polynomial regression (left) and CNN (right)

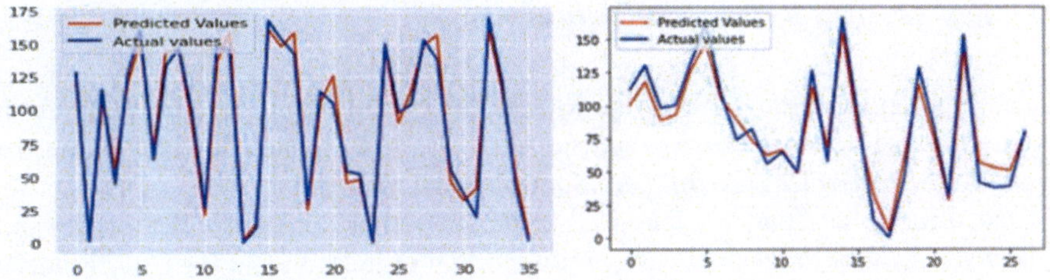

Fig. 8 Actual versus predicted values on applying polynomial regression (left) and CNN (right)

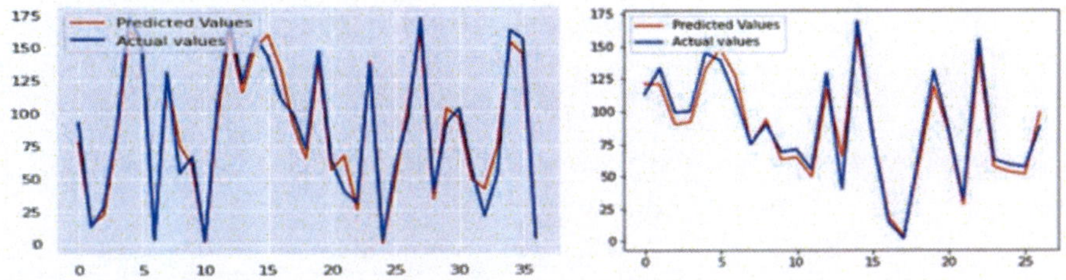

Fig. 9 Actual versus predicted values on applying polynomial regression (left) and CNN (right)

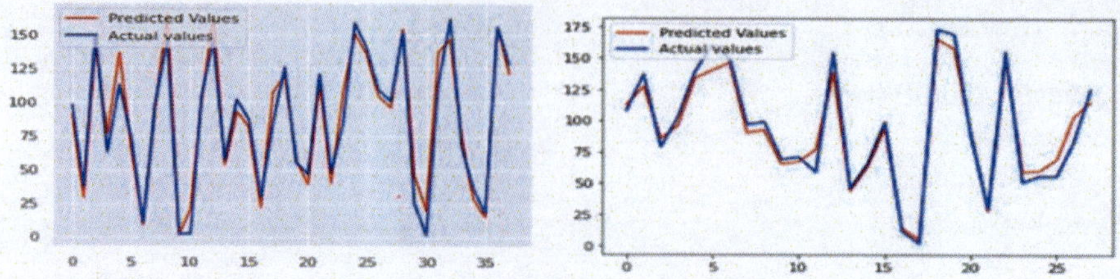

Fig. 10 Actual versus predicted values on applying polynomial regression (left) and CNN (right)

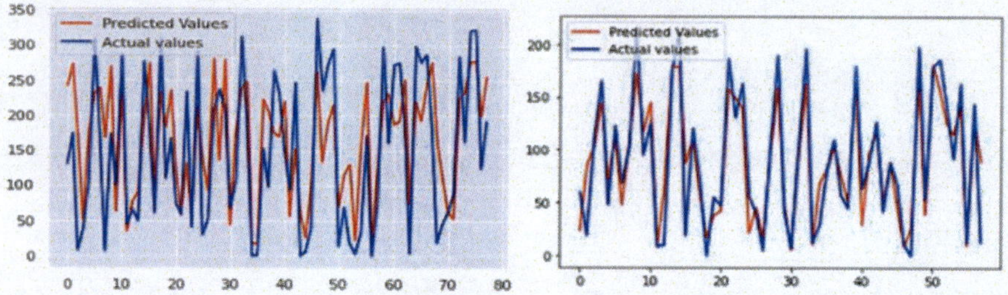

Fig. 11 Actual versus predicted values on applying polynomial regression (left) and CNN (right)

6.3 *RMSE and R-squared Values*

As can be seen from the Table 1, the RMSE values are lowest for Andhra Pradesh in case of polynomial regression and for Uttar Pradesh in case of CNN, while the highest values of RMSE are for Kerala for both the models.

Also, it can be seen that R-squared values are lowest for Kerala in both the cases and highest for Andhra Pradesh in case of polynomial regression and for Uttar Pradesh in case of CNN.

So, on an average, the RMSE value and R-squared values for polynomial regression model are 23.81 and 0.862, respectively. And on an average, the RMSE value and R-squared values for convolutional neural network model is 13.26 and 0.918, respectively.

Table 1 Name of the table that justify the values

States	Polynomial Regression		CNN	
	RMSE	*R*-Squared	RMSE	*R*-Squared
Maharashtra	12.31	0.95	11.55	0.92
Andhra Pradesh	9.36	0.97	9.82	0.94
Karnataka	11.03	0.95	9.48	0.94
Uttar Pradesh	11.73	0.94	9.22	0.96
Kerala	74.62	0.50	26.23	0.83

7 Conclusions

Experimental results show that among the top five states, Kerala will be having the highest number of cases in the next upcoming five days. And thus it can be concluded that Kerala requires more attention on preventive measures, so that the increasing number of cases could be controlled. The state will be requiring strict preventive measures to control the spread of Covid-19 in the state. Also, it can be concluded that if vaccine will be invented then Kerala will be requiring it first.

Further, it can be seen from the obtained RMSE values and R-squared values that CNN performed better in the analysis and prediction than the polynomial regression. And thus CNN proved out to be better model for prediction than polynomial regression.

References

1. Nadeski M (2019) Bringing machine learning to embedded systems. Texas Instruments
2. Dutt S, Chandramouli S, Das AK (2019) Machine learning. Pearson India
3. Ghosh P, Ghosh R, Chakraborty B (2020) COVID-19 in India: state-wise analysis and prediction. https://doi.org/10.1101/2020.04.24.20077792
4. Jia L, Li K, Jiang Y, Guo X et al. (2020) Prediction and analysis of coronavirus disease 2019. arXiv preprint arXiv:2003.05447
5. Wynants L, Van Calster B, Bonten MM, Collins GS, Debray TP, De Vos M et al. (2020) Systematic review and critical appraisal of prediction models for diagnosis and prognosis of COVID-19 infection. medRxiv
6. Chapter12-Regression-Polynomial Regression. Accessed on 29 Oct 2020. http://home.iitk.ac.in/~shalab/regression/Chapter12-Regression-PolynomialRegression.pdf
7. Kumari P, Kumar A A Study of polynomial regression towards machine learning. J Adv Sch Res Allied Educ 14(1)
8. Coronavirus outbreak in India. Accessed on 7 Sept 2020. https://www.covid19india.org/
9. Amar LA, Taha AA, Mohamed MY (2020) Prediction of the final size for COVID-19 epidemic using machine learning: a case study of Egypt. Infect Dis Model. https://doi.org/10.1016/j.idm.2020.08.008
10. Yan L, Zhang H, Goncalves J, Xiao Y, Wang M, Guo Y, Sun C, Tang X, Jing L, Zhang M, Huang X, Xiao Y, Cao H, Chen Y, Ren T, Wang F, Xiao Y, Huang S, Tan X, Huang N, Jiao B, Cheng C, Zhang Y, Luo A, Mombaerts L, Jin J, Cao Z, Li S, Xu H, Yuan Y (2020) An interpretable mortality prediction model for covid-19 patients. Nat Mach Intell 2:283–288
11. Bandyopadhyay SK, Dutta S (2020) Machine learning approach for confirmation of covid-19 cases: Positive, negative, death and release. medRxiv
12. Alimadadi A, Aryal S, Manandhar I, Munroe PB, Joe B, Cheng X (2020) Artificial intelligence and machine learning to fight COVID-19. Physiol Genomics
13. Huang C-J, Chen Y-H, Ma Y, Kuo P-H (2020) Multiple-input deep convolutional neural network model for COVID-19 forecasting in China. medRxiv
14. Haykin S Neural networks and learning machines, 3rd edn. Pearson Education Inc.
15. Convolutional Neural Network for Multi-Step time series forecasting. Accessed on 5 Sept 2020. https://machinelearningmastery.com/how-to-develop-convolutional-neural-networks-for-multi-step-time-series-forecasting/
16. Mehtab S, Sen J (2020).Stock price prediction using convolutional neural networks on a multivariate time series. In: Proceedings of the 3rd national conference on machine learning and artificial intelligence (NCMLAI2020)

17. Main Types of Neural Networks and its Applications—Tutorial, Towards AI. Accessed on 25 Oct 2020. https://medium.com/towards-artificial-intelligence/main-types-of-neural-networks-and-its-applications-tutorial-734480d7ec8e
18. Basic regression: Predict fuel efficiency | TensorFlow Core. Accessed on 5 Sept 2020. https://www.tensorflow.org/tutorials/keras/regression.
19. Yamashita R, Nishio M, Do RKG et al (2018) Convolutional neural networks: an overview and application in radiology. Insights Imaging 9:611–629. https://doi.org/10.1007/s13244-018-0639-9
20. An Intuitive Explanation of Convolutional Neural Networks—the data science blog. Accessed on 25 Oct 2020. https://ujjwalkarn.me/2016/08/11/intuitive-explanation-convnets/

Generation of One-Time Password for the Authentication of Software Requirements Using Secure Hash Algorithms

Javed Ahmad, Chaudhary Wali Mohammad, and Mohd. Sadiq

Abstract Security requirement is a non-functional requirement whose objective is to safeguard the data from misusers. Different algorithms have been developed to provide security in software by using various hash algorithms like message digest, secure hash algorithm, and race integrity primitives' evaluation message digest. Based on our literature review, we found that during the generation of one-time password (OTP) numeric values are mostly used for the authentication purpose. One of the limitations of such types of passwords is that they can be easily remembered or guessed by the misusers. Therefore, to address this, in this paper, we present a method for the generation of the OTPs in which alphanumeric values will be generated for authentication purpose. Finally, an example is given to explain the steps of the proposed method.

Keywords Security · Security requirements · Network security · Hash algorithms · Dynamic passwords and · One-time passwords

1 Introduction

Requirements engineering is a sub-research area of software engineering which is used to develop the secure system according to the need of the stakeholders having different views about the requirements. Stakeholders are the main source of the software requirements who participates in the requirements elicitation process. The outcome of the requirements elicitation process is the set of the requirements

J. Ahmad (✉) · C. W. Mohammad
Department of Applied Sciences and Humanities, Faculty of Engineering and Technology, Jamia Millia Islamia (A Central University), New Delhi 110025, India

C. W. Mohammad
e-mail: cmohammad@jmi.ac.in

Mohd. Sadiq
Department of Computer Science and Automation, Indian Institute of Science Bangalore, Bangalore, Karnataka 560012, India
e-mail: msadiq@jmi.ac.in

R. P. Mahapatra et al. (eds.), *Proceedings of International Conference on Recent Trends in Computing*, Lecture Notes in Networks and Systems 341,
https://doi.org/10.1007/978-981-16-7118-0_54

which are broadly divided into "functional requirements (FRs)" and "non-functional requirements (NFRs)" [1, 2]. It has been observed that during the requirements elicitation process, stakeholders are more concerned toward the elicitation of the FRs, whereas NFRs have received less attention during the elicitation process. Among various NFRs, a security requirement is considered as key NFR because it is used to safeguard the useful information and data from the misusers or unauthorized users [3]. Based on our review, we found that during the requirements elicitation process most of the focus is on the identification, authentication, and authorization of the software requirements. During the requirements, elicitation process less attention is given on how the identity of the end users will be authenticated during the transmission of the data using Internet [3, 4].

Security plays an important role in different areas like banking, e-commerce, social media, and institutions. because lots of data are transferred on the Internet. Security deals with the protection of information from unauthorized access or users. In [5], the authors have discussed different types of the security issues like information security, and network security. Among these security areas, network security is meant to provide security for information flowing on the network. Exchange of information between sender and receiver on open network such as Internet may pose some security attacks like masquerading and replay attacks which are threat to integrity. Network security deals with the protection of networks and their services from unauthorized access, alteration, disclosures, and destructive effect for users [6].

When messages are transmitted over network, there are chances of attacks by misusers to change the meaning of the message with the intention of providing the false information to the receiver. To save the information and data from the unauthorized users, OTPs are used in which numeric values are used to verify the authenticity of an end user. Based on our review [7–9], we found that during the generation of the OTPs, numeric values are used for the authentication purpose. One of the limitations of the OTPs based on numeric values is that it has limited combinations of the digits due to which it is easy to memorize, and as a result, it may be misused by unauthorized users [10]. Therefore, the objective of this paper is to develop an algorithm for the generation of OTPs based on alphanumeric values so that strong OTPs can be generated. The contributions of this paper are as follows:

1. An algorithm has been developed to generate the OTP from "*hash-based message authentication code*" (HMAC) based on alphanumeric values
2. The *N*-gram test has been used to show the randomness in the generated passwords

The remaining part of this paper is organized as follows: Related work is discussed in Sect. 2. An insight into selected hash algorithms is given in Sect. 3. The proposed method is discussed in Sect. 4. The explanation of the proposed method is discussed in Sect. 5. Finally, Sect. 6 concludes the paper and suggestion for future work in the area of OTP.

2 Related Work

For information to be secure, it must assure the following three goals of information security like “confidentiality” (C), “integrity” (I), and “availability” (A), which is also known as CIA triad. In CIA triad, the objective of the *confidentiality* is to hide the information from unauthorized users or access, whereas the term, *integrity* refers to protection of information from unauthorized change or alteration during transmission, and *availability* deals with providing the information at all times whenever needed by the authorized users. CIA is one of the key models to design and guide the policies for information security within an organization.

Based on data type, OTPs have been divided into two parts, i.e., (a) numeric OTP and (b) alphanumeric OTP. In *numeric OTPs*, numeric values like $0, 1, \ldots, 9$ are randomly selected from HMAC to generate the OTP. This type of OTPs is popularly used in various fields especially in the banking transactions, and it has also been observed that the length of the numeric OTP is of 4–6 digits based on the security requirements (SecR) of the system. On the other hand, *alphanumeric OTPs* include the combination of both numeral values and alphabets which is randomly extracted from HMAC. The uniqueness and time-based single use property of OTP provides an extra factor in authentication purposes. [11, 12]. In today’s scenario, there are many authentication-based protocols for OTP. In 1981, Leslie Lamport introduced the concept of OTP based on hash chains [13]. The OTP generation and verification algorithms have been divided into two types, i.e., statically generated OTPs and dynamically generated OTPs. In statically generated method, OTPs are stored in the form of string in the memory before the interaction between the two parties, i.e., client and server. This type of OTPs are not completely secure as they are stored which may be stolen by the attackers. Whereas, dynamically generated OTPs are more secure as the OTPs are generated during the authentication process, and it is valid only for one session. In real life applications, dynamically generated OTPs are widely used because a secret key is used during the communication.

Various methods have been developed for the generation of OTPs. For example, Raihi et al. [14, 15] focused on the generation of OTPs by means of timestamp and counter parameters. Bicakci and Baykal [16] proposed a *Bicakci protocol* for OTP generation and verification using asymmetric cryptography. A review on various authentication protocols are discussed in [17, 18]. In [19], the authors have generated alphanumeric OTPs with linear function that increases the degree of protection using automation theory. Saxena [20] has identified three important methods of OTPs which are used in enterprises like time synchronous technique, asynchronous challenge–response technique, and event synchronous technique. The clock time of client and server is synchronized, and an algorithm is used to generate OTP from synchronous times in time synchronous technique. In asynchronous challenge–response technique, the client application uses crypto-primitive for the generation of unique password from the challenge. The event synchronous technique uses the counter values for the authentication of the client by server. Client and server both generate an OTP from counter value and other inputs; if matching is found, then user is authenticated by the server. In 2020, Shukla et al. [21] proposed a method for OTP generation for the client and server applications. The proposed method was analyzed on the basis of the numeric OTP. Therefore, in this paper, an attempt has been made for the generation of the alphanumeric OTPs from the HMAC.

3 An Insight into Different Types of the Hash Algorithms

Hash algorithms, also known as cryptographic hash functions, provide the mechanism to secure the data by producing irreversible and unique hash value due to which a higher resistance can be achieved against attacks. Hash algorithms can be categorized into two types based on the secret key usage like "message authentication code" (MAC) and "manipulation detection code" (MDC). The MAC is also known as keyed hash function, whereas MDC is a keyless function. The MDC is further classified into two subtypes, i.e., "one way hash function" (OWHF) and "collision resistant hash function" (CRHF) [22], as exhibited in Fig. 1.

The OWHF was proposed by Merkle [23] which possesses the following properties: (a) the hash function should be publicly known, and there is no need of secret key; (b) the plaintext should be of any length, but the length of OWHF is fixed length having size $\geq$ 64 bits, and (c) the hash function must be one way and irreversible which means that it is very hard to find the plaintext from the hash code. On the other hand, CRHF was proposed by Damgard [24, 25] which possess the properties which are somehow quite similar to OWHF except that the hash code has a fixed length $\geq$ 128 bits and the hash function must be the collision resistant. The MAC provides a mechanism for checking the integrity of the message transmitted over insecure medium as it is an important requirement in the field of open computing and communications. Usually, MAC is used with a secret key shared between two parties for validation of messages exchanged between them. The cryptographic hash function along with MAC is known as hash-based message authentication code (HMAC). The input to hash algorithm is a variable length message (M), and a fixed length message is produced as an output, which is known as message digest (MD) or hash value or code represented as $H(M)$. The hash value created by hash algorithm is used to check the integrity of the message. The integrity and authenticity of the

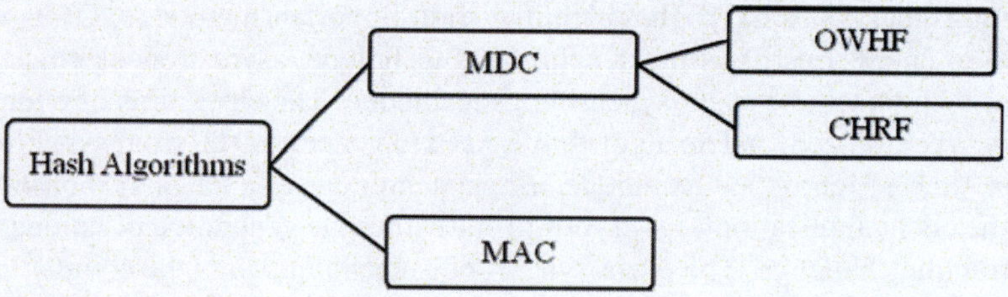

Fig. 1 Classification of hash algorithms

message are achieved through HMAC, which is created by using hash algorithm in combination of a shared secret key (K) as $H(K, M) \rightarrow$ HMAC.

There are various types of the hash algorithms like "SHA-0", "SHA-1", "SHA-2", "SHA-3", "MD-2", "MD-4", "MD-5", "MD-6", "RACE Integrity Primitives' Evaluation Message Digest"(RIPEMD), "RIPEMD-128", "RIPEMD-256", "RIPEMD-160", "RIPEMD-320", "BLAKE", "BLAKE2", "BLAKE3", "HAVAL", "Kangaroo Twelve", "Streebog", "Tiger", and "Whirlpool." In real life applications, the SHA and MDs algorithms have received much attention by research community [26]. In 1993, the National Institute of Standards and Technology (NIST) introduced a new algorithm called SHA-0 to overcome the shortcomings of MD4 and MD5 [27]. Therefore, in our work, we mainly focus on the SHA. A brief discussion about different types of the SHAs is given below:

- **SHA-0** produces 160 bit hash value. It was withdrawn after a short period and replaced by improved version due to undisclosed "significant flaw."
- **SHA-1** produces 160 bit MD based on the principles of MD4 and MD5. It is 40 digits long value represented in hexadecimal format.
- **SHA-2** incorporates significant modifications from its predecessor, i.e., SHA-1. The six hash functions, namely "SHA-224", "SHA-256", "SHA-384", "SHA-512", "SHA-512/224", and "SHA-512/256," are part of the SHA-2 family. SHA-256 and SHA-512, computed with 32-bit and 64-bit words, respectively, are novel hash functions [28, 29]. Both SHA-1 and SHA-2 follow the same *Merkle Demgard* structure.
- **SHA-3** was included as a draft of Federal Information Processing Standards (FIPS) 202 which was published by NIST. The standard "*Permutation-Based Hash and Extendable-Output Functions"* is used in SHA-3. The architectural design of SHA-3 uses a sponge construction [30].

4 Proposed Method

The objective of this section is to develop an algorithm for the generation of alphanumeric OTPs for different variants of SHA. The proposed algorithm for OTP generation is given below:

Step 1: Create a message based on the information of students and current time of the system

The aim of this step is to create string based on the information of the students which is stored in the student database of IES like name of the student, course name, and father's name. In addition to this information, the current time of the system/server will be used to form the message (M) which will be used as an input into SHA.

Step 2: Create HMAC from the message using SHA

In this step, the message (M) and the secret key will be used as an input into hash algorithm for the creation of the HMAC. The length of the HMAC will be stored into variable N, which will be considered as an upper bound for the generation of the random numbers.

Step 3: Generate random numbers
The objective of this step is to generate random numbers from 0 to N, where N is the total number of characters used in the SHA. If we want to generate 4 digits OTP, then four random numbers will be generated; and these numbers will be used as an input to extract the digits of the OTP.

Step 4: Extract the characters from the index of the HMAC for the generation of the OTP
The random numbers will be used to extract the characters from the index of the HMAC. For example, if the HMAC code is "$2497bdd9814d94fe70a116d684c5f12a82372481$" and suppose the random numbers are $24,\ 5,\ 38,$ and 2. Based on this information, the system will generate the following OTP:$8d89$.

The block diagram of the proposed method is exhibited in Fig. 2.

5 An Example

The objective of this section is to generate the OTP for the authentication of the requirements of an institute examination system (IES). The IES deals with the examination activities of the students like (i) to download the hall ticket, (ii) to submit the examination form, and (iii) to generate the end semester mark sheet [31]. In this study, we mainly focus on those requirements which need more security. For example, the hall ticket should be downloaded with proper security only for the bona fide students of an university/institution. The steps of the proposed method have been implemented using C language. According the method, the name and roll number of the student will be entered along with the secret key for the construction of the message. In this study, following information is used as an input into the program: $M =$ Javed Ahmad 16phdas1234 and $K = 13$ as a secret key. Based on hash function SHA-1, we got the following HMAC, i.e.,"$2497bdd9814d94fe70a116d684c5f12a82372481$". In our example, following random numbers are generated to extract the OTP from the HMAC, i.e., $12,\ 5,\ 9,\ 39,\ 4,$ and 11. Based on the given information, the program will generate the following OTP, i.e., $9d11ba$. In our work, we have analyzed different SHAs like SHA-1, SHA-2 variants like "SHA-224, SHA- 256, SHA-384, SHA-512, and SHA-3 512." The six characters are extracted randomly using the proposed algorithm from HMAC to generate OTP. The results are summarized in Table 1.

We have employed N-gram test to check the randomness in a password or a string. This test has also been used by Shukla et al. [21] to show the randomness in the generated passwords. The randomness depends on the frequency distribution of values in the string. The frequency distribution is shown in Table 2 (from 2a to 2f).

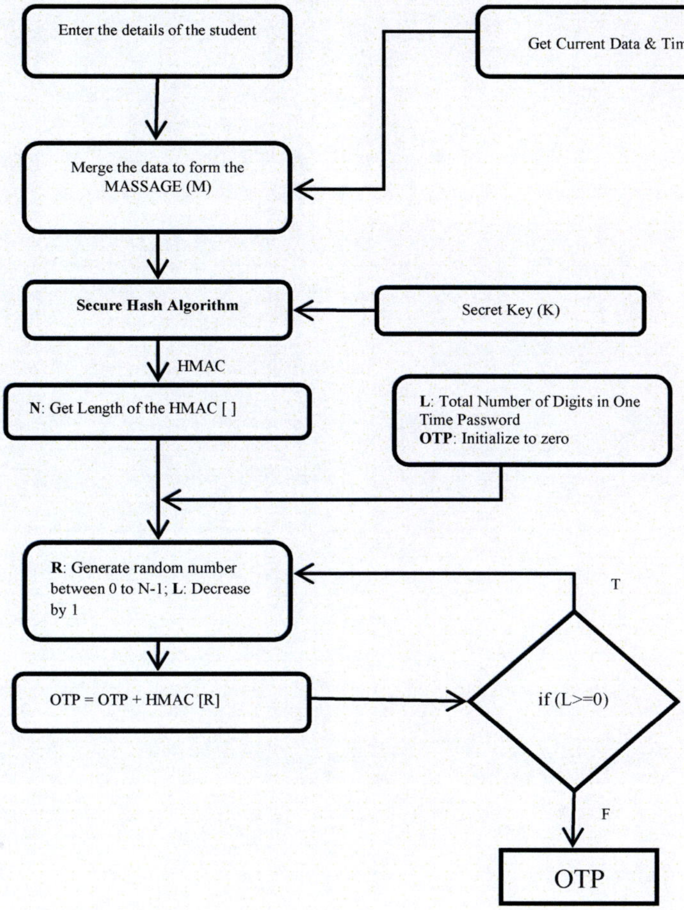

Fig. 2 Block diagram of the proposed method

The randomness of generated OTPs is represented graphically in Fig. 3 for all the SHA variants that we have used for analysis. The graph is drawn on the basis of value and their frequency of occurrence percentage (denoted by frequency %). The characters *a*, *b*, *c*, *d*, *e*, and *f* which are present in HMAC are represented by 10, 11, 12, 13, 14, and 15, respectively.

From Fig. 3, it is clear that there is high randomness in frequency distribution, so it is difficult to predict the OTPs based on the given input. We have compared the results of the proposed method with the selected methods for the generation of the OTPs, and the results are exhibited in Table 3.

From the results of Table 3, it is clear that proposed method generates a strong password by combining the alphabets and numeric values, i.e., $9d11ba$. On the other hand, the selected methods generate the OTPs based on the numeric values only.

Table 1 HMAC output and generated OTP

Message	Secret key	Hashing Function	HMAC	OTP
Javed Ahmad 16phdas1234	13	SHA 1	2497bdd9814d94fe70a116d684c5f12a82372481	9d11ba
		SHA 224	b26934758f00d0eb804d75acec3452c1c7e931012934373f2f303266	40efd6
		SHA 256	c3e57d1c5d753c6a7cd60f6de0b9e96a8974945ab95f98b34a668c05fde699f5	c73ba6
		SHA 384	9d33c4c6a6dca4a1b81636f31a519414a75c351fd350a0922280ae9de9bc61e87b8016eacd74f4d0ca6b1f4efa9d4f7a	ae85b9
		SHA 512	e2524c822a8208e36b6b238fdfa848bc2a3109377bd63b8bbab190de3bba729c3f3279c1d6f344babef0da4ac3e65209a1caf6f968c0fbeb32bfc4c2ac31383d	aef671
		SHA3 512	1e9f1834cc5e2e07f7e5d70ff18713897318416420110284faa471f32e635ae7b8477aa138a3d50d7e74fa01d3627f4ce3e11ea304c7ad4e09c345dc929d3f26	3ab816

Table 2 Frequency of occurrence of digits in descending order: **a** frequency of occurrence using SHA1, **b** frequency of occurrence using SHA224, **c** frequency of occurrence using SHA256, **d** frequency of occurrence using SHA384, **e** frequency of occurrence using SHA512, and **f** frequency of occurrence using SHA3512

Algorithm	Value	Frequency	Occurrence of Frequency in Percent (descending order) (%)
(a)			
SHA 1	4	5	12.50
	1	5	12.50
	2	4	10.00
	d	4	10.00
	8	3	07.50
	9	3	07.50
	7	2	07.50
	f	2	07.50
	a	2	07.50
	6	2	07.50
	b	1	02.50
	e	1	02.50
	0	1	02.50
	c	1	02.50
	5	1	02.50
	3	1	02.50
(b)			
SHA 224	3	8	14.29
	0	6	10.27
	2	5	08.93
	4	4	07.14
	7	4	07.14
	c	4	07.14
	6	3	05.36
	9	3	05.36
	5	3	05.36
	f	3	05.36
	e	3	05.36
	1	3	05.36
	b	2	03.57
	8	2	03.57
	d	2	03.57
	a	2	01.79

(continued)

Table 2 (continued)

Algorithm	Value	Frequency	Occurrence of Frequency in Percent (descending order) (%)
(*c*)			
SHA 256	9	8	12.50
	5	7	10.94
	6	7	10.94
	c	5	07.81
	d	5	07.81
	e	4	06.25
	7	4	06.25
	a	4	06.25
	f	4	06.25
	3	3	04.69
	0	3	04.69
	b	3	04.69
	8	3	04.69
	4	3	04.69
	1	1	01.56
	2	0	00.00
(*d*)			
SHA 384	a	11	11.46
	1	9	09.38
	4	8	08.33
	d	7	07.29
	c	7	07.29
	6	7	07.29
	9	6	06.25
	3	6	06.25
	f	6	06.25
	0	5	05.21
	e	5	05.21
	b	4	04.17
	8	4	04.17
	5	4	04.17
	7	4	04.17
	2	3	03.13

(continued)

Table 2 (continued)

Algorithm	Value	Frequency	Occurrence of Frequency in Percent (descending order) (%)
(e)			
SHA 512	b	15	11.72
	3	14	10.94
	2	12	09.38
	a	11	08.59
	c	10	07.81
	8	9	07.03
	f	9	07.03
	6	7	05.47
	e	6	04.69
	4	6	04.69
	0	6	04.69
	d	6	04.69
	9	6	04.69
	1	5	03.91
	7	4	03.13
	5	2	01.56
(f)			
SHA3 512	1	13	10.16
	7	13	10.16
	3	12	09.38
	e	11	08.59
	4	11	08.59
	f	9	07.03
	a	9	07.03
	0	8	06.25
	8	7	05.47
	2	7	05.47
	d	7	05.47
	c	6	04.69
	9	5	03.91
	5	5	03.91
	6	4	03.13
	b	1	00.78

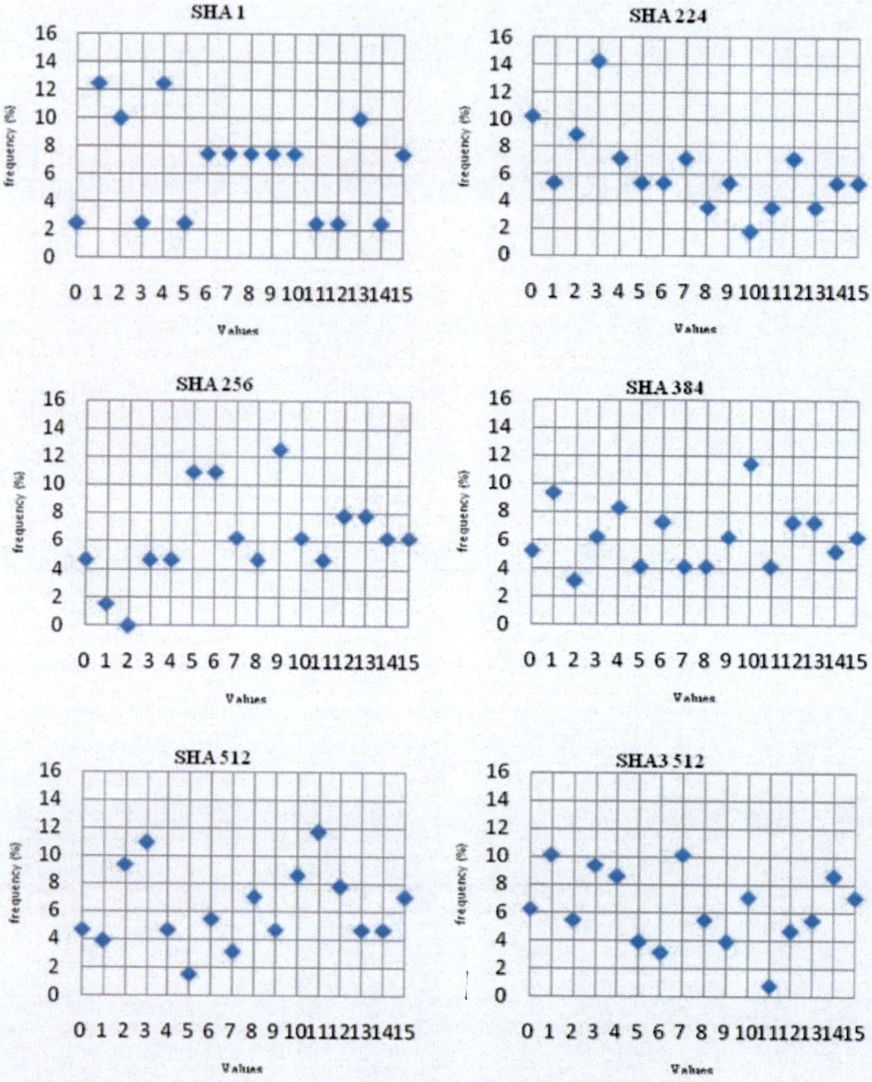

Fig. 3 Graphical representation of the randomness of all the generated OTP

Table 3 Comparative study between proposed method and other selected methods

S. No	Methods	Randomly generated digits	Generated OTP
1	Shukla et al. [21]	6	873,267
2	Parmar et al. [32]	8	67,592,317
3	Proposed method	6	9*d*11*ba*

6 Conclusion and Future Work

This paper presents a method for the generation of OTPs by considering the alphanumeric data for the authentication between client and server. Proposed method includes the following steps: (a) create a message based on the information of students and current time of the system, (b) create HMAC from the message using SHA, (c) generate random numbers, and (d) extract the characters from the index of the HMAC for the generation of the OTP. In our study, a message has been created based on the information received from the student. Then, HMAC was generated by taking the message and secret key value. Based on the random numbers, the OTPs were generated by retrieving the values which are present at the index of the HMAC. Finally, an alphanumeric OTP was generated for the given input.

References

1. Sadiq M, Jain SK (2014) Applying fuzzy preference relation for requirements prioritization in goal oriented requirements elicitation process. Int J Syst Assur Eng Manage 5:711–723
2. Arif M, Mohammad CW, Sadiq M (2020) Software requirements modeling: a systematic literature review. In:IEEE international conference on computing, power and communication technologies (GUCON), Greater Noida, India, pp 194–200
3. Glinz M (2007) On non-functional requirements. In: 15th IEEE international requirements engineering conference. pp 21–26
4. Firesmith D (2003) Engineering security requirements. J Object Technol 2(1):53–68
5. Parveen A, Khan ZH, Ahmad SN (2019) Block-based copy–move image forgery detection using DCT. Iran J Comput Sci 2:89–99
6. Pandey S (2011) Modern network security: issues and challenges. Int J Eng Sci Technol 3(5):4351–4357
7. Tzemos I, Fournaris PA, Sklavos N (2016) Security and efficiency analysis of one time password techniques. In: Proceedings of the 20th Pan-Hellenic conference on informatics (PCI '16), Association for Computing Machinery, New York, NY, USA, Article 67, pp 1–5
8. Ye X-J, Wu G-X (2002) Analysis and improvement of OTP authentication technology. Computer Engineering, pp 27–29
9. Huang Y, Huang Z, Zhao H, Lai X (2013) A new one-time password method. IERI Procedia 4:32–37
10. Yıldırım M, Mackie I (2019) Encouraging users to improve password security and memorability. Int J Inf Secur 18:741–759
11. Bakhtiari S, Safavi-Naini R, Pieprzyk J (1995) Cryptographic hash functions: a survey, technical report 95–09. University of Wollongong, Department of Computer Science
12. Aussel J (2007) Smart cards and digital identity. Telektronikk 3(4)
13. Lamport L (1981) Password authentication with insecure communication. Commun ACM 24(11):770–772
14. M' Raihi D, Bellare M, Hoornaert F, Naccache D, Ranen O (Dec 2005) HOTP: an HMAC-based one-time password algorithm. Internet RFC 4226
15. M'Raihi D, Bellare M, Hoornaert F, Naccache D, Ranen O (2011) TOTP: time-based one-time password algorithm. IETF RFC 6238 ser. Informational '11, pp 1–16
16. Bicakci K, Baykal N (2002) Infinite length hash chains and their applications. In: Eleventh IEEE international workshops on enabling technologies: infrastructure for collaborative enterprises, Pittsburgh, PA, USA, pp 57–61

17. Chefranov AG (2008) One-time password authentication with infinite hash chains. In: Novel algorithms and techniques in telecommunications, automation and industrial electronics, pp 283–286
18. Eldefrawy MH, Khan MK, Alghathbar K (2010) One-time password system with infinite nested hash chains. In: Kim T, Fang W, Khan MK, Arnett KP, Kang H, Ślęzak D (eds) Security technology, disaster recovery and business continuity. Communications in Computer and Information Science, vol 122. Springer, Berlin, Heidelberg, pp 161–170
19. Srivastava S, Sivasankar M (2016) On the generation of alphanumeric one time passwords. Int Conf Inventive Comput Technol 1–3
20. Saxena A (2008) Dynamic authentication: need than a choice. In: 3rd International conference on communication systems software and middleware and workshops
21. Shukla V, Chaturvedi A, Srivastava N (2019) A new one time password mechanism for client server applications. J Discrete Math Sci Crypt 22(8):1393–1406
22. Preneel B (2010) Cryptographic hash functions: theory and practice. In: Gong G, Gupta KC (eds) Progress in cryptology—INDOCRYPT 2010. INDOCRYPT 2010. Lecture Notes in Computer Science, vol 6498. Springer, Berlin, Heidelberg
23. Merkle RC (1990) One way hash functions and DES. In: Brassard G (eds) Advances in cryptology—CRYPTO' 89 Proceedings. LNCS, vol 435. Springer, New York, NY
24. Damgård IB (1988) Collision free hash functions and public key signature schemes. In: Chaum D, Price WL (eds) Advances in cryptology—EUROCRYPT' 87. LNCS, vol 304. Springer, Berlin, Heidelberg
25. Damgard IB (1988) The application of claw free functions in cryptography. PhD Thesis, Aarhus University, Mathematical Institute
26. Gupta P, Kumar S (2014) A comparative analysis of SHA and MD5 algorithm. Int J Comput Sci Inf Technol 5(3):4492–4495
27. Al-Odat ZA, Ali M, Abbas A, Khan SU (2020) Secure hash algorithms and the corresponding FPGA optimization techniques. ACM Comput Surv 53(5)
28. Glabb R, Imbert L, Jullien G, Tisserand A, Veyrat-Charvillon V (2006) Multi-mode operator for SHA-2 hash functions. J Syst Archit Elsevier 52(2–3):127–138
29. Debnath S, Chattopadhyay A, Dutta S (2017) Brief review on journey of secured hash algorithms. In: 4th International conference on opto-electronics and applied optics (Optronix), pp 1–5
30. Bayat-Sarmadi S, Mozaffari-Kermani M, Reyhani-Masoleh A (2014) Efficient and concurrent reliable realization of the secure cryptographic SHA-3 algorithm. In: IEEE transactions on computer-aided design of integrated circuits and systems 33(7):1105–1109
31. Nazim M, Mohammad CW, Sadiq M (2020) Generating datasets for software requirements prioritization research. In: 2020 IEEE international conference on computing, power and communication technologies (GUCON), Greater Noida, India, pp 344–349
32. Parmar H, Nainan N, Thaseen S (2012) Generation of secure one-time password based on image authentication. Computer Science & Information Technology, pp 195–206

QAudious: A Question Answering System for Academic Lecture Recordings

Dipti Pawade, Avani Sakhapara, Swati Pandey, Harsh Vasa, Kajal Shethia, and Sana Pagarkar

Abstract This paper puts forth a system for generating context from the media file and serving as a medium to answer questions posed by the user. The system accepts the meeting recordings and generates a textual transcript followed by preprocessing the transcript using embedding and data cleaning techniques of natural language processing. The transcript and the question are fed to the model to get the answer. The SQuAD 2.0 dataset is preprocessed for generation of a more rounded representation of the context, question, answer, span index, and mask. The pretrained models of BERT and Baseline BiDAF were implemented to generate an accurate answer for the given question. The system performance is measured through EM and $F1$ scores. Non-NLP methods were also executed that predicted the target sentence and not the answer, using supervised methods like fitting multinomial logistic regression, logistic regression with root match feature, random forest, and XGBoost and unsupervised methods like cosine similarity and Euclidean distance.

Keywords Question answering system · SQuAD 2.0 · BERT · Baseline BiDAF · Cosine similarity · Euclidean distance

D. Pawade (✉) · A. Sakhapara · S. Pandey (✉) · H. Vasa · K. Shethia · S. Pagarkar
Department of Information Technology, K. J. Somaiya College of Engineering, Vidyavihar, Mumbai, India
e-mail: diptipawade@somaiya.edu

S. Pandey
e-mail: swati.pandey@somaiya.edu

A. Sakhapara
e-mail: avanisakhapara@somaiya.edu

H. Vasa
e-mail: harsh.vasa@somaiya.edu

K. Shethia
e-mail: k.shethia@somaiya.edu

S. Pagarkar
e-mail: sana.p@somaiya.edu

R. P. Mahapatra et al. (eds.), *Proceedings of International Conference on Recent Trends in Computing*, Lecture Notes in Networks and Systems 341,
https://doi.org/10.1007/978-981-16-7118-0_55

1 Introduction

The COVID-19 pandemic has changed the world drastically. Looking at the lockdown conditions, almost every sector opted for online platforms to perform their operations. In the case of work from home scenarios, people prefer to conduct meetings on different applications like Zoom, Google Meet, Microsoft Teams etc. For future references, these meetings are recorded, so that if anyone has missed anything due to connectivity issues or other unforeseen reasons, then he/she can avail the ability of replaying the recorded sessions for catching up on the missed information. Let us consider the education sector for instance. Right from preschool to post graduation, all the classes are conducted online, and the recorded sessions are provided to the students so that they can later refer to it for revision/self-study. Now the problem that arises here is, if the student has understood the concept and is still unable to answer a particular question, he/she might have to go through the entire 45 min long video lecture, to find the answer, if the answer is at the end of video. Let us take an example of a classroom scenario where the topic covered was "World War II." During the class, the instructor explained the name of the belligerents involved, duration of the war, number of deaths incurred, the forces of the combatants, and other details. Now, the student was able to retain most of the information except he/she could not remember the date of this war. Thus by convention, the student would replay the video to find out the segment where the instructor may have mentioned this detail. This process, as a result, is taxing and time consuming. This aforementioned example is very simple and straightforward. Although, there can be more complicated questions too for which retrieving answers from recorded lectures would perhaps be difficult. This inspired us to build a Web application which serves as an NLP-based question answering system that answers questions based on the given audio file. For experimental purposes, we have restricted ourselves to recorded lectures belonging to the education sector only. The motivation behind this application is to ease the answer retrieval process and clear doubts of the student by procuring instant responses for which we intend to build a system that serves as a medium for the students to ask queries regarding the matter of discussion, as a result reducing the human effort of traversing through long hours of recorded conversations, and the time required for searching a said timeline and its corresponding answer. The proposed application makes use of independent pretrained models of BERT and Baseline BiDAF. A few non-NLP methods (supervised and unsupervised learning techniques) have also been implemented to predict the sentence that contains the answer and draw inferences from it.

2 Related Work

Researcher have discussed an answering system for SQuAD 2.0 dataset which includes steps like augmenting, randomly substituting with WordNet's synonyms, followed by paraphrasing using the larger PPDB database [1]. It is further enhanced by a novel GPT-2 model that generates completely new sentences, resulting in broader contexts. The author considered xxlarge and Baseline RoBERTa, XLNet, and GPT-2 models for their studies. Randomizing mini batches and Sentence Augmentation helped the system to achieve an EM/$F1$ score of 89.4/91.7 for dev set and 89.1/91.5 for test set, respectively. Wen Zhou et al. [2] proposed a question answering system based on the Baseline BERT model and significantly improved the single Baseline BERT model on SQuAD 2.0. They adopted a novel data augmentation approach and integrated linguistic knowledge achieving an EM/$F1$ score of 79.91/82.43 for dev set and 79.87/82.54 for test set. The addition of linguistic knowledge gave better performance results and data augmentation helped in adding diversity to the existing dataset preventing overfitting. Konrad Morzkowski et al. [3] implemented Hugging-Face's PyTorch using BERT2. To match the output with the output from the baseline model, a log-softmax function is used. The models used were Baseline BiDAF Model, BiDAF Model with character-level embeddings, BERT Model, attention over attention mechanism on BERT model, and an ensemble with BERT and BiDAF. It achieved an $F1$ score of 77.398 and EM score of 74.489. Zhen Qin et al. [4] applied a language representation model called BERT on SQuAD dataset to design a question answering system. They additionally applied L1 regularization, froze the first few layers of BERT, and used BERT embedding on BiDAF. For weighted Average Ensembling Model, EM/$F1$ scores for dev and test were 73.64/79.94 and 76.01/78.84, respectively. Danqi Chen et al. [5] developed a Wikipedia-based question answering system containing two modules namely Document Retriever and Document Reader. Document Retriever module used bigram hashing and TF-IDF matching to return a subset of relevant articles. The Document Reader module comprises of a multi-layer recurrent neural network machine comprehension model trained to detect answer spans. Four datasets namely SQuAD, CuratedTREC, WebQuestions, and WikiMovies were used. Liu et al. [6] proposed answering a model that had two components: a span detector and an unanswerable classifier. The model consists of two sets of layers: first the shared layers including a lexicon encoding layer, contextual encoding layer, and memory generation layer; second, the task-specific layers including the SAN answer module for span detection, and a binary classifier determining whether the question is unanswerable. $F1$ Score of this model was 72.66.

From the literature survey, it was concluded that most of the research papers focus on using either independent or an ensemble of different models with either Attention flow/AoA (Attention over Attention) mechanism, WordNet Synonym substitution, character- or word-level embedding etc. to give the most optimal results, as per their individual need-to-know basis. Thus, as observed, there is a vast scope of research in this field.

To acquire an overview about similar applications, we drew comparisons between the proposed system and existing applications like "START" [7], Intellexer [8], AllenNLP [9], and Quantum Stat [10]. Table 1 shows the comparison between these applications based on the following features: (1) Take media files as source, (2) Returns specific answers; (3) Returns answer from user context; (4) Detects unanswerable questions.

Table 1 Comparison of existing application

Application	Input	Feature			
		1	2	3	4
START [7]	Question	*x*	✓	*x*	✓
Intellexer [8]	Question	*x*	*x* (only URL)	*x*	✓
AllenNLP [9]	Question	*x*	✓	*x*	– (demo)
Quantum Stat [10]	Context + Question	*x*	✓	✓	✓
QAudious	Media files for context + Question	✓	✓	✓	✓

3 Methodology

The proposed approach includes three subsequent steps:

1. Processing the SQuAD dataset.
2. Input preprocessing.
3. Model training.

1. Processing the SQuAD Dataset

In the initial step of processing, the SquAD dataset was split into training and testing data, wherein each training example is broken up into context, question, answer, and span. Using Spacy tokenizer, these components were saved as space-separated tokens generated from JSON data. With the aforementioned data divided into separate files, a dictionary of words was built and the words were sorted by frequency, thereby associating each word with its corresponding index. Thereafter, the GloVe matrix (50, 100, 200, or 300 dimensional) was loaded and a dictionary of all the words representing keys and NumPy vectors as value was created, in order to build a matrix of embedding vectors. Instead of saving the procured set of experiment files containing context, question, answer and span, as tokens, the files contain indices into the embedding matrix for every word, wherein the context and question arrays are represented as NumPy arrays for easy loading.

2. Input Preprocessing

The next step pertains to the conversion of inputted media file into a transcript. File extensions matter while importing the audio file to the program. So on uploading the media file, the inputted file format was converted to a standard extension of.wav for convenient processing using the AudioSegment container from the Pydub module. The system splits the audio file into chunks and applies speech recognition on each chunk individually. These chunks were created on the basis of the splitting factor (700 ms) which is contingent upon the silences present in the intervals of the audio file. Here, 700 ms or more of interludes present in between the entire file were taken and split into chunks (multiple short audio files). After generating the overall unsegmented text from each chunk, Punctuator Model was used to create sentence boundaries within the specified boundary type. This inserted strings like ".PERIOD" into the raw text to denote a period. This was then converted from punctuation annotation symbols to readable and human interpreted punctuation marks, inserted throughout the text, so as to help to understand the tone, pauses and emphasis of the text. After this process, audio transcribing was completed by generating a clean and segmented transcript of the given media file.

3. Model Training

The final step is model training wherein three different approaches have been implemented to fetch the answers for questions posted by the user. In the first proposed method, a pretrained BERT Model was used. BERT is a very large model having 24 layers and an embedding size of 1024, for a total of 340 M parameters. Firstly, Bert-large-uncased-whole-word-masking-finetuned-squad which is already fine-tuned for question answering (version 1 of SQuAD) and tokenizers were loaded. After the question was asked, the BERT tokenizer was run against both the "question" and the "answer_text," which were actually concatenated together with the special token denoted as "SEP" in between them to distinguish the two. BERT has two special "Segment" embeddings, one for segment "A" and another for segment "B." Prior to the word embeddings entering the BERT layers, the segment "A" embeddings need to be added to the "question" tokens, and segment "B" embeddings to the "answer_text" tokens. For this, "0" or "1" was assigned for each token by the transformer library. This created a list of 0 s and 1 s, with every input token having a segment_id. These segment IDs help to differentiate "question" from "answer_text." Start and End scores were calculated to find the tokens with the highest score of both. The correct implementation is to pick the highest total score for which end $\geq$ start, as there is a possibility that it can predict an end word that is before the start word. Later, the input tokens were converted into their respective string versions. The desired answer for the query was formed by beginning with selecting the first token and then selecting the remaining answer tokens eventually joining them with the whitespace. Any word that got broken down in the process was reconstructed into subwords and regrouped with the previous token. Otherwise space was added before the token to store it as the final answer. This method gave accurate answers, although for questions that were impossible to answer, it did not output definitively whether the question is "impossible" to answer, which was then added in version 2 of SQuAD.

For the second method, Baseline BiDAF (Bidirectional Attention Flow) model was used. BiDAF is a multi-stage hierarchical process that represents the context at different levels of granularity. It uses a bidirectional attention flow mechanism to achieve a query-aware context representation without early summarization. It was introduced with PyTorch. Starting from the embedding layer, the word indices were embedded to get the word vectors. In the encoded layer, the embedded sequence was encoded. Later, attention mechanism was applied to the encoded sequence and in the model encoder layer, the encoding of the sequence was reiterated. In the embedding class, the embedding layer was used by BiDAF, without the character-level component. Also, the word-level embeddings were further refined using a 2-layer Highway Encoder (basically to encode an input sequence using a highway network). The embedding class used the following argument into consideration, pretrained word vectors, size of hidden activations and probability of zero-ing out activations. Bidirectional attention computes attention in two directions: the context attends to the query and the query attends to the context. The output of this layer is the concatenation of [context, c2q_attention, context × c2q_attention, context × q2c_attention]. This concatenation allows the attention vector at each timestep, along with the embeddings from previous layers, to flow through the attention layer to the modeling layer. The output has shape (batch_size, context_len, 8 × hidden_size). The "similarity matrix" between context and query was obtained. A naive implementation as described in BiDAF would concatenate the three vectors then project the result with a single weight matrix. This method is a more memory-efficient implementation of the same operation. Output layer used by BiDAF for question answering computes a linear transformation of the attention and modeling outputs, then takes the softmax of the result to get the start pointer. A bidirectional LSTM is applied to the modeling output to produce "mod_2." Then the linear + softmax of the attention output and "mod_2" is used to get the end pointer.

For the final approach, a few non-NLP methods were executed to compute the answers for our question answering system just to understand and analyze the results and compare them with the above-mentioned approaches. Traditionally, the approach that is conventionally followed is to average the vectors of all words in a sentence called the bag of words, but this medium did not perform as accurately because it neglected the consideration of the order of words. As opposed to the first step that involved creating word embeddings of the dataset, here, instead of dealing with individual words, we worked directly with individual sentences as in the case of large text, using only words would be limited by the information we can extract from the word embeddings. As word embedding techniques can fall short, sentence embedding techniques were introduced that represented entire sentences and their semantic information as vectors. This helped the machine in understanding the context, intention, and other nuances in the entire text. Facebook's "InferSent" technique which is a supervised sentence embedding method that is trained on Natural Language Inference data was also used. Firstly, the vocab was created from training data and was used to train the InferSent model. Textblob was used to break paragraphs into multiple sentences for sentence-wise processing. After that, the pair of sentences were encoded to generate the actual sentence embeddings. Therefore, the vector representation of each sentence and question was generated using the model. After these values were derived, the only objective was to return the most accurate result for the posed query, i.e., to extract the relations between the embeddings/sentence pairs, for which, two approaches were taken into account and accordingly the results were compared: The first method was using an unsupervised learning model in which the Euclidean distance was calculated to detect the sentence having minimum distance from the question. The accuracy of this model came to be around 45% and using Cosine similarity that takes alignment and angle between the vectors into account, enhanced the accuracy by almost 15%, i.e., 61%. The subsequent method was making use of supervised learning models. Here, for each sentence, one feature was built on cosine similarity and the maximum value being 1, another feature was Euclidean distance with maximum value of 60, and the third feature was root match. The paragraph length was restricted to 10 sentences. In case if the said paragraphs had a lesser number of sentences, its feature value was replaced with 1 making a total of 10 sentences for maintaining uniformity. Once the training data was created, multinomial and fitting logistic regression, random forest and XGBoost techniques were used. The XGBoost model delivered an accuracy of 68% and outperformed the other models in comparison.

4 Result and Discussion

For evaluation purposes, Exact Match Score and $F1$ score metrics were considered. Exact Match (EM) Score metric is a fairly strict binary measure (true/false) to calculate whether the system output exactly matches the ground truth answer exactly, i.e., word-to-word matching. If the predicted answer is exactly equal to the target answer then EM score is 1, else 0. The $F1$ Score metric is a less severe measure that calculates the score through a harmonic mean of precision and recall by formula 1.

$$F1 = (2 \times p \times r)/(p + r) \tag{1}$$

where p = precision, r = recall.

For greater $F1$ score, both precision and recall should be high. Precision and recall were calculated using Eqs. 2 and 3, respectively.

$$\text{precision} = \text{len (common\_ tokens) /len (predicted\_ tokens)} \tag{2}$$

$$\text{Recall} = \text{len (common\_ tokens) /len (target\_ tokens)} \tag{3}$$

where

len(common_tokens) = number of common tokens between the predicted and target answers.

len(predicted_tokens) = number of tokens in the predicted answers, and len(target_tokens) = number of tokens in the target answers.

Both Euclidean distance and cosine similarity [11] were used to predict the most similar sentence to the context of the question asked. Cosine similarity is better than Euclidean distance because the former takes into account the angle between the sentences. In the non-NLP methods, the sentence where the answer could be present was predicted. While calculating the EM and $F1$ score, the entire sentence was taken as the target rather than the original output in the SQuAD dataset. Tables 2 and 3 show the $F1$ and EM score for supervised and unsupervised learning methods.

In Table 4, the performance evaluation of BiDAF and BERT for exact answer prediction is given. In BiDAF, only attention is used, i.e., how much attention one

Table 2 Performance evaluation for supervised learning methods

Model name	Train accuracy		Test accuracy	
	EM	$F1$	EM	$F1$
Fitting multinomial logistic regression	0.636	0.634	0.632	0.631
Logistic-regression with root match feature	0.648	0.643	0.634	0.633
Random forest	0.726	0.737	0.648	0.644
XGBoost	0.715	0.694	0.681	0.662

Table 3 Performance evaluation for unsupervised learning methods

Algorithm for distance	EM	$F1$
Euclidean distance	0.458	0.476
Cosine similarity	0.616	0.638

Table 4 Performance evaluation for BiDAF and BERT

Model name	EM Score	$F1$ Score
BERT (BERT-large-uncased-whole-word-masking-finetuned-squad)	86.91	93.15
Baseline BiDAF	54.31	59.43

has to pay to other words in the sequence. In BERT, transformers are used that use self-attention, and this layer helps the encoder look at other words in the input sentence as it encodes a specific word. Transformer encoder reads the entire sequence of words at once. This characteristic allows the model to learn the context of a word based on all of its surroundings (left and right of the word).

5 Conclusion

To conclude, QAudious enabled the user to upload audio files, pose questions regarding the content of the uploaded file and retrieve answers based on the transcript and context generated by the system. The purpose of reducing the time consumed by the user to fetch the answer from a lecture recording or any other academic media was achieved through our system. To achieve good performance and better accuracy, NLP methods such as BERT and BiDAF were implemented. To measure the performance of these methods in our system, non-NLP methods based on supervised and unsupervised learning were explored. From the analysis of results, it was extrapolated that the NLP based method, BERT Model fine-tuned with SQuAD 2.0 dataset outperformed all the other methods. Our system can be extended further to be used with video files of conference sessions, conference meetings, expert sessions, etc. Also, support for multilingual content can be considered in further work.

References

1. Li Y (2020) Avengers: achieving superhuman performance for question answering on SQuAD 2.0 using multiple data augmentations, randomized mini-batch training and architecture ensembling. Stanford CS224N Final Project Reports
2. Zhou W, Zhang X, Jiang H (2019) Ensemble BERT with data augmentation and linguistic knowledge on SQuAD 2.0. Stanford CS224N Final Project Reports
3. Morzkowski K, Sun L, Hoovestol P (2019) SQuAD 2.0 project report. Stanford CS224N Final Project Reports
4. Qin Z, Mao W, Zhu Z (2019) Diverse ensembling with bert and its variations for question answering on SQuAD 2.0. Stanford CS224N Final Project Reports
5. Chen D, Weston J, Bordes A (2017) Reading wikipedia to answer open-domain questions. In: 2017 the 55th annual meeting of the association for computational linguistics
6. Liu X, Li W, Fang Y, Kim A, Duh K, Gao J (2018) Stochastic answer networks for SQuAD 2.0
7. http://start.csail.mit.edu/start-system.php. Retrieved on 27 June 2020
8. https://www.intellexer.com/. Retrieved on 27 June 2020
9. AllenNLP (https://allenai.org/). Retrieved on 27 June 2020
10. https://quantumstat.com/. Retrieved on 27 June 2020
11. Coenen A, Reif E, Yuan A, Kim B, Pearce A, Viégas F, Wattenberg M (2019) Visualizing and measuring the geometry of BERT. arXiv:1906.02715v2

Summarization of Tweets Related to Disaster

Avani Sakhapara, Dipti Pawade, Baiju Dodhia, Jigar Jain, Omkar Bhosale, and Onkar Chakrawar

Abstract Social media sites like Twitter have become a significant source of data during events such as disasters. A lot of important data is posted on Twitter during disasters. Twitter is one of the first places where users go to find information regarding any current event. When thousands of users tweet about any particular disaster, it then becomes a trend on twitter. But out of all these tweets pertaining to a particular disaster, usually, it is observed that the tweets containing important and relevant information about the disaster are often lost with other uninformative tweets on the topic. Another problem observed with tweets related to a particular disaster is regarding authenticity of disaster-related information extracted from these tweets. So to overcome these problems, in this paper, we have proposed and implemented a system to filter tweets pertaining to a particular disaster and generate summary of occurring disaster from these tweets. Two types of summaries are generated using the bidirectional autoregressive transformer (BART) model: one summary is generated from tweets of random user handles, and the other is generated from tweets of disaster management authorities and government Twitter handles. The system also provides information about prior occurred disaster events. Further in the paper, analysis of results is discussed, and inference is presented.

A. Sakhapara (✉) · D. Pawade · B. Dodhia (✉) · J. Jain · O. Bhosale · O. Chakrawar
Department of Information Technology, K. J. Somaiya College of Engineering, Vidyavihar, Mumbai, India
e-mail: avanisakhapara@somaiya.edu

B. Dodhia
e-mail: baiju.dodhia@somaiya.edu

D. Pawade
e-mail: diptipawade@somaiya.edu

J. Jain
e-mail: jain.ja@somaiya.edu

O. Bhosale
e-mail: o.bhosale@somaiya.edu

O. Chakrawar
e-mail: onkar.chakrawar@somaiya.edu

R. P. Mahapatra et al. (eds.), *Proceedings of International Conference on Recent Trends in Computing*, Lecture Notes in Networks and Systems 341,
https://doi.org/10.1007/978-981-16-7118-0_56

Keywords NLP · Twitter · Disaster · Sentiment analysis · Classification · Summarization · Machine learning · BART

1 Introduction

Social media platform such as twitter is used by large number of users to share information related to a disaster in real time, as the disaster is occurring. But, all this information communicated through tweets pertaining to the occurring disaster is unorganized on twitter. The tweets posted when disaster is occurring might be relevant to the disaster or they might not be relevant to the disaster. Usually, it is observed that the relevant tweets contain important information about occurring disaster such as the nature of the disaster, affected human life, relief measures, statistics, people's behavior, emotions, trends, and personal opinions. But, often, these tweets containing important information about the disaster are lost with other uninformative tweets. Also, it is observed that not all information exchanged through tweets about the occurring disaster is authentic. So, for an individual, going through these thousands of tweets to get precise information about the occurring disaster is difficult.

So, this motivated us to develop a system to summarize the widespread information from tweets related to occurring disaster and provide the necessary and important data to the concerned authorities and the general public of the occurring disaster. The system performs sentiment analysis on tweets and identifies disaster-related tweets. The system then retrieves the location information from tweets, such as state or city where the disaster is occurring. The system depicts this location on the map and presents two types of summaries for the disaster occurring. One summary is generated by retrieving tweets from the user handles, and the other summary is generated from disaster management authorities and government handles. The summary generated from disaster management authorities and government handles is used to verify the authenticity of information presented in the summary generated from user handles.

Further in the paper, in Sect. 2, literature survey of work carried out for summarization of text and for summary generation from tweets is discussed. In Sect. 3, the design and implementation of the proposed system are discussed. In Sect. 4, analysis and results are stated, and in Sect. 5, conclusion and future work is presented.

2 Literature Review

In this section, we have discussed about the work carried out by researchers to generate summary from text and to generate summary from tweets. K. Aziz et al. [1] classified tweets related to disaster into various sentiments like positive, neutral, and negative categories. Naive Bayes classifier was used for the classification of tweets, and the data were then plotted on a map to depict the affected regions. For this, first, they retrieved twitter data which included hashtags of ten natural disasters. Then, the tweets were preprocessed, cleaned, and filtered. After classification, the dominant disaster information was extracted using Word Cloud. S. Madichetty et al. [2] proposed a method based on the combination of CNN and ANN for detecting informative tweets during a disaster. K. Rudra et al. [3] classified tweets related to a disaster event and identified communal and anti-communal tweets. B. Shah et al. [4], retrieved tweets related to the Nepal earthquake, which were then preprocessed and analyzed. The paper showed measures and analysis of the situation using geolocation features for classifying crisis zones. Q. Zou [5] introduced a system for collecting Twitter data posted by people during a disaster and analyzing and presenting these data based on the geographical location. NLTK Naive Bayes classifier was used to perform sentiment analysis, and random forest was used to analyze social network data. S. Roy et al. [6] worked with the tweets related to the disaster and classified these tweets as informative or non-informative using an SVM classifier. The dataset of around 42,000 tweets was taken, and an accuracy of around 75% was obtained. The Fani cyclone dataset was used which contained tweets of that cyclone. They noted that informative tweets contained nouns, verbs, and hence, they performed speech tagging. The informative tweets were finally processed through Luhn and LSA summarization algorithms. Summary of tweets using LSTM-based approach was demonstrated by P. Goyal et al. [7], and then, a storyline was made from this summary. The authors proposed a novel approach called MYTHOS that detected events, subevents within an event, and generated abstract summary to provide different perspectives for an event. Mike Lewis et al. [8] introduced bidirectional autoregressive transformer (BART) which is an auto-encoder for pre-training sequence-to-sequence models. They found that BART was effective when it was fine-tuned to generate text. The model was trained by corrupting the documents and optimizing the loss between the original document and the decoder's output. The authors also discussed about experiments which replicate other training schemes in the framework of BART. This helped them to measure factors that influence end performance. A. K. Mohammad Masum et al. [9] generated an abstractive summary of a text document. They used a bidirectional RNN with LSTM in the encoding layer and attention model in the decoding layer and applied the sequence-to-sequence model to generate a summary. The amazon fine food reviews dataset was used for making a good summarizer. The main aim was to increase the efficiency and reduce the training loss of the sequence-to-sequence model to get a better abstractive text summarizer. They successfully reduced training loss with a value of 0.036 and created a summary of English text. The tensorflow1.12.0 sequence-to-sequence model was used. The input sentence for making a summary was taken from the dataset, and the summary length was defined randomly. K. Rudra et al. [10] proposed a classification–summarization method that divides tweets into different situational groups before summarizing them. For the summarization process, they suggested a two-step extractive–abstractive summarization method. It extracted a collection of relevant tweets from the entire set of knowledge in the first step, created a bigram-based word-graph from those tweets, and constructed paths by traversing the word-graph. Following that, it employed an integer linear programming (ILP) optimization technique to determine the most important tweets and paths based on various decision variables. The suggested technique is both time and memory effective, and it outperformed existing methods in terms of both quantitative and qualitative judgement. Koustav Rudra et al. [11] retrieved situational information from diverse quantities of emotions and views, then summarized the large amounts of real-time situational data. The approach was based on an interpretation as to how many ideas evolve in the context of a disaster on Twitter. This awareness aided in attaining high performance on English tweets as compared to state-of-the-art tweet classifiers and summarization techniques. A classification–summarization framework was used where both English and Hindi tweets were handled. They achieved 80% accuracy using SVM classifier. They further proposed SEMCOWTS methodology whose accuracy was 95% for generating a summary. K. Sarkar [12] generated an extractive summary of the text not just by using its information in itself but also by using the extractive summary of previously computed texts. Then by using similarity scores, the summaries generated using internal methods were compared with the summaries generated using external methods. For better text summarization results, the author proposed a hybrid summarization approach that combines internal and external methods. The effectiveness of the proposed summarization method was demonstrated by comparison to other state-of-the-art summarization systems. S. Shleifer et al. [13] demonstrated the working of CNN/Daily Mail and XSum datasets with various transformer-based models that are fine-tuned to smaller models for practical purposes. It analyzed the summarization results for 3 distillation methods, namely knowledge distillation, pseudo-labelling and shrinks, and fine-tunes. It also showed that making continuous small adjustments in the large models by changing the number of encoders and decoders for a specific task can result in a rapid generation of smaller models. It was inferred that the BART model gave better results for summarization. Hence, this model was used for summarizing the tweets. The summary of the literature review is presented in Table 1.

3 Implementation Details

In this section, the design and implementation of the proposed system are discussed. The overall working of the system is depicted in Fig. 1. Initially, the system connects to Twitter API. The Twitter API requires four API keys namely 'Consumer Key', 'Consumer Secret Key', 'Access Token Key', and 'Access Secret Token Key'. Then, the current Twitter trends for India are retrieved and filtered using a keywords list to identify disaster-related trends. If no disaster-related trend is identified, then, the system waits for 15 min to retrieve trends before continuing with the process again. If disaster-related trends are discovered, then, the system iterates over them one by one, retrieving two types of tweets for each trend that do not contain any retweets and contain a mix of tweets (recent and top), where the list of tweets may contain duplicate tweets.

The two types of tweets retrieved are one from random user handles and the other from disaster management authorities and government handles. Then, these both

Table 1 Summary of literature review

Ref No	Methodology	Dataset and results
[1]	Naive Bayes classifier for classification	32,117 tweets were used for the analysis. 68% of disaster-related tweets were negative
[2]	Combination of CNN and ANN for detecting informative tweets during the disaster	Real-time Twitter dataset Hurricane Harvey 2017 for analysis
[3]	Used seven different algorithms for classification available in scikit-learn including SVM, logistic regression and Naive Bayes	Dataset of five different disasters used. They achieved 94% precision and 91% recall in communal tweet detection
[4]	Retrieved twitter contents analysis based on keyword extraction from the dataset and statistical analysis to interpret the tweet content. To identify danger zones, a geolocation feature was used	40,236 tweets between the period of April and May, 2015 related to the Nepal earthquake were retrieved, processed, and analyzed
[5]	NLTK Naive Bayes classifier was used for sentiment analysis, and random forest was used to analyze social network data	17,748 tweets were retrieved for the flood keyword. 36% success rate achieved for retrieval of geographic information from tweets
[6]	SVM classifier for classification. Luhn and LSA algorithm for summarization	Dataset of 42,000 tweets and around 5000 meaningful tweets was obtained with an accuracy of 74.3%
[7]	LSTM-based approach used for a summary of tweets. The online clustering algorithm is used to preprocess tweets	CNN/Daily Mail dataset having a total of 92.5 k pairs of stories and summaries was used. Mythos framework generated a storyline for an event using summaries
[8]	BART model used for summary generation	The BART model was effective when it was fine-tuned to generate text and used for comprehension tasks
[9]	Generated a summary using bidirectional RNN with LSTM in encoding layer and attention model in decoding layer	Amazon fine food review datasets containing 20,000 data. The summary generated is not accurate for a few reviews
[10]	Summarizes disaster-related tweets using a combination of extractive and abstractive approaches	Developed an ILP-based summarization technique to generate a concise report of an event. COWEXABS performs better compared to APSAL and TSum4act by 23% and 26%, respectively, for summarization
[11]	Five algorithms are compared where SEMCOWTS outperforms others. Content word and semantic relation-based summarization approach (SEMCOWTS) used to summarize the situational tweet stream	Four disaster data is taken. SEMCOWTS had ROUGE-1, F-score better than COWTS, NAVTS and Sumblr
[12]	A novel approach for text summarization combining both internal and external information. A 100-word summary is generated	DUC 2001 dataset with 309 document summary pairs and DUC 2002 dataset with 567 document summary pairs are used

(continued)

Table 1 (continued)

Ref No	Methodology	Dataset and results
[13]	Demonstrated working of CNN and XSum datasets with various transformer-based models that are fine-tuned to smaller models for practical purposes. Analyses summarization results for 3 distillation methods	17,748 tweets were retrieved for the flood keyword. 36% success rate achieved for retrieval of geographic information from tweets

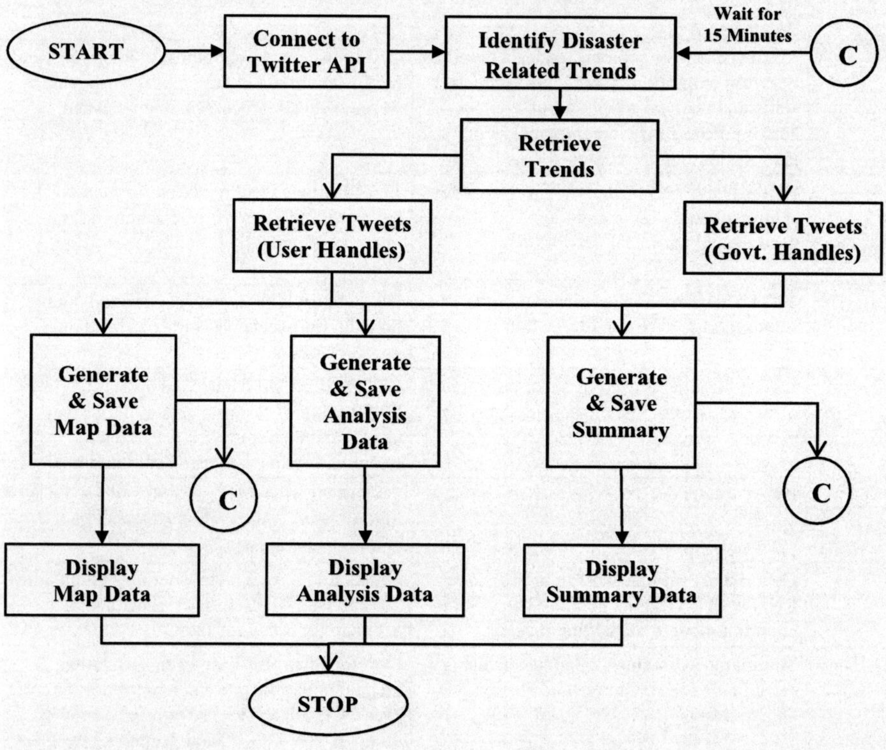

Fig. 1 Working of system

types of tweets are subjected to a series of processing steps to extract information separately.

To begin with the user handles tweets, three types of processing are done to extract information which are for map, analysis, and summary generation. Next, for government handles, tweets are processed to generate a summary.

Once all the information is extracted and generated, the map, analysis, and summary data are saved to further display it to the users.

3.1 Generation of Summary from User Handles

In this section, the methodology for classification of tweets and for generating summary from user handles are discussed as shown in Fig. 2. Here, the system starts by retrieving tweets for random user handles. Once the tweets are retrieved, they are subjected to a series of processing to extract information from them. Four types of processing are performed on the tweets to generate map, analysis, and summary data.

First, tweets are processed to identify if any location is present in the tweets or not. If the location is present, it is identified and saved to a list. In the end, the system identifies a list of locations that are present in the tweets which may contain duplicates. This list is then used to calculate the frequency of the locations present in the tweets, and then, the location with maximum frequency is identified and saved in the analysis data. The latitude and longitude for maximum frequency location are also identified, and this is saved as map data.

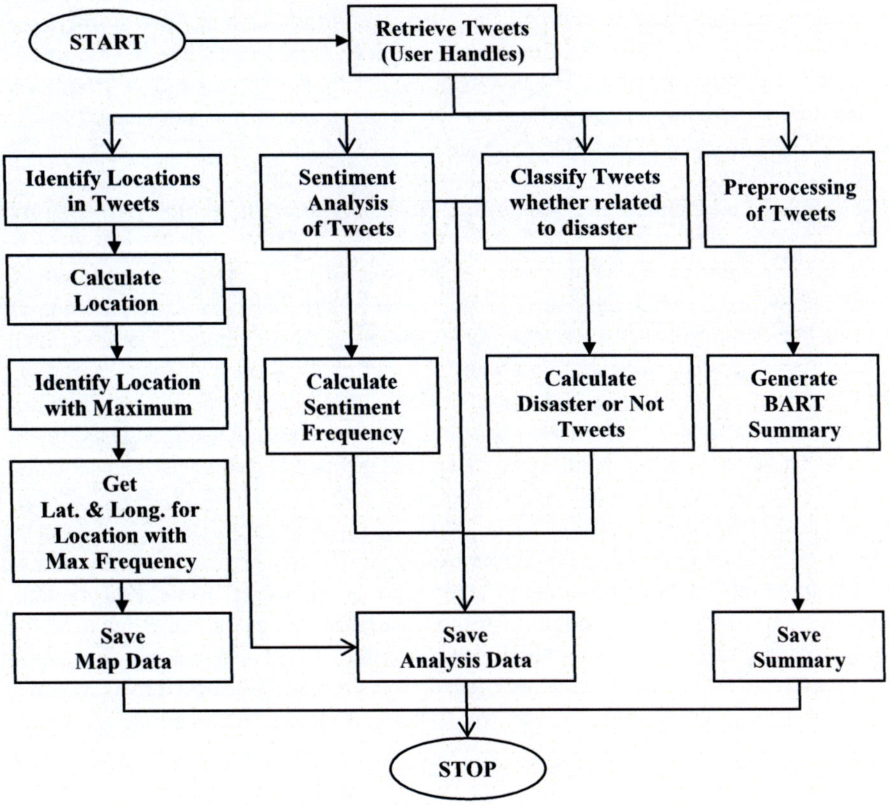

Fig. 2 Generation of summary from user handles

Second, sentiment analysis is performed to identify the sentiment of tweets and classify them as positive, neutral, or negative. A Python library 'TextBlob' is used for this task. TextBlob is a Python-based NLP library which is used for sentiment analysis. TextBlob provides two wrapper classes for sentiment analysis. One is based on the pattern library and the other is based on the Naive Bayes classifier. In this project, the default wrapper, i.e., the one based on the pattern library is used for sentiment analysis. The pattern library uses the movie reviews dataset for training the model to perform sentiment analysis. Once sentiment analysis is done, the frequency is also calculated and saved as in the analysis data.

Third, tweets are processed to classify them as 'related to disaster or not'. This is accomplished through the use of machine learning. Support Vector Machine (SVM) is a type of machine learning algorithm that can be used for classification. Scikit-learn Python library provides a wrapper for SVM with three classes that are SVC, NuSVC, and LinearSVC. Out of the three, NuSVC is used to train the model for the classification of disaster tweets with a human labeled dataset. The NuSVC model uses a radial basis function (RBF) kernel with a polynomial degree of 3. The RBF kernel uses normal curves around the data points and sums these so that the decision boundary can be defined by a type of topology condition such as curves where the sum is above a value of 0.5. The trained NuSVC model classifies the tweets as 'disaster' and 'not a disaster'. The results in the list are then processed for frequency calculation of tweets that are related to the disaster and that is not related to the disaster. This is saved in the analysis data.

Fourth, the tweets are processed to remove hashtags, URLs, and tagged accounts in the tweets, and the resulting tweets are used as inputs for the BART algorithm for the generation of summary. Two types of summaries are generated for tweets from random user handles: one which contains the duplicate tweets and another where duplicate tweets are removed from the input. The BART method is based on today's state-of-the-art approach, the transformer architecture. BART stands for bidirectional autoregressive transformer that combines bidirectional encoder and autoregressive decoder into a sequence-to-sequence transformer model. Its pre-training task includes token masking, token deletion, text infilling, sentence permutation, and document orientation. Once this pre-training is complete, the BART model is then adjusted for the text summarization task. In this, the input text is passed through the bidirectional encoder which scans the text from left-to-right and right-to-left. The output of this encoder is then passed to the autoregressive decoder which predicts the output based on the input text and the output tokens that are predicted so far. For practical purposes, this model is distilled into various smaller models which are then used for summarization tasks. One such model is the 'distilbart-cnn-12-6' which is fine-tuned with the CNN/Daily Mail dataset and is used. The summary generated by this model is saved in the summary data.

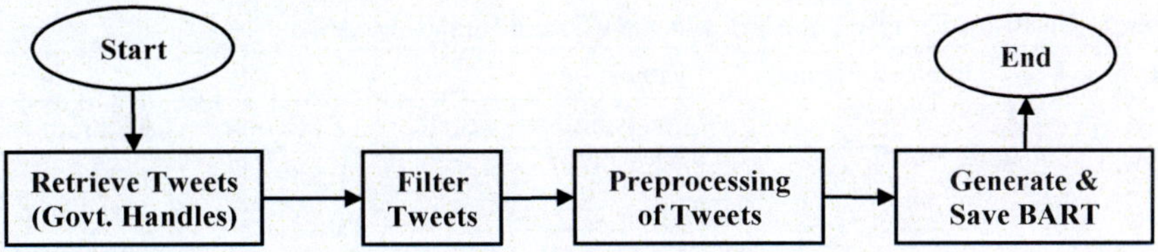

Fig. 3 Processing of tweets for government handles

3.2 Processing of Tweets from Government Handles

This section explains the methodology for generating the summary from disaster management authorities and government handle tweets such as NDRF, IMD, NDMA and. As shown in Fig. 3, the system starts by retrieving tweets from government handles. Once the tweets are retrieved, they are subjected to a series of processing to clean the tweets.

The processing includes removing hashtags, URLs, and tagged accounts from the tweet content. Then, the resulting tweets are used as inputs for the BART algorithm for the generation of summary. Two types of summaries are generated for tweets from government handles: one which contains the duplicate tweets and another where duplicate tweets are removed from the input. The summary generated by the BART model is saved in the summary data.

4 Results and Discussion

The results are discussed for disaster classification using the SVM-based classifiers and the summary generated using the BART models.

4.1 Disaster Classification Result

The dataset that is used for classification consist of 7613 tweets and includes 2 characteristics, viz. tweet and target. These tweets are classified into target values 0 or 1 such that 0 means the tweet is not related to disaster and 1 means the tweet is related to disaster. The 80% (6090) of this dataset is used to train three SVM-based classification models, viz. SVC, LinearSVC, and NuSVC, while 20% (1523) of the dataset is used for testing purpose. Table 2 shows the classification results of these 3 models. It includes calculating precision, recall, F1 score, and accuracy of the two classified values for these models.

Terminology for Classification Result:

Table 2 Classification result of SVC, LinearSVC, and NuSVC models

Model	0: Not a disaster 1: Disaster	Values			
		Precision	Recall	F1 Score	Accuracy
SVC	0	0.78	0.93	0.85	0.81
	1	0.87	0.63	0.73	
LinearSVC	0	0.80	0.87	0.83	0.80
	1	0.79	0.70	0.75	
NuSVC	0	0.79	0.92	0.85	0.82
	1	0.86	0.67	0.75	

- Accuracy: Accuracy is the most intuitive performance measure which is a ratio of correctly predicted observation and total observations. We can assume that our model is best if we have high accuracy. It is calculated as (TP + TN)/(TP + FP + FN + TN).
- Precision: Precision is the ratio of TP and (FP + TP). Here, FP is the number of false positives, and TP is the number of true positives. It is the ability of the classifier not to label a sample as positive that is negative.
- Recall: Recall is the ratio of TP and (TP + FN). Here, TP is the number of true positives, and FN is the number of false negatives. The classifier can identify all the positive samples.
- F1-Score: The harmonic mean between precision and recall is F1-Score.

Confusion Matrix

The confusion matrix is used to show the performance of the three classification models. It describes the model result over the testing dataset used and is used to compare actual values with the predicted values of the classification model. Table 3 shows the confusion matrix of three SVM-based classification models.

The acronyms in Table 3 are as follows:

- D: Disaster
- ND: Not a disaster
- AV: Actual values
- PV: Predicted values

Table 3 Confusion matrix for SVC, NuSVC, and LinearSVC models

Confusion matrix						
AV	PV					
	SVC		NuSVC		LinearSVC	
	N.D	D	N.D	D	N.D	D
N.D	826 (TP)	60 (FN)	818 (TP)	68 (FN)	768 (TP)	118 (FN)
D	234 (FP)	403 (TN)	213 (FP)	424 (TN)	118 (FP)	449 (TN)

Terminology for Confusion Matrix:

- True Positives (TP): These are cases in which the model predicted the tweet to be 'Not a Disaster', and it is actually 'Not a Disaster'.
- True Negatives (TN): These are cases in which the model predicted the tweet to be 'Disaster', and it is actually 'Disaster'.
- False Positives (FP): These are cases in which the model predicted the tweet to be 'Not a Disaster', but it is actually 'Disaster'.
- False Negatives (FN): These are cases in which the model predicted the tweet to be a 'Disaster', but it is actually 'Not a Disaster'.

4.2 Results for Summary Generation

For evaluation of results of summary generation, the following four different disasters were considered namely 'Cyclone Tauktae', 'Delhi Earthquake', 'Cyclone Nivar', and 'Assam Earthquake'. For each disaster, summary was generated using three different BART-based models, namely 'distilbart-cnn-12-6', 'distilbart-xsum-12-6', and 'bart-large-cnn (baseline)'. The summaries generated by system for 'Cyclone Nivar' using three different BART-based models are shown in Table 4.

In order to evaluate the results for the summaries generated using the three BART-based models, statistical analysis and human evaluation were performed. Four persons generated the summary manually from the tweets relevant to that particular disaster. Thereafter for statistical analysis, comparison of manually generated summary and system generated summary was carried out by calculating F1 score for each disaster. For calculation of F1-score, first precision and recall were calculated as shown in Eqs. (1) and (2)

$$\text{Precision} = \frac{\text{common statements in system and manually generated summary}}{\text{statements in system generated summary}} \quad (1)$$

$$\text{Recall} = \frac{\text{common statements in system and manually generated summary}}{\text{statements in manually generated summary}} \quad (2)$$

Table 4 Sample data for summary generated by different BART-based models for cyclone Nivar

Nivar cyclone	Summary generated by system
distilbart-cnn-12-6	Cyclone center to cross near Puducherry Coast in about 3 h. wind speed of 120–130 kmph. NDRF on ALERT. Teams already pre-deployed in Karaikal, Cuddalore, and Puducherry
distilbart-xsum-12-6	India's National Disaster Response Force (NDRF) is on alert for cyclone Nivar
bart-large-cnn (baseline)	Cyclone center to cross near Puducherry Coast. Teams already pre-deployed on alert-standby teams already on alert. The warning was issued by IMD

Table 5 F1-score of 3 different BART-based models

Disaster	Models		
	distilbart-cnn-12-6	distilbart-xsum-12-6	bart-large-cnn (baseline)
Cyclone Tauktae	0.82	0.33	0.66
Delhi Earthquake	0.79	0.28	0.60
Cyclone Nivar	0.80	0.29	0.63
Assam Earthquake	0.84	0.32	0.65
Average F1-score	*0.81*	*0.31*	*0.64*

Using the precision and recall values calculated for summary of each disaster, the corresponding F1-score was calculated as shown in Eq. (3)

$$\text{F1} - \text{score} = \frac{2 \times \text{precision} \times \text{recall}}{\text{precision} + \text{recall}} \tag{3}$$

The F1-score values computed for each disaster using the three BART-based models are shown in Table 5. From the average F1-score computed for each BART-based model, it was observed that the average F1-score for 'distilbart-cnn-12-6' was the highest. So, based on statistical analysis, we found performance of 'distilbart-cnn-12-6' to be better than other two BART models.

We also carried out human evaluation for the summaries generated by system for each disaster. For human evaluation, five human volunteers independently observed the summaries of four different disasters for three different BART-based models. Each of the summary generated was then evaluated out of 10, and the average was calculated for each BART-based model as shown in Table 6. Further in Table 7, the human evaluation scores of four different disaster summaries for three different BART-based models are given. From the results of human evaluation, we observed that the average score of summaries for 'distilbart-cnn-12-6' was higher than the other two models. Thus, from the results of statistical analysis and human evaluation, performance of 'distilbart-cnn-12-6' model was found to be better than other 2 BART models. Thus, to generate summary, we have used 'distilbart-cnn-12-6' model in our system.

Using 'distilbart-cnn-12-6' model, four types of summaries were generated by our system:

Table 6 Human evaluation scores for three different BART-based models for Cyclone Nivar summary

Nivar Cyclone summary	Human 1	Human 2	Human 3	Human 4	Human 5	Avg
distilbart-cnn-12-6	9	8	9	7	8	*8.20*
distilbart-xsum-12-6	2	3	3	4	2	*2.80*
bart-large-cnn (baseline)	7	7	6	8	7	*7.00*

Table 7 Human evaluation scores of different disaster summaries for three different BART-based models with average

Disaster	distilbart-cnn-12-6	distilbart-xsum-12-6	bart-large-cnn (baseline)
Cyclone Tauktae	8.90	2.90	7.30
Delhi Earthquake	8.60	3.60	6.80
Cyclone Nivar	8.20	2.80	7.00
Assam Earthquake	7.70	3.20	6.60
Average score	*8.35*	*3.13*	*6.93*

- Summary generated from government handles with duplicate tweets
- Summary generated from government handles without duplicate tweets
- Summary generated from random user handles with duplicate tweets
- Summary generated from random user handles without duplicate tweets

Sample of four different types of summaries generated for cyclone Nivar is shown in Table 8. From the results of all the summaries generated for four different disasters, we observed that the summaries generated without duplicate tweets were more informative than the summaries generated using duplicate tweets. Thus, for both random user handles and government handles, the summaries generated without duplicate tweets were considered.

Table 8 Four types of summaries generated for cyclone Nivar

Summary type	System generated summary
Government handles (With duplicates tweets)	Cyclone NIVAR to hit Tamilnadu with a very severe cyclonic storm during late evening of November 25. NDRF has 30 teams committed
Government handles (Without duplicates tweets)	Cyclone NIVAR to cross Tamilnadu coasts during late evening of November 25, 2020. Wind speed of 120–130 kmph. NDRF on ALERT:12 teams already pre-deployed, 18 on alert-standby, and 18 more on alert. Six teams reached Cuddalore, two in Puducherry, one in Karaikal, and one in Mamallapuram after cyclone warning issued by NDRF
User handles (With duplicates tweets)	The deep depression in the Bay of Bengal is likely to intensify further into a cyclonic storm by late Tuesday night. IMD issues red color coded warning for Southern TN and Kerala
User handles (Without duplicates tweets)	Cyclone Nivar is currently 50 km southwest of Chennai and is moving northwestward. Tamil Nadu may witness isolated extremely heavy rainfall on November 25, 2020 due to cyclone. The cyclone is expected to make a landfall tomorrow evening between Karaikal and Mamallapuram as an intense cyclonic storm of about 120 kmph is about to gust off

Table 9 Authenticity check performed for cyclone Nivar summary

Fact to be checked	Information in summary generated from government handles (without duplicate Tweets)	Information in summary generated from random user handles (without duplicate tweets)	Authentic (Yes/No)
Place	Tamilnadu, Cuddalore, Puducherry, Karaikal, Mamallapuram	Tamilnadu, Karaikal, Mamallapuram	Yes
Date	November 25, 2020	November 25, 2020	Yes
Wind speed	120–130 kmph	120 kmph	Yes

Further, to authenticate the summary information generated from random user handles, this summary information was compared with similar information present in summary generated from government handles. In case of discrepancy in the information, the information in summary generated from government handles was considered as authentic information. A sample of authenticity check performed for 'Cyclone Nivar Summary' is demonstrated in Table 9.The challenge we faced over here was, if there was no common information present in summary generated from random user handles and in summary generated from government handles, then the authenticity check could not be performed.

5 Conclusion and Future Work

In this paper, we have discussed the design and implementation of a system to generate two types of summaries related to a disaster. One summary is generated from random user handles, and another summary is generated from disaster management authority and government handles. For identifying disaster-related tweets, we have compared the classification results of SVC, Linear SVC, and NuSVC models. From the comparison, NuSVC model is found to give better results. Further for generating summary, we have considered three different BART-based models, viz, 'distilbart-cnn-12-6', 'distilbart-xsum-12-6', and 'bart-large-cnn' models. After comparing the results of these three models, we have found distilbart-cnn-12-6 to yield better results. So, we have used 'distilbart-cnn-12-6' model to generate summary from random user handles and government handles. Further, the authenticity of the information in the summary generated from random user handles is checked by verifying it against the information in the summary generated from government handles. The system can be extended further to provide multi-lingual support where the summaries can be generated in different regional languages of India. Further, the system can be expanded to be used for other countries.

References

1. Aziz K, Zaidouni D, Bellafkih M (2019) Social network analytics: natural disaster analysis through twitter. In: 2019 third international conference on intelligent computing in data sciences (ICDS), pp 1–7. https://doi.org/10.1109/ICDS47004.2019.8942337
2. Madichetty S, Sridevi M (2019) Detecting informative tweets during disaster using deep neural networks. In: 2019 11th international conference on communication systems networks (COMSNETS), pp 709–713. https://doi.org/10.1109/COMSNETS.2019.8711095
3. Rudra K, Sharma A, Ganguly N, Ghosh S (2018) Characterizing and countering communal microblogs during disaster events. IEEE Trans Comput Soc Syst 5(2):403–417. https://doi.org/10.1109/TCSS.2018.2802942
4. Shah B, Agarwal V, Dubey U, Correia S (2018) Twitter analysis for disaster management. In: 2018 fourth international conference on computing communication control and automation (ICCUBEA), pp 1–4. https://doi.org/10.1109/ICCUBEA.2018.8697382
5. Zou Q (2019) A prototype system using location-based twitter data for disaster management. In: IGARSS 2019—2019 IEEE international geoscience and remote sensing symposium, pp 9514–9517. https://doi.org/10.1109/IGARSS.2019.8898545
6. Roy S, Mishra S, Matam R (2020) Classification and summarization for informative tweets. In: 2020 IEEE international students' conference on electrical, electronics and computer science (SCEECS), pp 1–4. https://doi.org/10.1109/SCEECS48394.2020.128
7. Goyal P, Kaushik P, Gupta P, Vashisth D, Agarwal S, Goyal N (2020) Multilevel event detection, storyline generation, and summarization for tweet streams. IEEE Trans Comput Soc Syst 7(1):8–23. https://doi.org/10.1109/TCSS.2019.2954116
8. Lewis M, Liu Y, Goyal N, Ghazvininejad M, Mohamed A, Levy O, Stoyanov V, Zettlemoyer L (2019) BART: denoising sequence-to-sequence pre-training for natural language generation, translation, and comprehension. arXiv: 1910.13461 [cs.CL]
9. Masum AKM, Abujar S, Talukder MAI, Rabby AKMSA, Hossain SA (2019) Abstractive method of text summarization with sequence to sequence RNNs. In: 2019 10th international conference on computing, communication and networking technologies (ICCCNT), pp 1–5. https://doi.org/10.1109/ICCCNT45670.2019.8944620
10. Rudra K, Goyal P, Ganguly N, Imran M, Mitra P (2019) Summarizing situational tweets in crisis scenarios: an extractive-abstractive approach. IEEE Trans Comput Soc Syst 6(5):981–993. https://doi.org/10.1109/TCSS.2019.2937899
11. Rudra K, Ganguly N, Goyal P, Ghosh S (2018) Extracting and summarizing situational information from the twitter social media during disasters. ACM Trans. Web 12, 3, Article 17 (July 2018), 35 pages. https://doi.org/10.1145/3178541
12. Sarkar K (2018) Automatic text summarization using intenal and extemal information. In: 2018 fifth international conference on emerging applications of information technology (EAIT), pp 1–4https://doi.org/10.1109/EAIT.2018.8470412
13. Shleifer S, Rush AM (2020) Pre-trained summarization distillation. arXiv: 2010.13002 [cs.CL]

Analysing Effectiveness of ML Algorithms Used to Predict Diabetes Mellitus

Archit Sharma, Ashwani Raj, and Ramkumar Jayaraman

Abstract Diabetes mellitus is disease in disguise but is one of the deadliest diseases of all time. Its chronic nature is evident during the COVID-19 pandemic situation when the fatality rate was the most in the diabetes patient. India is home to the 77 million diabetes patient as hence called as the diabetes capital of the world. Diabetes is not a one-time disease but depends on various factors like lifestyle, age, high blood pressure, gene of ancestors, etc. Diabetes cannot be prevented but can be controlled when it is predicted at proper time. Thus, a proper prediction paradigm is the need of the hour. In our work, we will be comparing at least 10 machine learning classification algorithms like Zero-R, decision tree, naïve Bayes, random forest, ripper, gradient boosting, K-nearest neighbours, support vector machine, logistic regression, AdaBoost and artificial neural network too. We will be using PIMA Indian datasets which is one of the most widely datasets to predict diabetes and very simple. Hence, we did not see the use of feature selection to reduce the dimensionality. A few measurement metrics are compared for the algorithms used to predict. Our result shows the highest prediction accuracy of 80% with logistic regression classifier.

Keywords Machine learning · PIMA Indian Dataset · Diabetes prediction

1 Introduction

According to WHO, diabetes is a dreadful disease which occurs when enough insulin is not secreted by pancreas or the insulin is not effectively used by the body [1, 2]. Type 1 diabetes occurs during childhood only because of insulin deficiency. Diabetes

A. Sharma · A. Raj · R. Jayaraman (✉)
Department of Computing Technologies, SRM Institute of Science and Technology, Kattankulathur, Chennai, India

A. Sharma
e-mail: an4950@srmist.edu.in

A. Raj
e-mail: ar8293@srmist.edu.in

R. P. Mahapatra et al. (eds.), *Proceedings of International Conference on Recent Trends in Computing*, Lecture Notes in Networks and Systems 341,
https://doi.org/10.1007/978-981-16-7118-0_57

of type 2 is more common, and it occurs due to external factors mostly during the age of more than 30 [3]. In this case, the body cannot use the insulin effectively. The major culprit in this case is heavy weight and lazy life. Gestational diabetes is hyper glycaemia which can be seen during pregnancy [4]. It causes great health to the mother and the child. Diabetes is a big challenge in India with approximately 8.7% population of age 20–70 suffering from diabetes. The rising cases of diabetes is driven by a mixture of factors—fast urbanization, detrimental diets, inactive lifestyles, use of tobacco, and increasing life expectancy [5, 6]. Obesity is the greatest risk factor in rise of diabetes. Diabetes can be prevented to a great extent only by following healthy diet and having physical activity on a daily basis. During the coronavirus pandemic, diabetes patients are the most feared patients because they are much vulnerable to the deaths caused by the virus. Diabetes patients are more prone to the heart complications too. With the rapid change of lifestyle, diabetes is becoming common day by day [7, 8].

It is said that "if you can predict you can solve". Here comes machine learning to the rescue. The main objective of our project is to use various machine learning classification and ensemble methods to build a model to predict diabetes mellitus in an efficient way [9]. The most important thing in the prediction is the availability of data in good amount [10]. In our project, we will be using PIMA Indian Dataset which is sufficient for the classification and ensemble methods but will fail to predict in neural networks model due to insufficiency of data. We will introduce and discuss various methods and also evaluate with various measurement metrics which will be discussed in details. Apart from this, the methods will be compared in terms of the metrics. All the comparisons will be plotted with the help of tables and graphs. Finally, we will be discussing limitations and future aspects of machine learning models.

2 Literature Survey

Geetha and Prasad [11] proposed a theory that focused on creating hybrid models that is easy for doctors to treat diabetic patients. They have used only two ML algorithms, random forest and naïve Bayes. Performance metrics used for classification is confusion metrics. They compare the performance results of both algorithms. Shetty et al. [12] proposed a diabetes prediction system which is intelligent and gives analysis of diabetes using data mining [13]. In this system, they are using algorithms like KNN (K-nearest neighbour) and naïve Bayes to apply on diabetes patient's datasets and analyse them by taking some or all attributes of the dataset and hence predict the disease [14, 15]. Sisodia [16] designed a model which can predict the disease with maximum accuracy. They have used three classification algorithms namely decision tree, SVM and naïve Bayes. The tools used is WEKA and classification is done by splitting data into 70% of dataset for training. They verified the results using ROC curve. Zeki et al. [17] proposed a rule-based control statement system. They have used three modules for three stages, i.e. block diagram, decision tables and Mockler

Charts. After consideration of many factors and variables, this system provides a diagnosis of diabetics [18, 19].

Kumar et al. [20] implemented a system using Hadoop and MapReduce technique for processing and analysis of diabetes data. This model predicts the type of diabetes too. Their system is Hadoop based and provides a systematic way to treat the patients with availability and affordability. Nnamoko et al. [21] used five widely used classifiers that are employed for the ensembles and also aggregate the outputs with the help of meta-classifiers. The results are extracted and compared with other studies that is using same datasets. It is shown that with the help of the proposed method, diabetes can be predicted at initial phase with greater accuracy.

3 Problem Formulation

The existing system consists of the models that are using very few algorithms and hence they have less comparisons. For assisting health specialists, each machine learning algorithm presents different support systems. Accuracy is the best way to recognize the effectiveness of the decision support system. Therefore, our objective is to build a decision support system which can diagnose and predict the diabetes with higher accuracy and precision. The parameters taken into considerations in the existing system is very less and the use of evaluation metrics is minimal. Other disadvantage is that the systems ought to be connected very well to the related work. There is a dearth of an automated system in the existing system.

4 Proposed Work

The sole purpose of the project is to analyse the effectiveness of the machine learning algorithms used for predicting the occurrence of diabetes mellitus in a person. As we already understand the vulnerability of this disease, it has become the need of the hour to predict this early in a person. Especially in India, there are growing cases of diabetes. The symptoms of this disease are very common, and hence, most of the people ignore it and hence greater complications occur afterwards. In this project, we will be able to predict the diabetes using various classification-based algorithms and neural networks and also compare their effectiveness using various parameters in Fig. 1.

We are increasing the domain of the classification algorithms as well as the parameters used to evaluate the effectiveness of the algorithms. Most of the previous researches are based on few classification algorithms like decision tree, random forest and few of them uses the neural networks in the case of big datasets. But in this project, we will be using total of 11 classification techniques. We will also evaluate the results predicted by artificial neural networks. We also calculate the prediction results in various performance metrics.

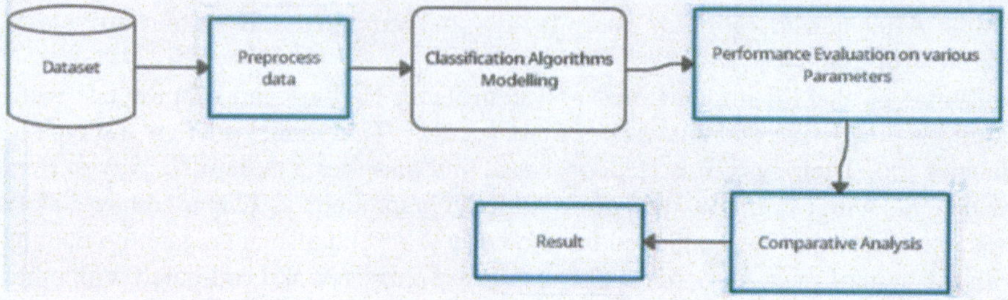

Fig. 1 Processes used in predicting diabetes

5 Methodology

5.1 Acquiring the Dataset

The dataset which we are using for prediction is PIMA Indian Dataset. It can be found in National Institute for Diabetes. There are 768 rows and 9 columns. Some of the input parameters used for attributes are BMI, Age, Blood Pressure, etc. There is 1 output parameter which gives the outcome.

The dataset which is in csv format is imported using panda library to the Jupyter Notebook. We are also using some other libraries like Numpy, Mathplotlib, etc for other purposes.

5.2 Pre-processing the Data

Data pre-processing is very important for datasets. Generally, data in health sectors contains lot of missing values and other complications. Data pre-processing helps in improving the quality and effectiveness of data. This process is essential for accurate result and successful prediction. Now, we need to check whether we can make some changes to the dataset. We have to do operations such as cleaning of data and initialization of variables. The dataset provides the information of patients with diabetes and patients who do not have.

- Diabetic people count: 268
- Non-diabetic people count: 500

In this way, there are 35% of patients who are afflicted with diabetes.

5.3 Algorithms Used

Zero-R: It is the first algorithm that was developed for classification. It is one of the simplest algorithms as it only performs the prediction of the majority class. It determines the standard performance and sets it for other classification methods. The confusion matrix of Zero-R for the diabetes prediction are as follows:

$N = 231$	Predicted NO:	Predicted YES:
Predicted No:	TN = 144	FP = 0
Predicted YES:	FN = 87	TP = 0

Naïve Bayes: Naïve Bayes classifier gives us the chance of the sample which we chose belongs to a particular category or not. Each character of a particular sample gives its independent offering and outputs the classification of that sample. This algorithm is well known for its ability to out-perform even highly sophisticated methods. The confusion matrix of Bernoulli naïve Bayes for the diabetes prediction are as follows:

$N = 231$	Predicted NO:	Predicted YES:
Predicted No:	TN = 130	FP = 21
Predicted YES:	FN = 38	TP = 36

The confusion matrix of Gaussian naïve Bayes for the diabetes prediction are as follows:

$N = 231$	Predicted NO:	Predicted YES:
Predicted No:	TN = 127	FP = 24
Predicted YES:	FN = 27	TP = 47

Decision Tree: This algorithm works very well for complex data. In this algorithm, we split the population into multiple homogenous sets. This is done on the basis of most significant attributes to make distinct groups. The confusion matrix of decision tree for the diabetes prediction are as follows:

$N = 231$	Predicted NO:	Predicted YES:
Predicted No:	TN = 117	FP = 27
Predicted YES:	FN = 42	TP = 45

Random Forest: Random forest is an ensemble of collection of decision trees. Random forests are build using a method called bagging in which decision trees are used as parallel estimators. Each tree gives a classification (like a vote) and then the forest chooses the classification which has the most votes. The confusion matrix of random forest + for the diabetes prediction are as follows:

$N = 231$	Predicted NO:	Predicted YES:
Predicted No:	TN = 132	FP = 12
Predicted YES:	FN = 65	TP = 22

Support Vector Machine: It is mostly used for classification tasks but can also be used for regression tasks. In simple terms, *I* could explain. SVM distinguishes classes by drawing a decision boundary. This decision boundary will split the data in two differently classified groups of data. The confusion matrix of support vector machine for the diabetes prediction are as follows:

$N = 231$	Predicted NO:	Predicted YES:
Predicted No:	TN = 138	FP = 13
Predicted YES:	FN = 34	TP = 40

KNN: It is used to classify the data. With the help of this algorithm, we can predict the category to which the input data belongs. This algorithm can also be used for regression, but mainly it is used for classification. It is called Lazy learning algorithm because to train the algorithms we need the input data, only then the algorithm can take any decision. The confusion matrix of KNN for the diabetes prediction are as follows:

$N = 231$	Predicted NO:	Predicted YES:
Predicted No:	TN = 122	FP = 29
Predicted YES:	FN = 34	TP = 40

AdaBoost: AdaBoost is the type of boosting technique in machine learning which use ensemble methods which is better than classifiers for classification. It is made by combining many classifiers which gives better accuracy than those classifiers. It uses the ensemble methods in repetition and makes a model with great accuracy. The confusion matrix of AdaBoost for the diabetes prediction are as follows:

$N = 231$	Predicted NO:	Predicted YES:
Predicted No:	TN = 120	FP = 24
Predicted YES:	FN = 40	TP = 47

The Ripper Algorithm: RIPPER is the algorithm which makes a model according to a rule which minimizes the errors to a great extent. The error in this case is the mispredictions which it shows frequently. The confusion matrix of the Ripper Algorithm for the diabetes prediction are as follows:

$N = 231$	Predicted NO:	Predicted YES:
Predicted No:	TN = 128	FP = 16
Predicted YES:	FN = 54	TP = 33

Logistic Regression: It is a classification algorithm, and it is not used in regression. Most commonly for binary classification tasks. It predicts the probability of occurrence of an event by fitting into a logit function which maps the probability. It underperforms when non-linear decision boundaries. The confusion matrix of the logistic regression for the diabetes prediction are as follows:

$N = 231$	Predicted NO:	Predicted YES:
Predicted No:	TN = 138	FP = 13
Predicted YES:	FN = 31	TP = 43

Gradient Boosting: Boosting is a technique of ensemble method in which we can predict sequentially instead of independently. Gradient boosting is an example of boosting. According to Wikipedia, we can combine the weak classification methods like decision tree and make a model which is boosted to make a more accurate model and here the birth of gradient boosting took place. The confusion matrix of gradient boosting for the diabetes prediction are as follows:

$N = 231$	Predicted NO:	Predicted YES:
Predicted No:	TN = 121	FP = 23
Predicted YES:	FN = 38	TP = 49

Artificial Neural Network: An artificial neural network (ANN) is the part of artificial intelligence that imitates human brain. It learns from the input that is provided and produces a desired output. The rule on which learning model of ANN works is backpropagation. ANN has a lot of practical applications in finance, industries, governance, entertainment, etc. We will be using three layers for the implementation of neural networks, i.e. Hidden layer1, Hidden layer2, and output layer. Keras will be used to build the architecture of the ML model.

5.4 Model Building

Model building is the phase which matters the most. All the algorithms discussed above will be implemented using various steps as follows.

Step1: Required libraries and diabetes dataset are imported.

Step2: Data is pre-processed for removing missing values.

Step3: dataset is divided in the ratio of 7:3 in training dataset and testing dataset.

Step4: All the Machine Learning algorithms discussed above are selected.

Step5: Training set is passed to the different classifier and ensemble methods.
Step6: Testing set is passed to the different classifier and ensemble methods.
Step7: Results obtained by all models is compared for evaluation.
Step8: The best performance algorithms based on the evaluation results are concluded.

6 Result

In this experiment, the effectiveness of each classification-based prediction model is calculated for predicting diabetes mellitus. In the following, we have shown the experimental results of classification techniques, such as Zero-R, random forest, support vector machine, decision tree, K-nearest neighbours, adaptive boosting, gradient boosting, repeated incremental pruning to produce error reduction, and logistic regression classifiers. Along with this, we also evaluate the results predicted by artificial neural networks. We calculate the prediction results in terms of accuracy, sensitivity, specificity, and Mathews correlation coefficient and also calculate F-measure, kappa, receiver operating characteristic value, precision, recall, and errors calculated by mean absolute error and root mean squared error of all those ensemble and classification-based models.

The values which are depicted from tables and Fig. 2 shows that logistic regression has the highest accuracy among all the classifiers and ensemble methods with an accuracy score of 80%. The ROC score of logistic regression is 0.86 which is also the highest. It shows that logistic regression has the best performance results.

Figure 3 shows the variations caused in performance metrics like precision, sensitivity, $F1$ Score, Kappa Score, MCC, and MAE. It depicts how the results are varying with respect to various classifiers. Variations in MAE is the least and the variations

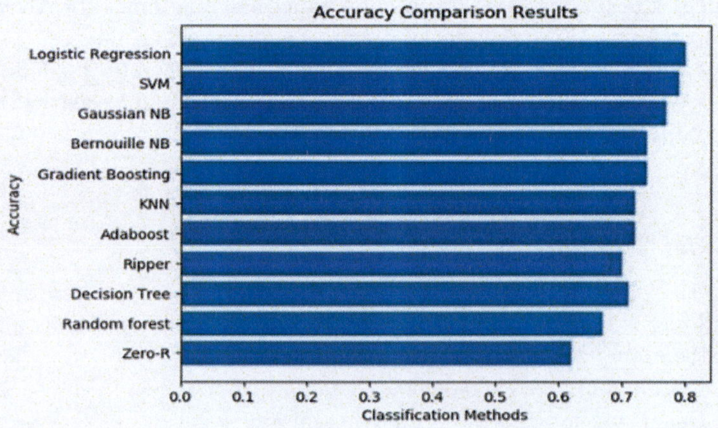

Fig. 2 Classification results based on accuracy

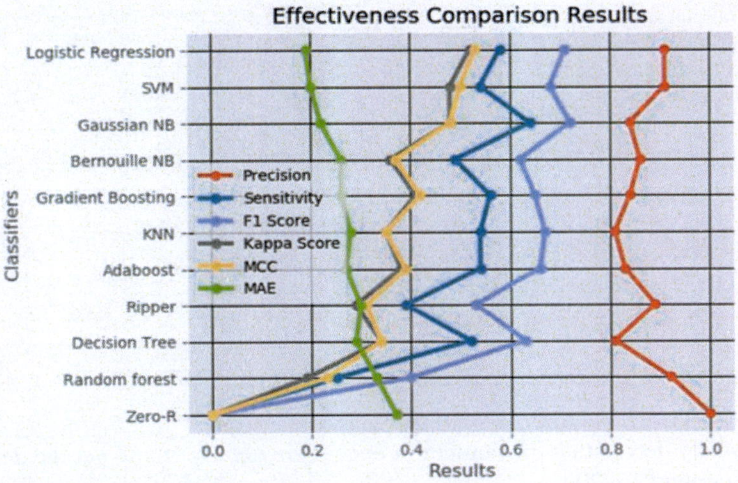

Fig. 3 Effectiveness comparison results

in $F1$ Score is the highest. Kappa Score and MCC has close variations. The values depicted by Zero-R are extreme (either 0.0 or 1.0). This shows its inefficiency.

Figure 4 shows the model performance of artificial neural network. Although the size of PIMA Indian Dataset is not sufficient to make a proper ANN model, but still accuracy is good in few epochs. Epochs refer to the iterations through the entire training dataset. It can also be understood as the number of times we feed the training set to the ANN model. In our model, we have implemented the model in 100 epochs. As we can see in Fig. 4 that the curve is increasing and the highest model accuracy is 0.76 when the number of epochs is 80.

If the average free energy is same for both valid and train dataset, then we can say that the model is not overfitting. The amount of overfitting can be seen by the gap between the valid and train dataset curves in the Fig. 5. So from the graph, we can get the value of epochs we need to run to get the best output. Looking at Fig. 5,

Fig. 4 Accuracy versus number of epochs

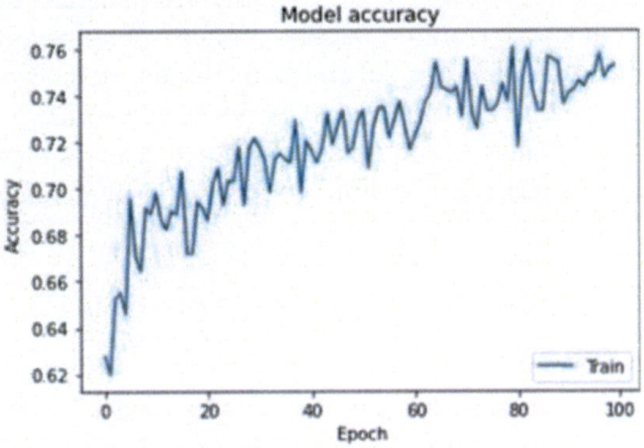

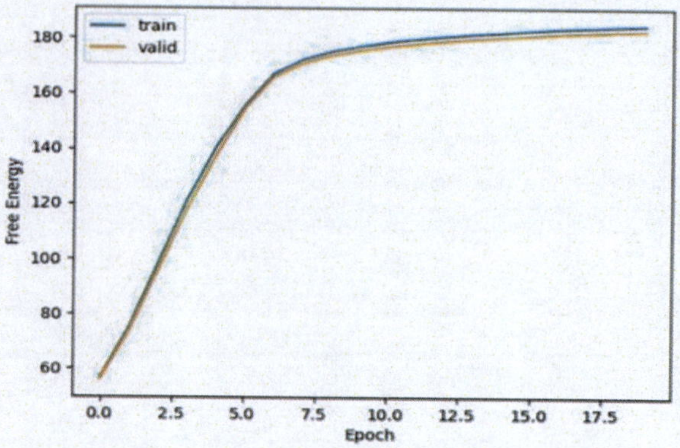

Fig. 5 This plot shows the free energy versus the epochs for training and the validation dataset

we can safely decide that the number of epochs we need to run to get the desirable output is around 15–20.

We used 20 epochs, and after the end, we got around 0.003 MAE(mean absolute error).

7 Conclusion

We have implemented a total of 11 classifiers and ensemble methods as well as artificial neural network to predict diabetes mellitus. We have also calculated various parameters to compare the performance results of all models. From all the observations, it is clear that Zero-R model is the poorest predictive model to predict diabetes, whereas logistic regression model is the best predictive model with an accuracy of 80%. Artificial neural network was also able to predict with an accuracy of 76% at most.

In the future, we are looking for a greater size of datasets and that too of Indian patients. It will give a clear perspective of predicting the diabetes. There are growing cases of diabetes in Indian people which is creating greater risks of death in COVID-19 pandemic. We are looking forward to collect datasets with more attributes where method like feature selection will be of more significance. We are also trying to find a more accurate model and also looking for the magic of deep learning once we get a large dataset with millions of rows. Diabetes prediction is the domain whose research will continue till eternity as diabetes cases is increasing in whooping numbers due to poor lifestyle. We have to com-bat it by predicting it before.

References

1. Zou Q, Qu K, Luo Y, Yin D, Ju Y, Tang H (2018) predicting diabetes mellitus with machine learning techniques. Front Genet 9:515
2. Pedregosa (2011) JMLR 12, pp 2825–2830
3. Hand DJ, Till RJ (2001) A simple generalisation of the area under the ROC curve for multiple class classification problems. Mach Learn 45(2):171–186
4. Ferri C, Hernandez-Orallo J, Modroiu R (2009) An experimental comparison of performance measures for classification. Pattern Recogn Lett 30:27–38
5. Provost F, Domingos P (2000) Well-trained PETs: Improving probability estimation trees (Section 6.2), CeDER Working Paper #IS-00-04, Stern School of Business, New York University
6. Fawcett T (2006) An introduction to ROC analysis. Pattern Recogn Lett 27(8):861–874
7. Fawcett T (2001) Using rule sets to maximize ROC performance In Data Mining, 2001. In: Proceedings IEEE international conference, pp 131–138
8. Everingham M, Van Gool L, Williams CKI, Winn J, Zisserman A (2010) The pascal visual object classes (VOC) challenge. IJCV
9. Davis J, Goadrich M (2006) The relationship between precision-recall and ROC curves. ICML
10. Flach PA, Kull M (2015) Precision-recall-gain curves: PR analysis done right. NIPS
11. Geeta G, Prasad (2020) Prediction of diabetics using machine learning. Int J Recent Technol Eng 8(5):1119–1124
12. Shetty D, Rit K, Shaikh S, Patil N (2017) Diabetes disease prediction using data mining. In: 2017 international conference on innovations in information, embedded and communication systems (ICIIECS), Coimbatore, India, pp 1–5
13. Sarwar MA, Kamal N, Hamid W, Shah MA (2018) Prediction of diabetes using machine learning algorithms in healthcare. In: 24th international conference on automation and computing (ICAC). Newcastle Upon Tyne, UK, pp 1–6
14. Mujumdar A, Vaidehi V (2019) Diabetes prediction using machine learning algorithms. Procedia Comput Sci 165:292–299
15. https://medium.com/mlreview/gradient-boosting-from-scratch-1e317ae4587d
16. Sisodia D, Sisodia D (2018) Prediction of diabetes using classification algorithms. Procedia Comput Sci 132:1578–1585
17. Zeki T, Malakooti MV, Ataeipoor Y, Tabibi ST (2012) An expert system for diabetes diagnosis. Am Acad Sch Res J 4
18. Manning CD, Raghavan P, Schütze H (2008) Introduction to information retrieval
19. Lai H, Huang H, Keshavjee K (2019) Predictive models for diabetes mellitus using machine learning techniques. BMC Endocr Disord 19:101
20. kumar N, Eswari T, Sampath P, Lavanya S (2015) Predictive methodology for diabetic data analysis in big data. Procedia Comput Sci 50:203–208
21. Nnamoko N, Hussain A, England D (2018) Predicting diabetes onset: an ensemble supervised learning approach, 2018 In: IEEE congress on evolutionary computation (CEC), Rio de Janeiro, Brazil, pp 1–7

Enhancing the Decentralized Application (Dapp) for E-commerce by Using the Ethereum Blockchain

P. Shamili and B. Muruganantham

Abstract Blockchain is the upcoming technology in creating the Dapp for different types of use cases. The transactions in the blockchain technology are transparent to all the nodes in the decentralized network. Most of the applications were built and trying to implement it with the Ethereum platform in blockchain for the security purpose. The nodes in the distributed network shares all the data to the other nodes without any modification of their data in the blockchain technology. The transactions done by the nodes are trusted through the digital signatures. In this paper, we built a decentralized web application for e-commerce by using the blockchain Ethereum platform. Our application is deployed by using the smart contracts, which are written in the solidity language and the front-end and back-end process are explained. All these data about the e-commerce details are stored into the blocks in blockchain. The Web application for online shopping will be secured and trust worthy for the nodes in the decentralized network. The other online shopping platforms like amazon are trying to deploy the blockchain for security of data and user's convenient. The issues like storage, scalability, and others can be considered for the future work.

Keywords Blockchain · Ethereum · Ethers · E-commerce · Smart contracts

1 Introduction

Decentralized applications are applications that run by using the blockchain Ethereum platform or in other words, the nodes in the peer-to-peer network or the nodes used to communicate with other nodes. The biggest thing is that the stored data are stored on a public ledger that records everything in a secure and transparent

P. Shamili (✉) · B. Muruganantham
Department of Engineering and Technology, SRM Institute of Science and Technology, Kattankulathur, Chennai, India
e-mail: shamilip@srmist.edu.in

B. Muruganantham
e-mail: muruganb@srmist.edu.in
URL: https://www.srmist.edu.in/

R. P. Mahapatra et al. (eds.), *Proceedings of International Conference on Recent Trends in Computing*, Lecture Notes in Networks and Systems 341,
https://doi.org/10.1007/978-981-16-7118-0_58

way and guarantees no one can manipulate it. The distributed database that stores all the records of the transactions among the nodes in the network are named as blockchain [1]. Each and every transaction is verified by the majority of the nodes in the network. It is the backbone for the digital crypto currency called bitcoin. One of the best examples for the blockchain is the bitcoin. This technology became popular through a paper named "Bitcoin: a peer-to-peer electronic cash system" published in the year 2008 by Satoshi Nakamoto [1]. In blockchain technology, the transactions are stored in the form of digital data and it is shared among all the nodes (users) in the distributed network. The data shared in the network are immutable. Some of the use cases in blockchain are land assets, supply chain management system, and so on.

In blockchain, the bitcoin uses the cryptographic proofs for trusting the nodes to execute their transactions. Every transaction by the nodes were trusted through the digital signatures. Here, there is no central control or the central system to keep track of other's data. Where, the nodes are connected to one another nodes in the distributed network and that the data shared here are immutable [2]. In blockchain, the data of one node is distributed among all the other nodes in the world through distributed peer-to-peer network. It is transparent to all the nodes that verify the data and are tamper proof [3, 4]. The transactions that are verified and stored in the blocks are done by miners. The main role of a miner is to solve the crypto puzzle to verify the transactions that occurs between the two nodes in the distributed network and the reward for the verification of a transaction is given to the miners. Just like the traditional centralized applications, Ethereum Dapps [3] also consists of back-end by smart contracts (the logic) and front-end by Web3 (the visual) [5]. The most significant difference is that the core of back-end is the blockchain technology.

Merkle Patricia tree The Patricia tree is a data structure and it is also known to be as the prefix tree, radix tree or the trie. Where it uses the key as the paths to share the same prefix and also it shares the same paths. The features of the Patricia tree are:

1. Fastest at finding the common preferences.
2. Simple to implement.
3. It also requires only small memory.

2 E-commerce

E-commerce is defined to be as the act of selling or buying the goods or the types of services by using the World Wide Web. The data and the amount are transferred for executing these transactions. Through Internet, the physical products are sold or purchased through the commercial transactions. Some of the solutions that blockchain brings to e-commerce are as follows: improved user experiences, optimized supply chain management, integration with the business processes, verified customer reviews, enhanced operational efficiency, dependable data security, trust and transparency in payment, personalized product offerings, fast transactions, and channel expansion. The layered architecture diagram for e-commerce in blockchain

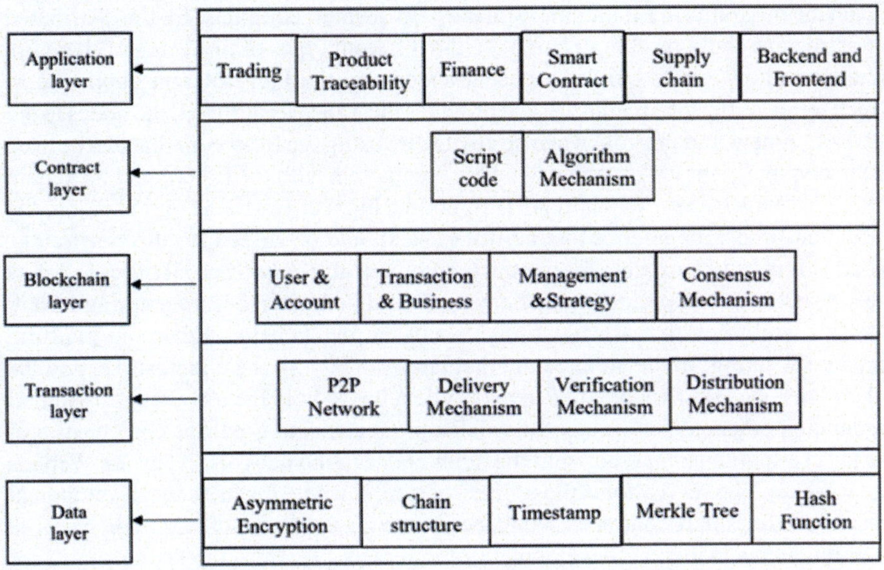

Fig. 1 Layered architecture diagram for e-commerce in blockchain technology

technology and the basic concepts that are explained below (see Fig. 1). The five layers are data layer, transaction layer, blockchain layer, contract layer, and the application layer.

- Data layer: It holds the functionalities of the hash function, merkle tree, time stamp, chain structure, and the asymmetric encryption.
- Transaction layer: In the transaction layer, data packets are transferred in the peer-to-peer network and it also follows the mechanisms of verification, distribution, and delivery of data.
- Blockchain layer: All the data are come under a mechanism of consensus and follows the management strategy, user their account and finally the transaction and business.
- Contract layer: The smart contracts are deployed in the contract layer by using the script codes and the algorithms.
- Application layer: Finally, the application layer plays a major role for front-end and back-end process for overall layers combined to perform by using the smart contract in areas like finance, supply chain, trading, etc.

3 Ethereum a Blockchain Platform

Blockchain allows developers to build many of the Dapps by using the different platforms. Here, we are going to build the decentralized application using the Ethereum platform [2]. One of the most important platforms to execute the arbitrary codes to run on the programs in Ethereum. With the help of smart contracts [6], the distributed infrastructure could be able to complete the different types of projects in Ethereum. Ethereum allows us to build the decentralized and the fault tolerant applications, which removes the intermediaries and provides the transparency to all the nodes in the network. A new tradable token can be created by using the Ethereum that can be used as the cryptocurrency. The wallets that are compatible with the Ethereum blockchain [5] are uses token, as a standard API. To build a blockchain-based organization, the nodes must me a member of that network and should be agreed for the constraints. Smart contracts will execute the contracts automatically only when all the constraints are satisfied. A clear concept of Ethereum should be known before entering into it. Now, we are going to learn few things about the blockchain—Ethereum platform and their working. In the architecture diagram (see Fig. 2), the client, server and the Ethereum nodes are explained. Where the client has to hold the e-wallet of particular amounts of ethers to sell or buy the needed products. And the front-end consists of the java script files to interact with the other nodes in the network. They are Web3.js and React.js. The server holds the process of truffle, where it acts as the development framework for smart contracts. Now, the Ethereum nodes which are main network as public network for the development of contracts [7]. Here the testnets used for connecting to the networks are deployed by using the smart contracts to the ganache (which holds the ethers) in the Ethereum platform.

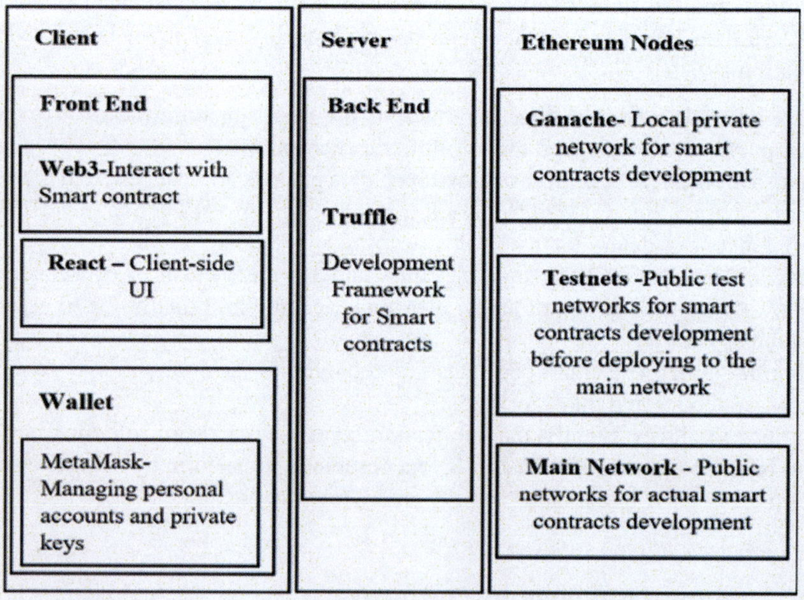

Fig. 2 Architecture and framework for Ethereum decentralized Web application

3.1 *Ethereum Virtual Machine (EVM)*

Ethereum virtual machine is provided by a platform called Ethereum [8], where the codes are executed by using the nodes in the distributed network. Here, the machine handles all the node's information like address, pertaining to balance, current gas price, and all about the block information. It is written in C++, Java, JavaScript, python, solidity, Node.js, [9] and many programming languages. By using this, anyone on the network can execute the codes that are guaranteed and are deterministic.

Computation and Turing completeness

The Ethereum virtual machine (EVM) is turing complete and it means that it can encode the data with any operation which is carried out, with infinite loops. There are two ways of loops. They are:

i The JUMP instruction that allows the program to jump back to the previous step in the code and JUMP instruction to do conditional jumping. For example, the statement like

$$X < 27; X = X * 2 \tag{1}$$

ii Contracts are allowed to call other contracts by allowing it for looping through recursion. Here comes the problem that the malicious users can essentially shut the miners and full nodes down by forcefully making them to enter into an infinite loop. It mainly arises because of the halting problem in computer science and there is no way to explain that whether or not a given program will halt or not

3.2 *Solidity Language*

For smart contract, the language used by Ethereum is the solidity language [10]. The programming language that is developed on top of the Ethereum virtual machine. The main purpose of solidity is to store all the states of the tokens, which are send and receive by the nodes. The programmers who are experienced in JavaScript, C++, and python could easily understand the solidity language also. If the solidity language is known well, then the program for the Ethereum smart contract could be written easily.

3.3 *Gas*

In Ethereum blockchain, the miners process the smart contracts and results in adding the new block to the Ethereum blockchain. The rewards given to the miners for their efforts, through the smart contracts on the Ethereum virtual machine requires some amounts of payments named as gas. The amount of gas needed to spend for the miners were created in their smart contracts. If the smart contract gets complicated, the gas fee also gets increased.

3.4 Algorithm

In blockchain, the newly created blocks will be added to the previous block, where it comes under the process of proof of work. While adding the blocks, the hash value of it would be generated and that hash value will be the value for the next block. For an example: the block 4 will be the hash value of block 3 and the block 5 will be the hash value of block 4 and so on. If suppose, any of the node tries to change or modify any of the data in the block, the hash value will be changed for it and the data will be updated. No one could modify it (i.e., data immutability).

Proof of Work (POW): If we want to add new block or modify the block, it has to be validated in the public blockchain without by validation we cannot add or modify the data in the blockchain network. When we do some processing to add a block in the blockchain, we have to go for the concept of POW [11], it will take minimum of 10 mins for any of the node to add the block in blockchain network. It will take more time to modify the data in the blockchain network. After the validation of the block, it is stored into multiple machines called nodes and also called distributed ledger database [DLD]. “Blockchain data will be stored on multiple machines” [12]. It is not a central server, and every node or machine in the blockchain network will have one copy of the data in the blockchain. Every node comes under the consensus to add a block. For that, we have the concept of consensus algorithm to add data in blockchain and also we have many consensus algorithms in blockchain.

POW consensus algorithm work mechanism

- Nodes need super computer (I e., computational power).
- 50% majority of the nodes in the blockchain network.
- Difficulty to solve the mathematical crypto puzzle.

3.5 Smart Contracts

The smart contracts are the agreements in which we follow some of the rules and restrictions while doing the transactions. The smart contract [10] that acts as a back-end storage and a logic is initialized by our distributed application [3] (see Fig. 3).

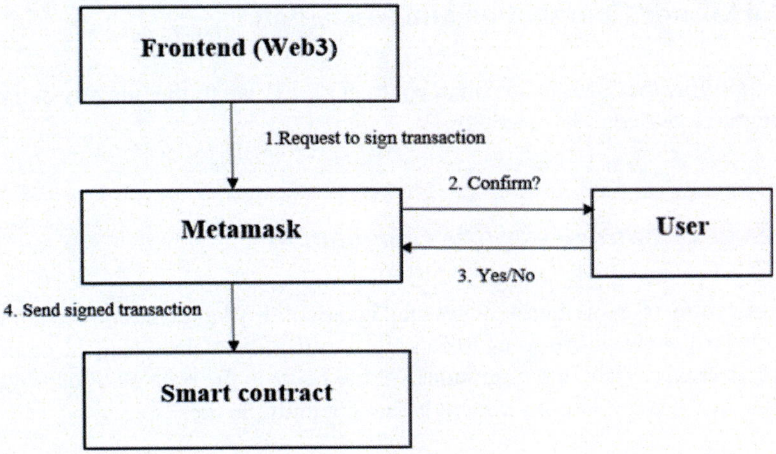

Fig. 3 Working of smart contract

3.5.1 Structure of the Directory

The truffle [13] directory structure contains the following:

1. Contract: which consists of all the source code for our smart contract.
2. Migration file: to handle the smart contract deployments, we use truffle migration and the changes will be traced.
3. Test/: it contains both the JavaScript and the solidity test files for the smart contract.
4. Truffle congif.js file: configuration file.

3.5.2 Setting up the Variables

The language used for writing the smart contracts is solidity programming language [10]. Means that the data types like integers, string, and array must be defined. The unique type in solidity is called as address. The address and the Ethereum address are stored as 20 byte values [3]. In the Ethereum blockchain, every smart contract address can send and receive the ethers from and to another addresses. Now, we are going to see how the codes should be suitable for our smart contract. Here, the pragma solidity defines the version for compiling the smart contract in the solidity language. Where the purchase defines the name of the contract in our application pragma solidity ˆ0.5.0;

Contract purchase { }

4 Back-End: Compilation and Migration

After modifying the codes in our smart contract [3, 14], we further move to compiling and migrating our contracts (see Fig. 3).

4.1 *Smart Contract—Truffle Compilation*

Here, the contracts are written and executed in the solidity language and the extension is .sol file [10]. Now we are going to compile a truffle file by providing the command as truffle compile. Then, the migrations.sol file helps in the process of deployment. If suppose we use truffle box, we will be having multiple files.

4.2 *Smart Contract—Truffle Migration*

Compiling [15] the truffle in our smart contract has done successfully. Now, we are moving to the migration of truffle in Ethereum blockchain network [13]. In Ethereum, the smart contracts are deployed by using the migrations files. The responsibility of doing this file is the task deployment and the assumptions over time to time. Our project evolves that the new migration scripts on the blockchain network [16]. As we compiled the truffle previously, now the command to run the migration is truffle migrate. The other migrations might move the data or replace it by new ones. The migration/directory: 1.initialmigration.js that handles the migration.sol file for deploying the contract [17]. To migrate, use the command truffle migrate to migrate the files.

1. Now, we have to create a new file called deploy contracts in the migration/ directory.
2. Add the following codes in the 2 deploy contracts.js file.

4.3 *Ganache*

In this paper, we used ganache [18] to deploy the contracts, develop the applications and to run the test in the Ethereum blockchain. We have already, downloaded ganache [19] (holds the default address for each ethers in 10 separate accounts) and the requisites as needed. It will generate a local blockchain running on the port of http://127.0.0.1:7545 (see Fig. 4).

The state of the block in ganache has been changed. After the computation, the block number is changed. In our side, it is changed to block 4 (see Fig. 5). Initially, the original account will be credited with 100 ethers. Now, it becomes less due to the transaction cost for the migrations.

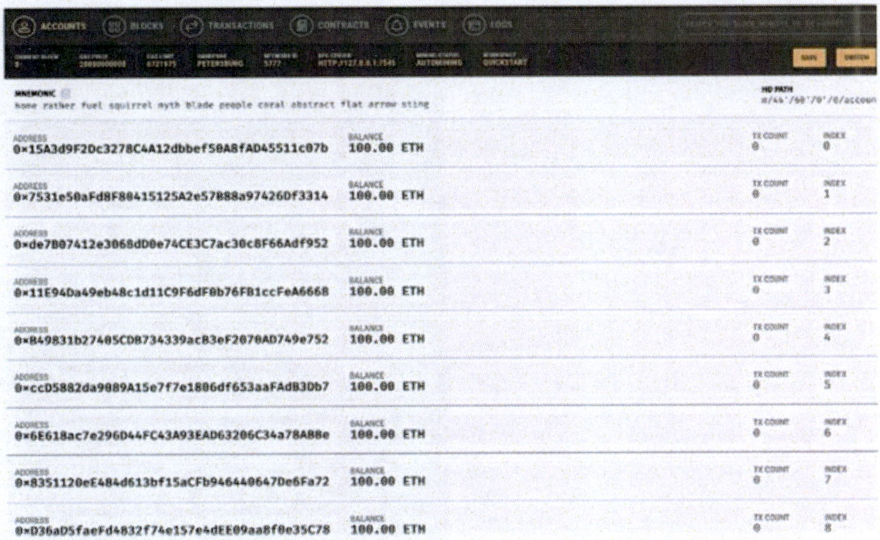

Fig. 4 Ganache after installation

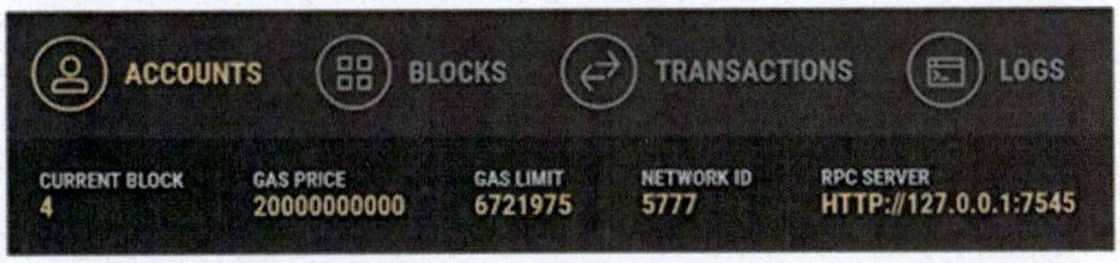

Fig. 5 Ganache after truffle migration

4.4 MetaMask

MetaMask is the simplest way to interact with the Dapp in the browser. We can install it through both the chrome extension and in firefox [6]. After the MetaMask is installed, create the user account and get the 12 seed phrase while doing the initial steps.

1. Click on the box "new network" and enter the link https://127.0.0.1:7545 and save it.
2. In ganache, each account is given with the 100 ethers. Because we used it, check that the ether will get reduced after we deployed our contract and when we run the test.

Now, we need to connect MetaMask to the blockchain through the ganache. By clicking the menu, we see the network is connected to "main network" and we have to select the custom RPC (see Fig. 6).

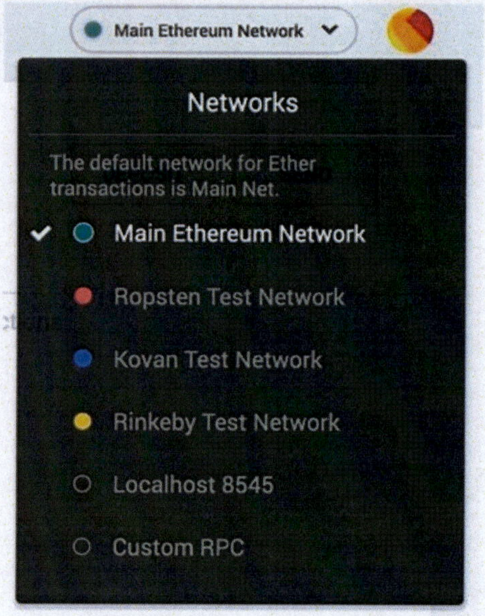

Fig. 6 Custom RPC

5 Decentralized Application in Blockchain

In this section, we built an application for the e-commerce by using the Ethereum platform for the users in the peer-to-peer network. Here, we used few of the functions for deploying the smart contract. It is as follows:

- By using the solidity programming language, the Ethereum smart contract is created.
- The tests for writing the smart contract in JavaScript.
- The smart contract is deployed to a blockchain.
- The Webpage for interaction between the user and the smart contract is created by using the Web3.JS and React.JS.

The smart contracts are being deployed by using the npm commands and few of the steps to be followed are listed and explained below.

5.1 Installing the Dependencies

We need to install few of the dependencies to run our smart contract and it is explained in the following subdivisions.

5.1.1 Ganache

It is a local development blockchain that can be used to mimic the behavior of a public blockchain. Where it allows us to deploy the smart contract, develop the applications and to run the test networks.

5.1.2 Node.JS

For deploying the smart contract, we need the node package manager or the npm, which comes along with node.js [9]. By running the npm commands, the requires files for our project will be included to the migration file, source file, and test file (truffle config.js, package lock.json, and package.json).

5.1.3 Truffle Framework

It provides a suite of the tools for development of Ethereum smart contract by using the solidity programming language.

5.1.4 MetaMask Ethereum Wallet

It is a browser extension which allows the users to connect to the blockchain. It turns our Web browser into the blockchain browser.

5.1.5 Web3.Js

Using the HTTP or IPC connection, Web3.js allows us to interact with the local or remote Ethereum nodes. It can also retrieve the user's account, all the transaction details, and also to interact with the smart contracts.

5.2 *Selling Products*

For selling the products to the customers, it allows us to list the items needed for sale. For that, a sample product with the struct is as follows, where the product name, id of the product and the address of owner who sells the product, and whether the product is purchased or not for the privacy of both the seller and receiver [20].

```
struct product {
uint id;
string name;
```

```
uint price;
address owner;
product purchased;
}
```

5.3 Buying Products

The products which are selected by the buyers has a function called the call function. If a user wants to buy the particular product, they have to submit the product id. They also have to send the Ethereum crypto currency [21] from their e-wallet to purchase the product, when the call function is used.

```
function purchaseproduct(uint id) public payable {//… }
```

5.4 Setting up the Web Page (Front-End)

The front-end is being run by the npm and the command for it is npm run start.

Here three of the main features are

i React.js to build the interface.
ii Bootstrap for creating the UI elements without writing CSS.
iii Web.js for connecting our app to the blockchain

Here, the React.js is used to manage the client-side behavior of our Web application and to store the data from the blockchain.

5.5 Deploying the Smart Contract

The smart contract is deployed and the Web application runs on a local host https://localhost:3000 and the page for e-commerce gets started running.

In this page, the product name, product price, and the details of particular purchased product gets displayed (see Fig. 7). It is done with the help of MetaMask and ethers in the e-wallet.

6 Front-End of Our Application

After a full node to connect with Ethereum main network [22], we want to pay some gas fee or transaction fee. Whenever we do transaction miner will get rewards to mine the block, before doing any transaction on the Ethereum network, we need ether

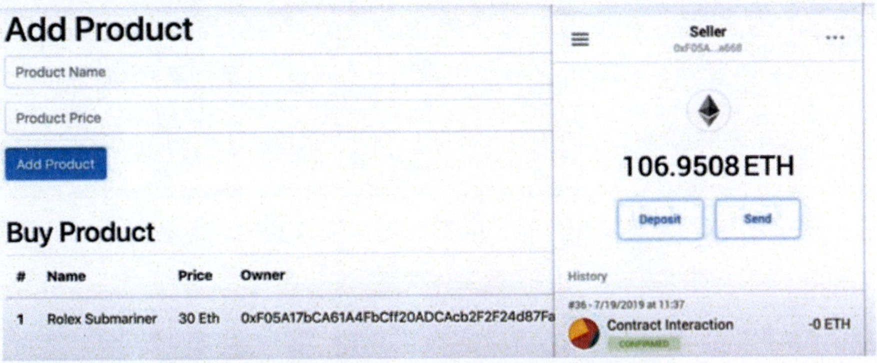

Fig. 7 Web page for deploying the smart contract by using MetaMask

(ETH). We have to buy the ethers. In this case, we are working on testing network called Rinkeby network. We have 18.75 ETH in our account to deploy transaction in Rinkey network and also, we allow to do transaction on Ethereum main network with 100 ETH. We have in Ethereum software test networks are [Ropsten, Kovan, and Rinkeby]. We have the browser, which working under the centralized system to make our browser to Dapp for that we already inject the MetaMask extension to our browser [get chrown ADDON], we built the app and deploy it with MetaMask after we have connected to main network our account address linked with truffle ganache, MetaMask which provider unique address. We can view the account details in MetaMask like QR code of address, we can also view etherscan.io [8] which is used track [23] all our activity with Ethereum blockchain (see Fig. 10).

1. When we are trying to interact with the distributed application, the popup message from the MetaMask will ask the confirmation request that is connected to our MetaMask wallet. If not, we will not be able to connect with the Dapp.
2. After confirmation, click on the Adopt button to select the pets.
3. Then, we will be directed to the MetaMask account to approve the transaction (see Fig. 8).
4. We will now see the success button after purchasing (see Fig. 9). After successfully submitted in MetaMask, in MetaMask [6] we will see the transaction [24] list (see Fig. 10)

7 Conclusion

In this paper, our main goal is to build the blockchain based decentralized application by using the Web3 library along with the ganache and MetaMask in our local host. It is to show that how the online shopping in the upcoming years will be secured and stored based upon the user's privacy concern [25]. To know the difference between each network in blockchain technology, it can also be implemented by using other

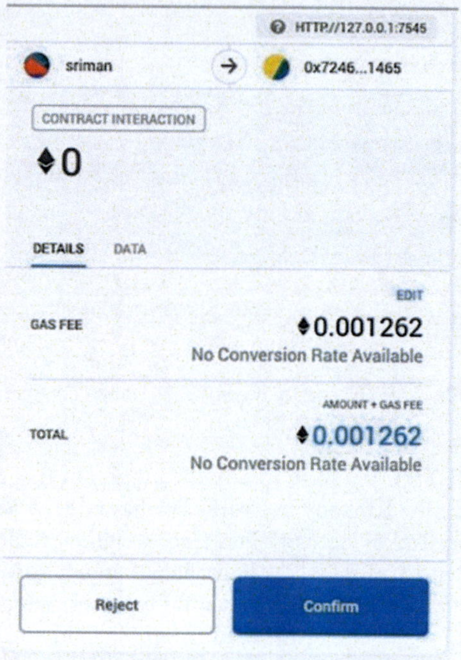

Fig. 8 Compound and uniswap

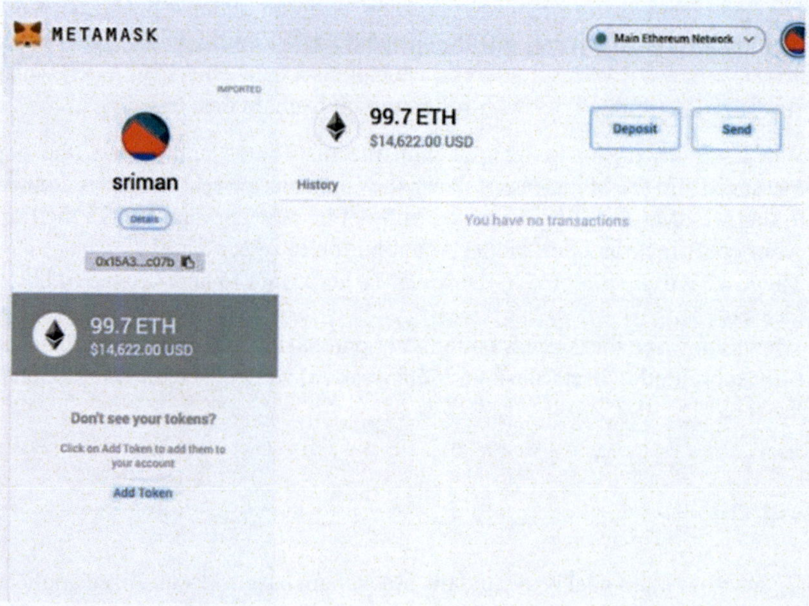

Fig. 9 Successfully purchased

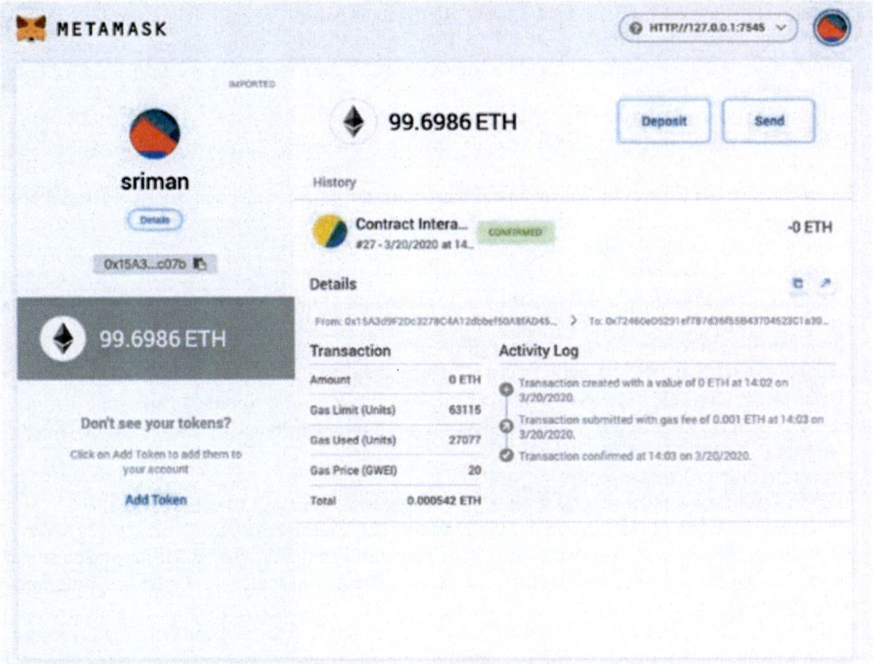

Fig. 10 MetaMask transaction history

networks like Rinkepy, Ropsten, etc. Our smart contract is made interacted with the Ethereum main network and the local host http://127.0.0.1:7454 RPC. Hereby, we conclude that, the implementation of the Dapp could done with the other test networks also for the efficient working of the nodes and the performance of the nodes with different test networks can be analyzed in future work. Also, the quantum computing [5] under blockchain is the current research for high computational power and to mine the mathematical crypto puzzle easily in future work.

References

1. Nakamoto S, Bitcoin: a peer-to-peer electronic cash system. https://bitcoin.org
2. Buterin V (2013) Ethereum white paper: a next generation smart contract decentralized application platform. Available https://github.com/Ethereum/wiki/wiki/White-Paper
3. Wood G (2014) Ethereum: a secure decentralized generalized transaction ledger. Ethereum project yellow paper 151
4. Shahriar MA, Bappy FH, Fakhrul Hossain AKM, Saikat DD, Ferdous MS, Chowdhury MJM, Bhuiyan MZA (2020) Modelling attacks in blockchain systems using petri nets. In: Published in: 2020 IEEE 19th international conference on trust, security and privacy in computing and communications (TrustCom). https://doi.org/10.1109/TrustCom50675.2020.00142

5. Azhar MT, Khan MB, Khan AUR (2019) Blockchain based Secure Crypto-currency system with quantum key distribution protocol. In: Published in: 2019 8th international conference on information and communication technologies (ICICT). https://doi.org/10.1109/ICICT47744.2019.9001979
6. Metamask (2020) Available https://metamask.io
7. Ethereum virtual machine [EVM] Ganache truffle framework. https://truffleframework.com/ganache
8. Ethereum (2018) Ethereum (ETH) blockchain explorer. https://etherscan.io/. Accessed 8 Nov 2018
9. Nodejs (2020) Available https://nodejs.org/en/
10. Solidity (2019) Available https://solidity.readthedocs.io/en/v0.5.11/
11. Kruglik S, Nazirkhanova K, Yanovich Y (2019) Challenges beyond blockchain: scaling, oracles and privacy preserving. In: Published in: 2019 XVI international symposium problems of redundancy in information and control systems (REDUNDANCY). https://doi.org/10.1109/REDUNDANCY48165.2019.9003331
12. Ekparinya P, Gramoli V, Jourjon G (2018) Impact of man-in-the-middle attacks on Ethereum. In: Published in: 2018 IEEE 37th symposium on reliable distributed systems (SRDS). https://doi.org/10.1109/SRDS.2018.00012
13. Ethereum. https://Ethereum.org
14. Web3ETHlibrary Available https://we b3js.readthedocs.io/en/v1.2.0/web3-eth.html/
15. Visual Studio Code (2019) Available https://code.visualstudio.com/
16. Gennaro R, Goldfeder S, Narayanan A (2016) Threshold-optimal DSA/ECDSA signatures and an application to bitcoin wallet security. In: International conference on applied cryptography and network security. Springer, pp 156–174
17. Goldfeder S, Bonneau J, Gennaro R, Narayanan A (2017) Escrow protocols for cryptocurrencies: How to buy physical goods using bitcoin. In: International conference on financial cryptography and data security. Springer, pp 321–339
18. Miers I, Garman C, Green M, Rubin AD (2013) Zerocoin: anonymous distributed e-cash from bitcoin. In: Security and privacy (SP), 2013 IEEE symposium. IEEE, pp 397–411
19. Sasson EB, Chiesa A, Garman C, Green M, Miers I, Tromer E, Virza M (2014) Zerocash: decentralized anonymous payments from bitcoin. In: Security and privacy (SP), 2014 IEEE symposium. IEEE, pp 459–474
20. Androulaki E, Karame GO, Roeschlin M, Scherer T, Capkun S (2013) Evaluating user privacy in bitcoin. In: International conference on financial cryptography and data security. Springer, pp 34–51
21. Ahuja A (2013) TensorFlip: a fast fully-decentralized computational lottery for cryptocurrency networks. In: 2021 13th International conference on communication systems and networks (COMSNETS). Published in IEEE. https://doi.org/10.1109/COMSNETS51098.2021.9352857
22. Green M, Miers I (2017) Bolt: anonymous payment channels for decentralized currencies. In: Proceedings of the 2017 ACM SIGSAC conference on computer and communications security. ACM, pp 473–48
23. Moschou K, Theodouli A, Terzi S, Votis K, Tzovaras D, Karamitros D, Diamantopoulos S (2020) Performance evaluation of different hyperledger sawtooth transaction processors for Blockchain log storage with varying workloads. In: Published in: 2020 IEEE international conference on Blockchain (Blockchain). https://doi.org/10.1109/Blockchain50366.2020.00069
24. Linoy S, Mahdikhani H, Ray S, Lu R, Stakhanova N, Ghorbani A (2019) Scalable privacy-preserving query processing over Ethereum blockchain. In: Published in: 2019 IEEE international conference on Blockchain (Blockchain). https://doi.org/10.1109/Blockchain.2019.00061
25. Linoy S, Mahdikhani H, Ray S, Lu R, Stakhanova N, Ghorbani A (2019) Scalable privacy-preserving query processing over Ethereum blockchain. In: Published in: 2019 IEEE international conference on blockchain (Blockchain). https://doi.org/10.1109/Blockchain.2019.00061

Efficient Machine Learning-Based Diagnosis System for Breast Cancer

M. Deva Priya, J. Sangeetha Priya, S. Sam Peter, Viraja Ravi, S. Karthik, and R. Karthik

Abstract Machine Learning (ML) algorithms find their application in the field of medicine especially in cancer prognosis enabling early diagnosis and treatment. Breast cancer is the predominant form of cancer occurring in women. It is a malignant growth or tumor resulting in uncontrolled division of cells that invades into the adjoining cells and also spreads to different parts of the body. It is the fifth leading cause of cancer mortality. In this paper, diverse ML algorithms are applied to the Wisconsin Diagnostic Breast Cancer (WDBC) dataset. Linear Discriminant Analysis (LDA) and Principal Component Analysis (PCA) are used for feature extraction, while Naive Bayes Classifier (NBC), Random Forest (RF), Neural Networks (NNs) and Support Vector Machine (SVM) are used for classification of data in the WDBC dataset. It is seen that LDA and NNs offer better results in terms of Accuracy, Sensitivity, Specificity and Kappa coefficient.

Keywords Machine learning · Feature selection · Classification · Breast cancer · Benign · Malignant

M. Deva Priya (✉) · J. Sangeetha Priya · S. Sam Peter · V. Ravi · R. Karthik
Department of Computer Science and Engineering, Sri Krishna College of Technology, Coimbatore, Tamil Nadu, India
e-mail: m.devapriya@skct.edu.in

J. Sangeetha Priya
e-mail: 19tpcs003@skct.edu.in

S. Sam Peter
e-mail: sampeter.s@skct.edu.in

V. Ravi
e-mail: virajaravi@skct.edu.in

R. Karthik
e-mail: karthik.r@skct.edu.in

S. Karthik
Department of Computer Science and Engineering, SRM Institute of Science and Technology, Delhi-NCR Campus, Ghaziabad, India

R. P. Mahapatra et al. (eds.), *Proceedings of International Conference on Recent Trends in Computing*, Lecture Notes in Networks and Systems 341,
https://doi.org/10.1007/978-981-16-7118-0_59

1 Introduction

Machine Learning (ML) deals with the growth, investigation and application of algorithm that enables computers to acquire, envisage and take decisions deprived of explicit programming for performing a task. Data mining is a domain that emphases on examining data using unsupervised learning. ML based algorithms find their effective application in the field of medicine, enabling the physician to predict and decide easily.

Breast cancer is found to be the most widespread kind of cancer found in women across the world. It is the main cause of death, and still there is no customary theory concerning the reasons and treatment of it. It is found that cancer is due to unrestrained growth of cells. Normal cells come with a life cycle wherein they grow and divide to form new cells and expire after suitable time. Any kind of irregularity in the life cycle may lead to cancer. The diagnostic methods like Fine-Needle Aspiration Cytology (FNAC) and mammography are usually used in cancer diagnosis. They are not very efficient. The main aim in classifying breast cancer is to forecast whether an unidentified tumor sample is benign or malignant, depending on the labeled training samples. Precise classification is indispensable and crucial in the field of medicine. An automated computer-based system is needed for classification.

Diverse expert systems and ML techniques are used in the classification of medical data as they have the ability to learn from experience, previous patterns and instances. A complete examination of medical data can be done in reduced time involving less number of errors. Nevertheless, unrelated and repeated features cannot be used in classification as they may lead to reduction of classification performance due to huge search space. Appropriate features should be chosen for classification so as to reduce the time taken for training by removing or dropping the inappropriate and repeated features.

2 Related Work

In this section, works done by authors in the application of ML algorithms on different datasets are discussed.

Akay [1] has come up with a breast cancer diagnosis system that is based on SVM combined with feature selection. They have applied the mechanism on the Wisconsin Breast Cancer Dataset (WBCD) and have shown a classification accuracy of 99.51% for five features. Maldonado et al. [2] have propounded a scheme to instantaneously choose appropriate features by considering the use of every feature in the formulation of Support Vector Machines (SVM). This mechanism called Kernel-Penalized SVM (KP-SVM) improves the shape of Radial Basis Function (RBF) kernel removing features that are of less relevance to the classifier. They have focused on real-world problems.

Chuang et al. [3] have designed an optimization algorithm called Catfish Binary Particle Swarm Optimization (CatfishBPSO) wherein catfish effect is applied to BPSO. New particles are introduced into the search space which replace particles that are of less fitness with the ones initialized at extreme points. This is done when the fitness of the global best particle is not made better for a number of successive iterations. Feature selection is eased with better classification accuracy, involving lesser features. Chen et al. [4] have propounded a Rough Set (RS)-based SVM classifier (RS-SVM) for breast cancer diagnosis. RS reduction algorithm is used for feature selection, and redundant features are removed to improve the diagnostic accuracy. Xue et al. [5] have designed two fitness functions in BPSO for feature selection. The dimensionality is reduced, thus improving the classification accuracy. In the first function, the relative prominence of classification and feature dimensionality are balanced by involving linearly increasing weight. The second function is a two-staged one, where classification is optimized and feature dimensionality is considered. Xue et al. [6] have proposed three initialization approaches and three personal best and global best updating schemes for effective feature selection. The main aim is to offer better classification, reduce the feature dimensionality and the computational time. The initialization strategy involves both forward and backward selections to reduce the feature dimensionality and thus the computational time. The updating scheme overcomes the drawbacks of existing mechanisms by considering the feature dimensionality.

Agrawal and Agrawal [7] have done a survey on the methods used in cancer diagnostics. This study will help the researchers to gain insight on the approaches that are suitable for clinicians to carry out effective diagnosis. Bhardwaj and Tiwari [8] have proposed Genetically Optimized Neural Network (GONN) algorithm for classifying breast cancer tumors, wherein a NN is used to optimize its architecture. Modified crossover and mutation operators are used to reduce their damaging nature. Nilashi et al. [9] have developed an intelligent system for breast cancer diagnosis. Expectation Maximization (EM) is used for clustering. This decision support system can assist physicians.

Huang et al. [10] have developed linear kernel-based SVM algorithm for diagnosing breast cancer. Statistical and ML mechanisms are used to develop diverse prediction models. SVM offers better results. It is seen that linear kernel-based SVM ensemble that uses bagging method and RBF kernel-based SVM ensemble using boosting method are better for a small-scale dataset, wherein feature selection can be done in the preprocessing phase itself. For a larger dataset, RBF kernel-based SVM ensemble that uses boosting seems effective. Nilashi et al. [9] have dealt with the multi-collinearity issue by using PCA. WDBC and mammographic mass datasets are considered, and it is seen that the prediction accuracy is improved. The proposed system acts as a clinical decision support system. Bhardwaj et al. [12] have proposed a concurrent feature selection and classification mechanism using Genetic Programming (GPsfsc) for diagnosis of breast cancer. The mechanism offers better classification accuracy involving less feature dimensionality. Devarriya et al. [13] have used Genetic Programming (GP) approach for breast cancer diagnosis. The system offers better recall. The fitness function designed for unstable data classification supports better classification.

3 Feature Extraction Mechanisms

Feature extraction considers original set of measured data and features that are planned to be useful and unique. It enables next learning and generalization resulting in improved human understandings. It may support dimensionality reduction, where raw variables are reduced to more adaptable groups, while precisely and wholly describing the existing dataset. Huge dataset can be reduced with reduced feature dimensionality. Subsets of data are anticipated to comprise of informative and unique input, so that the preferred tasks can be done on the reduced set. In this paper, Principal Component Analysis (PCA) and Linear Discriminant Analysis (LDA) are applied for carrying out feature extraction.

3.1 Principal Component Analysis (PCA)

Principal Component Analysis (PCA) is used for extracting features to reduce the dimensionality consisting of variables correlated to one another, while maintaining the difference in a dataset. The variables are transformed to a collection of variables referred to as Principal Components (PCs) and are orthogonal, well-ordered such that the retaining of variation among the variables drops. The first PC maintains extreme difference that is present in the components. The PCs are orthogonal eigen vectors of a covariance matrix. Prominently, the dataset is to be scaled, and the results are sensitive to relative scaling. Officially, a PC is a linear grouping of optimally weighted measured variables. PCs are basically the linear combinations of variables, and the weight vector is the eigen vector which consecutively satisfies the method of least squares. The variation in the PCs decreases from the first PC to the last one. The PCs of least importance are useful in regression, outlier detection, etc.

3.2 Linear Discriminant Analysis (LDA)

Linear Discriminant Analysis (LDA) is a customary statistical method that is used to reduce the dimensionality while conserving the class discriminatory information. It is assumed that the data are already available and the LDA feature space is found using the eigen decomposition of a suitable matrix. Data may be represented as a sequence, and the features are to be updated in stages by detecting the incoming samples. This dimensionality reduction tries to maximize the distance between the mean values of the classes and minimizes the spreading within the class.

4 Classification Mechanisms

Following classification algorithms are used in classification.

4.1 Naive Bayes Classifier (NBC)

A Naive Bayes Classifier (NBC) is a ML-based probabilistic technique. It is Bayes theorem-based, and the probability of an event happening given that another event has happened is considered. The event that has occurred is the evidence, and the happening event is the hypothesis. It is assumed that the predictors/features are independent of each other. It is termed as naïve because the presence of a feature does not affect the other. It can be applied to input with high dimensionality.

4.2 Random Forest (RF)

Random forest (RF), a supervised learning algorithm deals with building forests which are a collection of decision trees, typically trained using the "bagging" method. The common idea is that a collection of learning models improves the global result. It includes hyperparameters resembling a decision tree or a bagging classifier, where both need not be combined and the classifier class of RF can be used. Regression tasks can also be handled using RF. They add randomness to the model as trees are grown. Instead of finding the most significant feature while splitting a node, it finds the finest feature among the random subcategory of features. This leads to a wide variety that usually produces a better model. Random subset of features is considered for breaking a node. Random thresholds can be used with different features to generate more random trees rather than finding the finest thresholds. RF is versatile and can be employed in regression and classification. The relative significance of input features can be observed.

4.3 Neural Networks (NNs)

A Neural Network (NN) includes a sequence of algorithms that deals with recognizing the relationship that exists among data. They resemble the way the human brain functions. They refer to a system of neurons that may be biological or artificial. They are capable of adapting to varying input, thus producing the finest result without the need for redesigning the output criteria. A neuron mimics a mathematical function that gathers and categorizes information based on a particular architecture. The network mimics the statistical methods like curve fitting and regression analysis and includes layers of interrelated nodes. Every node is a perceptron resembling multiple linear regressions feeds the signal into an activation function that may be nonlinear. In a Multilayered Perceptron (MLP), the perceptions are organized in interconnected layers. The input layer gathers patterns, while the output layer includes classifications or signals to which the patterns map.

4.4 *Support Vector Machine (SVM)*

Support Vector Machines (SVMs), a supervised learning model includes related learning algorithms that examine data for both classification and regression study. Mainly, it is used for classification. The main aim is to find the decision boundary that can separate space of n-dimension into classes wherein the arriving data point can be put into. The extreme points or vectors that aid in finding the hyperplane are the support vectors.

5 Implementation

The WDBC dataset includes a record of 32 tumor features belonging to 569 subjects got from the digital image of breast Fine-Needle Aspirates (FNA). Thirty features are actual tumor features, while others represent the subject ID and the class label. The class label represents whether the tumor is benign or malignant. Digital images aid in getting the attributes of cell nuclei which include the radius, texture, concave points, fractal dimension, perimeter, area, concavity, symmetry, smoothness and compactness for the subjects. These measurements are of different scales and are necessary for normalizing the dataset before training. Figure 1 shows the framework.

In the preprocessing step, the missing values are handled. Exploratory data analysis is performed to recognize patterns, detect anomalies, perform hypothesis and check for assumptions using statistics and graphical illustrations. Figure 2 shows the missing values. In the given dataset, all features are complete, only "unnamed: 32" is completely null, and hence, the feature is dropped.

Figure 3 shows that 37.3% are malignant and 62.7% are benign. Figure 4 shows the relationship between the variables, wherein the values move in the same direction.

Figure 5 shows the independent attributes in the dataset. Figure 6 shows the relationship among the variables, in which they move in opposite directions from one another. Figures 7 and 8 show the results after applying PCA and LDA.

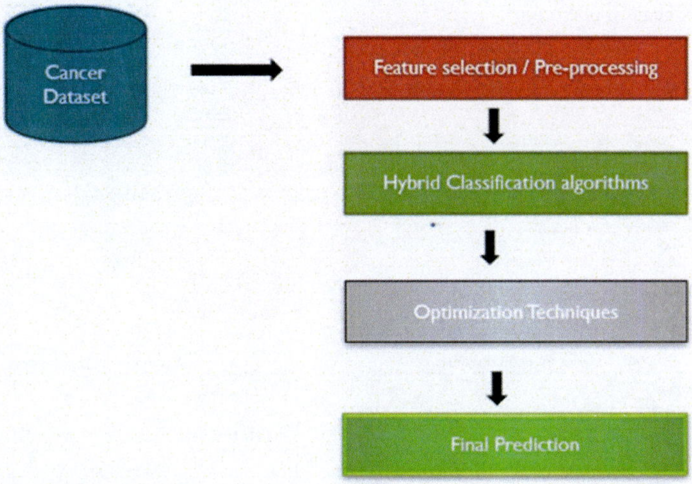

Fig. 1 Proposed Framework

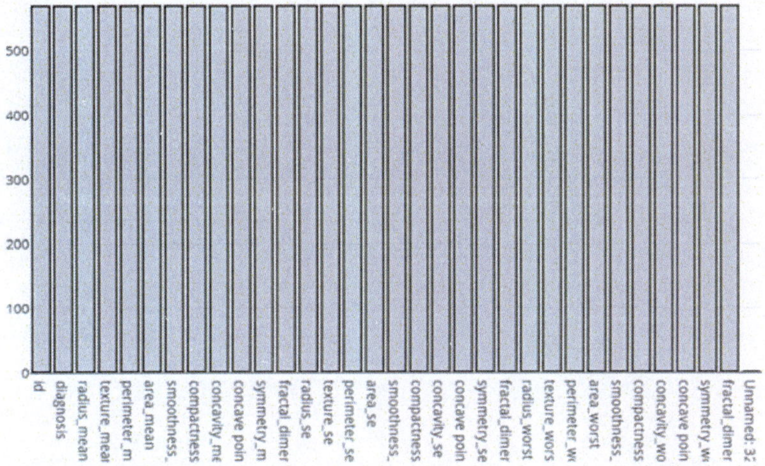

Fig. 2 Missing Value Check

6 Results and Discussion

A combination of algorithms is applied on the WDBC dataset, and the performance is analyzed. NBC, RF, NN and SVM are used for classification, while PCA and LDA are used in feature extraction. Accuracy, sensitivity, specificity and Kappa coefficients are measured.

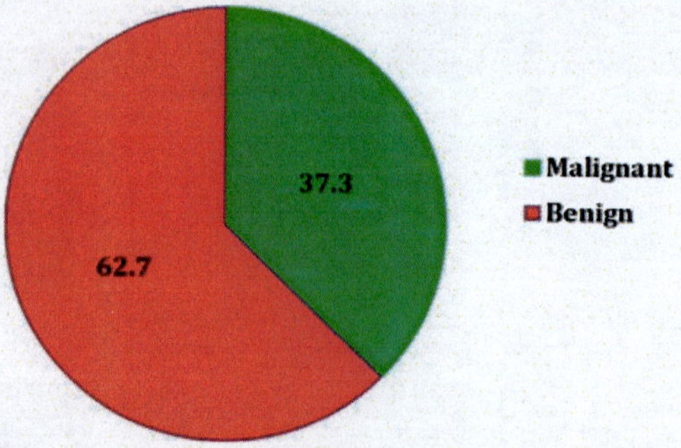

Fig. 3 Classification of Tumors

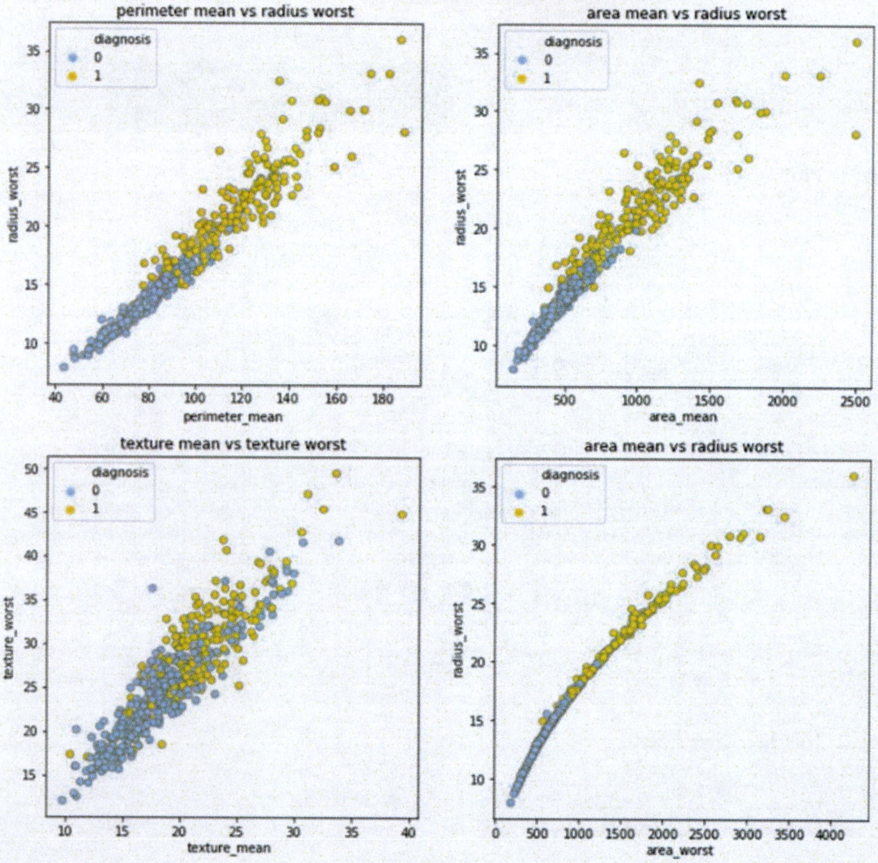

Fig. 4 Positive Correlated Features

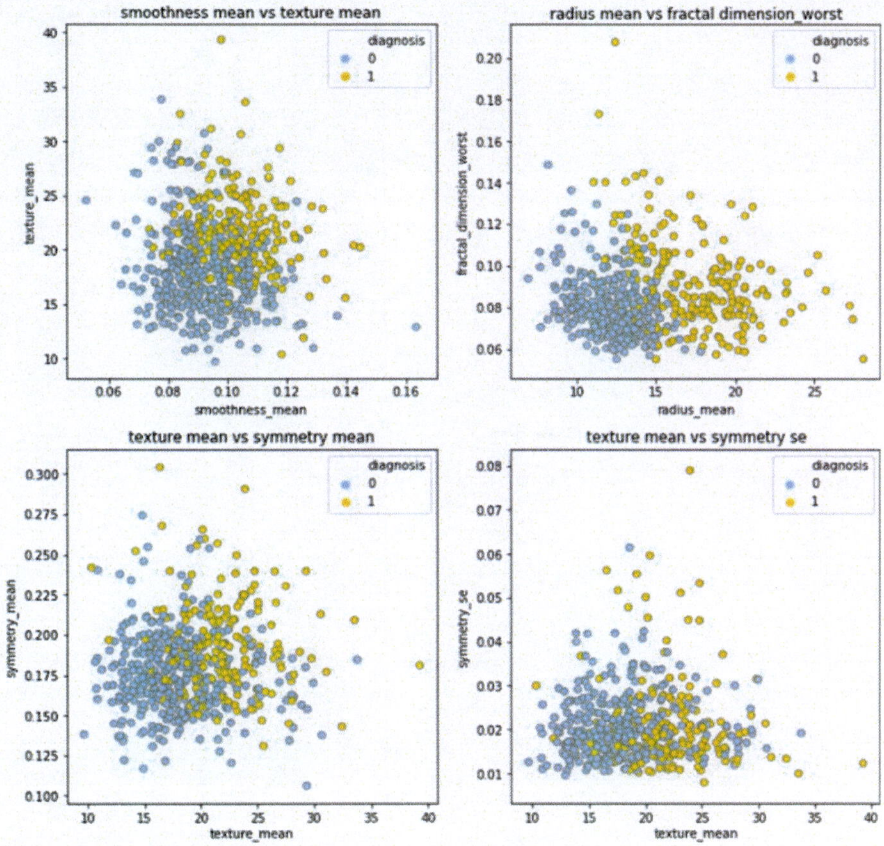

Fig. 5 Uncorrelated Features

Accuracy: It is the ratio of the sum of cases identified as TP and TN to the sum of cases identified as TP, TN, FP and FN.

- **True Positive (TP)**: Malignant tumor correctly recognized as malignant
- **False Positive (FP)**: Benign tumor incorrectly recognized as malignant
- **False Negative (FN)**: Malignant tumor incorrectly recognized as benign
- **True Negative (TN)**: Benign tumor correctly recognized as benign

Precision: It is the percentage of identified as TP to the sum of cases identified as TP and FP.

$$\text{Precision} = \frac{\text{TP}}{\text{TP} + \text{FP}} \tag{2}$$

Recall: It is the ratio of cases found as TP to the sum of cases identified as TP and FN.

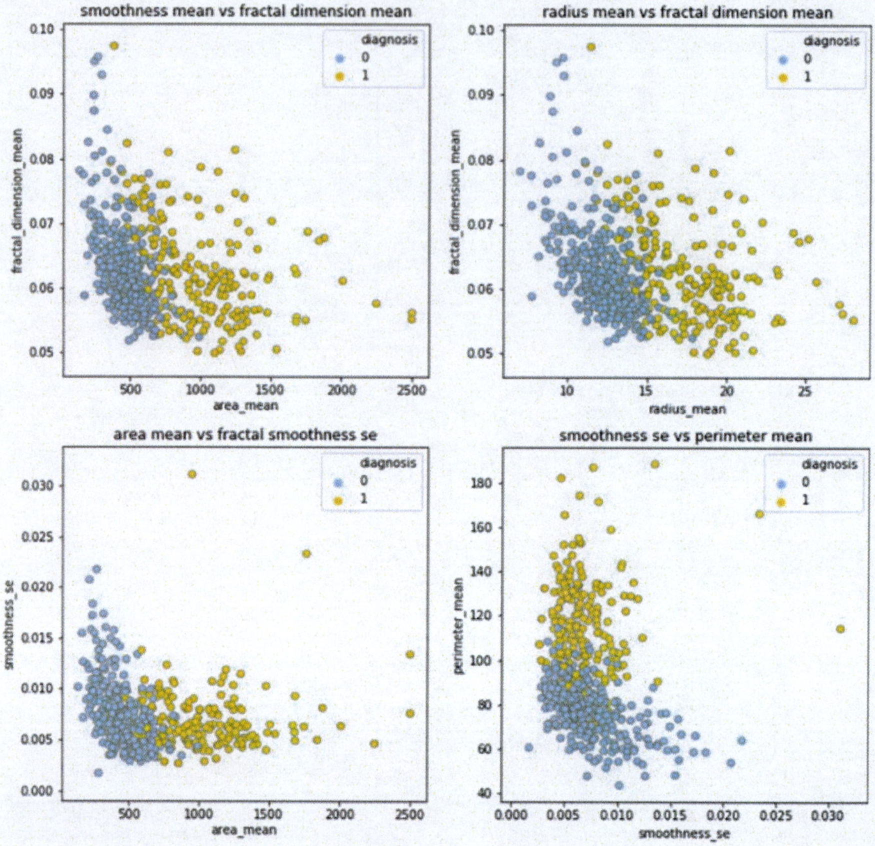

Fig. 6 Negative Correlated Features

$$\text{Recall} = \frac{\text{TP}}{\text{TP} + \text{FN}} \tag{3}$$

Specificity: It is the ratio of cases categorized into TN to the sum of TN and FP cases.

$$\text{Specificity} = \frac{\text{TN}}{\text{TN} + \text{FP}} \tag{4}$$

F-score: It is the weighted average computed based on precision and recall.

$$F\text{ - score} = 2 \times \frac{1}{\frac{1}{\text{Recall}} + \frac{1}{\text{Precision}}} \tag{5}$$

$$F\text{ - score} = 2 \times \frac{\text{Precision} \times \text{Recall}}{\text{Precision} + \text{Recall}} \tag{6}$$

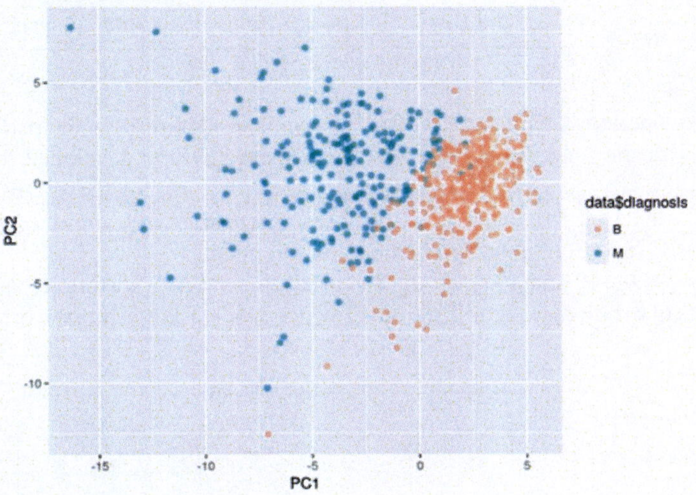

Fig. 7 Principal Component Analysis

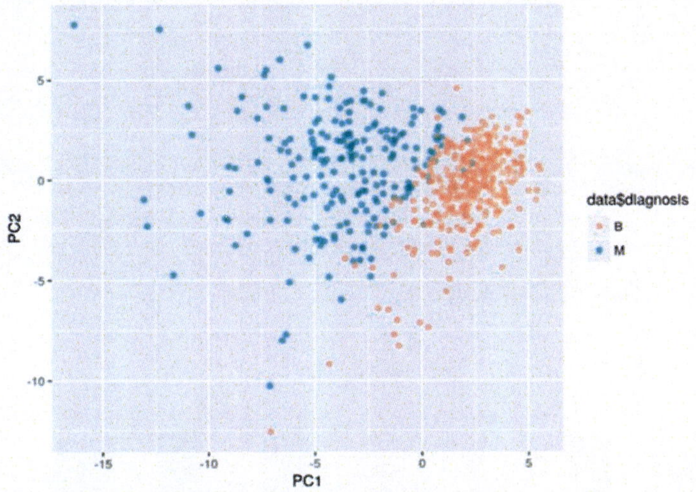

Fig. 8 Linear Discriminant Analysis

Kappa: It is the measure of inter-rater and intra-rater reliability for categorical items.

$$\text{Kappa} = \frac{P_o - P_e}{1 - P_e} \tag{7}$$

$$P_{\text{o}} = \frac{\text{TP} + \text{TN}}{\text{TP} + \text{TN} + \text{FP} + \text{FN}} \tag{8}$$

$$P_e = \frac{(TP + FN)*(TP + FP) + (FP + TN)*(FN + TN)}{(TP + TN + FP + FN)^2} \quad (9)$$

Receiver operator Characteristic (RoC) curve is used to display the performance of the classification algorithm (Fig. 9). The proposed system developed for cancer prediction is analyzed using Google Colaboratory platform for the WDBC dataset available in the UCI ML repository. Features obtained from a digitized image of breast mass FNA describe the appearances of cell nuclei.

Figures 10, 11, 12 and 13 show the Accuracy, Sensitivity, Specificity and Kappa coefficients of the diverse algorithms out of which NN + LDA offers the best results.

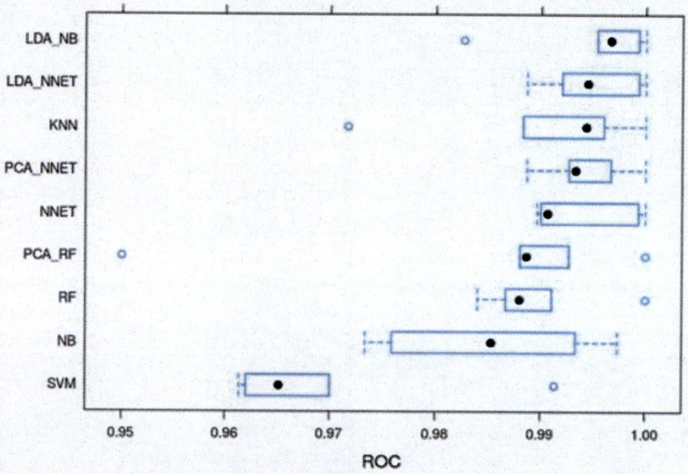

Fig. 9 RoC of various ML Classification Algorithms

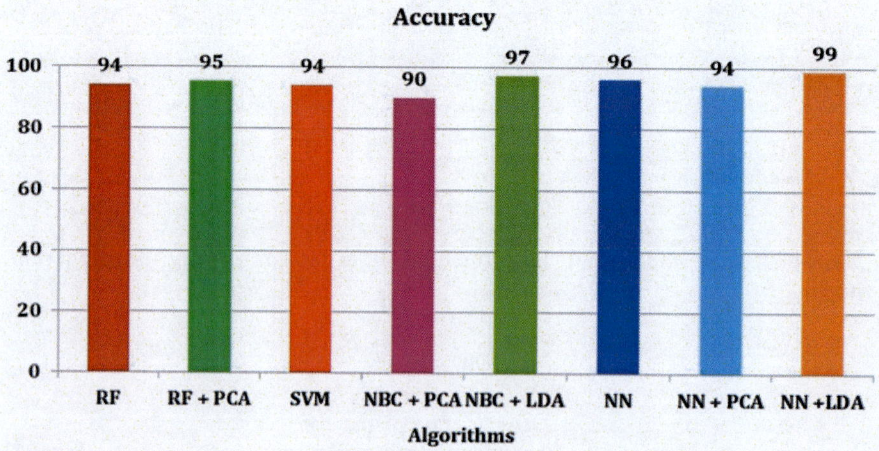

Fig. 10 Accuracy

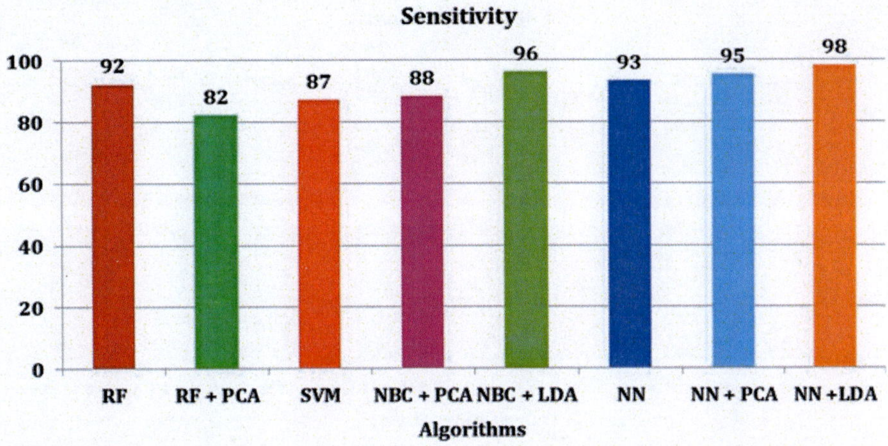

Fig. 11 Sensitivity

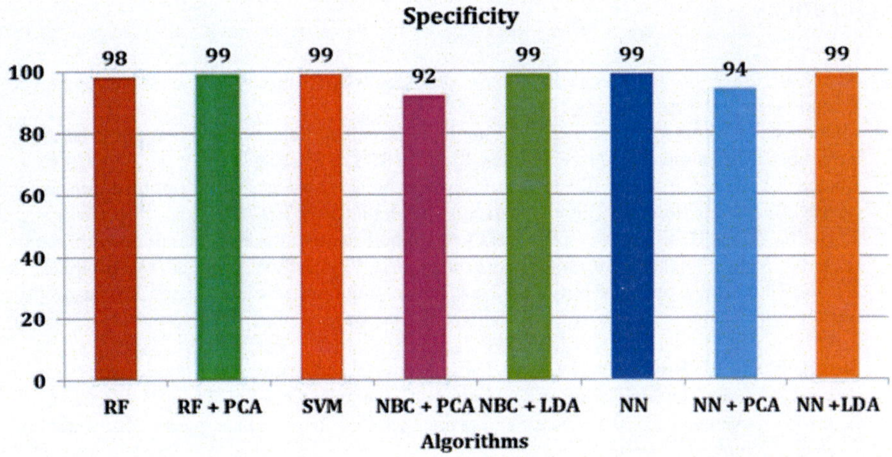

Fig. 12 Specificity

7 Conclusion

Linear Discriminant Analysis (LDA) and Principal Component Analysis (PCA) are used for feature extraction, while Naive Bayes Classifier (NBC), Random Forest (RF), Neural Networks (NNs) and Support Vector Machine (SVM) are used for classification of data in the WDBC dataset. LDA and NNs offer better results.

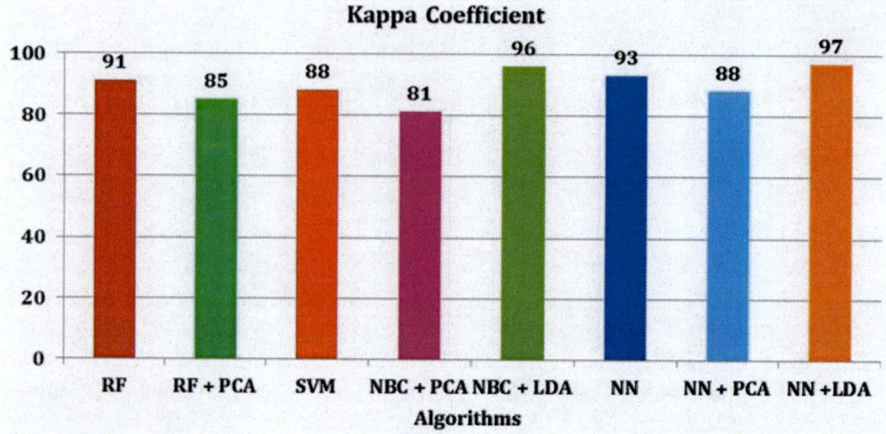

Fig. 13 Kappa coefficient

References

1. Akay MF (2009) Support vector machines combined with feature selection for breast cancer diagnosis. Expert Syst Appl 36(2):3240–3247
2. Maldonado S, Weber R, Basak J (2011) Simultaneous feature selection and classification using kernel-penalized support vector machines. Inf Sci 181(1):115–128
3. Chuang LY, Tsai SW, Yang CH (2011) Improved binary particle swarm optimization using catfish effect for feature selection. Expert Syst Appl 38(10):12699–12707
4. Chen HL, Yang B, Liu J, Liu DY (2011) A support vector machine classifier with rough set-based feature selection for breast cancer diagnosis. Expert Syst Appl 38(7):9014–9022
5. Xue B, Zhang M, Browne WN (2012) New fitness functions in binary particle swarm optimization for feature selection. In: 2012 IEEE congress on evolutionary computation. IEEE, pp 1–8
6. Xue B, Zhang M, Browne WN (2014) Particle swarm optimisation for feature selection in classification: novel initialisation and updating mechanisms. Appl Soft Comput 18:261–276
7. Agrawal S, Agrawal J (2015) Neural network techniques for cancer prediction: a survey. Procedia Comput Sci 60:769–774
8. Bhardwaj A, Tiwari A (2015) Breast cancer diagnosis using genetically optimized neural network model. Expert Syst Appl 42(10):4611–4620
9. Nilashi M, Ibrahim O, Ahmadi H, Shahmoradi L (2017) A knowledge-based system for breast cancer classification using fuzzy logic methods. Telematics Inform 34(4):133–144
10. Huang MW, Chen CW, Lin WC, Ke SW, Tsai CF (2017) SVM and SVM ensembles in breast cancer prediction. PLoS ONE 12(1):e0161501
11. Xiao Y, Wu J, Lin Z, Zhao X (2018) A deep learning-based multi-model ensemble method for cancer prediction. Comput Methods Prog Biomed 153:1–9
12. Bhardwaj H, Sakalle A, Tiwari A, Verma M, Bhardwaj A (2018) Breast cancer diagnosis using simultaneous feature selection and classification: a genetic programming approach. In: 2018 IEEE symposium series on computational intelligence (SSCI). IEEE, pp 2186–2192
13. Devarriya D, Gulati C, Mansharamani V, Sakalle A, Bhardwaj A (2020). Unbalanced breast cancer data classification using novel fitness functions in genetic programming

Deep Learning-based Stock Market Prediction

A. Christy Jeba Malar, M. Deva Priya, M. Kavin Kumar, S. Mangala Arunsankar, K. V. Bilal, and S. Karthik

Abstract Predictions on stock market and foreign currency exchange is a hot area of research. Stock market prediction is always challenging due to the non-linearity and instability nature of financial time series data. Finance experts and stock experts are continuously working on this analysis to predict future stock price which helps in deciding the buying or selling of stock for profit. Stock markets motivate investments by resource pooling, aiding corporations to make funds for enhancing their businesses. In Stock market, there are millions of resources available. As an investor, people fear to find the good resources from the pool of resources. The proposed stock price prediction model will help investors to price the resource. In this paper, an innovative stock market prediction management over time using deep learning approach is proposed.

Keywords Stock market prediction · Long short-term memory algorithm · Recurrent neural network

---

A. Christy Jeba Malar · M. Kavin Kumar · S. Mangala Arunsankar · K. V. Bilal
Department of Information Technology, Sri Krishna College of Technology, Coimbatore, Tamilnadu, India
e-mail: a.christyjebamalar@skct.edu.in

M. Kavin Kumar
e-mail: 17tuit032@skct.edu.in

S. Mangala Arunsankar
e-mail: 17tuit041@skct.edu.in

K. V. Bilal
e-mail: 17tuit010@skct.edu.in

M. Deva Priya (✉)
Department of Computer Science and Engineering, Sri Krishna College of Technology, Coimbatore, Tamilnadu, India
e-mail: m.devapriya@skct.edu.in

S. Karthik
Department of Computer Science and Engineering, Delhi-NCR Campus, SRM Institute of Science and Technology, Ghaziabad, India

R. P. Mahapatra et al. (eds.), *Proceedings of International Conference on Recent Trends in Computing*, Lecture Notes in Networks and Systems 341,
https://doi.org/10.1007/978-981-16-7118-0_60

1 Introduction

Stock market plays a dominant role in the economy by mobilizing domestic resources and directing them to fruitful investment. It acts as a support to countries that are moving toward economic integration with global economy. Increase in capital enables the organizations to enhance their businesses, operations and create job opportunities. Many investors try to forecast opening and closing prices along with trading capacity of favored stocks before investing in any stock. Maintaining assets regardless of time without forecasting may lead to drastic results. Forecasts in foreign exchange yields valuable decision-making information that increases return and reduces risk. Predicting stock price using appropriate prediction model offers the benefits of investors and raises the countries' economy. This work focusses on efficient stock price prediction problem that ensures guaranteed benefits to new investors. Deep learning techniques are applied on data, wherein increase in investment is encouraged by understanding the stock price variations under several situations. The proposed smart stock price prediction scheme lists out possibilities for investors and the results will be displayed in a Web application for easy access.

Feature selection can be done using a combination of Principle Component Analysis (PCA), Genetic Algorithm (GA) and Decision Tree (DT) [1] for better performance. A stock forecasting model integrates [2] wavelets transform, Multivariate Adaptive Regression Splines (MARS) and Support Vector Regression (SVR) to solve the problem of wavelet subseries. Sentiments of topics are incorporated as features in the prediction model [3]. Furthermore, a complete analysis with diverse dataset representations are carried out [4]. To recognize the potential changes in stock exchange, bitcoin autoregressive distributed lag model is used [5]. It also addresses the empirical studies on long-term relationship between the exchange values and the parameters that lead to fluctuations. Convolutional Neural Network (CNN) can be used to extract features from finance-related articles. Joint feature extraction method [6–9] is proposed which mines feature sets from both everyday trading and technical displays. Several online data sources with historical data, technical indicators and trends in Google searches are considered for building intelligent stock prediction systems. Predicting stock prices is an unwieldy task as it is not based on any specific pattern. Variations in stock prices are due to the supply and demand during a particular period. If a system is capable of keeping information in store for a longer duration, it will be useful for forecasting the stock prices. Prediction of a stock's price would yield significant profit for investors. Deep learning methods can be used in stock market prediction.

2 Methodology

Nowadays, people have started investing based on the results obtained from stock predictions systems. Since NNs are capable of identifying patterns in non-linear systems, they are used in predicting market directions more efficiently. To identify the pattern of stock price, deep learning techniques can be used. Various algorithms can be used for stock price prediction such as Decision Tree (DT), Support Vector Machine (SVM), NN etc., can be used.

Long Short-Term Memory (LSTM), a Recurrent NN (RNN) can be used for efficient series forecasting. It is capable of dealing with data over a longer period and is suitable for prediction. It uses large sets of non-linear and chaotic data. RNN-based methods are suitable for time-based data and are capable of processing the data in steps, thus preserving an internal state where information is stored.

Hybrid time series models are used by financial researchers and investors to create a model for stock markets based on historical stock data such as the Auto-Regressive Conditional Heteroscedasticity (ARCH) model. However, these models require more past data with normality postulates. With the development of these models, techniques like data mining, basic NN models and Back Propagation NN (BPNN) model [10] are used in economic analysis. Nevertheless, the Multi-Layer Perceptron (MLP) NN is extensively employed in developing predicting systems. They are incapable of restoring details of bygone events, taking much time to train as it includes non-linear mapping of global approximation. Basically, investors collaborate with more than one financial attribute to predict stock price. Multi-attribute data generates high-dimensional data and increases the computational complexity. Stock market analysts and fund managers apply numerous technical indicators to predict stock markets depending on personal experience which leads to wrong decisions on market signals. Finance-based multi-featured data statements generally produce high-dimensional data. To trade in stock market within an environment, economy moves from one phase of business cycle to the next. In contrast to the existing works, this paper analyzes the stock trading decisions using the technical conduct of trading patterns within the context of variable economic and business environments.

3 Related Deep Learning Methods

Deep learning model observes the features and trains the model to predict data for producing effective and continuous output predictions. In case of continuous data, the predicting values will never be discrete but a known collection of numbers or values. The input parameters include latest market data on stock price, major shareholders and many other minor and major parameters. Once the market data and information are downloaded, the deep learning model is trained and validated with that data. Further, data is normalized, time steps are added to the LSTM network and finally a deep learning model based on LSTM is created.

This work focuses on predicting the stock price and making investment estimation. The proposed methodology uses NN to predict the results efficiently. The accumulated data is analyzed to predict the stock price. Figure 1 shows the structure of the designed deep learning model. The agent performs various explorations during learning and computes the amount of stock trade as shown in Eq. 1.

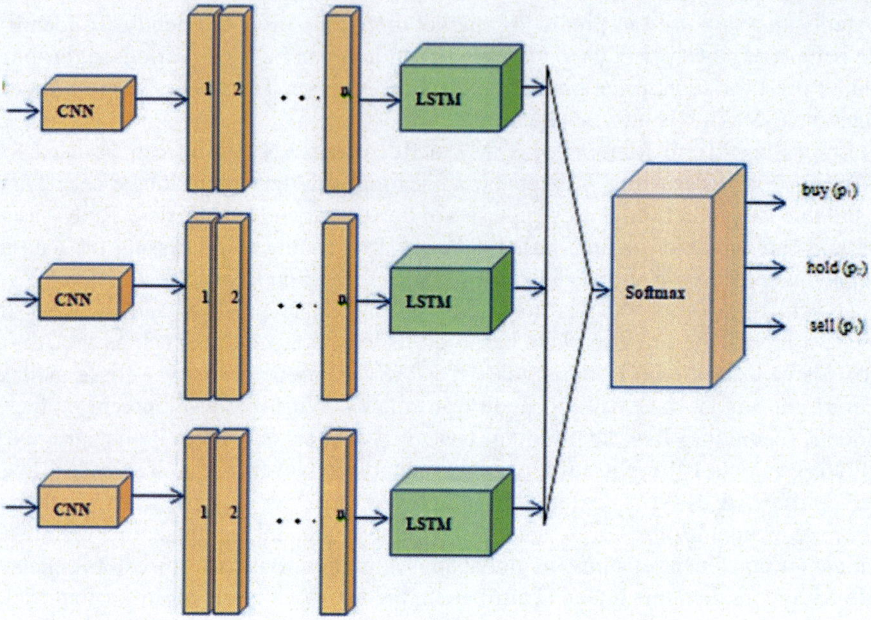

Fig. 1 Deep reinforcement learning structure for stock prediction

$$\mathrm{TU}_{(s,a)} = U_{\min} + \left| p(s, a) X (U_{\max} - U_{\min}) \right| \quad (1)$$

where
$\mathrm{TU}_{(s,a)}$—Number of stocks involved in trading for the selected action
$p(s, a)$—Probability of the selected action
$U_{\max}$—Maximum possible trading stocks
$U_{\min}$—Minimum possible trading stocks

3.1 Market Data and Information Acquisition

Initially, the dataset is prepared by collecting the latest stock market data for specific stock price. Python's yFinance library is used for data collection. The data in the form of a JSON document can be used to create security Master. It is necessary to store the data and keep track of the stock which the investors trade with. As there are chances for the data to come from multiple sources, this model stores it on a Data lake.

3.2 Market Data Exploration and Cleaning

One of the most significant steps in any ML project is data cleaning. There are several different methods of statistical analysis and data visualization techniques that can be used to explore data and identify the appropriate data cleaning operations needed. There are some simple data cleaning operations that can be applied on the datasets. Optimistic outcomes cannot be obtained if the models are not properly designed. The null values are cleaned to check whether all data types are valid.

Exploratory Data Analysis (EDA) is a very important phase in research for investigating various datasets and summarizing their significant characteristics, often using different methods of data visualization. It makes the job of a data analyst easy to obtain the repeated trends, spot anomalies, test theories and conclusions. This would help him to decide the way the data sources are monitored and get the results with greater precision.

3.3 Deep Learning Model Prediction

Recurrent Neural Network (RNN) belongs to a class of Artificial Neural Networks (ANNs), where the nodes are connected to form a directed graph. RNN is a well-known method that uses the internal state (memory) to process data making them applicable to tasks including non-linear, unsegmented and non-linear operations.

LSTM-based networks form a subset of RNN that evolves from the start, outperforming traditional models in certain applications. It is explicitly designed to overcome the long-term dependence issue in RNN. Information can be stored for a longer duration in LSTM. LSTMs are chain-like, but the repeating modules are of diverse structures.

4 Result and Discussion

In the pool of data obtained from yFinance, the training and testing data are separately defined. As stock prices could vary depending on the dates, currency and several other parameters, the function for the model requires three basic arguments as parameters.

- Ticker Symbol
- Start date which defines the date from which the data must be collected
- Validation date which defines to which date the data must be collected

As all the data cannot be used for prediction, the collected data must be normalized. To normalize data, it is scaled between 0 and 1 so that it can be taken on a common scale using a pre-processing tool called MinMaxScaler. Figure 2 shows the division between training and testing data.

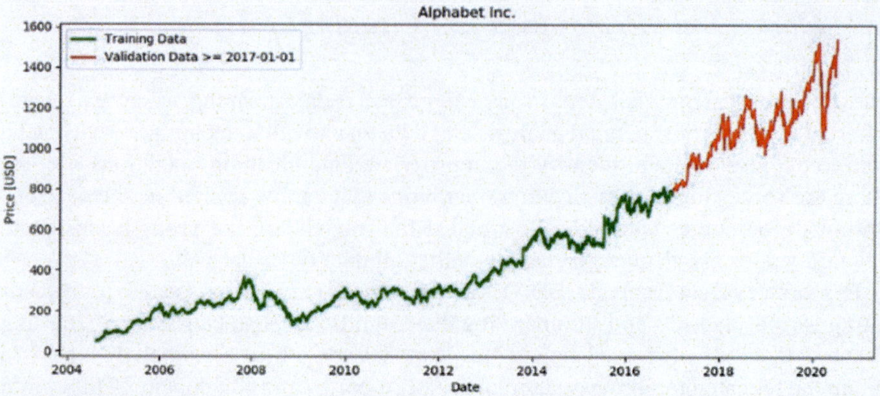

Fig. 2 Chart showing the difference of training and testing data

In a LSTM network, data is imported as a 3D array. To translate the obtained 2D array into 3D, a time step is used to loop through the data and generate minor partitions to be fed into the training model. The final array is restructured into training samples with 'n' time steps and 1 feature per step. In this model, 1 time step of 3 days is used. The model reads data of 3 former days to make prediction for the subsequent day. To create a deep learning LSTM model, the system is implemented using TensorFlow, TensorFlow-GPU and Keras. Once the model is defined, it has to be specified with suitable metrics to track the performance of the model. Figure 3 shows the graph for loss validation. The model is fine tuned to obtain the least validation loss. In

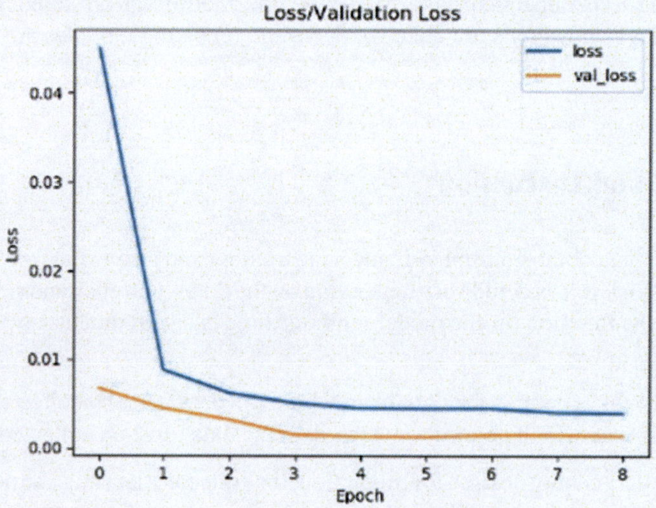

Fig. 3 Validation loss

this model, a validation loss of 0.14% is obtained with an MSE of 0.17% which is comparatively better. Figure 4 shows the Mean Square Error (MSE).

Once the data import is complete, the test data used for training with same time steps is taken. The model performs stock prediction based on the training done over the training data. A CSV file that includes the prediction and chart showing the real vs estimation is generated. It is made available to the investors on the client side of the Web application. The stock prices, their increase and decrease, and near-future prices are forecast (Fig. 5).

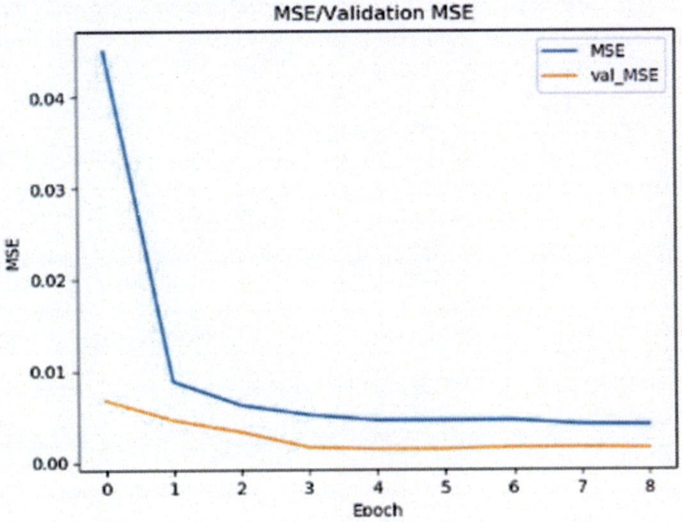

Fig. 4 Validation MSE

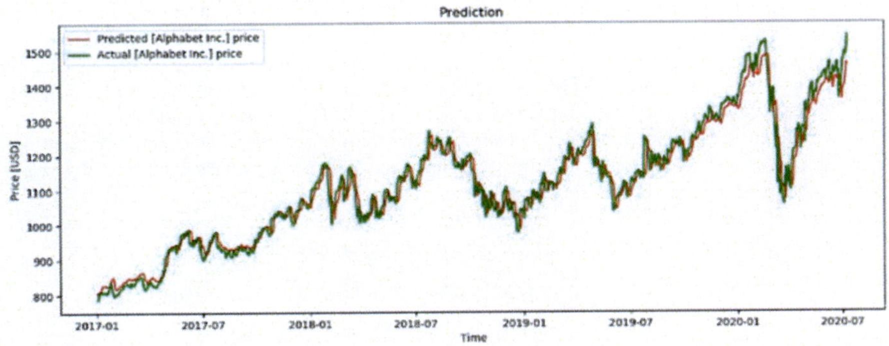

Fig. 5 Real versus the estimation value

5 Conclusion

In this paper, a deep learning-based model is proposed to predict the price of stocks in the near future. The data is presented to the user via a Web application. Efficient deep learning algorithms, NN-based models and technologies are used, and an overall user-friendly interface is provided to the investors. The acquired training datasets provide sample insights to forecast the required price and demand in the stock markets. Therefore, the scheme allows investors to make the right investments in the stock market. Long Short-Term Memory (LSTM)-based model is proposed. The model can be trained using currency deviation, company profit/loss history and can be made as an advanced AI model to predict the stock market's complete information and market price with more accuracy.

References

1. Tsai CF, Hsiao YC (2010) Combining multiple feature selection methods for stock prediction: union, intersection, and multi-intersection approaches. Decis Support Syst 50(1):258–269
2. Kao LJ, Chiu CC, Lu CJ, Chang CH (2013) A hybrid approach by integrating wavelet-based feature extraction with MARS and SVR for stock index forecasting. Decis Support Syst 54(3):1228–1244
3. Nguyen TH, Shirai K, Velcin J (2015) Sentiment analysis on social media for stock movement prediction. Expert Syst Appl 42(24):9603–9611
4. Chong E, Han C, Park FC (2017) Deep learning networks for stock market analysis and prediction: methodology, data representations, and case studies. Expert Syst Appl 83:187–205
5. Li X, Wang CA (2017) The technology and economic determinants of cryptocurrency exchange rates: the case of Bitcoin. Decis Support Syst 95:49–60
6. Liu Y, Zeng Q, Yang H, Carrio A (Aug 2018) Stock price movement prediction from financial news with deep learning and knowledge graph embedding. In: Pacific rim knowledge acquisition workshop, pp 102–113. Springer, Cham
7. Weng B, Lu L, Wang X, Megahed FM, Martinez W (2018) Predicting short-term stock prices using ensemble methods and online data sources. Expert Syst Appl 112:258–273
8. Shin HG, Ra I, Choi YH (Oct 2019) A deep multimodal reinforcement learning system combined with CNN and LSTM for stock trading. In: 2019 International conference on information and communication technology convergence (ICTC), pp 7–11. IEEE
9. Lee YS, Kim KH (2020) Experimental study on the short-term prediction of rebar price using bidirectional LSTM with data combination and deep learning related techniques. Korean J Constr Eng Manage 21(6):38–45
10. Cheon MJ (2020) Lee, OA Study on the stock price prediction and influence factors through NARX neural network optimization. J Korea Acad-Ind cooperation Soc 21(8):572–578

Neural Machine Translation Using Attention Mechanism

Sai Yashwanth Velpuri, Sonakshi Karanwal, and R. Anita

Abstract Neural Machine Translation (NMT) is the latest Machine Translation procedure that promotes exceptional upgrades compared to Rule-Based and Statistical Machine Translation procedures by conquering many shortcomings in conventional methods. NMT accepts preferences in a straightforward design and can catch lengthy relations in a sentence, demonstrating an immense possibility of turning into a different standard. Particularly for Indian Languages, NMT techniques show greater accuracy. The Attention Mechanism drastically improves the accuracy. The way to utilize the Attention Mechanism that guarantees the speed and accuracy of the translation together has likewise become a significant issue for scientists to explain. In this paper, we implement Attention Mechanism in the Machine Translation for Indian Languages and improve accuracy. Our primary objective is to test the model with the Hindi-English pair. Our NMT for this language pair showed a BLEU score of 0.8274. For future enhancement, we plan to work on other Indian Language pairs.

Keywords Natural language processing · Neural Machine Translation · Attention Mechanism · English · Hindi · Recurrent Neural Network · Long short-term memory

S. Y. Velpuri (✉) · S. Karanwal · R. Anita
Department of Computer Science and Engineering, SRM Institute of Science and Technology, Kattankulathur, Chennai, Tamil Nadu 603203, India
e-mail: ss6063@srmist.edu.in

S. Karanwal
e-mail: sa7544@srmist.edu.in

R. Anita
e-mail: anitar@srmist.edu.in

R. Mahapatra et al. (eds.), *Proceedings of International Conference on Recent Trends in Computing* , Lecture Notes in Networks and Systems 341,
https://doi.org/10.1007/978-981-16-7118-0_61

1 Introduction

1.1 Overview

Machine Translation (MT) is portrayed as the assignment of interpreting text or discourse starting with one common language then onto the next, with as minimal human exertion as could be expected under the circumstances. MT focuses on accomplishing standard translations that are semantically equal to input sentences along with linguistically accurate in objective language. MT executes essential replacement of phrase on a basic level, but that isn't enough, as an acknowledgment of the complete expressions and their nearest partners in the objective language are essential. It is a subfield in NLP that examines how to utilize computational power for translating the discourse or text from one language to another without human association. As the Machine Translation duty comprises a similar goal and the latest NLP and AI aim, i.e., to understand the human text (discourse) entirely at a semantic level, incredible consideration has been acquired. Besides the logical worth.

India is a country with a multilingual nation having individuals from various states communicating in different provincial dialects. Communication and data trade between individuals is essential for business purposes, and for individuals to share emotions, musings, assessments, and realities. As a country, we need to work in harmony with our nation's people to exchange information and ideas to have a better understanding. Machine Translation enables people from various cultural and linguistic backgrounds to communicate with each other easily using a mobile phone or a computer. Upgrading the accuracy of Machine Translation is a significant work that can bring substantial comfort and convenience for people worldwide. We look forward to contributing as much as possible to this problem.

1.2 Motivation

Our planet has people from different cultural backgrounds, and many of them have their communication language. When we want to connect and interact with people of different cultural backgrounds, we have a barrier, i.e., language. MT endeavors to limit the correspondence hole among individuals from other phonetic foundations. Not being acquainted with another person's tongue can make it hard to thoroughly and precisely communicate and makes the correspondence incapable. Development is blocked if the individual is not able to interact with associates on a more profound level. Interpreting dialects is essential for disseminating data, information, and thoughts that help in successful and compassionate communication between various societies and to beat linguistic hindrances to communication.

Researchers found that around 72% of the people give preference to products and websites in their language, 72.4% state they are bound to purchase an item with data in their language, and it has been discovered that people are more inclined to buy a product from a website inscribed in their language.

Though they are consistent in translating shorter sentences, the Machine Translation model is not yet fully reliable considering the accuracy of existing systems on comprehensive content and exceptional cases. There is a significant improvement required on these existing systems to make them completely reliable. Improving Machine Translation precision is critical work that can bring considerable solace and comfort for individuals worldwide. We are trying to contribute to this problem to help improve accuracy and reliability.

1.3 Problem Definition

Translation from one language to another may awaken various challenges based on structure, order, morphology, syntax, etc. Among these, syntactic and morphological differences are more challenging. The significant syntactic distinction is the word sequence. Subject-Object-Verb (SOV) and Subject-Verb-Object (SVO) are the two different word orders we can see. Likewise, numerous challenges arose during translation.

There are different ways to automate the process of Machine Translation. There are four leading approaches which showed remarkable results over the years, they are:

1. Rule-Based Machine Translation (RBMT)
2. Statistical Machine Translation (SMT)
3. Neural Machine Translation (NMT)
4. Hybrid Machine Translation (HMT).

RBMT uses pre-defined grammar rules; the sentence gets converted from the input language to the output language applying these rules. SMT utilizes measurable models from the examination of bilingual content corpora, as a rule finding a correspondence between a single word and another by making a probability-driven match. HMT usually employs two or more existing translation techniques by combining them to achieve better quality. NMT uses Neural Networks; this is a recent advancement that came to reality as we have more powerful GPUs and computers. NMT showed the best results in contrast with other existing methods. We discuss more NMT because our proposed methodology uses the same.

NMT has come out as a far unrivaled Machine Translation strategy in the previous 5–6 years. The severe intensity of GPUs and PCs has made Neural Networks more powerful and robust and increased the ascent of Neural Machine Translation is to a great extent because of the expanding fame of profound learning. We use bilingual parallel corpora to train and test. The dataset's quality greatly affects the accuracy of the model. Different lengths of sentences should be present with long-distance relations to enable the model to capture all the possibilities and actual relations. Our goal is to design an NMT with an Attention Mechanism that can perform well with short sentences and long sentences. Attention Mechanism encodes each word as a vector instead of each sentence as a vector. Each word gets a weight called "Attention". It describes how much attention must be given to a specific word. This enables the NMT to retain the past and map the future for translation. It is altogether making the NMT more accurate for comprehensive content.

The rest of the paper is organized as follows: Sect. 2 presents our literature survey on various research papers to understand the problem and its challenges. In Sect. 3, we describe the existing system and how it works. Section 4 describes our proposed work, the architecture of our model, and how the model works. Section 5 presents the implementation of the model, and how we achieved the proposed work in reality. In Sect. 6, we show the performance and evaluation of our model. Section 7 presents some results to show how our model performed in different scenarios. Section 8 consists of conclusion on the problem and future enhancements that can be done to achieve more promising results.

2 Literature Survey

Immense research has been done to understand different models which implement another mechanism to achieve Language Translation. Various techniques such as Rule-Based Machine Translation, Statistical Machine Translation, Hybrid Machine Translation, and Neural Machine Translation have been used to attain the model.

"Incorporating BERT into Neural Machine Translation" by Jinhua Zhu, Yingce Xia [1]; Introduced concerned training framework that integrated pre-trained model. Techniques used: Asymptotic distillation and dynamic switching gate. [1] "Hybrid Machine Translation by Combining Output from Multiple Machine Translation Systems" by Matīss Rikters [2]; improved accuracy by combining two HMT models. Techniques used: Hybrid Machine Translation Model that converts multiple languages. The LetsMT showed an accuracy of 21.92% [2] "Synchronous Bidirectional Neural Machine Translation" by Long Zhou [3]; Model used Bidirectional Neural Machine Translation. It not only relied upon its previously generated outputs but also on future context as it had synchronous Bidirectional decoding present inside an isolated model, which was used to decode from left-to-right and right-to-left. The model showed accuracy of 40.89 for first 4 token and 40.08 for last 4 tokens [3] "Neural-Based Machine Translation System: Outperforming Statistical Phrase-Based Machine for Low-Resource Languages" by Muskaan Singh, Ravinder Kumar [4]; this model delivered an important Deep Neural-Based Machine Translation model to translate text using Statistical and Neural MT. The model showed an accuracy of 60.4% outperforming the statistical phrase-based model by 2.4% [4]. "Attention Mechanism in Machine Translation" by Yuening Jia [5]; this model has used Attention Mechanism and CNN, which enabled to achieve better accuracy than conventional NMT [5]. "Machine

Translation in Indian Languages: Challenge & Resolution" by Raj Nath Patel [6]; This model converted English sentences to Indian Languages, such as, Hindi, Tamil, Malayalam, and Punjabi using SMT with suffix separation, which helped to improve the translation quality. It required less virtual space and gave accurate translation [6]. "Implementation of Hindi to English Idiom Translation System" by Himani Mishra, Chakrawarti RK (2018); Model used Hybrid MT consisting interlingual-based approach that achieved good accuracy of conversion of idioms where meaning was delivered effectively [7]. "Unsupervised Statistical Machine Translation" by Mikel Artetxe [8]; this model converted language using iterative back-translation and SMT to improve results. In which raw data trained a better backward model after each iteration. When the needed number of repetitions was reached, the final model was trained on the massive dataset after iterations [8]. "Neural Machine Translation of Indian Languages" by Karthik Revanuru [9]; Used Neural Machine Translation, Neural Networks. It had simpler architecture and was easy to train data that provides good accuracy. The model outperformed Google translate with a difference of 17, 29, 30 BLEU score for the language pairs Urdu-Hindi, Punjabi-Hindi, and Gujarati-Hindi, respectively [9]. "A Hybrid Approach for Hindi-English Machine Translation" by Omkar Dhariya [10]; Converted Hindi to English using HMT that gave the fluency and correct grammar. The model gave different accuracies for different types of sentences. The highest one is 94.74 for sentences with ambiguity [10]. "Hindi to English Machine Translation" by Kunal Sachdeva [11]; Used automatic systems to translate the text while maintaining the meaning of the source language using SMT [11]. "An efficient English to Hindi Machine Translation system using hybrid mechanism" by Jayashree Nair [12]; New rules were added for conversion and avoided most grammatical mistakes using HMT. Declension RBMT gave an accuracy of 96% [12]. "An analysis of Malayalam Machine Translation Systems" by Bijimol T.K [13]; Demanded vast knowledge of grammar, sentence formation, and its understanding of both input and output language. There was a quick generation of output with small dataset using RBMT and HMT [13].

3 Existing System

A Neural Network's (NN) primary objective is to discover the relations between the input features and target features. A dataset is distinguished into two groups, i.e., training and testing data, and every dataset will have input features and a target feature. A NN tries to find the association between them. However, it is not that straightforward in complex machine learning tasks like Machine Translation. In NMT, each word passes through an encoder, which encodes each sentence into a fixed-size vector. This vector is sent to the decoder, which decodes the input vector into a required target language sentence.

Before training, the data must be preprocessed. A typical NN-based machine learning task has the following steps:

1. Preprocessing of data, like encoding non-numeric data to data which is interpretable by computer.
2. Splitting of data into training data, validation data, and testing data.
3. Splitting the training data into small sets to make the training process feasible.

After the preparation of data, training of the model is performed. This is usually a four-step process.

1. Take input x_i for i belongs to [1, n] and predict the label $\hat{y}$.

$$(\hat{y} = softmax(Wx_i)) \tag{1}$$

2. Calculate a loss li from calculated $\hat{y}$.

$$(li = LossFunction(\hat{y}_i, y_i)) \tag{2}$$

3. Repeat (1) and (2) steps for all xi and calculate overall loss L.
4. As a final step, differentiate L with W.

With a process to modify numerical vectors from textual data, the duty of NMT will be started. The training set consists of a set of the input and the output for an RNN.

$$x_i = x_{it=1}, x_{it=2}, \ldots, x_{it=L_x} | L_x = length(input) \tag{3}$$

$$y_i = y_{it=1}, y_{it=2}, \ldots, y_{it=L_y} | L_y = length(output) \tag{4}$$

The undeniable contrast between this dataset and the sort utilized in the RNN area is the objective. y_i, is a variable-length vector instead of a solitary mark. To battle this problem, NMT uses the encoder–decoder model. Here, the initial RNN (the encoder) examines the input sentence and results from the last shrouded state, $h^E_{t=Lx,}$ onto the other RNN (the decoder) to utilize as its previously concealed state, $h^D_{t=0.}$.

During testing, the record with the most elevated incentive in forecast vector,$\hat{y}_{it=1,}$ figures out which word turns into the following info vector, $x_{it=1,}$ nonetheless, during preparing, a more regular strategy, alluded to as teacher forcing, is utilized to accelerate the preparation cycle. In instructor driving, instead of processing the anticipated word, the decoder uses the right word, $y_{it=1,}$ as the following info vector, $x_{it=2.}$ this cycle completes Ly times until an expectation vector is made for each and every word in the yield sentence. The misfortunes for every word are added to get a misfortune L for the whole sentence. L is separated regarding each weight grid also refreshes the weight of encoder and decoder.

4 Proposed System

4.1 Encoder

An encoder is used to provide the rendering of the input sentence. The input sentence is usually in the form of a sequence of words, for which we initially counsel the embedding matrix. As in the essential language model depicted beforehand, we process these words with an RNN. It develops in the hidden states, which converts every word to its left context. To likewise get the correct context, we additionally construct an RNN that runs from the right of the sentence to the left.

We are using a Bidirectional encoder where the output depends upon previous as well as future data. An LSTM will be inserted and connected between two hidden layers of the same production Bi-LSTM is a combination of two LSTM in which one LSTM takes the input in the forward direction while the other LSTM takes it backward. It is used to expand the input information that is accessible by the network. Bi-LSTM gives better prediction and also has access to future inputs without any time delay from the current state itself.

4.2 Decoder

The decoder or the Recurrent Neural Network produces the translation. It must not only have the complete information to predict the next word but also has to understand whether the next word is already translated or not. It combines input context, the previously hidden state, and output prediction word to generate a new decoder state which is hidden and new prediction output.

As shown in Fig. 1, when we combine previous hidden states, embedding of previous output and context input, we get sequence of hidden states.

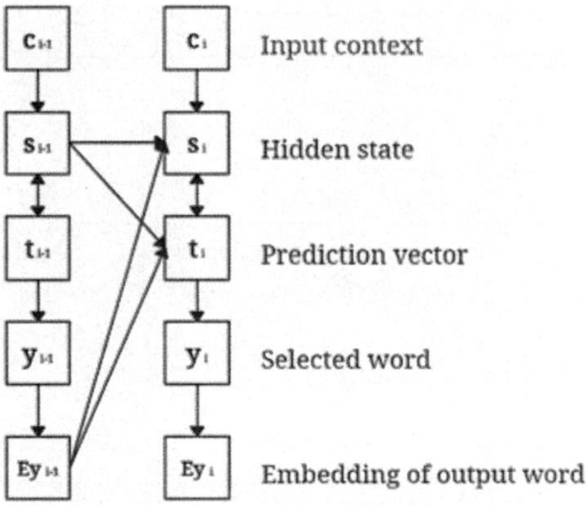

Fig. 1 Neural machine translation model

$$s_i = f(s_{i-1}, Ey_{i-1}, c_i) \tag{5}$$

We can now predict the output word from the hidden state as shown in Fig. 2. A probability distribution is added to the output. The prediction vector is obtained by the previous hidden state with the embedding of the previous output and input context.

$$t_i = softmax(W(Us_{i-1} + VEy_{i-1} + Cc_i)) \tag{6}$$

4.3 Attention Mechanism

The challenging part of the Machine Translation model is handling sentences and comprehensive content. RNNs are known for their capability of retaining the memory for predicting the output. RNNs were invented in 1986. In primitive developments of NMTs, RNNs showed promising results as they could hang on to the past results. An alternative of RNN named long short-term memory (LSTM) was introduced in 1997.

In contrast with RNN, LSTM showed greater accuracy in retaining the past knowledge, and they were more accurate in translating comprehensive content. Although LSTM can capture long-range relations better than the RNN, a recent advancement is the Attention Mechanism was put forward in 2014 by Bahdanau. This showed astounding results of the Machine Translation. According to Attention Mechanism, every term is encoded as a single vector instead of each sentence as a vector.

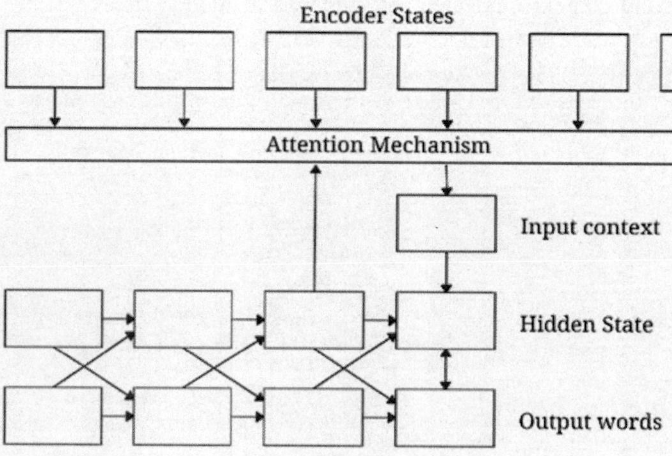

Fig. 2 Neural machine translational model with attention

We have a Bidirectional LSTM, which brings out a sequence of annotations for every input phrase. The vectors are succession forward and backward hidden states in the encoder. The weights are calculated using a feed-forward network.

$$C_j = \sum_{j=1}^{T_x} \alpha_{ij} \tag{7}$$

The weights α_{ij} are calculated using SoftMax function shown in the equation:

$$\alpha_{ij} = \frac{\exp(e_{ij})}{\sum_{k=1}^{T_x} \exp(e_{ik})} \tag{8}$$

The alignment model that links each word in source and target language is $e_{ij} = a(s_{ij}, h_j)$. The feed-forward neural network's output score expressed by function a is e_{ij} that tries capturing the positioning between input at j and output at *i*. T_x, number of annotations is obtained by the encoder. Each annotation has a dimension *d*, and the feed-forward network's input dimension is (T_x, 2*d*).

4.4 Architecture Diagram

The architecture diagram in Fig. 3 shows the working of NMT with Attention. The English sentence is taken as input till the < END > tag is encountered for encoder. Attention weights are assigned to all words while translation of each word. The output from encoder and attention weights are combined to get attention vector. The attention vector is used by decoder to decode the sentence and gives final translation.

5 Performance Evaluation

5.1 Human MT Evaluation (One Hour Translation)

This approach uses human translators, usually 20 per language. The human translators do not have any idea about the MT they are testing. They use two statistical tests to evaluate the inputs given by the testers.

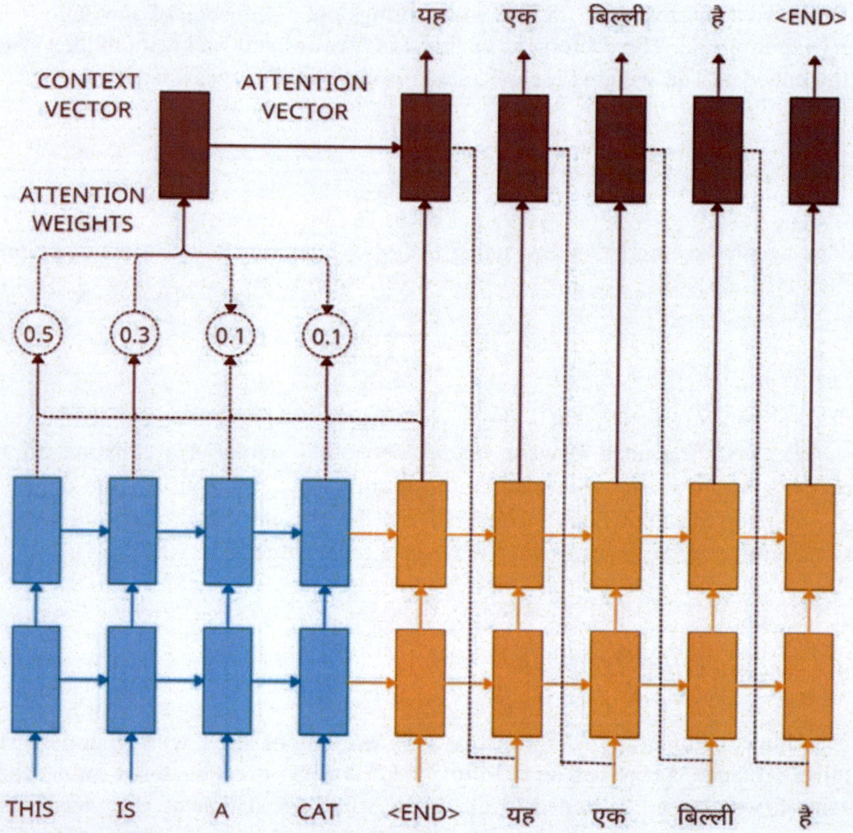

Fig. 3 Architecture diagram for NMT with attention mechanism

5.2 Automated MT Evaluation

The automated systems use segment-level similarity-based method. A translation from a human is compared with the predicted output of the MT to measure the closeness of the translation. Lesser the difference, the better the quality. We have used the Bilingual Evaluation Understudy (BLEU) evaluation method, which gives us a quality metric score for the translations of an MT. The BLEU score can be calculated by Python Natural Language Toolkit Library (NLTK) and is usually between 0 to 1. We have used a dataset with around 40,000 pairs of English and Hindi sentences.

5.3 Attention Plot

The method is specifically used for NMT with an Attention Mechanism. A graph is generated based on the amount of attention given to the parts of the sentence during translation. We use this method to calculate NMT's performance. We show the attention plot of the example sentence "This book contains many pictures" translated as "इस किताब में कई तस्वीरें हैं।" in Fig. 4. The colors show the amount of attention throughout the translation process.

6 Result Discussion

These are some samples of translations obtained after adding Attention Mechanism to NMT. Table 1 consists of correct Translations. In the second sentence in Table 2, for the word "back" there can be two translations possible "पीछे" and "वापसि".

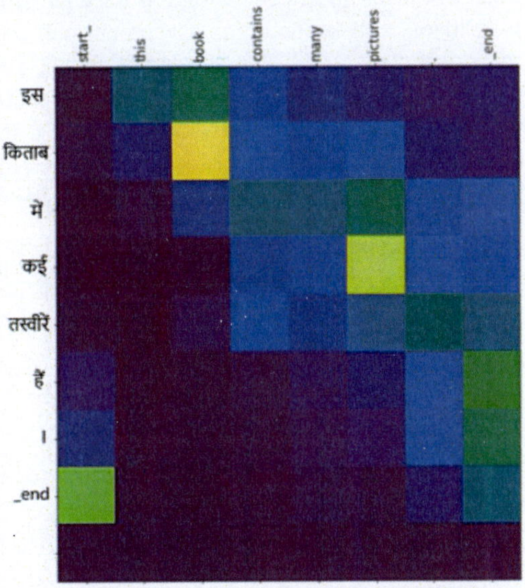

Fig. 4 Attention plot of our NMT

Table 1 Successful translations

Source	This book contains many pictures
Our NMT	इस किताब में कई तस्वीरें हैं ("iss kitaab mein kaee tasveeren hain"—"There are many pictures in this book")
Bing	इस किताब में कई तस्वीरें हैं ("iss kitaab mein kaee tasveeren hain"—"There are many pictures in this book")
Reference	इस किताब में बहुत सारी तस्वीरें हैं ("iss kitaab mein bahut sari tasveeren hain"—"This book contains many pictures")

Table 2 Failed translations

Source	He was taken back
Our NMT	उसे पीछे ले गयी थी ("use piche le gayi thi"—"he/she was taken behind")
Bing	उसे वापस ले जाया गया ("use wapis le jaya gaya"—"he/she was taken back")
Reference	उसे वापस ले लिया गया ("use wapis le liya gaya"—"he/she was taken back")

Though the appropriate translation would be "वापसि". The NMT failed to translate properly. These types of errors can be caused because of some irregularities in the dataset. The BLEU score of our model is 0.8274 based on the dataset containing 40,000 entries.

7 Conclusion and Future Work

In the proposed system, we designed a Neural Machine Translator to translate language with an added Attention Mechanism that encodes each word as a vector instead of each sentence as a vector. Each word gets a weight called "Attention" this describes how much attention should be paid to a particular word. A Bidirectional encoder has been used in which the result depends on previous as well as future data. We also implemented a Recurrent Neural Network which is a decoder to predict future words. It combines input context, previously hidden state, and output prediction word to generate a new decoder state which is hidden and new prediction output.

Our model mainly focused on Hindi-English pair. We foresee experimenting with this technique with different language pairs. Our model performed up to the mark in most cases, and there are some places where it can be improved, namely recognition of unknown words. Usage of two-layer LSTM can also help in achieving real-time translation. Combining NMT with other types of translation techniques may also perform better.

References

1. Xia Y, Zhu J, Wu L, He D, Qin T, Zhou W, Li H, Liu TY (2020) Incorporating BERT into neural machine translation. In: International conference on learning representations
2. Rikters M (2019) Hybrid machine translation by combining output from multiple machine translation systems
3. Zhou L, Zong C, Zhang J (2019) Synchronous bidirectional neural machine translation. CoRR
4. Singh M, Kumar R, Chana I (2019) Neural—based machine translation system: outperforming statistical phrase-based machine for low-resource languages. In: Twelfth international conference on contemporary computing
5. Jia Y (2019) Attention mechanism in machine translation. In: Journal of physics: conference series

6. Patel RN, Pimpale PB, Sasikumar M (2018) Machine translation in indian languages: challenge & resolution. J Intell Syst
7. Mishra H, Chakrawarti RK, Bansal P (2019) Implementation of hindi to english idiom translation system. In: International conference on advanced computing networking and informatics
8. Artetxe M, Labaka G, Agirre E (2018) Unsupervised statistical machine translation. In: Conference on empirical methods in natural language processing
9. Revanuru K, Turlapaty K, Rao S (2017) Neural machine translation of indian languages. In: Proceedings of the 10th annual ACM India compute conference
10. Dhariya O, Malviya S, Tiwary US (2017) A hybrid approach for hindi-english machine translation. In: International conference on information networking
11. Sachdeva K (2016) Hindi to english machine translation
12. Nair J, Krishnan KA, Deetha R (2016) An efficient English to Hindi machine translation system using hybrid mechanism. In: International Conference on Advances in Computing, Communications and Informatics
13. Professor BTK, Abraham J An analysis of Malayalam Machine Translation Systems. In: NCILC-14

Implementation of Automatic Detection of Text from Complex Images and Converting to Semantic Speech

Gorli Santoshi, Y. Srinivas, and Mullapudi Sri Harsha

Abstract The state of the art to recognize the printed English text character from any captured images containing misspelled text-to-speech signal which produces a voice with the correct pronunciation. The work uses an efficient preprocessing method to identify the region of interest (ROI) to detect the location of the text from the complex images having less resolution. The post-processing uses a Symspell correction model which provides a semantic text efficiently. The spell correction algorithms take minutes to process 127,000 words. The proposed model Symspell takes less than a minute to give the result. The innovation in the proposed algorithm produces an accuracy of 95.21%. The work is a cost-effective and truly efficient system that will be able to automatically identify and recite text aloud to visually challenged user.

Keywords OCR · TTS · Symspell correction

1 Introduction

Text to speech is an application which help in various areas such as driving assistance, narrating the stories for the young children, and most importantly, it creates a huge impact on the lives of visually disabled peoples. The motivation of the work is to support the listener to get the text from the complex natural images having less resolution and provide a sematic speech for the listeners to avoid the miscommunication. The printed text may have some spelling mistakes or the optical character recognition (OCR) used to convert the printed text have an accuracy of 93.42% which leads to misspells.

G. Santoshi · M. S. Harsha
ANITS Engineering College, Vishakhapatnam, India

Y. Srinivas (✉)
GITAM University, Vishakhapatnam, India
e-mail: syarrama@gitam.edu

R. P. Mahapatra et al. (eds.), *Proceedings of International Conference on Recent Trends in Computing*, Lecture Notes in Networks and Systems 341,
https://doi.org/10.1007/978-981-16-7118-0_62

The optical character recognition for the printed English text provides an outstanding output. In the recent days, various authors are working on reading the images and producing the voice output from the images using OCR. This work can provide the state of the art to the visually impaired persons. A knowledge extraction can be improved by just listening to sounds. Optical character recognition (OCR) is an aid for converting the image to text. The OCR can convert the handwritten text as well as the printed text of various languages into digital format which can be manipulated by the machine. Our work mainly concentrates on the Printed English text to a machine editable text. The OCR uses various classification algorithms like neural network [9], Bayesian, nearest neighborhood [15], adaptive classification [18], etc. The nearest neighborhood works perfect for the signboard detection, and neural network and Bayesian are complicated system and can be applied to the handwritten text detection. The adaptive classifier works perfectly for the printed text detection but can still needed improvisation for the skew estimation.

The existing works converting the images to speech uses preprocessing techniques, OCR-based speech synthesis [1], post-processing of spell correction [2], and text-to-speech synthesis algorithms as the basic modules. An extended version of OCR was also implemented which uses the HOG feature extraction [8] and SVM [7] for the classification of the characters. The state of the art of region of interest (ROI) [3] is not implemented by any author. After apply the region of interest, the input image is given to the adaptive classifiers OCR which reduces the complexity of processing the entire image and various spellcheck algorithms are compared and used the best algorithm to get the accurate output.

The popular spellcheck algorithms are LinSpell Linear Search spelling correction algorithm) which uses the Levenshtein distance [16] which finds the same substring many numbers of times which increases the time complexity of the algorithm. The BK tree (Burkhard–Keller tree) [17] uses the Levenshtein distance [16] method but needs to maintain a tree data structure which produces a better output. Norvig's spell correction [6] searches for the correct spelling which has high complexity.

Text-to-speech synthesis algorithms are classified into three broad categories. They are formant synthesis [14] which uses the formant frequency and its bandwidth. Articulatory synthesis [13] uses the natural speech production which is a high computational load and complex to implement. The last type is the concatenative synthesis [12] that produces a pre-recording unit of speech which requires huge dataset. Among all the three formants are an efficient and flexible method to execute the output in less time which is apt for the proposed method.

2 Proposed Work

The work mainly focuses on the processes of image preprocessing and post-processing as the spellcheck algorithm. The images are captured directly by the camera and given for the preprocessing. The image consists of all the data extra information along with the text data. The proposed model can automatically filter the text region from the image using the region of interest algorithm which identifies the region having the text briefly explained in Sect. 2.1 followed by the adaptive thresholding algorithm [4] for preprocessing, explained in Sect. 2.2. The output of preprocessing is a binary image which is segmented from text to character. The OCR uses an adaptive classifier to classify the character which may produce misspell words. The post-processing of spell correction which is carried out by Symspell is based on the approach which takes a word of a sentence and word dictionary at a time and processes the semantic or meaningful words mentioned in Sect. 2.3.

2.1 Region of Interest

The natural image contains lot of other information along with the text. Some images also have less text. The time complexity increases if the complete image is provided for the feature extraction. The image is having a quality of 72dpi work accurately by applying the scene text detection (STD) module for identifying the region containing the text. The algorithm also detects the images having multiple text regions and provides all the data using slicing.

The steps for the algorithm are as follows.

1. Convert the RGB image to grayscale.
2. Remove noise from the grayscale image by applying a Gaussian blurring as it keeps the edges sharp.
3. Dilation is applied to the image over multiple iterations about 4 times using a rectangular structure of (9, 9) matrix to combine the image portion containing text into a single contour.
4. Now we can extract the contours in this image and select only those contours which have the rectangular or square shape with the area above 10,000 thresholds.

2.2 Preprocessing and OCR

The region of interest extracted from scene text detection algorithm is converted into grayscale image. This process removes all color information, leaving only the luminance of each pixel. The proposed work uses k-means with adaptive thresholding algorithm [10] to convert the grayscale image into a binary image.

Image segmentation is an important image processing step, and it is used everywhere if we want to analyze what is inside the image. Image segmentation basically provides the meaningful objects of the image. Segmentation is the process of partitioning a digital image into multiple segments. The aim of the segmentation process is to change the representation of an image into more meaningful and easier to analyze. The steps use the histogram detection method for line segmentation, word segmentation, and character segmentation.

The adaptive classifier [18] is based on structural analysis method which analyzes the character structure like curves and lines in character. The baseline/*x*-height normalization makes it easier to distinguish upper- and lower-case characters as well as improves immunity to noise speckles.

The algorithm compares the output of ROI, preprocessing, and OCR and considers the best result among all to get an accuracy of 95.21%. For all the images only, ROI or preprocessing does not give the accurate output.

2.3 Post-processing (Symspell)

Symspell is a Symmetric Delete Spelling correction algorithm used to find all strings by using the maximum edit distance from a list of strings in less than 1 s as proposed by Reacheal Tatman [11].

The algorithm has the following steps.

(1) Split the sentences into tokens and remove all the stop words.
(2) Check the words in the english_dictionary of 1.3 Megabytes in less the 1 s and collects all the misspelled words.
(3) Symmetric delete spelling correction algorithm generates all the candidate keys deleting the word up to 2 edit distances and identifies the word having least edit distance and an appropriate work is replaces.

3 Experimental Results and Discussions

The work uses an open-source python library with pytesseract tool as the adaptive classifier as the OCR classification. We used eng.tessdata as training dataset for classification of the characters in an OCR. The output generated by the OCR is a string that is written to the text file as shown in Table 1. The output from the OCR may contain errors which are further given as input to the Symspell algorithm to correct the misspell works. The Symspell uses a predefined dataset of ½ million most frequency English words by Reacheal Tatman [11] available on the web. The Symspell algorithm gives an accurate and efficient output as discussed in Sect. 3.2. The last step of the module uses a formant speech synthesis method to convert the text to speech, which generates a sine wave for a particular phonetics and produces a machine voice-based system. As no human voice training is not used resulting in converting the text to speech using less time and space.

3.1 *Comparative Matrix of OCR and OCR with STD Algorithm as ROI*

Table 1 shows the experimental outputs of various complex images producing output after using the ROI and without applying scene text detection as ROI as the existing system. Some of the images having complex output do not produce an accurate output in the existing system without using the STD algorithm before providing the image to OCR. About 100 complex images are tested proved a better output compared to the existing system. The efficiency of the system is increased and gives an accurate output. The accuracy was calculated based on 100 images containing over 5366 original characters detection using Eq. 1.

$$\text{ACCURACY} = \text{Tr} / \text{Ti X } 100 \tag{1}$$

where Tr = total no of characters in result.

Ti = total no of characters in input images.

Table 1 Comparison of proposed work with ROI and without ROI as existing system

Images	Proposed work	Existing system
	SECUTIRY	NO OUTPUT
	PRIVATE CAR PARK LION WALK CHURCH	PRIVATE + CAR PARK LION WALK: CHURCH
	PLEASE TAKE NOTHING BUT PICTURES LEAVE NOTHING BUT FOOT PRINTS	NO OUTPUT
	POLICEY HEADCUWATER	_.POLICE Y HEADCUWATER

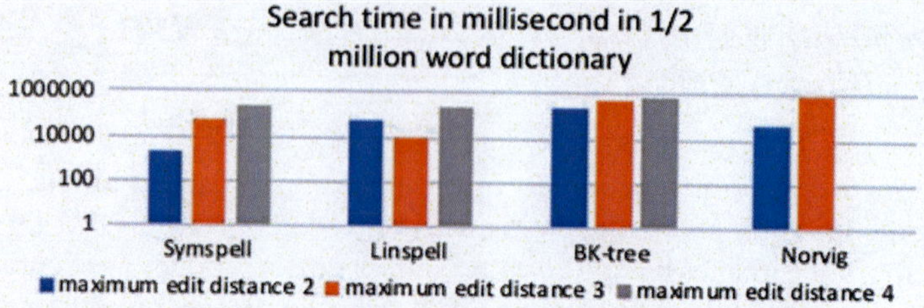

Fig. 1 Different searching algorithms compared with the proposed algorithm

The accuracy of the proposed work is 95.21%, whereas the existing system produces an accuracy of 93.42% producing more accurate output compared with the only OCR.

3.2 *Symspell Model Accuracy*

The Symspell model uses a fast and accurate output for an edit distance of 3 as explained in Fig. 1. Various algorithms like LinSpell [16], BK tree [17], and Norvig's spell correction. The graphical representation shows how the Symspell algorithm work to search 1000 words in ½ million word dictionary for various edit distance. The algorithm produces approximately 5000 words for an edit distance of 290,000 words for an edit distance of 3 and 227,000 for an edit distance of 4. In the proposed model only of edit distance of 2 is used which can produces a 5000 words and can be searched in about 1 seconds. As the model used a printed text which have a clear text and even though OCR output is wrong then, only an edit distance of 2 can solve the issue.

3.3 *Experimental Result and Discussions*

The paper uses total of six modules to implement the proposed works. Table 1 gives the detail experimental steps to get the speech for the complex text images. The time taken to execute the complex image depends on the complexity of model and number of words in the image. The output for the image in Table 1 takes 82 s to get the voice output. Figure 2a is the original complex image which is cropped to get the exact location of the text as presented in Fig. 2b. The character in the image is having various fonts but the preprocessing steps of adaptive thresholding filter give the accurate output in 8 s as in Fig. 2c. The OCR model produces a wrong output in 43 s as in Fig. 2d with an edit distance for deletion is 2. A Symspell algorithm is applied to the text output to convert it to a sematic sentence as proposed in the

a. Original Image

b. ROI after applying STD

c. Adaptive thresholding

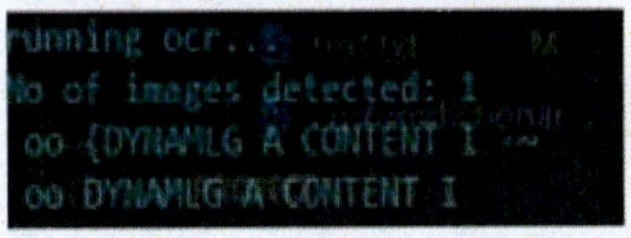

d. OCR

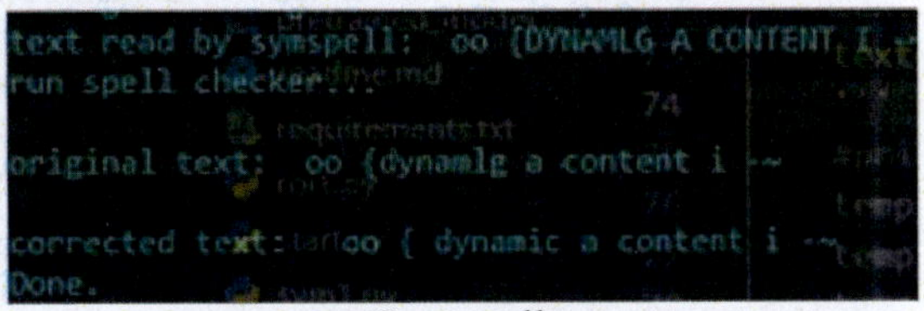

e. Symspell

Fig. 2 Steps of execution and the experimental outputs

paper as there are 2 words in the time complex 13 s. The total time taken to produce a semantic text from a complex image is 82 s.

4 Conclusion

The modified ROI and the application of the Symspell to the text-to-speech conversion system using OCR increase the accuracy of the model. The automation of cropping the images and providing the input to the preprocessing techniques give a good quality output and increase the efficiency of the system. The cropping of the images is not cropped every time correctly which leads to an inaccuracy in the system. The proposed work compares the output of the OCR with ROI and without OCR to produce an accurate output. In the future, the application can be improvised by applying the feature extraction method to differentiate the images which are needed to be cropped to get more accurate results in less time. The spellcheck works perfectly for all the dictionary words, but not for the naïve words. The work can be extended in these areas to get more accurate outputs.

References

1. Sawant NK, Borkar S (2018) Devanagari printed text to speech conversion using OCR. In:

2018 2nd international conference on I-SMAC (IoT in social, mobile, analytics and cloud) (I-SMAC)I-SMAC (IoT in social, mobile, analytics and cloud) (I-SMAC), pp 504–507. https://doi.org/10.1109/I-SMAC.2018.8653685
2. Mullani JJ, Sankar M, Khade PS, Sonalkar SH, Patil NL (2018) OCR based speech synthesis system using labview : text to speech conversion system using OCR. In: 2018 second international conference on computing methodologies and communication (ICCMC), pp 7–14
3. Lin H, Si J, Abousleman GP (2007) Region-of-interest detection and its application to image segmentation and compression. In: 2007 International conference on integration of knowledge intensive multi-agent systems, pp 306–311. https://doi.org/10.1109/KIMAS.2007.369827
4. Roy P, Dutta S, Dey N, Dey G, Chakraborty S, Ray R (2014) Adaptive thresholding: a comparative study. In: 2014 International conference on control, instrumentation, communication and computational technologies (ICCICCT), pp 1182–1186
5. Jurafsky D, Martin JH (2019) Speech and language processing—an introduction to natural language processing, computational linguistics, and speech recognition, 3rd edn. Prentice Hall Series in Artificial Intelligence, Prentice Hall: Englewood Cliffs, NJ, USA
6. Guo J, Sainath TN, Weiss RJ (2019) A spelling correction model for end-to-end speech recognition. In: Proceedings of the 2019 IEEE international conference on Acoustics, speech and signal processing (ICASSP), Brighton, UK, 12–17 May 2019, pp 5651–5655
7. Santoshi G, Mishra SR (2015) Pedestrian with direction detection using the combination of decision tree learning and SVM. In: Satapathy S, Govardhan A, Raju K, Mandal J (eds) Emerging ICT for bridging the future—proceedings of the 49th annual convention of the computer society of India (CSI) Volume 1. Advances in Intelligent Systems and Computing, vol 337. Springer, Cham
8. Santoshi G, Parwekar P, Gowri Pushpa G, Kranthi T (2021) Multiple hand gestures for cursor movement using convolution neural networks. In: Satapathy S, Bhateja V, Janakiramaiah B, Chen YW (eds) Intelligent system design. advances in intelligent systems and computing, vol 1171. Springer, Singapore. https://doi.org/10.1007/978-981-15-5400-1_77
9. Kranthi T, Mishra TK, Deepthi KS, Santhoshi G (Mar 2019) A multi-font OCR for Telugu characters using convolutional neural networks. Int J Res Anal Rev 6(1)
10. Trivedi S, Nandwana B, Khunteta DK, Narayan S (2017) K-means clustering with adaptive threshold for segmentation of hand images. In: 2017 7th international conference on communication systems and network technologies (CSNT), pp 183–187. https://doi.org/10.1109/CSNT.2017.8418534
11. https://www.kaggle.com/rtatman/english-word-frequency dataset for english_dictionary
12. Khan RA, Chitode JS (Feb 2016) Concatenative speech synthesis: a review. Int J Comput Appl (0975–8887) 136(3):1–4
13. Kröger BJ, Birkholz P (2009) Articulatory synthesis of speech and singing: state of the art and suggestions for future research. Multimodal Signals: Cognitive and Algorithmic Issues. pp 306–319
14. Lemmetty S (1999) Review of speech synthesis technology. Helsinki University of Technology Department of Electrical and Communications Engineering. 30 Mar 1999
15. Shen H, Coughlan JM (2012) Towards a real time system for finding and reading signs for visually impaired users. Computers Helping People with Special Needs. Springer International Publishing, Linz, Austria
16. Levenshtein VI (1966) Binary codes capable of correcting deletions, insertions, and reversals. Soviet Phys Doklady. 10(8):707–710
17. Burkhard WA, Keller RM (Apr 1973) Some approaches to best-match file searching. Commun ACM 16(4):230–236. https://doi.org/10.1145/362003.362025
18. Nguyen TT (2002) Adaptive classifier construction: an approach to handwritten digit recognition. In: Alpigini JJ, Peters JF, Skowron A, Zhong N (eds) Rough sets and current trends in computing. RSCTC 2002. Lecture Notes in Computer Science, vol 2475. Springer, Berlin, Heidelberg

Various Security Problems and Its Solving for Future Dynamic IoT-Based Smart Home Automation

S. Karthick, N. Gomathi, R. P. Mahapatra, Anitha Rajakumari, and Pritee Parwekar

Abstract Nowadays, Internet of things (IoT) is becoming a huge research platform because of IoT reducing human regular works. IoT provides lot of facilities with different components and different engineering designs. This fast-growing field is having some issues with securities. IoT is having embedded area, so it gets both software and hardware attacks. Through this proposed work firstly, it summarizes the security threats to current and future smart homes. Secondly, with this information as a basis, it provides a contribution to manage the security overview of a smart home environment. The contribution is in the form of a security module prototype which focuses on detecting malicious behavior and reports it to the end user. This research work introduced an effective security solution for all kind of Internet of things. Through smart hub concept, this proposed system detects evil twin attacks, botnet device, some malicious activities detections, etc. The proposed security system will be shown the result, if the smart hub is a viable location for such a mechanism to reside in both concerning resource usage and detection rate with smart home security system.

1 Motivation

The growth of Internet of Things (IoT) and the spread of smart homes are steadily increasing. In recent years, the market has boomed up with new devices, such as smart locks, heating, ventilation and air conditioning (HVAC) systems. Gartner has

S. Karthick (✉) · N. Gomathi
Department of Computer Science and Engineering, Vel Tech Rangarajan Dr. Sagunthala R&D Institute of Science and Technology, Chennai, Tamil Nadu, India

N. Gomathi
e-mail: gomathin@veltech.edu.in

R. P. Mahapatra · A. Rajakumari · P. Parwekar
Department of Computer Science and Engineering, SRM Institute of Science and Technology, Delhi-NCR Campus, Ghaziabad, India
e-mail: mahapatra.rp@srmist.edu.in

R. P. Mahapatra et al. (eds.), *Proceedings of International Conference on Recent Trends in Computing*, Lecture Notes in Networks and Systems 341,
https://doi.org/10.1007/978-981-16-7118-0_63

predicted number of IoT devices to reach as many as 25 billion in 2020. Many of these devices will be inside the homes of everyday people, bringing technology closer than ever.

However, there have been many setbacks for smart homes regarding their security. There have been several cases where actual devices have been compromised, displaying the importance of high security within this field. Fernandes et al. exploited several vulnerabilities in Samsung SmartThings with accompanying apps to, e.g., disable certain functionality and cause fake fire alarms. Schwartz et al. performed black-box testing on 16 smart home devices and recovered passwords from eight of them. Furthermore, there have been additional reports of actual users and companies being hacked. These attacks range from gaining access to baby monitors to accessing casino backend servers through an aquarium thermometer.

While these reports make the need for security apparent, developing a smart home product with security in mind can be difficult. The resource-constrained hardware means that old conventional methods, such as computationally heavy encryption, do not work effectively on these devices. Even though there are several initiatives around to provide security to resource-constrained IoT devices, much remains to be done.

Intrusion detection systems (IDS) is a traditional field of security still in its incipiency for smart homes. Several new intrusion detection methodologies for resource-restricted devices have been proposed by researchers. However, their focus is mostly on wireless sensor networks (WSN) or general IoT devices, and only a few bring up the case of the smart home. As far as we know, none of these methodologies take the smart hub, a central entity in many smart homes, into account. Exploring detection of threats in the hub may contribute to the overall security situation in smart homes since it increases awareness of threats among users.

1.1 Problem Description

Security in smart homes is a new and complex field. This section brings up the main problems and challenges with security in smart homes.

Heterogeneity: Smart homes comprise a large number of devices which differ in multiple ways. Firstly, the communication protocols differ between devices, ranging from Wi-Fi and Bluetooth to Zigbee and Z-Wave and many more. Furthermore, the data protocols used vary from device to device and include message queueing telemetry transfer (MQTT) and constrained application protocol (CoAP) among others. A general security solution would need to understand and adapt to every protocol. Overall, heterogeneity makes it hard to find one solution that fits every need—especially when numerous new products enter the market every year.

User awareness: The target audience of a smart home is the average person. This target group generally have little awareness of the risks with technology and how to ensure their devices remain uncompromised. Due to this lack of security awareness, customers have little interest in paying more for a more secure product.

Consequently, the market is not particularly motivated to increase the security in smart home technology.

Position of defense: The positioning of defense mechanisms is troublesome in smart homes since we want to secure the network not only from external threats but internal threats as well. A network provider will never notice whether a device sends malicious data locally on an internal network. A router cannot detect if a Z-Wave node abuses the multi-hop feature, in which a node can act as an intermediate router, to alter or drop messages. A specific node does not have the capability to run advanced detection due to its resource-restricted properties. From these examples, we see that no obvious position for a strong defense mechanism that solves every problem is apparent. We use a smart hub as the basis for our defense mechanism. The smart hub provides a position within the local network, relatively close to the end nodes, allowing for detailed knowledge of the network data flows. In addition, the smart hub can provide information on the devices' states.

Resource restriction: Even though smart hubs are generally connected to wall sockets, their computational power and memory usage are resource restricted. Consequently, it is not possible to keep a large database over all known devices and how they communicate. Neither is it possible to run computationally heavy algorithms to detect every behavior of the device and find anomalies. To adapt to the resource restriction and use our resources in the most effective way, we need to use a combination of these two approaches.

2 Related Works

Even though the security of smart homes is a rather young research area, many studies on the subject exists. Here presents research relevant to this thesis and explain how it relates to this work.

2.1 Risk Analyses from Other Researchers

Risk analyses have long been used to evaluate risks and threats in a system, and several methods on how to perform these analyzes exist on the market. However, Nurse et al. claim there is a need for new approaches for conducting risk analyzes within IoT systems as many of the traditional risk assessment methodologies were developed prior to the pervasive cyber-physical networks currently deployed. Consequently, these methodologies have weaknesses when assessing modern IoT environments. Due to these factors, we focused on risk analyses that specifically targeted IoT or smart homes.

In 2016, Jacobsson et al. conducted a smart home project for which they did a rigorous risk analysis focused on IoT using the ISRA risk analysis method. The analysis resulted in a wide range of possible issues, several applicable to smart homes.

Furthermore, other studies investigating the security in smart home applications and devices mainly present vulnerabilities in the systems and propose solutions on how to mitigate these vulnerabilities. They sometimes touch on the risks within the system, but this topic is not their primary focus. Our focus is to understand a large variety of risks to provide an extendable platform on which prevention for threats against smart homes can be implemented.

2.2 Security Management in Smart Homes

Regarding security management in smart homes, Batalla et al. published an extensive paper on smart home security challenges in which they propose a solution where network providers implement a security layer between Internet service providers (ISP) and home networks. They propose to do this via a multi-functional home gateway residing in the home of the end user and with a home area network management system. Sivaraman et al. in turn propose a similar solution with a security management provider. However, contrary to Batalla et al., they argue that this device does not necessarily need to be operated by the network provider. Their justification for the decoupling is that it opens up for more actors on the market of security solutions which allows for competition within the security field, ideally leading to more secure solutions. In our work, we instead explore the viability of placing a security mechanism in a smart hub since it has additional information on the smart home state.

2.3 Intrusion Detection in IoT Devices

Multiple works on intrusion detection for IoT exists. However, the topic is still in its incipiency and many problems need to be solved before the technology can be considered mature. Many of these works on intrusion detection focus on distributed solutions in sensor networks, where each node in the network contains an instance of or part of an IDS. Drawbacks of a distributed IDS in the smart home environment include reluctance from manufacturers to agree on a common system, and the heterogeneous nature of smart homes comprises several devices using different technologies. Our solution instead investigates a centralized system for detecting security breaches. Sforzin et al. developed a project called RPiDS in which they provided a portable Raspberry Pi 2, a small single-board computer, running the well-known general-purpose intrusion detection system (IDS) Snort. The authors highlighted two main features. Firstly, due to its portability and ease of use, deploying the system on any desired location would be easy. Secondly, the architecture allowed multiple Raspberry Pis to work together to increase detection rates and reduce false positives. However, they found that a single Raspberry Pi had issues with high network loads.

Our research differs from these studies on how our focus is on the smart hub and how to use the data it provides. Using intelligent filters and data from the smart hub, the amount of network traffic necessary to examine can be minimized. Furthermore, smart hub data could help create more situationally adapted detection filters than otherwise possible, leading to increased efficiency and decreased CPU/RAM load.

2.4 Resource Restriction

Smart home devices often have restrictions in their resources. Some devices may be battery operated, giving them a restriction on their power supply. Other devices have restrictions in CPU, memory or bandwidth. These restrictions give way to challenges that are not present for less resource-restricted devices such as workstations and laptops. The main challenges can be divided into two categories: communication issues and computational issues. A common communication challenge is that battery-operated devices may periodically go offline in order to save power. Furthermore, there might be a need to combine data since the data is transmitted less often. As these changes affect the prerequisites for communication, it also creates a need for new or updated communication protocols where transmission only happens during specific periods of time with concatenated packets.

Computational challenges are affected both by CPU- and memory restrictions. Low CPU resources may make traditional solutions for computation and security impossible to implement. For example, some cryptographic functions need a substantial amount of CPU resources to be fast enough to be practical. As such, these are currently not an alternative for CPU-restricted devices. However, there exist projects to find low demanding cryptographic algorithms for resource-restricted environments.

3 System Models

3.1 Communication Protocols

The IoT has brought with it many new communication technologies, the development of which has often been driven by the challenges caused by resource restrictions and new use cases. As an example, many smart home devices broadcast their wireless traffic on 433 MHz or 900 MHz since a lower frequency penetrates walls better. Some technologies used in smart homes also utilize multi-hop routing in order to reach even further with the help of other nodes. This section brings up some of the most common protocols and communication technologies within smart home devices and compares it to more traditional technology.

Table 1 Common protocols in traditional technology in terms of the OSI model

Name	Protocols
Physical layer	Ethernet, IEEE 802.11, USB, Bluetooth, DSL
Data link layer	Ethernet, IEEE 802.11, PPP, ARP, NDP, USB
Network layer	IPv4, IPv6, ICMP, OSPF, RIP, NAT
Application layer	HTTP, DNS, SMTP, POP3, UPnP, SOAP

3.2 Common Network Protocols

In traditional networking and the Internet at large, the layered structure of the OSI model are mostly used. Layers 3 and 4 are dominated by the TCP/IP communication stack, while either Ethernet or IEEE 802.11 (Wi-Fi) usually inhabits the two lowest layers. In layer 5 and 6, we find important encryption protocols such as SSL and TSL. Layer 7, the application layer, contains a large variety of protocols including well-known ones such as HTTP, domain name system (DNS) and SMTP. Table 1 presents common protocols in terms of the OSI 4 layers.

3.3 Network Protocols in the Smart Some

In order to be easy to set up and use, smart home devices are almost always wireless, and hence, they almost exclusively use network protocols designed for wireless transfers. Some of the most common network protocols used are described in this section.

Wi-Fi is frequently used in smart home devices since it makes the devices compatible with smartphones and other already existing equipment, which in turn makes it easy for the user to get started without any additional one. The introduction of Wi-Fi mesh networks means better coverage which in turns enables smart home devices by weakening the demand on the power of their radios.

Bluetooth is another traditional technology which has made its way into the smart home. Many users are already used to using Bluetooth headphones, speakers and keyboards with smartphones and tablets. All this makes it easy for the user to incorporate new devices that use the same technology. There also exists a more energy-efficient version of Bluetooth called Bluetooth Low Energy (BLE) which makes battery driven devices last longer.

Z-Wave is a relatively young technology which broadcasts on frequencies around 900 MHz. It is specifically designed to be used in smart homes and is therefore energy efficient, uses multi-hop routing and focuses on low latency and reliability instead of maximizing data throughput. Z-Wave is a proprietary protocol owned by the Z-Wave Alliance, which means little open research has been done on its security aspects. However, in 2016, a mandatory security standard, called the S2 standard, was issued by the Z-Wave Alliance. All products incorporating Z-Wave must follow this standard to receive a Z-Wave certificate.

Zigbee, on the other hand, is a non-proprietary protocol developed by the Zigbee Alliance. Aside from this, the hardware is designed to be as cheap as possible in order to enable product owners to lower their costs and bring consumers cheaper technology. Over the history of Zigbee, there have been multiple reports of breaches and security issues, but these have been resolved and nowadays Zigbee has grown into a more secure protocol.

3.4 Application Protocols in the Smart Home

CoAP, defined in RFC 7252, is a Web transfer protocol on top of UDP which is designed for constrained devices and low-power, lossy networks [1]. MQTT is a messaging protocol based on the publish/subscribe pattern, a pattern which is very suitable for a highly generic environment such as the smart home. The protocol requires a central server, called message broker, through which all messages are routed. One key feature of MQTT is its low network overhead which enables resource-restricted devices such as low-power sensors and actuators to use it [2].

3.5 Security Risks

In the literature for security in smart homes, the terms risk, threat and vulnerability are often used. Sometimes, they are used interchangeably, and it can be difficult to pinpoint what the words mean at a given moment. In this thesis report, we are using the following definitions as defined in:

- A threat is anything that has the potential to damage a system.
- A vulnerability is an existing weakness in a system, which can be exploited.
- A risk is the probability of a threat, coupled with the cost of the consequence of said threat.

Jacobsson et al. published an extensive risk analysis on IoT, where many of the found risks fit into the profile of smart homes as well. We have highlighted some of the most pressing security risks for smart homes from this report, namely those of a mean risk value higher than 8, and present them. In summary, these risks fall into two categories, inadequate security mechanisms in all levels of the system, and bad behavior from users. These risks are often brought up as pressing risks in the field.

The Open Web Application Security Project (OWASP) is a well-known foundation which have compiled a list of the top 10 vulnerabilities for IoT devices in 2014. Even though the fields of IoT and smart homes do not entirely overlap, the main security issues of IoT can be applied to a smart home setting as well. Insecure control interfaces, whether it is a Web-based architecture or a smartphone app, are important attack surfaces to secure. If an attacker is able to gain access to said interface, they can operate the device to send and retrieve data as the original user. Furthermore, they might be able to reconfigure the device entirely, detaching the original owners and leaving them with an uncontrollable device. Hence, securing both virtual and physical access to the device is of great importance.

Denning et al. published a study which not only brings up risks with smart homes but also the end consequences of a possible breach. Their reasoning describes the process in three steps:

- Low-level mechanisms are exploitable security risks in the smart home. Example: Hack a smart lock.
- Intermediary goals are the action(s) the attacker is trying to accomplish. Example: Enable physical entry to the apartment. Viewing private data.
- High-level goals are the end-goal for the attacker. This step provides the actual benefit of the entire process for the attacker. Example: Blackmailing.

They argue that by recognizing possible end goals for attackers, new risks for smart homes can be discovered that were not previously considered.

3.6 *Attacks on Smart Homes*

There have been a number of attacks against devices that can be found in smart homes. Some have been documented in research as proof-of-concept attacks in an attempt to motivate further development of security countermeasures, while others are real-life breaches which can mostly be found in online articles, on news sites, and blogs. Some of the most recent academic papers on attacks against smart home devices are presented in Table 2.

3.7 *Proof-of-Concept Attacks in Research*

Ling et al. published a paper in 2017 on the Edimax SP-2101 W smart plug in which they are able to retrieve full control of the smart plug with little or no access to the network it is residing in. Even if the end user tries to use a password to protect the device, the infrastructure of the system allows an attacker to stealthily acquire this password in multiple ways. Furthermore, the authors show the possibility to upload malicious firmware to the device in order to gain total control of it.

Hernandez et al. evaluated security in the Nest Thermostat and found lacking hardware protection. They showed that an attacker with physical access can overwrite the firmware, using only a standard USB-cable, to gain root access on the device. This technique only required 15 s alone with the device to perform the attack. In a paper by Lei et al., the security properties of the voice assistant Alexa is studied. The focus of the authors was specifically put on protection from acoustic attacks, where

Table 2 OWASP Top 5 IoT vulnerabilities

Rank	Vulnerability	Examples of issues	Impact
1	Insecure Web Interface	Use of default passwords, no lockout on multiple failed logins, valid usernames can be determined	Severe
2	Insufficient Authentication Insufficient Authorization	Insecure passwords are allowed, credentials transmitted in clear text, no separation in levels of privilege	Severe
3	Insecure Network Services	Unused ports are open, services vulnerable to buffer overflow, abnormal traffic not filtered out	Moderate
4	Lack of transport encryption integrity verification	Use of unencrypted protocols, encryption protocols are out of date, packet integrity is not checked	Severe
5	Privacy concerns	Arbitrary data is collected, personal data not properly secured Not using a data retention policy	Severe

the attacker plays a sound and tricks the voice assistant into believing a human is issuing the command, e.g., purchasing products online. The authors developed two proof-of-concept attacks which show that Alexa is susceptible to these kinds of attacks.

3.8 Real-World Attacks on Smart Homes

One attack type that can especially benefit from the increased number of devices due to the growth of smart homes is distributed denial of service (DDoS) attacks. This was clearly shown by the Mirai worm, a worm that has affected hundreds of thousands of IoT devices, including many found in smart homes. Mirai was discovered in 2016 and has been the cause of several big DDoS attacks. One of these attacks had a traffic peak at 1.1Tbps, magnitudes above what most sites on the Internet can handle. Mirai exploits naive vulnerabilities in IoT devices, mainly the use of common default passwords, but has the potential to severely impact the infrastructure of the internet. This was shown by the October 2016 attack against the service provider Dyn which caused several large sites on the internet to become unavailable.

There have been multiple news reports concerning a few minor single-target attacks as well. In 2014, a man hacked a connected baby monitor and screamed "Wake up baby" multiple times at a sleeping baby. In 2016, hackers attacked the thermostats in two buildings in Finland which resulted in non-working heating in the facilities. At the time, the outside temperature was around 5∘C. One attack in 2018, which went rather viral, concerned not an attack against a private home but against a casino. The attackers managed to get their hands on the high-roller database by entering the network through a smart fish tank thermostat.

3.9 Traditional Security Measures

Traditional network security consists of several techniques which are often bundled together to provide a secure environment. This section brings up a few of the most common techniques. One of the most used techniques is encryption. Modern protocols, in multiple stages of data transfer, use public-key cryptography and intricate hand-shaking protocols to set up secure connections. Encryption is often seen as a base premise for confidentiality. Authorization is the process of deciding whether a certain entity is allowed to perform a certain protected data. A prerequisite for authorization is authentication. Traditionally, authentication can be done in various ways: passwords, passcodes, biometric validations and near field communication (NFC) among other. Two-factor-authentication (2FA), where users need to authenticate themselves in two different ways, is becoming increasingly popular.

One of the oldest protective security measures is firewalls. Nowadays, most homes are equipped with a home Wi-Fi router in order to supply a wireless Internet connection for portable devices such as smartphones, tablets and laptops. These routers usually include a basic firewall. Modern operating systems for computers, such as Microsoft Windows, Mac OS and various Linux distributions, provide a basic firewall by default as well. The existence of zero-day exploits means that not all attacks can be prevented from entering a system. However, it is important to detect when such a breach has occurred. Traditionally, intrusion detection systems and intrusion prevention systems (IPS) have been used to fill this task. IDSs only monitor traffic passively while an IPS also has the possibility to take action against the threat.

Another way to improve the chance that attacks do not go undetected is extensive logging of all activity on the device. These logs help system administrators analyze system actions in order to find anomalies, understand the passage of events and use this data to improve the security. Penetration testing is often used by companies to make sure their product or office network is impregnable. These tests are run by security experts and quality of the outcome relies heavily on the skill and experience of the testers. Common Criteria (CC) is a certificate standard which reassures that a product has undergone significant security review by an expert third-party certifier. It is an international recognized standard which assesses security functions, performs penetration tests and evaluates the development process of the product to assure security compliance. Certificates need to be re-issued if any hardware of software changes occur.

3.9.1 Smart Home Security Measures

This section brings up issues with traditional security measures based on the environment of smart homes and the current risk situation. Furthermore, it presents a set of security techniques and approaches adapted to a smart home setting.

3.9.2 Issues with Traditional Security Measures

One general issue with traditional security in smart homes is the expectations from the user. In a research study by Sandström about smart home users, he found that users are less willing to use a smart home device if it is harder to access. This means that if security measures of a smart home make it noticeably more difficult or more time-consuming for a user to use services of the smart home, the user is less likely to use them as much. As a consequence, it is important to provide security solutions which does not affect the user productivity in the system. As an example, new ways to authenticate a user might be needed to ease the process while still keeping high security.

The non-technical nature of the user, or rather the lack of knowledgeable system administrators, renders some security countermeasures unusable. This includes logging, where the user generally neither has the knowledge nor the interest to assess device actions via data logs. Furthermore, evaluation of penetration tests would not be feasible either. In addition to the above issues, the heterogeneous and resource-restrained nature of smart homes can make it difficult or impossible to apply traditional security measures. Hossain et al. pinpointed a number of important constraints and how they make applying traditional security measures difficult. The limitations on resources in smart home devices mean that computationally heavy cryptographic algorithms are not applicable, and security solutions that have not specifically been designed with memory and storage limitations in mind may not work. For example, local firewalls or IDSs on smart home devices may not be possible due to lack of processing power.

The heterogeneity of smart home networks also come with problems. Not only do smart home networks contain a lot of devices, making scalability of the security solution an issue, these devices may also be mobile, joining and leaving the network at any time, something that existing security models may not be able to cope with. In addition, the heterogeneity when it comes to communication protocols can also leave traditional security measures lacking. Hossain et al. also point out that the slimmed, embedded operating systems on IoT devices often have a thin network protocol stack that might lack the security measures of traditional systems. In addition, the OS or protocol stack on the device may not be able to integrate new code, making it very difficult to patch discovered vulnerabilities.

4 New Proposed Techniques and Approaches

Even though the field of security in smart homes is relatively young, some new techniques and approaches have been proposed. This section brings up a few of them as a contrast and extension to the section on traditional approaches. All approaches either address the problem of resource restriction or the lack of technical experience among users.

One of the most used encryption layers, transport layer security (TLS), operates on top of TCP in the TCP/IP stack. However, a similar protocol called datagram transport layer security (DTLS) has been developed to provide support for encryption over UDP and can therefore be used with the resource-efficient protocol CoAP. This enables some nodes to introduce encryption into data transmissions, leading to higher security. However, the devices which are most resource restricted will still be unable to use this feature due to the complexity of the encryption algorithms themselves.

There is ongoing work to secure MQTT, a frequently used protocol within smart homes. In a study from 2015, researchers used a lightweight attribute-based encryption (ABE) algorithm to establish secure communications in an MQTT broadcast domain and specified the protocol SMQTT. The technique was also adapted for MQTT-SN, MQTT for sensor nodes, in order to provide support for power constrained devices. Several new ways to authenticate users have been explored. Voice authentication has been a concept for a long time, and patents dates back all the way to 2000. However, this technique was not commonly used until a few years ago, and nowadays, companies like Nuance, VoicePin and Google offer this service on a large scale. Zhang et al. extended this idea to involve both voice and gesture authentication, in order to prevent an attacker from carrying out a replay attack by using a pre-recorded audio clip. Lei et al. suggested that voice features only should be available if that particular user is present at home.

IDSs cannot stop vulnerabilities from being exploited, but detection of an attack can provide the end user with information concerning potential breaches of particular devices in their smart home setup. This information could help the users, even those with little to no technological knowledge, to take action by either sending the device to a service shop for repairs or by replacing it. When an infected device only gets a small-time window to transfer the virus which infected it, there is less chance it will succeed and hence virus spread will be mitigated in that particular network. The work in this thesis is focused on this approach.

4.1 Design

As previously stated, we use a smart hub as the location for the Security Supervisor. We compared several smart hubs and chose Home Assistant due to its popularity, availability, and openness. Thus, we designed our Security Supervisor to fit into the Home Assistant architecture even though the general principles of the software are adaptable to any smart hub system. A key design feature was that the Security Supervisor should be easily extendable. This section describes the architecture of both Home Assistant and the Security Supervisor and shows how they interface with one another. The Security Supervisor is designed to be able to integrate with any smart hub, but the implementation described in this thesis interfaces with Home Assistant.

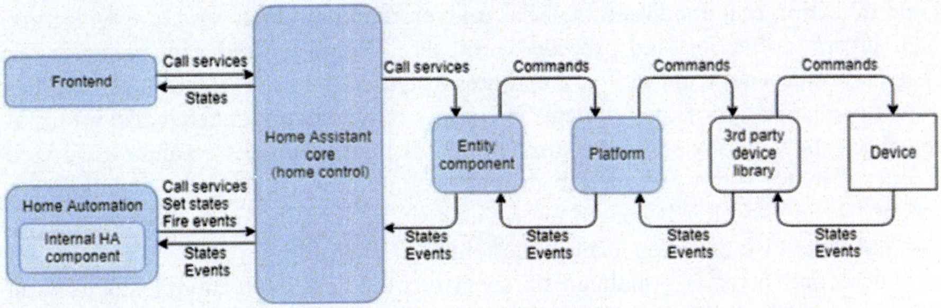

Fig. 1 Home automation embedded design

4.2 Architecture of Home Assistant

Home Assistant has a straightforward and modular software architecture. It consists of a central core, a user interface, home automation and components to communicate with smart devices in the home. In Fig. 1, the shaded parts of the Home Assistant source code while the non-shaded parts are external libraries and physical devices.

The core of Home Assistant consists of an EventBus and a StateMachine. The EventBus allows other parts of Home Assistant to fire and listen to events in order to communicate changes to each other. The StateMachine holds the current state of the smart home, and changes to the states are facilitated through the events sent over the EventBus.

Communication with smart home devices is handled through entity components and platforms. The entity components are components which handle communication of a type of device, i.e., lights or switches. They contain shared functionality common to a type of device. The platforms expand the entity components so that they are compatible with certain brands of devices. The actual communication with the device is in the shape of external third-party libraries. The platforms communicate commands, states and events with these libraries through API calls. The modularity allowed by the entity components and platforms makes it easy for developers to add support to more devices.

The final part of Home Assistant is the home automation. This is controlled by user configurations and internal components that use triggers from events together with information from the core to activate commands. An example of an automation would be to turn on a light when the user comes home and it is dark outside.

4.3 Architecture of Security Supervisor

The threat detection is added to Home Assistant as an internal component. As a component, it has access to the EventBus and internal knowledge of the hub, allowing it to factor details of the smart home setup into its threat detection mechanisms. The threat detection component in itself has a layered and modular structure with three main layers: collection and preprocessing, threat analysis and data presentation. These layers are depicted in Fig. 2 and are explained below. The first layer handles collection and preprocessing of data. This is not only network data but also information about states of devices in the smart home. The data is processed and categorized in order to match filters provided from analysis modules in layer two. All data which matches a module filter is passed on to that specific module for further processing. The first layer also provides utility functionality for profiling.

The second layer is populated by security modules. Each module focuses on detecting a particular threat and provides, as previously mentioned, a filter to match to general characteristics of this threat. Once data enters the module, an internal algorithm evaluates the data on a deeper level and alarms layer three if any malicious activity is found.

The third layer acts as an aggregator and summarizes all information from security modules in layer two in order to present the data to the end user. Even though the design of this layer might affect the outcome of the end product, since users prioritize easy-to-use tools, this layer has not been the focus of the thesis and will not be presented further.

The layered structure of the Security Supervisor not only fits in with the general layered structure of Home Assistant, it also allows for easy extension of the Security Supervisor. With the fast pace of the market and the always evolving cyber-attacks, easy extendibility is a key feature for software which focuses on detecting security breaches.

Layer 1: Data collection and preprocessing.

The main purpose of layer one is to act as a utility layer and provide easy and fast access to data gathered from the device for the analysis modules in layer two. The layer itself can be divided into three separate parts: reading input data, communicating

Fig. 2 The three-layered architecture of the Security Supervisor

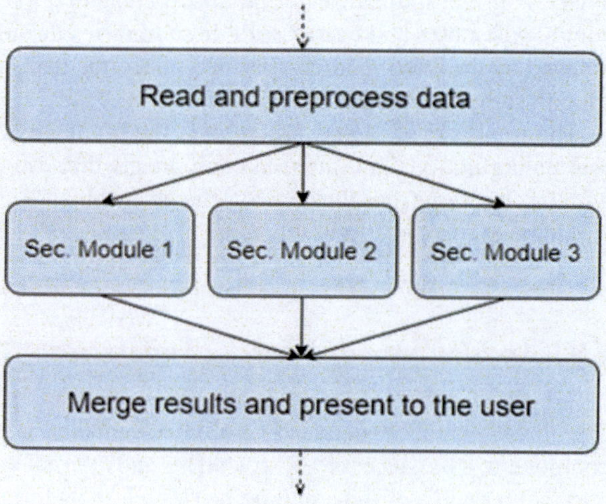

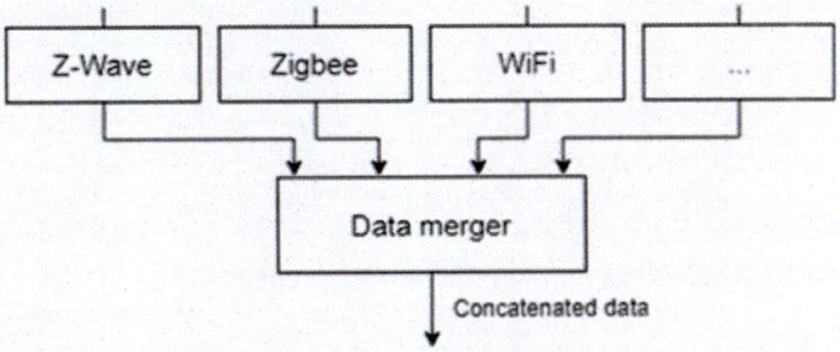

Fig. 3 The traffic concatenation module provides generic interface for multiple technologies

with the smart hub and profiling. These parts will be explained in detail below. In order to scan for threats, the Security Supervisor needs to be able to intercept network traffic and process it. Smart homes consist of devices with multiple communication protocols; it is important to provide a generic interface for input data. This interface component is responsible to gather data from multiple sources and merge it into one single data stream, a process which is depicted in Fig. 3. The concatenated data from the data merger will be profiled and processed by each analysis module for which the data match the announced filter of corresponding analysis module.

One big advantage with locating threat detection within a smart hub is that it gains access to the data of the smart hub. Layer one is responsible for gathering updated information about the state of the smart home and to communicate any changes to the analysis modules. It also provides the network capture drivers with information about smart device states in order to limit the amount of network traffic which need to be processed. Instead of listening to all devices on the network, including PCs streaming movie content or performing other high network load tasks, the Security Supervisor needs only to process smart home device data. With less data to be processed, the amount of dropped packets decrease. The packet drop rate was mentioned both by Sforzin et al. and Kyaw et al. as a main issue with threat detection on resource-restricted devices.

The Security Supervisor uses profiling as a central concept. Layer one provides a set of high-level functions to maintain profiles for each device. Once network data has been captured and preprocessed, it is run through a set of profiling functions provided by the analysis modules. By giving the responsibility of profiling functions to the analysis modules, we assure that the data needed for the modules to work is guaranteed to be profiled. However, a set of default profiling functions for common protocols, such as IPv4, TCP and UDP, are provided by layer one by default in order to reduce redundant profiling functions. The profiling process and how it affects the other layers in the Security Supervisor are depicted in Fig. 4. Especially notice how layer two is separated into two sections to adhere to the profiling—one part to perform the profiling and one part to run the actual analysis.

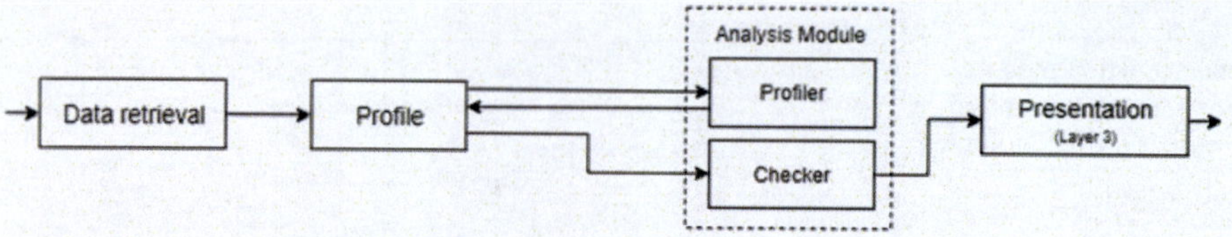

Fig. 4 Data flows through the profiling functionality

5 Results

This section presents results from the experiments on the Security Supervisor. Results concerning resource usage and focuses on what level of detection rate the analysis modules provide.

5.1 CPU and RAM Usage

Two different network libraries were considered—Scapy and Pypacker. Performance in terms of CPU and RAM usage differed a lot between these two libraries. A comparison between the number of packets each of them are able to process per second is presented in Table 3.

Results from running the Security Supervisor with Scapy are presented in Fig. 5. When exposed to a low network load with normal traffic from smart devices and a single AP, CPU usage averaged 14% and RAM usage did not differ from the baseline. However, when exposed to a higher network load with 6–8 nearby APs, Scapy was too slow to process data which led to a constant increase in RAM and a system crash within 29 min. APs generally send 10 beacon frames per second resulting in 60–80 frames per second for 6–8 APs, which exceeds the processing capacity of 57 packets per second for Scapy.

Pypacker was able to handle a high load without any noticeable issues. When exposed to the higher network load, CPU usage averaged at 4% and RAM usage did not differ from the baseline. No tests were carried out for the easier case with lower network load due to the high performance of the library. Results from the Pypacker tests are presented in Fig. 6. Note that these tests were carried out in periods of only 30 min instead of two hours due to two main reasons—the large expenditure of time

Table 3 Comparison between Scapy and Pypacker performance-wise

Library	Packets per second	
	Pypacker test	Our testbed
Scapy	726	57
Pypacker	17 938	15 941

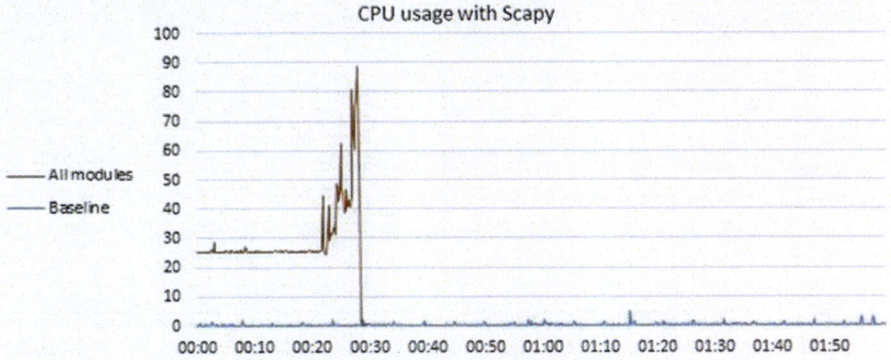

a) CPU usage (%), high network load

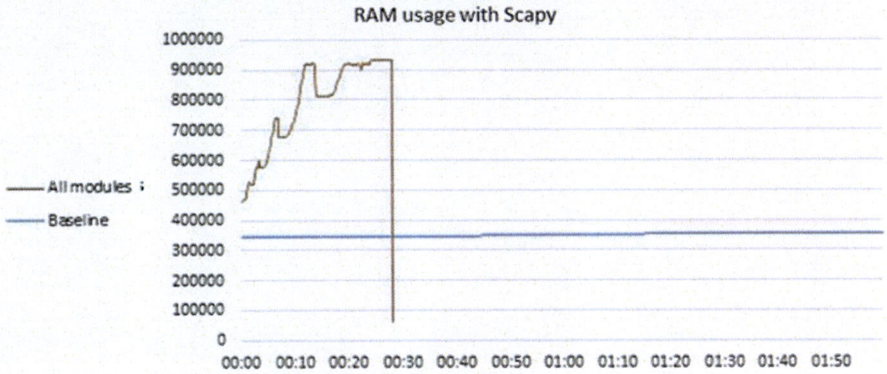

b) RAM usage (KB), high network load

Fig. 5 CPU and RAM usage over time when using Security Supervisor with Scapy, each with low and high network traffic load

to carry out tests and the fact that resource impacts are noticeable already early in the process.

5.2 *Permanent Storage*

Figure 7 shows an average of the results of the iostat runs for the tests. Both 7a and 7b show a comparison between the three cases of tests, namely the base version with only Home Assistant running, the botnet version where only the botnet detection module is running, and the full instance of the Security Supervisor. While all datasets show some significant variance in their values, it is clear from the charts that both the number of write and the data written to the memory cards is the lowest in the base case and the highest in the full version. From the values collected during the tests,

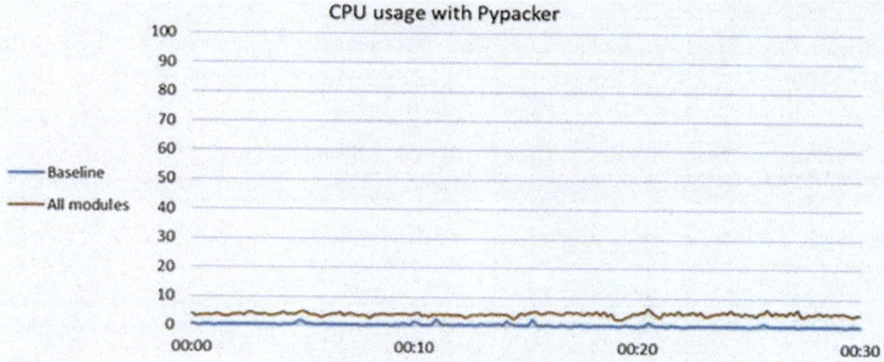

a) CPU usage (%), high networkload

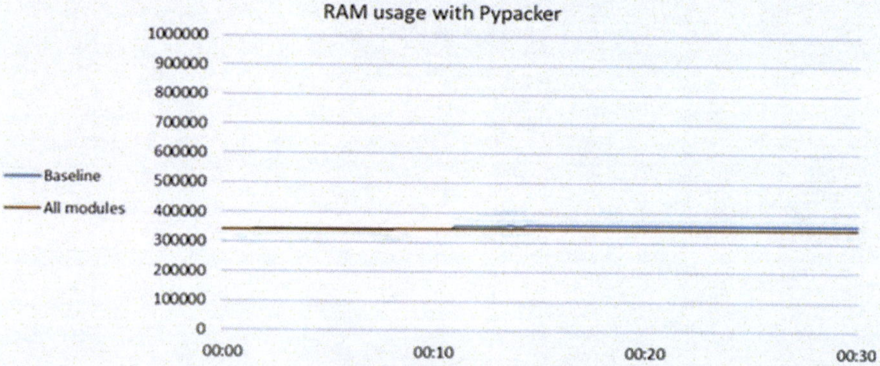

b) RAM usage (KB), high network load.

Fig. 6 CPU and RAM usage over time when using the Security Supervisor with Pypacker

we can make a rough estimate of the expected lifetime of the memory card in the different versions tested. To make this estimate, we need to establish some values.

We use a 16GiB memory card. Out of these 16 GiB, 4GiB is taken up by the card's internal functionality, as well as the operative system of the Raspberry Pi. As such, there are 12GiB available for Home Assistant and the Security Supervisor. The file system write block size (wbs) of our operating system is 4KiB, and our memory card has an erase block size (ebs) of 4MiB. This gives us 3146 sectors (ws) available to use in the card. Assuming up to 3000 write cycles (wc) per sector we can theoretically do 9.4 $*$ 106 4Kib writes in total to our memory card. From the iostat data, we discovered the average number of writes per second (wps), as well as the average size of these writes (s). Using this data, we can construct a formula that gives us the lifetime of the card in years. The results of this formula for each test case can be found in Table 4. The derivation of this formula can be seen below.

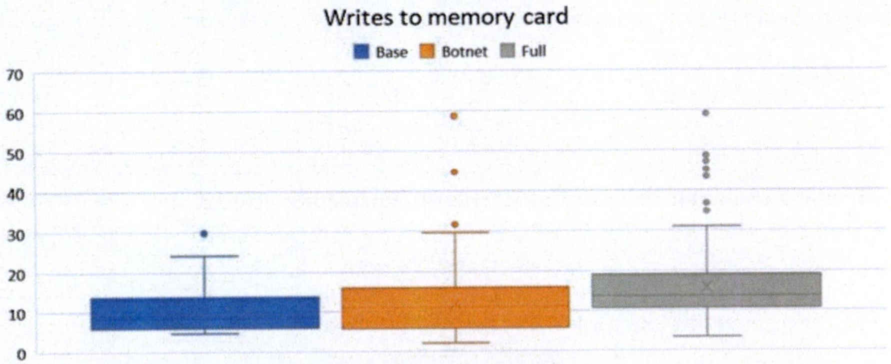

(a) Memory writes over time.

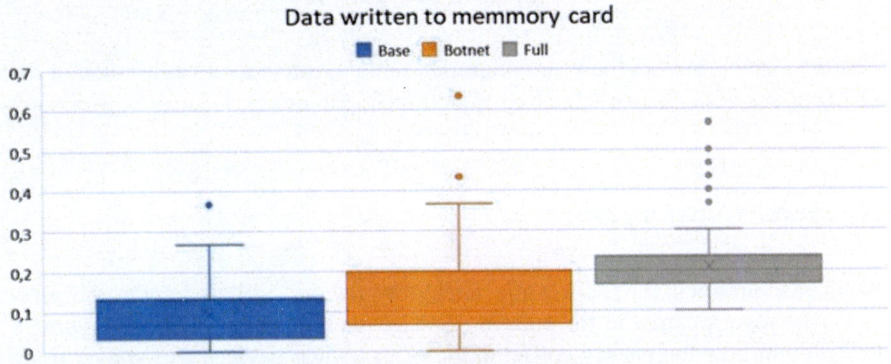

(b) Data written to memory over time.

Fig. 7 The number of writes to memory over time, as well as the amount of data written to memory over time, for three different cases

Table 4 Estimated lifetime for the memory card given the three test cases. The average writes per second from our test data can be found in the w/s column, while the KiB/w denotes the measured average size per write

Version	w/s	KiB/w	wb/w	w/sector	w	y
Only Home Assistant	1.1	9.8	3	333	$3.1e^9$	91
Only botnet module	1.2	11.4	3	333	$3.1e^9$	81
Full Security Supervisor	1.8	14.0	4	250	$2.4e^9$	42

Find the amount of write blocks per average write:

$$\left(\frac{\omega b}{\omega}\right) = \begin{cases} \frac{s}{\omega bs} & \text{if } s \bmod \omega bs = 0 \\ \frac{s}{\omega bs} + 1 & \text{otherwise} \end{cases}$$

Find the number of writes per write sector:

$$\omega p\omega s = \frac{\frac{\omega b}{\omega}}{\frac{ebs}{\omega bs \times \frac{\omega b}{\omega}}} = \frac{\omega bs \times \left(\frac{\omega b}{\omega}\right)^2}{ebs}$$

Find the amount of total available writes to the memory card:

$$t\omega = \omega p\omega s \times \omega s$$

Find expected lifetime for the memory card (in years):

$$\text{years} = \frac{t\omega}{\omega ps} \times (60 \times 60 \times 24 \times 365)^{-1}$$

As the results show in the year column of Table 4, the theoretical lifetime of the card decreases with 49 years due to the additional load of the Security Supervisor.

5.3 *Power Consumption*

Power consumption in the setup is affected by two factors: current draw by the wireless USB network card and the current draw by the Raspberry Pi itself. Figures 8 and 9 present how the energy consumption of these two devices varied over time. The energy consumption of the wireless network card was not affected by the Security Supervisor at all, while the Raspberry Pi experienced a small increase in power consumption. The Raspberry Pi used slightly more energy with the Security Supervisor running than when only Home Assistant core modules were active.

Since the voltage for both the Raspberry Pi as well as the wireless network card was 5 V at all times, we can compare energy usage based on the amount of current used. Average current usage for the Raspberry Pi increased from 0.43A to 0.48A when using the Security Supervisor. The wireless network card averaged at 0.14A. Table 5 summarizes the energy usage measurements for both of these devices.

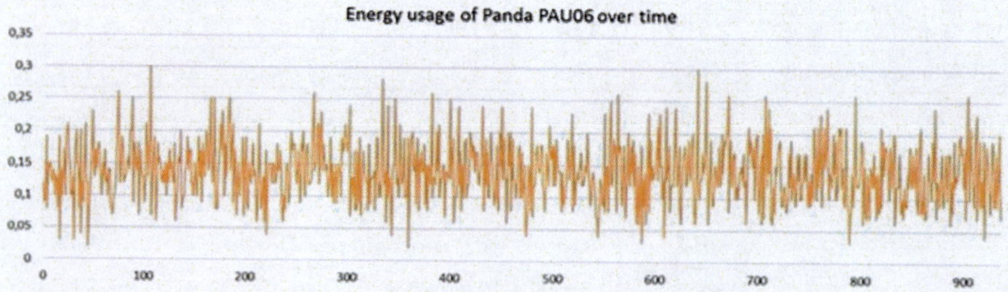

Fig. 8 Energy usage over time of wireless network card Panda PAU06 in terms of current, measured in ampere

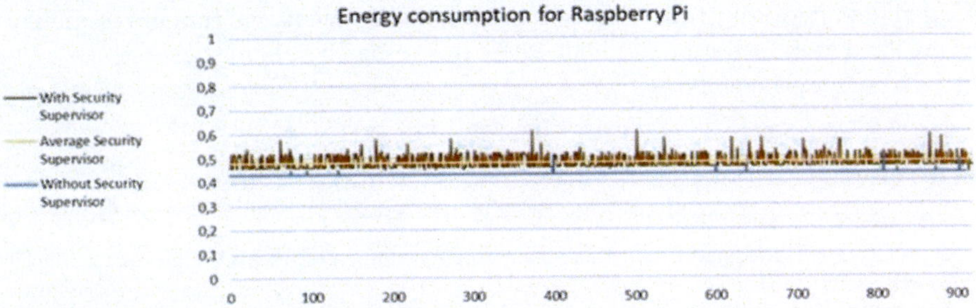

Fig. 9 Energy usage over time of Raspberry Pi running Home Assistant in terms of current, measured in ampere

Table 5 Energy usage without and with the Security Supervisor running

Device	Baseline	With Sec. Sup	Increased usage		
			Current	Cost/month*	CO^2/month
Raspberry Pi	0.43A	0.48A	0.05A	0.12SEK	0.133 kg
Panda PAU06	0A	0.14A	0.14A	0.36SEK	0.385 kg
Total	0.43A	0.62A	0.19A	0.60SEK	0.518 kg

* Cost is based on 0.70SEK/kWh according to statistics from Vattenfall and a voltage of 5 V

6 Conclusion

It is very clear that smart homes are, and will continue to be an appealing target in cybercrime. Security breaches in smart homes can be used as steppingstones for cybercrime that can have far-reaching and severe consequences for both society and individuals. Smart homes with their multitude of devices are an appealing target for obtaining DDoS slaves. These DDoS attacks could potentially target important societal infrastructures, causing severe damage. Smart homes also make it possible for attackers to collect sensitive information about individuals, which could cause severe economic damage or even physical harm to the individual. In this thesis work, we have focused on developing and evaluating a security solution for the smart home that can mitigate security threats against individuals and society. Solutions to mitigate two specific attacks have been implemented. The attacks in question are the use of smart home devices in botnets, and the evil twin attack, where an adversary tries to steal devices from an access point to steal information.

The current Security Supervisor is not able to be deployed at this time. Further improvement is necessary. This is because there are significant weaknesses in the functionality of the security modules. A large amount of additional security modules needs to be developed as well to offer a wider security solution. From a resource usage perspective, we believe that it is possible to place an IDS for the smart home on a smart hub. It is our opinion that a smart hub of a similar capacity to the Raspberry Pi has enough hardware and processing capability to handle a limited IDS while still being able to perform its function as a smart hub. We encourage further research in the area and look forward to find out where this research field may evolve.

References

1. Light RA (2017) Mosquitto: server and client implementation of the MQTT pro-tocol. J Open Source Softw 2(13)
2. Wilson C, Hargreaves T, Hauxwell-Baldwin R (2015) Smart homes and their users: a systematic analysis and key challenges. Personaland Ubiquitous Comput 19(2):463–476

Machine Vision for Intelligent Vehicles in India

Vatsal Verma, Srishti Ahlawat, and Darpan Khanna

Abstract This paper consists of a cross-examination on previous exploration done on machine vision for intelligent vehicles with respect to our new technical ideas which can be implemented for the betterment in intelligent vehicles. We have stated about various technical advancements in machine vision in the past few decades in various parts of world which can be considered in India. We have proposed a model with the advancement of autonomous vehicles which can be productive for use in India. We stated the current environmental conditions which impacts vehicle and its performance. Depth and optical flow has been acknowledged in the vision algorithms, which is used for betterment in image processing with minimal computation power and special instruments/hardware in the autonomous vehicles for efficiency. We demonstrated routine tasks, mapping and localization processes, which play an important role for detection of objects. Map data often influences high-profile verdicts of intelligent vehicles for better accuracy in meters. The results seem to be promising, showing that the machine vision system is useful in enhancing the vision algorithms, and the proposed ideas are robust for acknowledgment of complex road environments conditions in India. Hereafter, challenges of the machine vision in intelligent vehicles are also conferred.

Keywords Machine vision · Automatic vehicle guidance · Intelligent machines · Intelligent transportation system · Vision algorithm · Multivariate decision trees · Classification

---

V. Verma (✉) · S. Ahlawat · D. Khanna
Department of Computer Science and Engineering, SRM Institute of Science and Technology, Modinagar, India
e-mail: vv7871@srmist.edu.in

S. Ahlawat
e-mail: sa3467@srmist.edu.in

D. Khanna
e-mail: dk9420@srmist.edu.in

R. P. Mahapatra et al. (eds.), *Proceedings of International Conference on Recent Trends in Computing*, Lecture Notes in Networks and Systems 341,
https://doi.org/10.1007/978-981-16-7118-0_64

1 Introduction

1.1 Motivation

Since 1980s, a lot of research has been done on machine vision for intelligent vehicles in many different countries like Germany, USA and Japan. But still the concept of autonomous vehicles is still far from becoming a reality on Indian roads. This is mainly because of problems like lack of research and awareness, unavailability of proper road infrastructure and carelessness of majority of Indian drivers toward traffic rules. But, the benefits of these autonomous vehicles should also be considered and development should be done in this sector too. These autonomous vehicles can help in reducing the deaths caused by road accidents, economical loss due to accidents, and it will make vehicles safer, more efficient, and more environment-friendly. It will travel a lot more easier for the driver along with ensuring safety. Development in this sector can bring a major profitable change in Indian transport system. With the help of some previous research work and new ideas, these autonomous vehicles can be brought in India in near future. This paper provides review of some previous research work done on machine vision for intelligent vehicles along with the problems these intelligent vehicles may face on Indian roads with their probable solutions.

1.2 Previous Work

1980s was the year in which many countries had been seen to go ahead with working on machine vision for autonomous vehicles [1]. In 1987, the UBM test vehicle VaMoRs set the first milestone in developing an autonomous vehicle by the guidance of computer vision after demonstrating the capability to drive autonomously on a highway for more than 20 km at a speed of 96 km/h [1]. This triggered many countries like Japan to also start working on computer vision for autonomous vehicle guidance [1]. In Japan, MITI along with few other top companies of Japan went ahead with the project 'Personal Vehicle System' [1].

During 1921, black/white CCD cameras were used [1]. Then, color machine vision came into picture, as it was seen that it could help in recognizing characteristics of the road which earlier could not be detected in black/white vision. It was shown that under extreme weather conditions or even the normal outdoor environment, the change in the color can be shown with the help of characteristics distributions in the RGB space and that they can be 'learned' with the help of multivariate decision trees [2]. MDTs are mostly used for (PLNFE) piecewise linear nonparametric function estimation. This is done to get a better understanding about the color of the object ahead by using training samples, detecting them by classifying pixels according to the approximation function. This approach had appeared to be a very useful in recognizing different colors. [2]. It also showed that color is an essential feature in computer vision for intelligent vehicles.

Three main technologies generally used by intelligent transport system's machine vision applications are: *image-acquisition hardware, real-time image processors, and algorithms* [3]. These help in the major challenges that are object detection and lane detection. Early autonomous vehicles included systems which gave audio warnings to the driver whenever the vehicle used to leave its lane beyond a certain threshold [1]. They also had longitudinal control to some extent, wherein the driver is alarmed if an obstacle is very close to it [1]. There were 'Autonomous Stop and Go' vehicles too. These vehicles were designed for autonomous driving in crowded areas. Vision Technology Application (VITA) of Daimler–Benz in 1991 was one of the first vehicles to demonstrate these capabilities in public [1]. These types of vehicles help the driver in difficult tasks like stopping and accelerating frequently in congested places [4]. Here, a congested place refers to a place where all the lanes on the road are occupied by vehicles and the speed limit is very low [4]. Another functionality of the early autonomous vehicles was lane change systems wherein the assistance systems tell the driver when it is safe to change the lane [1]. This was depicted by the 'Prometheus' project of the UBM group [5] on AutoRoute 1 near Paris [1].

The test vehicle NavLab-S Carnegie Mellon University (CMU) having a simple vision system which could only recognize the road's horizontal curvature and also detect the lateral position of the vehicle in the lane covered a distance of 98% of more than 5000 km autonomously from the East coast to the West coast of the USA in 1995. Driver of the vehicle was only handling longitudinal control [1, 6]. After this, VAMP of UBM demonstrated fully autonomous driving for about 95% of a total of 1600 km from Munich to Odense, Denmark [7]. This time the vehicle had both lateral and longitudinal controls [1].

After this, collision warning systems came into picture. These systems had various features like forward collision warnings, lane-departure warning, intersection-collision warning, and pedestrian detection [3]. It also has a special category which is driver monitoring, and it detects it the driver is drowsy or some other issue with the driver which prevents him/her from safely operating the vehicle [3]. Also, if the driver does not respond to warnings, this system might take control of the steering, brakes, or throttle to keep the vehicle and driver safe [3]. After that the focus was on detecting "negative" obstacles. In order to handle this challenge, UniBwM evolved a Multifocal, Saccadic (EMS) vision [7].

Recently, autonomous vehicles have been divided into two categories, one with "confirmation-type" vision and the other with "scout-type" vision [7]. Autonomous vehicles have started using GPS, so because of the availability of these precise maps machine vision was mainly being used to assure the presence of motionless obstacles in the knowledge base that is the sort of vision which is known as "confirmation-type" [7]. "Bertha Benz" is an example of this type, whose performance was demonstrated in 2013 [7]. Scout-type vision is used by autonomous vehicles to drive in the absence of any sort of help in places that are new to the driver [7]. The scout type of vision has a high growth potential as compared to confirmation type as it has a large data set [7]. So, it can be said that these machine vision technologies for autonomous vehicles are advancing rapidly. Further, the paper will be talking about how we can use these technologies and some of our ideas to bring the concept of autonomous vehicles in India.

1.3 Methodology

In this paper, we have proposed many models and ideas dealing with the advancement of autonomous vehicles for efficient use in India stating the current complex environment like roads, weather, and shading, etc.

1.4 Vision Algorithm

We aim to develop an algorithm that deals with the complex sections dealing with streets and unstable roads and environmental conditions like weather and lighting, etc. [8]. We will be using depth and optical flow for betterment in image feature detection systems to track the detected objects like car, truck and human, etc.

We will be extracting patterns from an image, and this might change with slight variation due to outside conditions. Depth and optical flow will acknowledge the complex and multiple scenes. We will be using depth as it gives us a virtual representation of three-dimensional world and environment which helps in recognition. Moving objects will be accounted for under optical flow which is a major component to detect a moving object.

1.5 Concept Implementation

As optical flow and depth are really expensive in terms of high computing tasks and no computer has this ability based on the idea we shared in the above section, so we need an optimal system to implement this algorithm.

We need to focus on developing hardware tools which can extract images like depth maps, edges, and optical flow. We had a focus on speed and efficiency to extract as it is nearly impossible to extract images using bottom-up process [8] (Fig. 1).

1.6 Multivariate Decision Trees (MDT) for Indian Highway

Multivariate decision trees (MDT) is used in many parts of the world as part of the automated highway system project [9] for improvised lanes and objects. In MDT, each decision node creates space by recursively dividing into two with a hyperplane [10]. The binary test is also known as linear threshold units (LTU), and a single LTU

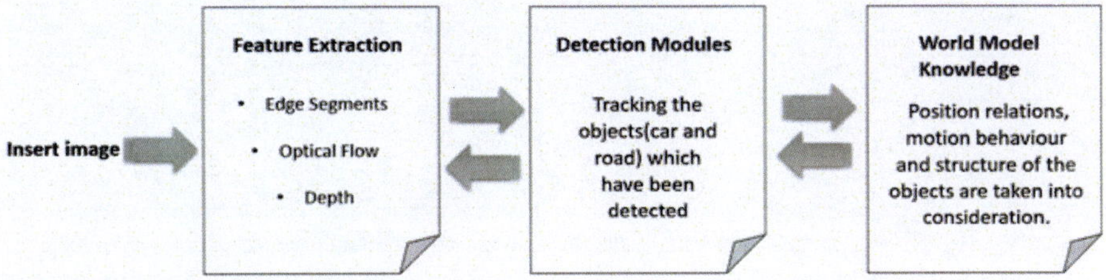

Fig. 1 Overview of vision algorithm [8]

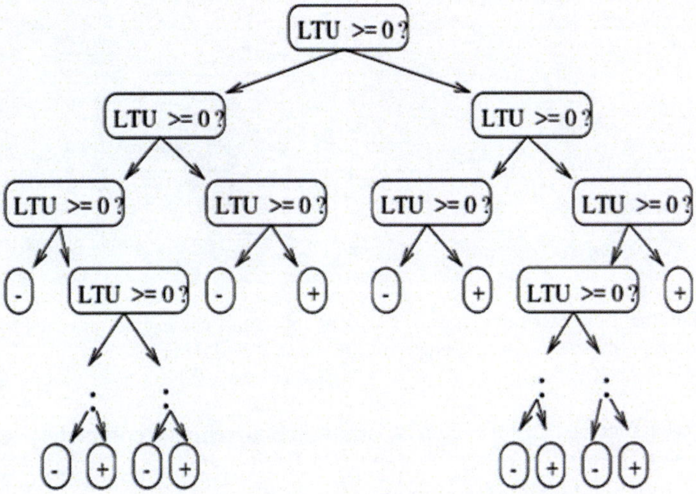

Fig. 2 Targets and LTUs of an MDT showing target (+) and background (−) [9]

can separate two subsets are linearly separable, and the MDT consists of only one node else to separate instances to the extent, and LTU will divide space linearly [9]. We will be using motion techniques to eliminate the obstacle by identifying that lie above the plane. The distributions in RGB space can also be learned using these to help in analyzing variation in apparent color of the environment during different times of the day [2]. Also, they have been used in lane as well as obstacle detection on highways [2] (Fig. 2).

1.7 Localizations and Maps

Virtual sensor which is nowadays known as digital map has become an important component which is used as driver assistance [11]. Maps which are used in these systems today are composed by culminating an image coming from a satellite consisting of street-level information with a good accuracy in meters. More than 95%

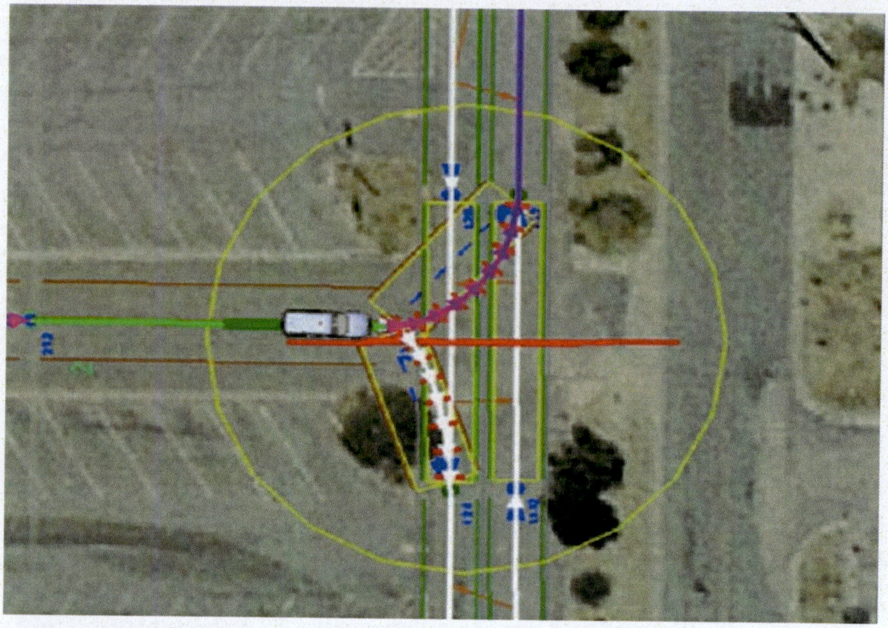

Fig. 3 RNDF (the route network definition file) map represents center lines of roads by polygons of geo-positions permitting a directed graph for the road network [11]

of a route was successfully traveled by autonomous vehicles completely relying on onboard sensor information [12, 13], and according to references provided in [12, 14, 15], a more enhanced map information is allowed for automation over long mileage. Route network definition file (RNDF) of a two-dimensional map in Fig. 3.

This has been given in the DARPA Challenge 2007 [16]. The vehicle must limit itself within the map to employ a map for planning the vehicle in terms of behavior and trajectory. The localization action of the six-dimension pose in an automated drive requires the identification and association of at least three landmarks per frame [17, 18].

1.8 Classification and Recognition

In machine vision, selecting the area of notice is the first and primary processing step to aim. The major focus should be to distract big portions of the image with minimal computation. In the field of machine vision, such things may employ arrival cues that are sensitive to symmetry, shadows, local texture or color gradients, as these encode attributes of vehicles and other traffic objects [19–21]. Optical flow is another way to find objects as it jointly integrates information on motion and geometry.

Fig. 4 (SLGT) Semantic labeling ground truth from the Cityscapes dataset [24] which shows classes for road, sky, pathways, sidewalk, humans, buildings, trees, and sky

Two modes came into this area which have earned popularity and importance since the last few years which are: (SLCIS) Semantic labeling combines image segmentation and classification which order to assign labels with pixel resolution as represented in Fig. 4. Corresponding approaches employ classical methods such as support vector machines [22] or conditional random fields [23].

The surveys [25, 26] differentiate between the circumstances of fatigue and distraction and represent both non-visual and visual features and attributes for their detection. This formerly includes eye and facial moments like (closing and blinking) and yawning, which can be recognized with classical techniques and some general facial expressions esoteric which involves neural networks.

2 Conclusion

Machine vision has made remarkable progress over the latest decennary. An overview of the previous research work done on machine vision for intelligent vehicles has been presented in this paper. The paper focuses on important issues, challenges, and metrics associated with evolution of vision-based systems for intelligent vehicles in India. The favorable results gained in the initial phases of the research on intelligent vehicles complete cyber-nation of hold-up which is proficiently feasible. The experimental response reflected the figure of merit and feasibility of the solution.

The algorithms and the equipment addressed in this paper have already been tried-out on board of the model vehicle accessible to all the research units. Autonomous vehicles succeeded in traveling some good portion of a route completely depending on onboard sensor information, elaborated map information. The obstacle detection system operates in the time period to excavate obstructions in the visual precept from 5 to 20 m ahead of the vehicle. MDT motion styles are adopted to eliminate the obstruction by discovering that lie above the plane. Difficulties come forth when multitude of vehicles coincide or in city-bred scenes, where road infrastructures, sign-posts and shadows make the scene too colonial. It is essential to idealize the aspects of the driving circumstances, for example weather conditions, the type of road, and traffic density. Optical flow integrates information on motion and geometry, thus proving to be a commending way to trace objects.

It is essential to set-up more norms and execution judgment on environment notion for intelligent vehicles. Hereafter, research of this project will focus on how to accurately trace the lanes when the circumstances on the road are too complicated. Such state-of-the-art research commands will be dealt within the years to come, unitedly with the new evolutionary push that, although compelling a complex parametric quantity calibrating stage, seems to present high qualitative execution.

References

1. Dickmanns E (nd) The development of machine vision for road vehicles in the last decade. Intelligent Vehicle Symposium, 2002. IEEE. https://doi.org/10.1109/ivs.2002.1187962
2. Buluswar SD, Draper BA (1998) Color machine vision for autonomous vehicles. Eng Appl Artif Intell 11(2):245–256. https://doi.org/10.1016/s0952-1976(97)00079-1
3. Bishop R (2000) Intelligent vehicle applications worldwide. IEEE Intell Syst 15(1):78–81. https://doi.org/10.1109/5254.820333
4. Yi K et al (2001) A vehicle-to-vehicle distance control algorithm for stop-and-go cruise control. ITSC 2001. 2001 IEEE intelligent transportation systems. Proceedings (Cat. No.01TH8585). https://doi.org/10.1109/itsc.2001.948704
5. Kujawski C (1995) Deciding the behavior of an autonomous road vehicle in complex traffic situations. In: Proceedings of 2nd EAC conference on intelligent autonomous vehicles, Espw (Helsinki)
6. Pomerleau DA (1995) Ralph Rapidly adapting lateral position handler. In: Proceedings of IEEE symposium on intelligent, Vehicles'95, Detroit, MI, USA
7. Dickmanns ED (2017) Developing the sense of vision for autonomous road vehicles at UniBwM. Computer 50(12):24–31. https://doi.org/10.1109/MC.2017.4451214
8. Ninomiya Y, Matsuda S, Ohta M, Harata Y, Suzuki T (1995) A real-time vision for intelligent vehicles. In: Proceedings of the intelligent vehicles '95. Symposium, 1995, pp 315–320. https://doi.org/10.1109/IVS.1995.528300
9. Buluswar SD, Draper BA (1998) Color machine vision for autonomous vehicles. Eng Appl Artif Intell 11(2):245–256. ISSN 0952-1976
10. Brodley CE, Utgoff PE (1995) Multivariate decision trees. Mach Learn 19(1):45–77. https://doi.org/10.1023/A:1022607123649
11. Ranft B, Stiller C (2016) The role of machine vision for intelligent vehicles. IEEE Trans Intell Veh 1(1):8–19. https://doi.org/10.1109/TIV.2016.2551553
12. Urmson C, Anhalt J, Bagnell D, Baker C, Bittner R, Clark MN, Dolan J, Duggins D, Gittleman M, Harbaugh S, Wolkowicki Z, Ziglar J, Bae H, Brown T, Demitrish D, Sadekar V, Zhang W,

Struble J, Taylor M, Darms M, Ferguson D (2008) Autonomous driving in urban environments: boss and the urban challenge. J Field Robot: Spec Issues 2007 DARPA Urban Challenge 425–466
13. Maurer M, Behringer R, Fürst S, Thomanek F, Dickmanns E (1996) A compact vision system for road vehicle guidance. In: 13th International conference on pattern recognition, pp 313–317
14. Thrun S, Montemerlo M, Dahlkamp H, Stavens D, Aron A, Diebel J, Fong P, Gale J, Halpenny M, Hoffmann G, Lau K, Oakley C, Palatucci M, Pratt V, Stang P, Strohband S, Dupont C, Jendrossek L-E, Koelen C, Markey C, Rummel C, van Niekerk J, Jensen E, Alessandrini P, Bradski G, Davies B, Ettinger S, Kaehler A, Nefian A, Mahoney P (2006) Stanley: the robot that won the darpa grand challenge. J Field Robot 23(9):661–692
15. Ziegler J, Bender P, Schreiber M, Lategahn H, Strauss T, Stiller C, Dang T, Franke U, Appenrodt N, Keller C, Kaus E, Herrtwich R, Rabe C, Pfeiffer D, Lindner F, Stein F, Erbs F, Enzweiler M, Knöppel C, Hipp J, Haueis M, Trepte M, Brenk C, Tamke A, Ghanaat M, Braun M, Joos A, Fritz H, Mock H, Hein M, Zeeb E (2014) Making Bertha drive —an autonomous journey on a historic route. IEEE Intell Transp Syst Mag 6(2):8–20
16. Dickmanns E, Behringer R, Dickmanns D, Hildebrandt T, Maurer M, Thomanek F, Schiehlen J (1994) The seeing passenger car 'VaMoRs-P'. In: IEEE intelligent vehicles symposium (IV), pp 68–73
17. Lategahn H, Stiller C (2014) Vision-only localization. IEEE Trans Intell Transp Syst 15(3):1246–1257
18. Ziegler J, Lategahn H, Schreiber M, Keller CG, Knöppel C, Hipp J, Haueis M, Stiller C (2014) Video based localization for Bertha. In: IEEE intelligent vehicles symposium, June 2014, pp 1231–1238
19. Bertozzi M, Broggi A, Castelluccio S (1997) A real-time oriented system for vehicle detection. J Syst Architect 43(1):317–325
20. Kalinke T, Tzomakas C, Seelen WV (1998) A texture-based object detection and an adaptive model-based classification. In: IEEE international conference on intelligent vehicles. Proceedings of the 1998 IEEE international conference on intelligent vehicles, vol 1
21. Hoffmann C, Dang T, Stiller C (2004) Vehicle detection fusing 2D visual features. In: Proceedings of IEEE intelligent vehicles symposium, Parma, Italy, June 2004, pp 280–285
22. Scharwächter T, Enzweiler M, Franke U, Roth S (2013) Efficient multi-cue scene segmentation. In: Pattern recognition. Springer, Berlin, pp 435–445
23. Verbeek J, Triggs W (2008) Scene segmentation with crfs learned from partially labeled images. In: Advances in neural information processing systems, 2008, pp 1553–1560
24. Cordts M, Omran M, Ramos S, Scharwächter T, Enzweiler M, Benenson R, Franke U, Roth S, Schiele B (2015) The cityscapes dataset. In: Conference on computer vision and pattern recognition, workshop on the future of datasets in vision
25. Kaplan S, Guvensan A, Yavuz MG, Karalurt Y (2015) Driver behavior analysis for safe driving: a survey. Trans Intell Transp Syst 16(6):3017–3032
26. Dong Y, Hu Z, Uchimura K, Murayama N (2011) Driver inattention monitoring system for intelligent vehicles: a review. Trans Intell Transp Syst 12(2):596–614

An Efficient Contour Detection Approach for Extracting Rim from Wheel Images

D. Karthik, P. Mirunalini, R. Priyadharsini, and T. T. Mirnalinee

Abstract In the automobile industry, the wheel alignment process plays a major role in ensuring that the wheels are placed at proper angles, and pointing straight for proper road contact and safe drive. Identifying the location of the rim in the wheel plays a major role in aligning wheels. There are many state of art image processing techniques available for rim detection from wheel images which includes circle and ellipse detection algorithms. When the captured wheel images are skewed (angulary/translationally), then the existing techniques fail to detect the rim. Hence, we propose a contour-based rim detection method which can detect the position of the rim from the images of automobile wheels even when they are angularly or translationally skewed. The rim detection is done as a two step process which encompasses detection of contours on the image and filtering the irrelevant contours. If the rim is detected with an anomalous contour, our proposed methodology will correct it using K-means clustering algorithm by considering the distance from center of the contour. The performance of the proposed methodology has been analyzed using 50 images having various viewpoint variations, and it is observed that it is extracting the rim perfectly.

Keywords Rim detection · Contour method · Wheel alignment

D. Karthik (✉) · P. Mirunalini · R. Priyadharsini · T. T. Mirnalinee
Sri Sivasubramaniya Nadar College of Engineering, Chennai, Tamil Nadu, India
e-mail: karthik19047@cse.ssn.edu.in

P. Mirunalini
e-mail: miruna@ssn.edu.in

R. Priyadharsini
e-mail: priyadharsinir@ssn.edu.in

T. T. Mirnalinee
e-mail: mirnalineett@ssn.edu.in

R. P. Mahapatra et al. (eds.), *Proceedings of International Conference on Recent Trends in Computing*, Lecture Notes in Networks and Systems 341,
https://doi.org/10.1007/978-981-16-7118-0_65

1 Introduction

Maintaining proper wheel alignment is essential to avoid unnecessary wear on automobile tyres, steering, suspension, and brakes. Misalignment of the wheel may lead to accidents. Accurate wheel alignment optimizes driving stability, maximizes tire life and improves the vehicle's overall performance. During manufacturing and also maintenance of the vehicle, wheel alignment checking is a major step which involves rim detection. The rim of a wheel is the outer edge of the metallic portion, holding the tire. There are many state of art shape detection techniques available for object detection such as Hough transform for circle, ellipse, and any arbitrary shape. Even though there are proven techniques available for circular object detection, the challenges in rim detection lies in the orientation of the wheel position and translation parameters such as height of the wheel from the ground surface.

In literature, to count vehicle axles in a video sequence for use in automatic vehicle classifiers, fast Hough transforms [6] have been used. The method also has been extended for allowing systematic false positive wheel defections. To meet the same objective of vehicle axle detection and classification, circular Hough transform-assisted CNN [5] has been proposed. The method has used MIO dataset for training the network and has been tested for five axle classes. A video-based method for the detection of truck axles [3] has used a sobel edge detection for detecting the true edges from the video frames, and the Hough transform has been applied on those images. The peak in the accumulator array of the Hough transform has been used for detecting the axle. In order to detect the vehicle and wheel, a Single Shot MultiBox Detector [9] has been used with multiple concatenation modules aiming to improve the performance of small object detection. The authors have used the Kitti dataset for training and testing the model. The method had categorized the vehicle into truck, pickup, tractor, car, and wheel. These object detection methods involve feature extraction and semantic segmentation [4].

Boosted cascade of feature classifiers have been used to extract Haar like features in [8] for rapid object detection. The system which proposed wheel detection using random forest classifiers [7] can be used for vehicle counting and classification. Using the images convolved by the difference of Gaussian filter bank and the precomputed set of principle components [1, 2], the wheels are detected and tracked. Wheel being circular in shape has been amplified using Pseudo wavelet filters [11] for detecting the rim from the wheel.

Contour detection methods have also been applied in [10] for detecting the defective complex structural chips. Pixels with same intensity values are divided into subregions which are coded by two pass algorithm. Incomplete edges are corrected using interpolation and the interference contours were eliminated on shape features. Convolution-based neural network has been proposed in [12] to connect low and high-resolution features. VGG net was used as encoding part, and then, feature maps are combined with up or down sample methods for getting specific resolution. From literature, it is evident that objects in an image can be detected using shape-based, region-based, and contour-based approaches. In this work, a contour detection approach is used to extract the rim from wheel images.

2 Proposed Contour-Based Rim Detection Method

We propose a contour-based image processing methodology to detect the position of the rim from the automobile wheel images. The input images are of different orientations due to different zoom levels, angular skew, and translational skew. The rim of the wheel is the outermost concentric circle on the wheel which needs to be identified. The rim detection for the wheel images is performed by detecting all the contours and filtering the imperfect contours. A collection of several closely spaced pixels on the image can be used to reconstruct a near-accurate boundary of the rim. A set of such closely spaced pixels are called contours. The number of pixels and their closeness with each other will parameterize the accuracy of the said boundary. The proposed methodology involves two steps to detect rim from wheel images: extracting the possible contours from the image and filtering the contours to isolate the rim boundary and rectification of anomalous rim boundary if needed. The process of extracting the rim is illustrated in Algorithm 1.

2.1 Detection of All Contours

In order to detect the contours of the image, a Gaussian blur filter is first applied to the image to reduce random noise. This process smoothens the image by minimizing the brightness and hue variations. The blur filter helps to eliminate contours that may be detected due to inherent noise in the image. The image is then converted to a binary image by image thresholding using Otsu's binarization. Otsu's method helps to identify a single intensity threshold that separate pixels into two classes

To detect the contours present in the image, a pixel-based approach was used. The algorithm parses the image pixels row-wise until the bottom right of the image is reached. Each time a nonzero pixel is found, it further scans the neighboring pixels to find another nonzero value. This is done recursively until the first pixel is reached back, thereby tracing a closed boundary of nonzero pixels—a contour. The algorithm also distinguishes between outer boundaries and hole boundaries. While parsing, a transition of pixels from 0 to 1 indicates an outer boundary, whereas a transition from 1 to 0 indicates a hole boundary. The above process is repeated to detect all the contours present in the image. This algorithm also maintains a hierarchy among the detected contours by assigning an incremental sequential value to the contour pixels during the neighbors scan.

Thus, several contours were detected in the input image using the algorithm. Furthermore, the obtained contours were ordered hierarchically. The inessential contours were eliminated from the obtained contours using image processing techniques to isolate the rim.

Algorithm 1: Detection of rim from wheel images

```
Input: Wheel image
Output: Image with rim detected
Apply Guassian Blur
Convert to binary image by thresholding
//(To detect all contours)
Initialize ctr ← 0
for image-pixels from top-left to bottom-right do
    if non-zero pixel found then
        foreach neighbor of pixel do
            if non-zero pixel then
                mark as ctr
                ctr++
            else
                continue
            end
        end

end
//(To perform Hierarchy-Test)
Initialize array candidates containing all contours
foreach contour in candidates do
    if no sub-contours within candidate contour then
        Eliminate corresponding candidate contour
    else
        Add contour's area to areas array
        Retain the candidate contour
    end
end
//(To retain only convex-apeirogon shaped contours)
Initialize empty array dual-loops
foreach contour in candidates do
    Compute perimeter-ratio
    Compute area-ratio
    if area-ratio=1 and area-ratio=1 then
        Retain the candidate contour
    else
        Add contour area to dual-loops array
        Retain the candidate contour
    end
    Eliminate corresponding candidate contour
end
```

```
//(To find the contour with largest image relative area)
Initialize max − ratio ← 1.1
foreach contour in candidates do
    Compute Image-relative area-ratio
    if area-ratio > 1 then
        Eliminate corresponding candidate contour
    else if area-ratio > max-ratio then
        max − ratio ← area − ratio
        Mark the corresponding candidate contour as the Rim
    else
        Eliminate corresponding candidate contour
    end
end
//(To rectify anamolous contours )
Set rim − center ← centerofenclosingcircle
if rim in dual-loops then
    Store distance of each pixel from rim-center in array distances
    Sort the distances array
    Perorm K-Means Clustering on distances with clusters=2
    Compute average of means of the two clusters
    Increment this average by a small value and store as threshold; foreach pixel in contour
    do
        if distance from rim-center < threshold then
            Eliminate corresponding candidate pixel
        else
            Retain corresponding candidate pixel
        end
    end
else
    Leave contour unchanged
end
Rim is detected
```

2.2 Filtering the Contours

A wheel consists of distinctive parts such as the rim, wheel center, spokes, and lug holes. The contour detection algorithm identified each part of the wheel as a separate contour. In some cases, the regions around the wheel may also be detected as contours, due to aribitrary pixel variations in the image. The first step of filtering involves removing the contours found outside the rim.

Regions around the wheel are not characterized by as many contours. This is due to relatively less number of objects with a distinct boundary in these regions. This trait is used to perform a hierarchy test, wherein the contours not enclosing any other contours are eliminated. Consequently, contours that are outside the wheel but similar in size to the wheel are eliminated from the list of potential rim boundaries. Among the contours that qualify this test, the one encompassing the largest enclosing circle is chosen as the Region of Interest (ROI).

At this stage, there may still be some lower area contours present outside the ROI. We perform a point-polygon test and radius comparison between the ROI contour and enclosing circles of all other contours. A contour will pass this test only if it lies entirely within the ROI. Hence, through this test, all the remaining contours outside the ROI are eliminated. Thus, the ROI has been detected along with other contours inside it.

The different parts of the wheel are detected as arbitrary-shaped boundaries by the proposed algorithm. However, the real shape of a wheel rim is always circular. But due to different orientations of the input wheel image on account of different levels of scaling as well as angular and translational skewing, the detected contours of the ROI can be any convex apeirogon—a convex polygon with infinitely many sides. In order to retain only convex apeirogon-shaped contours, we have performed tests based on perimeter and area ratios between the contours and their enclosing circles, using Eqs. 1 and 2.

$$AreaRatio = ContourArea/EnclosingCircleArea \tag{1}$$

$$PerimeterRatio = ContourPerimeter/EnclosingCirclePerimeter \tag{2}$$

A convex apeirogon-shaped contour must have both perimeter and area ratio close to unit value. If either of the ratio tests fail, the obtained contour does not describe the rim of the wheel.

After removing all the arbitrary-shaped contours inside the wheel, we obtained only convex apeirogon-shaped contours within the ROI. Since the input images are of different resolutions, all area measurements will be linearly parameterized by the image area. Hence, the relative areas are found as the ratio of the area of contours' enclosing circles to that of the image.

Area of the image is defined as the product of the first two dimensions of its shape—height and width. The ROI is known, and all contours outside of it have already been eliminated. Furthermore, the wheel will occupy a large portion of the image, but cannot be larger than the image itself. Hence, the contour with the largest relative area, not greater than 1 is isolated as the contour of the rim. This is referred to as the rim boundary.

In over 30% of the input images, the detected rim boundary was characterized by a dual-loop. Unlike an ordinary boundary, they form a thin circular strip as shown in Fig. 1. This happens in images where there may be a small object such as an air-valve present on the periphery, causing a discontinuity in the occurrence of non-zero pixels. A dual-loop boundary does not allow accurate measurement of parameters of the rim such as radius, circumference, and position of lateral and vertical extremes. Such parameters are often required in applications such as automated wheel-alignment systems and wear-out detection. Hence, this is an anomalous contour which is corrected by following the steps in Sect. 2.3.

2.3 Rectification of Anomalous Rim Boundaries

In the proposed methodology, a specific type of anomalous contours herein referred to as dual-loop contours is rectified. All dual-loop contours will have a perimeter ratio close to 2. Hence, they were detected during the perimeter ratio test performed in the previous step. The following steps are then adopted to retain only the outer loop of the contour.

The basis for discarding the pixels on the inner loop of the contour stems from the fact that there is a characteristic distinction in the distances of the pixels on the inner loop and those on the outer loop from the center of the contour's enclosing circle. However, the impediment in setting a strict number to filter out these pixels arises from the fact that each image can have a different orientation. As a result, the contours are not perfect circles and might be flattened along one of the diameters. The consequent characteristic variation of the distances of points on a dual-loop contour is illustrated in Fig. 1. Two distinct sets of pixels as well as convergence regions can be noticed. The characteristic hump shape stems from the angular skewing of images, causing a deviation from a perfectly circular rim shape.

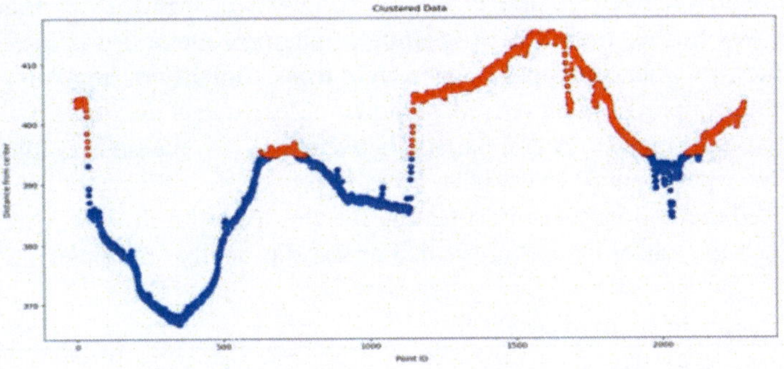

Fig. 1 Distances of pixels on the inner and outer loops, clustered before sorting

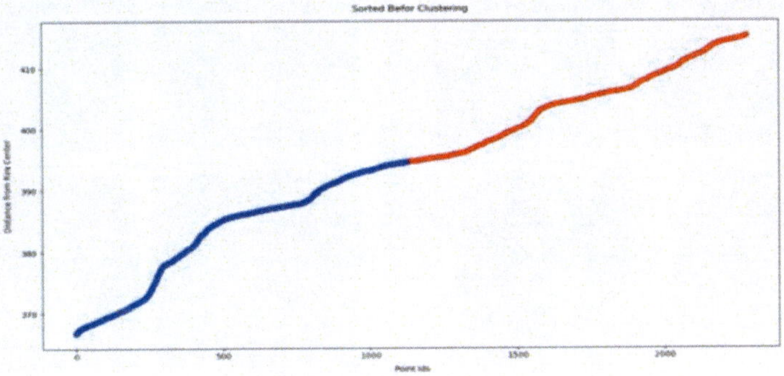

Fig. 2 Distances of pixels on the inner and outer loops, clustered after sorting

The distances of each contour pixel from the center of its enclosing circle is found. The obtained distances are then sorted, and K-means clustering is applied on this data to detect two distinct clusters. The result of clustering can be seen in Fig. 2. The mean of average distances from each of the clusters is found and shifted up by a small value to arrive at a threshold. This shift ensures that no pixels on the inner loop are retained. The threshold is then used to retain only the outer-loop pixels, whose distances will lie on the higher side of the threshold.

3 Experimental Results

We have applied the proposed contour-based rim detection algorithm on several wheel images of different orientations collected from publicly available sources. We also applied the Hough transform circle-based detection algorithm on these images to compare their abilities to detect contours, particularly in the skewed images where the rims may not be perfectly circular. About 50 images of car wheels were randomly collected from the Internet. The proposed algorithm was able to accurately detect the rim on different orientations of 43 of these images. A subset of the results for angularly or translationally skewed, scaled as well as straight wheel images is illustrated in Table 1. It was further found that the Hough transform method was ineffective in detecting the rim when the input images were translationally or angularly skewed. However, the proposed contour-based method is able to detect the rim in both skewed as well as straight images. A comparison between the two methods is illustrated in Table 2. The experimental results have visually proven the efficacy of the proposed contour-based method in not only detecting the rim from the straight wheel images but also when the wheel images are scaled or translationally or angularly skewed.

Table 1 Results of the proposed algorithm

Skewed Input	Detected Rim	Straight Input	Detected Rim

Table 2 Comparison of proposed algorithm and Hough transform method

Input Image	Hough Transform	Proposed Method

4 Conclusion

Rim detection, which is an important step in the automated wheel alignment process, has been performed using contour based methods. On finding the rims using contour detection method, we analyzed that many potential boundaries were generated. Hence, the inessential contours were filtered out to retain only the rim. In some cases when the rim was detected with a dual-loop, it was rectified to retain only the outer boundary. This method works very well for rim detection compared to the state-of-the-art Hough transform which many researchers have been using for circular object detection. This is because the wheel rims may not be perfectly circular in the image due to different orientations, skewing, and scaling. As illustrated in Sect. 3, the proposed algorithm is able to detect the rim in straight wheel images. Furthermore, it performs better than the existing Hough transform method, particularly when detecting the rim from scaled and translationally or angularly skewed images. As a future work, after detecting the rim in the wheel, measurements to correct different points on the wheel can be made to achieve automated wheel alignment.

References

1. Achler O, Trivedi M (2004) Vehicle wheel detector using 2d filter banks. In: IEEE intelligent vehicles symposium, pp 25–30. https://doi.org/10.1109/IVS.2004.1336350
2. Achler O, Trivedi M (2004) Camera based vehicle detection, tracking, and wheel baseline estimation approach, pp 743–748. https://doi.org/10.1109/ITSC.2004.1398995
3. Frenze J (2006) A video-based method for the detection of truck axles
4. Girshick R, Donahue J, Darrell T, Malik J (2014) Rich feature hierarchies for accurate object detection and semantic segmentation. In: Proceedings of the IEEE conference on vomputer Vision and Pattern Recognition (CVPR)
5. Gothankar N, Kambhamettu C, Moser P (2019) Circular hough transform assisted cnn based vehicle axle detection and classification. In: 2019 4th International Conference on Intelligent Transportation Engineering (ICITE), pp 217–221. https://doi.org/10.1109/ICITE.2019.8880232
6. Grigoryev A, Bocharov D, Terekhin A, Nikolaev D (2015) Vision-based vehicle wheel detector and axle counter. In: ECMS
7. Hultström K (2013) Image based wheel detection using random forest classification
8. Jiayue F, Cairong Z, Xia Y, Wenbin L (2002) Vehicle and wheel detection: a novel ssd-based approach and associated large-scale benchmark dataset. Multimedia Tools Appl 1:I–I. https://doi.org/10.1007/s11042-019-08523-y
9. Lienhart R, Maydt J (2020) An extended set of haar-like features for rapid object detection. Proc Int Conf Image Process 79:12615–12634. https://doi.org/10.1109/ICIP.2002.1038171
10. Lin B, Wang J, Yang X, Tang Z, Li X, Duan C, Zhang X (2021) Defect contour detection of complex structural chips. Math Problems Eng 2021:1–11. https://doi.org/10.1155/2021/5518675
11. OBrien EJ, Caprani CC, Blacoe S, Guo D, Malekjafarian A (2018) Detection of vehicle wheels from images using a pseudo-wavelet filter for analysis of congested traffic. IET Image Process 12(12):2222–2228. https://doi.org/10.1049/iet-ipr.2018.5369, https://ietresearch.onlinelibrary.wiley.com/doi/abs/10.1049/iet-ipr.2018.5369
12. Xu G, Lin C, Cheng Y (2021) Dense connection decoding network for crisp contour detection. IET Image Process 15(4):956–963. https://doi.org/10.1049/ipr2.12076

Design of GaN HEMT Based Down-Convertor Passive Mixer from S Band to X Band

Manoj Kumar Vishnoi and Satya Sai Srikant

Abstract The essential components of transmitter and receiver systems fabricated using Gallium-Nitride are being gradually accepted. Gallium Arsenide based MMICs are being replaced Gallium Nitride MMIC. Gallium Nitride transmit/receive switches, power amplifiers as well as low noise amplifiers, provide significantly better performance than GaAs components. In this work, the Gallium Nitride high electron mobility transistors (HEMT) is used to design the passive mixer. The large bandwidth is achieved at the IF stage by providing the input RF signal to mixing stage with inductor. The conversion gain at the output for a frequency range of 2–10 GHz is achieved as 22 dB, while the noise figure is 17 dB. The mixer achieve large bandwidth by shunting the mixing stage. This mixer is designed using the UMS GH_25_10 HEMT foundry for down convertor system.

Keywords Passive · Mixing stage · TR switches · Shunting

1 Introduction

Recently, the wide bandgap semiconductor material have drawn the attention of the researcher to design the monolithic microwave integrated circuit (MMIC). The GaN devices are ideal for high-power circuits, due to their high breakdown voltage. These high-power microwave circuits can be utilized in military systems, satellite communications and cellular base stations, among other uses. As the demand for wireless communication grows, so does the need for technology to evolve. GaN-based devices have been shown to handle a lot of power at high frequencies. Circuits based on CMOS technology have previously demonstrated large bandwidths of several tens of GHz [1–5]. AlGaN/GaN HEMTs based monolithic, hybrid and flip chip circuits have been demonstrated using these unique performance features [6]. Broadband amplifiers [7–9], low-noise amplifiers [10, 11], active filter [12] mixer, and voltage-controlled oscillators [13] are examples of these circuits. The number of viable circuit

M. K. Vishnoi (✉) · S. S. Srikant
SRM Institute of Science and Technology, Delhi-NCR Campus, Modinagar, Ghaziabad 201204, India

R. P. Mahapatra et al. (eds.), *Proceedings of International Conference on Recent Trends in Computing*, Lecture Notes in Networks and Systems 341,
https://doi.org/10.1007/978-981-16-7118-0_66

applications grows as the technology improves. Wide-bandgap devices have been recognised with lot of advantages in the design of TR switches [14]. AlGaN/GaN HEMT based switches are capable of handling higher power at RF frequency and produce low insertion loss [15]. The main disadvantage of using GaN technology for the frequency mixers is that the requisite lumped components, such as baluns, take up a significant portion of total chip area. The ability to create a completely GaN HEMT based receiver is one of the benefits of technology upgradation. Other benefits include the ability to quickly integrate a local oscillator, amplifier and other components. The GaN HEMT also provide robustness for mechanical circumstances. A double-balanced passive mixer MMIC has been constructed using UMS 0.25 μm AlGaN/GaN technology to investigate the capabilities and limitations of a GaN mixer (GH25-10 technology).

2 Passive Mixer Circuit Design

RF Mixers are active or passive devices with three ports shown in Fig. 1. When two different input frequencies are introduced into the other two ports, they are designed to produce both a sum and a difference frequency at a single output port.

When this input signal is sent, it can be converted to a higher IF frequency ($f_{RF} + f_{LO}$); the mixer is then referred to as an up converter and is involved in the modulation process in the transmitter. It can also be down converted to a lower frequency ($f_{RF} + f_{LO}$), making it a down converter in the receiver demodulation process. If mixer circuit is providing the signal of less than 25 kHz bandwidth, the mixer is called the narrow band converter. If the output signal is greater than 25 kHz, then this is called wide band conversion.

The double-balanced mixer is a frequently utilized circuit for millimeter (mm)-wave applications with high port to port isolation, linearity and low noise. The passive mixer using the GaN HEMT is shown in Fig. 2. The capacitive coupling is used between the input stage and the mixing stage. The passive frequency mixer is simply made up of six HEMTs that have a resistance (R_{on}).

At the mixing stage, the input signal is routed to the source of HEMT through C_1 and C_2. The output is achieved by combining the input RF signal with local oscillator signal. The inductor L is introduced to the mixing stage that improve the bandwidth

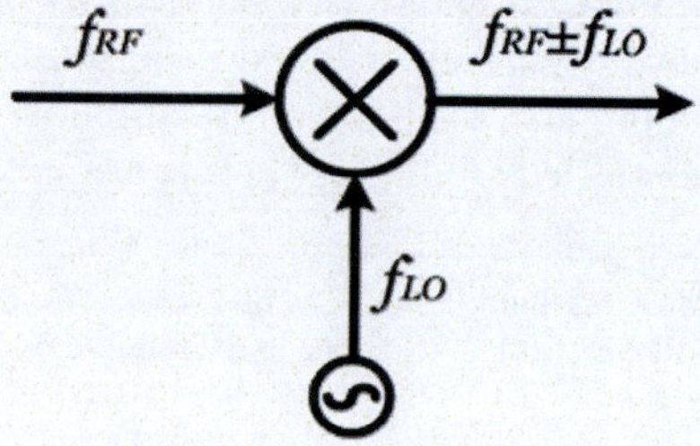

Fig. 1 Fundamental mixer block diagram

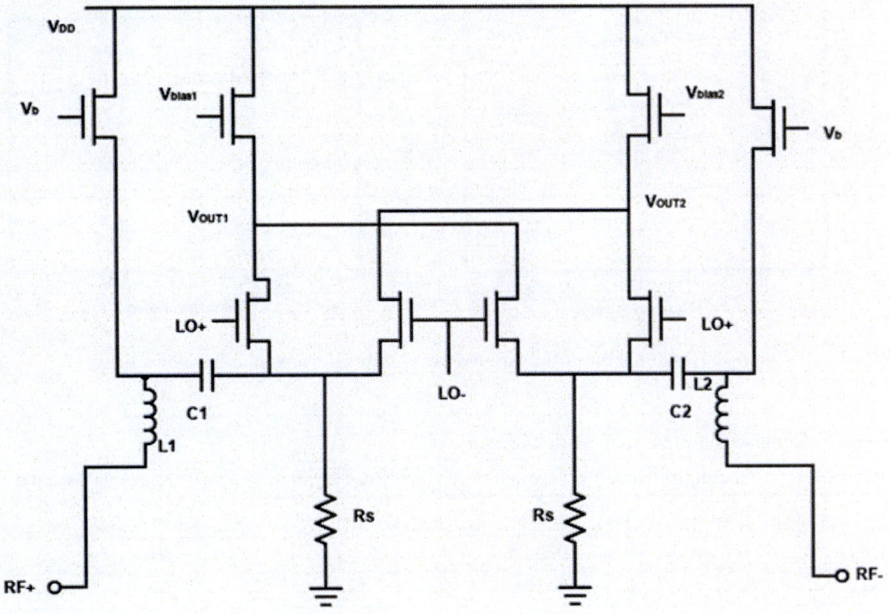

Fig. 2 GaN HEMT based passive mixer circuit

of the mixer. This improved shunt peaking technique is introduced in passive mixer to make the large transfer function at the input to mixing stage, so as to provide the large bandwidth.

3 Results and Discussion

The passive mixer is designed and simulated with UMS GH_25_10 HEMT in microwave electronic design automation (EDA) tools called Applied Wave Research (AWR); developed by AWR corporation. The voltage conversion gain and noise figure for proposed mixer are shown in Fig. 3a, b, respectively. The voltage conversion gain in the frequency range of 2–10 GHz provides 22 dB approximately whereas its noise figure shows 17 dB. The gain and noise figure obtained from designed GaN HEMT based passive mixer can be stable at S Band (2–4 GHz), C Band (4–8 GHz) and some portion of X band (i.e. 8–10 GHz) and hence this mixer is mostly used for down convertors for wireless applications, ranging from S band to some extent of X Band. The comparison of mixer design using various technology with proposed work is shown in Table. 1.

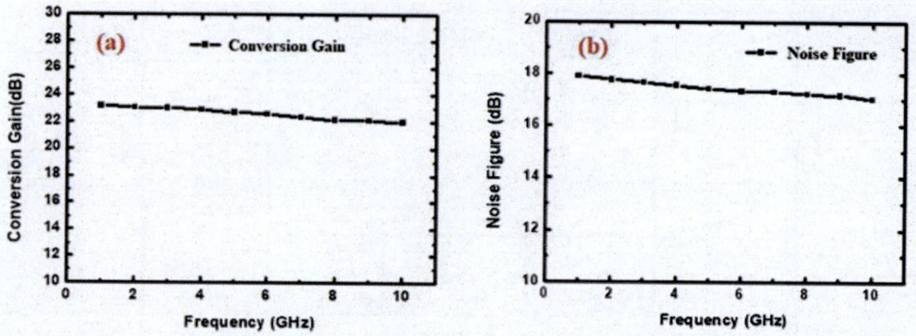

Fig. 3 **a** Voltage conversion gain of the passive mixer with frequency, and **b** noise figure

Table 1 Comparison of various work with our work

Ref. No.	Technology used	Frequency (GHz)	Conversion gain (dB)	Noise figure (dB)
Present work	0.25 μm HEMT	2–10	22	17
[1]	0.18-μm CMOS	0.3–25	11	NA
[4]	0.13-μm CMOS	2–10	9–24	NA
[5]	0.18-μm CMOS	8.9–5.5	−17	NA
[6]	65 nm RF CMOS	0.1–5	22–31	8–14.2

4 Conclusion

A wideband downconverter passive mixer was designed for wireless communication from S band to some portion of X band (i.e., from 2 to 10 GHz). However, it can also be used for transmitter. The GaN HEMT (UMS foundry) with 0.25 μm length and 100 μm width having 8 gate fingers is used in mixer design. The obtained results show that the passive mixer in terms of conversion gain and noise figure is a perfect option for receiver and other wireless communication application as a down convertor receiver.

References

1. Tsai M-D, Wang H (2004) A 0.3–25-GHz ultra-wideband mixer using commercial 0.18-um CMOS technology. IEEE MWC Lett 14:522–524
2. Vidojkovic V, van der Tang J, Leeuwenburgh A, van Roermund A (2005) A low-voltage folded-switching mixer in 0.18-μm CMOS. IEEE J SSC 40:1259–1264
3. Poobuapheun N, Chen W-H, Boos Z, Niknejad AM (2006) A 1.5 V 0.7–2.5 GHz CMOS quadrature demodulator for multi-band direct-conversion receivers. IEEE CICC, pp 797–800
4. Wang M, Carlos E (2011) Saavedra. Reconfigurable broadband mixer with variable conversion gain reconfigurable broadband mixer with variable conversion gain. IEEE MTTS, pp 1–4

5. Kaper VS, Thompson RM, Prunty TR, Shealy JR (2005) Signal generation, control, and frequency conversion AlGaN/GaN HEMT MMICs. IEEE Trans Microw Theory Tech 53:1
6. Gupta N, Dutta A, Singh SG (2015) A low/high band highly linearized reconfigurable down conversion mixer in 65 nm CMOS process. NORCAS
7. Green BM, Tilak V, Lee S, Kim H, Smart JA, Webb KJ, Shealy JR, Eastman LF (2001) High-power broad-band AlGaN/GaN HEMT MMICs on SiC substrates. IEEE Trans Microwave Theory Tech 49(12):2486–2493
8. Xu JJ, Keller S, Parish G, Heikman S, Mishra UK, York RA (2000) A 3–10-GHz GaN-based flip-chip integrated broad-band power amplifier. IEEE Trans Microwave Theory Tech 48(12):2573–2578
9. Greensboro (2003) RF Micro Devices announces GaN process technology milestone and availability of first GaN power amplifiers. RFMD, NC, US
10. Mishra M, Tomar SK, Sharma S, Mittal A (2014) Design and analysis of three stages pHEMT LNA at K-band. IJRASET 2:202–205
11. Ellis G, Moon J, Wong D, Micovic M, Hashimoto P, Hu M (2004) Wideband AlGaN/GaN HEMT MMIC low noise amplifier. IEEE MTT-S international microwave symposium digits, pp 153–156
12. Sharma SS, Pandey AK, Tiwary AK (2018) New method of analysis and design of frequency and bandwidth reconfigurable active filter. Int J RF Microwave CAD: MMCE 21556:1–10
13. Shealy JB, Smart JA, Shealy JR (2001) Low-phase noise AlGaN/GaN FET-based voltage controlled oscillators (VCOs). IEEE Microw Wirel Compon Lett 11(6):244–245
14. Shealy JB et al (2002) Gallium nitride (GaN) HEMT's: progress and potential for commercial applications. Gallium arsenide integrated circuit symposium digits, pp 243–246
15. Caverly RH, Drozdovski NV, Joye C, Quinn M (2002) Gallium nitride: use in high power control application. Gallium arsenide integrated circuit symposium digits, pp 3499–3509

Psychological Impacts of Covid-19 on Human Health

Md Adnan Baig and Ihtiram Raza Khan

Abstract Recently, the world is going through a global pandemic COVID-19 caused by SARS—COV2. Which has raised huge and difficult challenges in front of world for every layer of society and from every aspect, be it economy crisis, temporary unemployment, starvation, collision of fundamental services, educational pause, it has affected every sector globally. This disease is hitting the medical system of almost 213 countries brutally. If we talk about health issues raised by global pandemic, so at one side where it is drastically dangerous for physical health of a person, at the same time, there are many psychological impacts of COVID-19 on human health. Its highly spreading nature is the reason why restrictions have been made to cope against its growth throughout the world, due to which almost everyone whether they are student, teachers, business persons, is locked inside to stop the outbreak of this virus. Now this situation leads to the mental health issues in people sitting far apart from each other since long, a feeling of loneliness, anxiety, depressive symptoms, and other negative emotions of mental illness can be seen largely in people during pandemic around the globe. There could be many reasons, like sitting at home for so long, fear of being infected by the virus, bereavement, rumors and myths in the society about diseases and what not. The main aim of this work is to implement the dataset gathered in the direction of getting useful information and generating results out of it, to study the psychological impacts of COVID-19 on human health, what are the different symptoms, what could be the reasons, and a general solution roadmap to the respective problem.

Keywords COVID-19 · Pandemic · Mental health · Anxiety · Human health

1 Introduction

Wuhan, China at December 31, 2019, reported initial cases of deadly pandemic in a cluster having symptoms of pneumonia. On January 20, 2020, official actions were

M. A. Baig (✉) · I. R. Khan
Department of Computer Science, Jamia Hamdard, Delhi, India

R. P. Mahapatra et al. (eds.), *Proceedings of International Conference on Recent Trends in Computing*, Lecture Notes in Networks and Systems 341,
https://doi.org/10.1007/978-981-16-7118-0_67

started taken to fight against it after being declared as B-type infectious disease by National Health Commission [1]. And till March 11, 2020, only after a month later WHO declared corona virus as Global pandemic [2]. Conditions gets worsen as people were traveling in the initial days before the declaration of pandemic and restrictions were made. This made approximately 213 countries to suffer. COVID-19 has raised difficult challenges in front of whole world in all aspects. During the pandemic economies of most of the wealthy countries fell by 7% [3]. Similarly if we talk in terms of human health, it has drastically shaken the primary medical system of almost every country. With almost 117M cases till date and 2.59M deaths around the globe [4]. It usually spreads when someone comes in closed contact with infected person (less than 1 meter) [5]. On one hand where people are fighting from physical health issues raised by infection, there are several psychological impacts too of COVID-19 on human health. State of well being, where an individual can easily cope with the situation of stress, realizes own abilities, while making contributions to his/her community [6]. When addressing COVID-19 as global pandemic, WHO has already warned that this will raise the fear, anxiety, stress and other mental illness emotions in people which may result in negative psychological impacts of it on human health [7]. COVID-19 has socially and emotionally affected people's health that has negative but not yet estimated consequences [8].

Along with all these things restrictions made by the governments to cope the situation locked the population indoors for months which leads to a serious mental pressure and sometimes depression also. These changes are extremely complex and challenging for everyone, it caused temporary unemployment, working from home, online education for students, and a huge responsibility for health professionals, these are the situations we are not used to, and hence chances of being mentally ill during pandemic are high. Its initial symptoms could be depression and anxiety which can lead to major mental breakdown later [1]. There is a constant fear of being infected, stress about our friend's and families' health. Changes in eating habit, sleeping disorder, irritation, etc., are some of the symptoms. There are several ways to fight against this traumatized phase like using social media but to a limited extent, proper sleeping, healthy eating habit, meditation, yoga, etc. We will discuss about this later in this chapter.

According to Mark Schaller, there is a psychological mechanism that helps organisms to detect infectious parasites in environment and then to fight against it, so there are more chances of growing emotions like anxiety, anger, etc., while accounting the situation raised [9].

2 COVID-19: Physical Health

As discussed earlier, this is an infectious diseases caused by SARS-CoV-2 Noble Corona Virus. This disease spreads from one person to another person when someone comes in closed contact with infected person. It transmits through direct contacts, like with respiratory droplets. Its spike proteins function as key lock mechanism in the human body to due to which it remains undetected for a long time, unlike other flu [10]. Estimated results in different research works show that estimated average incubation period of COVID-19 is 11.5 days from the day of infection caught them. Which means a person will get infected within 12 days after been infected from this virus, and hence symptoms will be developed during this period [11]. In some cases, long term effects of Covid-19 can be seen in patients. COVID-19 mainly targets organs like, lungs, liver, kidneys to shut them down. It blocks the respiratory organs due to which problem of shortness of breathing can be seen in majority of the cases. According to WHO, symptoms of Covid-19 have been categorized as common, less common and few serious symptoms [12]. Common symptoms include fever, dry cough and fatigue. Loss of smell and taste senses, headache, joint pain and loss of appetite, shortness of breath are some symptoms included in less common and serious symptoms of COVID-19, respectively.

3 COVID-19: Mental Health

As we have already discussed, global pandemic has affected almost every person physically, socially and emotionally around the world, but mental trauma for every person is of different kind due to the different circumstances. Like, not so managed remote education system for students, not so desirable working environment for people working from home, business collapse, fear of being infected in health workers and social workers, doctors and so on, all these things resulted in mental distress. There are various different reasons of mental illness among people during pandemic but misinformation or false information about virus which gives birth to myths and rumors in the society could be a big reason which enforced people to develop symptoms of melancholy in population [13]. If we talk about patients of COVID-19, a high stress can be seen among them caused by fear of disease. An early study in China shows that constant fear of virus among population due to its unpredictable and uncertain nature and period, it can cause mental disorders [14]. And also a huge exposure to negative or false news is also a reason of stress and anxiety in people [15]. How and due to which reasons different people are mentally affected are listed below.

3.1 Students

This portion covers majority of youth, whether they are school kids, university or college going students. In the beginning of March 2020, when restrictions were put up as a prevention of COVID-19 outbreak all the university, school and colleges stand closed for almost 8 to 9 months and still closed. This situation becomes a reason of educational pause at beginning, but as soon as officials realized that pandemic will take more of 2020, remote educational plan came into play. Improperly managed remote educational system becomes a reason of stress and anxiety in students. Virtual classes, frequent assignments and social isolation from friends also need to be considered when talking about psychological impacts of COVID-19 on students. And the worst part is mental illness in students can affect the ability to concentrate shown by a study [16].

3.2 *People Working from Home*

If we see from occupation's point of view, where health professionals are at higher risk in terms of physical and mental health both due to various reasons during COVID-19 period with increased workload one of the reason, people working in other sectors are also at risk of being mentally ill during the same. Adaptation to a whole new environment, which disconnects you from people out there physically, can cause several mental issues. Also not all people have proper setup to work on with them at home. Like, laptop, working internet connection. According to Dominique (Professor of people, organizations, and society, at Grenoble School of business, France). He states that regular practice of virtual meetings frequently on daily bases can give birth to mental fatigue in employees [17]. Not getting proper work environment at home and increased workload play an important role in disturbing mental status of a person, working from home. That's why we can see anxiety, stress, anger, irritation and sometimes even other depressive symptoms in them.

3.3 *Patients of Covid-19*

A rapid development of symptoms of mental disorder can be seen among the patients of COVID-19. Improper and incomplete information about nature of virus is one reason in early stages of pandemic. As in early stage of pandemic, there were no proper medicines available for cure and different vaccines were also under trial. In addition, recommendation made by WHO for COVID-19 patients that an isolation period of 14 days is must for COVID-19 patients, was salt in the wound. This raised the cases of significant level of anxiety, confusion, stress and anger among them [18]. While fear of death, becomes constant among them. When compared to people not suffering from COVID-19, patients of COVID-19 show higher level of PTSD according to studies made in this area [19]. Aside from it, there are patients who were already fighting with some other diseases like asthma, diabetes, etc., and some prior mental disorder problems too.

3.4 *Health Professionals*

The biggest reason among health workers and doctors to develop negative emotions of mental disorder like, stress, anxiety, irritation, anger and other is the huge exposure to the disease which increase the chance of being infected by COVID-19 at a rapid rate as compared to other people. When every other service whether they are public or private are in idle state, medical sector is the one where people are working even more due to the increase workload. There are doctors, nurses, cleaning staff at hospitals, lab technicians, ambulance staff, and many other health professionals who are in direct contact with patients of COVID-19. A study on 1563 health professionals shows, depressive symptoms in more than 50.7%, while symptoms of anxiety and sleeping issues were 44.7% and 36.1%, respectively [20]. On other hand, global supply chain of medical equipments, medicines, protective kits and other essential equipments had also been questioned in many countries during pandemic. For example, if talk about US medical system, it was estimated that there will be not enough ventilators for patients in upcoming time [21]. So all these short comes becomes a reason of mental disturbance in health professionals.

3.5 *Elders*

In elders, fear is more as compared to youth. Fighting against a disease is all about what a strong immune system do but in elder people lack of fighting capability from diseases becomes dangerous for physical as well as for mental health. Also it can be seen that even before pandemic mental distress problems are more in elders than others. Risk of being infected increases with age [22]. All these factors are enough to create negative psychological impacts in them in the time of global pandemic COVID-19.

4 Method

A systematic dataset has been taken for the implementation process and to generate results out of it. Dataset contains information about a short population, approximately 500 people consist of nearly 45.7% males and 54.3% females. Dataset contains information like, gender, age group they belongs to (focused on a particular age group of 18 to 30 years), their occupation, symptoms of mental trauma encountered by them, reasons of it, and the ways adopted by people to tackle the situation during pandemic. It also includes rating given by people to their mental situation before and during pandemic, which will be helpful to estimate a rise or drop in this parameter. For implementation, Tableau software has been used. This is a data visualization and analytic tool.

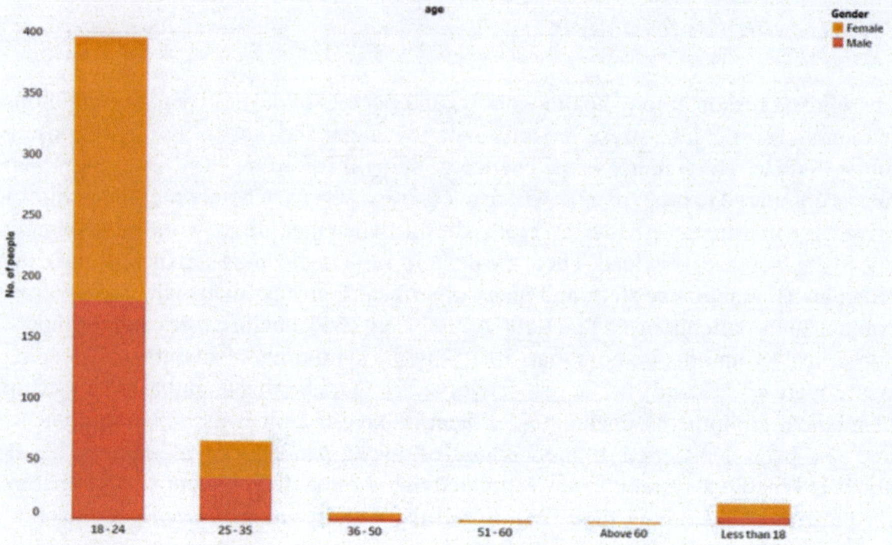

5 Results

Results generated as a part of survey include symptoms, reasons and solution suggested by our participants to handle this mental trauma during pandemic. So first if have a look over symptoms so, almost 45.9% is the response rate for reduced ability to concentrate, 56.9% for stress, 37.8%, 37.6% and 35.6% for fear/worries, anger and change in eating habit, respectively. Whereas anxiety is responded as a symptom at a rate of 46.1%. Results also include few but dangerous suicidal thoughts in almost 11.9% responders.

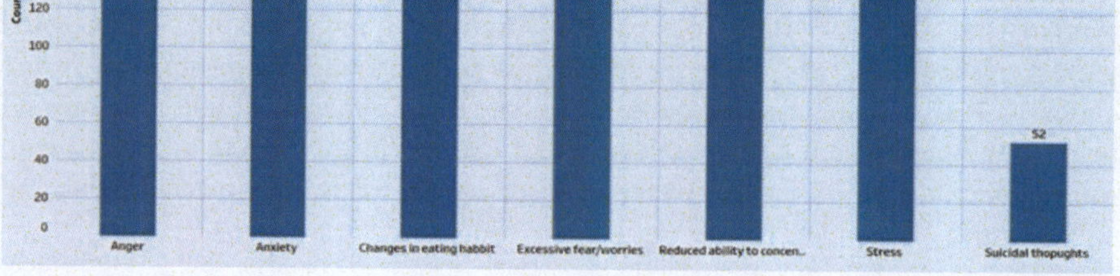

Now if we see reasons of being mentally affected during COVID-19, as expected it shows highest that is 56.9% respond rate for, being at home for a long time due to the various restrictions made by governments to reduce the spread of infection. One of them is lockdown. It also shows 33.8% response by the people concerned about their nearest and dearest people to get infected. 30% for educational pressure on students and people working from home. Economic instability and fear of getting infected also contributed toward disturbing psychological health of people during pandemic with respond rate of 23.5% and 22.9%, respectively. While with this, there were almost 16.1% people who developed these negative emotions because they were directly affected from this disease.

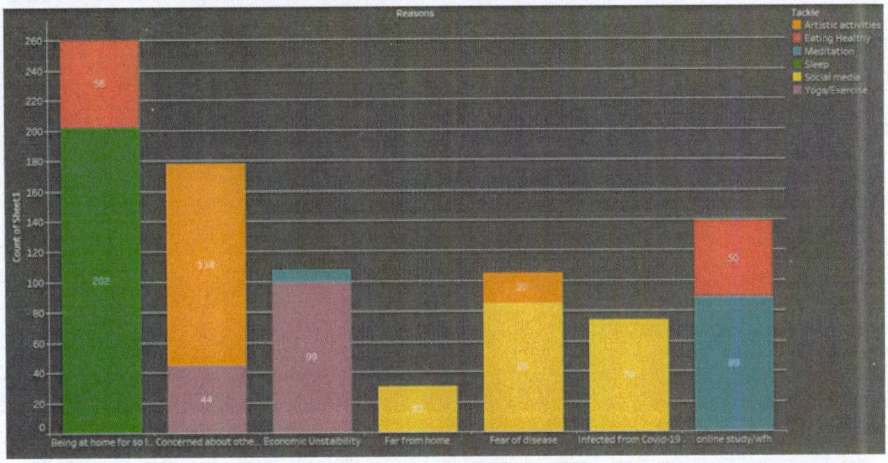

As we have already discussed dataset has been framed, so to estimate the mental situation of population before and during pandemic. For which dataset includes rating on a scale of 1 to 5 in both phases. After implementation, results shown are like this. Comparison between these two graphs can be analyzed easily. We can see the graph of rating of peoples mental health before COVID-19 is maximum at 197 (mental health rated as 5), and 192 (mental health rated as 4). But if we see graph showing rating of peoples mental health during COVID-19 it's not so systematic symmetry show that its highest at rating 3, rated by almost 157 people. So it can be concluded that people's mental health has been at stacked during pandemic in comparison with non pandemic era.

Following is the implementation analysis of the same, which shows a comparison between average count of rating of mental health before and during pandemic. Here, average rating has been calculated which is 4.11 before pandemic and 3.01 during pandemic. It clearly shows a drop in mental health as committed by people.

Rating of mental health (x)	Beforte Covid-19 f1(x)	During Covid-19 f2(x)	xf1(x)	xf2(x)
1	17	54	17	54
2	14	109	28	218
3	62	157	186	471
4	192	101	768	404
5	197	61	985	305
	482	482	1984	1452
		Average of rating of mental health :	4.11618257	3.01244813

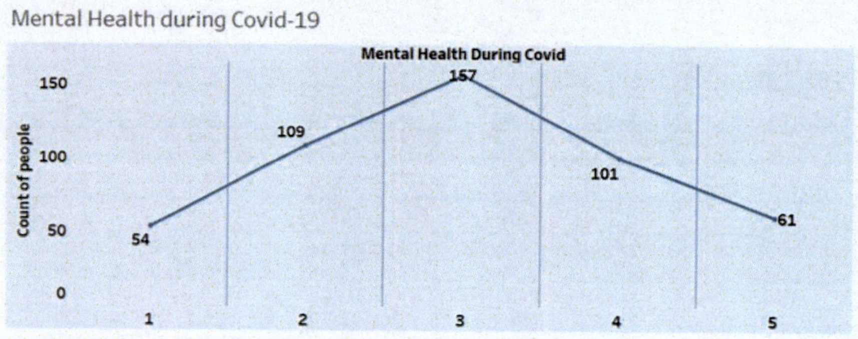

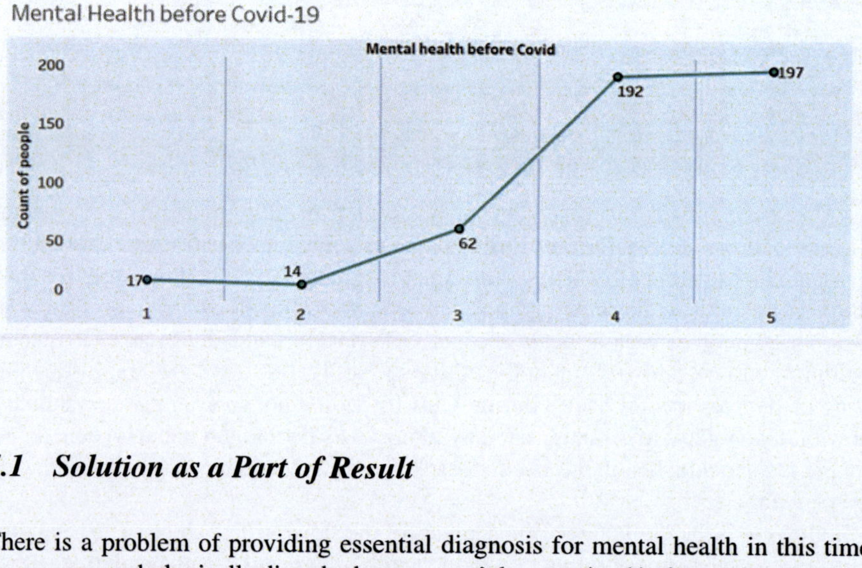

5.1 Solution as a Part of Result

There is a problem of providing essential diagnosis for mental health in this time because a psychologically disturbed person can't be examined in face to face because it increases the chances of transmission of virus rapidly [23]. WHO also recommended that mental health services should also be a part of recovery plan for COVID-19 [24]. A work suggests use of PFA (psychological first aid) immediately after such mental crisis [25]. PFA is first psychological treatment, and it includes family members, friends and other people around you [26]. Some ways executed by people to fight against this situation, which were part of survey conducted, are given below.

5.1.1 Awareness About Mental Health

One of the biggest problems behind this is the lack of awareness about mental health. Either they don't prioritize their mental health or are not aware of its consequences. Problem of lack of awareness can be seen largely in elders and in children. A study in Bangladesh on 2425 adults aged from 18–90 shows that only 0.7% people were aware about mental health conditions and almost 56.28% were not aware of MHCs at all [27]. This could be a reason of not getting proper psychological treatment at needed time. So as a solution we can move toward online awareness campaigns for people to talk about a healthy mind. So that people can identify the early attacks and cope with the situation as soon as possible.

5.1.2 Psychology as a Profession

Pandemic situation is a reason of mental disorder and a reason for not getting proper treatment as well. According to WHO, mental health services have been disrupted due to pandemic in many countries [24]. Especially in most low and middle income countries there are very slow mental health service delivery [28]. Along with this, there is a shortage of psychiatrists for treatment. In India, there are only 3 psychiatrists per 1 M people, and that is a huge shortage [23]. Also in this situation, face to face treatment is not possible, and hence, many hospitals have adapted online counseling sessions for treatment [20]. So it's the need of time to take psychology as serious profession for future assessments. Teletherapy is also a solution; "Bharosa helpline" by UNION HRD Minister, Mr. Ramesh pokhriyal is the best example of teletherapy started for university students mainly [29].

5.1.3 Artistic Activities

Artistic activities like dancing, singing, painting, writing, etc., play important role in maintaining a healthy mental situation especially in the time of pandemic. Results show that almost 33.3% results shows that it's a possible solution. There are enough evidences of arts and creativity being a positive approach toward overcoming the emotions of anxiety and depression, also it is a cost-effective method [30]. It is also known as creative therapy. Our implementation also shows 33.1% people committed artistic activities as a solution to this problem.

5.1.4 Yoga/Meditation

Yoga and meditation are very helpful in attaining mental peace in lockdown situation at home. Results generated after implementation on dataset targeted shows that 28.4% results contributes toward Yoga and meditation as one solution. It can reduce the mental stress. Different forms of yoga like, Yogasana, Sahajyoga, had shown positive results, in the treatment of depression [31]. It can also be seen that mediation, exercise/yoga as a healthy remedy with 21.1% and 30.8% respond rate.

5.1.5 Other Solutions

There are many other solutions also present like maintaining a healthy and a positive lifestyle, and for this many ways are present. Like, maintaining a proper sleeping schedule which is responded at a very high rate of 43.7% among other solutions. Eating healthy food is also important reflects its results that is 23.2%. Though using a lot more social media is not a part of healthy lifestyle but as it was a survey focused on youth so it came up as a solution with highest respond rate of 44.5%.

6 Conclusion

Conclusion can be made that COVID-19 pandemic has raised many mental disorder problems in people on a large scale throughout the world. Reasons could be financial instability, increased workload, and social detachment from loved ones and so on. So there are several solutions suggested, like we have discussed about awareness, campaigns, healthy lifestyle and an optimistic approach when living in between of a pandemic. Also there is a need of taking psychology as a serious profession to fulfill the future demands of this area.

References

1. Li S, Wang Y, Xue J, Zhao N, Zhu T (2020) The impact of COVID-19 Epidemic declaration on psychological consequences: a study on active weibo users. Int J Environ Res Publ Health 17:2032
2. WHO (2020) WHO response timeline of corona virus, https://www.who.int/emergencies/diseases/novel-coronavirus-2019/interactive-timeline
3. Sunitha S, Sudha S (2020) The continental shift in global economy during COVID-19 pandemic, World Bank report, 18.21.716954795.005
4. WIKIPEDIA (2020) COVID-data: Worldwide. https://en.m.wikipedia.org/wiki/template:COVID-19_pandemic_data
5. WHO (2020) How COVID spread between 2 people. https://www.who.int/news-room/q-a-detail/coronavirus-disease-covid-19-how-is-it-transmitted

6. Vigo D, Thornicroft G, Atun R (2016) Estimating all the global burden of mental health illness. Lancet Psychiatry 3(2):171–178
7. WHO Europe (2020) Impacts of pandemic on mental health. https://www.euro.who.int/en/health-topics/health-emergencies/coronavirus-covid-19/publications-and-technical-guidance/noncommunicable-diseases/mental-health-and-covid-19
8. Pfefferbaun B, North CS (2020) Mental health and covid-19 pandemic, 383(6):510–512
9. Schaller M, Park JH (2011) The behavioral immune system (and why it matters). SAGE J 20(2):99–103
10. The Hindu (2020) Covid-19; Impact on immune system. https://www.google.com/amp/s/www.thehindu.com/sci-tech/health/coronavirus-how-does-the-immune-system-respond-to-a-coronavirus-attack/article31319716.ece/amp/
11. Dutta SS (2020) Covid-19 Incubation period, NEWS Medical life science. https://www.news-medical.net/amp/health/Coronavirus-Incubation-Period.aspx
12. WHO (2020) Symptoms of Covid-19. https://www.who.int/emergencies/diseases/novel-coronavirus-2019/question-and-answers-hub/q-a-detail/coronavirus-disease-covid-19
13. Salari N, Hosseinian-far A, Jalali R, Vaisi-Raygani A, Rasoulpoor S, Mohammadi M, Raoulpoor S, Khaledi-Paveh B (2020) Prevalence general population during the COVID-19 pandemic: a systematic review and Meta analysis. Springer Nature. https://doi.org/10.1186/s12992-020-00589-w
14. Shigemura J, Ursano RJ, Morganstein JC, Kurosawa M, Bendek DM, (2020) Public responses to novel 2019 Coronavirus in Japan: mental; health consequences and target populations. Psychiatry Clin neurosci 74(4):281
15. Moghanibhashi-Mansourieh A (2020) Assessing the anxiety level of Iranian general population during COVID-19 outbreak. Asian J Psychiatry 51:102076
16. Son C, Hegde S, Smith A, Wanmg X, Sasangohar F (2020) Effects of Covid-19 on college students' mental health in the United States: Interview survey study. Jf Med Internet Res. https://www.jmir.org
17. Business Because Article (2021) Mental health of people working from Home. https://www.google.com/amp/s/amp.businessbecause.com/news/insights/7244/working-from-home-mental-health
18. Brooks SK, Webster RK, Smith LE, Woodland L, Wessely S, Greenberg N et al. (2020) The psychological impact of quarantine and how to reduce it: rapid review of the evidence. Lancet 14; 395(10227):912–20
19. Guo Q, Zheng Y, Shi J, Wang J, Li G, Fromson JA, Xu Y, Liu X, Xu H, Zhang T, Lu Y, Chen X, Hu H, Tang Y, Yang S, Zhuo H, Wang X, Chen H, Wang Z, Yang Z (2020) Immediate psychological distress in quarantined patients with COVID-19 and its association with peripheral inflammation: a mixed-method study. Brain Behav Immune
20. Ho CS, Chee CY, Ho RC (2020) Mental health strategies to combat the psychological impact of COVID-19 beyond paranoia and panic. Ann Acad Med Singapore 49:01–03
21. Ranney ML, Griffeth V, Jha AK (2020) Critical supply shortages-The need for ventilators and personal protective equipment during the COVID-19 pandemic. New England J Med 382:e41
22. Centers for Disease Control and Prevention (2021) https://www.cdc.gov/coronavirus/2019
23. Simonetti G, Iosco C, Taruschio G (2020) Mental health and COVID-19: an action plan, preprints. www.preprints.org
24. WHO Bulletin (2020) Mental Health services in different countries. https://www.who.int/news/item/05-10-2020-covid-19-disrupting-mental-health-services-in-most-countries-who-survey
25. Jacobs GA, Meyer DL, Laura B, Robert JS (2006) Psychological interventions in times of crisis. In: Community based psychological first aid for oncology professionals. Springer Publishing Company, New York, pp 57–71
26. Presentation by WHO (2016) Psychological First Aid for all. https://www.who.int/mental_health/world-mental-health-day/ppt.pdf
27. Uddin MN, Bhar S, Islam FA (2019) Assessment of awareness of mental health conditions and its association with socio-demographic characteristics: a cross sectional study in a rural district in Bangladesh. BMC Health Serv Res 19:562

28. Srivastava K, Chatterjee K (2016) Pookala shivaram Bhat, 2016 Mental Health awareness: The Indian Scenario. Ind Psychiatry J 25(2):131–134
29. Zee News Article (2020) Bharosa Helpline. https://www.google.com/amp/s/zeenews.india.com/india/bharosa-helpline-launched-in-odhisa-to-help-students-relieve-their-distress-amid-covid-19-2282806.html/amp
30. Bungay H (2018) How prescription and creativity can improve mental and physical health, Medical Xpress. https://medicalxpress.com/news/2018-04-prescription-creativity-mental-physical-health.html
31. Cramer H, Lauche R, Langhorst J, Dobos G (2013) Yoga for depression: a systematic review and meta-analysis. Depress Anxiety 2013(30):1068–1083

Improved Symbiosis Organism Search Algorithm-Based Clustering Scheme for Enhancing Longevity in Wireless Sensor Networks (WSNs)

Sengathir Janakiraman and Bandi Rambabu

Abstract Wireless sensor networks (WSNs) possess the merits of low cost, low power consumption and flexible communication that makes it suitable and ideal for diversified applications such as military operations, weather forecasting, health monitoring and network surveillance. Dynamic cluster head selection schemes from the dimension of WSN clustering algorithm are considered to handle the issues of unreasonable cluster head selection. Impotent cluster head selection results in unbalanced energy consumption and overlapping coverage during the process of cluster communication. Thus, an optimal clustering scheme that sustains energy stability and energy consumptions by discovering the reliable oath without hindering the reliability in transmission is a highly challenging task. In this paper, Improved Symbiosis Organism Search Algorithm-based Clustering Scheme (ISOSACS) is proposed for predominant cluster head selection that aids in maintaining energy stability with enhancing longevity in WSNs. This ISOSACS is proposed with the benefits of self-adaptive superior factor and three-way phase of mutualism, such that frequent selection of cluster head is completely prevented in the network. It included the benefits of new control operator and coefficient of random weighted reflection for potential selection of cluster heads, such that unnecessary wastage of energy is eliminated. Simulation results of ISOSACS confirmed better performance in throughput of 18.24% with minimized energy consumptions of 15.68% and reduced delay of 16.38% on par with the competitive clustering schemes.

Keywords Improved symbiosis organism search algorithm (ISOSACS) · Cluster head selection · Three-Way mutualism phase · Self-adaptive superior factor · Wireless sensor networks (WSNs)

S. Janakiraman · B. Rambabu (✉)
Hyderabad, India

R. P. Mahapatra et al. (eds.), *Proceedings of International Conference on Recent Trends in Computing*, Lecture Notes in Networks and Systems 341,
https://doi.org/10.1007/978-981-16-7118-0_68

1 Introduction

Wireless sensor networks (WSNs) collect, examine and use data obtained from sensors that were used for diverse applications like smart homes, air purifiers and disaster monitoring owing to the enhanced performance, easiness of use and reduced price [1]. Sensors may be positioned in dangerous environments deterring their replacement. Enhancing the battery life of a node raises the costs. Network lifetime and steadiness have to be enhanced through a diverse collection of protocols [2]. The low-energy adaptive clustering hierarchy (LEACH) protocol increases energy efficacy through clustering. When data is sent from a sensor to the base station (BS), the amount of energy consumed depends on the distance between them. Clustering aims at reducing the distance of nodes that are not cluster heads (CHs). Selection of appropriate CHs enables efficient use of energy [3].

In LEACH, nodes are uniformly and probabilistically chosen as CHs, ignoring the state and features of chosen nodes like residual energy, anticipated energy consumption and number of neighbors [4]. Information regarding nodes' energy is used at the BS for choosing CHs. Nevertheless, it is demanding to concurrently obtain this information during transmission. In LEACH-C [5], to get the information regarding residual energy, synchronization is performed through time-division multiple access (TDMA). It involves more computing power at the BS when compared to other nodes [6]. Clustering using swarm intelligence is a precise scheme that is extensively applied to optimization protocols. This scheme is included in protocols based on particle swarm optimization (PSO), bee colony optimization [6] and ant colony optimization [7].

In this paper, Improved Symbiosis Organism Search Algorithm-based Clustering Scheme (ISOSACS) is proposed for predominant cluster head selection that aids in maintaining energy stability with enhancing longevity in WSNs. This ISOSACS is proposed with the benefits of self-adaptive superior factor and three-way phase of mutualism, such that frequent selection of cluster head is completely prevented in the network. It included the benefits of new control operator and coefficient of random weighted reflection for potential selection of cluster heads, such that unnecessary wastage of energy is eliminated. Simulation experiments of the proposed ISOSACS and the competitive cluster head selection schemes such as SOSACS, BCOCS and COSOCS schemes are conducted with the evaluation metrics of throughput, mean energy consumptions, network latency and network lifetime with respect to different sensor nodes.

2 Related Work

Cuckoo-based particle swarm optimization scheme [8] is proposed for energy-effective computation to focus on optimizing the network performance. This method is propounded for offering energy efficacy by including a multi-modal objective function. Once CHs are selected, the aggregated data is sent to the BS using a particle-based algorithm. This meta-heuristic optimization scheme is used to collect data. Nodes are categorized into normal nodes and CHs based on the amount of remaining energy. The CH is chosen based on the best fitness value determined by the search agents involved in the cuckoo search process. Energy is balanced among nodes based on the remaining energy.

GA-based hierarchical clustering approach [9] is propounded to focus on the transmission involving long distances between the nodes and the BS. GA is combined with hierarchical clustering for lessening the distance, which in turn reduces the amount of energy consumed to the maximized level. In this scheme, position of the BS, size of the population and number of runs are determined by applying GA is applied on the network. The CHs that are in proximity to the BS collect the data from the CHs and forward it to the sink. It computed the fitness of nodes using genetic operators that are assigned to nodes, and the one with maximum energy is chosen as the CH. The trusts of nodes are calculated for the whole network. The sensed data is sent to the CHs through multi-hop communication, which improves the network lifespan.

Multipath Routing Protocol (MRP) is proposed for clustering based on ACO [10]. It uses a clustering algorithm, wherein a node with more remaining energy, a greater number of adjacent nodes and more RSSI are elected as CHs. An enhanced ACO algorithm is used to find several paths with reduced energy. Search ANTs (SANTs) are involved in finding paths and collecting information along the paths to the sink, while backward ANTs (BANTs) are used for finding the cost of paths found by the SANTs, and the abnormal ANTs (AANTs) aid in circumventing failure. New CHs are chosen when the remaining energy goes below 50% of the mean energy of the nodes.

MPSICA is a smart and speedy routing recovery technique based on PSO [11], which is proposed for a heterogeneous network to preserve disjoint paths from every source to the nearest CH and the accessible path from CHs to the BS. It is extremely faulted forbearing to the malfunctioning of paths of normal and CHs, as outmoded retransmissions can be reduced, and the dependability can be enhanced with reduced energy consumption and lengthier lifespan. To proficiently reduce energy consumption, this centralized approach is applied to CH nodes with better resources than simple nodes. Each node is a particle where nodes form a particle sequence relating to a path to the BS. The propounded scheme optimizes every node's particle sequence through the optimal path by using an ideal fitness function on the CH.

BCO based on artificial bee colony (ABC) algorithm is proposed [12]. ABC is similar to the smart foraging nature of honey-bee swarms. The onlookers deal with choosing the food source, while scouts are involved in performing random searches, and employer bees go in search of the source. The clustering scheme of the ABC-C is based on LEACH, where CHs are involved in aggregation. The CHs with unrestricted energy are selected using the ABC algorithm. Nodes are allocated to clusters based on the distance to the CHs. On computing the cross-distances, the nodes send the values to the BS.

3 Proposed Improved Symbiosis Organism Search Algorithm-Based Clustering Scheme (ISOSACS)

The proposed Improved Symbiosis Organism Search Algorithm-based Clustering Scheme (ISOSACS) is formulated based on the inspiration derived from the primitive nature-inspired from the cooperative association determined between the organisms existing in the ecosystem. The inherited SOSA algorithm [13] comprises of three potential steps that includes initialization, mutualism, commensalism and parasitism. In this context, solutions are considered to be identical to the organisms of the ecosystem. Moreover, the objective function value of each solution represents the adaptation degree of the necessitated objective.

In the initialization phase of SOSA, each solution (organism) is generated in the search space randomly. In this context, organism (feasible solution) existing in the ecosystem (search space) is generated based on Eq. (1).

$$\mathrm{Sol}_{\mathrm{Org(i,m)}} = \mathrm{LB_m} + \mathrm{rand}(0, 1) + (\mathrm{UB_m} - \mathrm{LB_m}) \tag{1}$$

where $1 \leq \mathrm{i} \leq \mathrm{NP}$ and $1 \leq \mathrm{m} \leq \mathrm{d}$

where NP is the number of possible solutions (number of organisms considered in the ecosystem) with as the optimization problems' dimension. This feasible solution is controlled by the upper $\mathrm{UB_m}$ and lower $\mathrm{LB_m}$ bounds associated with the search agent.

In the phase of mutualism, the two solutions (organisms) mutually cooperate and derive their benefits during the interaction process. In this mutualism processes, each organism $\mathrm{Sol}_{\mathrm{Org(i,m)}}$ cooperates with the randomly selected organism $\mathrm{Sol}_{\mathrm{Org(j,m)}}$ interacting in the ecosystem. Moreover, the solutions try to enhance themselves based on the derivation of merits of mutual survival that they achieve through the interaction in the ecosystem. The solution and under cooperation or interaction is represented based on Eq. (2) and (3).

$$\mathrm{Sol}^{\mathrm{New}}_{\mathrm{(i,m)}} = \mathrm{Sol}_{\mathrm{(i,m)}} + \mathrm{rand}(0.1) + (\mathrm{Sol}_{\mathrm{(Best)}} - \mathrm{V}_{\mathrm{Mutual}} + \mathrm{FB}_1) \tag{2}$$

$$\mathrm{Sol}^{\mathrm{New}}_{\mathrm{(j,m)}} = \mathrm{Sol}_{\mathrm{(j,m)}} + \mathrm{rand}(0.1) + (\mathrm{Sol}_{\mathrm{(Best)}} - \mathrm{V}_{\mathrm{Mutual}} + \mathrm{FB}_2) \tag{3}$$

$$\mathrm{V}_{\mathrm{Mutual}} = \frac{\mathrm{Sol}_{\mathrm{(i,m)}} + \mathrm{Sol}_{\mathrm{(j,m)}}}{2} \tag{4}$$

The aforementioned solutions are updated, when the fitness value of the newly generated solution is predominant over the existing solution depending on their values of their fitness. In this context, $\mathrm{Sol}_{\mathrm{(Best)}}$ highlights the best solution determined based on the value of the fitness corresponding to factors of benefits FB_1 and FB_2, which are utilized randomly as 1 or 2. This factor of benefits portrays whether the solution is gaining any advantage fully or partially from other solutions existing in the search space.V_{Mutual} is responsible for representing the association features that exists between $Sol^{New}_{(i,m)}$ and$Sol^{New}_{(j,m)}$, respectively.

In the phase of commensalism, the two solutions (organisms) are mutually cooperative, but only the first solution derives the benefits from another solution. While, the second solution is not benefited in any aspects by the first solution. The first solution $Sol^{New}_{(i,m)}$ derives the merits from the second solution $Sol^{New}_{(j,m)}$, which increases the higher adaptation degree. The new solution is updated through the position in which they are existing in the search space based on Eq. (5)

$$Sol^{New}_{(i,m)} = Sol_{(i,m)} + rand(-1.1) + (Sol_{(Best)} - Sol_{(j,m)}) \tag{5}$$

This solution updating process is always attained, when the fitness value of new solution is predominant than its associated fitness value identified in the previous iteration.

In the final parasitism phase, the two solutions (organisms) mutually cooperative, but only the first solution derives the benefits from second solution. While the second solution is negatively influenced or impacted by the first solution. In this case, the vector of parasitism is attained through the following steps, such as:

Choose a single dimension from the complete set of dimensions that are considered for evaluating the potential of a solution (organism).

Change the chosen randomly selected dimensions and replace the dimension position into $Sol^{New}_{(i,m)}$ for generating the vector of parasitism $V_{Parasite}$ based on Eq. (6)

$$Sol_{Org(i,p)} = LB_p + rand(0, 1) + \left(UB_p - LB_p\right) \tag{6}$$

where $1 \leq i \leq NP$ and $1 \leq m \leq d$

In this context, the vector of parasite plays a role of a host in order to facilitate the random selection of any other solution in order to derive its benefits and impose negative influence in the search space. If the objective function value of $V_{Parasite}$ of $Sol_{Org(i,p)}$ is better than the $V_{Parasite}$ of $Sol_{Org(j,p)}$, then the solution $Sol_{Org(j,p)}$ is completely replaced by the first $Sol_{Org(i,p)}$ solution.

In the primitive SOSA algorithm, the phenomenon of mutualism generally happens between any two solutions, but three solutions also may interact in the search space; hence, the method of three-phase mutualism is essential for attaining the better solution (position of cluster head to be selected). In addition, factors of benefits FB_1 and FB_2, which are utilized randomly are considered as 1 or 2. This value of 1 or 2 clearly depicts that each solution is deriving the benefits from another solution or not is represented in the phase of mutualism. This improvement of SOSA aids in enhancing the advantage of survival in order to determine the best solution that targets at the process of minimizing the rate of convergence with improved quality of solution. This objective targets on including a new factor of benefit for improving the capability of searching based on the inclusion of boundary control operators and coefficient of random weighted reflection.

In the first mutualism phase of the proposed Improved Symbiosis Organism Search Algorithm-based Clustering Scheme (ISOSACS), three-way mutualism is included based on the steps, such as:

Three selected solutions in the search space is randomly selected and made to improve themselves in the search space.

New solutions in the search space are determined based on mutual symbiosis associated between as specified in (7), (8) and (9), respectively.

$$\text{Imp\_Sol}^{\text{New}}_{(\text{i,m})} = \text{Sol}_{(\text{i,m})} + \text{rand}(0.1) + (\text{Sol}_{(\text{Best})} - V_{\text{Mutual}} + \text{FB}_1) \quad (7)$$

$$\text{Imp\_Sol}^{\text{New}}_{(\text{j,m})} = \text{Sol}_{(\text{j,m})} + \text{rand}(0.1) + (\text{Sol}_{(\text{Best})} - V_{\text{Mutual}} + \text{FB}_2) \quad (8)$$

$$\text{Imp\_Sol}^{\text{New}}_{(\text{k,m})} = \text{Sol}_{(\text{k,m})} + \text{rand}(0.1) + (\text{Sol}_{(\text{Best})} - V_{\text{Mutual}} + \text{FB}_3) \quad (9)$$

where the mutual vector determined in this three-way mutualism is presented in Eq. (10)

$$V_{\text{Mutual}} = \frac{(\text{Imp\_Sol}_{(\text{i,m})} + \text{Imp\_Sol}_{(\text{j,m})} + \text{Imp\_Sol}_{(\text{k,m})})}{3} \quad (10)$$

In this proposed ISOSACS scheme, factor of self-adaptive advantage [14] is used for preventing the problem inherent with the factors of benefits FB_1 and FB_2 that randomly varies with the value of 1 and 2. When the value of FB_1 and FB_2 is equal to 1, the rate of convergence is slow. On the other hand, when the value of FB_1 and FB_2 is equal to 2, the rate of convergence speeds up but minimizes the capability of searching in the search space. Thus, three new factors of self-adaptive advantage defined in Eqs. (11), (12) and (13) are used for balancing the search with improved capability of convergence.

$$F_{\text{SAA(i)}} = 1 + \text{rand}(0, 1) + \frac{\text{fn}_{(\text{i,m})}}{F_{\text{Max(i,m)}}}, \text{ where } F_{\text{Max(i,m)}} \neq 0 \quad (11)$$

$$F_{\text{SAA(j)}} = 1 + \text{rand}(0, 1) + \frac{\text{fn}_{(\text{j,,m})}}{F_{\text{Max(j,m)}}}, \text{ where } F_{\text{Max(j,m)}} \neq 0 \quad (12)$$

$$F_{\text{SAA(k)}} = 1 + \text{rand}(0, 1) + \frac{\text{fn}_{(\text{k,m})}}{F_{\text{Max(k,m)}}}, \text{ where } F_{\text{Max(k,m)}} \neq 0 \quad (13)$$

Where $\text{fn}_{(\text{i,m})}$, $\text{fn}_{(\text{j,,m})}$ and $\text{fn}_{(\text{k,m})}$ refers to the three solutions considered for cluster head selection in the phase of mutualism with $f_{\text{Max(m)}}$ as the maximum fitness value associated with a solution in the entire search space. Moreover, this factor of self-adaptive advantage is determined automatically in the process of improving the quality of the solution. Moreover, the coefficient of weighted reflection proposed in [15] was utilized for improving the mechanism of commensalism phase in the primitive SOSA algorithm. This coefficient of weighted reflection is made to vary between 0.5 and 1 as defined in Eq. (14).

$$\text{Coeff}_{\text{RWR}} = 0.5 + (1 + \text{R}), \text{ Where r} \in [0, 1] \tag{14}$$

In the successive step of mutualism, the coefficient of weighted reflection is adopted in the proposed ISOSACS scheme during the phase of commensalism is used for updating the solution based on Eq. (15).

$$\text{Sol}_{(\text{i,d})}^{\text{New}} = \text{Sol}_{(\text{i,m})} + \text{Coeff}_{\text{RWR}} * \left(\text{Sol}_{(\text{Best})} - \text{Sol}_{(\text{j,d})}\right) \tag{15}$$

Finally, the same process of parasitism inherent with the fundamental SOSA algorithm is included in the proposed ISOSACS scheme for randomly selecting the position of the solution. This process of cluster head selection is adopted, until the condition of termination is satisfied. This proposed ISOSACS scheme is thus beneficial in selecting the cluster head in order to improve the energy stability and minimize the energy consumptions in the network.

4 Simulation Results and Discussion

The simulation experiments of the proposed ISOSACS scheme and the competitive cluster head selection schemes such as SOSACS, BCOCS and COSOCS schemes are conducted based on MATLAB R2018 [12, 16–19]. This experimental investigation is conducted with the evaluation metrics of throughput, mean energy consumptions, network latency and network lifetime with respect to different sensor nodes. Figures 1 and 2 present the mean throughput and mean energy consumptions of the proposed ISOSACS scheme with different sensor nodes.

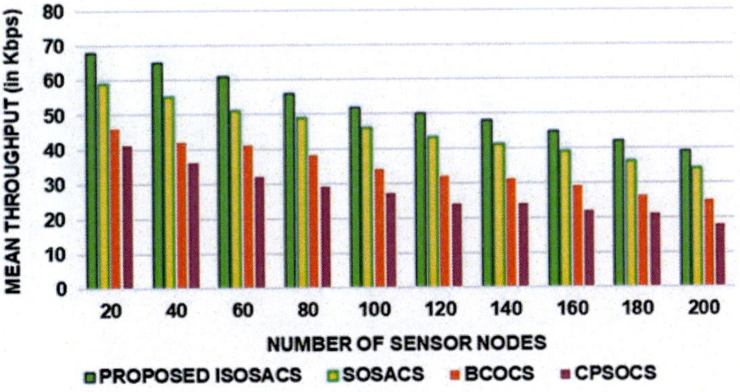

Fig. 1 Proposed ISOSACS: mean throughput with increasing sensor nodes

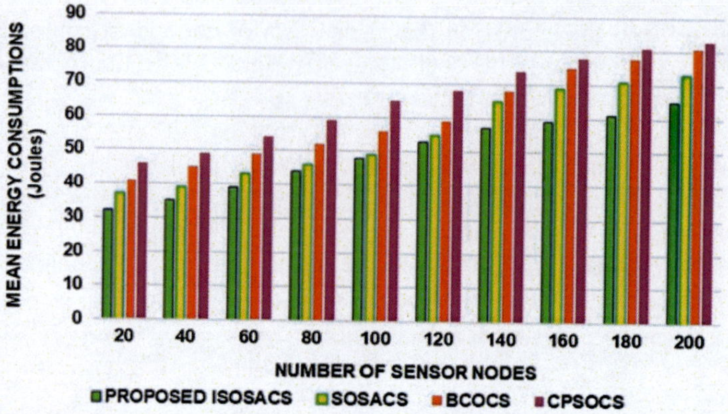

Fig. 2 Proposed ISOSACS: mean energy consumptions with increasing sensor nodes

The mean throughput of the proposed ISOSACS scheme is determined to increase with different sensor nodes, since the inclusion of self-adaptive advantage factors attributed toward better delivery of reliable packets in the network. On the other hand, the mean energy consumptions of the proposed ISOSACS scheme is minimized during the factors of three different factors of benefits. The proposed ISOSACS scheme improved mean throughput by 6.21, 5.64 and 5.28%, better than the competitive approaches. Moreover, mean energy consumptions of the proposed ISOSACS scheme is minimized by 5.12, 4.56 and 3.98%, better than the competitive approaches.

In addition, Figs. 3 and 4 present the network latency and network lifetime of the proposed ISOSACS scheme with different sensor nodes. The network latency of the proposed ISOSACS scheme is identified to be minimized with different sensor nodes, since the inclusion of the coefficient of weighted reflection included into the phase of commensalism. In addition, the network lifetime of the proposed ISOSACS scheme is improved due to the inclusion of this three-way mutualism. The proposed

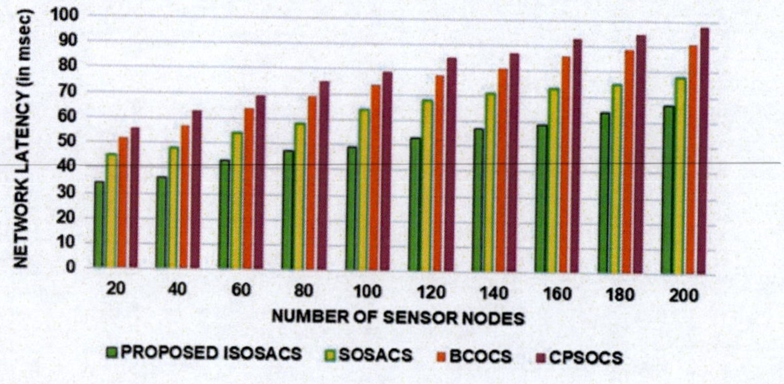

Fig. 3 Proposed ISOSACS: network latency with increasing sensor nodes

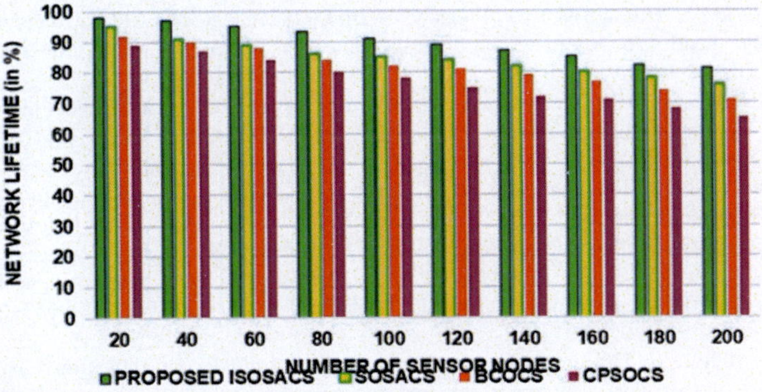

Fig. 4 Proposed ISOSACS: network lifetime with increasing sensor nodes

ISOSACS scheme minimized network latency by 5.94, 4.98 and 4.12% better than the competitive approaches. Moreover, network lifetime of the proposed ISOSACS scheme is improved by 4.21, 3.98 and 3.49% better than the competitive approaches.

5 Conclusion

In this paper, ISOSACS was proposed for attaining significant cluster head selection that aids in maintaining energy stability with enhancing longevity in WSNs. This ISOSACS is proposed with the benefits of self-adaptive superior factor and three-way phase of mutualism, such that frequent selection of cluster head is completely prevented in the network. It included the benefits of new control operator and coefficient of random weighted reflection for potential selection of cluster heads, such that unnecessary wastage of energy is eliminated. Simulation results of ISOSACS confirmed better performance in throughput of 18.24% with minimized energy consumptions of 15.68% and reduced delay of 16.38% on par with the competitive clustering schemes. Furthermore, mean energy consumptions of the proposed ISOSACS scheme is minimized by 5.12, 4.56 and 3.98% better than the competitive approaches. Moreover, network lifetime of the proposed ISOSACS scheme is improved by 4.21, 3.98 and 3.49% better than the competitive approaches. As the part of the future scope, it is planned to devise a Salp swarm optimization algorithm-based clustering scheme and compare with the proposed ISOSACS scheme.

References

1. Ni Q, Pan Q, Du H, Cao C, Shai Y (2017) A novel cluster head selection algorithm based on fuzzy clustering and particle swarm optimization. In: IEEE/ACM transactions on computational biology and bioinformatics, vol 14, no 1, pp 76–84, 1 Jan–Feb 2017

2. Ali H, Tariq UU, Hussain M, Lu L, Panneerselvam J, Zhai X, ARSH-FATI a novel metaheuristic for cluster head selection in wireless sensor networks. IEEE Syst J https://doi.org/10.1109/JSYST.2020.2986811
3. Kang SH, Nguyen T (2012) Distance Based Thresholds for Cluster Head Selection in Wireless Sensor Networks. IEEE Commun Lett 16(9):1396–1399. https://doi.org/10.1109/LCOMM.2012.073112.120450
4. Umbreen S, Shehzad D, Shafi N, Khan B, Habib U (2020) An energy-efficient mobility-based cluster head selection for lifetime enhancement of wireless sensor networks. IEEE Access 8:207779–207793. https://doi.org/10.1109/ACCESS.2020.3038031
5. Choi H-H, Muy S, Lee J-R (2021) Geometric analysis-based cluster head selection for sectorized wireless powered sensor networks. IEEE Wirel Commun Lett 10(3):649–653. https://doi.org/10.1109/LWC.2020.3044902
6. Lata S, Mehfuz S, Urooj S, Alrowais F (2020) Fuzzy clustering algorithm for enhancing reliability and network lifetime of wireless sensor networks. IEEE Access 8:66013–66024. https://doi.org/10.1109/ACCESS.2020.2985495
7. Leu J, Chiang T, Yu M, Su K (2015) Energy efficient clustering scheme for prolonging the lifetime of wireless sensor network with isolated nodes. IEEE Commun Lett 19(2):259–262. https://doi.org/10.1109/LCOMM.2014.2379715
8. Vijayalakshmi K, Anandan P (2020) Global Levy flight of cuckoo search with particle swarm optimization for effective cluster head selection in wireless sensor network. Intell Autom Soft Comput 2(1):45–58
9. Maheswari AU, Pushpalatha MS (2014) Cluster head selection based on genetic algorithm using AHYMN approaches in WSN. Inte J Comput Organ Trends 5(2):63–66
10. Chen Y, Wang H (2017) Evolutionary energy balanced ant colony algorithm based on WSNs. Clust Comput 22(S1):609–621
11. Shankar T, Shanmugavel S, Rajesh A (2016) Hybrid HSA and PSO algorithm for energy efficient cluster head selection in wireless sensor networks. Swarm Evol Comput 30(3):1–10
12. Janakiraman S (2018) A hybrid ant colony and artificial bee colony optimization algorithm-based cluster head selection for IoT. Procedia Comput Sci 143(2):360–366
13. Küçükuğurlu B, Gedikli E (2020) Symbiotic organisms search algorithm for multilevel thresholding of images. Expert Syst Appl 147(2):113210
14. Rambabu B, Reddy AV, Janakiraman S (2019) A hybrid artificial bee colony and bacterial foraging algorithm for optimized clustering in wireless sensor network. IJTEEE-2186-2190
15. Satapathy CS, Naik A (2012) Improved teaching learning-based optimization for global function optimization. Decis Sci Lett 4(2):23–34
16. Rambabu B, Venugopal Reddy A, Janakiraman S (2019) Hybrid artificial bee colony and monarchy butterfly optimization algorithm (HABC-mboa)-based cluster head selection for WSNs. J King Saud Univ Comput Inf Sci 2(1):45–56
17. Janakiraman S, Priya M, Devi S, Sandhya G, Nivedhitha G, Padmavathi S (2018) A Markov process-based opportunistic trust factor estimation mechanism for efficient cluster head selection and extending the lifetime of wireless sensor networks. EAI Endorsed Trans Energy Web 2(1):168093
18. Janakiraman S, Deva Priya M (2020) An energy-proficient clustering-inspired routing protocol using improved BKD-tree for enhanced node stability and network lifetime in wireless sensor networks. Int J Commun Syst e4575
19. Bandi R, Ananthula VR, Janakiraman S (2021) Self adapting differential search strategies improved artificial bee colony algorithm-based cluster head selection scheme for WSNs. Wireless Pers Commun 1–22

A Novel 3D Facial Recognition for Digital Payments

K. C. Prabu Shankar and M. Hema

Abstract In this research paper, we propose a system that envisions to revolutionize the way digital payments work across the globe. With over six-hundred million users transacting regularly in India alone, the landscape of the global market is potentially huge. But when we talk about the security of these, a big question arises as the current methods are either breachable or complex. The majority of them use age authentication mechanisms, or the ones which have advanced to use new technologies are too slow and memory inefficient. All of which is not very convenient, simplified, and reliable for an end-user. Considering all these issues and to effectively solve them, we propose a 3D facial recognition system backed with a series of preprocessing algorithms, a point cloud neural network along with a VoxelNet layer to reconstruct a textured full 3D face model out from a single image. This system will be further implemented on applications like 3D facial payments.

Keywords 3D · Point cloud · VRN · DeFa · PRNet

1 Introduction

Over the past few years, digital payments have tremendously evolved in countries like us. While it had existed in the market for a long time, however the adoption of it has taken a boom in the recent years. Arguably, the demonetization five years back acted as a catalyst for its adoption, and inception of new systems, and likewise this pandemic had acted as a growth fueler of it. Today, over 643.9 M users in India regularly transact via digital ways resulting in transaction value projection of ₹10,355,332 M. Not just that, this landscape in India is witnessing a year-on-year growth rate of 13.6%. This revolution of how we see the payments would not have

K. C. Prabu Shankar (✉) · M. Hema
Department of Computer Science and Engineering, SRM Institute of Science and Technology, Chennai, Tamil Nadu, India
e-mail: prabushc@srmist.edu.in

M. Hema
e-mail: hemam@srmist.edu.in

R. P. Mahapatra et al. (eds.), *Proceedings of International Conference on Recent Trends in Computing*, Lecture Notes in Networks and Systems 341,
https://doi.org/10.1007/978-981-16-7118-0_69

been possible if the government had not taken the initiatives to support the existing players in the market, create better policies and create various of it is own verticals within, all of which contributed [1].

While, we have multiple options available which covers a wide spectrum of features, but when we talk about the security of these, it boils down to primarily two options: fingerprint and 2 factor authentication. 2 FA, which is still used in most of the systems despite being the weakest security feature. Every now and then, we come across the news of password breaches in the news. To combat this, the biometric authentication like fingerprints, first came into existence but sooner it was fooled by using gelatin-based sweets. Technologists advanced to use methods like facial recognition systems to make the system more secure. It has been adopted over multiple applications across the world, be it payments at kiosks, restaurants etc. In India, it is majorly being used in biometric attendance only whereas countries like China have implemented it even over the payments system. Alipay and WeChat are the dominant players of the payment's ecosystem in countries like China who have been using this technology since a while now, and are heavily investing in improving it [2, 3].

Today, a user or consumer in India has various modes to digitally pay or receive money. Each of them associates certain advantages and disadvantages with it. Below, we have compared all of them on certain no. of factors as provided in Table 1.

Though such systems already exist in some means, but the drawback of it is the way it works. They have been implemented in such a manner that it has created two different processes based on the user's camera hardware. If a user has a device, which has a 3D camera, then the application runs a 3D facial recognition algorithm, and otherwise a 2D recognition algorithm on the rest of the devices. With only a few devices which come with three-dimensional cameras, it questions the bias of the system which is proportional to the security of this system, and the transactions which run over it [3]. A simple and viable solution to it is to use an algorithm that takes 2D image as input, converts it into a 3D image and then runs through the recognition algorithm. The question to this is how it could be done efficiently-and here the technology of point clouds comes into picture.

Point cloud is a 3D structure of an object presented in the form of points arranged in a three-dimensional space. With a very high density of points one can accurately represent a 3D object in terms of a point cloud. If represented properly using deep learning techniques, one can represent complex objects like human body and space. Previously, there have been methods where the 2D facial scan which had normal pixel matrices were used for verification and matching purposes. These are not very much accurate because there is no depth sensing in these types of images. To improve the depth of the image, we use point cloud technology. Using this technology gives our picture a sense of depth which makes the picture of the face feature rich and very difficult to clone as not all crest and troughs of the face cannot be cloned. Advantages of point clouds is that they represent relatively complex objects with a finite number of elements (points) and are very accurate. Security-wise as the technology advances, we would need more and more accurate images but beyond a threshold of resolution and we will not be able to achieve that accuracy in 2D images [4, 5]. Therefore, it

Table 1 Comparison of various payment systems

Factors	Unstructured supplementary service data	Aadhaar enabled payment system	Unified payments interface	Phone wallets	Phone banking	Net banking		
						NEFT	RTGS	IMPS
Service availability	24 × 7	Business Hours	24 × 7	24 × 7	24 × 7	8.00 to 6.30	8.00 to 4.30	24 × 7
Transaction time	Instantaneous	Real Time	Instantaneous	Instantaneous	Instantaneous	Same day	Real time	Instantaneous
Limit	INR 5000/ transaction	Depends on Bank	INR 1,00,000/ transaction	INR 20,000/ month	INR 50,000/ day	No Limit	INR 2,00,000 to no limit	INR 10,000 to 2,00,000
Required information	Aadhar number/MPIN/ AC number/ IFSC	Bank name and aadhar number	MPIN and VPA	Login for payment	Aadhar number/MPIN/ AC number/ IFSC	AC number and IFSC		
Authentication	2FA	Thumb impression	2FA					
Tech. requirement	Smartphone	PoS device	Smartphone with Net access		WAP and Net access	Net access		

becomes very necessary to use point clouds data structures for more secure facial verification systems. It gives another dimension to the pictures which is difficult to recreate if a hacker wants to do it.

2 Review of Literature

This section elaborates on the study of various research papers and outlines their advantages and disadvantages.

Li et al. (2020) proposed an exhaustive point cloud geometrical relational network concentrating three-dimensional object identification, called geometric relation network (GRNet) which depends upon the points of objects as an intra-object features and the association amongst varied objects as the inter-object feature. There are disservices in assessing and finding 3D bounding boxes because of the inadequate surface, void item place, and different sizes of point cloud objects. As an improvement, usage of shapes with geometry, topological structure, and association of objects is done to draw out boxes with features and reasoning information. Application of object relation learning module is done so as to estimate parameters of the bounding box. To further improve the performance of this method, RGB features of geometrically fragile objects should be exploited and the association of objectness score and a study of the accuracies of the estimated bounding boxes should be done [1].

Suo et al. (2020) proposed a novel deep neural network, LPD-AE, to get significant neighborhood and context oriented highlights for the age of idle space from 3D point cloud straightforwardly. Computationally perplexing assignments depending on point mists, similar to relocalization, circle conclusion location, and guide pressure reproduction are led easily on inactive space [2].

Liang et al. (2020) proposed an approach related to regression to rebuild full 3D models about the faces having pictures from multiple close-up views. There are disadvantages in effectively utilizing these multiple viewed close-up shots by the conventional techniques. The proposed method reconstructs three-dimensional facial geometry via linear or non-linear regressions. The texture of these close-up pictures is stitched to the rebuilt 3D facial geometry by using a flawless texture recovery system. This reduces the 3D face recognition error significantly as well as improves the recognition accuracy of face matchers. A major refinement of the recognition accuracy may be attained by fine-tuning generated multiple angled face pictures [3].

Contreras et al. (2020) proposed a classifying framework which can accommodate large 3D territorial data. There are disadvantages due to a large number of points that make the data, the proposed method segments the scene into meaningful regions, and then classifies these segments utilizing an amalgamation of PointNet and geometric deep learning. In addition, surface normalization techniques are applied and features are enriched with geometric attributes. Ameliorating of accuracy and avg cover of union can be seen in semantic 3D standard tests [4].

Zhai et al. (2020) proposed a multiscale dynamic GCN model for point clouds classification. Due to the irregularity of the info format, ancient convolutional neural networks can not be applied to purpose clouds directly. Within the projected methodology, a farthest purpose sampling methodology is applied to expeditiously cowl the complete purpose set. It achieves a far better performance on classification accuracy and model quality than different progressive models [5].

Luo et al. (2020) proposed a sequential slice representation with an attention embedding network which is new, named RSSNet. There are disadvantages like humongous computational resources and inaccurate representations. The proposed method uses a sequential slice module to build 3D point clouds, for encoding territorial data of these sequential matrices, a GRU is must, and to refine architectural differentiation. A better performance is achieved than the other recognition and information retrieval algorithms. To further improve, making the network model less complex and fast as well as ameliorating the classification accuracy is required [6].

Lu et al. (2020) proposed a novel end-to-end deep learning network known as PointNGCNN which will consume purpose clouds for 3D visual perception and segmentation tasks. There is an obstacle that 3D point clouds can not be applied to the standard convolutional neural networks attributable to irregularities in information. The proposed method obtains better global features by extracting two spaces simultaneously, is efficient in semisupervised tasks such as citation networks and node classification, and can extract more neighborhood geometric information [7].

Boulch (2020) proposed a generalization of distinct convolutional neural networks (CNNs) so as to affect point clouds by commuting distinct kernels by continuous ones. There is a drawback that ancient machine learning approaches developed for image can not be directly transferred to point clouds. The proposed method helps in simple formulation of neural networks and is agnostic to object scales. To further improve, extra features such as normals or curvatures should be feeded to the network, variations of network architectures should be explored, use a fixed-radius neighborhood instead of a fixed number of neighbors [8].

Mukhaimar (2019) proposed a vigorous composition so as to make it immune to data variations like disintegrated outliers, congregated outliers, and noise and missing points and detect objects accurately. It is composed of planar patches and transferred to neural net. It is an uncomplicated system but requires manual fine-tuning. Lots of objects can not be transformed into planars. Future work is to make the process of parameter tuning automatic [9].

Lee et al. (2019) proposed a deep learning network with extended Laplacian filter (ELF) which makes feature extraction possible on irregular and unordered point set of a point cloud. Properties of ELF-it is a 2-state filter mistreatment two filter matrices (one for a middle purpose and therefore the different for neighboring points), it employs a scalar weight function to predict the relative vital neighboring points. The proposed system is efficient and effective in classification, memory, and time [10].

Cai et al. (2019) proposed a speedy 3D face recognition approach which is immuned to errors and consists of three major technologies: a swift three-dimensional scan preprocessing, multiple data augmentation, and a method dependent upon facial component patches. There is a disadvantage that the conventional 3D facial identification methods are expensive computationally. The proposed method uses only three facial landmarks, improves supervision, and has low computational cost. To further improve, a more efficacious facial stance augmentation method can be used [11].

Zhou et al. (2019) proposed a multi-view PointNet (MVPointNet) which takes multiple angles of vision of a point cloud as input. This method achieves better performance than traditional methods because of more number of features extracted from the middle of a centroid and local neighborhood points (norm, edge, and angle).To further improve, the author proposes a more appropriate blend of strategies to enable the network to avail the information from multiple angles [12].

Yu et al. (2019) proposed a stiff registration method using surface resampling and deblurring. There is a disadvantage that the traditional methods have poor accuracy in experimental results. The proposed methods minimize the effect on registration residuals, significantly improving accuracy. To further improve, more efficacious facial stance correction methods can be used [13].

Arshad et al. (2019) proposed a deep neural network which consists of a cascaded combination of 3D point-based residual networks for coincident linguistics scene segmentation and object classification to accommodate the unstructured point cloud information. The projected design incorporates a easy style, easier implementation, and higher performance. Point-based convolution, network architecture design, neighborhood selection (grid cell size), hybrid features (deep + hand crafted features) are the design parameters to be explored. To further improve the system design parameters like–point-based convolution, network architecture design, neighborhood selection (grid cell size), hybrid features (deep + hand crafted features) are to be explored [14].

Luo et al. (2019) proposed a novel framework integrating the BoW model with the JDA model to classify three-dimensional objects in diverging point clouds. There is a disadvantage in traditional methods of acquiring supervised information as it is assiduous and tedious. The proposed method effectively describes three dimensional objects in point clouds, alleviating the occlusions, interclass discrepancies, and intraclass discrepancies. To further improve, new methods can be developed to consider more contextual information, amalgamate other BoW based methods with domain adaptation techniques [15].

Wang et al. (2019) proposed classification algorithm for large-scale 3D point clouds using deep learning models with Feature Description Matrices (FDM) to acquire a high classification accuracy. This method tackles the challenge of unordered 3D points which can not be fed to the deep learning pipelines. To further improve, author proposes to accelerate the feature extraction process, study the principle of the arrangement for the FDM, explore mesh models, and deep learning models [16].

Song et al. (2019) proposed an effective coding technique that encodes every purpose into a one-dimensional feature vector. There are disadvantages in traditional deep learning networks as it encounters difficulties in utilizing the local information because of its irregular format. The Proposed technique improves the classification accuracy of the easy deep learning network. To further improve, hyperparameters for the number of neighborhood points can be removed, and classification accuracy of individual categories can be improved [17].

Yao et al. (2019) proposed OG-PointNet++ (Octree-Grouping-PointNet++) to deal with heavier computational overhead of PointNet++.This method determines point density of point cloud through an unbalanced octree and groups the points according to the density. It assigns these groups to different layers uses PointNet++ to extract features. Performs better in classification and semantic segmentation tasks [18].

An et al. (2019) proposed a face alignment method for pose invariant face recognition, called adaptive pose alignment. There is a disadvantage in traditional methods of noise in the alignment process, especially in unconstrained settings. The proposed method corrects it and greatly reduces the intraclass difference as well [19].

Liu et al. (2018) proposed a joint face alignment and 3D face reconstruction methodology to at the same time solve these 2 issues for 2nd face pictures of absolute poses and expressions by applying 2 cascaded regressors. These two were previously considered to be a separate task of alignment and 3D face reconstruction. It naturally creates both posture and demeanor standardized and expressive 3D face shapes, limit both noticeable and undetectable 2D landmarks, detect milestones and reproduce 3D countenances continuously, additional face alignment methods not required (only single face image with arbitrary poses) [20].

Zaganidis et al. (2018) proposed a complete pipeline for semantic assisted registration of point clouds using SE-NDT, SE-GICP, and PointNet. There is a disadvantage in traditional methods as it cannot learn real semantic labels from data. The proposed method demonstrates the improvement of the registration in terms of robustness, precision, and speed. To further improve, point feature vector can be used before the output layer as input to the semantic registration and test against the probabilistic versions of the algorithms [21].

Ben-Shabat et al. (2018) proposed a novel, intuitively interpretable, 3D point cloud representation called 3D modified Fisher Vectors (3DmFV) so as to overcome the challenge of severe accuracy vs memory size tradeoffs in voxelization. It is hybrid of a coarse discrete matrix structure with consistent summed up Fisher vectors, using the lattice to plan another CNN engineering for continuous point cloud grouping. It is robust to data corruptions, 3DmFV shows high exactness for low thickness point clouds, and quick run time for higher thickness point clouds [22].

Judith Leo and Suchitra (2018) proposed a face recognition system using an efficient combination of 3D principal component analysis and SVM. There is a disadvantage in traditional methods that it has poor results in recognition under various expressions. The proposed method improves the registration rate on near frontal 3D faces consisting of rich facial expressions. To further improve, face segmentation into multiple regions can be done along with registration of partial data [23].

Zou et al. (2017) proposed a new voxel-based approach to classify the point clouds. The method involves individual tree (object) extraction dependent on the point cloud density; low-level component portrayal through voxel-based rasterization; and order of tree (object) by a deep learning model. To improve this technique, effective ways to represent the objects need to be explored in the future [24].

Ding and Tao (2017) proposed an exceptionally effective posture standardization approach (HPN) that depends on homography. There are impediments in conventional strategies like loss of semantic correspondence, nonstraight facial surface twisting, and impediment. The proposed method has high efficient normalization and high quality synthesized frontal face images. To further improve, evaluation of the ability of HPN to recognize unconstrained large-pose face images needs to be done [25].

3 Proposed Architecture

The proposed architecture consists of three main modules namely, preprocessing stage, point cloud generation stage, and finally training, classification, and annotation stage. The three stages or modules again have their subprocesses where the image is firstly preprocessed by 3 subprocesses, then a point cloud is extracted after which the point cloud is again preprocessed and is made available for training and classification. The pictures accurately describe the processes [6, 7].

The flow diagram accurately describes the main processes and outline of the whole process in a more self-explanatory way. We start, capture an image, preprocess it, and fine tune. Using the preprocessed image, we send this to our proposed network and obtain a point cloud. This point cloud is filtered and voxelized. VoxelNet is trained on our point cloud and we finally are able to classify faces (Figs. 1 and 2).

The functions of each of the elements of the architecture is elaborated below:

Equalization of Histogram: The system has to capture images very first. The number of images captured is 1 which makes this system less memory consuming and faster. After capturing these images will go to the preprocessing phase where we resize, adjust contrast, intensity for fine-tuning and remove the texture from the frontal image so that there are no dark patches due to invariant lighting conditions and finally extract saturation information from the images to obtain maximum features which will help us achieve good accuracy.

$$''\text{cv} :: \text{equalizeHist}'' \tag{1}$$

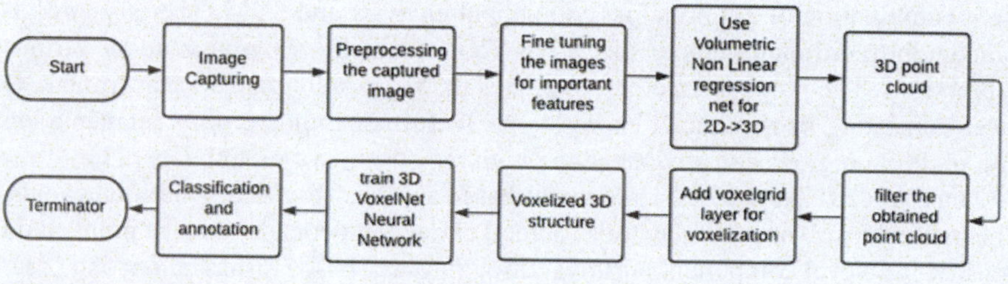

Fig. 1 Flow diagram of the proposed architecture

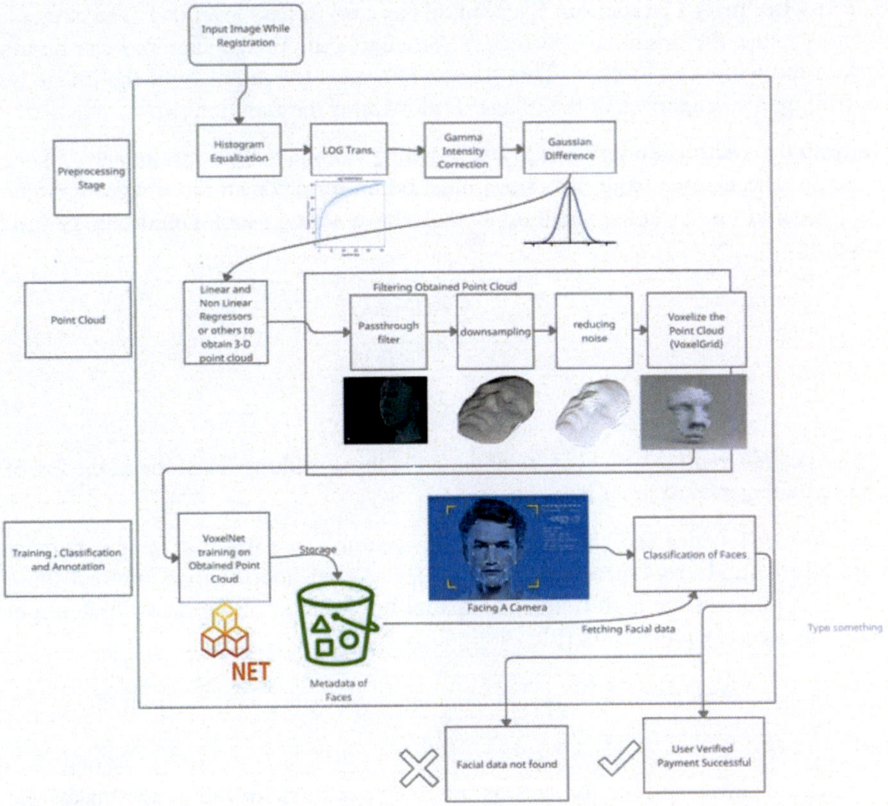

Fig. 2 Architecture of 3D point clouds

Log Transformation: In log transformation, we replace the pixel values with that of its log value. The dark pixels are made to expand compared to its higher counterparts and we get an enhanced image.

$$s = c * \log(r + 1) \tag{2}$$

$$c = 255/(\log(1 + \max\_\text{input\_pixel\_value}) \tag{3}$$

where,

s is output image's pixel value.

r is input image's pixel value.

c is the constant.

These calculations can be made using functions in the Numpy library.

Gamma Intensity Correction: Every pixel has a brightness level 0–1. The cameras do not capture the brightness correctly. Sometimes the image captured can be too dark or too white or bleached. This function corrects the intensity of the image by controlling the brightness of the image. Following is the function [8]:

Gamma Correction Function (x): = x^γ, where gamma is the constant and X is the intensity. The gamma value we chose must be uniform for all the images and one can see better results in the value 0.5–0.9. To have a closer look behind the gamma function.

$$\gamma = \frac{\delta \log V_{\text{out}}}{\delta \log V_{\text{in}}} \tag{4}$$

$$V_{\text{out}} = V_{\text{in}}^{\gamma} \tag{5}$$

The desired gamma value can be obtained in this by taking the slope of the log of required intensity and input intensity.

Gaussian Difference: To enhance the image features, we use two images namely more blurred and less blurred using a frequency filter and then we subtract these images. It removes the high frequency spatial information of the face which might cause unnecessary processing [9].

$$G(x) = \frac{1}{\sqrt{2\pi\sigma^2}} \mathrm{e}^{\frac{x^2}{2\sigma^2}} \tag{6}$$

Here, the Gaussian function acts as the low pass filter for the image makes the image blurred. Similarly, here the two Gaussian filters with two different variances $\sigma 1 < \sigma 2$ we get two different images-one is blurred and other is of better quality. The final output obtained will be-

$$\Gamma_{\sigma_{1,\sigma_2}(x) = I * \frac{1}{\sigma_{1\sqrt{2\pi}}} \mathrm{e}^{\frac{x^2}{2\sigma_1{}^2}} - I * \frac{1}{\sigma_{1\sqrt{2\pi}}} \mathrm{e}^{\frac{x^2}{2\sigma_2{}^2}}} \tag{7}$$

where I is the image convolved with Gaussian variances and has highlighted features of the image. Finally, it preserves the spatial information between the high pass and low pass frequency filter.

Regression Point Cloud formation technique: We will use multi linear regression with convolutional neural network so that we can predict the projection of the third dimension of the point cloud by fusing various predetermined orientations of the faces. It will be a single image but we can change its orientation using various tools like skewing. Also, we can use AutoCAD for 2D to 3D conversion which is a very simple process but it is not open source. For this, we are going to use a volumetric non-linear regression network (VNLRN). For the training of the VNLRN, we will use softmax cross entropy loss function. We get a very clean formula after amalgamation of softmax and cross entropy loss.

$$\frac{\delta L}{\delta \ln} = \widehat{y_n} - y_n \tag{8}$$

Softmax function is given as

$$y_{\text{whd}} = \frac{e^l \text{whd}}{\sum e^l} \tag{9}$$

Cross entropy loss is given by Eq. (9)

$$L_{\text{whd}} = -\sum_w \sum_h \sum_d y_{\text{whd}} \log(\widehat{y_{\text{whd}}}) \tag{10}$$

By derivation and substitution, we get-

$$\frac{\delta L}{\delta l_{\text{whd}}} = \widehat{y_{\text{whd}}} - y_{\text{whd}} \tag{11}$$

After we preprocess the 2D image, it is fed into VNLRN CNN and trained on the 2D image using the softmax cross entropy loss function. By regression we are able to predict the 3D facial structure.

Pass through filter: After obtaining the 3D point cloud from the previous step we use a passthrough point cloud filter and take only the point cloud space we require. To achieve that we will use a pass through filter to get a point cloud which only contains the face of the person. We will feed the xyz coord to the point cloud library functions [9].

DownSampling: A point cloud of face with only important points and areas is very efficient for the process of training. To achieve this, we use a critical point layer (CPL) which only takes the point which is very critical for our facial point cloud recognition. By adding this layer to our network we downsample the point clouds. It filters out not important points and only saves the important ones, according to a point's level of contribution to the global max-pooling (max-reduced feature vector). The CPL is efficient and robust as it does not require further processing to find the nearest neighbors. We are able to preserve the important information unlike others as CPL is adaptive in nature [10].

Reducing Noise: We first use outlier detection on the above obtained point cloud. Outliers can be found by traditional methods of graph plotting. But a more advanced method is that of neural networks. In point CleanNET, a two stage process where it first removes the outliers and in the second, removes the noise in the point cloud. This method is autonomous and does not require any parameters where the neural net automatically detects the curvature and dense points filled with important information. Point CleanNET uses the PointNet [QSMG17] and PCPNet [GKOM18] to automatically to find and delete the outliers in the point cloud [11].

Voxelization: We will convert the point cloud obtained to voxels. Voxels will expedite the process as it is more structured than the point clouds. We will use PyntCloud library. PyntCloud is a python library which can be used to voxelize and process point cloud. PyntCloud can be integrated with libraries like open3d (openCV for 3D images). PyntCloud library has a voxel grid function (VoxelGrid) which will find a centroid for a particular number of points () and will form a voxel on that centroid. By this method there will be a significant decrease in the number of points in the point cloud which will result in faster processing and less space. For example, this is a function with "voxelgrid" as the point cloud and x, y, z variables for the dimensions of the voxel/cube to be formed on the centroid.

$$\begin{aligned} \text{voxelgrid\_id} &= \text{cloud.add\_structure}(''\text{voxelgrid}'', n\_x \\ &= 32, n\_y = 32, n\_z = 32) \end{aligned} \tag{12}$$

VoxelNet: After the voxelization process, we get voxelized point cloud where 2–3 points form a voxel. This voxel is fed into VoxelNet neural network with a softmax loss function,3D convolution Net, regional proposal network. The training is done using feature extraction. After feature extraction we get a feature map in the form of tensors and these tensors are then stored in a MetaData from where we fetch the features for detection and annotation of the face, which is then fed to the pipeline of the middle layers of 3D ConvNet. The feature map obtained from the middle layers is then fed to the regional proposal network which consists of 3 convnet layers. RPN is very well known for the object detection purposes. RPN convnet layer downsamples, ReLu is applied, layer up samples to a high resolution. We obtain a probability score map and a regression map. The highest number in the probability score map represents the face detected [12].

3.1 Architecture Blocks Description

Capturing and Preprocessing: The system has to capture images very first. The amount of images captured is 1 which makes this system less memory consuming and faster. After capturing these images will go to the preprocessing phase where we resize, adjust contrast, intensity for fine-tuning and remove the texture from the frontal image so that there are no dark patches due to invariant lighting conditions and finally extract saturation information from the images to obtain maximum features which will help us achieve good accuracy through various preprocessing stages like histogram equalization, log transformation, gamma intensity correction, and difference of Gaussian surfaces as described in the architecture diagram. By this step, we get a feature rich 2D image which is ready to be fed to the network for 2D to 3D conversion.

Conversion of 2D Images to 3D Point Cloud: These 2D images after the preprocessing phase are ready for the conversion to 3D point cloud which we require for the facial recognition part. This preprocessed 2D frontal face image is fed into a pipeline of neural networks which we call as volumetric non-linear regression network (VNLRN) to reconstruct a 3D point cloud from the preprocessed 2D image. The softmax loss function in the regression network helps us to predict the third dimension of the 2D image which we orient it in various directions to get a complete three dimensional structure in the form of a point cloud. Here, we not only get the complete frontal face but also some less observed features. This complete point cloud 3D structure of the face is then fed to the filtering engine of the system. We use the pass through filter, then use downsampling to highlight the features and finally we reduce the noise of the obtained point cloud. This obtained point cloud is then fed to the next layer VoxelGrid layer of the PyntCloud library.

Training, Classification, and Annotation: Through the VoxelGrid layer we voxelized our obtained point clouds by turning the points into cubes. As we are combining 3–4 points in the point cloud to a single voxel/cube we are thereby reducing the data points required for processing. Also we make our point cloud more structured which was randomized/unstructured previously due to the random points. Finally after voxelization we get a point voxel 3D structure of the face. Now we train our 3D VoxelNet (neural network) model on the point voxel obtained after voxelization. The VoxelNet has three separate components in it feature extraction to get a feature map, ConvNet middle layers and a regional proposal network (RPN) for Object detection. After the training phase ends, we classify the point clouds using RPN and annotate them according to their faces and in this way, we complete the face verification for the facial payments.

4 Results and Discussion

For this project we have used python 3.7 and also some bash and python scripts. The python script makes our job easier and tries to ease the job of creating a 3D point cloud. This project has been completely developed on the google colab environment with hardware acceleration (GPU cuda cores).The hardware acceleration is mainly needed during the phase of face recognition as the face recognition model tries to extract features from the 3D features of the face.

We are using pretrained weights as we expect to limit the time taken to train the model to make the process faster. This training was done with the help of AFLW2000-3D and florence dataset which are standardized datasets and give the best results till date. For testing we tried and tested our own images having high resolutions and got a pretty decent point cloud mesh structure which contained only our facial features. The process of point cloud filtering was done inside the model itself. Also for the initial preprocessing phase of image we are using normal openCV function which again gave us some descent image with highlighted facial features (Fig. 3).

Here, the images (a, e, i) are the normal input images. The images (b, f, j) are the ones we get after doing the preprocessing. The images (c, g, k) are the ones where we get our point cloud. Finally, we get a labelled image (d, h, l) with the name of the user. This is the output which we have got using our model for face recognition. This gave us a clean point cloud diagram with minimal and important points or important facial landmarks.

Also, In Table 2 we compare different 2D images to 3D point clouds models which are based on deep learning methods. Here, we can see that the first three methods require 3 images. Here, F is frontal; L means Left; and R means Right. This unnecessarily increases the processing and training time. Our proposed architecture

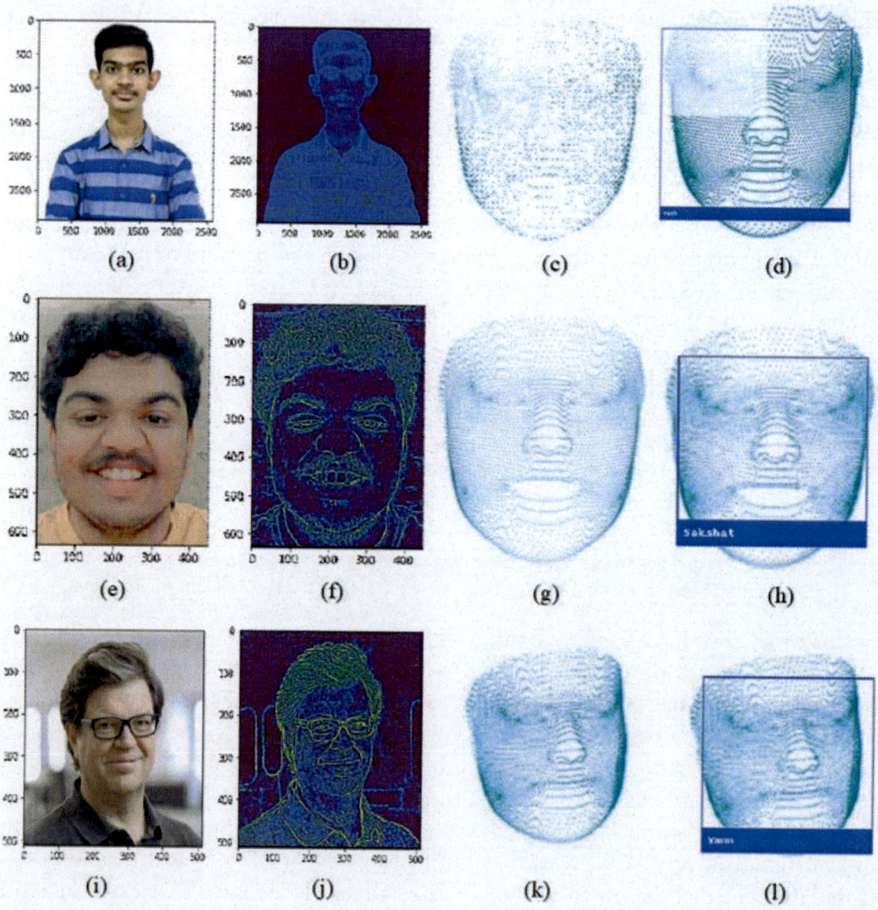

Fig. 3 Architecture of 3D point clouds

Table 2 Input image comparison of existing methods vs proposed method

Method	Input image number	Input image type
Fast and robust auto face detection	3	F + L + R
3D Face mugshots method (Linear)	3	F + L + R
3D Face mugshots method (Non-Linear)	3	F + L + R
Proposed architecture	1	F

(VNLRN) requires just one frontal image which reduces processing and training time.

In Table 3, we compare 4 very widely used techniques for 3D point cloud reconstruction especially for facial data. One can see that the proposed method is very fast as the run time can be very low and reach upto 3.0 ms as we are using only a single image for the whole process (Fig. 4).

From this system we infer that the system we have built is more advanced, fast and secure than the 2D face recognition payment system. Also here we do not require multiple images to form a point cloud as the traditional system uses 3 different images to stack them up for a point cloud whereas we are using just a single image to form a complete point cloud which is more than enough to get an accurate result for the face

Table 3 Run time comparison of existing methods vs proposed method

Method	Run time (ms)
VRN	23.2
DeFA	11.8
PRNet	9.8
Proposed architecture	4.1

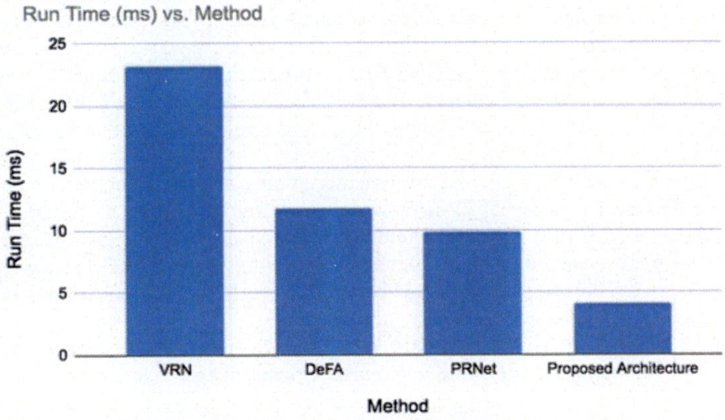

Fig. 4 Graphical representation of runtime of various methods

recognition part. Finally, we do not have any specialized hardware requirement like LiDAR, high res cameras, etc. Our system only requires a clear image which ideally should have a face covering at least 65–70% for better feature learning process. Our system is better than others as our aim is to eliminate heavy duty costs of hardware specs like LiDAR and make this system available to every shop and household with normal smartphones.

5 Conclusion

In this work, we have successfully created a novel system that envisions to revolutionize the way digital payments work across the globe. This system has a potential to replace the old complex traditional systems. The majority of face scanning techniques use age-authentication mechanisms, or the ones which have advanced to use new technologies are too slow and memory inefficient. All of which is not very convenient, simplified, and reliable for an end-user. Considering all these issues and to effectively solve them, our proposed 3D facial recognition system which is backed with a series of preprocessing algorithms, a point cloud neural network along with a VoxelNet layer to reconstruct a textured full 3D face model out from a single image. This system will be further implemented on applications like 3D facial payments.

References

1. Li Y, Ma L, Tana W, Sun C, Cao D, Li J (2020) GRNet: Geometric relation network for 3D object detection from point clouds. ISPRS J Photogrammetry Remote Sens 165:43–53. https://doi.org/10.1016/j.isprsjprs.2020.05.008
2. Suo C, Liu Z, Mo L, Liu Y (2020) LPD-AE: latent space representation of large-scale 3D point cloud. IEEE Access 8:108402–108417. https://doi.org/10.1109/ACCESS.2020.2999727
3. Liang J, Tu H, Liu F, Zhao Q, Jain AK (2020) 3D face reconstruction from mugshots: Application to arbitrary view face recognition. Neurocomputing. 410:12–27. https://doi.org/10.1016/j.neucom.2020.05.076
4. Contreras J, Sickert S, Denzle J (2020) Region-Based edge convolutions with geometric attributes for the semantic segmentation of large-scale 3-D point clouds. IEEE J Sel Top Appl Earth Observations Remote Sens 13:2598–2609. https://doi.org/10.1109/JSTARS.2020.2998037
5. Zhai Z, Zhang X, Yao L (2020) Multi-Scale dynamic graph convolution network for point clouds classification. IEEE Access 8:65591–65598. https://doi.org/10.1109/ACCESS.2020.2985279
6. Luoa Z, Liua D, Li J, Chena Y, Xiaoa Z, Junior JM, Gonçalves WN, Wanga C (2020) Learning sequential slice representation with an attention-embedding network for 3D shape recognition and retrieval in MLS point clouds. ISPRS J Photogrammetry Remote Sens 161:147–163. https://doi.org/10.1016/j.isprsjprs.2020.01.003.
7. Lu, Chen C, Xie W, Luo Y (2020) PointNGCNN: Deep convolutional networks on 3D point clouds with neighborhood graph filters. Comput Graph 86:42–51. https://doi.org/10.1016/j.cag.2019.11.005
8. Boulch S (2020) ConvPoint: continuous convolutions for point cloud processing. Comput Graph 88:24–34. https://doi.org/10.1016/j.cag.2020.02.005

9. Mukhaimar A, Tennakoon R, Lai CY, Hoseinnezhad R, Bab-Hadiashar A (2019) PL-Net3d: robust 3D object class recognition using geometric models. IEEE Access 7:163757–163766. https://doi.org/10.1109/ACCESS.2019.2952638
10. Lee SH, Kim HU, Kim CS (2019) ELF-Nets: deep learning on point clouds using extended laplacian filter. IEEE Access 7:156569–156581. https://doi.org/10.1109/ACCESS.2019.2949785
11. Cai Y, Lei Y, Yanga M, Youa Z, Shane S (2019) A fast and robust 3D face recognition approach based on deeply learned face representation. Neurocomputing 363:375–397. https://doi.org/10.1016/j.neucom.2019.07.047
12. Zhou W, Jiang X, Liu YH (2019) MVPointNet: multi-view network for 3D object based on point cloud. IEEE Sens J 19(24):12145–12152. https://doi.org/10.1109/JSEN.2019.2937089
13. Yu Y, Da F, Guo Y () Sparse ICP with resampling and denoising for 3D face verification. IEEE Trans Inf Forensics Secur 14(7):1917–1927. https://doi.org/10.1109/TIFS.2018.2889255
14. Arshad S, Shahzad M, Riaz Q, Fraz MM (2019) DPRNet: deep 3D point based residual network for semantic segmentation and classification of 3D point clouds. IEEE Access 7:68892–68904. https://doi.org/10.1109/ACCESS.2019.2918862
15. Luo H, Wang C, Wen Y, Guo W (2019) 3-D object classification in heterogeneous point clouds via bag-of-words and joint distribution adaption. IEEE Geoscience Remote Sens Lett 7(12):1909–1913. https://doi.org/10.1109/LGRS.2019.2911200
16. Wang L, Meng W, Xi R, Zhang Y, Ma C, Lu L, Zhang X (2019) 3D Point cloud analysis and classification in large-scale scene based on deep learning. IEEE Access 7:55649–55658. https://doi.org/10.1109/ACCESS.2019.2909742
17. Song Y, Gao L, Li X, Pan QK (2019) An effective encoding method based on local information for 3D point cloud classification. IEEE Access 7:39369–39377. https://doi.org/10.1109/ACCESS.2019.2905595.
18. Yao X, Guo, J Hu J, Cao Q (2019) Using deep learning in semantic classification for point cloud data. IEEE Access 7:37121–37130. https://doi.org/10.1109/ACCESS.2019.2905546
19. An Z, Deng W, Hu J, Zhong Y, Zhao Y (2019) APA: adaptive pose alignment for pose-invariant face recognition. IEEE Access 7:14653–14670. https://doi.org/10.1109/ACCESS.2019.2894162
20. Liu F, Zhao G, Liu X, Zeng D (2018) Joint face alignment and 3D face reconstruction with application to face recognition. IEEE Trans Pattern Analy Mach Intell 42(3): 664–678. https://doi.org/10.1109/TPAMI.2018.2885995, December 2018
21. Zaganidis A, Sun L, Duckett T, Cielniak G (2018) Integrating deep semantic segmentation Into 3-D point cloud registration", IEEE Robotics Autom Lett 3(4):2942–2949. https://doi.org/10.1109/LRA.2018.2848308
22. Ben-Shabat Y, Lindenbaum M, Fischer A (2018) 3DmFV: three-dimensional point cloud classification in real-time using convolutional neural networks. IEEE Robotics Autom Lett 3(4):3145–3152. https://doi.org/10.1109/LRA.2018.2850061
23. Judith Leo M, S. Suchitra (2018) SVM based expression-invariant 3D face recognition system. Procedia Comput Sci 143:619–625. https://doi.org/10.1016/j.procs.2018.10.441
24. Zou X, Cheng M, Wang C, Xia Y, Jonathan L (2017) Tree classification in complex forest point clouds based on deep learning. IEEE Geoscience Remote Sens Lett 14(12):2360–2364. https://doi.org/10.1109/LGRS.2017.2764938
25. Ding C, Tao D (2017) Pose-invariant face recognition with homography-based normalization. Pattern Recogn 66:144–152. https://doi.org/10.1016/j.patcog.2016.11.024

Denoising of Thermal Images Using Deep Neural Network

Ruchika Thukral, A. S. Arora, Ashwani Kumar, and Gulshan

Abstract In the recent scenario, there are various methods for imaging the human body for medical diagnostics, viz. X-ray radiology, magnetic resonance imaging (MRI), ultrasound, computerized axial tomography (CAT), positron emission tomography (PET) scanning, etc. Imaging the human body using IR is non-invasive as it is not ionizing in the tissue, nevertheless, it suffers from the limitation of less penetration power in the tissue and poor resolution. Thermal imaging is one of the imaging techniques, which has shown a promising result in the diagnosis of cancer. The use of thermal radiations for cancer diagnosis is safe as compared to many other imaging modalities. With the advancements in image processing routines, it has become feasible to make use of non-invasive nature of thermal radiations for cancer diagnosis. The time for the capturing of thermal image, pre-processing, and analysis of image data has improved from last few decades with recent advancements in sensor technology and analysis tools. Thus, it has become an important task to remove noise from the thermal image and restore a high-quality image in order for better thermographic assessment of images. This paper presents the denoising of thermal images using deep neural network by adding the different types of the noises to the original thermal image. The quantitative analysis is done using the three metrices which are peak signal to noise ratio, structural similarity index measurement, and mean square error in which the deep neural network shows a promising result and remove a lot of Gaussian noise and also improve the image quality than normal filtering techniques.

Keywords Infrared thermal imaging · Image denoising · Deep learning · Noise · Filtering techniques

1 Introduction

Metastatic cancer or advanced cancer is a stage when the cancer is spread from one part of the body or from original site to other parts of the body. It is a process when

R. Thukral (✉) · A. S. Arora · A. Kumar · Gulshan
Department of Electrical and Instrumentation Engineering, Sant Longowal Institute of Engineering and Technology, Longowal, Punjab 148106, India

R. P. Mahapatra et al. (eds.), *Proceedings of International Conference on Recent Trends in Computing*, Lecture Notes in Networks and Systems 341,
https://doi.org/10.1007/978-981-16-7118-0_70

the cancer cells are spreading through the bloodstream or through the lymph system to different parts of the body. A secondary tumor is formed from the cancer cells that have spread during the metastasis process in other parts of the body. The cancer cells that have spread to the nearby areas from the primary site of origin are called as regional metastasis, and the cancer cells spreading to other body parts that are farther away from the original site are called as distant metastasis. The primary aim for the treatment of the metastatic cancer is to slow down the growth and speed of spreading the cancerous cells.

Now a days, the most common type of cancer that spread in women and becomes the most significant cause of death is breast cancer. In women, infection in the breast may not be felt or diagnosed by pre-existing medical imaging modalities until a lump becomes sizeable. With the time passes, a lump grows in noticeable size, and it usually reached to stage 2 [1]. Due to this, the doctors recommended performing the early-stage cancer screening so that the survival rate of the patient will become higher [2, 3]. Breast awareness, annual screening, and clinical breast examination are the protocols of early screening [4].

The survival chance of breast cancer patient is depending on the early detection as the early detection increases the rate of survival and also provide the possibility of complete cures. The survival rate of a breast cancer patients has a span of ten-year such as for Stage 0-95% chances of survival; Stage I-88% chances of survival; Stage II-66% chances of survival; Stage III-36%; and Stage IV-7% [4]. Diagnosis of large size of tumors is also associated with decreased chances of survival [5].

The early detection of cancer by screening may provide significant benefits if the cancer was more specific and sensitive. For the improvement in the rates of early detection, a large variety of the latest technologies, involving alternative imaging modalities, and upgrading in x-ray mammography, is being investigated. Various types of imaging modalities can be used for screening of breast, like X-ray, ultra-sonography, MRI, computed tomography (CT), ultrasound, PET scans, digital mammography (FFDM), and lastly infrared thermography, which is most popular imaging technology nowadays.

During the process of thermal image acquisition, often, the image becomes degraded due to numerous reasons. The main objective of denoising of thermal images is to estimate the original acquired thermal image from the noise contaminated version of the image by removing or suppressing the noise [6]. Now a days, the image denoising becomes the most important part that is actively used for image reconstruction, image processing, and computer vision.

This paper is mainly focused to denoise the thermal images using deep neural network using MATLAB version R2019a. The benefit of using this CNN model over traditional filtering techniques is that it is optimizing continuously, and hence, improving the weights of convolutional kernel during training of network whereas traditional filtering techniques have fixed parameters and cannot be adjusted during filtering. By using quantitative analysis parameters such as MSE, PSNR, and SSIM Index [7, 8], the denoised image was compared with the original image.

2 Image Acquisition and Denoising of Image

Thermal imaging camera E60, FLIR® Systems, USA has been used to acquire the thermograms of subjects. The participant group was aged between 40 and 60 years. The passive thermograms of all participants have been captured after the acclimatization period of 15 min. The data acquisition procedure was explained to all the participants; then they read and signed the consent form prior to imaging acquisition. The imaging procedure was decided in such a way that the participants do not feel discomfort. Furthermore, it ensures the accuracy and high reproducibility in data acquisition by standardizing the imaging at various levels. The imaging was performed in an investigation room which exhibits the properties, such as isolated from external radiations, noise free, complete darkness, 25 °C ambient temperature, and 60% humidity.

To avoid the motion artifacts, the camera was mounted on a photographic tripod. Furthermore, the camera settings were kept constant such as 0.98 emissivity value and 20 reflected temperatures. Prior the imaging, the camera was kept power ON for 10 min for the stabilization of radiometric performance. To attain the thermal equilibration, the participant should be seated in the ambient condition of investigation room for 15 min before the imaging.

The thermal image denoising process is carried out to minimize the distortions and extract the noise from the thermal image to make it free from noise. Image denoising is always carried out prior to the pre-processing of the thermal image. The enhancement of a thermal image is widely used in research areas. A quiet variation is found in the problem of denoising due to a large number of variety of noises and distorted images. In few important coefficients, the SNR ratio is significantly rising up, and the image is highlighted in this domain. The noise is not highlighted in this domain.

Thermal image can be easily distorted by the noises, and distortions that are created randomly and also not associated with the thermal image which can be approximating by the help of a source of Gaussian noise [9]. The denoising is done using pretrained denoising convolutional neural network, in MATLAB R2019a software after adding the different type of noises in the grayscale image.

The PSNR and MSE are calculated as:

$$\text{PSNR} = 10\log\sum_{i=1}^{N}(2^n - 1)^2/\text{MSE}) \tag{1}$$

The PSNR as stated in Eq. (1), PSNR is an engineering-related term that is defined as a ratio of the highest power of signal or image to the corrupting or distorting noise power that alters the quality of its representation. The PSNR is calculated between two images, and the value is represented in decibels. The value of PSNR indicates the quality of measurement among the original image and the compression image. The PSNR value is mainly used to measure the quality of the compressed image that have been reconstructed. The greater the value of PSNR, the superior or fine will be the quality of the compressed, or reconstructed image. The MSE and PSNR are mainly used to differentiate the quality of compressed image. The MSE indicates the cumulative squared error amongt the original thermal image and the denoised thermal image, while PSNR indicates the peak error measurement [10, 11, 14]. The error will be minor if the MSE value is lower. The SSIM is a perceptual metric that quantifies image quality degradation caused by processing like compression of data or by data loss during transmission [12, 13].

3 Results and Discussion

The experimental testing of the deep neural network for denoising is performed on various acquired thermal images using MATLAB software, version: R2019a. But in this paper, the results are presented on the thermal image acquired of "Diabetic foot" shown in Fig. 1. The deep neural network-based denoising is tested after adding various noises Gaussian noise (variance of 0.01), Poisson noise, salt and pepper noise, and speckle noise shown in Fig. 2 [15].

The denoising results are also assessed using certain metrices like MSE, PSNR, and SSIM Index. These parameters are for image quality assessment [16–20]. It is observed that denoising results are much better in case of Gaussian noise, Poisson

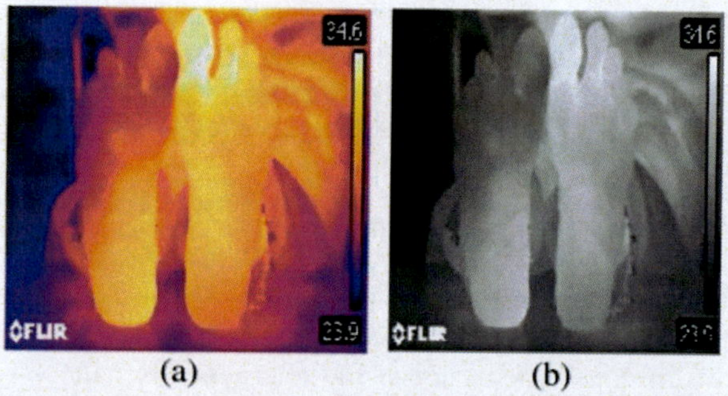

Fig. 1 Thermal image of diabetic foot **a** Original Image, **b** Converted RGB thermal image into gray image

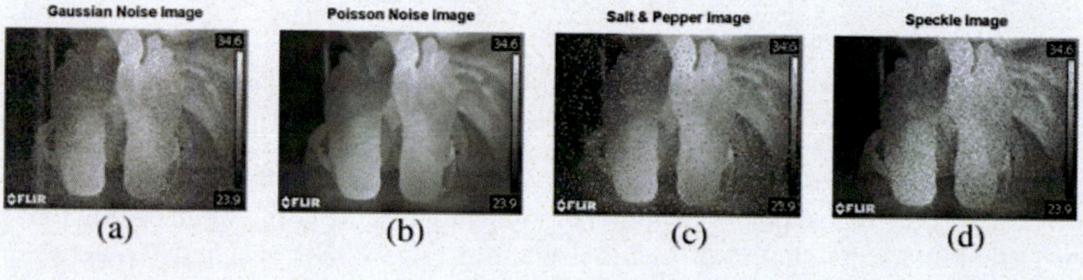

Fig. 2 Thermal image of diabetic foot after adding noise **a** Gaussian noise, **b** Poisson noise, **c** Salt and pepper noise, **d** Speckle noise

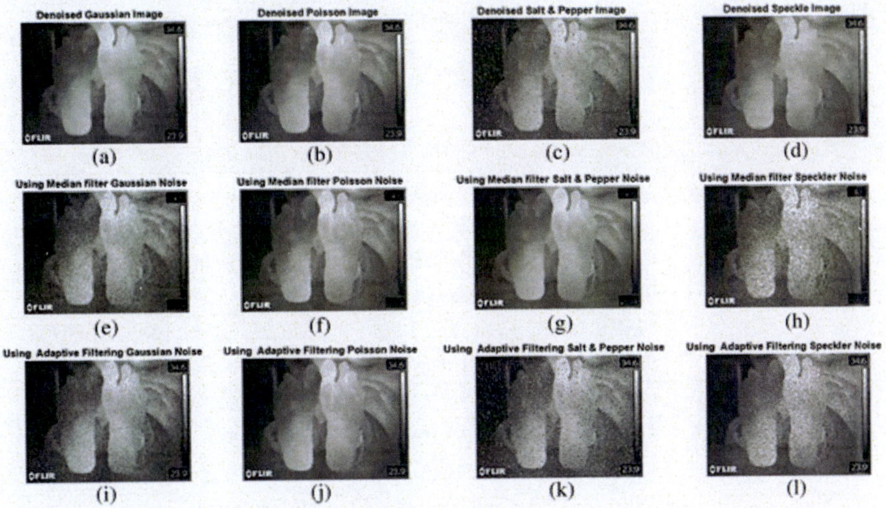

Fig. 3 Denoised thermal image after **a-d** Deep neural network, **e–h** Median filter, **i–l** Wiener filtering

noise and also for speckle noise but for the case of Salt and pepper noise results are not as good as many objects as in this case shown in Fig. 3.

The denoising results can be more analyzed by zooming the results. Histogram of all denoising results helps to analyze the results comparatively after adding various type of noises. If the curves of the denoised image and reference image are closed to each other or they are overlapped, then that represents effective denoising results. Deep neural network shows better smoothness in the uniform regions as there is no blurriness in the homogeneous part of the image. In terms of deep neural network-based denoising method shows the best edge and fine detail preservation scheme, as here in this method's result, the corner of the object is well preserved, while in the other method's results the edge and corners are not well preserved.

The quantitative analysis is performed using metrics like MSE, PSNR, and SSIM in Table 1; that clearly indicates the overall best performance of CNN-based denoising method over other filtering methods. MSE helps to assess the similarity rate amongt the denoised thermal image and the reference image. The PSNR assess overall performance of the denoising method. SSIM measures the similarity amongt denoised image against reference image. The luminance, contrast, and structural term are the terms on which it depends. The parameter values of CNN method are better than all other compared method that shows overall best performance in denoising method due to highest overlapping.

Table 1 Quantitative analysis of different thermal image denoising techniques

S. No.	Technique used	Image quality assessment parameters			
		Gaussian noise	Poisson noise	Salt and Pepper noise	Speckle noise
1	Without using technique	PSNR:20.3118	PSNR:27.7156	PSNR:18.0812	PSNR:19.3603
		SNR:14.3491	SNR:21.7530	SNR:12.1186	SNR:13.3977
		MSE:605.2054	MSE:110.0318	MSE:1011.4828	MSE:753.4359
2	Deep neural network	**PSNR:34.0712**	**PSNR:38.445**	PSNR:23.9278	**PSNR:32.2120**
		SNR:28.1086	**SNR:32.4818**	SNR:17.9652	**SNR:26.2493**
		MSE:25.4657	**MSE:9.3032**	MSE:263.2081	**MSE:39.0734**
		SSIM:0.87355	**SSIM:0.94094**	SSIM:0.5589	**SSIM:0.88284**
3	Using median filter	PSNR:26.7712	PSNR:31.5868	**PSNR:32.9184**	PSNR:23.9552
		SNR:20.8085	SNR:25.6242	**SNR:26.9557**	SNR:17.9935
		MSE:136.7605	MSE:45.1230	**MSE:33.2078**	MSE:261.559
		SSIM:0.53503	SSIM:0.83988	**SSIM:0.96844**	SSIM:0.51033
4	Using Weiner filtering	PSNR:28.1022	PSNR:35.4492	PSNR:20.7462	PSNR:25.0846
		SNR:22.1395	SNR:29.4865	SNR:14.7836	SNR:19.1219
		MSE:100.6613	MSE:18.5423	MSE:547.5900	MSE:201.6608
		SSIM:0.60573	SSIM:0.89091	SSIM:0.43158	SSIM:0.60178

4 Conclusion and Future Scope

The paper studies about the deep neural network-based image denoising, and it also critically analyses its results and compare with median filtering, Wiener filtering. Thus, it has been concluded that the median filter is far superior at preserving the sharp edges in compare to the mean filter. Adaptive iltering approach often shows superior results rather than the linear filtering. The adaptive filter is much more preferable rather than the comparable linear filter, preserving edges, and other high-frequency parts of a thermal image. Adaptive filtering shows superior results, when the noise is of constant-power, i.e., white or additive noise, like Gaussian noise. Noisy images have a smooth distribution of pixel intensity as compared to the original images. On analyzing the quantitative parameters like MSE, PSNR, and SSIM, it is observed that deep learning-based denoising method shows best performance than all compared method.

The residual encoder-decoder networks (REDNet) shows much better results in denoising of thermal images when clubbed with multi-level wavelet CNN (MWCNN) and pyramid real image denoising network (PRIDNet).

References

1. Lipari, CA, Head JF(1997) Advanced infrared image processing for breast cancer risk assessment. In: Proceedings of the 19th annual international conference of the IEEE engineering in medicine and biology society. 'magnificent milestones and emerging opportunities in medical engineering'(Cat. No. 97CH36136), vol 2, ppp 673–676. IEEE
2. Tomohiro, KS, Onodera JI (2001) United nations scientific committee on the effects of atomic radiation 2000 report. Jap J Health Phys 36(2):149–158
3. Zhou Y, Chen J, Li Q, Huang W, Lan H, Jiang H (2015) Association between breastfeeding and breast cancer risk: evidence from a meta-analysis. Breastfeed Med 10(3):175–182
4. Gram IT, Bremnes Y, Ursin G, Maskarinec G, Bjurstam N, Lund NE (2005) Percentage density, Wolfe's and Tabar's mammographic patterns: agreement and association with risk factors for breast cancer. Breast Cancer Res 7(5):R854
5. Michaelson JS, Silverstein M, Wyatt J, Weber G, Moore R, Halpern E, Kopans DB, Hughes K (2002) Predicting the survival of patients with breast carcinoma using tumor size. Cancer: Interdisciplinary Int J Am Cancer Soc 95(4)):713–723
6. Liu Z, Yan WQ, Yang ML (2018) Image denoising based on a CNN model. In:2018 4th International conference on control, automation and robotics (ICCAR), pp.389–393. IEEE
7. He Y, Li X, Li R, Wang J, Jing X (2020) A Deep-Learning Method for Radar Micro-Doppler Spectrogram Restoration. Sensors 20(17):5007
8. AlHajjar B (2013) Occupational stress among hospital nurses in Gaza-Palestine. University of Manchester
9. Thukral R, Kumar A, Arora AS, Gulshan (2019) Effect of different thresholding techniques for denoising of EMG signals by using different wavelets. In: 2019 2nd International conference on intelligent communication and computational techniques (ICCT), 2019, pp 161–165. https://doi.org/10.1109/ICCT46177.2019.8969036
10. Ganesh L, Chaitanya SK, Rao JD, Kumar MNVSS (2014) Development of image fusion algorithm for impulse noise removal in digital images using the quality assessment in spatial domain. International Journal of Engineering Research and Applications (IJERA) 1:786–792
11. Singh P, Shankar A (2021) A novel optical image denoising technique using convolutional neural network and anisotropic diffusion for real-time surveillance applications. J Real-Time Image Proc 1–18
12. Singh AP, Kamal TS, Kumar S (2006) Development of a virtual linearizer for correcting transducer static nonlinearity. ISA Trans 45(3):319–328
13. Sezan MI, Lagendijk RL (eds) (2012) Motion analysis and image sequence processing (Vol 220). Springer Science & Business Media
14. Singh P, Shree R (2020) A new homomorphic and method noise thresholding based despeckling of SAR image using anisotropic diffusion. J King Saud Univ Comput Inf Sci 32(1):137–148
15. Chakre R, Thapa P, Siddharth R, Baruah U (2013) An experimental study on image de-noising filters. Int J Emerging Technol Adv Eng 3(2):549–553
16. Lee D et al (2015) Learning speed improvement using multi-GPUs on DNN-Based acoustic model training in Korean Intelligent personal assistant. In: Natural language dialog systems and intelligent assistants. Springer, Cham, pp 263–271
17. Singh P, Shree R (2017) A new homomorphic and method noise thresholding based despeckling of SAR image using anisotropic diffusion. J King Saud Univ Comput Inf Sci
18. Singh P, Shree R (2016) Statistical modelling of log transformed speckled image. Int J Comput Sci Inf Secur (IJCSIS). 14(8)
19. Singh P, Shree R (2018) A new SAR image despeckling using directional smoothing filter and method noise thresholding. Eng Sci Tech Int J
20. Singh P et al (2018) A new SAR image despeckling using correlation-based fusion and method noise thresholding. J King Saud Univ Comput Inf Sci

Author Index

A
Abhishek, K. G., 99
Adithya Narayan, K. A., 293
Agarwal, Divyanshu, 549
Agarwal, Srishti, 337
Aggarwal, Daksh, 81
Ahlawat, Srishti, 761
Ahmad, Javed, 627
Alam, M. Afshar, 9
Amirthavarsini, G., 201
Anand, Anjali, 73
Anantha Prabha, P., 237, 403, 415
Angu Kaushika, S., 201
Anita, R., 717
Arathi, P., 503
Arjuna Rao, A., 139
Arjun, K. P., 549
Arjun, N., 415
Arora, A. S., 827
Arora, Deepak, 431, 443
Arunima, 73
Ayush, 91

B
Baig, Md Adnan, 787
Baliyan, Priyanshi, 91
Baranidharan, B., 391
Baweja, Asis Kaur, 227
Bhatia, Hansika, 81
Bhosale, Omkar, 651
Bhushan Singh, Shashi, 569
Bilal, K. V., 709
Biswas, Siddhartha Sankar, 9

C
Chakrawar, Onkar, 651
Chauhan, Divyansh, 81
Chauhan, Shekhar, 381
Chaurasia, Divyansh, 569
Christy Jeba Malar, A., 709
Chudhey, Arshdeep Singh, 325

D
Dave, Poornima, 569
Dayal, Mukul, 121
Deeptheshanmathie, K., 201
Deva Priya, M., 201, 213, 237, 359, 695, 709
Dhanalakshmi, Samiappan, 483, 589
Dinesh Kumar, Y., 139
Divya Prasanth, S., 415
Dixit, Ashutosh, 317, 347
Dixit, Vibhu, 337
Dodhia, Baiju, 651

G
Gangwar, Mohit, 527, 537
Garg, Apoorv, 121
Garg, Jyoti, 517
Ghosh, Soumalya, 559
Goel, Diksha, 91
Goel, Mohak, 121
Goel, Naman, 371
Gogul, J., 415
Gokulakrishnan, K., 237
Gomathi, N., 147, 739
Gulshan, 827

R. P. Mahapatra et al. (eds.), *Proceedings of International Conference on Recent Trends in Computing*, Lecture Notes in Networks and Systems 341,
https://doi.org/10.1007/978-981-16-7118-0

Gupta, Aman, 493
Gupta, Ankush, 111
Gupta, Megha, 227
Gupta, Utkarsh, 111
Gupta, Vatsal, 381

H
Harikumar, Sandhya, 293
Hari Prakash, Shilpa, 293
Harsha, Mullapudi Sri, 731
Hema, M., 809
Hithyshi, K., 161

I
Iftekhar, Nida, 9

J
Jadhav, Kishor P., 527
Jain, Jatin, 549
Jain, Jigar, 651
Janakiraman, Sengathir, 799
Jayaraman, Ramkumar, 469, 667
Joe Raymond, A., 237
Joga, S. Ramana Kumar, 263
Jorvekar, Ganesh N., 537

K
Karanwal, Sonakshi, 717
Karimisetty, Sujatha, 139
Karthickraja, T., 237
Karthick, S., 147, 739
Karthik, D., 771
Karthik, R., 695
Karthik, S., 695, 709
Kaur, Gurneet, 81
Kavin Kumar, M., 709
Khan, Ihtiram Raza, 787
Khanna, Darpan, 761
Khan, Samia, 9
Kumar, Amit, 263
Kumar, Anil, 443
Kumar, Ashwani, 827
Kumar, Avneesh, 569
Kumar, R., 589
Kumar, Rajeev, 317
Kumar, Sunil, 81
Kumar, Vinesh, 61
Kumar, Yogesh, 251
Kumudini Devi, P., 139

L
Lakshmanan, Madan Kumar, 275
Lavanya, S., 99

M
Mahadevan, Balaji, 275
Mahapatra, R. P., 739
Maharana, Manoj Kumar, 263
Mahesh, Gadiraju, 51
Maheswara Rao, V. V. R., 51
Malhotra, Harshit, 559
Malhotra, Kritika, 613
Malik, Abhishek, 131
Mangala Arunsankar, S., 709
Mani Tripathi, Aanjey, 73
Mirnalinee, T. T., 771
Mirunalini, P., 771
Mishra, Arpit, 251
Mittal, Mahima, 161
Mittal, Punit, 91
Mohammad, Chaudhary Wali, 627
Monika, R., 589
Muruganantham, B., 679

N
Nair, Gouri S., 293
Navya, V., 307
Nayak, Sweta, 493
Nehra, Rahul, 391

P
Pagarkar, Sana, 641
Pal, Abhisikta, 391
Paliwal, Shweta, 131
Pandey, Garima, 559
Pandey, Swati, 641
Parwekar, Pritee, 739
Pathak, Ayush Shankar, 559
Patil, Kavita Avinash, 307
Pawade, Dipti, 641, 651
Payal, Ritu, 251
Pooja Shree, H. R., 161
Prabu Shankar, K. C., 809
Prasanna, M. S. M., 161
Priyadharsini, R., 771

R
Raj, Aditi, 61
Rajakumari, Anitha, 739
Raj, Ashwani, 667
Rajesh, T. M., 99